INTEGRALS (AN ARBITRARY CONSTANT MAY BE ADDED TO EACH INTEGRAL.)

1. $\int x^n \, dx = \dfrac{1}{n+1} x^{n+1} \quad (n \neq -1)$

2. $\int \dfrac{1}{x} \, dx = \log |x|$

3. $\int e^x \, dx = e^x$

4. $\int a^x \, dx = \dfrac{a^x}{\log a}$

5. $\int \sin x \, dx = -\cos x$

6. $\int \cos x \, dx = \sin x$

7. $\int \tan x \, dx = -\log |\cos x|$

8. $\int \cot x \, dx = \log |\sin x|$

9. $\int \sec x \, dx = \log |\sec x + \tan x| = \log |\tan (\tfrac{1}{2}x + \tfrac{1}{4}\pi)|$

10. $\int \csc x \, dx = \log |\csc x - \cot x| = \log |\tan \tfrac{1}{2}x|$

11. $\int \arcsin \dfrac{x}{a} \, dx = x \arcsin \dfrac{x}{a} + \sqrt{a^2 - x^2} \quad (a > 0)$

12. $\int \arccos \dfrac{x}{a} \, dx = x \arccos \dfrac{x}{a} - \sqrt{a^2 - x^2} \quad (a > 0)$

13. $\int \arctan \dfrac{x}{a} \, dx = x \arctan \dfrac{x}{a} - \dfrac{a}{2} \log (a^2 + x^2) \quad (a > 0)$

14. $\int \sin^2 mx \, dx = \dfrac{1}{2m} (mx - \sin mx \cos mx)$

15. $\int \cos^2 mx \, dx = \dfrac{1}{2m} (mx + \sin mx \cos mx)$

16. $\int \sec^2 x \, dx = \tan x$

17. $\int \csc^2 x \, dx = -\cot x$

18. $\int \sin^n x \, dx = -\dfrac{\sin^{n-1} x \cos x}{n} + \dfrac{n-1}{n} \int \sin^{n-2} x \, dx$

19. $\int \cos^n x \, dx = \dfrac{\cos^{n-1} x \sin x}{n} + \dfrac{n-1}{n} \int \cos^{n-2} x \, dx$

20. $\int \tan^n x \, dx = \dfrac{\tan^{n-1} x}{n-1} - \int \tan^{n-2} x \, dx \quad (n \neq 1)$

21. $\int \cot^n x \, dx = -\dfrac{\cot^{n-1} x}{n-1} - \int \cot^{n-2} x \, dx \quad (n \neq 1)$

22. $\int \sec^n x \, dx = \dfrac{\tan x \sec^{n-2} x}{n-1} + \dfrac{n-2}{n-1} \int \sec^{n-2} x \, dx \quad (n \neq 1)$

23. $\int \csc^n x \, dx = -\dfrac{\cot x \csc^{n-2} x}{n-1} + \dfrac{n-2}{n-1} \int \csc^{n-2} x \, dx \quad (n \neq 1)$

(Continued at the back of the book)

Vector Calculus

VECTOR CALCULUS

Fourth Edition

JERROLD E. MARSDEN

California Institute of Technology, Pasadena

ANTHONY J. TROMBA

University of California, Santa Cruz

 W. H. FREEMAN AND COMPANY / NEW YORK

Politics is for the moment.
An equation is for eternity.

A. EINSTEIN

Some calculus tricks are quite easy.
Some are enormously difficult. The fools
who write the textbooks of
advanced mathematics seldom take the trouble
to show you how easy the easy
calculations are.

SILVANUS P. THOMPSON *Calculus Made Easy*, Macmillan (1910)

Cover photo: False-color photograph of Comet Bennett, 1970. Copyright Photo Researchers.

LIBRARY OF CONGRESS CATALOGING-IN-PUBLICATION DATA

Marsden, Jerrold E.
    Vector calculus / Jerrold E. Marsden, Anthony J. Tromba. — 4th ed.
        p.   cm.
    Includes index.
    ISBN 0-7167-2432-4 (hc)
    1. Calculus.   2. Vector analysis.   I. Tromba, Anthony.
    II. Title.
    QA303.M338   1996
    515′.63—dc20   95-37696
                                                                              CIP

Printed in the United States of America

Second printing 1996, RRD

CONTENTS

PREFACE

This text is intended for a one-semester course in the calculus of functions of several variables and vector analysis taught at the sophomore or junior level. Sometimes the course is preceded by a beginning course in linear algebra, but this is not an essential prerequisite. We require only the bare rudiments of matrix algebra, and the necessary concepts are developed in the text. However, we do assume a knowledge of the fundamentals of one-variable calculus—differentiation and integration of the standard functions.

The text includes much of the basic theory as well as many concrete examples and problems. Some of the technical proofs for theorems in Chapters 2 and 5 are given in the optional Sections 2.7 and 5.5. Section 2.2, on limits and continuity, is designed to be treated lightly and is deliberately brief. More sophisticated theoretical topics, such as compactness and delicate proofs in integration theory, have been omitted, because they usually belong to a more advanced course.

Computational skills and intuitive understanding are important at this level, and we have tried to meet this need by making the book concrete and student-oriented. For example, although we formulate the definition of the derivative correctly, it is done by using matrices of partial derivatives rather than abstract linear transformations. Also, we include a number of physical illustrations from such areas of physics as fluid mechanics, gravitation, and electromagnetic theory, and from economics as well, although prior knowledge of these subjects is not assumed.

A special feature of the text is the early introduction of vector fields, divergence, and curl in Chapter 4, before integration. Vector analysis usually suffers in a course of this type, and the present arrangement is designed to offset this tendency. To go even further, one might consider teaching Chapter 3 (Taylor's theorem, maxima and minima, Lagrange multipliers) after Chapter 8 (the integral theorems of vector analysis).

This fourth edition was completely reset, but retains the balance between theory, applications, optional material, and historical notes that was present in earlier editions.

The main changes in this edition and in its supplements are as follows:

1. Summary boxes to emphasize the main points and help guide the student.

2. An *Instructor's Guide* that contains supplementary examples and practice exams was prepared.

3. The *Student Guide* by Pao and Soon was reworked and reset.

PREREQUISITES AND NOTATION

We assume that students have studied the calculus of functions of a real variable, including analytic geometry in the plane. Some students may have had some exposure to matrices as well, although what we shall need is given in Sections 1.3 and 1.5.

We also assume that students are familiar with functions of elementary calculus, such as $\sin x$, $\cos x$, e^x, and $\log x$ (we write $\log x$ or $\ln x$ for the natural logarithm, which is sometimes denoted $\log_e x$). Students are expected to know, or to review as the course proceeds, the basic rules of differentiation and integration for functions of one variable, such as the chain rule, the quotient rule, integration by parts and so forth.

We now summarize the notations to be used later. Students can read through these quickly now, then refer to them later if the need arises.

The collection of all real numbers is denoted $\mathbb{R}$. Thus $\mathbb{R}$ includes the *integers*, ..., $-3, -2, -1, 0, 1, 2, 3, \ldots$; the *rational numbers*, p/q, where p and q are integers ($q \neq 0$); and the *irrational numbers*, such as $\sqrt{2}$, π, and e. Members of $\mathbb{R}$ may be visualized as points on the real-number line, as shown in Figure P.1.

Figure P.1   The geometric representation of points on the real number line.

When we write $a \in \mathbb{R}$ we mean that a is a member of the set $\mathbb{R}$, in other words, that a is a real number. Given two real numbers a and b with $a < b$ (that is, with a less than b), we can form the *closed interval* $[a, b]$, consisting of all x such that $a \leq x \leq b$, and the *open interval* (a, b), consisting of all x such that $a < x < b$. Similarly, we may form half-open intervals $(a, b]$ and $[a, b)$ (Figure P.2).

Figure P.2   The geometric representation of the intervals $[a, b]$, (c, d), and $[e, f)$.

The *absolute value* of a number $a \in \mathbb{R}$ is written $|a|$ and is defined as

$$|a| = \begin{cases} a & \text{if } a \geq 0 \\ -a & \text{if } a < 0. \end{cases}$$

For example, $|3| = 3$, $|-3| = 3$, $|0| = 0$, and $|-6| = 6$. The inequality $|a + b| \leq |a| + |b|$ always holds. The *distance from a to b* is given by $|a - b|$. Thus, the distance from 6 to 10 is 4 and from -6 to 3 is 9.

If we write $A \subset \mathbb{R}$, we mean A is a *subset* of $\mathbb{R}$. For example, A could equal the set of integers $\{\ldots, -3, -2, -1, 0, 1, 2, 3, \ldots\}$. Another example of a subset of $\mathbb{R}$ is the set $\mathbb{Q}$ of rational numbers. Generally, for two collections of objects (that is, sets) A and B, $A \subset B$ means A is a subset of B; that is, every member of A is also a member of B.

The symbol $A \cup B$ means the **union** of A and B, the collection whose members are members of either A or B. Thus

$$\{\ldots, -3, -2, -1, 0\} \cup \{-1, 0, 1, 2, \ldots\} = \{\ldots, -3, -2, -1, 0, 1, 2, \ldots\}.$$

Similarly, $A \cap B$ means the **intersection** of A and B; that is, this set consists of those members of A and B that are in both A and B. Thus the intersection of the two sets above is $\{-1, 0\}$.

We shall write $A \backslash B$ for those members of A that are not in B. Thus

$$\{\ldots, -3, -2, -1, 0\} \backslash \{-1, 0, 1, 2, \ldots\} = \{\ldots, -3, -2\}.$$

We can also specify sets as in the following examples:

$$\{a \in \mathbb{R} \mid a \text{ is an integer}\} = \{\ldots, -3, -2, -1, 0, 1, 2, \ldots\}$$
$$\{a \in \mathbb{R} \mid a \text{ is an even integer}\} = \{\ldots, -2, 0, 2, 4, \ldots\}$$
$$\{x \in \mathbb{R} \mid a \leq x \leq b\} = [a, b].$$

A **function** $f \colon A \to B$ is a rule that assigns to each $a \in A$ one specific member $f(a)$ of B. We call A the **domain** of f and B the **target** of f. The set $\{f(x) \mid x \in A\}$ consisting of all the values of $f(x)$ is called the **range** of f. Denoted by $f(A)$, the range is a subset of the target B. It may be all of B, in which case f is said to be **onto** B. The fact that the function f sends a to $f(a)$ is denoted by $a \mapsto f(a)$. For example, the function $f(x) = x^3/(1 - x)$ that assigns the number $x^3/(1 - x)$ to each $x \neq 1$ in $\mathbb{R}$ can also be defined by the rule $x \mapsto x^3/(1 - x)$. Functions are also called **mappings, maps,** or **transformations**. The notation $f \colon A \subset \mathbb{R} \to \mathbb{R}$ means that A is a subset of $\mathbb{R}$ and that f assigns a value $f(x)$ in $\mathbb{R}$ to each $x \in A$. The **graph** of f consists of all the points $(x, f(x))$ in the plane (Figure P.3).

graph of f

$(x, f(x))$

$A =$ domain

Figure P.3   The graph of a function with the half-open interval A as domain.

The notation $\sum_{i=1}^{n} a_i$ means $a_1 + \cdots + a_n$, where $a_1, \ldots, a_n$ are given numbers. The sum of the first n integers is

$$1 + 2 + \cdots + n = \sum_{i=1}^{n} i = \frac{n(n+1)}{2}.$$

The *derivative* of a function $f(x)$ is denoted $f'(x)$ or

$$\frac{df}{dx},$$

and the *definite integral* is written

$$\int_a^b f(x)\, dx.$$

If we set $y = f(x)$, the derivative is also denoted by

$$\frac{dy}{dx}.$$

Readers are assumed to be familiar with the chain rule, integration by parts, and other basic facts from the calculus of functions of one variable. In particular, they should know how to differentiate and integrate exponential, logarithmic, and trigonometric functions. Short tables of derivatives and integrals, which are adequate for the needs of this text, are printed at the front and back of the book.

The following notations are used synonymously: $e^x = \exp x$, $\ln x = \log x$, and $\sin^{-1} x = \arcsin x$.

The end of a proof is denoted by the symbol ■, while the end of an example or remark is denoted by the symbol ▲. Optional material is preceded by a star: *.

ACKNOWLEDGMENTS

Many colleagues and students in the mathematical community have made valuable contributions and suggestions since this book was begun. An early draft of the book was written in collaboration with Ralph Abraham. We thank him for allowing us to draw upon his work. It is impossible to list all those who assisted with this book, but we wish especially to thank Michael Hoffman and Joanne Seitz for their help on earlier editions. We also received valuable comments from Mary Anderson, John Ball, Gerald Edgar, Frank Gerrish, Jenny Harrison, David Knudson, Richard Koch, Andrew Lenard, Gordon McLean, David Merriell, Jeanette Nelson, Dan Norman, Keith Phillips, Anne Perleman, Kenneth Ross, Ray Sachs, Diane Sauvageot, Joel Smoller, Melvyn Tews, Ralph and Bob Tromba, Steve Wan, Alan Weinstein, John Wilker, and Peter Zvengrowski. The students and faculty of Austin Community College deserve a special note of thanks.

We are grateful to the following instructors who provided detailed reviews of the manuscript for this edition: Richard Laugesen, University of Michigan; John Lott, University of

Michigan; Joan Rand Moschovakis, Occidental College; and A. Shadi Tahvildar-Zadeh, Princeton University.

A final word of thanks goes to those who helped in the preparation of the manuscript and the production of the book. We especially thank Connie Calica, Nora Lee, Marnie McElhiney, Rosemarie Stampful, Ruth Suzuki, Ikuko Workman, and Esther Zack for their excellent typing of various versions and revisions of the manuscript; Herb Holden of Gonzaga University and Jerry Kazdan of the University of Pennsylvania for suggesting and preparing the computer-generated figures; Jerry Lyons and Holly Hodder for their roles as our mathematics editors; Christine Hastings for editorial supervision; and Trumbull Rogers for his expert copyediting.

We will be maintaining an up-to-date list of corrections and suggestions concerning this fourth edition. We will be happy to mail this list to any user of the text. Please send your request to either Jerrold Marsden at Control and Dynamical Systems 104-44, California Institute of Technology, Pasadena, CA 91125 (marsden@cds.caltech.edu) or Anthony Tromba at the Department of Mathematics, University of California, Santa Cruz, CA 95064 (tromba@cats.ucsc.edu).

1 THE GEOMETRY OF EUCLIDEAN SPACE

Quaternions came from Hamilton ... and have been an unmixed evil
to those who have touched them in any way. Vector is a useless
survival ... and has never been of the slightest use to any creature.

LORD KELVIN

In this chapter we consider the basic operations on vectors in two- and three-dimensional space: vector addition, scalar multiplication, and the dot and cross products. In Section 1.5 we generalize some of these notions to n-space and review properties of matrices that will be needed in Chapters 2 and 3.

1.1 VECTORS IN TWO- AND THREE-DIMENSIONAL SPACE

Points P in the plane are represented by ordered pairs of real numbers (a_1, a_2); the numbers a_1 and a_2 are called the ***Cartesian coordinates of*** P. We draw two perpendicular lines, label them as the x and y axes, and then drop perpendiculars from P to these axes, as in Figure 1.1.1. After designating the intersection of the x and y axes as the origin and choosing units on these axes, we produce two signed distances a_1 and a_2 as shown in the figure; a_1 is called the x ***component*** of P, and a_2 is called the y ***component***.

Points in space may be similarly represented as ordered triples of real numbers. To construct such a representation, we choose three mutually perpendicular lines that meet at a point in space. These lines are called x ***axis***, y ***axis***, and z ***axis***, and the point at which they meet is called the ***origin*** (this is our reference point). We choose a scale on these axes, as shown in Figure 1.1.2.

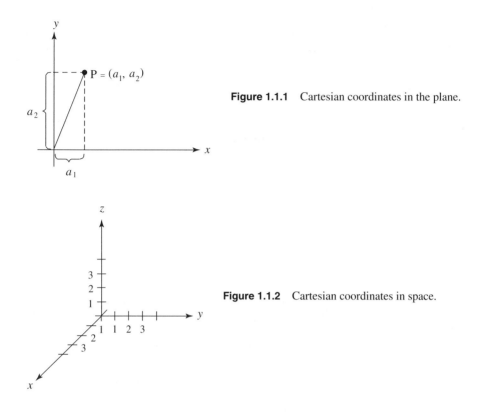

Figure 1.1.1   Cartesian coordinates in the plane.

Figure 1.1.2   Cartesian coordinates in space.

The triple $(0, 0, 0)$ corresponds to the origin of the coordinate system, and the arrows on the axes indicate the positive directions. For example, the triple $(2, 4, 4)$ represents a point 2 units from the origin in the positive direction along the x axis, 4 units in the positive direction along the y axis, and 4 units in the positive direction along the z axis (Figure 1.1.3).

Figure 1.1.3   Geometric representation of the point $(2, 4, 4)$ in Cartesian coordinates.

Because we can associate points in space with ordered triples in this way, we often use the expression "the point (a_1, a_2, a_3)" instead of the longer phrase "the point P that corresponds to the triple (a_1, a_2, a_3)." We say that a_1 is the x *coordinate* (or first coordinate), a_2 is the y *coordinate* (or second coordinate), and a_3 is the z *coordinate* (or third coordinate) of P. It is also common to denote points in space with the letters x, y, and z in place of a_1, a_2, and a_3. Thus the triple (x, y, z) represents a point whose first coordinate is x, second coordinate is y, and third coordinate is z.

We employ the following notation for the line, the plane, and three-dimensional space:

(i)   The real number line is denoted $\mathbb{R}^1$ or simply $\mathbb{R}$.

(ii) The set of all ordered pairs (x, y) of real numbers is denoted $\mathbb{R}^2$.

(iii) The set of all ordered triples (x, y, z) of real numbers is denoted $\mathbb{R}^3$.

When speaking of $\mathbb{R}^1$, $\mathbb{R}^2$, and $\mathbb{R}^3$ simultaneously, we write $\mathbb{R}^n$, where $n = 1, 2$, or 3; or $\mathbb{R}^m$, where $m = 1, 2, 3$. Starting in Section 1.5 we will also study $\mathbb{R}^n$ for $n = 4, 5, 6, \ldots$, but the cases $n = 1, 2, 3$ are closest to our geometric intuition and will be stressed throughout the book.

Vector Addition and Scalar Multiplication

The operation of addition can be extended from $\mathbb{R}$ to $\mathbb{R}^2$ and $\mathbb{R}^3$. For $\mathbb{R}^3$, this is done as follows. Given the two triples (a_1, a_2, a_3) and (b_1, b_2, b_3), we define their *sum* to be

$$(a_1, a_2, a_3) + (b_1, b_2, b_3) = (a_1 + b_1, a_2 + b_2, a_3 + b_3).$$

EXAMPLE 1

$$(1, 1, 1) + (2, -3, 4) = (3, -2, 5),$$

$$(x, y, z) + (0, 0, 0) = (x, y, z),$$

$$(1, 7, 3) + (a, b, c) = (1 + a, 7 + b, 3 + c). \quad \blacktriangle$$

The element $(0, 0, 0)$ is called the *zero element* (or just *zero*) of $\mathbb{R}^3$. The element $(-a_1, -a_2, -a_3)$ is the *additive inverse* (or *negative*) of (a_1, a_2, a_3), and we will write $(a_1, a_2, a_3) - (b_1, b_2, b_3)$ for $(a_1, a_2, a_3) + (-b_1, -b_2, -b_3)$.

There are several important product operations that we will define on $\mathbb{R}^3$. One of these, called the *inner product*, assigns a real number to each pair of elements of $\mathbb{R}^3$. We shall discuss it in detail in Section 1.2. Another product operation for $\mathbb{R}^3$ is called *scalar multiplication* (the word "scalar" is a synonym for "real number"). This product combines scalars (real numbers) and elements of $\mathbb{R}^3$ (ordered triples) to yield elements of $\mathbb{R}^3$ as follows: Given a scalar α and a triple (a_1, a_2, a_3), we define the *scalar multiple* by

$$\alpha(a_1, a_2, a_3) = (\alpha a_1, \alpha a_2, \alpha a_3).$$

EXAMPLE 2

$$2(4, e, 1) = (2 \cdot 4, 2 \cdot e, 2 \cdot 1) = (8, 2e, 2),$$

$$6(1, 1, 1) = (6, 6, 6),$$

$$1(u, v, w) = (u, v, w),$$

$$0(p, q, r) = (0, 0, 0). \quad \blacktriangle$$

Addition and scalar multiplication of triples satisfy the following properties:

(i) $(\alpha\beta)(a_1, a_2, a_3) = \alpha[\beta(a_1, a_2, a_3)]$ (associativity)

(ii) $(\alpha + \beta)(a_1, a_2, a_3) = \alpha(a_1, a_2, a_3) + \beta(a_1, a_2, a_3)$ (distributivity)

(iii) $\alpha[(a_1, a_2, a_3) + (b_1, b_2, b_3)] = \alpha(a_1, a_2, a_3) + \alpha(b_1, b_2, b_3)$ (distributivity)

(iv) $\alpha(0, 0, 0) = (0, 0, 0)$ (property of zero)

(v) $0(a_1, a_2, a_3) = (0, 0, 0)$ (property of zero)

(vi) $1(a_1, a_2, a_3) = (a_1, a_2, a_3)$ (property of the unit element)

The identities are proven directly from the definitions of addition and scalar multiplication. For instance,

$$(\alpha + \beta)(a_1, a_2, a_3) = ((\alpha + \beta)a_1, (\alpha + \beta)a_2, (\alpha + \beta)a_3)$$
$$= (\alpha a_1 + \beta a_1, \alpha a_2 + \beta a_2, \alpha a_3 + \beta a_3)$$
$$= \alpha(a_1, a_2, a_3) + \beta(a_1, a_2, a_3).$$

For $\mathbb{R}^2$, addition and scalar multiplication are defined just as in $\mathbb{R}^3$, with the third component of each vector dropped off. All the properties above still hold.

EXAMPLE 3 Interpret the chemical equation $2NH_2 + H_2 = 2NH_3$ as a relation in the algebra of ordered pairs.

SOLUTION We think of the molecule N_xH_y (x atoms of nitrogen, y atoms of hydrogen) as represented by the ordered pair (x, y). Then the chemical equation given is equivalent to $2(1, 2) + (0, 2) = 2(1, 3)$. Indeed, both sides are equal to $(2, 6)$. $\quad \blacktriangle$

Let us turn to the geometry of these operations in $\mathbb{R}^2$ and $\mathbb{R}^3$. For the moment, we define a **vector** to be a directed line segment beginning at the origin, that is, a line segment with specified magnitude and direction, and initial point at the origin. Figure 1.1.4 shows several vectors, drawn as arrows beginning at the origin. In print, vectors are usually denoted by boldface letters such as **a**. By hand, we usually write them as $\vec{a}$ or simply as a, possibly with a line or wavy line under it.

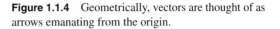

Figure 1.1.4   Geometrically, vectors are thought of as arrows emanating from the origin.

Using this definition of a vector, we associate with each vector **a** the point (a_1, a_2, a_3) where **a** terminates, and conversely, with each point (a_1, a_2, a_3) in space we can associate a vector **a**. Thus, we shall identify **a** with (a_1, a_2, a_3) and write $\mathbf{a} = (a_1, a_2, a_3)$. For this reason, the elements of $\mathbb{R}^3$ not only are ordered triples of real numbers, but are also regarded as vectors. The triple $(0, 0, 0)$ is denoted **0**. We call a_1, a_2, and a_3 the ***components*** of **a**, or when we think of **a** as a point, its ***coordinates***.

Two vectors $\mathbf{a} = (a_1, a_2, a_3)$ and $\mathbf{b} = (b_1, b_2, b_3)$ are equal if and only if $a_1 = b_1$, $a_2 = b_2$, and $a_3 = b_3$. Geometrically this means that **a** and **b** have the same direction and the same length (or "magnitude").

Geometrically, we define vector addition as follows. In the plane containing the vectors $\mathbf{a} = (a_1, a_2, a_3)$ and $\mathbf{b} = (b_1, b_2, b_3)$ (see Figure 1.1.5), form the parallelogram having **a** as one side and **b** as its adjacent side. The sum $\mathbf{a} + \mathbf{b}$ is the directed line segment along the diagonal of the parallelogram.

Figure 1.1.5   The geometry of vector addition.

This geometric view of vector addition is useful in many physical situations, as we shall see in the next section. For an easily visualized example, consider a bird or an airplane flying through the air with velocity $\mathbf{v}_1$, but in the presence of a wind with velocity $\mathbf{v}_2$. The resultant velocity, $\mathbf{v}_1 + \mathbf{v}_2$, is what one sees; see Figure 1.1.6.

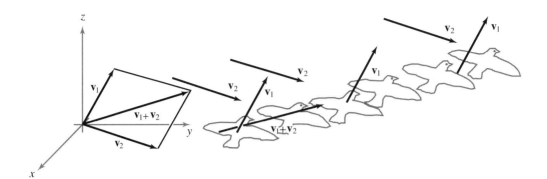

Figure 1.1.6   A physical interpretation of vector addition.

To show that our geometric definition of addition is consistent with our algebraic definition, we demonstrate that $\mathbf{a} + \mathbf{b} = (a_1 + b_1, a_2 + b_2, a_3 + b_3)$. We shall prove this result in the plane and leave the proof in three-dimensional space to the reader. Thus, we wish to show that if $\mathbf{a} = (a_1, a_2)$ and $\mathbf{b} = (b_1, b_2)$, then $\mathbf{a} + \mathbf{b} = (a_1 + b_1, a_2 + b_2)$.

In Figure 1.1.7 let $\mathbf{a} = (a_1, a_2)$ be the vector ending at the point A, and let $\mathbf{b} = (b_1, b_2)$ be the vector ending at point B. By definition, the vector $\mathbf{a} + \mathbf{b}$ ends at the vertex C of parallelogram OBCA. To verify that $\mathbf{a} + \mathbf{b} = (a_1 + b_1, a_2 + b_2)$, it suffices to show that the coordinates of C are $(a_1 + b_1, a_2 + b_2)$. The sides of the triangles OAD and BCG are parallel, and the sides OA and BC have equal lengths, which we write as OA = BC. The triangles are congruent, so BG = OD; since BGFE is a rectangle, EF = BG. Furthermore, OD = a_1 and OE = b_1. Hence, EF = BG = OD = a_1. Since OF = EF + OE, it follows that OF = $a_1 + b_1$. This shows that the x coordinate of $\mathbf{a} + \mathbf{b}$ is $a_1 + b_1$. The proof that the y coordinate is $a_2 + b_2$ is analogous. This argument assumes A and B to be in the first quadrant, but similar arguments hold for the other quadrants.

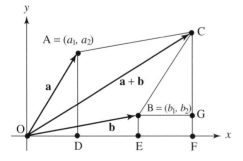

Figure 1.1.7   The construction used to prove that $(a_1, b_1) + (a_2, b_2) = (a_1 + b_1, a_2 + b_2)$.

Figure 1.1.8(a) illustrates another way of looking at vector addition: in terms of triangles rather than parallelograms. That is, we translate (without rotation) the directed line segment representing the vector $\mathbf{b}$ so that it begins at the end of the vector $\mathbf{a}$. The endpoint of the resulting directed segment is the endpoint of the vector $\mathbf{a} + \mathbf{b}$. We note that when $\mathbf{a}$ and $\mathbf{b}$ are collinear, the triangle collapses to a line segment, as in Figure 1.1.8(b).

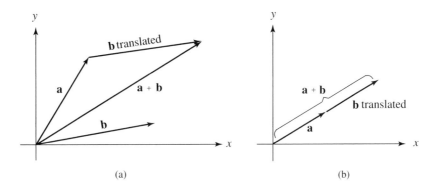

Figure 1.1.8 (a) Vector addition may be visualized in terms of triangles as well as parallelograms. (b) The triangle collapses to a line segment when $\mathbf{a}$ and $\mathbf{b}$ are collinear.

In Figure 1.1.8 we have placed $\mathbf{a}$ and $\mathbf{b}$ *head to tail*. That is, the tail of $\mathbf{b}$ is placed at the head of $\mathbf{a}$, and the vector $\mathbf{a} + \mathbf{b}$ goes from the tail of $\mathbf{a}$ to the head of $\mathbf{b}$. If we do it in the other order, $\mathbf{b} + \mathbf{a}$, we get the same vector by going around the parallelogram the other way. Consistent with this figure, it is useful to let vectors "glide" or "slide," keeping the same magnitude and direction. We want, in fact, to regard two vectors as the *same* if they have the same magnitude and direction. When we insist on vectors beginning at the origin, we will say that we have ***bound vectors***. If we allow vectors to begin at other points, we will speak of ***free vectors*** or just ***vectors***.

VECTORS Vectors (also called free vectors) are directed line segments in [the plane or] space represented by directed line segments with a beginning (tail) and an end (head). Line segments obtained from each other by parallel translation (but not rotation) represent the same vector.

The components (a_1, a_2, a_3) of $\mathbf{a}$ are the (signed) lengths of the projections of $\mathbf{a}$ along the three coordinate axes; equivalently, they are defined by placing the tail of $\mathbf{a}$ at the origin and letting the head be the point (a_1, a_2, a_3). We write $\mathbf{a} = (a_1, a_2, a_3)$.

Two vectors are added by placing them head to tail and drawing the vectors from the tail of the first to the head of the second, as in Figure 1.1.8.

Scalar multiplication of vectors also has a geometric interpretation. If α is a scalar and $\mathbf{a}$ a vector, we define $\alpha\mathbf{a}$ to be the vector that is $|\alpha|$ times as long as $\mathbf{a}$, with the same direction as $\mathbf{a}$ if $\alpha > 0$, but with the opposite direction if $\alpha < 0$. Figure 1.1.9 illustrates several examples.

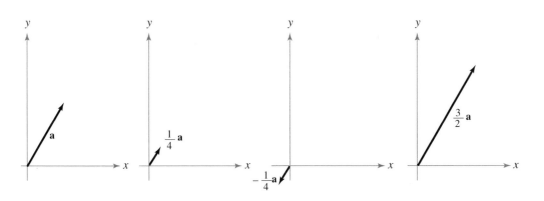

Figure 1.1.9   Some scalar multiples of a vector **a**.

Using an argument based on similar triangles, one finds that if $\mathbf{a} = (a_1, a_2, a_3)$, and α is a scalar, then

$$\alpha\mathbf{a} = (\alpha a_1, \alpha a_2, \alpha a_3).$$

That is, the geometric definition coincides with the algebraic one.

Given two vectors **a** and **b**, how do we represent the vector $\mathbf{b} - \mathbf{a}$ geometrically, that is, what is the geometry of vector subtraction? Since $\mathbf{a} + (\mathbf{b} - \mathbf{a}) = \mathbf{b}$, we see that $\mathbf{b} - \mathbf{a}$ is the vector that one adds to **a** to get **b**. In view of this, we may conclude that $\mathbf{b} - \mathbf{a}$ is the vector parallel to, and with the same magnitude as, the directed line segment beginning at the endpoint of **a** and terminating at the endpoint of **b** when **a** and **b** begin at the same point (see Figure 1.1.10).

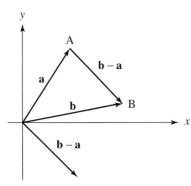

Figure 1.1.10   The geometry of vector subtraction.

EXAMPLE 4   Let **u** and **v** be the vectors shown in Figure 1.1.11. Draw the two vectors $\mathbf{u} + \mathbf{v}$ and $-2\mathbf{u}$. What are their components?

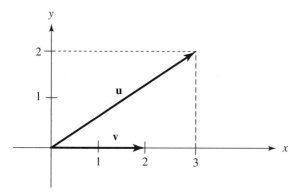

Figure 1.1.11 Find $\mathbf{u} + \mathbf{v}$ and $-2\mathbf{u}$.

SOLUTION Place the tail of $\mathbf{v}$ at the tip of $\mathbf{u}$ to obtain the vector shown in Figure 1.1.12.

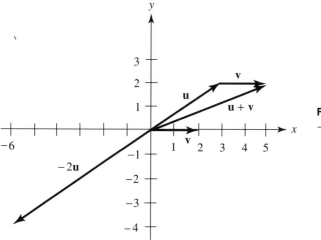

Figure 1.1.12 Computing $\mathbf{u} + \mathbf{v}$ and $-2\mathbf{u}$.

The vector $-2\mathbf{u}$, also shown, has length twice that of $\mathbf{u}$ and points in the opposite direction. From the figure, we see that the vector $\mathbf{u} + \mathbf{v}$ has components $(5, 2)$ and $-2\mathbf{u}$ has components $(-6, -4)$. ▲

EXAMPLE 5

(a) Sketch $-2\mathbf{v}$, where $\mathbf{v}$ has components $(-1, 1, 2)$.

(b) If $\mathbf{v}$ and $\mathbf{w}$ are any two vectors, show that $\mathbf{v} - \frac{1}{3}\mathbf{w}$ and $3\mathbf{v} - \mathbf{w}$ are parallel.

SOLUTION

(a) The vector $-2\mathbf{v}$ is twice as long as $\mathbf{v}$ but points in the opposite direction (see Figure 1.1.13).

(b) $\mathbf{v} - \frac{1}{3}\mathbf{w} = \frac{1}{3}(3\mathbf{v} - \mathbf{w})$; vectors that are multiples of one another are parallel. ▲

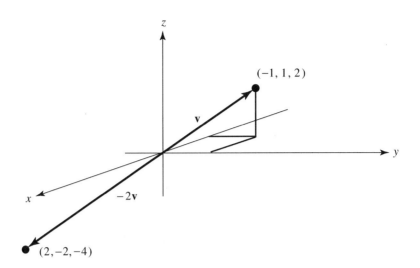

Figure 1.1.13 Multiplying $(-1, 1, 2)$ by -2.

The Standard Basis Vectors

To describe vectors in space, it is convenient to introduce three special vectors along the x, y, and z axes:

i: the vector with components $(1, 0, 0)$

j: the vector with components $(0, 1, 0)$

k: the vector with components $(0, 0, 1)$.

These *standard basis vectors* are illustrated in Figure 1.1.14. In the plane one has the standard basis **i** and **j** with components $(1, 0)$ and $(0, 1)$.

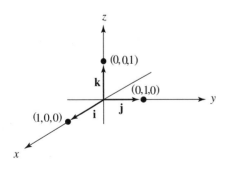

Figure 1.1.14 The standard basis vectors.

Let **a** be any vector, and let (a_1, a_2, a_3) be its components. Then

$$\mathbf{a} = a_1\mathbf{i} + a_2\mathbf{j} + a_3\mathbf{k},$$

since the right-hand side is given in components by

$$a_1(1, 0, 0) + a_2(0, 1, 0) + a_3(0, 0, 1) = (a_1, 0, 0) + (0, a_2, 0) + (0, 0, a_3)$$

$$= (a_1, a_2, a_3).$$

Thus, we can express every vector as a sum of scalar multiples of $\mathbf{i}$, $\mathbf{j}$, and $\mathbf{k}$.

THE STANDARD BASIS VECTORS

1. The vectors $\mathbf{i}$, $\mathbf{j}$, and $\mathbf{k}$ are unit vectors along the three coordinate axes, as shown in Figure 1.1.14.

2. If $\mathbf{a}$ has components (a_1, a_2, a_3), then

$$\mathbf{a} = a_1\mathbf{i} + a_2\mathbf{j} + a_3\mathbf{k}.$$

EXAMPLE 6

Express the vector whose components are $(e, \pi, -\sqrt{3})$ in the standard basis.

SOLUTION

$$\mathbf{v} = e\mathbf{i} + \pi\mathbf{j} - \sqrt{3}\mathbf{k}. \quad \blacktriangle$$

EXAMPLE 7 The vector $(2, 3, 2)$ equals $2\mathbf{i} + 3\mathbf{j} + 2\mathbf{k}$, and the vector $(0, -1, 4)$ is $-\mathbf{j} + 4\mathbf{k}$. Figure 1.1.15 shows $2\mathbf{i} + 3\mathbf{j} + 2\mathbf{k}$; the student should draw in the vector $-\mathbf{j} + 4\mathbf{k}$. $\quad \blacktriangle$

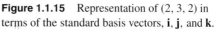

Figure 1.1.15 Representation of $(2, 3, 2)$ in terms of the standard basis vectors, $\mathbf{i}$, $\mathbf{j}$, and $\mathbf{k}$.

Addition and scalar multiplication may be written in terms of the standard basis vectors as follows:

$$(a_1\mathbf{i} + a_2\mathbf{j} + a_3\mathbf{k}) + (b_1\mathbf{i} + b_2\mathbf{j} + b_3\mathbf{k}) = (a_1 + b_1)\mathbf{i} + (a_2 + b_2)\mathbf{j} + (a_3 + b_3)\mathbf{k}$$

and

$$\alpha(a_1\mathbf{i} + a_2\mathbf{j} + a_3\mathbf{k}) = (\alpha a_1)\mathbf{i} + (\alpha a_2)\mathbf{j} + \alpha(a_3)\mathbf{k}.$$

The Vector Joining Two Points

To apply vectors to geometric problems, it is useful to assign a vector to a *pair* of points in the plane or in space, as follows. Given two points P and P' we can draw the vector $\mathbf{v}$ with tail P and head P', as in Figure 1.1.16, where we write $= \overrightarrow{PP'}$ for $\mathbf{v}$.

Figure 1.1.16 The vector from P to P' is denoted $\overrightarrow{PP'}$.

If P = (x, y, z) and P' = (x', y', z'), then the vectors from the origin to P and P' are $\mathbf{a} = x\mathbf{i} + y\mathbf{j} + z\mathbf{k}$ and $\mathbf{a}' = x'\mathbf{i} + y'\mathbf{j} + z'\mathbf{k}$, respectively, so the vector $\overrightarrow{PP'}$ is the difference $\mathbf{a}' - \mathbf{a} = (x' - x)\mathbf{i} + (y' - y)\mathbf{j} + (z' - z)\mathbf{k}$. (See Figure 1.1.17.)

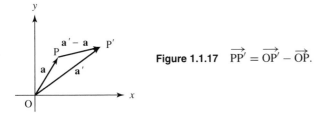

Figure 1.1.17 $\overrightarrow{PP'} = \overrightarrow{OP'} - \overrightarrow{OP}$.

> **THE VECTOR JOINING TWO POINTS** If the point P has coordinates (x, y, z) and P' has coordinates (x', y', z'), then the vector $\overrightarrow{PP'}$ from the tip of P to the tip of P' has components $(x' - x, y' - y, z' - z)$.

EXAMPLE 8

(a)   Find the components of the vector from $(3, 5)$ to $(4, 7)$.

(b)   Add the vector **v** from $(-1, 0)$ to $(2, -3)$ and the vector **w** from $(2, 0)$ to $(1, 1)$.

(c)   Multiply the vector **v** in (b) by 8. If the resulting vector is represented by the directed line segment from $(5, 6)$ to Q, what is Q?

SOLUTION

(a)   As in the preceding box, we subtract the ordered pairs: $(4, 7) - (3, 5) = (1, 2)$. Thus the required components are $(1, 2)$.

(b)   The vector **v** has components $(2, -3) - (-1, 0) = (3, -3)$ and **w** has components $(1, 1) - (2, 0) = (-1, 1)$. Therefore, the vector **v** + **w** has components $(3, -3) + (-1, 1) = (2, -2)$.

(c)   The vector 8**v** has components $8(3, -3) = (24, -24)$. If this vector is represented by the directed line segment from $(5, 6)$ to Q, and Q has coordinates (x, y), then $(x, y) - (5, 6) = (24, -24)$, so $(x, y) = (5, 6) + (24, -24) = (29, -18)$.   ▲

EXAMPLE 9   Let $P = (-2, -1), Q = (-3, -3)$, and $R = (-1, -4)$ in the xy plane.

(a)   Draw these vectors: **v** joining P to Q; **w** joining Q to R; **u** joining R to P.

(b)   What are the components of **v**, **w**, and **u**?

(c)   What is **v** + **w** + **u**?

SOLUTION

(a)   See Figure 1.1.18.

Figure 1.1.18   The vector **v** joins P to Q; **w** joins Q to R; and **u** joins R to P.

(b)
$$\mathbf{v} = (-3, -3) - (-2, -1) = (-1, -2),$$
$$\mathbf{w} = (-1, -4) - (-3, -3) = (2, -1),$$
$$\mathbf{u} = -(-1, -4) + (-2, -1) = (-1, 3).$$

(c) $\mathbf{v} + \mathbf{w} + \mathbf{u} = (-1, -2) + (2, -1) + (-1, 3) = (0, 0).$ ▲

Geometry Theorems by Vector Methods

Many of the theorems of plane geometry can be proved by vector methods. Here is one example.

EXAMPLE 10   Use vectors to prove that the diagonals of a parallelogram bisect each other.

SOLUTION   Let OPRQ be the parallelogram, with two adjacent sides represented by the vectors $\mathbf{a} = \overrightarrow{OP}$ and $\mathbf{b} = \overrightarrow{OQ}$. Let M be the midpoint of the diagonal OR, N the midpoint of the other, PQ. (See Figure 1.1.19.)

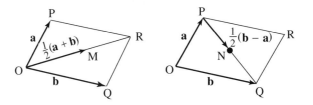

Figure 1.1.19   The midpoints M and N coincide, so the diagonals OR and PQ bisect each other.

Then $\overrightarrow{OR} = \overrightarrow{OP} + \overrightarrow{OQ} = \mathbf{a} + \mathbf{b}$ by the parallelogram rule for vector addition, so $\overrightarrow{OM} = \frac{1}{2}\overrightarrow{OR} = \frac{1}{2}(\mathbf{a} + \mathbf{b})$. On the other hand,

$$\overrightarrow{PQ} = \overrightarrow{OQ} - \overrightarrow{OP} = \mathbf{b} - \mathbf{a}, \text{ so } \overrightarrow{PN} = \frac{1}{2}\overrightarrow{PQ} = \frac{1}{2}(\mathbf{b} - \mathbf{a}),$$

and hence

$$\overrightarrow{ON} = \overrightarrow{OP} + \overrightarrow{PN} = \mathbf{a} + \frac{1}{2}(\mathbf{b} - \mathbf{a}) = \frac{1}{2}(\mathbf{a} + \mathbf{b}).$$

Since $\overrightarrow{OM}$ and $\overrightarrow{ON}$ are equal vectors, the points M and N coincide, so the diagonals bisect each other.   ▲

Equations of Lines

Planes and lines are geometric objects that can be represented by equations. We shall defer until Section 1.3 a study of equations representing planes. However, using the geometric interpretation of vector addition and scalar multiplication, we will now find the *equation of a line l that passes through the endpoint of the vector* **a**, *with the direction of a vector* **v** (see Figure 1.1.20).

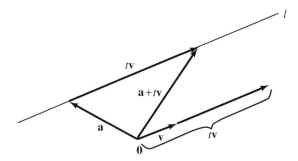

Figure 1.1.20 The line l, parametrically given by $\mathbf{l}(t) = \mathbf{a} + t\mathbf{v}$, lies in the direction **v** and passes through the tip of **a**.

As t varies through all real values, the points of the form $t\mathbf{v}$ are all scalar multiples of the vector **v**, and therefore exhaust the points of the line *passing through the origin* in the direction of **v**. Since every point on l is the endpoint of the diagonal of a parallelogram with sides **a** and $t\mathbf{v}$ for some real value of t, we see that all the points on l are of the form $\mathbf{a} + t\mathbf{v}$. Thus, the line l may be expressed by the equation $\mathbf{l}(t) = \mathbf{a} + t\mathbf{v}$. We say that l is expressed ***parametrically***, with t the parameter. At $t = 0, \mathbf{l}(t) = \mathbf{a}$. As t increases, the point $\mathbf{l}(t)$ moves away from **a** in the direction of **v**. As t decreases from $t = 0$ through negative values, $\mathbf{l}(t)$ moves away from **a** in the direction of $-\mathbf{v}$.

POINT-DIRECTION FORM OF A LINE The equation of the line l through the tip of **a** and pointing in the direction of the vector **v** is $\mathbf{l}(t) = \mathbf{a} + t\mathbf{v}$, where the parameter t takes on all real values. In coordinate form, the equations are

$$x = x_1 + at,$$
$$y = y_1 + bt,$$
$$z = z_1 + ct,$$

where $\mathbf{a} = (x_1, y_1, z_1)$ and $\mathbf{v} = (a, b, c)$. For lines in the xy plane, one simply drops the z component.

EXAMPLE 11   Determine the equation of the line l passing through $(1, 0, 0)$ in the direction **j**. See Figure 1.1.21.

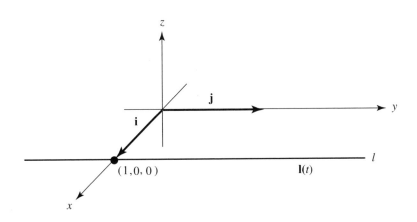

Figure 1.1.21   The line l passes through the tip of **i** in the direction **j**.

SOLUTION   The desired line can be expressed parametrically as $\mathbf{l}(t) = \mathbf{i} + t\mathbf{j}$. In terms of coordinates,

$$\mathbf{l}(t) = (1, 0, 0) + t(0, 1, 0) = (1, t, 0).\quad \blacktriangle$$

EXAMPLE 12

(a)   Find the equations of the line in space through the point $(3, -1, 2)$ in the direction $2\mathbf{i} - 3\mathbf{j} + 4\mathbf{k}$.

(b)   Find the equation of the line in the plane through the point $(1, -6)$ in the direction of $5\mathbf{i} - \pi\mathbf{j}$.

(c)   In what direction does the line $x = -3t + 2$, $y = -2(t - 1)$, $z = 8t + 2$ point?

SOLUTION

(a)   Here $\mathbf{a} = (3, -1, 2) = (x_1, y_1, z_1)$ and $\mathbf{v} = 2\mathbf{i} - 3\mathbf{j} + 4\mathbf{k}$, so $a = 2, b = -3$, and $c = 4$. From the box above, the equations are

$$x = 3 + 2t, \qquad y = -1 - 3t, \qquad z = 2 + 4t.$$

(b)   Here $\mathbf{a} = (1, -6)$ and $\mathbf{v} = 5\mathbf{i} - \pi\mathbf{j}$, so the line is

$$\mathbf{l}(t) = (1, -6) + (5t, -\pi t) = (1 + 5t, -6 - \pi t),$$

or

$$x = 1 + 5t, \qquad y = -6 - \pi t.$$

(c)   Using the preceding box, we construct the direction $\mathbf{v} = a\mathbf{i} + b\mathbf{j} + c\mathbf{k}$ from the coefficients of t: $a = -3, b = -2, c = 8$. Thus the line points in the direction of $\mathbf{v} = -3\mathbf{i} - 2\mathbf{j} + 8\mathbf{k}$. $\blacktriangle$

EXAMPLE 13 Do the two lines $(x, y, z) = (t, -6t + 1, 2t - 8)$ and $(x, y, z) = (3t + 1, 2t, 0)$ intersect?

SOLUTION If the lines intersect, there must be numbers t_1 and t_2 such that the corresponding points are equal:

$$(t_1, -6t_1 + 1, 2t_1 - 8) = (3t_2 + 1, 2t_2, 0);$$

that is, all three of the following equations hold:

$$t_1 = 3t_2 + 1,$$
$$-6t_1 + 1 = 2t_2,$$
$$2t_1 - 8 = 0.$$

From the third equation, $t_1 = 4$. The first equation then becomes $4 = 3t_2 + 1$ or $t_2 = 1$. We must check whether these values satisfy the middle equation:

$$-6t_1 + 1 \overset{?}{=} 2t_2.$$

Since $t_1 = 4$ and $t_2 = 1$, this reads

$$-24 + 1 \overset{?}{=} 2,$$

which is false, so the lines do not intersect. ▲

Notice that there can be many equations of the same line. Some may be obtained by choosing instead of **a** a different point on the given line, and forming the parametric equation of the line beginning at that point and in the direction of **v**. For example, the endpoint of $\mathbf{a} + \mathbf{v}$ is on the line $\mathbf{l}(t) = \mathbf{a} + t\mathbf{v}$, and thus, $\mathbf{l}_1(t) = (\mathbf{a} + \mathbf{v}) + t\mathbf{v}$ represents the same line. Still other equations may be obtained by observing that if $\alpha \neq 0$, the vector $\alpha\mathbf{v}$ has the same (or opposite) direction as **v**. Thus $\mathbf{l}_2(t) = \mathbf{a} + t\alpha\mathbf{v}$ is another equation of $\mathbf{l}(t) = \mathbf{a} + t\mathbf{v}$.

For example, both $\mathbf{l}(t) = (1, 0, 0) + (t, t, 0)$ and $\mathbf{l}_1(s) = (0, -1, 0) + (s, s, 0)$ represent the same line since both are in the direction $\mathbf{i} + \mathbf{j}$ and both pass through the point $(1, 0, 0)$; **l** passes through $(1, 0, 0)$ at $t = 0$ and $\mathbf{l}_1$ passes through $(1, 0, 0)$ at $s = 1$.

Therefore the equation of a line is not uniquely determined. Nevertheless, it is customary to use the term "the" equation. Keeping this in mind, let us derive *the equation of a line passing through the endpoints of two given vectors* **a** *and* **b**. Since the vector $\mathbf{b} - \mathbf{a}$ is parallel to the directed line segment from **a** to **b**, we calculate the parametric equation of the line passing through **a** in the direction of $\mathbf{b} - \mathbf{a}$ (Figure 1.1.22). Thus

$$\mathbf{l}(t) = \mathbf{a} + t(\mathbf{b} - \mathbf{a}); \qquad \text{that is,} \qquad \mathbf{l}(t) = (1 - t)\mathbf{a} + t\mathbf{b}.$$

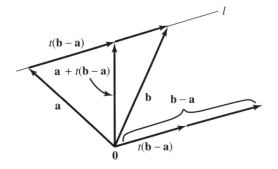

Figure 1.1.22   The line l, parametrically given by $\mathbf{l}(t) = \mathbf{a} + t(\mathbf{b} - \mathbf{a}) = (1 - t)\mathbf{a} + t\mathbf{b}$, passes through the tips of $\mathbf{a}$ and $\mathbf{b}$.

As t increases from 0 to 1, $t(\mathbf{b} - \mathbf{a})$ starts as the zero vector and increases in length (remaining in the direction of $\mathbf{b} - \mathbf{a}$) until at $t = 1$ it is the vector $\mathbf{b} - \mathbf{a}$. Thus for $\mathbf{l}(t) = \mathbf{a} + t(\mathbf{b} - \mathbf{a})$, as t increases from 0 to 1, the vector $\mathbf{l}(t)$ moves from the endpoint of $\mathbf{a}$ to the endpoint of $\mathbf{b}$ along the directed line segment from $\mathbf{a}$ to $\mathbf{b}$.

If $P = (x_1, y_1, z_1)$ is the tip of $\mathbf{a}$ and $Q = (x_2, y_2, z_2)$ is the tip of $\mathbf{b}$, then $\mathbf{v} = (x_2 - x_1)\mathbf{i} + (y_2 - y_1)\mathbf{j} + (z_2 - z_1)\mathbf{k}$, and so the equations of the line are

$$x = x_1 + (x_2 - x_1)t,$$

$$y = y_1 + (y_2 - y_1)t,$$

$$z = z_1 + (z_2 - z_1)t.$$

By eliminating t, these can be written as

$$\frac{x - x_1}{x_2 - x_1} = \frac{y - y_1}{y_2 - y_1} = \frac{z - z_1}{z_2 - z_1}.$$

PARAMETRIC EQUATION OF A LINE: POINT-POINT FORM
The parametric equations of the line l through the points $P = (x_1, y_1, z_1)$ and $Q = (x_2, y_2, z_2)$ are

$$x = x_1 + (x_2 - x_1)t,$$

$$y = y_1 + (y_2 - y_1)t,$$

$$z = z_1 + (z_2 - z_1)t,$$

where (x, y, z) is the general point of l, and the parameter t takes on all real values.

EXAMPLE 14   Find the equation of the line through $(2, 1, -3)$ and $(6, -1, -5)$.

SOLUTION Using the preceding box, we choose $(x_1, y_1, z_1) = (2, 1, -3)$ and $(x_2, y_2, z_2) = (6, -1, -5)$, so the equations are

$$x = 2 + (6 - 2)t = 2 + 4t,$$
$$y = 1 + (-1 - 1)t = 1 - 2t,$$
$$z = -3 + (-5 - (-3))t = -3 - 2t. \quad \blacktriangle$$

EXAMPLE 15 Find the equation of the line passing through $(-1, 1, 0)$ and $(0, 0, 1)$ (see Figure 1.1.23).

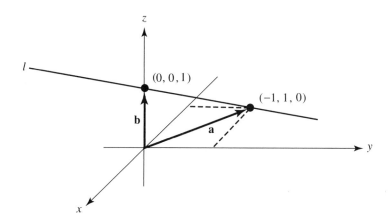

Figure 1.1.23 Finding the equation of the line through two points.

SOLUTION Letting $\mathbf{a} = -\mathbf{i} + \mathbf{j}$ and $\mathbf{b} = \mathbf{k}$ represent the given points, we have

$$\mathbf{l}(t) = (1 - t)(-\mathbf{i} + \mathbf{j}) + t\mathbf{k} = -(1 - t)\mathbf{i} + (1 - t)\mathbf{j} + t\mathbf{k}.$$

The equation of this line may thus be written as

$$\mathbf{l}(t) = (t - 1)\mathbf{i} + (1 - t)\mathbf{j} + t\mathbf{k},$$

or, equivalently, if $\mathbf{l}(t) = x\mathbf{i} + y\mathbf{j} + z\mathbf{k}$,

$$x = t - 1, \qquad y = 1 - t, \qquad z = t. \quad \blacktriangle$$

The description of a line *segment* requires that the domain of the parameter t be restricted.

EXAMPLE 16 Find the equation of the line segment between $(1, 1, 1)$ and $(2, 1, 2)$.

SOLUTION   The *line* through $(1, 1, 1)$ and $(2, 1, 2)$ is described in parametric form by $(x, y, z) = (1 + t, 1, 1 + t)$, as t takes on all real values. When $t = 0$, the point (x, y, z) is $(1, 1, 1)$, and when $t = 1$, the point (x, y, z) is $(2, 1, 2)$. Thus, the point (x, y, z) lies between $(1, 1, 1)$ and $(2, 1, 2)$ when $0 \leq t \leq 1$, so the line *segment* is described by the equations

$$x = 1 + t,$$

$$y = 1,$$

$$z = 1 + t,$$

together with the inequalities $0 \leq t \leq 1$.   ▲

We can also give parametric descriptions of geometric objects other than lines.

EXAMPLE   17   Describe the points that lie within the parallelogram whose adjacent sides are the vectors **a** and **b** based at the origin ("within" includes points on the edges of the parallelogram).

SOLUTION   Consider Figure 1.1.24. If P is any point within the given parallelogram and we construct lines l_1 and l_2 through P parallel to the vectors **a** and **b**, respectively, we see that l_1 intersects the side of the parallelogram determined by the vector **b** at some point $t\mathbf{b}$, where $0 \leq t \leq 1$. Likewise, l_2 intersects the side determined by the vector **a** at some point $s\mathbf{a}$, where $0 \leq s \leq 1$.

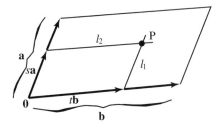

Figure 1.1.24   Describing points within the parallelogram formed by vectors **a** and **b**, with vertex **0**.

Note that P is the endpoint of the diagonal of a parallelogram having adjacent sides $s\mathbf{a}$ and $t\mathbf{b}$; hence, if **v** denotes the vector $\overrightarrow{OP}$, we see that $\mathbf{v} = s\mathbf{a} + t\mathbf{b}$. Thus, all the points in the given parallelogram are endpoints of vectors of the form $s\mathbf{a} + t\mathbf{b}$ for $0 \leq s \leq 1$ and $0 \leq t \leq 1$. Reversing our steps we see that all vectors of this form end within the parallelogram.   ▲

Since two different lines through the origin determine a plane through the origin, so do two nonparallel vectors. If we apply the same reasoning as in Example 17, we see that the entire plane formed by two nonparallel vectors **v** and **w** consists of all points of the form $s\mathbf{v} + t\mathbf{w}$ where s and t can be any real numbers, as in Figure 1.1.25.

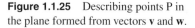
Figure 1.1.25 Describing points P in the plane formed from vectors **v** and **w**.

We have thus described the points in the plane by two parameters. For this reason, we say the plane is ***two-dimensional***. Similarly, a line is called ***one-dimensional*** whether it lies in the plane or in space or is the real number line itself.

The plane determined by **v** and **w** is called the plane ***spanned by*** **v** and **w**. When **v** is a scalar multiple of **w** and $\mathbf{w} \neq \mathbf{0}$, then **v** and **w** are parallel and the plane degenerates to a straight line. When $\mathbf{v} = \mathbf{w} = \mathbf{0}$ (that is, both are zero vectors), we obtain a single point.

There are three particular planes that arise naturally in a coordinate system and which will be useful to us later. We call the plane spanned by vectors **i** and **j** the xy plane, the plane spanned by **j** and **k** the yz plane, and the plane spanned by **i** and **k** the xz plane. These planes are illustrated in Figure 1.1.26.

Figure 1.1.26 The three coordinate planes.

HISTORICAL NOTE

Until around 1900, many scientists resisted the use of vectors in favor of more complicated objects called quaternions. The book that popularized vector methods was *Vector Analysis* by E. B. Wilson (reprinted by Dover in 1960), which was based on lectures delivered by J. W. Gibbs at Yale in 1899 and 1900. Wilson was reluctant to take Gibbs' course, since he had just completed a full-year course in quaternions at Harvard under J. M. Pierce, a champion of quaternionic methods, but was forced by a dean to add the course to his program. (For more details, see M. J. Crowe, *A History of Vector Analysis*, Dover, 1967, 1985.)

EXERCISES

(Exercises with shaded boxes are solved in the Study Guide.)

Complete the computations in Exercises 1 to 4.

1. $(-21, 23) - (?, 6) = (-25, ?)$

2. $3(133, -0.33, 0) + (-399, 0.99, 0) = (?, ?, ?)$

3. $(8a, -2b, 13c) = (52, 12, 11) + \frac{1}{2}(?, ?, ?)$

4. $(2, 3, 5) - 4\mathbf{i} + 3\mathbf{j} = (?, ?, ?)$

In Exercises 5 to 8, sketch the given vectors $\mathbf{v}$ and $\mathbf{w}$. On your sketch, draw in $-\mathbf{v}$, $\mathbf{v} + \mathbf{w}$, and $\mathbf{v} - \mathbf{w}$.

5. $\mathbf{v} = (2, 1)$ and $\mathbf{w} = (1, 2)$

6. $\mathbf{v} = (0, 4)$ and $\mathbf{w} = (2, -1)$

7. $\mathbf{v} = (2, 3, -6)$ and $\mathbf{w} = (-1, 1, 1)$

8. $\mathbf{v} = (2, 1, 3)$ and $\mathbf{w} = (-2, 0, -1)$

9. What restrictions must be made on x, y, and z so that the triple (x, y, z) will represent a point on the y axis? On the z axis? In the xz plane? In the yz plane?

10. (a) Generalize the geometric construction in Figure 1.1.7 to show that if $\mathbf{v}_1 = (x, y, z)$ and $\mathbf{v}_2 = (x', y', z')$ then $\mathbf{v}_1 + \mathbf{v}_2 = (x + x', y + y', z + z')$.
(b) Using an argument based on similar triangles, prove that $\alpha \mathbf{v} = (\alpha x, \alpha y, \alpha z)$ when $\mathbf{v} = (x, y, z)$.

In Exercises 11 to 17, use set theoretic or vector notation or both to describe the points that lie in the given configurations.

11. The plane spanned by $v_1 = (2, 7, 0)$ and $v_2 = (0, 2, 7)$

12. The plane spanned by $v_1 = (3, -1, 1)$ and $v_2 = (0, 3, 4)$

13. The line passing through $(-1, -1, -1)$ in the direction of j

14. The line passing through $(0, 2, 1)$ in the direction of $2i - k$

15. The line passing through $(-1, -1, -1)$ and $(1, -1, 2)$

16. The line passing through $(-5, 0, 4)$ and $(6, -3, 2)$

17. The parallelogram whose adjacent sides are the vectors $i + 3k$ and $-2j$

18. Find the points of intersection of the line $x = 3 + 2t, y = 7 + 8t, z = -2 + t$, that is, $l(t) = (3 + 2t, 7 + 8t, -2 + t)$, with the coordinate planes.

19. Show that there are no points (x, y, z) satisfying $2x - 3y + z - 2 = 0$ and lying on the line $v = (2, -2, -1) + t(1, 1, 1)$.

20. Show that every point on the line $v = (1, -1, 2) + t(2, 3, 1)$ satisfies $5x - 3y - z - 6 = 0$.

In Exercises 21 to 23, use vector methods to describe the given configurations.

21. The parallelepiped with edges the vectors a, b, and c emanating from the origin.

22. The points within the parallelogram with one corner at (x_0, y_0, z_0) whose sides extending from that corner are equal in magnitude and direction to vectors a and b.

23. The plane determined by the three points (x_0, y_0, z_0), (x_1, y_1, z_1), and (x_2, y_2, z_2).

Prove the statements in Exercises 24 to 26.

24. The line segment joining the midpoints of two sides of a triangle is parallel to and has half the length of the third side.

25. If PQR is a triangle in space and $b > 0$ is a number, then there is a triangle with sides parallel to those of PQR and side lengths b times those of PQR.

26. The medians of a triangle intersect at a point, and this point divides each median in a ratio of $2:1$.

Problems 27 and 28 require some knowledge of chemical notation.

27. Write the chemical equation $CO + H_2O = H_2 + CO_2$ as an equation in ordered triples (x_1, x_2, x_3) where x_1, x_2, x_3 are the number of carbon, hydrogen, and oxygen atoms, respectively, in each molecule.

28. (a) Write the chemical equation $pC_3H_4O_3 + qO_2 = rCO_2 + sH_2O$ as an equation in ordered triples with unknown coefficients p, q, r, and s.

(b) Find the smallest positive integer solution for p, q, r, and s.

(c) Illustrate the solution by a vector diagram in space.

29. Find a line that lies in the set defined by the equation $x^2 + y^2 - z^2 = 1$.

1.2 THE INNER PRODUCT, LENGTH, AND DISTANCE

In this section and the next we shall discuss two products of vectors: the inner product and the cross product. These are very useful in physical applications and have interesting geometric interpretations. The first product we shall consider is called the *inner product*. The name *dot product* is often used instead.

Suppose we have two vectors **a** and **b** in $\mathbb{R}^3$ (Figure 1.2.1) and we wish to determine the angle between them, that is, the smaller angle subtended by **a** and **b** in the plane that

Figure 1.2.1 θ is the angle between the vectors **a** and **b**.

they span. The inner product enables use to do this. Let us first develop the concept formally and then prove that this product does what we chaim. Let $\mathbf{a} = a_1\mathbf{i} + a_2\mathbf{j} + a_3\mathbf{k}$ and $\mathbf{b} = b_1\mathbf{i} + b_2\mathbf{j} + b_3\mathbf{k}$. We define the *inner product* of **a** and **b**, written $\mathbf{a} \cdot \mathbf{b}$, to be the real number

$$\mathbf{a} \cdot \mathbf{b} = a_1b_1 + a_2b_2 + a_3b_3.$$

Note that the inner product of two vectors is a scalar quantity. Sometimes the inner product is denoted $\langle \mathbf{a}, \mathbf{b} \rangle$. This is often done for typographic reasons. *Thus $\langle \mathbf{a}, \mathbf{b} \rangle$ and $\mathbf{a} \cdot \mathbf{b}$ mean exactly the same thing.*

EXAMPLE 1

(a) If $\mathbf{a} = 3\mathbf{i} + \mathbf{j} - 2\mathbf{k}$ and $\mathbf{b} = \mathbf{i} - \mathbf{j} + \mathbf{k}$, calculate $\mathbf{a} \cdot \mathbf{b}$.

(b) Calculate $(2\mathbf{i} + \mathbf{j} - \mathbf{k}) \cdot (3\mathbf{k} - 2\mathbf{j})$.

SOLUTION

(a) $\mathbf{a} \cdot \mathbf{b} = 3 \cdot 1 + 1 \cdot (-1) + (-2) \cdot 1 = 3 - 1 - 2 = 0.$

(b) $(2\mathbf{i} + \mathbf{j} - \mathbf{k}) \cdot (3\mathbf{k} - 2\mathbf{j}) = (2\mathbf{i} + \mathbf{j} - \mathbf{k}) \cdot (0\mathbf{i} - 2\mathbf{j} + 3\mathbf{k}) = 2 \cdot 0 - 1 \cdot 2 - 1 \cdot 3 = -5.$ ▲

Certain properties of the inner product follow from the definition. If $\mathbf{a}, \mathbf{b}$, and $\mathbf{c}$ are vectors in $\mathbb{R}^3$ and α and β are real numbers, then

(i) $\mathbf{a} \cdot \mathbf{a} \geq 0$;

   $\mathbf{a} \cdot \mathbf{a} = 0$ if and only if $\mathbf{a} = \mathbf{0}$.

(ii) $\alpha\mathbf{a} \cdot \mathbf{b} = \alpha(\mathbf{a} \cdot \mathbf{b})$ and $\mathbf{a} \cdot \beta\mathbf{b} = \beta(\mathbf{a} \cdot \mathbf{b})$.

(iii) $\mathbf{a} \cdot (\mathbf{b} + \mathbf{c}) = \mathbf{a} \cdot \mathbf{b} + \mathbf{a} \cdot \mathbf{c}$ and $(\mathbf{a} + \mathbf{b}) \cdot \mathbf{c} = \mathbf{a} \cdot \mathbf{c} + \mathbf{b} \cdot \mathbf{c}$.

(iv) $\mathbf{a} \cdot \mathbf{b} = \mathbf{b} \cdot \mathbf{a}$.

To prove the first of these properties, observe that if $\mathbf{a} = a_1\mathbf{i} + a_2\mathbf{j} + a_3\mathbf{k}$, then $\mathbf{a} \cdot \mathbf{a} = a_1^2 + a_2^2 + a_3^2$. Since a_1, a_2, and a_3 are real numbers, we know $a_1^2 \geq 0, a_2^2 \geq 0, a_3^2 \geq 0$. Thus $\mathbf{a} \cdot \mathbf{a} \geq 0$. Moreover, if $a_1^2 + a_2^2 + a_3^2 = 0$, then $a_1 = a_2 = a_3 = 0$; therefore $\mathbf{a} = \mathbf{0}$ (zero vector). The proofs of the other properties of the inner product are also easily obtained.

It follows from the Pythagorean theorem that the **length** of the vector $\mathbf{a} = a_1\mathbf{i} + a_2\mathbf{j} + a_3\mathbf{k}$ is $\sqrt{a_1^2 + a_2^2 + a_3^2}$ (see Figure 1.2.2). The length of the vector $\mathbf{a}$ is denoted by $\|\mathbf{a}\|$. This quantity is often called the **norm** of $\mathbf{a}$. Since $\mathbf{a} \cdot \mathbf{a} = a_1^2 + a_2^2 + a_3^2$, it follows that

$$\|\mathbf{a}\| = (\mathbf{a} \cdot \mathbf{a})^{1/2}.$$

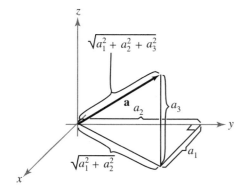

Figure 1.2.2   The length of the vector $\mathbf{a} = (a_1, a_2, a_3)$ is given by the Pythagorean formula: $\sqrt{a_1^2 + a_2^2 + a_3^2}$.

Vectors with norm 1 are called **unit vectors**. For example, the vectors $\mathbf{i}, \mathbf{j}, \mathbf{k}$ are unit vectors. Observe that for any nonzero vector $\mathbf{a}$, $\mathbf{a}/\|\mathbf{a}\|$ is a unit vector; when we divide $\mathbf{a}$ by $\|\mathbf{a}\|$, we say that we have **normalized** $\mathbf{a}$.

EXAMPLE 2

(a)   Normalize $\mathbf{v} = 2\mathbf{i} + 3\mathbf{j} - \frac{1}{2}\mathbf{k}$.

(b)   Find unit vectors $\mathbf{a}$, $\mathbf{b}$, and $\mathbf{c}$ in the plane such that $\mathbf{b} + \mathbf{c} = \mathbf{a}$.

SOLUTION

(a)   We have $\|\mathbf{v}\| = \sqrt{2^2 + 3^2 + (1/2)^2} = (1/2)\sqrt{53}$, so the normalization of $\mathbf{v}$ is

$$\mathbf{u} = \frac{1}{\|\mathbf{v}\|}\mathbf{v} = \frac{4}{\sqrt{53}}\mathbf{i} + \frac{6}{\sqrt{53}}\mathbf{j} - \frac{1}{\sqrt{53}}\mathbf{k}.$$

(b)   Since all three vectors are to have length 1, a triangle with sides $\mathbf{a}$, $\mathbf{b}$, and $\mathbf{c}$ must be equilateral as in Figure 1.2.3. Orienting the triangle as in the figure, we take $\mathbf{a} = \mathbf{i}$, then necessarily $\mathbf{b} = \frac{1}{2}\mathbf{i} + (\sqrt{3}/2)\mathbf{j}$, and $\mathbf{c} = \frac{1}{2}\mathbf{i} - (\sqrt{3}/2)\mathbf{j}$. You should check that $\|\mathbf{a}\| = \|\mathbf{b}\| = \|\mathbf{c}\| = 1$ and that $\mathbf{b} + \mathbf{c} = \mathbf{a}$.   ▲

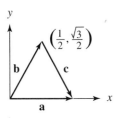

Figure 1.2.3   The vectors $\mathbf{a}$, $\mathbf{b}$, and $\mathbf{c}$ are represented by the sides of an equilateral triangle.

In the plane, define the vector $\mathbf{i}_\theta = (\cos\theta)\mathbf{i} + (\sin\theta)\mathbf{j}$, which is the unit vector making an angle θ with the x axis (see Figure 1.2.4).

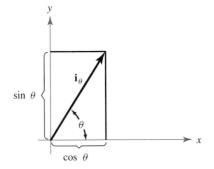

Figure 1.2.4   The coordinates of $\mathbf{i}_\theta$ are $\cos\theta$ and $\sin\theta$; it is a unit vector since $\cos^2\theta + \sin^2\theta = 1$.

If $\mathbf{a}$ and $\mathbf{b}$ are vectors, we have seen that the vector $\mathbf{b} - \mathbf{a}$ is parallel to and has the same magnitude as the directed line segment from the endpoint of $\mathbf{a}$ to the endpoint of $\mathbf{b}$. It follows that the distance from the endpoint of $\mathbf{a}$ to the endpoint of $\mathbf{b}$ is $\|\mathbf{b} - \mathbf{a}\|$ (see Figure 1.2.5).

Figure 1.2.5 The distance between the tips of $\mathbf{a}$ and $\mathbf{b}$ is $\|\mathbf{b} - \mathbf{a}\|$.

INNER PRODUCT, LENGTH, AND DISTANCE

Letting $\mathbf{a} = a_1\mathbf{i} + a_2\mathbf{j} + a_3\mathbf{k}$ and $\mathbf{b} = b_1\mathbf{i} + b_2\mathbf{j} + b_3\mathbf{k}$,

$$\mathbf{a} \cdot \mathbf{b} = a_1b_1 + a_2b_2 + a_3b_3,$$

$$\|\mathbf{a}\| = \sqrt{a_1^2 + a_2^2 + a_3^2}.$$

To **normalize** a vector $\mathbf{a}$, form the vector

$$\frac{\mathbf{a}}{\|\mathbf{a}\|}.$$

The **distance between** the endpoints of $\mathbf{a}$ and $\mathbf{b}$ is $\|\mathbf{a} - \mathbf{b}\|$, and the **distance between** P and Q is $\|\overrightarrow{PQ}\|$.

EXAMPLE 3 Find the distance from the endpoint of the vector $\mathbf{i}$, that is, the point $(1, 0, 0)$, to the endpoint of the vector $\mathbf{j}$, that is, the point $(0, 1, 0)$.

SOLUTION $\|\mathbf{j} - \mathbf{i}\| = \sqrt{(0 - 1)^2 + (1 - 0)^2 + (0 - 0)^2} = \sqrt{2}.$ ▲

The Angle Between Two Vectors

Let us now show that the inner product does indeed measure the angle between two vectors.

THEOREM 1 *Let $\mathbf{a}$ and $\mathbf{b}$ be two vectors in $\mathbb{R}^3$ and let θ, where $0 \leq \theta \leq \pi$, be the angle between them (Figure 1.2.6). Then*

$$\mathbf{a} \cdot \mathbf{b} = \|\mathbf{a}\|\|\mathbf{b}\|\cos\theta.$$

It follows from this vector that if **a** and **b** are nonzero, we may express the angle between them as

$$\theta = \cos^{-1}\left(\frac{\mathbf{a}\cdot\mathbf{b}}{\|\mathbf{a}\|\,\|\mathbf{b}\|}\right).$$

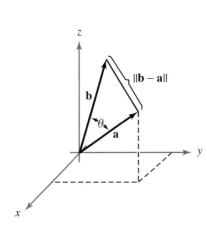

Figure 1.2.6 The vectors **a**, **b** and the angle θ between them; the geometry for Theorem 1 and its proof.

PROOF If we apply the law of cosines from trigonometry to the triangle with one vertex at the origin and adjacent sides determined by the vectors **a** and **b** (as in the figure), it follows that

$$\|\mathbf{b}-\mathbf{a}\|^2 = \|\mathbf{a}\|^2 + \|\mathbf{b}\|^2 - 2\|\mathbf{a}\|\,\|\mathbf{b}\|\cos\theta.$$

Since $\|\mathbf{b}-\mathbf{a}\|^2 = (\mathbf{b}-\mathbf{a})\cdot(\mathbf{b}-\mathbf{a})$, $\|\mathbf{a}\|^2 = \mathbf{a}\cdot\mathbf{a}$, and $\|\mathbf{b}\|^2 = \mathbf{b}\cdot\mathbf{b}$, we can rewrite the above equation as

$$(\mathbf{b}-\mathbf{a})\cdot(\mathbf{b}-\mathbf{a}) = \mathbf{a}\cdot\mathbf{a} + \mathbf{b}\cdot\mathbf{b} - 2\|\mathbf{a}\|\,\|\mathbf{b}\|\cos\theta.$$

Now

$$\begin{aligned}
(\mathbf{b}-\mathbf{a})\cdot(\mathbf{b}-\mathbf{a}) &= \mathbf{b}\cdot(\mathbf{b}-\mathbf{a}) - \mathbf{a}\cdot(\mathbf{b}-\mathbf{a})\\
&= \mathbf{b}\cdot\mathbf{b} - \mathbf{b}\cdot\mathbf{a} - \mathbf{a}\cdot\mathbf{b} + \mathbf{a}\cdot\mathbf{a}\\
&= \mathbf{a}\cdot\mathbf{a} + \mathbf{b}\cdot\mathbf{b} - 2\mathbf{a}\cdot\mathbf{b}.
\end{aligned}$$

Thus,

$$\mathbf{a}\cdot\mathbf{a} + \mathbf{b}\cdot\mathbf{b} - 2\mathbf{a}\cdot\mathbf{b} = \mathbf{a}\cdot\mathbf{a} + \mathbf{b}\cdot\mathbf{b} - 2\|\mathbf{a}\|\,\|\mathbf{b}\|\cos\theta.$$

That is,

$$\mathbf{a}\cdot\mathbf{b} = \|\mathbf{a}\|\,\|\mathbf{b}\|\cos\theta. \quad \blacksquare$$

EXAMPLE 4 Find the angle between the vectors $\mathbf{i}+\mathbf{j}+\mathbf{k}$ and $\mathbf{i}+\mathbf{j}-\mathbf{k}$ (Figure 1.2.7).

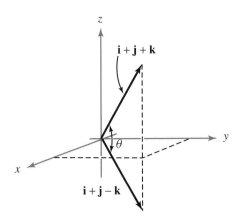

Figure 1.2.7 Finding the angle between $\mathbf{a} = \mathbf{i} + \mathbf{j} + \mathbf{k}$ and $\mathbf{b} = \mathbf{i} + \mathbf{j} - \mathbf{k}$.

SOLUTION   Using Theorem 1, we have

$$(\mathbf{i} + \mathbf{j} + \mathbf{k}) \cdot (\mathbf{i} + \mathbf{j} - \mathbf{k}) = \|\mathbf{i} + \mathbf{j} + \mathbf{k}\| \|\mathbf{i} + \mathbf{j} - \mathbf{k}\| \cos\theta$$

and so

$$1 + 1 - 1 = (\sqrt{3})(\sqrt{3})\cos\theta.$$

Hence,

$$\cos\theta = \tfrac{1}{3}.$$

That is,

$$\theta = \cos^{-1}(\tfrac{1}{3}) \approx 1.23 \text{ radians } (71°). \quad \blacktriangle$$

The Cauchy-Schwarz Inequality

Theorem 1 shows that the inner product of two vectors is the product of their lengths times the cosine of the angle between them. This relationship is often of value in problems of a geometric nature. An important consequence of Theorem 1 is:

COROLLARY: CAUCHY-SCHWARZ INEQUALITY   *For any two vectors* $\mathbf{a}$ *and* $\mathbf{b}$*, we have*

$$|\mathbf{a} \cdot \mathbf{b}| \le \|\mathbf{a}\| \|\mathbf{b}\|$$

with equality if and only if $\mathbf{a}$ *is a scalar multiple of* $\mathbf{b}$*, or one of them is* $\mathbf{0}$*.*

PROOF   If **a** is not a scalar multiple of **b**, then $|\cos\theta| < 1$ and so the inequality holds; in fact, if **a** and **b** are both nonzero, *strict* inequality holds in this case. When **a** is a scalar multiple of **b**, then $\theta = 0$ or π and $|\cos\theta| = 1$, so equality holds in this case.   ■

EXAMPLE 5   Verify the Cauchy-Schwarz inequality for $\mathbf{a} = -\mathbf{i} + \mathbf{j} + \mathbf{k}$ and $\mathbf{b} = 3\mathbf{i} + \mathbf{k}$.

SOLUTION   The dot product is $\mathbf{a} \cdot \mathbf{b} = -3 + 0 + 1 = -2$, so $|\mathbf{a} \cdot \mathbf{b}| = 2$. Also, $\|\mathbf{a}\| = \sqrt{1+1+1} = \sqrt{3}$ and $\|\mathbf{b}\| = \sqrt{9+1} = \sqrt{10}$, and it is true that $2 \leq \sqrt{3} \cdot \sqrt{10}$ because $\sqrt{3} \cdot \sqrt{10} > \sqrt{3} \cdot \sqrt{3} = 3 \geq 2$.   ▲

If **a** and **b** are nonzero vectors in $\mathbb{R}^3$ and θ is the angle between them, we see that $\mathbf{a} \cdot \mathbf{b} = 0$ if and only if $\cos\theta = 0$. Thus, *the inner product of two nonzero vectors is zero if and only if the vectors are perpendicular*. Hence, the inner product provides us with a convenient method for determining whether two vectors are perpendicular. Often we say that perpendicular vectors are *orthogonal*. The standard basis vectors **i**, **j**, and **k** are mutually orthogonal and of length 1; any such system is called *orthonormal*. We shall adopt the convention that the zero vector is orthogonal to all vectors.

EXAMPLE 6   The vectors $\mathbf{i}_\theta = (\cos\theta)\mathbf{i} + (\sin\theta)\mathbf{j}$ and $\mathbf{j}_\theta = -(\sin\theta)\mathbf{i} + (\cos\theta)\mathbf{j}$ are orthogonal, since

$$\mathbf{i}_\theta \cdot \mathbf{j}_\theta = -\cos\theta \sin\theta + \sin\theta \cos\theta = 0$$

(see Figure 1.2.8).   ▲

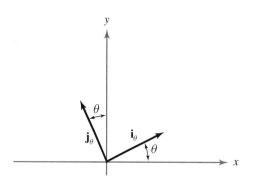

Figure 1.2.8   The vectors $\mathbf{i}_\theta$ and $\mathbf{j}_\theta$ are orthogonal and of unit length, that is, they are orthonormal.

EXAMPLE 7   Let **a** and **b** be two nonzero orthogonal vectors. If **c** is a vector in the plane spanned by **a** and **b**, then there are scalars α and β such that $\mathbf{c} = \alpha\mathbf{a} + \beta\mathbf{b}$. Use the inner product to determine α and β (see Figure 1.2.9).

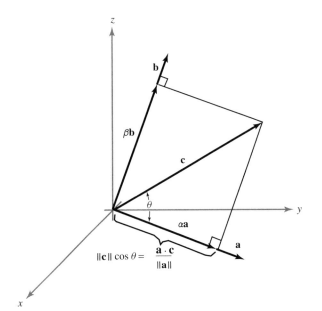

Figure 1.2.9 The geometry for finding α and β where $\mathbf{c} = \alpha\mathbf{a} + \beta\mathbf{b}$.

SOLUTION   Taking the inner product of $\mathbf{a}$ and $\mathbf{c}$, we have

$$\mathbf{a} \cdot \mathbf{c} = \mathbf{a} \cdot (\alpha\mathbf{a} + \beta\mathbf{b}) = \alpha\mathbf{a} \cdot \mathbf{a} + \beta\mathbf{a} \cdot \mathbf{b}.$$

Since $\mathbf{a}$ and $\mathbf{b}$ are orthogonal, $\mathbf{a} \cdot \mathbf{b} = 0$, and so

$$\alpha = \frac{\mathbf{a} \cdot \mathbf{c}}{\mathbf{a} \cdot \mathbf{a}} = \frac{\mathbf{a} \cdot \mathbf{c}}{\|\mathbf{a}\|^2}.$$

Similarly,

$$\beta = \frac{\mathbf{b} \cdot \mathbf{c}}{\mathbf{b} \cdot \mathbf{b}} = \frac{\mathbf{b} \cdot \mathbf{c}}{\|\mathbf{b}\|^2}. \quad \blacktriangle$$

Orthogonal Projection

In the preceding example, the vector $\alpha\mathbf{a}$ is called the *projection* of $\mathbf{c}$ along $\mathbf{a}$, and $\beta\mathbf{b}$ is its *projection* along $\mathbf{b}$. Let us formulate this idea more generally. If $\mathbf{v}$ is a vector, and l is the line through the origin in the direction of a vector $\mathbf{a}$, then the ***orthogonal projection*** of $\mathbf{v}$ ***on a*** is the vector whose tip is obtained by dropping a perpendicular line to l from the tip of $\mathbf{v}$, as in Figure 1.2.10.

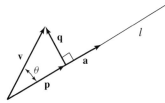

Figure 1.2.10 $\mathbf{p}$ is the orthogonal projection of $\mathbf{v}$ on $\mathbf{a}$.

Referring to the figure, we see that **p** is a multiple of **a** and that **v** is the sum of **p** and a vector **q** perpendicular to **a**. Thus,

$$\mathbf{v} = c\mathbf{a} + \mathbf{q},$$

where $\mathbf{p} = c\mathbf{a}$ and $\mathbf{a} \cdot \mathbf{q} = 0$. Taking the dot product of **a** with both sides of $\mathbf{v} = c\mathbf{a} + \mathbf{q}$, we find $\mathbf{a} \cdot \mathbf{v} = c\mathbf{a} \cdot \mathbf{a}$, so $c = (\mathbf{a} \cdot \mathbf{v})/(\mathbf{a} \cdot \mathbf{a})$, and hence

$$\mathbf{p} = \frac{\mathbf{a} \cdot \mathbf{v}}{\|\mathbf{a}\|^2}\, \mathbf{a}.$$

The length of **p** is

$$\|\mathbf{p}\| = \frac{|\mathbf{a} \cdot \mathbf{v}|}{\|\mathbf{a}\|^2}\, \|\mathbf{a}\| = \frac{|\mathbf{a} \cdot \mathbf{v}|}{\|\mathbf{a}\|} = \|\mathbf{v}\|\cos\theta.$$

ORTHOGONAL PROJECTION   The *orthogonal projection* of **v** on **a** is the vector

$$\mathbf{p} = \frac{\mathbf{a} \cdot \mathbf{v}}{\|\mathbf{a}\|^2}\, \mathbf{a}.$$

EXAMPLE 8   Find the orthogonal projection of $\mathbf{i} + \mathbf{j}$ on $\mathbf{i} - 2\mathbf{j}$.

SOLUTION   With $\mathbf{a} = \mathbf{i} - 2\mathbf{j}$ and $\mathbf{v} = \mathbf{i} + \mathbf{j}$, the orthogonal projection of **v** on **a** is

$$\frac{\mathbf{a} \cdot \mathbf{v}}{\mathbf{a} \cdot \mathbf{a}}\, \mathbf{a} = \frac{1-2}{1+4}\,(\mathbf{i} - 2\mathbf{j}) = -\frac{1}{5}\,(\mathbf{i} - 2\mathbf{j})$$

(see Figure 1.2.11).   ▲

Orthogonal projection of **v** on **a**

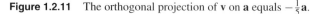

Figure 1.2.11   The orthogonal projection of **v** on **a** equals $-\frac{1}{5}\mathbf{a}$.

The Triangle Inequality

A useful consequence of the Cauchy-Schwarz inequality, called the *triangle inequality*, relates the lengths of vectors **a** and **b** and of their sum $\mathbf{a} + \mathbf{b}$.

Geometrically, it says that the length of any side of a triangle is no greater than the sum of the lengths of the other two (see Figure 1.2.12).

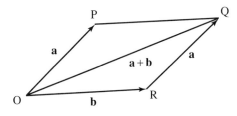

Figure 1.2.12 This geometry shows that $\|OQ\| \leq \|OR\| + \|RQ\|$, or, in vector notation, that $\|\mathbf{a} + \mathbf{b}\| \leq \|\mathbf{a}\| + \|\mathbf{b}\|$, which is the triangle inequality.

THEOREM 2: TRIANGLE INEQUALITY *For vectors **a** and **b** in space,*

$$\|\mathbf{a} + \mathbf{b}\| \leq \|\mathbf{a}\| + \|\mathbf{b}\|.$$

PROOF   We consider the square of the left-hand side:

$$\|\mathbf{a} + \mathbf{b}\|^2 = (\mathbf{a} + \mathbf{b}) \cdot (\mathbf{a} + \mathbf{b}) = \|\mathbf{a}\|^2 + 2\mathbf{a} \cdot \mathbf{b} + \|\mathbf{b}\|^2.$$

By the Cauchy-Schwarz inequality, we have

$$\|\mathbf{a}\|^2 + 2\mathbf{a} \cdot \mathbf{b} + \|\mathbf{b}\|^2 \leq \|\mathbf{a}\|^2 + 2\|\mathbf{a}\|\|\mathbf{b}\| + \|\mathbf{b}\|^2 = (\|\mathbf{a}\| + \|\mathbf{b}\|)^2.$$

Thus,

$$\|\mathbf{a} + \mathbf{b}\|^2 \leq (\|\mathbf{a}\| + \|\mathbf{b}\|)^2;$$

taking square roots proves the result.   ∎

EXAMPLE 9

(a)   Verify the triangle inequality for $\mathbf{a} = \mathbf{i} + \mathbf{j}$ and $\mathbf{b} = 2\mathbf{i} + \mathbf{j} + \mathbf{k}$.

(b)   Prove that $\|\mathbf{u} - \mathbf{v}\| \leq \|\mathbf{u} - \mathbf{w}\| + \|\mathbf{w} - \mathbf{v}\|$ for any vectors $\mathbf{u}$, $\mathbf{v}$, and $\mathbf{w}$. Illustrate with a figure in which $\mathbf{u}$, $\mathbf{v}$, and $\mathbf{w}$ have the same base point.

SOLUTION

(a)   We have $\mathbf{a} + \mathbf{b} = 3\mathbf{i} + 2\mathbf{j} + \mathbf{k}$, so $\|\mathbf{a} + \mathbf{b}\| = \sqrt{9 + 4 + 1} = \sqrt{14}$. On the other hand, $\|\mathbf{a}\| = \sqrt{2}$ and $\|\mathbf{b}\| = \sqrt{6}$, so the triangle inequality asserts that $\sqrt{14} \leq \sqrt{2} + \sqrt{6}$. The numbers bear this out: $\sqrt{14} \approx 3.74$, while $\sqrt{2} + \sqrt{6} \approx 1.41 + 2.45 = 3.86$.

(b)   We find that $\mathbf{u} - \mathbf{v} = (\mathbf{u} - \mathbf{w}) + (\mathbf{w} - \mathbf{v})$, so the result follows from the triangle inequality with $\mathbf{a}$ replaced by $\mathbf{u} - \mathbf{w}$ and $\mathbf{b}$ replaced by $\mathbf{w} - \mathbf{v}$. Geometrically, we are considering the shaded triangle in Figure 1.2.13.   ▲

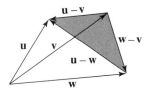

Figure 1.2.13   Illustrating the inequality $\|\mathbf{u} - \mathbf{v}\| \le \|\mathbf{u} - \mathbf{w}\| + \|\mathbf{w} - \mathbf{v}\|$.

Physical Applications of Vectors

A simple example of a physical quantity represented by a vector is a displacement. Suppose that, on a part of the earth's surface small enough to be considered flat, we introduce coordinates so that the x axis points east, the y axis points north, and the unit of length is the kilometer. If we are at a point P and wish to get to a point Q, the ***displacement vector*** $\mathbf{d} = \overrightarrow{PQ}$ joining P to Q tells us the direction and distance we have to travel. If x and y are the components of this vector, the displacement of P to Q is "x kilometers east, y kilometers north."

EXAMPLE 10   Suppose that two navigators who cannot see one another but can communicate by radio wish to determine the relative position of their ships. Explain how they can do this if they can each determine their displacement vector to the same lighthouse.

SOLUTION   Let P_1 and P_2 be the positions of the ships, and let Q be the position of the lighthouse. The displacement of the lighthouse from the ith ship is the vector $\mathbf{d}_i$ joining P_i to Q. The displacement of the second ship from the first is the vector $\mathbf{d}$ joining P_1 to P_2. We have $\mathbf{d} + \mathbf{d}_2 = \mathbf{d}_1$ (Figure 1.2.14), and so $\mathbf{d} = \mathbf{d}_1 - \mathbf{d}_2$. That is, the displacement from one ship to the other is the difference between the displacements from the ships to the lighthouse.   ▲

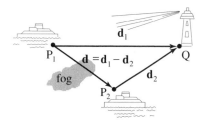

Figure 1.2.14   Vector methods can be used to locate objects.

We can also represent the velocity of a moving object as a vector. For the moment, we will consider only objects moving at uniform speed along straight lines. Suppose, for example, that a boat is steaming across a lake at 10 kilometers per hour (km/h) in the northeast direction. After 1 hour of travel, the displacement is $(10/\sqrt{2}, 10/\sqrt{2}) \approx (7.07, 7.07)$; see Figure 1.2.15.

position after 1 h

10

initial
position

45°

$\dfrac{10}{\sqrt{2}}$

$\dfrac{10}{\sqrt{2}}$

Figure 1.2.15 If an object moves northeast at 10 km/h, its velocity vector has components $(10/\sqrt{2}, 10/\sqrt{2}) = 10(1/\sqrt{2}, 1/\sqrt{2})$ where $(1/\sqrt{2}, 1/\sqrt{2})$ is the unit vector in the northeast direction.

The vector whose components are $(10/\sqrt{2}, 10/\sqrt{2})$ is called the *velocity vector* of the boat. In general, if an object is moving uniformly along a straight line, *its **velocity vector** is the displacement vector from the position at any moment to the position 1 unit of time later.* If a current appears on the lake, moving due eastward at 2 km/h, and the boat continues to point in the same direction with its engine running at the same rate, its displacement after 1 hour will have components given by $(10/\sqrt{2} + 2, 10/\sqrt{2})$; see Figure 1.2.16. The new velocity vector, therefore, has components $(10/\sqrt{2} + 2, 10/\sqrt{2})$. We note that this is the sum of the original velocity vector $(10/\sqrt{2}, 10/\sqrt{2})$ of the boat and the velocity vector $(2, 0)$ of the current.

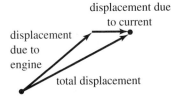

displacement due
to current

displacement
due to
engine

total displacement

Figure 1.2.16 The total displacement is the sum of the displacements due to the engine and the current.

DISPLACEMENT AND VELOCITY If an object has a (constant) velocity vector **v**, then in t units of time the resulting displacement vector of the object is $\mathbf{d} = t\mathbf{v}$; thus, after time $t = 1$, the displacement vector equals the velocity vector. See Figure 1.2.17.

v

$\mathbf{d} = t\mathbf{v}$

displacement in time t

Figure 1.2.17 Displacement = time × velocity.

EXAMPLE 11 A bird is flying in a straight line with velocity vector $10\mathbf{i} + 6\mathbf{j} + \mathbf{k}$ (in kilometers per hour). Suppose that (x, y) are its coordinates on the ground and z is its height above the ground.

(a) If the bird is at position $(1, 2, 3)$ at a certain moment, where is it 1 hour later? 1 minute later?

(b) How many seconds does it take the bird to climb 10 meters?

SOLUTION (a) The displacement vector from $(1, 2, 3)$ after 1 hour is $10\mathbf{i} + 6\mathbf{j} + \mathbf{k}$, so the new position is $(1, 2, 3) + (10, 6, 1) = (11, 8, 4)$. After 1 minute, the displacement vector from $(1, 2, 3)$ is $\frac{1}{60}(10\mathbf{i}, 6\mathbf{j}, \mathbf{k}) = \frac{1}{6}\mathbf{i} + \frac{1}{10}\mathbf{j} + \frac{1}{60}\mathbf{k}$, and so the new position is $(1, 2, 3) + (\frac{1}{6}, \frac{1}{10}, \frac{1}{60}) = (\frac{7}{6}, \frac{21}{10}, \frac{181}{60})$.

(b) After t seconds $(= t/3600 \text{ h})$, the displacement vector from $(1, 2, 3)$ is $(t/3600)$ $(10\mathbf{i} + 6\mathbf{j} + \mathbf{k}) = (t/360)\mathbf{i} + (t/600)\mathbf{j} + (t/3600)\mathbf{k}$. The increase in altitude is the z component $t/3600$. This will equal 10 m $(= \frac{1}{100} \text{ km})$ when $t/3600 = \frac{1}{100}$, that is, when $t = 36$ s. ▲

EXAMPLE 12 Physical forces have magnitude and direction and may thus be represented by vectors. If several forces act at once on an object, the resultant force is represented by the sum of the individual force vectors. Suppose that forces $\mathbf{i} + \mathbf{k}$ and $\mathbf{j} + \mathbf{k}$ are acting on a body. What third force $\mathbf{F}$ must we impose to counteract the two—that is, to make the total force equal to zero?

SOLUTION The force $\mathbf{F}$ should be chosen so that $(\mathbf{i} + \mathbf{k}) + (\mathbf{j} + \mathbf{k}) + \mathbf{F} = \mathbf{0}$; that is, $\mathbf{F} = -(\mathbf{i} + \mathbf{k}) - (\mathbf{j} + \mathbf{k}) = -\mathbf{i} - \mathbf{j} - 2\mathbf{k}$. (Recall that $\mathbf{0}$ is the *zero vector*, the vector whose components are all zero.) ▲

EXERCISES

1. Calculate $(3\mathbf{i} + 2\mathbf{j} + \mathbf{k}) \cdot (\mathbf{i} + 2\mathbf{j} - \mathbf{k})$.

2. Calculate $\mathbf{a} \cdot \mathbf{b}$ where $\mathbf{a} = 2\mathbf{i} + 10\mathbf{j} - 12\mathbf{k}$ and $\mathbf{b} = -3\mathbf{i} + 4\mathbf{k}$.

3. Find the angle between $7\mathbf{j} + 19\mathbf{k}$ and $-2\mathbf{i} - \mathbf{j}$ (to the nearest degree).

4. Compute $\mathbf{u} \cdot \mathbf{v}$, where $\mathbf{u} = \sqrt{3}\mathbf{i} - 315\mathbf{j} + 22\mathbf{k}$ and $\mathbf{v} = \mathbf{u}/\|\mathbf{u}\|$.

5. Is $\|8\mathbf{i} - 12\mathbf{k}\| \cdot \|6\mathbf{j} + \mathbf{k}\| - |(8\mathbf{i} - 12\mathbf{k}) \cdot (6\mathbf{j} + \mathbf{k})|$ equal to zero? Explain.

In Exercises 6 to 11, compute $\|\mathbf{u}\|$, $\|\mathbf{v}\|$, *and* $\mathbf{u} \cdot \mathbf{v}$ *for the given vectors in* $\mathbb{R}^3$.

6. $\mathbf{u} = 15\mathbf{i} - 2\mathbf{j} + 4\mathbf{k}, \mathbf{v} = \pi\mathbf{i} + 3\mathbf{j} - \mathbf{k}$

7. $\mathbf{u} = 2\mathbf{j} - \mathbf{i}, \mathbf{v} = -\mathbf{j} + \mathbf{i}$

8. $\mathbf{u} = 5\mathbf{i} - \mathbf{j} + 2\mathbf{k}, \mathbf{v} = \mathbf{i} + \mathbf{j} - \mathbf{k}$

9. $\mathbf{u} = -\mathbf{i} + 3\mathbf{j} + \mathbf{k}$, $\mathbf{v} = -2\mathbf{i} - 3\mathbf{j} - 7\mathbf{k}$

10. $\mathbf{u} = -\mathbf{i} + 3\mathbf{k}$, $\mathbf{v} = 4\mathbf{j}$

11. $\mathbf{u} = -\mathbf{i} + 2\mathbf{j} - 3\mathbf{k}$, $\mathbf{v} = -\mathbf{i} - 3\mathbf{j} + 4\mathbf{k}$

12. Normalize the vectors in Exercises 6 to 8. (Only the solution corresponding to Exercise 7 is in the Student Guide.)

13. Find the angle between the vectors in Exercises 9 to 11. If necessary, express your answer in terms of $\cos^{-1}$.

14. Find the projection of $\mathbf{u} = -\mathbf{i} + \mathbf{j} + \mathbf{k}$ onto $\mathbf{v} = 2\mathbf{i} + \mathbf{j} - 3\mathbf{k}$.

15. Find the projection of $\mathbf{v} = 2\mathbf{i} + \mathbf{j} - 3\mathbf{k}$ onto $\mathbf{u} = -\mathbf{i} + \mathbf{j} + \mathbf{k}$.

16. What restrictions must be made on b so that the vector $2\mathbf{i} + b\mathbf{j}$ is orthogonal to (a) $-3\mathbf{i} + 2\mathbf{j} + \mathbf{k}$ and (b) $\mathbf{k}$?

17. Find two nonparallel vectors both orthogonal to $(1, 1, 1)$.

18. Find the line through $(3, 1, -2)$ that intersects and is perpendicular to the line $x = -1 + t$, $y = -2 + t$, $z = -1 + t$. [HINT: If (x_0, y_0, z_0) is the point of intersection, find its coordinates.]

19. A ship at position $(1, 0)$ on a nautical chart (with north in the positive y direction) sights a rock at position $(2, 4)$. What is the vector joining the ship to the rock? What angle θ does this vector make with due north? (This is called the *bearing* of the rock from the ship.)

20. Suppose that the ship in Exercise 19 is pointing due north and traveling at a speed of 4 knots relative to the water. There is a current flowing due east at 1 knot. The units on the chart are nautical miles; 1 knot = 1 nautical mile per hour.
   (a) If there were no current, what vector $\mathbf{u}$ would represent the velocity of the ship relative to the sea bottom? (b) If the ship were just drifting with the current, what vector $\mathbf{v}$ would represent its velocity relative to the sea bottom? (c) What vector $\mathbf{w}$ represents the total velocity of the ship? (d) Where would the ship be after 1 hour? (e) Should the captain change course? (f) What if the rock were an iceberg?

21. An airplane is located at position $(3, 4, 5)$ at noon and traveling with velocity $400\mathbf{i} + 500\mathbf{j} - \mathbf{k}$ kilometers per hour. The pilot spots an airport at position $(23, 29, 0)$.
   (a)   At what time will the plane pass directly over the airport? (Assume that the earth is flat and that the vector $\mathbf{k}$ points straight up.)
   (b)   How high above the airport will the plane be when it passes?

22. The wind velocity $\mathbf{v}_1$ is 40 miles per hour (mi/h) from east to west while an airplane travels with air speed $\mathbf{v}_2$ of 100 mi/h due north. The speed of the airplane relative to the earth is the vector sum $\mathbf{v}_1 + \mathbf{v}_2$.
   (a)   Find $\mathbf{v}_1 + \mathbf{v}_2$. (b)   Draw a figure to scale.

23. A force of 50 lb is directed $50°$ above horizontal, pointing to the right. Determine its horizontal and vertical components. Display all results in a figure.

24. Two persons pull horizontally on ropes attached to a post, the angle between the ropes being $60°$. Person A pulls with a force of 150 lb, while B pulls with a force of 110 lb.

(a) The resultant force is the vector sum of the two forces in a conveniently chosen coordinate system. Draw a figure to scale which graphically represents the three forces.

(b) Using trigonometry, determine formulas for the vector components of the two forces in a conveniently chosen coordinate system. Perform the algebraic addition, and find the angle the resultant force makes with A.

25. A 1-kilogram (1-kg) mass located at the origin is suspended by ropes attached to the two points $(1, 1, 1)$ and $(-1, -1, 1)$. If the force of gravity is pointing in the direction of the vector $-\mathbf{k}$, what is the vector describing the force along each rope? [HINT: Use the symmetry of the problem. A 1-kg mass weighs 9.8 newtons (N).]

26. Suppose that an object moving in direction $\mathbf{i} + \mathbf{j}$ is acted on by a force given by the vector $2\mathbf{i} + \mathbf{j}$. Express this force as a sum of a force in the direction of motion and a force perpendicular to the direction of motion.

27. A force of 6 N (newtons) makes an angle of $\pi/4$ radian with the y axis, pointing to the right. The force acts against the movement of an object along the straight line connecting $(1, 2)$ to $(5, 4)$. (a) Find a formula for the force vector $\mathbf{F}$. (b) Find the angle θ between the displacement direction $\mathbf{D} = (5 - 1)\mathbf{i} + (4 - 2)\mathbf{j}$ and the force direction $\mathbf{F}$. (c) The *work* done is $\mathbf{F} \cdot \mathbf{D}$, or equivalently, $\|\mathbf{F}\| \|\mathbf{D}\| \cos\theta$. Compute the work from both formulas and compare.

1.3 MATRICES, DETERMINANTS, AND THE CROSS PRODUCT

In Section 1.2 we defined a product of vectors that was a scalar. In this section we shall define a product of vectors that is a vector; that is, we shall show how, given two vectors $\mathbf{a}$ and $\mathbf{b}$, we can produce a third vector $\mathbf{a} \times \mathbf{b}$, called the *cross product* of $\mathbf{a}$ and $\mathbf{b}$. This new vector will have the pleasing geometric property that it is perpendicular to the plane spanned (determined) by $\mathbf{a}$ and $\mathbf{b}$. The definition of the cross product is based on the notions of the matrix and the determinant, and so these are developed first. Once this has been accomplished, we can study the geometric implications of the mathematical structure we have built.

We define a 2×2 *matrix* to be an array

$$\begin{bmatrix} a_{11} & a_{12} \\ a_{21} & a_{22} \end{bmatrix}$$

where a_{11}, a_{12}, a_{21}, and a_{22} are four scalars. For example,

$$\begin{bmatrix} 2 & 1 \\ 0 & 4 \end{bmatrix}, \qquad \begin{bmatrix} -1 & 0 \\ 1 & 1 \end{bmatrix}, \qquad \text{and} \qquad \begin{bmatrix} 13 & 7 \\ 6 & 11 \end{bmatrix}$$

are 2×2 matrices. The *determinant*

$$\begin{vmatrix} a_{11} & a_{12} \\ a_{21} & a_{22} \end{vmatrix}$$

of such a matrix is the real number defined by the equation

$$\begin{vmatrix} a_{11} & a_{12} \\ a_{21} & a_{22} \end{vmatrix} = a_{11}a_{22} - a_{12}a_{21}. \tag{1}$$

E X A M P L E   1

$$\begin{vmatrix} 1 & 1 \\ 1 & 1 \end{vmatrix} = 1 - 1 = 0; \qquad \begin{vmatrix} 1 & 2 \\ 3 & 4 \end{vmatrix} = 4 - 6 = -2; \qquad \begin{vmatrix} 5 & 6 \\ 7 & 8 \end{vmatrix} = 40 - 42 = -2. \quad \blacktriangle$$

A 3×3 *matrix* is an array

$$\begin{bmatrix} a_{11} & a_{12} & a_{13} \\ a_{21} & a_{22} & a_{23} \\ a_{31} & a_{32} & a_{33} \end{bmatrix}$$

where, again, each a_{ij} is a scalar; a_{ij} denotes the entry in the array that is in the ith row and the jth column. We define the ***determinant*** of a 3×3 matrix by the rule

$$\begin{vmatrix} a_{11} & a_{12} & a_{13} \\ a_{21} & a_{22} & a_{23} \\ a_{31} & a_{32} & a_{33} \end{vmatrix} = a_{11}\begin{vmatrix} a_{22} & a_{23} \\ a_{32} & a_{33} \end{vmatrix} - a_{12}\begin{vmatrix} a_{21} & a_{23} \\ a_{31} & a_{33} \end{vmatrix} + a_{13}\begin{vmatrix} a_{21} & a_{22} \\ a_{31} & a_{32} \end{vmatrix}. \tag{2}$$

Without some mnemonic device, formula (2) would be difficult to memorize. The rule to learn is that you move along the first row, multiplying a_{1j} by the determinant of the 2×2 matrix obtained by canceling out the first row and the jth column, and then you add these up, remembering to put a minus in front of the a_{12} term. For example, the determinant multiplied by the middle term of formula (2), namely,

$$\begin{vmatrix} a_{21} & a_{23} \\ a_{31} & a_{33} \end{vmatrix}$$

is obtained by crossing out the first row and the second column of the given 3×3 matrix:

$$\begin{bmatrix} \cancel{a_{11}} & \cancel{a_{12}} & \cancel{a_{13}} \\ a_{21} & \cancel{a_{22}} & a_{23} \\ a_{31} & \cancel{a_{32}} & a_{33} \end{bmatrix}.$$

E X A M P L E   2

$$\begin{vmatrix} 1 & 0 & 0 \\ 0 & 1 & 0 \\ 0 & 0 & 1 \end{vmatrix} = 1\begin{vmatrix} 1 & 0 \\ 0 & 1 \end{vmatrix} - 0\begin{vmatrix} 0 & 0 \\ 0 & 1 \end{vmatrix} + 0\begin{vmatrix} 0 & 1 \\ 0 & 0 \end{vmatrix} = 1.$$

$$\begin{vmatrix} 1 & 2 & 3 \\ 4 & 5 & 6 \\ 7 & 8 & 9 \end{vmatrix} = 1\begin{vmatrix} 5 & 6 \\ 8 & 9 \end{vmatrix} - 2\begin{vmatrix} 4 & 6 \\ 7 & 9 \end{vmatrix} + 3\begin{vmatrix} 4 & 5 \\ 7 & 8 \end{vmatrix} = -3 + 12 - 9 = 0. \quad \blacktriangle$$

Properties of Determinants

An important property of determinants is that interchanging two rows or two columns results in a change of sign. For 2×2 determinants, this is a consequence of the definition as follows: For rows, we have

$$\begin{vmatrix} a_{11} & a_{12} \\ a_{21} & a_{22} \end{vmatrix} = a_{11}a_{22} - a_{21}a_{12} = -(a_{21}a_{12} - a_{11}a_{22}) = -\begin{vmatrix} a_{21} & a_{22} \\ a_{11} & a_{12} \end{vmatrix}$$

and for columns,

$$\begin{vmatrix} a_{11} & a_{12} \\ a_{21} & a_{22} \end{vmatrix} = -(a_{12}a_{21} - a_{11}a_{22}) = -\begin{vmatrix} a_{12} & a_{11} \\ a_{22} & a_{21} \end{vmatrix}$$

We leave it to the reader to verify this property for the 3×3 case.

A second fundamental property of determinants is that *we can factor scalars out of any row or column.* For 2×2 determinants, this means

$$\begin{vmatrix} \alpha a_{11} & a_{12} \\ \alpha a_{21} & a_{22} \end{vmatrix} = \begin{vmatrix} a_{11} & \alpha a_{12} \\ a_{21} & \alpha a_{22} \end{vmatrix} = \alpha \begin{vmatrix} a_{11} & a_{12} \\ a_{21} & a_{22} \end{vmatrix} = \begin{vmatrix} \alpha a_{11} & \alpha a_{12} \\ a_{21} & a_{22} \end{vmatrix} = \begin{vmatrix} a_{11} & a_{12} \\ \alpha a_{21} & \alpha a_{22} \end{vmatrix}.$$

Similarly, for 3×3 determinants we have

$$\begin{vmatrix} \alpha a_{11} & \alpha a_{12} & \alpha a_{13} \\ a_{21} & a_{22} & a_{23} \\ a_{31} & a_{32} & a_{33} \end{vmatrix} = \alpha \begin{vmatrix} a_{11} & a_{12} & a_{13} \\ a_{21} & a_{22} & a_{23} \\ a_{31} & a_{32} & a_{33} \end{vmatrix} = \begin{vmatrix} a_{11} & \alpha a_{12} & a_{13} \\ a_{21} & \alpha a_{22} & a_{23} \\ a_{31} & \alpha a_{32} & a_{33} \end{vmatrix}$$

and so on. These results follow from the definitions. In particular, if any row or column consists of zeros, then the value of the determinant is zero.

A third fundamental fact about determinants is the following: *If we change a row (or column) by adding another row (or respectively, column) to it, the value of the determinant remains the same.* For the 2×2 case this means that

$$\begin{vmatrix} a_1 & a_2 \\ b_1 & b_2 \end{vmatrix} = \begin{vmatrix} a_1 + b_1 & a_2 + b_2 \\ b_1 & b_2 \end{vmatrix} = \begin{vmatrix} a_1 & a_2 \\ b_1 + a_1 & b_2 + a_2 \end{vmatrix}$$

$$= \begin{vmatrix} a_1 + a_2 & a_2 \\ b_1 + b_2 & b_2 \end{vmatrix} = \begin{vmatrix} a_1 & a_1 + a_2 \\ b_1 & b_1 + b_2 \end{vmatrix}.$$

For the 3×3 case, this means

$$\begin{vmatrix} a_1 & a_2 & a_3 \\ b_1 & b_2 & b_3 \\ c_1 & c_2 & c_3 \end{vmatrix} = \begin{vmatrix} a_1 + b_1 & a_2 + b_2 & a_3 + b_3 \\ b_1 & b_2 & b_3 \\ c_1 & c_2 & c_3 \end{vmatrix} = \begin{vmatrix} a_1 + a_2 & a_2 & a_3 \\ b_1 + b_2 & b_2 & b_3 \\ c_1 + c_2 & c_2 & c_3 \end{vmatrix}$$

and so on. Again, this property can be proved using the definition of the determinant.

EXAMPLE 3 Suppose

$$\mathbf{a} = \alpha\mathbf{b} + \beta\mathbf{c}; \quad \text{that is, } \mathbf{a} = (a_1, a_2, a_3) = \alpha(b_1, b_2, b_3) + \beta(c_1, c_2, c_3).$$

Show that

$$\begin{vmatrix} a_1 & a_2 & a_3 \\ b_1 & b_2 & b_3 \\ c_1 & c_2 & c_3 \end{vmatrix} = 0.$$

SOLUTION We shall prove the case $\alpha \neq 0, \beta \neq 0$. The case $\alpha = 0 = \beta$ is trivial, and the case where exactly one of α, β is zero is a simple modification of the case we prove. Using the fundamental properties of determinants, the determinant in question is

$$\begin{vmatrix} \alpha b_1 + \beta c_1 & \alpha b_2 + \beta c_2 & \alpha b_3 + \beta c_3 \\ b_1 & b_2 & b_3 \\ c_1 & c_2 & c_3 \end{vmatrix}$$

$$= -\frac{1}{\alpha} \begin{vmatrix} \alpha b_1 + \beta c_1 & \alpha b_2 + \beta c_2 & \alpha b_3 + \beta c_3 \\ -\alpha b_1 & -\alpha b_2 & -\alpha b_3 \\ c_1 & c_2 & c_3 \end{vmatrix}$$

$$\text{(factoring } -1/\alpha \text{ out of the second row)}$$

$$= \left(-\frac{1}{\alpha}\right)\left(-\frac{1}{\beta}\right) \begin{vmatrix} \alpha b_1 + \beta c_1 & \alpha b_2 + \beta c_2 & \alpha b_3 + \beta c_3 \\ -\alpha b_1 & -\alpha b_2 & -\alpha b_3 \\ -\beta c_1 & -\beta c_2 & -\beta c_3 \end{vmatrix}$$

$$\text{(factoring } -1/\beta \text{ out of the third row)}$$

$$= \frac{1}{\alpha\beta} \begin{vmatrix} \beta c_1 & \beta c_2 & \beta c_3 \\ -\alpha b_1 & -\alpha b_2 & -\alpha b_3 \\ -\beta c_1 & -\beta c_2 & -\beta c_3 \end{vmatrix} \quad \text{(adding the second row to the first row)}$$

$$= \frac{1}{\alpha\beta} \begin{vmatrix} 0 & 0 & 0 \\ -\alpha b_1 & -\alpha b_2 & -\alpha b_3 \\ -\beta c_1 & -\beta c_2 & -\beta c_3 \end{vmatrix} \quad \text{(adding the third row to the first row)}$$

$$= 0. \quad \blacktriangle$$

HISTORICAL NOTE

Determinants appear to have been invented and first used by Leibniz in 1693, in connection with solutions of linear equations. Maclaurin and Cramer developed their properties between

1729 and 1750; in particular, they showed that the solution of the system of equations

$$a_{11}x_1 + a_{12}x_2 + a_{13}x_3 = b_1$$
$$a_{21}x_1 + a_{22}x_2 + a_{23}x_3 = b_2$$
$$a_{31}x_1 + a_{32}x_2 + a_{33}x_3 = b_3$$

is

$$x_1 = \frac{1}{\Delta} \begin{vmatrix} b_1 & a_{12} & a_{13} \\ b_2 & a_{22} & a_{23} \\ b_3 & a_{32} & a_{33} \end{vmatrix}, \qquad x_2 = \frac{1}{\Delta} \begin{vmatrix} a_{11} & b_1 & a_{13} \\ a_{21} & b_2 & a_{23} \\ a_{31} & b_3 & a_{33} \end{vmatrix}, \qquad x_3 = \frac{1}{\Delta} \begin{vmatrix} a_{11} & a_{12} & b_1 \\ a_{21} & a_{22} & b_2 \\ a_{31} & a_{32} & b_3 \end{vmatrix}$$

where

$$\Delta = \begin{vmatrix} a_{11} & a_{12} & a_{13} \\ a_{21} & a_{22} & a_{23} \\ a_{31} & a_{32} & a_{33} \end{vmatrix},$$

a fact now known as *Cramer's rule*. Later, Vandermonde (1772) and Cauchy (1812), treating determinants as a separate topic worthy of special attention, developed the field more systematically, with contributions by Laplace, Jacobi, and others. Formulas for volumes of parallelepipeds in terms of determinants are due to Lagrange (1775). We shall study these later in this section. Although during the nineteenth century mathematicians studied matrices and determinants, the subjects were considered separate. For the full history up to 1900, see T. Muir, *The Theory of Determinants in the Historical Order of Development*, reprinted by Dover, New York, 1960.

Cross Products

Now that we have established the necessary properties of determinants and discussed their history, we are ready to proceed with the cross product of vectors.

DEFINITION: THE CROSS PRODUCT Let $\mathbf{a} = a_1\mathbf{i} + a_2\mathbf{j} + a_3\mathbf{k}$ and $\mathbf{b} = b_1\mathbf{i} + b_2\mathbf{j} + b_3\mathbf{k}$ be vectors in $\mathbb{R}^3$. The *cross product* or *vector product* of $\mathbf{a}$ and $\mathbf{b}$, denoted $\mathbf{a} \times \mathbf{b}$, is defined to be the vector

$$\mathbf{a} \times \mathbf{b} = \begin{vmatrix} a_2 & a_3 \\ b_2 & b_3 \end{vmatrix} \mathbf{i} - \begin{vmatrix} a_1 & a_3 \\ b_1 & b_3 \end{vmatrix} \mathbf{j} + \begin{vmatrix} a_1 & a_2 \\ b_1 & b_2 \end{vmatrix} \mathbf{k},$$

or, symbolically,

$$\mathbf{a} \times \mathbf{b} = \begin{vmatrix} \mathbf{i} & \mathbf{j} & \mathbf{k} \\ a_1 & a_2 & a_3 \\ b_1 & b_2 & b_3 \end{vmatrix}.$$

Even though we only defined determinants for arrays of *real* numbers, this formal expression involving *vectors* is a useful memory aid for the cross product.

EXAMPLE 4   Find $(3\mathbf{i} - \mathbf{j} + \mathbf{k}) \times (\mathbf{i} + 2\mathbf{j} - \mathbf{k})$.

SOLUTION

$$(3\mathbf{i} - \mathbf{j} + \mathbf{k}) \times (\mathbf{i} + 2\mathbf{j} - \mathbf{k}) = \begin{vmatrix} \mathbf{i} & \mathbf{j} & \mathbf{k} \\ 3 & -1 & 1 \\ 1 & 2 & -1 \end{vmatrix} = -\mathbf{i} + 4\mathbf{j} + 7\mathbf{k}. \quad \blacktriangle$$

Certain algebraic properties of the cross product follow from the definition. If $\mathbf{a}$, $\mathbf{b}$, and $\mathbf{c}$ are vectors and α, β, and γ are scalars, then

(i) $\mathbf{a} \times \mathbf{b} = -(\mathbf{b} \times \mathbf{a})$

(ii) $\mathbf{a} \times (\beta\mathbf{b} + \gamma\mathbf{c}) = \beta(\mathbf{a} \times \mathbf{b}) + \gamma(\mathbf{a} \times \mathbf{c})$ and $(\alpha\mathbf{a} + \beta\mathbf{b}) \times \mathbf{c} = \alpha(\mathbf{a} \times \mathbf{c}) + \beta(\mathbf{b} \times \mathbf{c})$

Note that $\mathbf{a} \times \mathbf{a} = -(\mathbf{a} \times \mathbf{a})$, by property (i). Thus, $\mathbf{a} \times \mathbf{a} = \mathbf{0}$. In particular,

$$\mathbf{i} \times \mathbf{i} = \mathbf{0}, \qquad \mathbf{j} \times \mathbf{j} = \mathbf{0}, \qquad \mathbf{k} \times \mathbf{k} = \mathbf{0}.$$

Also,

$$\mathbf{i} \times \mathbf{j} = \mathbf{k}, \qquad \mathbf{j} \times \mathbf{k} = \mathbf{i}, \qquad \mathbf{k} \times \mathbf{i} = \mathbf{j},$$

which can be remembered by cyclicly permuting $\mathbf{i}, \mathbf{j}, \mathbf{k}$ like this:

To give a geometric interpretation of the cross product, we first introduce the triple product. Given three vectors $\mathbf{a}$, $\mathbf{b}$, and $\mathbf{c}$, the real number

$$(\mathbf{a} \times \mathbf{b}) \cdot \mathbf{c}$$

is called the ***triple product*** of $\mathbf{a}$, $\mathbf{b}$, and $\mathbf{c}$ (in that order). To obtain a formula for it, let $\mathbf{a} = a_1\mathbf{i} + a_2\mathbf{j} + a_3\mathbf{k}$, $\mathbf{b} = b_1\mathbf{i} + b_2\mathbf{j} + b_3\mathbf{k}$, and $\mathbf{c} = c_1\mathbf{i} + c_2\mathbf{j} + c_3\mathbf{k}$. Then

$$(\mathbf{a} \times \mathbf{b}) \cdot \mathbf{c} = \left(\begin{vmatrix} a_2 & a_3 \\ b_2 & b_3 \end{vmatrix} \mathbf{i} - \begin{vmatrix} a_1 & a_3 \\ b_1 & b_3 \end{vmatrix} \mathbf{j} + \begin{vmatrix} a_1 & a_2 \\ b_1 & b_2 \end{vmatrix} \mathbf{k} \right) \cdot (c_1\mathbf{i} + c_2\mathbf{j} + c_3\mathbf{k})$$

$$= \begin{vmatrix} a_2 & a_3 \\ b_2 & b_3 \end{vmatrix} c_1 - \begin{vmatrix} a_1 & a_3 \\ b_1 & b_3 \end{vmatrix} c_2 + \begin{vmatrix} a_1 & a_2 \\ b_1 & b_2 \end{vmatrix} c_3.$$

This is the expansion by minors of the third row of the determinant, so

$$(\mathbf{a} \times \mathbf{b}) \cdot \mathbf{c} = \begin{vmatrix} a_1 & a_2 & a_3 \\ b_1 & b_2 & b_3 \\ c_1 & c_2 & c_3 \end{vmatrix}.$$

If $\mathbf{c}$ is a vector in the plane spanned by the vectors $\mathbf{a}$ and $\mathbf{b}$, then the third row in the determinant expression for $(\mathbf{a} \times \mathbf{b}) \cdot \mathbf{c}$ is a linear combination of the first and second rows and therefore $(\mathbf{a} \times \mathbf{b}) \cdot \mathbf{c} = 0$. In other words, *the vector $\mathbf{a} \times \mathbf{b}$ is orthogonal to any vector in the plane spanned by $\mathbf{a}$ and $\mathbf{b}$, in particular to both $\mathbf{a}$ and $\mathbf{b}$.*

Next, we calculate the length of $\mathbf{a} \times \mathbf{b}$. Note that

$$\|\mathbf{a} \times \mathbf{b}\|^2 = \begin{vmatrix} a_2 & a_3 \\ b_2 & b_3 \end{vmatrix}^2 + \begin{vmatrix} a_1 & a_3 \\ b_1 & b_3 \end{vmatrix}^2 + \begin{vmatrix} a_1 & a_2 \\ b_1 & b_2 \end{vmatrix}^2$$

$$= (a_2 b_3 - a_3 b_2)^2 + (a_1 b_3 - b_1 a_3)^2 + (a_1 b_2 - b_1 a_2)^2.$$

If we expand the terms in the last expression, we can recollect them to give

$$(a_1^2 + a_2^2 + a_3^2)(b_1^2 + b_2^2 + b_3^2) - (a_1 b_1 + a_2 b_2 + a_3 b_3)^2,$$

which equals

$$\|\mathbf{a}\|^2 \|\mathbf{b}\|^2 - (\mathbf{a} \cdot \mathbf{b})^2 = \|\mathbf{a}\|^2 \|\mathbf{b}\|^2 - \|\mathbf{a}\|^2 \|\mathbf{b}\|^2 \cos^2 \theta = \|\mathbf{a}\|^2 \|\mathbf{b}\|^2 \sin^2 \theta$$

where θ is the angle between $\mathbf{a}$ and $\mathbf{b}$, $0 \le \theta \le \pi$. Taking square roots and using $\sqrt{k^2} = |k|$, we find that $\|\mathbf{a} \times \mathbf{b}\| = \|\mathbf{a}\| \|\mathbf{b}\| |\sin \theta|$.

Combining our results, we conclude that $\mathbf{a} \times \mathbf{b}$ is a *vector perpendicular to the plane $\mathcal{P}$ spanned by $\mathbf{a}$ and $\mathbf{b}$ with length $\|\mathbf{a}\| \|\mathbf{b}\| |\sin \theta|$.* We see from Figure 1.3.1 that this length is also the area of the parallelogram (with base $\|\mathbf{a}\|$ and height $\|\mathbf{b} \sin \theta\|$) spanned by $\mathbf{a}$ and $\mathbf{b}$. There are still two possible vectors that satisfy these conditions, because there are two choices of direction that are perpendicular (or normal) to $\mathcal{P}$. This is clear from Figure 1.3.1, which shows the two choices $\mathbf{n}_1$ and $-\mathbf{n}_1$ perpendicular to $\mathcal{P}$, with $\|\mathbf{n}_1\| = \|-\mathbf{n}_1\| = \|\mathbf{a}\| \|\mathbf{b}\| |\sin \theta|$.

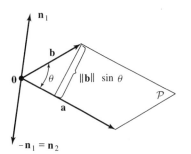

Figure 1.3.1 $\mathbf{n}_1$ and $\mathbf{n}_2$ are the two possible vectors orthogonal to both $\mathbf{a}$ and $\mathbf{b}$, and with norm $\|\mathbf{a}\| \|\mathbf{b}\| |\sin \theta|$.

Which vector represents $\mathbf{a} \times \mathbf{b}$, $\mathbf{n}_1$ or $-\mathbf{n}_1$? The answer is $\mathbf{n}_1$. Try a few cases such as $\mathbf{k} = \mathbf{i} \times \mathbf{j}$ to verify this. The following "right-hand rule" determines the direction of $\mathbf{a} \times \mathbf{b}$ in general. Take your right hand and place it so your fingers curl from $\mathbf{a}$ toward $\mathbf{b}$ through the *acute* angle θ, as in Figure 1.3.2. Then your thumb points in the direction of $\mathbf{a} \times \mathbf{b}$.

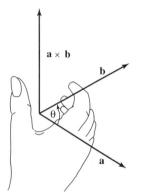

Figure 1.3.2   The right-hand rule for determining in which of the two possible directions $\mathbf{a} \times \mathbf{b}$ points.

THE CROSS PRODUCT

Geometric definition: $\mathbf{a} \times \mathbf{b}$ is the vector such that:

(1) $\|\mathbf{a} \times \mathbf{b}\| = \|\mathbf{a}\|\,\|\mathbf{b}\| \sin\theta$, the area of the parallelogram spanned by $\mathbf{a}$ and $\mathbf{b}$ (θ is the angle between $\mathbf{a}$ and $\mathbf{b}$; $0 \le \theta \le \pi$); see Figure 1.3.3.

(2) $\mathbf{a} \times \mathbf{b}$ is perpendicular to $\mathbf{a}$ and $\mathbf{b}$, and the triple $(\mathbf{a}, \mathbf{b}, \mathbf{a} \times \mathbf{b})$ obeys the right-hand rule.

Component formula:

$$(a_1\mathbf{i} + a_2\mathbf{j} + a_3\mathbf{k}) \times (b_1\mathbf{i} + b_2\mathbf{j} + b_3\mathbf{k}) = \begin{vmatrix} \mathbf{i} & \mathbf{j} & \mathbf{k} \\ a_1 & a_2 & a_3 \\ b_1 & b_2 & b_3 \end{vmatrix}$$

$$= (a_2 b_3 - a_3 b_2)\mathbf{i} + (a_3 b_1 - a_1 b_3)\mathbf{j} + (a_1 b_2 - a_2 b_1)\mathbf{k}.$$

Algebraic rules:
1. $\mathbf{a} \times \mathbf{b} = \mathbf{0}$ if and only if $\mathbf{a}$ and $\mathbf{b}$ are parallel or $\mathbf{a}$ or $\mathbf{b}$ is zero.
2. $\mathbf{a} \times \mathbf{b} = -\mathbf{b} \times \mathbf{a}$.
3. $\mathbf{a} \times (\mathbf{b} + \mathbf{c}) = \mathbf{a} \times \mathbf{b} + \mathbf{a} \times \mathbf{c}$.
4. $(\mathbf{a} + \mathbf{b}) \times \mathbf{c} = \mathbf{a} \times \mathbf{c} + \mathbf{b} \times \mathbf{c}$
5. $(\alpha\mathbf{a}) \times \mathbf{b} = \alpha(\mathbf{a} \times \mathbf{b})$.

Multiplication table:

		Second factor		
	$\times$	$\mathbf{i}$	$\mathbf{j}$	$\mathbf{k}$
	$\mathbf{i}$	$\mathbf{0}$	$\mathbf{k}$	$-\mathbf{j}$
First	$\mathbf{j}$	$-\mathbf{k}$	$\mathbf{0}$	$\mathbf{i}$
factor	$\mathbf{k}$	$\mathbf{j}$	$-\mathbf{i}$	$\mathbf{0}$

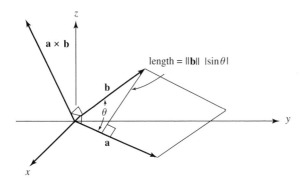

Figure 1.3.3   The length of $\mathbf{a} \times \mathbf{b}$ is the area of the parallelogram formed by $\mathbf{a}$ and $\mathbf{b}$.

EXAMPLE 5   Find the area of the parallelogram spanned by the two vectors $\mathbf{a} = \mathbf{i} + 2\mathbf{j} + 3\mathbf{k}$ and $\mathbf{b} = -\mathbf{i} - \mathbf{k}$.

SOLUTION   We calculate the cross product of $\mathbf{a}$ and $\mathbf{b}$ by applying the component or determinant formula, with $a_1 = 1, a_2 = 2, a_3 = 3, b_1 = -1, b_2 = 0, b_3 = -1$:

$$\mathbf{a} \times \mathbf{b} = [(2)(-1) - (3)(0)]\mathbf{i} + [(3)(-1) - (1)(-1)]\mathbf{j} + [(1)(0) - (2)(-1)]\mathbf{k}$$

$$= -2\mathbf{i} - 2\mathbf{j} + 2\mathbf{k}.$$

Thus the area is

$$\|\mathbf{a} \times \mathbf{b}\| = \sqrt{(-2)^2 + (-2)^2 + (2)^2} = 2\sqrt{3}. \quad \blacktriangle$$

EXAMPLE 6   Find a unit vector orthogonal to the vectors $\mathbf{i} + \mathbf{j}$ and $\mathbf{j} + \mathbf{k}$.

SOLUTION   A vector perpendicular to both $\mathbf{i} + \mathbf{j}$ and $\mathbf{j} + \mathbf{k}$ is the vector

$$(\mathbf{i} + \mathbf{j}) \times (\mathbf{j} + \mathbf{k}) = \begin{vmatrix} \mathbf{i} & \mathbf{j} & \mathbf{k} \\ 1 & 1 & 0 \\ 0 & 1 & 1 \end{vmatrix} = \mathbf{i} - \mathbf{j} + \mathbf{k}.$$

Since $\|\mathbf{i} - \mathbf{j} + \mathbf{k}\| = \sqrt{3}$, the vector

$$\frac{1}{\sqrt{3}}(\mathbf{i} - \mathbf{j} + \mathbf{k})$$

is a unit vector perpendicular to $\mathbf{i} + \mathbf{j}$ and $\mathbf{j} + \mathbf{k}$. $\blacktriangle$

EXAMPLE 7 Derive an identity relating the dot and cross products from the formulas

$$\|\mathbf{u} \times \mathbf{v}\| = \|\mathbf{u}\| \|\mathbf{v}\| \sin\theta \qquad \text{and} \qquad \mathbf{u} \cdot \mathbf{v} = \|\mathbf{u}\| \|\mathbf{v}\| \cos\theta$$

by eliminating θ.

SOLUTION Seeing $\sin\theta$ and $\cos\theta$ multiplied by the same expression suggests squaring the two formulas and adding the results. We get

$$\|\mathbf{u} \times \mathbf{v}\|^2 + (\mathbf{u} \cdot \mathbf{v})^2 = \|\mathbf{u}\|^2 \|\mathbf{v}\|^2 (\sin^2\theta + \cos^2\theta) = \|\mathbf{u}\|^2 \|\mathbf{v}\|^2,$$

so

$$\|\mathbf{u} \times \mathbf{v}\|^2 = \|\mathbf{u}\|^2 \|\mathbf{v}\|^2 - (\mathbf{u} \cdot \mathbf{v})^2.$$

This identity is interesting because it establishes a link between the dot and cross products. ▲

Geometry of Determinants

Using the cross product, we may obtain a basic geometric interpretation of 2×2 and 3×3 determinants. Let $\mathbf{a} = a_1\mathbf{i} + a_2\mathbf{j}$ and $\mathbf{b} = b_1\mathbf{i} + b_2\mathbf{j}$ be two vectors in the plane. If θ is the angle between $\mathbf{a}$ and $\mathbf{b}$, we have seen that $\|\mathbf{a} \times \mathbf{b}\| = \|\mathbf{a}\| \|\mathbf{b}\| |\sin\theta|$ is the area of the parallelogram with adjacent sides $\mathbf{a}$ and $\mathbf{b}$. The cross product as a determinant is

$$\mathbf{a} \times \mathbf{b} = \begin{vmatrix} \mathbf{i} & \mathbf{j} & \mathbf{k} \\ a_1 & a_2 & 0 \\ b_1 & b_2 & 0 \end{vmatrix} = \begin{vmatrix} a_1 & a_2 \\ b_1 & b_2 \end{vmatrix} \mathbf{k}.$$

Thus the area $\|\mathbf{a} \times \mathbf{b}\|$ is the absolute value of the determinant

$$\begin{vmatrix} a_1 & a_2 \\ b_1 & b_2 \end{vmatrix} = a_1 b_2 - a_2 b_1.$$

GEOMETRY OF 2 × 2 DETERMINANTS The absolute value of the determinant $\begin{vmatrix} a_1 & a_2 \\ b_1 & b_2 \end{vmatrix}$ is the area of the parallelogram whose adjacent sides are the vectors $\mathbf{a} = a_1\mathbf{i} + a_2\mathbf{j}$ and $\mathbf{b} = b_1\mathbf{i} + b_2\mathbf{j}$. The sign of the determinant is $+$ when, rotating in the counterclockwise direction, the angle from $\mathbf{a}$ to $\mathbf{b}$ is less than π.

EXAMPLE 8 Find the area of the triangle with vertices at the points $(1, 1)$, $(0, 2)$, and $(3, 2)$ (Figure 1.3.4).

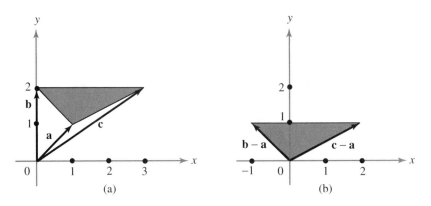

Figure 1.3.4 Problem (a): Find the area A of the shaded triangle. Solution: Express the sides as vector differences (b) to get $A = \|(\mathbf{b} - \mathbf{a}) \times (\mathbf{c} - \mathbf{a})\|/2$.

SOLUTION   Let $\mathbf{a} = \mathbf{i} + \mathbf{j}, \mathbf{b} = 2\mathbf{j}$, and $\mathbf{c} = 3\mathbf{i} + 2\mathbf{j}$. It is clear that the triangle whose vertices are the endpoints of the vectors $\mathbf{a}, \mathbf{b}$, and $\mathbf{c}$ has the same area as the triangle with vertices at $\mathbf{0}, \mathbf{b} - \mathbf{a}$, and $\mathbf{c} - \mathbf{a}$ (Figure 1.3.4). Indeed, the latter is merely a translation of the former triangle. Since the area of this translated triangle is one-half the area of the parallelogram with adjacent sides $\mathbf{b} - \mathbf{a} = -\mathbf{i} + \mathbf{j}$, and $\mathbf{c} - \mathbf{a} = 2\mathbf{i} + \mathbf{j}$, we find that the area of the triangle with vertices $(1, 1)$, $(0, 2)$, and $(3, 2)$ is the absolute value of

$$\frac{1}{2}\begin{vmatrix} -1 & 1 \\ 2 & 1 \end{vmatrix} = -\frac{3}{2},$$

that is, 3/2.   ▲

There is an interpretation of determinants of 3×3 matrices as volumes that is analogous to the interpretation of determinants of 2×2 matrices as areas.

GEOMETRY OF 3×3 DETERMINANTS   The absolute value of the determinant

$$D = \begin{vmatrix} a_1 & a_2 & a_3 \\ b_1 & b_2 & b_3 \\ c_1 & c_2 & c_3 \end{vmatrix}$$

is the volume of the parallelepiped whose adjacent sides are the vectors

$$\mathbf{a} = a_1\mathbf{i} + a_2\mathbf{j} + a_3\mathbf{k}, \qquad \mathbf{b} = b_1\mathbf{i} + b_2\mathbf{j} + b_3\mathbf{k}, \qquad \text{and} \qquad \mathbf{c} = c_1\mathbf{i} + c_2\mathbf{j} + c_3\mathbf{k}.$$

To prove the statement in the box above, we follow Figure 1.3.5 and note that $\|\mathbf{a} \times \mathbf{b}\|$ is the area of the parallelogram with adjacent sides $\mathbf{a}$ and $\mathbf{b}$. Moreover, $(\mathbf{a} \times \mathbf{b}) \cdot \mathbf{c} = \|\mathbf{a} \times \mathbf{b}\| \|\mathbf{c}\| \cos \psi$, where ψ is the angle that $\mathbf{c}$ makes with the normal to the plane spanned by **a** and **b**. Since the volume of the parallelepiped with adjacent sides **a**, **b**, and **c** is the product of the area of the base $\|\mathbf{a} \times \mathbf{b}\|$ and the altitude $\|\mathbf{c}\| |\cos \psi|$, it follows that the volume is $|(\mathbf{a} \times \mathbf{b}) \cdot \mathbf{c}|$. We saw earlier that $(\mathbf{a} \times \mathbf{b}) \cdot \mathbf{c} = D$, so the volume equals the absolute value of D.

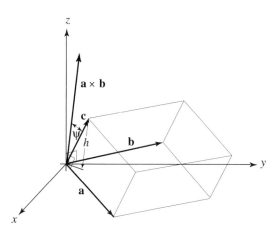

Figure 1.3.5 The volume of the parallelepiped spanned by **a**, **b**, **c** is the absolute value of the determinant of the 3×3 matrix having **a**, **b**, **c** as its rows.

EXAMPLE 9 Find the volume of the parallelepiped spanned by the three vectors $\mathbf{i} + 3\mathbf{k}$, $2\mathbf{i} + \mathbf{j} - 2\mathbf{k}$, and $5\mathbf{i} + 4\mathbf{k}$.

SOLUTION The volume is the absolute value of

$$\begin{vmatrix} 1 & 0 & 3 \\ 2 & 1 & -2 \\ 5 & 0 & 4 \end{vmatrix}.$$

If we expand this determinant by minors of the second column, the only nonzero term is

$$\begin{vmatrix} 1 & 3 \\ 5 & 4 \end{vmatrix} (1) = -11.$$

so the volume equals 11. ▲

Equations of Planes

Let $\mathcal{P}$ be a plane in space, $P_0 = (x_0, y_0, z_0)$ a point in the plane, and $\mathbf{n} = A\mathbf{i} + B\mathbf{j} + C\mathbf{k}$ a vector normal to the plane (see Figure 1.3.6). Let $P = (x, y, z)$ be a point in $\mathbb{R}^3$. Then P lies on the plane $\mathcal{P}$ if and only if the vector $\overrightarrow{P_0 P} = (x - x_0)\mathbf{i} + (y - y_0)\mathbf{j} + (z - z_0)\mathbf{k}$ is perpendicular to **n**, that is, $\overrightarrow{P_0 P} \cdot \mathbf{n} = 0$, or

$$(A\mathbf{i} + B\mathbf{j} + C\mathbf{k}) \cdot [(x - x_0)\mathbf{i} + (y - y_0)\mathbf{j} + (z - z_0)\mathbf{k}] = 0.$$

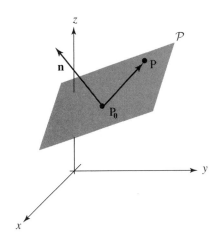

Figure 1.3.6   The points P of the plane through P_0 and perpendicular to **n** satisfy the equation $\overrightarrow{P_0P} \cdot \mathbf{n} = 0$.

Thus,

$$A(x - x_0) + B(y - y_0) + C(z - z_0) = 0.$$

EQUATION OF A PLANE IN SPACE   The equation of the plane $\mathcal{P}$ through (x_0, y_0, z_0) that has a normal vector $\mathbf{n} = A\mathbf{i} + B\mathbf{j} + C\mathbf{k}$ is

$$A(x - x_0) + B(y - y_0) + C(z - z_0) = 0;$$

that is, $(x, y, z) \in \mathcal{P}$ if and only if

$$Ax + By + Cz + D = 0$$

where $D = -Ax_0 - By_0 - Cz_0$.

The four numbers $A, B, C,$ and D are not determined uniquely by the plane $\mathcal{P}$. To see this, note that (x, y, z) satisfies the equation $Ax + By + Cz + D = 0$ if and only if it also satisfies the relation

$$(\lambda A)x + (\lambda B)y + (\lambda C)z + (\lambda D) = 0$$

for any constant $\lambda \neq 0$. Furthermore, if A, B, C, D and A', B', C', D' determine the same plane $\mathcal{P}$, then $A = \lambda A', B = \lambda B', C = \lambda C', D = \lambda D'$ for a scalar λ. Consequently, A, B, C, D are *determined by $\mathcal{P}$ up to a scalar multiple*.

EXAMPLE 10   Determine an equation for the plane that is perpendicular to the vector $\mathbf{i} + \mathbf{j} + \mathbf{k}$ and contains the point $(1, 0, 0)$.

SOLUTION Using the general form $A(x - x_0) + B(y - y_0) + C(z - z_0) = 0$, the plane is $1(x - 1) + 1(y - 0) + 1(z - 0) = 0$; that is, $x + y + z = 1$. ▲

EXAMPLE 11 Find an equation for the plane containing the three points $(1, 1, 1)$, $(2, 0, 0)$, and $(1, 1, 0)$.

SOLUTION *Method 1.* This is a "brute force" method which you can use if you have forgotten the vector methods. The equation for any plane is of the form $Ax + By + Cz + D = 0$. Since the points $(1, 1, 1)$, $(2, 0, 0)$, and $(1, 1, 0)$ lie in the plane, we have

$$
\begin{aligned}
A + B + C + D &= 0, \\
2A \qquad\quad + D &= 0, \\
A + B \qquad + D &= 0.
\end{aligned}
$$

Proceeding by elimination, we reduce this system of equations to the form

$$
\begin{aligned}
2A + D &= 0 \quad \text{(second equation)} \\
2B + D &= 0 \quad (2 \times \text{third} - \text{second}), \\
C &= 0 \quad \text{(first} - \text{third)}.
\end{aligned}
$$

Since the numbers A, B, C, and D are determined only up to a scalar multiple, we can fix the value of one of them, say $A = 1$, and then the others will be determined uniquely. We get $A = 1$, $D = -2$, $B = 1$, $C = 0$. Thus an equation of the plane that contains the given points is $x + y - 2 = 0$.

Method 2. Let $P = (1, 1, 1)$, $Q = (2, 0, 0)$, $R = (1, 1, 0)$. Any vector normal to the plane must be orthogonal to the vectors $\overrightarrow{QP}$ and $\overrightarrow{RP}$, which are parallel to the plane, since their endpoints lie on the plane. Thus, $\mathbf{n} = \overrightarrow{QP} \times \overrightarrow{RP}$ is normal to the plane. Computing the cross product, we have

$$
\mathbf{n} = \begin{vmatrix} \mathbf{i} & \mathbf{j} & \mathbf{k} \\ -1 & 1 & 1 \\ 0 & 0 & 1 \end{vmatrix} = \mathbf{i} + \mathbf{j}.
$$

Since the point $(2, 0, 0)$ lies on the plane, we conclude that the equation is given by $(x - 2) + (y - 0) - 0 \cdot (z - 0) = 0$, or $x + y - 2 = 0$. ▲

Distance: Point to Plane

Let us now determine the distance from a point $E = (x_1, y_1, z_1)$ to the plane $\mathcal{P}$ determined by the equation $A(x - x_0) + B(y - y_0) + C(z - z_0) = Ax + By + Cz + D = 0$. To do so, consider the unit normal vector

$$\mathbf{n} = \frac{A\mathbf{i} + B\mathbf{j} + C\mathbf{k}}{\sqrt{A^2 + B^2 + C^2}},$$

which is a unit vector normal to the plane. Drop a perpendicular from E to the plane and construct the triangle REQ shown in Figure 1.3.7. The distance $d = \|\overrightarrow{EQ}\|$ is the length of the projection of $\mathbf{v} = \overrightarrow{RE}$ (the vector from R to E) onto $\mathbf{n}$; thus,

$$\text{distance} = |\mathbf{v} \cdot \mathbf{n}| = |[(x_1 - x_0)\mathbf{i} + (y_1 - y_0)\mathbf{j} + (z_1 - z_0)\mathbf{k}] \cdot \mathbf{n}|$$
$$= \frac{|A(x_1 - x_0) + B(y_1 - y_0) + C(z_1 - z_0)|}{\sqrt{A^2 + B^2 + C^2}}.$$

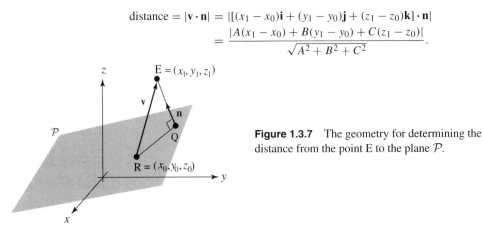

Figure 1.3.7   The geometry for determining the distance from the point E to the plane $\mathcal{P}$.

If the plane is given in the form $Ax + By + Cz + D = 0$, then for any point (x_0, y_0, z_0) on it, $D = -(Ax_0 + By_0 + Cz_0)$. Substitution into the previous formula gives the following:

DISTANCE FROM A POINT TO A PLANE   The distance from (x_1, y_1, z_1) to the plane $Ax + By + Cz + D = 0$ is

$$\text{distance} = \frac{|Ax_1 + By_1 + Cz_1 + D|}{\sqrt{A^2 + B^2 + C^2}}.$$

EXAMPLE 12   Find the distance from $Q = (2, 0, -1)$ to the plane $3x - 2y + 8z + 1 = 0$.

SOLUTION   We substitute into the formula in the preceding box the values $x_1 = 2, y_1 = 0, z_1 = -1$ (the point) and $A = 3, B = -2, C = 8, D = 1$ (the plane)to give

$$\text{distance} = \frac{|3 \cdot 2 + (-2) \cdot 0 + 8(-1) + 1|}{\sqrt{3^2 + (-2)^2 + 8^2}} = \frac{|-1|}{\sqrt{77}} = \frac{1}{\sqrt{77}}. \quad \blacktriangle$$

EXERCISES

1. Verify that interchanging the first two rows of the 3×3 determinant

$$\begin{vmatrix} 1 & 2 & 1 \\ 3 & 0 & 1 \\ 2 & 0 & 2 \end{vmatrix}$$

changes the sign of the determinant.

2. Evaluate

(a) $\begin{vmatrix} 2 & -1 & 0 \\ 4 & 3 & 2 \\ 3 & 0 & 1 \end{vmatrix}$

(b) $\begin{vmatrix} 36 & 18 & 17 \\ 45 & 24 & 20 \\ 3 & 5 & -2 \end{vmatrix}$

(c) $\begin{vmatrix} 1 & 4 & 9 \\ 4 & 9 & 16 \\ 9 & 16 & 25 \end{vmatrix}$

(d) $\begin{vmatrix} 2 & 3 & 5 \\ 7 & 11 & 13 \\ 17 & 19 & 23 \end{vmatrix}$

3. Compute $\mathbf{a} \times \mathbf{b}$, where $\mathbf{a} = \mathbf{i} - 2\mathbf{j} + \mathbf{k}$, $\mathbf{b} = 2\mathbf{i} + \mathbf{j} + \mathbf{k}$.

4. Compute $\mathbf{a} \cdot (\mathbf{b} \times \mathbf{c})$, where $\mathbf{a}$ and $\mathbf{b}$ are as in Exercise 3 and $\mathbf{c} = 3\mathbf{i} - \mathbf{j} + 2\mathbf{k}$.

5. Find the area of the parallelogram with sides the vectors $\mathbf{a}$ and $\mathbf{b}$ given in Exercise 3.

6. A triangle has vertices $(0, 0, 0)$, $(1, 1, 1)$, and $(0, -2, 3)$. Find its area.

7. What is the volume of the parallelepiped with sides $2\mathbf{i} + \mathbf{j} - \mathbf{k}$, $5\mathbf{i} - 3\mathbf{k}$, and $\mathbf{i} - 2\mathbf{j} + \mathbf{k}$?

8. What is the volume of the parallelepiped with sides $\mathbf{i}$, $3\mathbf{j} - \mathbf{k}$, and $4\mathbf{i} + 2\mathbf{j} - \mathbf{k}$?

In Exercises 9 to 12, describe all unit vectors orthogonal to the given vectors.

9. $\mathbf{i}, \mathbf{j}$

10. $-5\mathbf{i} + 9\mathbf{j} - 4\mathbf{k}, 7\mathbf{i} + 8\mathbf{j} + 9\mathbf{k}$

11. $-5\mathbf{i} + 9\mathbf{j} - 4\mathbf{k}, 7\mathbf{i} + 8\mathbf{j} + 9\mathbf{k}, \mathbf{0}$

12. $2\mathbf{i} - 4\mathbf{j} + 3\mathbf{k}, -4\mathbf{i} + 8\mathbf{j} - 6\mathbf{k}$

13. Compute $\mathbf{u} + \mathbf{v}, \mathbf{u} \cdot \mathbf{v}, \|\mathbf{u}\|, \|\mathbf{v}\|$, and $\mathbf{u} \times \mathbf{v}$ where $\mathbf{u} = \mathbf{i} - 2\mathbf{j} + \mathbf{k}, \mathbf{v} = 2\mathbf{i} - \mathbf{j} + 2\mathbf{k}$.

14. Repeat Exercise 13 for $\mathbf{u} = 3\mathbf{i} + \mathbf{j} - \mathbf{k}, \mathbf{v} = -6\mathbf{i} - 2\mathbf{j} - 2\mathbf{k}$.

15. Find an equation for the plane that
(a) is perpendicular to $\mathbf{v} = (1, 1, 1)$ and passes through $(1, 0, 0)$.
(b) is perpendicular to $\mathbf{v} = (1, 2, 3)$ and passes through $(1, 1, 1)$.
(c) is perpendicular to the line $\mathbf{l}(t) = (5, 0, 2)t + (3, -1, 1)$ and passes through $(5, -1, 0)$.
(d) is perpendicular to the line $\mathbf{l}(t) = (-1, -2, 3)t + (0, 7, 1)$ and passes through $(2, 4, -1)$.

16. Find an equation for the plane that passes through (a) $(0, 0, 0), (2, 0, -1)$, and $(0, 4, -3)$; (b) $(1, 2, 0), (0, 1, -2)$, and $(4, 0, 1)$; and (c) $(2, -1, 3), (0, 0, 5)$, and $(5, 7, -1)$.

17. (a)   Prove the two triple-vector-product identities $(\mathbf{A} \times \mathbf{B}) \times \mathbf{C} = (\mathbf{A} \cdot \mathbf{C})\mathbf{B} - (\mathbf{B} \cdot \mathbf{C})\mathbf{A}$ and $\mathbf{A} \times (\mathbf{B} \times \mathbf{C}) = (\mathbf{A} \cdot \mathbf{C})\mathbf{B} - (\mathbf{A} \cdot \mathbf{B})\mathbf{C}$.
   (b)   Prove $(\mathbf{u} \times \mathbf{v}) \times \mathbf{w} = \mathbf{u} \times (\mathbf{v} \times \mathbf{w})$ if and only if $(\mathbf{u} \times \mathbf{w}) \times \mathbf{v} = \mathbf{0}$.
   (c)   Also prove that $(\mathbf{u} \times \mathbf{v}) \times \mathbf{w} + (\mathbf{v} \times \mathbf{w}) \times \mathbf{u} + (\mathbf{w} \times \mathbf{u}) \times \mathbf{v} = \mathbf{0}$ (the *Jacobi identity*).

18. (a)   Prove, without recourse to geometry, that

$$\mathbf{u} \cdot (\mathbf{v} \times \mathbf{w}) = \mathbf{v} \cdot (\mathbf{w} \times \mathbf{u}) = \mathbf{w} \cdot (\mathbf{u} \times \mathbf{v}) = -\mathbf{u} \cdot (\mathbf{w} \times \mathbf{v})$$
$$= -\mathbf{w} \cdot (\mathbf{v} \times \mathbf{u}) = -\mathbf{v} \cdot (\mathbf{u} \times \mathbf{w}).$$

(b)   Prove

$$(\mathbf{u} \times \mathbf{v}) \cdot (\mathbf{u}' \times \mathbf{v}') = (\mathbf{u} \cdot \mathbf{u}')(\mathbf{v} \cdot \mathbf{v}') - (\mathbf{u} \cdot \mathbf{v}')(\mathbf{u}' \cdot \mathbf{v}) = \begin{vmatrix} \mathbf{u} \cdot \mathbf{v}' & \mathbf{u} \cdot \mathbf{v}' \\ \mathbf{u}' \cdot \mathbf{v} & \mathbf{v} \cdot \mathbf{v}' \end{vmatrix}.$$

[HINT: Use part (a) and Exercise 17(a).]

19. Verify Cramer's rule, presented in the Historical Note in this section.

20. Find an equation for the plane that passes through the point $(2, -1, 3)$ and is perpendicular to the line $\mathbf{v} = (1, -2, 2) + t(3, -2, 4)$.

21. Find an equation for the plane that passes through the point $(1, 2, -3)$ and is perpendicular to the line $\mathbf{v} = (0, -2, 1) + t(1, -2, 3)$.

22. Find the equation of the line that passes through the point $(1, -2, -3)$ and is perpendicular to the plane $3x - y - 2z + 4 = 0$.

23. Find an equation for the plane containing the two (parallel) lines

$$\mathbf{v}_1 = (0, 1, -2) + t(2, 3, -1) \qquad \text{and} \qquad \mathbf{v}_2 = (2, -1, 0) + t(2, 3, -1).$$

24. Find the distance from the point $(2, 1, -1)$ to the plane $x - 2y + 2z + 5 = 0$.

25. Find an equation for the plane that contains the line $\mathbf{v} = (-1, 1, 2) + t(3, 2, 4)$ and is perpendicular to the plane $2x + y - 3z + 4 = 0$.

26. Find an equation for the plane that passes through $(3, 2, -1)$ and $(1, -1, 2)$ and that is parallel to the line $\mathbf{v} = (1, -1, 0) + t(3, 2, -2)$.

27. Redo Exercises 19 and 20 of Section 1.1 using the dot product and what you know about normals to planes.

28. Given vectors $\mathbf{a}$ and $\mathbf{b}$, do the equations $\mathbf{x} \times \mathbf{a} = \mathbf{b}$ and $\mathbf{x} \cdot \mathbf{a} = \|\mathbf{a}\|$ determine a unique vector $\mathbf{x}$? Argue both geometrically and analytically.

29. Determine the distance from the plane $12x + 13y + 5z + 2 = 0$ to the point $(1, 1, -5)$.

30. Find the distance to the point $(6, 1, 0)$ from the plane through the origin that is perpendicular to $\mathbf{i} - 2\mathbf{j} + \mathbf{k}$.

31. (a)   In mechanics, the ***moment*** M *of a force* $\mathbf{F}$ *about a point* O is defined to be the magnitude of $\mathbf{F}$ times the perpendicular distance d from O to the line of action of $\mathbf{F}$. The ***vector moment*** $\mathbf{M}$ is the vector of magnitude M whose direction is perpendicular to the plane of O and $\mathbf{F}$, determined by the right-hand rule. Show that $\mathbf{M} = \mathbf{R} \times \mathbf{F}$, where $\mathbf{R}$ is any vector from O to the line of action of $\mathbf{F}$. (See Figure 1.3.8.)

Figure 1.3.8   Moment of a force.

(b)   Find the moment of the force vector $\mathbf{F} = \mathbf{i} - \mathbf{j} + 2\mathbf{k}$ Newtons about the origin if the line of action is $x = 1 + t, y = 1 - t, z = 2t$.

32. Show that the plane that passes through the three points $A = (a_1, a_2, a_3)$, $B = (b_1, b_2, b_3)$, and $C = (c_1, c_2, c_3)$ consists of the points $P = (x, y, z)$ given by

$$\begin{vmatrix} a_1 - x & a_2 - y & a_3 - z \\ b_1 - x & b_2 - y & b_3 - z \\ c_1 - x & c_2 - y & c_3 - z \end{vmatrix} = 0.$$

[HINT: Write the determinant as a triple product.]

33. Two media with indices of refraction n_1 and n_2 are separated by a plane surface perpendicular to the unit vector $\mathbf{N}$. Let $\mathbf{a}$ and $\mathbf{b}$ be unit vectors along the incident and refracted rays, respectively, their directions being those of the light rays. Show that $n_1(\mathbf{N} \times \mathbf{a}) = n_2(\mathbf{N} \times \mathbf{b})$ by using *Snell's law*, $\sin \theta_1 / \sin \theta_2 = n_2/n_1$, where θ_1 and θ_2 are the angles of incidence and refraction, respectively. (See Figure 1.3.9.)

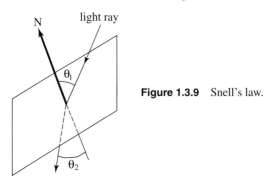

Figure 1.3.9   Snell's law.

34. Justify the steps in the following computation:

$$\begin{vmatrix} 1 & 2 & 3 \\ 4 & 5 & 6 \\ 7 & 8 & 10 \end{vmatrix} = \begin{vmatrix} 1 & 2 & 3 \\ 0 & -3 & -6 \\ 7 & 8 & 10 \end{vmatrix} = \begin{vmatrix} 1 & 2 & 3 \\ 0 & -3 & -6 \\ 0 & -6 & -11 \end{vmatrix} = \begin{vmatrix} -3 & -6 \\ -6 & -11 \end{vmatrix} = 33 - 36 = -3.$$

35. Show that adding a multiple of the first row of a matrix to the second row leaves the determinant unchanged; that is,

$$\begin{vmatrix} a_1 & b_1 & c_1 \\ a_2 + \lambda a_1 & b_2 + \lambda b_1 & c_2 + \lambda c_1 \\ a_3 & b_3 & c_3 \end{vmatrix} = \begin{vmatrix} a_1 & b_1 & c_1 \\ a_2 & b_2 & c_2 \\ a_3 & b_3 & c_3 \end{vmatrix}.$$

[In fact, adding a multiple of any row (column) of a matrix to another row (column) leaves the determinant unchanged.]

1.4 CYLINDRICAL AND SPHERICAL COORDINATES

A standard way to represent a point in the plane $\mathbb{R}^2$ is by means of rectangular coordinates (x, y). However, as the reader has probably learned in elementary calculus, polar coordinates in the plane can be extremely useful. As portrayed in Figure 1.4.1, the coordinates (r, θ) are related to (x, y) by the formulas

$$x = r \cos \theta \qquad \text{and} \qquad y = r \sin \theta,$$

where we usually take $r \geq 0$ and $0 \leq \theta < 2\pi$.

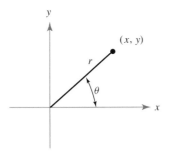

Figure 1.4.1 The polar coordinates of (x, y) are (r, θ).

Readers not familiar with polar coordinates are advised to study the relevant section of their calculus texts. We now set forth two ways of representing points in space other than by using rectangular Cartesian coordinates (x, y, z). These alternative coordinate systems are particularly well suited for certain types of problems, such as the evaluation of integrals using a change of variables.

Cylindrical Coordinates

DEFINITION The *cylindrical coordinates* (r, θ, z) of a point (x, y, z) are defined by (see Figure 1.4.2)

$$x = r \cos \theta, \qquad y = r \sin \theta, \qquad z = z. \tag{1}$$

Figure 1.4.2 Representing a point (x, y, z) in terms of its cylindrical coordinates r, θ, and z.

To express r, θ, and z in terms of x, y, and z, and to ensure that θ lies between 0 and 2π, we can write

$$r = \sqrt{x^2 + y^2}, \qquad \theta = \begin{cases} \tan^{-1}(y/x) & \text{if } x > 0 \text{ and } y \geq 0 \\ \pi + \tan^{-1}(y/x) & \text{if } x < 0 \\ 2\pi + \tan^{-1}(y/x) & \text{if } x > 0 \text{ and } y < 0, \end{cases} \qquad z = z,$$

where $\tan^{-1}(y/x)$ is between $-\pi/2$ and $\pi/2$. The requirement that $0 \leq \theta < 2\pi$ uniquely determines θ and $r \geq 0$ for a given x and y. If $x = 0$, then $\theta = \pi/2$ for $y > 0$ and $3\pi/2$ for $y < 0$. If $x = y = 0$, θ is undefined.

In other words, for any point (x, y, z), we represent the first and second coordinates in terms of polar coordinates and leave the third coordinate unchanged. Formula (1) shows that, given (r, θ, z), the triple (x, y, z) is completely determined, and vice versa, if we restrict θ to the interval $[0, 2\pi)$ (sometimes the range $(-\pi, \pi]$ is convenient) and require that $r > 0$.

To see why we use the term "cylindrical coordinates," note that if $0 \leq \theta < 2\pi$, $-\infty < z < \infty$, and $r = a$ is some positive constant, then the locus of these points is a cylinder of radius a (see Figure 1.4.3).

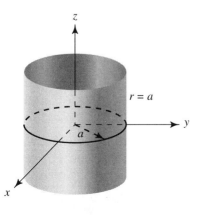

Figure 1.4.3 The graph of the points whose cylindrical coordinates satisfy $r = a$ is a cylinder.

EXAMPLE 1   (a) Find and plot the cylindrical coordinates of (6, 6, 8).   (b) If a point has cylindrical coordinates $(8, 2\pi/3, -3)$, what are its Cartesian coordinates? Plot.

SOLUTION For part (a), we have $r = \sqrt{6^2 + 6^2} = 6\sqrt{2}$ and $\theta = \tan^{-1}(6/6) = \tan^{-1}(1) = \pi/4$. Thus the cylindrical coordinates are $(6\sqrt{2}, \pi/4, 8)$. This is point P in Figure 1.4.4. For part (b), we have

$$x = r \cos\theta = 8 \cos\frac{2\pi}{3} = -\frac{8}{2} = -4$$

and

$$y = r \sin\theta = 8 \sin\frac{2\pi}{3} = 8\frac{\sqrt{3}}{2} = 4\sqrt{3}.$$

Thus the Cartesian coordinates are $(-4, 4\sqrt{3}, -3)$. This is point Q in the figure.   ▲

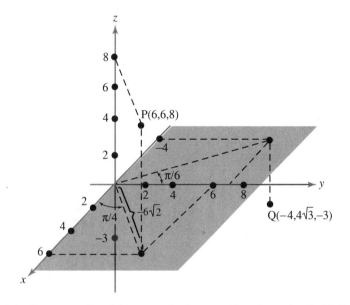

Figure 1.4.4 Some examples of the conversion between Cartesian and cylindrical coordinates.

Spherical Coordinates

Cylindrical coordinates are not the only possible generalization of polar coordinates to three dimensions. Recall that in two dimensions the magnitude of the vector $x\mathbf{i} + y\mathbf{j}$ (that is, $\sqrt{x^2 + y^2}$) is the r in the polar coordinate system. For cylindrical coordinates, the length of the vector $x\mathbf{i} + y\mathbf{j} + z\mathbf{k}$, namely,

$$\rho = \sqrt{x^2 + y^2 + z^2},$$

is not one of the coordinates of the system—instead, we used the magnitude $r = \sqrt{x^2 + y^2}$, the angle θ, and the "height" z.

We now modify this by introducing the *spherical coordinate* system, which uses ρ as a coordinate. Spherical coordinates are often useful for problems that possess spherical symmetry (symmetry about a point), whereas cylindrical coordinates can be applied when cylindrical symmetry (symmetry about a line) is involved.

Given a point $(x, y, z) \in \mathbb{R}^3$, let

$$\rho = \sqrt{x^2 + y^2 + z^2}$$

and represent x and y by polar coordinates in the xy plane:

$$x = r\cos\theta, \quad y = r\sin\theta \tag{2}$$

where $r = \sqrt{x^2 + y^2}$ and θ is given by formula (1). The coordinate z is given by

$$z = \rho\cos\phi,$$

where ϕ is the angle (between 0 and π, inclusive) that the radius vector $\mathbf{v} = x\mathbf{i} + y\mathbf{j} + z\mathbf{k}$ makes with the positive z axis, in the plane containing the vector $\mathbf{v}$ and the z axis (see Figure 1.4.5). Using the dot product, we can express ϕ as follows:

$$\cos\phi = \frac{\mathbf{v}\cdot\mathbf{k}}{\|\mathbf{v}\|}, \qquad \text{that is,} \qquad \phi = \cos^{-1}\left(\frac{\mathbf{v}\cdot\mathbf{k}}{\|\mathbf{v}\|}\right).$$

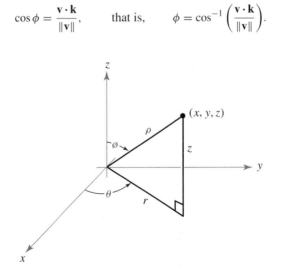

Figure 1.4.5 Spherical coordinates (ρ, θ, ϕ); the graph of points satisfying $\rho = a$ is a sphere.

We take as our coordinates the quantities ρ, θ, ϕ. Since

$$r = \rho \sin \phi,$$

we can use formula (2) to find x, y, z in terms of the spherical coordinates ρ, θ, ϕ.

DEFINITION   The *spherical coordinates* of (x, y, z) are defined as follows:

$$x = \rho \sin \phi \cos \theta, \qquad y = \rho \sin \phi \sin \theta, \qquad z = \rho \cos \phi \qquad (3)$$

where

$$\rho \geq 0, \qquad 0 \leq \theta < 2\pi, \qquad 0 \leq \phi \leq \pi.$$

Notice that the spherical coordinates θ and ϕ are similar to the geographic coordinates of longitude and latitude if we take the earth's axis to be the z axis. There are differences, though: The geographical longitude is $|\theta|$ and is called east or west longitude according to whether θ is positive or negative measure from the Greenwich meridian; the geographical latitude is $|\pi/2 - \phi|$ and is called north or south latitude according to whether $\pi/2 - \phi$ is positive or negative.

Note that in spherical coordinates the equation of the sphere of radius a centered at the origin takes on the particularly simple form: $\rho = a$.

EXAMPLE 2

(a) Find the spherical coordinates of $(1, -1, 1)$ and plot.

(b) Find the Cartesian coordinates of $(3, \pi/6, \pi/4)$ and plot.

(c) Let a point have Cartesian coordinates $(2, -3, 6)$. Find its spherical coordinates and plot.

(d) Let a point have spherical coordinates $(1, \pi/2, \pi/4)$. Find its Cartesian coordinates and plot.

SOLUTION

(a) $\rho = \sqrt{x^2 + y^2 + z^2} = \sqrt{1^2 + (-1)^2 + 1^2} = \sqrt{3},$

$\theta = \pi + \tan^{-1}\left(\frac{y}{x}\right) = \pi + \tan^{-1}\left(\frac{-1}{1}\right) = \pi - \frac{\pi}{4} = \frac{3\pi}{4}$

$\phi = \cos^{-1}\left(\frac{z}{\rho}\right) = \cos^{-1}\left(\frac{1}{\sqrt{3}}\right) \approx 0.955 \approx 54.74°.$

See Figure 1.4.6(a) and the formula for θ following equation (1).

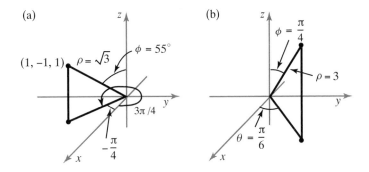

Figure 1.4.6 Finding (a) the spherical coordinates of the point $(1, -1, 1)$ and (b) the Cartesian coordinates of $(3, \pi/6, \pi/4)$.

(b) $x = \rho \sin\phi \cos\theta = 3\sin\left(\dfrac{\pi}{4}\right)\cos\left(\dfrac{\pi}{6}\right) = 3\left(\dfrac{1}{\sqrt{2}}\right)\dfrac{\sqrt{3}}{2} = \dfrac{3\sqrt{3}}{2\sqrt{2}}$,

$y = \rho \sin\phi \sin\theta = 3\sin\left(\dfrac{\pi}{4}\right)\sin\left(\dfrac{\pi}{6}\right) = 3\left(\dfrac{1}{\sqrt{2}}\right)\left(\dfrac{1}{2}\right) = \dfrac{3}{2\sqrt{2}}$,

$z = \rho \cos\phi = 3\cos\left(\dfrac{\pi}{4}\right) = \dfrac{3}{\sqrt{2}} = \dfrac{3\sqrt{2}}{2}$.

See Figure 1.4.6(b).

(c) $\rho = \sqrt{x^2 + y^2 + z^2} = \sqrt{2^2 + (-3)^2 + 6^2} = \sqrt{49} = 7$,

$\theta = 2\pi + \tan^{-1}\left(\dfrac{y}{x}\right) = 2\pi + \tan^{-1}\left(\dfrac{-3}{2}\right) \approx 5.3002 \text{ radians} \approx 303.69°$,

$\phi = \cos^{-1}\left(\dfrac{z}{\rho}\right) = \cos^{-1}\left(\dfrac{6}{7}\right) \approx 0.541 \approx 31.0°$.

See Figure 1.4.7(a).

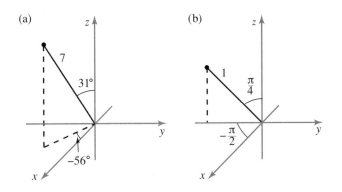

Figure 1.4.7 Finding (a) the spherical coordinates of $(2, -3, 6)$ and (b) the Cartesian coordinates of $(1, -\pi/2, \pi/4)$.

(d) $x = \rho \sin \phi \cos \theta = 1 \sin \left(\dfrac{\pi}{4} \right) \cos \left(\dfrac{\pi}{2} \right) = \left(\dfrac{\sqrt{2}}{2} \right) \cdot 0 = 0,$

$y = \rho \sin \phi \sin \theta = 1 \sin \left(\dfrac{\pi}{4} \right) \sin \left(\dfrac{\pi}{2} \right) = \left(\dfrac{\sqrt{2}}{2} \right)(-1) = -\dfrac{\sqrt{2}}{2},$

$z = \rho \cos \phi = 1 \cos \left(\dfrac{\pi}{4} \right) = \dfrac{\sqrt{2}}{2}.$

See Figure 1.4.7(b)   ▲

EXAMPLE 3   Express (a) the surface $xz = 1$ and (b) the surface $x^2 + y^2 - z^2 = 1$ in spherical coordinates.

SOLUTION   From formula (3), $x = \rho \sin \phi \cos \theta$, $z = \rho \cos \phi$, and so the surface (a) consists of all (ρ, θ, ϕ) such that

$$\rho^2 \sin \phi \cos \theta \cos \phi = 1, \quad \text{that is,} \quad \rho^2 \sin 2\phi \cos \theta = 2.$$

For part (b) we can write

$$x^2 + y^2 - z^2 = x^2 + y^2 + z^2 - 2z^2 = \rho^2 - 2\rho^2 \cos^2 \phi,$$

so that the surface is $\rho^2 (1 - 2 \cos^2 \phi) = 1$, or $-\rho^2 \cos (2\phi) = 1$.   ▲

Associated with cylindrical and spherical coordinates are unit vectors that are the counterparts of $\mathbf{i}, \mathbf{j},$ and $\mathbf{k}$ for rectangular coordinates. They are shown in Figure 1.4.8. For example, $\mathbf{e}_r$ is the unit vector parallel to the xy plane and in the radial direction, so that $\mathbf{e}_r = (\cos \theta)\mathbf{i} + (\sin \theta)\mathbf{j}$. Similarly, in spherical coordinates, $\mathbf{e}_\phi$ is the unit vector tangent to the curve parametrized by the variable ϕ with the variables ρ and θ held fixed. We shall use these unit vectors later when we use cylindrical and spherical coordinates in vector calculations.

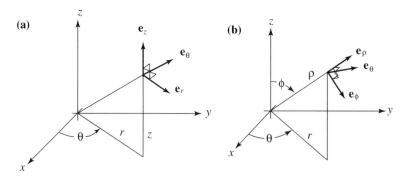

Figure 1.4.8   (a) Orthonormal vectors $\mathbf{e}_r$, $\mathbf{e}_\theta$, and $\mathbf{e}_z$ associated with cylindrical coordinates. The vector $\mathbf{e}_r$ is parallel to the line labeled r. (b) Orthonormal vectors $\mathbf{e}_\rho$, $\mathbf{e}_\theta$, and $\mathbf{e}_\phi$ associated with spherical coordinates.

EXERCISES

1. (a) The following points are given in cylindrical coordinates; express each in rectangular coordinates and spherical coordinates: $(1, 45°, 1)$, $(2, \pi/2, -4)$, $(0, 45°, 10)$, $(3, \pi/6, 4)$, $(1, \pi/6, 0)$, and $(2, 3\pi/4, -2)$. (Only the first point is solved in the Study Guide.)

(b) Change each of the following points from rectangular coordinates to spherical coordinates and to cylindrical coordinates: $(2, 1, -2)$, $(0, 3, 4)$, $(\sqrt{2}, 1, 1)$, $(-2\sqrt{3}, -2, 3)$. (Only the first point is solved in the Study Guide.)

2. Describe the geometric meaning of the following mappings in cylindrical coordinates:

(a) $(r, \theta, z) \mapsto (r, \theta, -z)$

(b) $(r, \theta, z) \mapsto (r, \theta + \pi, -z)$

(c) $(r, \theta, z) \mapsto (-r, \theta - \pi/4, z)$

3. Describe the geometric meaning of the following mappings in spherical coordinates:

(a) $(\rho, \theta, \phi) \mapsto (\rho, \theta + \pi, \phi)$

(b) $(\rho, \theta, \phi) \mapsto (\rho, \theta, \pi - \phi)$

(c) $(\rho, \theta, \phi) \mapsto (2\rho, \theta + \pi/2, \phi)$

4. (a) Describe the surfaces $r = $ constant, $\theta = $ constant, and $z = $ constant in the cylindrical coordinate system.

(b) Describe the surfaces $\rho = $ constant, $\theta = $ constant, and $\phi = $ constant in the spherical coordinate system.

5. Show that to represent each point in $\mathbb{R}^3$ by spherical coordinates it is necessary to take only values of θ between 0 and 2π, values of ϕ between 0 and π, and values of $\rho \geq 0$. Are coordinates unique if we allow $\rho \leq 0$?

6. Using cylindrical coordinates and the orthonormal (orthogonal normalized) vectors $\mathbf{e}_r$, $\mathbf{e}_\theta$, and $\mathbf{e}_z$ (see Figure 1.4.8),

(a) express each of $\mathbf{e}_r$, $\mathbf{e}_\theta$, and $\mathbf{e}_z$ in terms of $\mathbf{i}, \mathbf{j}, \mathbf{k}$ and (x, y, z); and

(b) calculate $\mathbf{e}_\theta \times \mathbf{j}$ both analytically, using part (a), and geometrically.

7. Using spherical coordinates and the orthonormal (orthogonal normalized) vectors $\mathbf{e}_\rho$, $\mathbf{e}_\theta$, and $\mathbf{e}_\phi$ (see Figure 1.4.8(b)),

(a) express each of $\mathbf{e}_\rho$, $\mathbf{e}_\theta$, and $\mathbf{e}_\phi$ in terms of $\mathbf{i}, \mathbf{j}, \mathbf{k}$ and (x, y, z); and

(b) calculate $\mathbf{e}_\theta \times \mathbf{j}$ and $\mathbf{e}_\phi \times \mathbf{j}$ both analytically and geometrically.

8. Express the plane $z = x$ in (a) cylindrical and (b) spherical coordinates.

9. Show that in spherical coordinates:

(a) ρ is the length of $x\mathbf{i} + y\mathbf{j} + z\mathbf{k}$.

(b) $\phi = \cos^{-1} (\mathbf{v} \cdot \mathbf{k}/\|\mathbf{v}\|)$, where $\mathbf{v} = x\mathbf{i} + y\mathbf{j} + z\mathbf{k}$.

(c) $\theta = \cos^{-1} (\mathbf{u} \cdot \mathbf{i}/\|\mathbf{u}\|)$, where $\mathbf{u} = x\mathbf{i} + y\mathbf{j}$.

10. Two surfaces are described in spherical coordinates by the two equations $\rho = f(\theta, \phi)$ and $\rho = -2f(\theta, \phi)$, where $f(\theta, \phi)$ is a function of two variables. How is the second surface obtained geometrically from the first?

11. A circular membrane in space lies over the region $x^2 + y^2 \leq a^2$. The maximum z component of points in the membrane is b. Assume that (x, y, z) is a point on the deflected membrane. Show that the corresponding point (r, θ, z) in cylindrical coordinates satisfies the conditions $0 \leq r \leq a, 0 \leq \theta \leq 2\pi, |z| \leq b$.

12.   A tank in the shape of a right-circular cylinder of radius 10 ft and height 16 ft is half filled and lying on its side. Describe the air space inside the tank by suitably chosen cylindrical coordinates.

13.   A vibrometer is to be designed which withstands the heating effects of its spherical enclosure of diameter d, which is buried to a depth $d/3$ in the earth, the upper portion being heated by the sun (assume the earth is flat). Heat conduction analysis requires a description of the buried portion of the enclosure in spherical coordinates. Find it.

14.   An oil filter cartridge is a porous right-circular cylinder inside which oil diffuses from the axis to the outer curved surface. Describe the cartridge in cylindrical coordinates, if the diameter of the filter is 4.5 in, the height is 5.6 in, and the center of the cartridge is drilled (all the way through) from the top to admit a $\frac{5}{8}$-in-diameter bolt.

15.   Describe the surface given in spherical coordinates by $\rho = \cos 2\theta$.

1.5 *n*-DIMENSIONAL EUCLIDEAN SPACE

Vectors in *n*-space

In Sections 1.1 and 1.2 we studied the spaces $\mathbb{R} = \mathbb{R}^1, \mathbb{R}^2$, and $\mathbb{R}^3$ and gave geometric interpretations to them. For example, a point (x, y, z) in $\mathbb{R}^3$ can be thought of as a geometric object, namely, the directed line segment or vector emanating from the origin and ending at the point (x, y, z). We can therefore think of $\mathbb{R}^3$ in either of two ways:

(i)   Algebraically, as a set of triples (x, y, z) where x, y, and z are real numbers

(ii)   Geometrically, as a set of directed line segments

These two ways of looking at $\mathbb{R}^3$ are equivalent. For generalization it is easier to use definition (i). Specifically, we can define $\mathbb{R}^n$, where n is a positive integer (possibly greater than 3), to be the set of all ordered n-tuples $(x_1, x_2, \ldots, x_n)$, where the x_i are real numbers. For instance, $(1, \sqrt{5}, 2, \sqrt{3}) \in \mathbb{R}^4$.

The set $\mathbb{R}^n$ so defined is known as ***Euclidean n-space***, and its elements, which we write as $\mathbf{x} = (x_1, x_2, \ldots, x_n)$, are known as ***vectors*** or ***n-vectors***. By setting $n = 1, 2$, or 3, we recover the line, the plane, and three-dimensional space, respectively.

We launch our study of Euclidean n-space by introducing several algebraic operations. These are analogous to those introduced in Section 1.1 for $\mathbb{R}^2$ and $\mathbb{R}^3$. The first two, addition and scalar multiplication, are defined as follows:

(i) $(x_1, x_2, \ldots, x_n) + (y_1, y_2, \ldots, y_n) = (x_1 + y_1, x_2 + y_2, \ldots, x_n + y_n)$;

and

(ii)   for any real number α,

$$\alpha(x_1, x_2, \ldots, x_n) = (\alpha x_1, \alpha x_2, \ldots, \alpha x_n).$$

The geometric significance of these operations for $\mathbb{R}^2$ and $\mathbb{R}^3$ was discussed in Section 1.1.

The n vectors $\mathbf{e}_1 = (1, 0, 0, \ldots, 0)$, $\mathbf{e}_2 = (0, 1, 0, \ldots, 0)$, $\ldots$, $\mathbf{e}_n = (0, 0, \ldots, 0, 1)$ are called the ***standard basis vectors*** of $\mathbb{R}^n$, and they generalize the three mutually orthogonal unit vectors $\mathbf{i}, \mathbf{j}, \mathbf{k}$ of $\mathbb{R}^3$. The vector $\mathbf{x} = (x_1, x_2, \ldots, x_n)$ can then be written as $\mathbf{x} = x_1 \mathbf{e}_1 + x_2 \mathbf{e}_2 + \cdots + x_n \mathbf{e}_n$.

For two vectors $\mathbf{x} = (x_1, x_2, x_3)$ and $\mathbf{y} = (y_1, y_2, y_3)$ in $\mathbb{R}^3$, we defined the *dot* or *inner product* $\mathbf{x} \cdot \mathbf{y}$ to be the real number $\mathbf{x} \cdot \mathbf{y} = x_1 y_1 + x_2 y_2 + x_3 y_3$. This definition easily extends to $\mathbb{R}^n$; specifically, for $\mathbf{x} = (x_1, x_2, \ldots, x_n)$, $\mathbf{y} = (y_1, y_2, \ldots, y_n)$, we define the ***inner product*** of $\mathbf{x}$ and $\mathbf{y}$ to be $\mathbf{x} \cdot \mathbf{y} = x_1 y_1 + x_2 y_2 + \cdots + x_n y_n$. In $\mathbb{R}^n$, the notation $\langle \mathbf{x}, \mathbf{y} \rangle$ is often used in place of $\mathbf{x} \cdot \mathbf{y}$ for the inner product.

Continuing the analogy with $\mathbb{R}^3$, we are led to define the abstract notion of the ***length*** or ***norm*** of a vector $\mathbf{x}$ by the formula

$$\text{length of } \mathbf{x} = \|\mathbf{x}\| = \sqrt{\mathbf{x} \cdot \mathbf{x}} = \sqrt{x_1^2 + x_2^2 + \cdots + x_n^2}.$$

If $\mathbf{x}$ and $\mathbf{y}$ are two vectors in the plane ($\mathbb{R}^2$) or in space ($\mathbb{R}^3$), then we know that the angle θ between them is given by the formula

$$\cos \theta = \frac{\mathbf{x} \cdot \mathbf{y}}{\|\mathbf{x}\| \|\mathbf{y}\|}.$$

The right side of this equation can be defined in $\mathbb{R}^n$ as well as in $\mathbb{R}^2$. It still represents the cosine of the angle between $\mathbf{x}$ and $\mathbf{y}$; this angle is geometrically well defined, since $\mathbf{x}$ and $\mathbf{y}$ lie in a two-dimensional subspace of $\mathbb{R}^n$ (the plane determined by $\mathbf{x}$ and $\mathbf{y}$).

It will be useful to have available some algebraic properties of the inner product. These are summarized in the next theorem [see properties (i), (ii), (iii), and (iv) of Section 1.2].

THEOREM 3   *For* $\mathbf{x}, \mathbf{y}, \mathbf{z} \in \mathbb{R}^n$ *and* α, β*, real numbers, we have*

(i) $(\alpha \mathbf{x} + \beta \mathbf{y}) \cdot \mathbf{z} = \alpha(\mathbf{x} \cdot \mathbf{z}) + \beta(\mathbf{y} \cdot \mathbf{z})$.

(ii) $\mathbf{x} \cdot \mathbf{y} = \mathbf{y} \cdot \mathbf{x}$.

(iii) $\mathbf{x} \cdot \mathbf{x} \geq 0$.

(iv) $\mathbf{x} \cdot \mathbf{x} = 0$ *if and only if* $\mathbf{x} = \mathbf{0}$.

PROOF   Each of the four assertions can be proved by a simple computation. For example, to prove property (i) we write

$$
\begin{aligned}
(\alpha \mathbf{x} + \beta \mathbf{y}) \cdot \mathbf{z} &= (\alpha x_1 + \beta y_1, \alpha x_2 + \beta y_2, \ldots, \alpha x_n + \beta y_n) \cdot (z_1, z_2, \ldots, z_n) \\
&= (\alpha x_1 + \beta y_1) z_1 + (\alpha x_2 + \beta y_2) z_2 + \cdots + (\alpha x_n + \beta y_n) z_n \\
&= a x_1 z_1 + \beta y_1 z_1 + a x_2 z_2 + \beta y_2 z_2 + \cdots + \alpha x_n z_n + \beta y_n z_n \\
&= \alpha(\mathbf{x} \cdot \mathbf{z}) + \beta(\mathbf{y} \cdot \mathbf{z}).
\end{aligned}
$$

The other proofs are similar.   ■

In Section 1.2 we proved an interesting property of dot products, called the Cauchy-Schwarz inequality (sometimes called the Cauchy-Bunyakovskii-Schwarz inequality, or simply the CBS inequality, because it was independently discovered in special cases by the French mathematician Cauchy, the Russian mathematician Bunyakovskii, and the German mathematician Schwarz). For $\mathbb{R}^2$ our proof required the use of the law of cosines. For $\mathbb{R}^n$ we could also use this method, by confining our attention to a plane in $\mathbb{R}^n$. However, we can also give a direct, completely algebraic proof.

THEOREM 4: CAUCHY-SCHWARZ INEQUALITY IN $\mathbb{R}^n$ *Let* $\mathbf{x}, \mathbf{y}$ *be vectors in* $\mathbb{R}^n$. *Then*

$$|\mathbf{x} \cdot \mathbf{y}| \le \|\mathbf{x}\| \|\mathbf{y}\|.$$

PROOF   Let $a = \mathbf{y} \cdot \mathbf{y}$ and $b = -\mathbf{x} \cdot \mathbf{y}$. If $a = 0$ the theorem is clearly valid, since then $\mathbf{y} = \mathbf{0}$ and both sides of the inequality reduce to 0. Thus we may suppose $a \ne 0$. By Theorem 2 we have

$$0 \le (a\mathbf{x} + b\mathbf{y}) \cdot (a\mathbf{x} + b\mathbf{y}) = a^2 \mathbf{x} \cdot \mathbf{x} + 2ab\mathbf{x} \cdot \mathbf{y} + b^2 \mathbf{y} \cdot \mathbf{y}$$
$$= (\mathbf{y} \cdot \mathbf{y})^2 \mathbf{x} \cdot \mathbf{x} - (\mathbf{y} \cdot \mathbf{y})(\mathbf{x} \cdot \mathbf{y})^2.$$

Dividing by $\mathbf{y} \cdot \mathbf{y}$ gives $0 \le (\mathbf{y} \cdot \mathbf{y})(\mathbf{x} \cdot \mathbf{x}) - (\mathbf{x} \cdot \mathbf{y})^2$, that is, $(\mathbf{x} \cdot \mathbf{y})^2 \le (\mathbf{x} \cdot \mathbf{x})(\mathbf{y} \cdot \mathbf{y}) = \|\mathbf{x}\|^2 \|\mathbf{y}\|^2$. Taking square roots on both sides of this inequality yields the desired result. ∎

There is a useful consequence of the Cauchy-Schwarz inequality in terms of lengths. The triangle inequality is geometrically clear in $\mathbb{R}^3$ and was discussed in Section 1.2. The *analytic* proof of the triangle inequality that we gave in Section 1.2 works exactly the same in $\mathbb{R}^n$ and proves the following:

COROLLARY: TRIANGLE INEQUALITY IN $\mathbb{R}^n$ *Let* $\mathbf{x}, \mathbf{y}$ *be vectors in* $\mathbb{R}^3$. *Then*

$$\|\mathbf{x} + \mathbf{y}\| \le \|\mathbf{x}\| + \|\mathbf{y}\|.$$

If Theorem 4 and its corollary are written out algebraically, they become the following useful inequalities:

$$\left| \sum_{i=1}^{n} x_i y_i \right| \le \left(\sum_{i=1}^{n} x_i^2 \right)^{1/2} \left(\sum_{i=1}^{n} y_i^2 \right)^{1/2};$$
$$\left(\sum_{i=1}^{n} (x_i + y_i)^2 \right)^{1/2} \le \left(\sum_{i=1}^{n} x_i^2 \right)^{1/2} + \left(\sum_{i=1}^{n} y_i^2 \right)^{1/2}.$$

EXAMPLE 1   Let $\mathbf{x} = (1, 2, 0, -1)$ and $\mathbf{y} = (-1, 1, 1, 0)$. Verify Theorem 4 and its corollary in this case.

SOLUTION

$$\|\mathbf{x}\| = \sqrt{1^2 + 2^2 + 0^2 + (-1)^2} = \sqrt{6}$$
$$\|\mathbf{y}\| = \sqrt{(-1)^2 + 1^2 + 1^2 + 0^2} = \sqrt{3}$$
$$\mathbf{x} \cdot \mathbf{y} = 1(-1) + 2 \cdot 1 + 0 \cdot 1 + (-1)0 = 1$$
$$\mathbf{x} + \mathbf{y} = (0, 3, 1, -1)$$
$$\|\mathbf{x} + \mathbf{y}\| = \sqrt{0^2 + 3^2 + 1^2 + (-1)^2} = \sqrt{11}.$$

We compute $\mathbf{x} \cdot \mathbf{y} = 1 \le 4.24 \approx \sqrt{6}\sqrt{3} = \|\mathbf{x}\|\,\|\mathbf{y}\|$, which verifies Theorem 4. Similarly, we can check its corollary:

$$\|\mathbf{x} + \mathbf{y}\| = \sqrt{11} \approx 3.32$$
$$\le 4.18 = 2.45 + 1.73 \approx \sqrt{6} + \sqrt{3} = \|\mathbf{x}\| + \|\mathbf{y}\|. \quad \blacktriangle$$

By analogy with $\mathbb{R}^3$, we can define the notion of distance in $\mathbb{R}^n$; namely, if $\mathbf{x}$ and $\mathbf{y}$ are points in $\mathbb{R}^n$, the ***distance between x and y*** is defined to be $\|\mathbf{x} - \mathbf{y}\|$, or the length of the vector $\mathbf{x} - \mathbf{y}$. We do not attempt to define the cross product on $\mathbb{R}^n$ except for $n = 3$.[*]

General Matrices

Generalizing 2×2 and 3×3 matrices (see Section 1.3), we can consider $m \times n$ matrices, which are arrays of mn numbers:

$$A = \begin{bmatrix} a_{11} & a_{12} & \cdots & a_{1n} \\ a_{21} & a_{22} & \cdots & a_{2n} \\ \vdots & \vdots & & \vdots \\ a_{m1} & a_{m2} & \cdots & a_{mn} \end{bmatrix}$$

We shall also write A as $[a_{ij}]$. We define addition and multiplication by a scalar componentwise, just as we did for vectors. Given two $m \times n$ matrices A and B, we can add them to obtain a new $m \times n$ matrix $C = A + B$, whose ijth entry c_{ij} is the sum of a_{ij} and b_{ij}. It is clear that $A + B = B + A$.

EXAMPLE 2

(a) $\begin{bmatrix} 2 & 1 & 0 \\ 3 & 4 & 1 \end{bmatrix} + \begin{bmatrix} -1 & 1 & 3 \\ 0 & 0 & 7 \end{bmatrix} = \begin{bmatrix} 1 & 2 & 3 \\ 3 & 4 & 8 \end{bmatrix}.$

(b) $[1 \quad 2] + [0 \quad -1] = [1 \quad 1].$

---

[*]There is a vector product with most of the properties of the cross product in $\mathbb{R}^7$; this involves a number system called the "octonians." The nonexistence of a cross product in other dimensions is a deep result that goes beyond the scope of this book. For further information see W. S. Massey, *American Mathematical Monthly* 90 (1983): 697.

(c) $\begin{bmatrix} 2 & 1 \\ 1 & 2 \end{bmatrix} - \begin{bmatrix} 1 & 0 \\ 0 & 1 \end{bmatrix} = \begin{bmatrix} 1 & 1 \\ 1 & 1 \end{bmatrix}.$ ▲

Given a scalar λ and an $m \times n$ matrix A, we can multiply A by λ to obtain a new $m \times n$ matrix $\lambda A = C$, whose ijth entry c_{ij} is the product λa_{ij}.

EXAMPLE 3

$$3 \begin{bmatrix} 1 & -1 & 2 \\ 0 & 1 & 5 \\ 1 & 0 & 3 \end{bmatrix} = \begin{bmatrix} 3 & -3 & 6 \\ 0 & 3 & 15 \\ 3 & 0 & 9 \end{bmatrix}.$$ ▲

Next we turn to matrix multiplication. If $A = [a_{ij}]$, $B = [b_{ij}]$ are $n \times n$ matrices, then the product $AB = C$ has entries given by

$$c_{ij} = \sum_{k=1}^{n} a_{ik} b_{kj},$$

which is the dot product of the ith row of A and the jth column of B:

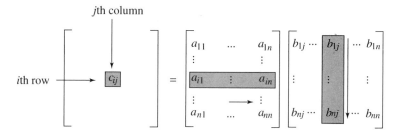

EXAMPLE 4   Let

$$A = \begin{bmatrix} 1 & 0 & 3 \\ 2 & 1 & 0 \\ 1 & 0 & 0 \end{bmatrix} \quad \text{and} \quad B = \begin{bmatrix} 0 & 1 & 0 \\ 1 & 0 & 0 \\ 0 & 1 & 1 \end{bmatrix}$$

Then

$$AB = \begin{bmatrix} 0 & 4 & 3 \\ 1 & 2 & 0 \\ 0 & 1 & 0 \end{bmatrix} \quad \text{and} \quad BA = \begin{bmatrix} 2 & 1 & 0 \\ 1 & 0 & 3 \\ 3 & 1 & 0 \end{bmatrix}.$$

Observe that $AB \neq BA$. ▲

Similarly, we can multiply an $m \times n$ matrix (m rows, n columns) by an $n \times p$ matrix (n rows, p columns) to obtain an $m \times p$ matrix (m rows, p columns) by the same rule. Note that for AB to be defined, the number of *columns* of A must equal the number of *rows* of B.

EXAMPLE 5   Let

$$A = \begin{bmatrix} 2 & 0 & 1 \\ 1 & 1 & 2 \end{bmatrix} \quad \text{and} \quad B = \begin{bmatrix} 1 & 0 & 2 \\ 0 & 2 & 1 \\ 1 & 1 & 1 \end{bmatrix}.$$

Then

$$AB = \begin{bmatrix} 3 & 1 & 5 \\ 3 & 4 & 5 \end{bmatrix},$$

and BA is not defined.   ▲

EXAMPLE 6   Let

$$A = \begin{bmatrix} 1 \\ 2 \\ 1 \\ 3 \end{bmatrix} \quad \text{and} \quad B = [2 \quad 2 \quad 1 \quad 2].$$

Then

$$AB = \begin{bmatrix} 2 & 2 & 1 & 2 \\ 4 & 4 & 2 & 4 \\ 2 & 2 & 1 & 2 \\ 6 & 6 & 3 & 6 \end{bmatrix} \quad \text{and} \quad BA = [13]. \quad ▲$$

Any $m \times n$ matrix A determines a mapping of $\mathbb{R}^n$ to $\mathbb{R}^m$ defined as follows: Let $\mathbf{x} = (x_1, \ldots, x_n) \in \mathbb{R}^n$; consider the $n \times 1$ column matrix associated with $\mathbf{x}$, which we shall *temporarily* denote $\mathbf{x}^T$:

$$\mathbf{x}^T = \begin{bmatrix} x_1 \\ \vdots \\ x_n \end{bmatrix}$$

and multiply A by $\mathbf{x}^T$ (considered to be an $n \times 1$ matrix) to get a new $m \times 1$ matrix:

$$A\mathbf{x}^T = \begin{bmatrix} a_{11} & \cdots & a_{1n} \\ \vdots & & \vdots \\ a_{m1} & \cdots & a_{mn} \end{bmatrix} \begin{bmatrix} x_1 \\ \vdots \\ x_n \end{bmatrix} = \begin{bmatrix} y_1 \\ \vdots \\ y_m \end{bmatrix} = \mathbf{y}^T$$

corresponding to the vector $\mathbf{y} = (y_1, \ldots, y_m)$.*   Thus, although it may cause some

---

*To use a matrix A to get a mapping from vectors $\mathbf{x} = (x_1, \ldots, x_n)$ to vectors $\mathbf{y} = (y_1, \ldots, y_n)$ according to the equation $A\mathbf{x}^T = \mathbf{y}^T$, we write the vectors in the column form $\mathbf{x}^T$ instead of the row form $(x_1, \ldots, x_n)$. This sudden switch from writing $\mathbf{x}$ as a row to writing $\mathbf{x}$ as a column is necessitated by standard conventions on matrix multiplication.

confusion, we will write $\mathbf{x} = (x_1, \ldots, x_n)$ and $\mathbf{y} = (y_1, \ldots, y_m)$ as column matrices

$$\mathbf{x} = \begin{bmatrix} x_1 \\ \vdots \\ x_n \end{bmatrix}, \qquad \mathbf{y} = \begin{bmatrix} y_1 \\ \vdots \\ y_m \end{bmatrix}$$

when dealing with matrix multiplication; that is, we will *identify* these two forms of writing vectors. Thus, we will delete the T on $\mathbf{x}^T$ and view $\mathbf{x}^T$ and $\mathbf{x}$ as the same; that is, $\mathbf{x} = \mathbf{x}^T$.

Thus, $A\mathbf{x} = \mathbf{y}$ will "really" mean the following: Write $\mathbf{x}$ as a column matrix, multiply it by A, and let $\mathbf{y}$ be the vector whose components are those of the resulting column matrix. The rule $\mathbf{x} \mapsto A\mathbf{x}$ therefore defines a mapping of $\mathbb{R}^n$ to $\mathbb{R}^m$. This mapping is linear; that is, it satisfies

$$A(\mathbf{x} + \mathbf{y}) = A\mathbf{x} + A\mathbf{y}$$

$$A(\alpha\mathbf{x}) = \alpha(A\mathbf{x}), \qquad \alpha \text{ a scalar,}$$

as may be easily verified. One learns in a linear algebra course that, conversely, any linear transformation of $\mathbb{R}^n$ to $\mathbb{R}^m$ is representable in this way by an $m \times n$ matrix.

If $A = [a_{ij}]$ is an $m \times n$ matrix and $\mathbf{e}_j$ is the jth standard basis vector of $\mathbb{R}^n$, then $A\mathbf{e}_j$ is a vector in $\mathbb{R}^m$ with components the same as the jth column of A. That is, the ith component of $A\mathbf{e}_j$ is a_{ij}. In symbols, $(A\mathbf{e}_j)_i = a_{ij}$.

EXAMPLE 7   If

$$A = \begin{bmatrix} 1 & 0 & 3 \\ -1 & 0 & 1 \\ 2 & 1 & 2 \\ -1 & 2 & 1 \end{bmatrix},$$

then $\mathbf{x} \mapsto A\mathbf{x}$ of $\mathbb{R}^3$ to $\mathbb{R}^4$ is the mapping defined by

$$\begin{bmatrix} x_1 \\ x_2 \\ x_3 \end{bmatrix} \mapsto \begin{bmatrix} x_1 + 3x_3 \\ -x_1 + x_3 \\ 2x_1 + x_2 + 2x_3 \\ -x_1 + 2x_2 + x_3 \end{bmatrix}. \quad \blacktriangle$$

EXAMPLE 8   The following illustrates what happens to a specific point when mapped by a 4×3 matrix:

$$A\mathbf{e}_2 = \begin{bmatrix} 4 & 2 & 9 \\ 3 & 5 & 4 \\ 1 & 2 & 3 \\ 0 & 1 & 2 \end{bmatrix} \begin{bmatrix} 0 \\ 1 \\ 0 \end{bmatrix} = \begin{bmatrix} 2 \\ 5 \\ 2 \\ 1 \end{bmatrix} = \text{2nd column of } A. \quad \blacktriangle$$

Properties of Matrices

Matrix multiplication is not, in general, ***commutative***: If A and B are $n \times n$ matrices, then generally

$$AB \neq BA,$$

as Examples 4, 5, and 6 show.

An $n \times n$ matrix is said to be ***invertible*** if there is an $n \times n$ matrix B such that

$$AB = BA = I_n,$$

where

$$I_n = \begin{bmatrix} 1 & 0 & 0 & \cdots & 0 \\ 0 & 1 & 0 & \cdots & 0 \\ 0 & 0 & 1 & \cdots & 0 \\ \vdots & \vdots & \vdots & & \vdots \\ 0 & 0 & 0 & \cdots & 1 \end{bmatrix}$$

is the $n \times n$ identity matrix: I_n has the property that $I_n C = C I_n = C$ for any $n \times n$ matrix C. We denote B by A^{-1} and call A^{-1} the ***inverse*** of A. The inverse, when it exists, is unique.

EXAMPLE 9   If

$$A = \begin{bmatrix} 2 & 4 & 0 \\ 0 & 2 & 1 \\ 3 & 0 & 2 \end{bmatrix}, \quad \text{then} \quad A^{-1} = \frac{1}{20} \begin{bmatrix} 4 & -8 & 4 \\ 3 & 4 & -2 \\ -6 & 12 & 4 \end{bmatrix},$$

since $AA^{-1} = I_3 = A^{-1}A$, as may be checked by matrix multiplication. $\blacktriangle$

Methods of computing inverses are learned in linear algebra; we won't require these methods in this book. If A is invertible, the equation $A\mathbf{x} = \mathbf{y}$ can be solved for the vector $\mathbf{x}$ by multiplying both sides by A^{-1} to obtain $\mathbf{x} = A^{-1}\mathbf{y}$.

In Section 1.3 we defined the determinant of a 3×3 matrix. This can be generalized by induction to $n \times n$ determinants. We illustrate here how to write the determinant of a 4×4 matrix in terms of the determinants of 3×3 matrices:

$$\begin{vmatrix} a_{11} & a_{12} & a_{13} & a_{14} \\ a_{21} & a_{22} & a_{23} & a_{24} \\ a_{31} & a_{32} & a_{33} & a_{34} \\ a_{41} & a_{42} & a_{43} & a_{44} \end{vmatrix} = a_{11} \begin{vmatrix} a_{22} & a_{23} & a_{24} \\ a_{32} & a_{33} & a_{34} \\ a_{42} & a_{43} & a_{44} \end{vmatrix} - a_{12} \begin{vmatrix} a_{21} & a_{23} & a_{24} \\ a_{31} & a_{33} & a_{34} \\ a_{41} & a_{43} & a_{44} \end{vmatrix}$$

$$+ a_{13} \begin{vmatrix} a_{21} & a_{22} & a_{24} \\ a_{31} & a_{32} & a_{34} \\ a_{41} & a_{42} & a_{44} \end{vmatrix} - a_{14} \begin{vmatrix} a_{21} & a_{22} & a_{23} \\ a_{31} & a_{32} & a_{33} \\ a_{41} & a_{42} & a_{43} \end{vmatrix}$$

[see formula (2) of Section 1.3; the signs alternate $+, -, +, -$].

The basic properties of 3×3 determinants reviewed in Section 1.3 remain valid for $n \times n$ determinants. In particular, we note the fact that if A is an $n \times n$ matrix and B is the matrix formed by adding a scalar multiple of one row (or column) of A to another row (or, respectively, column) of A, then the determinant of A is equal to the determinant of B (see Example 10 below).

A basic theorem of linear algebra states that an $n \times n$ matrix A is invertible if and only if the determinant of A is not zero. Another basic property is that $\det(AB) = (\det A)(\det B)$. In this text we shall not make use of many details of linear algebra, and so we shall leave these assertions unproved.

EXAMPLE 10   Let

$$A = \begin{bmatrix} 1 & 0 & 1 & 0 \\ 1 & 1 & 1 & 1 \\ 2 & 1 & 0 & 1 \\ 1 & 1 & 0 & 2 \end{bmatrix}.$$

Find det A. Does A have an inverse?

SOLUTION   Adding $(-1) \times$ first column to the third column, we get

$$\det A = \begin{vmatrix} 1 & 0 & 0 & 0 \\ 1 & 1 & 0 & 1 \\ 2 & 1 & -2 & 1 \\ 1 & 1 & -1 & 2 \end{vmatrix} = 1 \begin{vmatrix} 1 & 0 & 1 \\ 1 & -2 & 1 \\ 1 & -1 & 2 \end{vmatrix}.$$

Adding $(-1) \times$ first column to the third column of this 3×3 determinant gives

$$\det A = \begin{vmatrix} 1 & 0 & 0 \\ 1 & -2 & 0 \\ 1 & -1 & 1 \end{vmatrix} = \begin{vmatrix} -2 & 0 \\ -1 & 1 \end{vmatrix} = -2.$$

Thus, $\det A = -2 \neq 0$, and so A has an inverse.   ▲

If we have three matrices A, B, and C such that the products AB and BC are defined, then the products $(AB)C$ and $A(BC)$ are defined and are in fact equal (that is, matrix multiplication is *associative*). We call this the *triple product* of matrices and denote it by ABC.

EXAMPLE 11   Let

$$A = \begin{bmatrix} 3 \\ 5 \end{bmatrix}, \qquad B = [1 \quad 1], \qquad \text{and} \qquad C = \begin{bmatrix} 1 \\ 2 \end{bmatrix}.$$

Then

$$ABC = A(BC) = \begin{bmatrix} 3 \\ 5 \end{bmatrix}[3] = \begin{bmatrix} 9 \\ 15 \end{bmatrix}. \quad \blacktriangle$$

EXAMPLE 12

$$\begin{bmatrix} 2 & 0 \\ 0 & 1 \end{bmatrix}\begin{bmatrix} 1 & 1 \\ 1 & 1 \end{bmatrix}\begin{bmatrix} 0 & -1 \\ 1 & 1 \end{bmatrix} = \begin{bmatrix} 2 & 0 \\ 0 & 1 \end{bmatrix}\begin{bmatrix} 1 & 0 \\ 1 & 0 \end{bmatrix} = \begin{bmatrix} 2 & 0 \\ 1 & 0 \end{bmatrix}. \quad \blacktriangle$$

EXERCISES

1. Calculate the dot product of $\mathbf{x} = (1, -1, 0, 2) \in \mathbb{R}^4$ and $\mathbf{y} = (1, 2, 3, 4) \in \mathbb{R}^4$.

2. In $\mathbb{R}^n$ show that
    (a) $2\|\mathbf{x}\|^2 + 2\|\mathbf{y}\|^2 = \|\mathbf{x} + \mathbf{y}\|^2 + \|\mathbf{x} - \mathbf{y}\|^2$ (This is known as the *parallelogram law*.)
    (b) $\|\mathbf{x} - \mathbf{y}\|\|\mathbf{x} + \mathbf{y}\| \leq \|\mathbf{x}\|^2 + \|\mathbf{y}\|^2$
    (c) $4\langle \mathbf{x}, \mathbf{y} \rangle = \|\mathbf{x} + \mathbf{y}\|^2 - \|\mathbf{x} - \mathbf{y}\|^2$ (This is called the *polarization identity*.)

Interpret these results geometrically in terms of the parallelogram formed by $\mathbf{x}$ and $\mathbf{y}$.

Verify the CBS inequality and the triangle inequality for the vectors in Exercises 3 to 6.

3. $\mathbf{x} = (2, 0, -1)$, $\mathbf{y} = (4, 0, -2)$

4. $\mathbf{x} = (1, 0, 2, 6)$, $\mathbf{y} = (3, 8, 4, 1)$

5. $\mathbf{x} = (1, -1, 1, -1, 1)$, $\mathbf{y} = (3, 0, 0, 0, 2)$

6. $\mathbf{x} = (1, 0, 0, 0, 1)$, $\mathbf{y} = (-1, 0, 0, 1)$

7. Compute AB, det A, det B, det (AB), and det $(A + B)$ for

$$A = \begin{bmatrix} 1 & -1 & 0 \\ 0 & 3 & 2 \\ 3 & 1 & 1 \end{bmatrix} \qquad \text{and} \qquad B = \begin{bmatrix} -2 & 0 & 2 \\ -1 & 1 & -1 \\ 1 & 4 & 3 \end{bmatrix}.$$

8. Compute AB, $\det A$, $\det B$, $\det (AB)$, and $\det (A + B)$ for

$$A = \begin{bmatrix} 3 & 0 & 1 \\ 1 & 2 & -1 \\ 1 & 0 & 1 \end{bmatrix} \quad \text{and} \quad B = \begin{bmatrix} 1 & 0 & -1 \\ 2 & 0 & 1 \\ 0 & 1 & 0 \end{bmatrix}.$$

9. Use induction on k to prove that if $\mathbf{x}_1, \ldots, \mathbf{x}_k \in \mathbb{R}^n$, then

$$\|\mathbf{x}_1 + \cdots + \mathbf{x}_k\| \leq \|\mathbf{x}_1\| + \cdots + \|\mathbf{x}_k\|.$$

10. Prove using algebra, the *identity of Lagrange*: For real numbers $x_1, \ldots, x_n$ and $y_1, \ldots, y_n$.

$$\left(\sum_{i=1}^{n} x_i y_i \right)^2 = \left(\sum_{i=1}^{n} x_i^2 \right) \left(\sum_{i=1}^{n} y_i^2 \right) - \sum_{i<j} (x_i y_j - x_j y_i)^2.$$

Use this to give another proof of the Cauchy-Schwarz inequality in $\mathbb{R}^n$,

11. Prove that if A is an $n \times n$ matrix, then
   (a) $\det (\lambda A) = \lambda^n \det A$; and
   (b)   if B is a matrix obtained from A by multiplying any row or column by a scalar λ, then $\det B = \lambda \det A$.

In Exercises 12 to 14, A, B, and C denote $n \times n$ matrices.

12. Is $\det (A + B) = \det A + \det B$? Give a proof or counterexample.

13. Does $(A + B)(A - B) = A^2 - B^2$?

14. Assuming the law $\det (AB) = (\det A)(\det B)$, prove that $\det (ABC) = (\det A)(\det B)(\det C)$.

15. (This exercise assumes a knowledge of integration of continuous functions of one variable.) Note that the proof of the CBS inequality (Theorem 4) depends only on the properties of the inner product listed in Theorem 1. Use this observation to establish the following inequality for continuous functions $f, g: [0, 1] \to \mathbb{R}$:

$$\left| \int_0^1 f(x)g(x)\, dx \right| \leq \sqrt{\int_0^1 [f(x)]^2\, dx} \sqrt{\int_0^1 [g(x)]^2\, dx}.$$

Do this by
   (a)   verifying that the space of continuous functions from $[0, 1]$ to $\mathbb{R}$ forms a vector space; that is, we may think of functions f, g abstractly as "vectors" that can be added to each other and multiplied by scalars.
   (b)   introducing the inner product of functions

$$f \cdot g = \int_0^1 f(x)g(x)\, dx$$

and verifying that it satisfies conditions (i) to (iv) of Theorem 3.

16. Define the transpose A^T of an $n \times n$ matrix A as follows: the ijth element of A^T is a_{ji} where a_{ij} is the ijth entry of A. Show that A^T is characterized by the following property: For all $\mathbf{x}, \mathbf{y}$ in $\mathbb{R}^n$,

$$(A^T \mathbf{x}) \cdot \mathbf{y} = \mathbf{x} \cdot (A\mathbf{y}).$$

17. Verify that the inverse of

$$\begin{bmatrix} a & b \\ c & d \end{bmatrix} \quad \text{is} \quad \frac{1}{ad - bc} \begin{bmatrix} d & -b \\ -c & a \end{bmatrix}.$$

18. Use your answer in Exercise 17 to show that the solution of the system

$$ax + by = e$$
$$cx + dy = f$$

is

$$\begin{bmatrix} x \\ y \end{bmatrix} = \frac{1}{ad - dc} \begin{bmatrix} d & -b \\ -c & a \end{bmatrix} \begin{bmatrix} e \\ f \end{bmatrix}.$$

19. Assuming the law $\det (AB) = (\det A)(\det B)$, verify that $(\det A)(\det A^{-1}) = 1$ and conclude that if A has an inverse, then $\det A \neq 0$.

REVIEW EXERCISES FOR CHAPTER 1

1. Let $\mathbf{v} = 3\mathbf{i} + 4\mathbf{j} + 5\mathbf{k}$ and $\mathbf{w} = \mathbf{i} - \mathbf{j} + \mathbf{k}$. Compute $\mathbf{v} + \mathbf{w}$, $3\mathbf{v}$, $6\mathbf{v} + 8\mathbf{w}$, $-2\mathbf{v}$, $\mathbf{v} \cdot \mathbf{w}$, $\mathbf{v} \times \mathbf{w}$. Interpret each operation geometrically by graphing the vectors.

2. Repeat Exercise 1 with $\mathbf{v} = 2\mathbf{j} + \mathbf{k}$ and $\mathbf{w} = -\mathbf{i} - \mathbf{k}$.

3. (a) Find the equation of the line through $(-1, 2, -1)$ in the direction of $\mathbf{j}$.
  (b) Find the equation of the line passing through $(0, 2, -1)$ and $(-3, 1, 0)$.
  (c) Find the equation for the plane perpendicular to $(-2, 1, 2)$ and passing through $(-1, 1, 3)$.

4. (a) Find the equation of the line through $(0, 1, 0)$ in the direction of $3\mathbf{i} + \mathbf{k}$.
  (b) Find the equation of the line passing through $(0, 1, 1)$ and $(0, 1, 0)$.
  (c) Find an equation for the plane perpendicular to $(-1, 1, -1)$ and passing through $(1, 1, 1)$.

5. Compute $\mathbf{v} \cdot \mathbf{w}$ for the following sets of vectors:
  (a) $\mathbf{v} = -\mathbf{i} + \mathbf{j}$; $\mathbf{w} = \mathbf{k}$.
  (b) $\mathbf{v} = \mathbf{i} + 2\mathbf{j} - \mathbf{k}$; $\mathbf{w} = 3\mathbf{i} + \mathbf{j}$.
  (c) $\mathbf{v} = -2\mathbf{i} - \mathbf{j} + \mathbf{k}$; $\mathbf{w} = 3\mathbf{i} + 2\mathbf{j} - 2\mathbf{k}$.

6. Compute $\mathbf{v} \times \mathbf{w}$ for the vectors in Exercise 5. [Only part (b) is solved in the Study Guide.]

7. Find the cosine of the angle between the vectors in Exercise 5. [Only part (b) is solved in the Study Guide.]

8. Find the area of the parallelogram spanned by the vectors in Exercise 5. [Only part (b) is solved in the Study Guide.]

9. Use vector notation to describe the triangle in space whose vertices are the origin and the endpoints of vectors $\mathbf{a}$ and $\mathbf{b}$.

10. Show that three vectors $\mathbf{a}, \mathbf{b}, \mathbf{c}$ lie in the same plane through the origin if and only if there are three scalars α, β, γ, not all zero, such that $\alpha\mathbf{a} + \beta\mathbf{b} + \gamma\mathbf{c} = \mathbf{0}$.

11. For real numbers $a_1, a_2, a_3, b_1, b_2, b_3$, show that

$$(a_1b_1 + a_2b_2 + a_3b_3)^2 \leq (a_1^2 + a_2^2 + a_3^2)(b_1^2 + b_2^2 + b_3^2).$$

12. Let $\mathbf{u}, \mathbf{v}, \mathbf{w}$ be unit vectors that are orthogonal to each other. If $\mathbf{a} = \alpha\mathbf{u} + \beta\mathbf{v} + \gamma\mathbf{w}$, show that

$$\alpha = \mathbf{a} \cdot \mathbf{u}, \qquad \beta = \mathbf{a} \cdot \mathbf{v}, \qquad \gamma = \mathbf{a} \cdot \mathbf{w}.$$

Interpret the results geometrically.

13. Let $\mathbf{a}, \mathbf{b}$ be two vectors in the plane, $\mathbf{a} = (a_1, a_2)$, $\mathbf{b} = (b_1, b_2)$, and let λ be a real number. Show that the area of the parallelogram determined by $\mathbf{a}$ and $\mathbf{b} + \lambda\mathbf{a}$ is the same as that determined by $\mathbf{a}$ and $\mathbf{b}$. Sketch. Relate this result to a known property of determinants.

14. Find the volume of the parallelepiped determined by the vertices $(0, 1, 0), (1, 1, 1), (0, 2, 0), (3, 1, 2)$.

15. Given nonzero vectors $\mathbf{a}$ and $\mathbf{b}$ in $\mathbb{R}^3$, show that the vector $\mathbf{v} = \|\mathbf{a}\|\mathbf{b} + \|\mathbf{b}\|\mathbf{a}$ bisects the angle between $\mathbf{a}$ and $\mathbf{b}$.

16. Use vector methods to prove that the distance from the point (x_1, y_1) to the line $ax + by = c$ is

$$\frac{|ax_1 + by_1 - c|}{\sqrt{a^2 + b^2}}.$$

17. Verify that the direction of $\mathbf{b} \times \mathbf{c}$ is given by the right-hand rule, by choosing $\mathbf{b}, \mathbf{c}$ to be two of the vectors $\mathbf{i}, \mathbf{j}$, and $\mathbf{k}$.

18. (a) Suppose $\mathbf{a} \cdot \mathbf{b} = \mathbf{a}' \cdot \mathbf{b}$ for all $\mathbf{b}$. Show that $\mathbf{a} = \mathbf{a}'$.
　　 (b) Suppose $\mathbf{a} \times \mathbf{b} = \mathbf{a}' \times \mathbf{b}$ for all $\mathbf{b}$. Is it true that $\mathbf{a} = \mathbf{a}'$?

19. (a) Using vector methods, show that the distance between two nonparallel lines l_1 and l_2 is given by

$$d = \frac{|(\mathbf{v}_2 - \mathbf{v}_1) \cdot (\mathbf{a}_1 \times \mathbf{a}_2)|}{\|\mathbf{a}_1 \times \mathbf{a}_2\|},$$

where $\mathbf{v}_1, \mathbf{v}_2$ are any two points on l_1 and l_2, respectively, and $\mathbf{a}_1$ and $\mathbf{a}_2$ are the directions of l_1 and l_2. [HINT: Consider the plane through l_2 which is parallel to l_1. Show that $(\mathbf{a}_1 \times \mathbf{a}_2)/\|\mathbf{a}_1 \times \mathbf{a}_2\|$ is a unit normal for this plane; now project $\mathbf{v}_2 - \mathbf{v}_1$ onto this normal direction.]
　　 (b) Find the distance between the line l_1 determined by the points $(-1, -1, 1)$ and $(0, 0, 0)$ and the line l_2 determined by the points $(0, -2, 0)$ and $(2, 0, 5)$.

20. Show that two planes given by the equations $Ax + By + Cz + D_1 = 0$ and $Ax + By + Cz + D_2 = 0$ are parallel, and that the distance between them is

$$\frac{|D_1 - D_2|}{\sqrt{A^2 + B^2 + C^2}}.$$

21. (a) Prove that the area of the triangle in the plane with vertices (x_1, y_1), (x_2, y_2), (x_3, y_3) is the absolute value of

$$\frac{1}{2} \begin{vmatrix} 1 & 1 & 1 \\ x_1 & x_2 & x_3 \\ y_1 & y_2 & y_3 \end{vmatrix}.$$

(b) Find the area of the triangle with vertices $(1, 2)$, $(0, 1)$, $(-1, 1)$.

22. Convert the following points from Cartesian to cylindrical and spherical coordinates and plot:
(a) $(0, 3, 4)$ (b) $(-\sqrt{2}, 1, 0)$
(c) $(0, 0, 0)$ (d) $(-1, 0, 1)$
 (e) $(-2\sqrt{3}, -2, 3)$

23. Convert the following points from cylindrical to Cartesian and spherical coordinates and plot:
(a) $(1, \pi/4, 1)$ (b) $(3, \pi/6, -4)$
(c) $(0, \pi/4, 1)$ (d) $(2, -\pi/2, 1)$
(e) $(-2, -\pi/2, 1)$

24. Convert the following points from spherical to Cartesian and cylindrical coordinates and plot:
(a) $(1, \pi/2, \pi)$ (b) $(2, -\pi/2, \pi/6)$
(c) $(0, \pi/8, \pi/35)$ (d) $(2, -\pi/2, -\pi)$
(e) $(-1, \pi, \pi/6)$

25. Rewrite the equation $z = x^2 - y^2$ using cylindrical and spherical coordinates.

26. Using spherical coordinates, show that

$$\phi = \cos^{-1}\left(\frac{\mathbf{u} \cdot \mathbf{k}}{\|\mathbf{u}\|}\right)$$

where $\mathbf{u} = x\mathbf{i} + y\mathbf{j} + z\mathbf{k}$. Interpret geometrically.

27. Verify the Cauchy-Schwarz and triangle inequalities for

$$\mathbf{x} = (3, 2, 1, 0) \quad \text{and} \quad \mathbf{y} = (1, 1, 1, 2).$$

28. Multiply the matrices

$$A = \begin{bmatrix} 3 & 0 & 1 \\ 2 & 0 & 1 \\ 1 & 0 & 1 \end{bmatrix} \quad \text{and} \quad B = \begin{bmatrix} 1 & 0 & 1 \\ 1 & 1 & 1 \\ 0 & 0 & 1 \end{bmatrix}.$$

Does $AB = BA$?

29. (a) Show that for two $n \times n$ matrices A and B, and $\mathbf{x} \in \mathbb{R}^n$,

$$(AB)\mathbf{x} = A(B\mathbf{x}).$$

(b) What does the equality in part (a) imply about the relationship between the composition of the mappings $\mathbf{x} \mapsto B\mathbf{x}$, $\mathbf{y} \mapsto A\mathbf{y}$ and matrix multiplication?

30. Find the volume of the parallelepiped spanned by the vectors

$$(1, 0, 1), \quad (1, 1, 1), \quad \text{and} \quad (-3, 2, 0).$$

31. Verify that a linear mapping T of $\mathbb{R}^n$ to $\mathbb{R}^n$ is determined by an $n \times n$ matrix.

32. Find an equation for the plane that contains $(3, -1, 2)$ and the line $\mathbf{v} = (2, -1, 0) + t(2, 3, 0)$.

33. The work W done in moving an object from $(0, 0)$ to $(7, 2)$ subject to a constant force $\mathbf{F}$ is $W = \mathbf{F} \cdot \mathbf{r}$, where $\mathbf{r}$ is the vector with head at $(7, 2)$ and tail at $(0, 0)$. The units are feet and pounds.
   (a) Suppose the force $\mathbf{F} = 10 \cos \theta \mathbf{i} + 10 \sin \theta \mathbf{j}$. Find W in terms of θ.
   (b) Suppose the force $\mathbf{F}$ has magnitude of 6 lb and makes an angle of $\pi/6$ rad with the horizontal, pointing right. Find W in foot-pounds.

34. If a particle with mass m moves with velocity $\mathbf{v}$, its *momentum* is $\mathbf{p} = m\mathbf{v}$. In a game of marbles, a marble with mass 2 grams (g) is shot with velocity 2 meters per second (m/s), hits two marbles with mass 1 g each, and comes to a dead halt. One of the marbles flies off with a velocity of 3 m/s at an angle of 45° to the incident direction of the larger marble as in Figure 1.R.1. Assuming that the total momentum before and after the collision is the same (according to the law of conservation of momentum), at what angle and speed does the second marble move?

Figure 1.R.1   Momentum and marbles.

35. Show that for all x, y, z,

$$\begin{vmatrix} x+2 & y & z \\ z & y+1 & 10 \\ 5 & 5 & 2 \end{vmatrix} = - \begin{vmatrix} y & x+2 & z \\ 1 & z-x-2 & 10-z \\ 5 & 5 & 2 \end{vmatrix}.$$

36. Show that

$$\begin{vmatrix} 1 & x & x^2 \\ 1 & y & y^2 \\ 1 & z & z^2 \end{vmatrix} \neq 0$$

if x, y, and z are all different.

37. Show that

$$\begin{vmatrix} 66 & 628 & 246 \\ 88 & 435 & 24 \\ 2 & -1 & 1 \end{vmatrix} = \begin{vmatrix} 68 & 627 & 247 \\ 86 & 436 & 23 \\ 2 & -1 & 1 \end{vmatrix}.$$

38. Show that

$$
\begin{vmatrix}
n & n+1 & n+2 \\
n+3 & n+4 & n+5 \\
n+6 & n+7 & n+8
\end{vmatrix}
$$

has the same value no matter what n is. What is this value?

39. The volume of a tetrahedron with concurrent edges $\mathbf{a}, \mathbf{b}, \mathbf{c}$ is given by $V = \frac{1}{6}\mathbf{a} \cdot (\mathbf{b} \times \mathbf{c})$.
(a) Express the volume as a determinant.
(b) Evaluate V when $\mathbf{a} = \mathbf{i} + \mathbf{j} + \mathbf{k}, \mathbf{b} = \mathbf{i} - \mathbf{j} + \mathbf{k}, \mathbf{c} = \mathbf{i} + \mathbf{j}$.

*Use the following definition for problems 40 and 41: Let $\mathbf{r}_1, \ldots, \mathbf{r}_n$ be vectors in $\mathbb{R}^3$ from 0 to the masses $m_1, \ldots, m_n$. The **center of mass** is the vector*

$$
\mathbf{c} = \frac{\sum_{i=1}^{n} m_i \mathbf{r}_i}{\sum_{i=1}^{n} m_i}.
$$

40. A tetrahedron sits in xyz coordinates with one vertex at $(0, 0, 0)$, and the three edges concurrent at $(0, 0, 0)$ are coincident with the vectors $\mathbf{a}, \mathbf{b}, \mathbf{c}$.
(a) Draw a figure and label the heads of the vectors $\mathbf{a}, \mathbf{b}, \mathbf{c}$.
(b) Find the center of mass of each of the four triangular faces of the tetrahedron if a unit mass is placed at each vertex.

41. Show that for any vector $\mathbf{r}$, the center of mass of a system satisfies

$$
\sum_{i=1}^{n} m_i \|\mathbf{r} - \mathbf{r}_i\|^2 = \sum_{i=1}^{n} m_i \|\mathbf{r}_i - \mathbf{c}\|^2 + m\|\mathbf{r} - \mathbf{c}\|^2,
$$

where $m = \sum_{i=1}^{n} m_i$ is the total mass of the system.

In Exercises 42 to 47, find a unit vector which has the given property.

42. Parallel to the line $x = 3t + 1, y = 16t - 2, z = -(t + 2)$.

43. Orthogonal to the plane $x - 6y + z = 12$.

44. Parallel to both the planes $8x + y + z = 1$ and $x - y - z = 0$.

45. Orthogonal to $\mathbf{i} + 2\mathbf{j} - \mathbf{k}$ and to $\mathbf{k}$.

46. Orthogonal to the line $x = 2t - 1, y = -t - 1, z = t + 2$, and the vector $\mathbf{i} - \mathbf{j}$.

47. At an angle of $30°$ to $\mathbf{i}$ and making equal angles with $\mathbf{j}$ and $\mathbf{k}$.

2 DIFFERENTIATION

I turn away with fright and horror from the lamentable evil of
functions which do not have derivatives.

CHARLES HERMITE, in a letter to Thomas Jan Stieltjes

This chapter extends the principles of differential calculus for functions of one variable
to functions of several variables. We begin in Section 2.1 with the geometry of real-
valued functions and study the graphs of these functions as an aid in visualizing them.
Section 2.2 gives some basic definitions relating to limits and continuity. This subject is
treated briefly, since it requires time and mathematical maturity to develop fully and is
therefore best left to a more advanced course. Fortunately, a complete understanding of
all the subtleties of the limit concept is not necessary for our purposes; the student who
has difficulty with this section should bear this in mind. However, we hasten to add that
the notion of a limit is central to the definition of the derivative, but not to the computation
of most derivatives in specific problems, as we already know from one-variable calculus.
Sections 2.3 and 2.4 deal with the definition of the derivative, and establish some basic
rules of calculus: namely, how to differentiate a sum, product, quotient, or composition.
In Section 2.5 we study directional derivatives and tangent planes, relating these ideas to
those in Section 2.1. Finally, the optional Section 2.7 gives some of the technical proofs.

In generalizing calculus from one dimension to several, it is often convenient to use the
language of matrix algebra. What we shall need has been summarized in Section 1.5.

2.1 THE GEOMETRY OF REAL-VALUED FUNCTIONS

We launch our investigation of real-valued functions by developing methods for visualizing
them. In particular, we introduce the notions of a graph, a level curve, and a level surface
of such functions.

Let f be a function whose domain is a subset A of $\mathbb{R}^n$ and with range contained in $\mathbb{R}^m$. By this we mean that to each $\mathbf{x} = (x_1, \ldots, x_n) \in A$, f assigns a value $f(\mathbf{x})$, an m-tuple in $\mathbb{R}^m$. Such functions f are called ***vector-valued functions***[*] if $m > 1$, and ***scalar-valued functions*** if $m = 1$. For example, the scalar-valued function $f(x, y, z) = (x^2 + y^2 + z^2)^{-3/2}$ maps the set A of $(x, y, z) \neq (0, 0, 0)$ in $\mathbb{R}^3$ ($n = 3$ in this case) to $\mathbb{R}$ ($m = 1$). To denote f we sometimes write

$$f: (x, y, z) \mapsto (x^2 + y^2 + z^2)^{-3/2}.$$

Note that in $\mathbb{R}^3$ we often use the notation (x, y, z) instead of (x_1, x_2, x_3). In general, the notation $\mathbf{x} \mapsto f(\mathbf{x})$ is useful for indicating the value to which a point $\mathbf{x} \in \mathbb{R}^n$ is sent. We write $f: A \subset \mathbb{R}^n \to \mathbb{R}^m$ to signify that A is the domain of f (a subset of $\mathbb{R}^n$) and the range is contained in $\mathbb{R}^m$. We also use the expression f maps A *into* $\mathbb{R}^m$. Such functions f are called ***functions of several variables*** if $A \subset \mathbb{R}^n, n > 1$.

As another example we can take the vector-valued function $g: \mathbb{R}^6 \to \mathbb{R}^2$ defined by the rule

$$g(\mathbf{x}) = g(x_1, x_2, x_3, x_4, x_5, x_6) = \left(x_1 x_2 x_3 x_4 x_5 x_6, \sqrt{x_1^2 + x_6^2} \right).$$

The first coordinate of the value of g at $\mathbf{x}$ is the product of the coordinates of $\mathbf{x}$.

Functions from $\mathbb{R}^n$ to $\mathbb{R}^m$ are not just mathematical abstractions, they arise naturally in problems studied in all the sciences. For example, to specify the temperature T in a region A of space requires a function $T: A \subset \mathbb{R}^3 \to \mathbb{R}$ ($n = 3, m = 1$); thus $T(x, y, z)$ is the temperature at the point (x, y, z). To specify the velocity of a fluid moving in space requires a map $\mathbf{V}: \mathbb{R}^4 \to \mathbb{R}^3$, where $\mathbf{V}(x, y, z, t)$ is the velocity vector of the fluid at the point (x, y, z) in space at time t (see Figure 2.1.1). To specify the reaction rate of a solution consisting of six reacting chemicals A, B, C, D, E, F in proportions x, y, z, w, u, v requires a map $\sigma: U \subset \mathbb{R}^6 \to \mathbb{R}$, where $\sigma(x, y, z, w, u, v)$ gives the rate when the chemicals are in the indicated proportions. To specify the cardiac vector (the vector giving the magnitude and direction of electric current flow in the heart) at time t requires a map $\mathbf{c}: \mathbb{R} \to \mathbb{R}^3, t \mapsto \mathbf{c}(t)$.

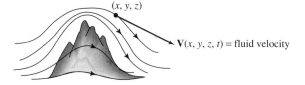
(x, y, z)

$\mathbf{V}(x, y, z, t) = $ fluid velocity

Figure 2.1.1 A fluid in motion defines a vector field $\mathbf{V}$ by specifying the velocity of the fluid particles at each point in space and time.

---

[*]Some mathematicians would write such an f in boldface, using the notation $\mathbf{f}(\mathbf{x})$, since it is vector-valued. We did not do so, as a matter of personal taste. We use boldface primarily for mappings that are vector fields, introduced later. The notion of function was developed over many centuries, with the definition extended to cover more cases as they arose. For example, in 1667 James Gregory defined a function as "a quantity obtained from other quantities by a succession of algebraic operations or by any other operation imaginable." In 1755 Euler gave the following definition: "If some quantities depend on others in such a way as to undergo variation when the latter are varied then the former are called functions of the latter."

When $f\colon U \subset \mathbb{R}^n \to \mathbb{R}$, we say that f is a ***real-valued function of n variables with domain*** U. The reason we say "n variables" is simply that we regard the coordinates of a point $\mathbf{x} = (x_1, \ldots, x_n) \in U$ as n variables, and $f(\mathbf{x}) = f(x_1, \ldots, x_n)$ depends on these variables. We say "real-valued" because $f(x_1, \ldots, x_n)$ is a real number. A good deal of our work will be with real-valued functions, so we give them special attention.

Graphs of Functions

For $f\colon U \subset \mathbb{R} \to \mathbb{R}$ ($n = 1$), the ***graph*** of f is the subset of $\mathbb{R}^2$ consisting of all points $(x, f(x))$ in the plane, for x in U. This subset can be thought of as a curve in $\mathbb{R}^2$. In symbols, we write this as

$$\operatorname{graph} f = \{(x, f(x)) \in \mathbb{R}^2 \mid x \in U\},$$

where the curly braces mean "the set of all" and the vertical bar is read "such that." Drawing the graph of a function of one variable is a useful device to help visualize how the function actually behaves. (See Figure 2.1.2.) It will be helpful to generalize the idea of a graph to functions of several variables. This leads to the following definition:

DEFINITION: GRAPH OF A FUNCTION

Let $f\colon U \subset \mathbb{R}^n \to \mathbb{R}$. Define the **graph** of f to be the subset of $\mathbb{R}^{n+1}$ consisting of all the points $(x_1, \ldots, x_n, f(x_1, \ldots, x_n))$ in $\mathbb{R}^{n+1}$ for $(x_1, \ldots, x_n)$ in U. In symbols,

$$\operatorname{graph} f = \{(x_1, \ldots, x_n, f(x_1, \ldots, x_n)) \in \mathbb{R}^{n+1} \mid (x_1, \ldots, x_n) \in U\}.$$

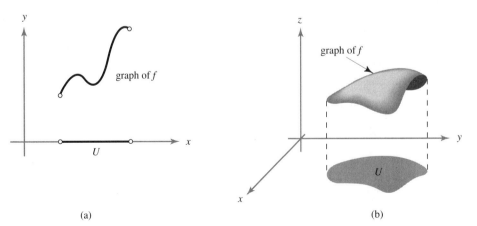

(a)   (b)

Figure 2.1.2   The graphs of (a) a function of one variable and (b) a function of two variables.

For the case $n = 1$ the graph is a curve in $\mathbb{R}^2$, while for $n = 2$ it is a surface in $\mathbb{R}^3$ (see Figure 2.1.2). For $n = 3$ it is difficult to visualize the graph, because, since we are humans living in a three-dimensional world, it is hard for us to envisage sets in $\mathbb{R}^4$. To help overcome this handicap, we introduce the idea of a level set.

Level Sets

Suppose $f(x, y, z) = x^2 + y^2 + z^2$. A *level set* is a subset of $\mathbb{R}^3$ on which f is constant; for instance, the set where $x^2 + y^2 + z^2 = 1$ is a level set for f. This we can visualize: it is just a sphere of radius 1 in $\mathbb{R}^3$. Formally, a level set is the set of (x, y, z) such that $f(x, y, z) = c$, c a constant. The behavior or structure of a function is determined in part by the shape of its level sets; consequently, understanding these sets aids us in understanding the function in question. Level sets are also useful for understanding functions of two variables $f(x, y)$, in which case we speak of *level curves* or *level contours*.

The idea is similar to that used to prepare contour maps, where one draws lines to represent constant altitudes; walking along such a line would mean walking on a level path. In the case of a hill rising from the xy plane, a graph of all the level curves gives us a good idea of the function $h(x, y)$, which represents the height of the hill at point (x, y) (see Figure 2.1.3).

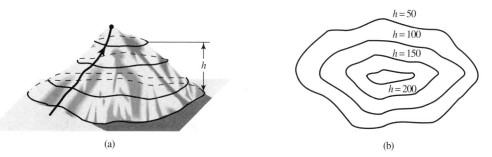

(a) (b)

Figure 2.1.3 Level contours of a function are defined in the same manner as contour lines for a topographical map.

EXAMPLE 1   The constant function $f \colon \mathbb{R}^2 \to \mathbb{R}$, $(x, y) \mapsto 2$, that is, the function $f(x, y) = 2$, has as its graph the horizontal plane $z = 2$ in $\mathbb{R}^3$. The level curve of value c is empty if $c \neq 2$, and is the whole xy plane if $c = 2$.   ▲

EXAMPLE 2   The function $f \colon \mathbb{R}^2 \to \mathbb{R}$, defined by $f(x, y) = x + y + 2$ has as its graph the inclined plane $z = x + y + 2$. This plane intersects the xy plane ($z = 0$) in the line $y = -x - 2$ and the z axis at the point $(0, 0, 2)$. For any value $c \in \mathbb{R}$, the level curve of value c is the straight line $y = -x + (c - 2)$; or in symbols, the set

$$L_c = \{(x, y) \mid y = -x + (c - 2)\} \subset \mathbb{R}^2.$$

We indicate a few of the level curves of the function in Figure 2.1.4. This is a contour map of the function f.

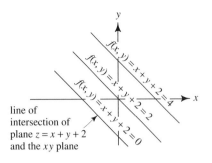

Figure 2.1.4 The level curves of $f(x, y) = x + y + 2$ show the behavior of this function.

From level curves labeled with the value or "height" of the function, the shape of the graph may be inferred by mentally elevating each level curve to the appropriate height, without stretching, tilting, or sliding it. If this procedure is visualized for all level curves L_c, that is, for all values $c \in \mathbb{R}$, they will assemble to give the entire graph of f, as indicated by the shaded plane in Figure 2.1.5. If the graph is visualized using a finite number of level curves, a contour model is produced. If f is a smooth function, its graph will be a smooth surface, and so the contour model, mentally smoothed over, gives a good impression of the graph. ▲

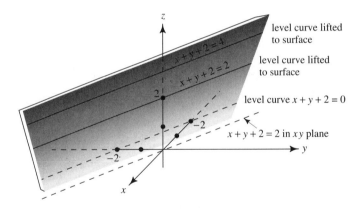

Figure 2.1.5 Relationship of level curves of Figure 2.1.4 to the graph of the function $f(x, y) = x + y + 2$, which is the plane $z = x + y + 2$.

DEFINITION: LEVEL CURVES AND SURFACES

Let $f: U \subset \mathbb{R}^n \to \mathbb{R}$ and let $c \in \mathbb{R}$. Then the *level set of value* c is defined to be the set those points $\mathbf{x} \in U$ at which $f(\mathbf{x}) = c$. If $n = 2$, we speak of a *level curve* (of value c); and if $n = 3$, we speak of a *level surface*. In symbols, the level set of value c is written

$$\{\mathbf{x} \in U \mid f(\mathbf{x}) = c\} \subset \mathbb{R}^n.$$

Note that the level set is always in the domain space.

EXAMPLE 3 Describe the graph of the quadratic function

$$f: \mathbb{R}^2 \to \mathbb{R}, (x, y) \mapsto x^2 + y^2.$$

SOLUTION The graph is the ***paraboloid of revolution*** $z = x^2 + y^2$, oriented upward from the origin, around the z axis. The level curve of value c is empty for $c < 0$; for $c > 0$ the level curve of value c is the set $\{(x, y) \mid x^2 + y^2 = c\}$, a circle of radius $\sqrt{c}$ centered at the origin. Thus, raised to height c above the xy plane, the level set is a circle of radius $\sqrt{c}$, indicating a parabolic shape (see Figures 2.1.6 and 2.1.7). ▲

Figure 2.1.6 Some level curves for the function $f(x, y) = x^2 + y^2$.

Figure 2.1.7 Level curves in Figure 2.1.6 raised to the graph.

The Method of Sections

By a ***section*** of the graph of f we mean the intersection of the graph and a (vertical) plane. For example, if P_1 is the xz plane in $\mathbb{R}^3$, defined by $y = 0$, then the section of f

in Example 3 is the set

$$P_1 \cap \text{graph } f = \{(x, y, z) \mid y = 0, z = x^2\},$$

which is a parabola in the xz plane. Similarly, if P_2 denotes the yz plane, defined by $x = 0$, then the section

$$P_2 \cap \text{graph } f = \{(x, y, z) \mid x = 0, z' = y^2\}$$

is a parabola in the yz plane (see Figure 2.1.8). It is usually helpful to compute at least one section to complement the information given by the level sets.

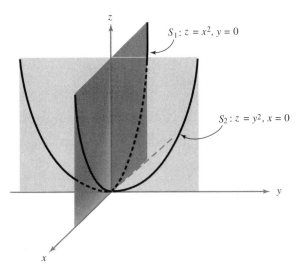

$S_1: z = x^2, y = 0$

$S_2: z = y^2, x = 0$

Figure 2.1.8 Two sections of the graph of $f(x, y) = x^2 + y^2$.

EXAMPLE 4   The graph of the quadratic function $f: \mathbb{R}^2 \to \mathbb{R}, (x, y) \mapsto x^2 - y^2$ is called a ***hyperbolic paraboloid***, or ***saddle***, centered at the origin. Sketch the graph.

SOLUTION   To visualize this surface, we first draw the level curves. To determine the level curves, we solve the equation $x^2 - y^2 = c$. Consider the values $c = 0, \pm1, \pm4$. For $c = 0$, we have $y^2 = x^2$, or $y = \pm x$, so that this level set consists of two straight lines through the origin. For $c = 1$, the level curve is $x^2 - y^2 = 1$, or $y = \pm\sqrt{x^2 - 1}$, which is a hyperbola that passes vertically through the x axis at the points $(\pm1, 0)$ (see Figure 2.1.9). Similarly, for $c = 4$, the level curve is defined by $y = \pm\sqrt{x^2 - 4}$, the hyperbola passing vertically through the x axis at $(\pm2, 0)$. For $c = -1$, we obtain the curve $x^2 - y^2 = -1$, that is, $x = \pm\sqrt{y^2 - 1}$, the hyperbola passing horizontally through the y axis at $(0, \pm1)$. And for $c = -4$, the hyperbola through $(0, \pm2)$ is obtained. These level curves are shown in Figure 2.1.9. Since it is not easy to visualize the graph of f

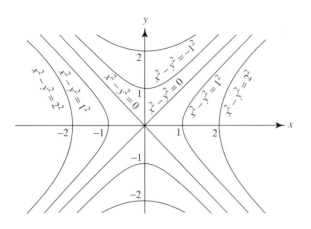

Figure 2.1.9 Level curves for the function $f(x, y) = x^2 - y^2$.

from these data alone, we shall compute two sections, as in the previous example. For the section in the xz plane, we have

$$P_1 \cap \text{ graph of } f = \{(x, y, z) \mid y = 0, z = x^2\},$$

which is a parabola opening upward; and for the yz plane,

$$P_2 \cap \text{ graph } f = \{(x, y, z) \mid x = 0, z = -y^2\},$$

which is a parabola opening downward. The graph may now be visualized by lifting the level curves to the appropriate heights and smoothing out the resulting surface. Their placement is aided by computing the parabolic sections. This procedure generates the hyperbolic saddle indicated in Figure 2.1.10. Compare this with the computer-generated graphs in Figure 2.1.11 (note that the orientation of the axes has been changed). ▲

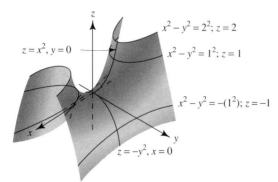

$z = x^2, y = 0$

$x^2 - y^2 = 2^2; z = 2$

$x^2 - y^2 = 1^2; z = 1$

$x^2 - y^2 = -(1^2); z = -1$

$z = -y^2, x = 0$

Figure 2.1.10 Some level curves on the graph of $f(x, y) = x^2 - y^2$.

EXAMPLE 5 Describe the level sets of the function

$$f: \mathbb{R}^3 \to \mathbb{R}, (x, y, z) \mapsto x^2 + y^2 + z^2.$$

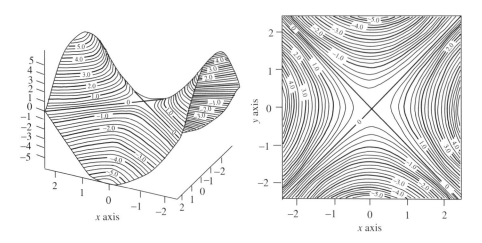

Figure 2.1.11   The graph of $z = x^2 - y^2$ and its level curves.

SOLUTION   This is the three-dimensional analogue of Example 3. In this context, level sets are surfaces in the three-dimensional domain $\mathbb{R}^3$. The graph, in $\mathbb{R}^4$, cannot be visualized directly, but sections can nevertheless be computed.

The level set with value c is the set

$$L_c = \{(x, y, z) \mid x^2 + y^2 + z^2 = c\},$$

which is the sphere centered at the origin with radius $\sqrt{c}$ for $c > 0$, is a single point at the origin for $c = 0$, and is empty for $c < 0$. The level sets for $c = 0, 1, 4,$ and 9 are indicated in Figure 2.1.12.   ▲

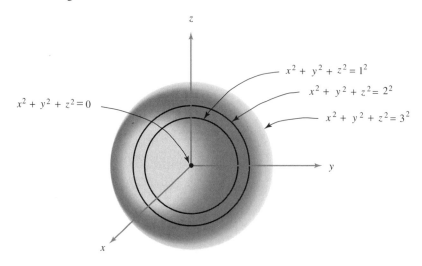

Figure 2.1.12   Some level surfaces for $f(x, y, z) = x^2 + y^2 + z^2$.

EXAMPLE 6   Describe the graph of the function $f\colon \mathbb{R}^3 \to \mathbb{R}$ defined by $f(x, y, z) = x^2 + y^2 - z^2$, which is the three-dimensional analogue of Example 4, and is also called a *saddle.*

SOLUTION   The level surfaces are defined by

$$L_c = \{(x, y, z) \mid x^2 + y^2 - z^2 = c\}.$$

For $c = 0$, this is the cone $z = \pm\sqrt{x^2 + y^2}$ centered on the z-axis. For c negative, say, $c = -a^2$, we obtain $z = \pm\sqrt{x^2 + y^2 + a^2}$, which is a hyperboloid of two sheets around the z axis, passing through the z axis at the points $(0, 0, \pm a)$. For c positive, say, $c = b^2$, the level surface is the **single-sheeted hyperboloid of revolution** around the z axis defined by $z = \pm\sqrt{x^2 + y^2 - b^2}$, which intersects the xy plane in the circle of radius $|b|$. These level surfaces are sketched in Figure 2.1.13.

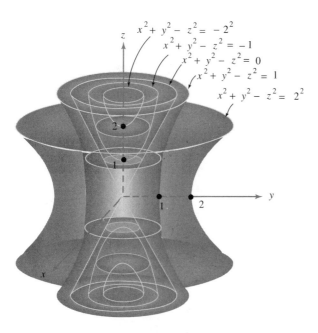

Figure 2.1.13 Some level surfaces of the function $f(x, y, z) = x^2 + y^2 - z^2$.

Another view of the graph may be obtained from a section. For example, the subspace $S_{y=0} = \{(x, y, z, t) \mid y = 0\}$ intersects the graph in the section

$$S_{y=0} \cap \text{graph } f = \{(x, y, z, t) \mid y = 0, t = x^2 - z^2\},$$

that is, the set of points of the form $(x, 0, z, x^2 - z^2)$, which may be considered to be a surface in xzt space (see Figure 2.1.14).   ▲

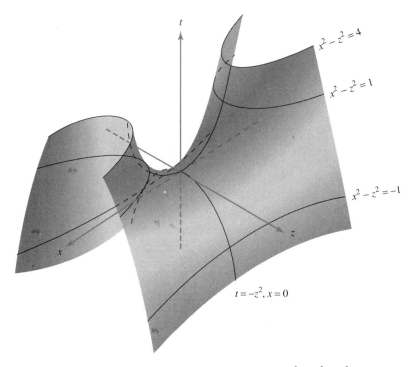

Figure 2.1.14   The $y = 0$ section of the graph of $f(x, y, z) = x^2 + y^2 - z^2$.

(a)

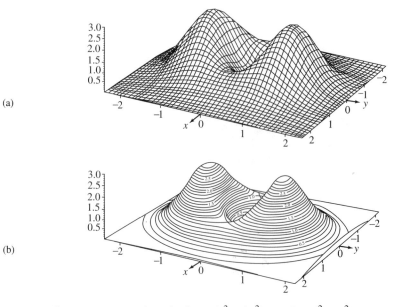

(b)

Figure 2.1.15   Computer-generated graph of $z = (x^2 + 3y^2) \exp(1 - x^2 - y^2)$ represented in four ways: (a) by sections, (b) by level curves on graph, (c) by level curves in the xy plane seen in perspective, and (d) by level curves in the xy plane seen from above.

(Continues)

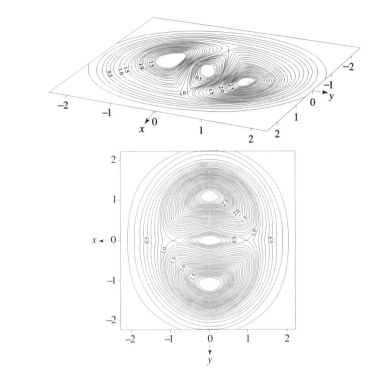

(c)

(d)

Figure 2.1.15 (*Continued*)

We have seen how the methods of sections and level sets can be used to understand the behavior of a function and its graph; these techniques can be quite useful to people who desire comprehensive visualization of complicated data. There are many computer programs available to do this, and we show the results of one such program in Figure 2.1.15.

EXERCISES

1. Sketch the level curves and graphs of the following functions:

 (a) $f: \mathbb{R}^2 \to \mathbb{R}, \ (x, y) \mapsto x - y + 2$ **(b)** $f: \mathbb{R}^2 \to \mathbb{R}, \ (x, y) \mapsto x^2 + 4y^2$

     (c) $f: \mathbb{R}^2 \to \mathbb{R}, \ (x, y) \mapsto -xy$

2. Describe the behavior, as c varies, of the level curve $f(x, y) = c$ for each of these functions:

     (a) $f(x, y) = x^2 + y^2 + 1$ **(b)** $f(x, y) = 1 - x^2 - y^2$ **(c)** $f(x, y) = x^3 - x$

3. For the functions in Examples 2, 3, and 4, compute the section of the graph defined by the plane

$$S_\theta = \{(x, y, z) \mid y = x \tan \theta\}$$

for a given constant θ. Do this by expressing z as a function of r, where $x = r \cos \theta$, $y = r \sin \theta$. Determine which of these functions f have the property that the shape of $S_\theta \cap$ graph f is independent of θ. (The solution for Example 3 only is in the Study Guide.)

In Exercises 4 to 10, draw the level curves (in the xy plane) for the given function f and specified values of c. Sketch the graph of $z = f(x, y)$.

4. $f(x, y) = 4 - 3x + 2y, c = 0, 1, 2, 3, -1, -2, -3$

5. $f(x, y) = (100 - x^2 - y^2)^{1/2}, c = 0, 2, 4, 6, 8, 10$

6. $f(x, y) = (x^2 + y^2)^{1/2}, c = 0, 1, 2, 3, 4, 5$

7. $f(x, y) = x^2 + y^2, c = 0, 1, 2, 3, 4, 5$

8. $f(x, y) = 3x - 7y, c = 0, 1, 2, 3, -1, -2, -3$

9. $f(x, y) = x^2 + xy, c = 0, 1, 2, 3, -1, -2, -3$

10. $f(x, y) = x/y, c = 0, 1, 2, 3, -1, -2, -3$

In Exercises 11 to 13, sketch or describe the level surfaces and a section of the graph of each function.

11. $f: \mathbb{R}^3 \to \mathbb{R}, (x, y, z) \mapsto -x^2 - y^2 - z^2$

12. $f: \mathbb{R}^3 \to \mathbb{R}, (x, y, z) \mapsto 4x^2 + y^2 + 9z^2$

13. $f: \mathbb{R}^3 \to \mathbb{R}, (x, y, z) \mapsto x^2 + y^2$

In Exercises 14 to 18, describe the graph of each function by computing some level sets and sections.

14. $f: \mathbb{R}^3 \to \mathbb{R}, (x, y, z) \mapsto xy$

15. $f: \mathbb{R}^3 \to \mathbb{R}, (x, y, z) \mapsto xy + yz$

16. $f: \mathbb{R}^3 \to \mathbb{R}, (x, y, z) \mapsto xy + z^2$

17. $f: \mathbb{R}^2 \to \mathbb{R}, (x, y) \mapsto |y|$

18. $f: \mathbb{R}^2 \to \mathbb{R}, (x, y) \mapsto \max(|x|, |y|)$

Sketch or describe the surfaces in $\mathbb{R}^3$ of the equations presented in Exercises 19 to 31.

19. $4x^2 + y^2 = 16$ **20.** $x + 2z = 4$

21. $z^2 = y^2 + 4$

22. $x^2 + y^2 - 2x = 0$

23. $\dfrac{x}{4} = \dfrac{y^2}{4} + \dfrac{z^2}{9}$

24. $\dfrac{y^2}{9} + \dfrac{z^2}{4} = 1 + \dfrac{x^2}{16}$

25. $z = x^2$

26. $y^2 + z^2 = 4$

27. $z = \dfrac{y^2}{4} - \dfrac{x^2}{9}$

28. $y^2 = x^2 + z^2$

29. $4x^2 - 3y^2 + 2z^2 = 0$

30. $\dfrac{x^2}{9} + \dfrac{y^2}{12} + \dfrac{z^2}{9} = 1$

31. $x^2 + y^2 + z^2 + 4x - by + 9z - b = 0$, where b is a constant

32. Using polar coordinates, describe the level curves of the function defined by

$$f(x, y) = 2xy/(x^2 + y^2) \text{ if } (x, y) \neq (0, 0) \text{ and } f(0, 0) = 0.$$

33. Let $f: \mathbb{R}^2 \backslash \{\mathbf{0}\} \to \mathbb{R}$ be given in polar coordinates by $f(r, \theta) = (\cos 2\theta)/r^2$. Sketch a few level curves in the xy plane. Here, $\mathbb{R}^2 \backslash \{\mathbf{0}\} = \{\mathbf{x} \in \mathbb{R}^2 \mid \mathbf{x} \neq \mathbf{0}\}$.

34. In Figure 2.1.15(d), the level "curve" $z = 3$ appears to consist of two points. Prove this algebraically.

2.2 LIMITS AND CONTINUITY

This section develops the concepts of open sets, limits, and continuity; open sets are needed to understand limits, and limits are in turn needed to understand continuity and differentiability.

As in elementary calculus, it is not necessary to completely master the limit concept in order to work problems in differentiation. For this reason, instructors may treat the following material with varying degrees of rigor. The student should consult with the instructor about the depth of understanding required.

We begin formulating the concept of an open set by defining an open disk. Let $\mathbf{x}_0 \in \mathbb{R}^n$ and let r be a positive real number. The ***open disk*** (or ***open ball***) of radius r and center $\mathbf{x}_0$ is defined to be the set of all points $\mathbf{x}$ such that $\|\mathbf{x} - \mathbf{x}_0\| < r$. This set is denoted $D_r(\mathbf{x}_0)$, and is the set of points $\mathbf{x}$ in $\mathbf{R}^n$ whose distance from $\mathbf{x}_0$ is less than r. Notice that we include only those $\mathbf{x}$ for which *strict* inequality holds. The disk $D_r(\mathbf{x}_0)$ is illustrated in Figure 2.2.1 for $n = 1, 2, 3$. For the case $n = 1$ and $x_0 \in \mathbb{R}$, the open disk $D_r(x_0)$ is the open interval $(x_0 - r, x_0 + r)$, which consists of all numbers $x \in \mathbb{R}$ *strictly* between $x_0 - r$ and $x_0 + r$. For the case $n = 2, \mathbf{x}_0 \in \mathbb{R}^2$, $D_r(\mathbf{x}_0)$ is the "inside" of the disk of radius r centered at $\mathbf{x}_0$. For the case $n = 3, \mathbf{x}_0 \in \mathbb{R}^3$, $D_r(\mathbf{x}_0)$ is the part strictly "inside" of the ball of radius r centered at $\mathbf{x}_0$.

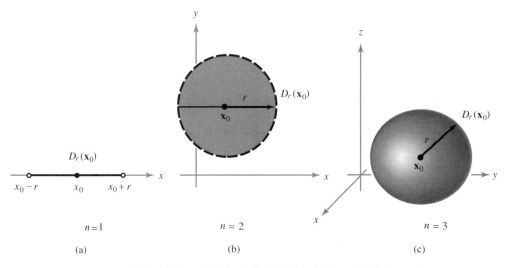

Figure 2.2.1   What disks $D_r(\mathbf{x}_0)$ look like in (a) 1, (b) 2, and (c) 3 dimensions.

DEFINITION: OPEN SETS   Let $U \subset \mathbb{R}^n$ (that is, let U be a subset of $\mathbb{R}^n$). We call U an ***open set*** when for every point $\mathbf{x}_0$ in U there exists some $r > 0$ such that $D_r(\mathbf{x}_0)$ is contained within U; symbolically, we write $D_r(\mathbf{x}_0) \subset U$ (see Figure 2.2.2).

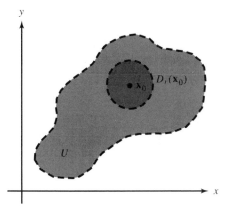

Figure 2.2.2   An open set U is one that completely encloses some disk $D_r(\mathbf{x}_0)$ about each of its points $\mathbf{x}_0$.

The number $r > 0$ can depend on the point $\mathbf{x}_0$, and generally r will shrink as $\mathbf{x}_0$ gets closer to the "edge" of U. Intuitively speaking, a set U is open when the "boundary" points of U do not lie in U. In Figure 2.2.2, the dashed line is *not* included in U.

We establish the convention that the empty set $\emptyset$ (the set consisting of no elements) is open.

We have defined an open disk and an open set. From our choice of terms it would seem that an open disk should also be an open set. A little thought shows that this fact requires some proof.

THEOREM 1 *For each* $\mathbf{x}_0 \in \mathbb{R}^n$ *and* $r > 0$, $D_r(\mathbf{x}_0)$ *is an open set.*

PROOF Let $\mathbf{x} \in D_r(\mathbf{x}_0)$, that is, let $\|\mathbf{x} - \mathbf{x}_0\| < r$. According to the definition of open set we must find an $s > 0$ such that $D_s(\mathbf{x}) \subset D_r(\mathbf{x}_0)$. Referring to Figure 2.2.3, we see that $s = r - \|\mathbf{x} - \mathbf{x}_0\|$ is a reasonable choice; note that $s > 0$ but that s becomes smaller if $\mathbf{x}$ is nearer the edge of $D_r(\mathbf{x}_0)$.

$$d = \|\mathbf{x} - \mathbf{x}_0\|$$
$$s = r - \|\mathbf{x} - \mathbf{x}_0\|$$

Figure 2.2.3 The geometry of the proof that an open disk is an open set.

To prove that $D_s(\mathbf{x}) \subset D_r(\mathbf{x}_0)$, let $\mathbf{y} \in D_s(\mathbf{x})$; that is, let $\|\mathbf{y} - \mathbf{x}\| < s$. We want to prove that $\mathbf{y} \in D_r(\mathbf{x}_0)$ as well. Proving this, in view of the definition of an r-disk, entails showing that $\|\mathbf{y} - \mathbf{x}_0\| < r$. This is done by using the triangle inequality for vectors in $\mathbb{R}^n$:

$$\|\mathbf{y} - \mathbf{x}_0\| = \|(\mathbf{y} - \mathbf{x}) + (\mathbf{x} - \mathbf{x}_0)\| \le \|\mathbf{y} - \mathbf{x}\| + \|\mathbf{x} - \mathbf{x}_0\| < s + \|\mathbf{x} - \mathbf{x}_0\| = r.$$

Hence $\|\mathbf{y} - \mathbf{x}_0\| < r$. ∎

The following example illustrates some techniques that are useful in establishing the openness of sets.

EXAMPLE 1 Prove that $A = \{(x, y) \in \mathbb{R}^2 \mid x > 0\}$ is an open set.

SOLUTION The set is pictured in Figure 2.2.4.

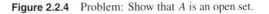

Figure 2.2.4 Problem: Show that A is an open set.

Intuitively, this set is open, since no points on the "boundary," $x = 0$, are contained in the set. Such an argument will often suffice after one becomes acustomed to the concept of openness. At first, however, we should give details. To prove that A is open, we show that for every point $(x, y) \in A$ there exists an $r > 0$ such that $D_r(x, y) \subset A$. If $(x, y) \in A$, then $x > 0$. Choose $r = x$. If $(x_1, y_1) \in D_r(x, y)$, we have

$$|x_1 - x| = \sqrt{(x_1 - x)^2} \leq \sqrt{(x_1 - x)^2 + (y_1 - y)^2} < r = x,$$

and so $x_1 - x < x$ and $x - x_1 < x$. The latter inequality implies $x_1 > 0$, that is, $(x_1, y_1) \in A$. Hence $D_r(x, y) \subset A$, and therefore A is open (see Figure 2.2.5).   ▲

Figure 2.2.5   The construction of a disk about a point in A that is completely enclosed in A.

It is useful to have a special name for an open set containing a given point $\mathbf{x}$, since this idea arises often in the study of limits and continuity. Thus, by a ***neighborhood*** of $\mathbf{x} \in \mathbb{R}^n$ we merely mean an open set U containing the point $\mathbf{x}$. For example, $D_r(\mathbf{x}_0)$ is a neighborhood of $\mathbf{x}_0$ for any $r > 0$. The set A in Example 1 is a neighborhood of the point $\mathbf{x}_0 = (3, -10)$.

Let us formally introduce the concept of a boundary point, which we alluded to in Example 1.

DEFINITION: BOUNDARY POINTS   Let $A \subset \mathbb{R}^n$. A point $\mathbf{x} \in \mathbb{R}^n$ is called a ***boundary point*** of A if every neighborhood of $\mathbf{x}$ contains at least one point in A and at least one point not in A.

In this definition, $\mathbf{x}$ may or may not be in A; if $\mathbf{x} \in A$, then $\mathbf{x}$ is a boundary point if every neighborhood of $\mathbf{x}$ contains at least one point *not* in A (it already contains a point of A, namely, $\mathbf{x}$). Similarly, if $\mathbf{x}$ is not in A, it is a boundary point if every neighborhood of $\mathbf{x}$ contains at least one point of A.

We shall be particularly interested in boundary points of open sets. By the definition of an open set, no point of an open set A can be a boundary point of A. Thus *a point* $\mathbf{x}$ *is a boundary point of an open set* A *if and only if* $\mathbf{x}$ *is not in* A *and every neighborhood of* $\mathbf{x}$ *has a nonempty intersection with* A.

This expresses in precise terms the intuitive idea that a boundary point of A is a point just on the "edge" of A. In many examples it is perfectly clear what the boundary points are.

EXAMPLE 2 (a) Let $A = (a, b)$ in $\mathbb{R}$. Then the boundary points of A consist of the points a and b. A consideration of Figure 2.2.6 and the definition will make this clear. [The reader will be asked to prove this in Exercise 2(c).]

boundary points

Figure 2.2.6 The boundary points of the interval (a, b).

(b) Let $A = D_r(x_0, y_0)$ be an r-disk about (x_0, y_0) in the plane. The boundary consists of points (x, y) with $(x - x_0)^2 + (y - y_0)^2 = r^2$ (Figure 2.2.7).

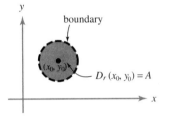

boundary

Figure 2.2.7 The boundary of A consists of points on the edge of A.

$D_r(x_0, y_0) = A$

(c) Let $A = \{(x, y) \in \mathbb{R}^2 \mid x > 0\}$. Then the boundary of A consists of all points on the y axis (the student should draw a figure).

(d) Let A be $D_r(\mathbf{x}_0)$ minus the point $\mathbf{x}_0$ (a "punctured" disk about $\mathbf{x}_0$). Then $\mathbf{x}_0$ is a boundary point of A. ▲

We now turn our attention to the concept of a limit. *Throughout the following discussions the domain of definition of the function f will be an open set A.* We are interested in finding the limit of f as $\mathbf{x} \in A$ approaches either a point of A or a boundary point of A.

The reader should appreciate the fact that the limit concept is a basic and useful tool for the analysis of functions; it enables us to study derivatives, and hence maxima and minima, asymptotes, improper integrals, and other important features of functions, as well as being useful for infinite series and sequences. We will present a theory of limits for functions of several variables that includes the theory for functions of one variable as a special case.

In one-variable calculus, the student has encountered the notion of $\displaystyle\lim_{x \to x_0} f(x) = l$ for a function $f \colon A \subset \mathbb{R} \to \mathbb{R}$ from a subset A of the real numbers to the real numbers. Intuitively, this means that as x gets closer and closer to x_0, the values $f(x)$ get closer and closer to (the limiting value) l. To put this intuitive idea on a firm, mathematical foundation, either the "epsilon (ε) and delta (δ) method" or the "neighborhood method" is usually introduced. The same is true for functions of several variables. In what follows we develop the neighborhood approach to limits. The epsilon-delta approach is left for optional study at the end of this section.

DEFINITION: LIMIT Let $f: A \subset \mathbb{R}^n \to \mathbb{R}^m$, where A is an open set. Let $\mathbf{x}_0$ be in A or be a boundary point of A, and let N be a neighborhood of $\mathbf{b} \in \mathbb{R}^m$. We say f is *eventually in N as $\mathbf{x}$ approaches* $\mathbf{x}_0$ if there exists a neighborhood U of $\mathbf{x}_0$ such that $\mathbf{x} \neq \mathbf{x}_0, \mathbf{x} \in U$, and $\mathbf{x} \in A$ imply $f(\mathbf{x}) \in N$. [The geometric meaning of this assertion is illustrated in Figure 2.2.8; note that $\mathbf{x}_0$ need not be in the set A, so that $f(\mathbf{x}_0)$ is not necessarily defined.] We say $f(\mathbf{x})$ *approaches* $\mathbf{b}$ as $\mathbf{x}$ approaches $\mathbf{x}_0$, or, in symbols,

$$\underset{\mathbf{x} \to \mathbf{x}_0}{\text{limit}}\, f(\mathbf{x}) = \mathbf{b} \quad \text{or} \quad f(\mathbf{x}) \to \mathbf{b} \quad \text{as} \quad \mathbf{x} \to \mathbf{x}_0,$$

when, given *any* neighborhood N of $\mathbf{b}$, f is eventually in N as $\mathbf{x}$ approaches $\mathbf{x}_0$ [that is, "$f(\mathbf{x})$ is close to $\mathbf{b}$ if $\mathbf{x}$ is close to $\mathbf{x}_0$.]" It may be that as $\mathbf{x}$ approaches $\mathbf{x}_0$ the values $f(\mathbf{x})$ do not get close to any particular number. In this case we say that $\underset{\mathbf{x} \to \mathbf{x}_0}{\text{limit}}\, f(\mathbf{x})$ *does not exist*.

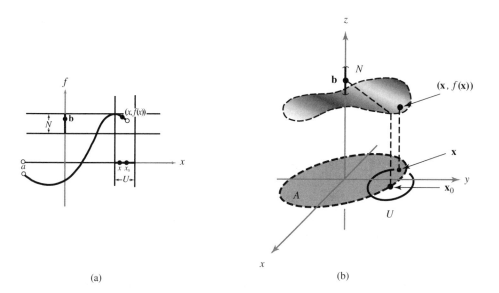

(a)   (b)

Figure 2.2.8 Limits in terms of neighborhoods; if $\mathbf{x}$ is in U, then $f(\mathbf{x})$ will be in N. [The little open circles denote the fact that $(a, f(a))$ and $(x_0, f(x_0))$ do not lie on the graph.] (a) $f: A = (a, x_0) \to \mathbb{R}$. (b) $f: A = \{(x, y) \mid x^2 + y^2 < 1\} \to \mathbb{R}$. (The dashed line is not in the graph of f.)

Henceforth, whenever we consider the notion $\underset{\mathbf{x} \to \mathbf{x}_0}{\text{limit}}\, f(\mathbf{x})$, we shall always assume that $\mathbf{x}_0$ either belongs to some open set on which f is defined or is on the boundary of such a set.

One reason we insist on $\mathbf{x} \neq \mathbf{x}_0$ in the definition of limit will become clear if we remember from one-variable calculus that we want to be able to define the derivative

$f'(x_0)$ of a function f at a point x_0 by

$$f'(x_0) = \lim_{x \to x_0} \frac{f(x) - f(x_0)}{x - x_0},$$

and this expression is not defined at $x = x_0$.

EXAMPLE 3   (a)   This example illustrates a limit that does not exist. Consider the function $f \colon \mathbb{R} \to \mathbb{R}$ defined by

$$f(x) = \begin{cases} 1 & \text{if } x > 0 \\ -1 & \text{if } x \leq 0. \end{cases}$$

The limit $f(x)$ does not exist, since there are points x_1 arbitrarily close to 0 with $f(x_1) = 1$ and also points x_2 arbitrarily close to 0 with $f(x_2) = -1$; that is, there is no single number that f is close to when x is close to 0 (see Figure 2.2.9). If f is restricted to the domain $(0, 1)$ or $(-1, 0)$, then the limit does exist. Can you say why?

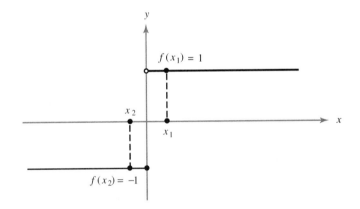

Figure 2.2.9   The limit of this function as $x \to 0$ does not exist.

(b)   This example illustrates a function whose limit does exist but whose limiting value does not equal its value at the limiting point. Define $f \colon \mathbb{R} \to \mathbb{R}$ by

$$f(x) = \begin{cases} 0 & \text{if } x \neq 0 \\ 1 & \text{if } x = 0. \end{cases}$$

It is true that $\lim\limits_{x \to 0} f(x) = 0$, since for any neighborhood U of 0, $x \in U$ and $x \neq 0$ implies that $f(x) = 0$. One sees from the graph in Figure 2.2.10 that f approaches 0 as $x \to 0$; we do not care that f happens to take on some other value at 0.   ▲

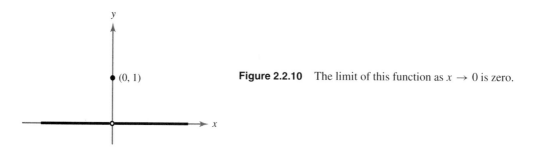

Figure 2.2.10 The limit of this function as $x \to 0$ is zero.

EXAMPLE 4 Use the definition to verify that the "obvious" limit $\mathbf{x} = \mathbf{x}_0$ holds, where $\mathbf{x}$ and $\mathbf{x}_0 \in \mathbb{R}^n$.
$\underset{\mathbf{x} \to \mathbf{x}_0}{}$

SOLUTION Let f be the function defined by $f(\mathbf{x}) = \mathbf{x}$, and let N be any neighborhood of $\mathbf{x}_0$. We must show that $f(\mathbf{x})$ is eventually in N as $\mathbf{x} \to \mathbf{x}_0$. Thus, by the definition we must find a neighborhood U of $\mathbf{x}_0$ with the property that if $\mathbf{x} \neq \mathbf{x}_0$ and $\mathbf{x} \in U$, then $f(\mathbf{x}) \in N$. Pick $U = N$. If $\mathbf{x} \in U$, then $\mathbf{x} \in N$; since $\mathbf{x} = f(\mathbf{x})$, it follows that $f(\mathbf{x}) \in N$. Thus we have shown that $\lim_{\mathbf{x} \to \mathbf{x}_0} \mathbf{x} = \mathbf{x}_0$. In similar way we have

$$\lim_{(x,y) \to (x_0, y_0)} x = x_0, \quad \text{etc.} \quad \blacktriangle$$

In what follows, the student may assume, without proof, the validity of limits from one-variable calculus. For example, $\lim_{x \to 1} \sqrt{x} = \sqrt{1} = 1$ and $\lim_{\theta \to 0} \sin \theta = \sin 0 = 0$ may be used.

EXAMPLE 5 (This example demonstrates another case in which the limit cannot simply be "read off" from the function.) *Find* $\lim_{x \to 1} g(x)$ *where*

$$g: x \mapsto \frac{x - 1}{\sqrt{x} - 1}.$$

SOLUTION This function is graphed in Figure 2.2.11.

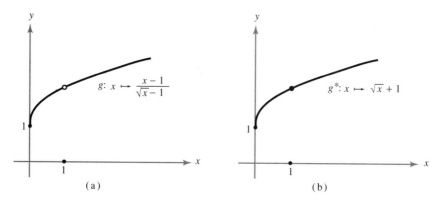

(a)   (b)

Figure 2.2.11 These graphs are the same except that in part (a), g is undefined at $x = 1$, while in part (b), g^* is defined for all $x \geq 0$.

We see that $g(1)$ is not defined, since division by zero is not defined. However, if we multiply the numerator and denominator of $g(x)$ by $\sqrt{x} + 1$, we find that for all x in the domain of g we have

$$g(x) = \frac{x - 1}{\sqrt{x} - 1} = \sqrt{x} + 1, \qquad x \neq 1.$$

The expression $g^*(x) = \sqrt{x} + 1$ is defined and takes the value 2 at $x = 1$; from one-variable calculus, $g^*(x) \to 2$ as $x \to 1$. But since $g^*(x) = g(x)$ for all $x \geq 0, x \neq 1$, we must have as well that $g(x) \to 2$ as $x \to 1$. ▲

To properly speak of *the* limit, we should establish that f can have *at most one* limit as $\mathbf{x} \to \mathbf{x}_0$. This is intuitively clear and we now state it formally. (See Section 2.7 for the proof.)

THEOREM 2: UNIQUENESS OF LIMITS

If $\lim\limits_{\mathbf{x} \to \mathbf{x}_0} f(\mathbf{x}) = \mathbf{b}_1$ *and* $\lim\limits_{\mathbf{x} \to \mathbf{x}_0} f(\mathbf{x}) = \mathbf{b}_2$, *then* $\mathbf{b}_1 = \mathbf{b}_2$.

To carry out practical computations with limits, we require some rules for limits, for example, that the limit of a sum is the sum of the limits. These rules are summarized in the following theorem (see Section 2.7 for the proof).

THEOREM 3: PROPERTIES OF LIMITS *Let* $f: A \subset \mathbb{R}^n \to \mathbb{R}^m$, $g: A \subset \mathbb{R}^n \to \mathbb{R}^m$, $\mathbf{x}_0$ *be in* A *or be a boundary point of* $A, \mathbf{b} \in \mathbb{R}^m$, *and* $c \in \mathbb{R}$; *then*

(i) *If* $\lim\limits_{\mathbf{x} \to \mathbf{x}_0} f(\mathbf{x}) = \mathbf{b}$, *then* $\lim\limits_{\mathbf{x} \to \mathbf{x}_0} cf(\mathbf{x}) = c\mathbf{b}$, *where* $cf: A \to \mathbb{R}^m$ *is defined by* $\mathbf{x} \mapsto (c)(f(\mathbf{x}))$.

(ii) *If* $\lim\limits_{\mathbf{x} \to \mathbf{x}_0} f(\mathbf{x}) = \mathbf{b}_1$ *and* $\lim\limits_{\mathbf{x} \to \mathbf{x}_0} g(\mathbf{x}) = \mathbf{b}_2$, *then* $\lim\limits_{\mathbf{x} \to \mathbf{x}_0} (f + g)(\mathbf{x}) = \mathbf{b}_1 + \mathbf{b}_2$, *where* $(f + g): A \to \mathbb{R}^m$ *is defined by* $\mathbf{x} \mapsto f(\mathbf{x}) + g(\mathbf{x})$.

(iii) *If* $m = 1, \lim\limits_{\mathbf{x} \to \mathbf{x}_0} f(\mathbf{x}) = b_1$, *and* $\lim\limits_{\mathbf{x} \to \mathbf{x}_0} g(\mathbf{x}) = b_2$, *then* $\lim\limits_{\mathbf{x} \to \mathbf{x}_0} (fg)(\mathbf{x}) = b_1 b_2$, *where* $(fg): A \to \mathbb{R}$ *is defined by* $\mathbf{x} \mapsto f(\mathbf{x})g(\mathbf{x})$.

(iv) *If* $m = 1, \lim\limits_{\mathbf{x} \to \mathbf{x}_0} f(\mathbf{x}) = b \neq 0$, *and* $f(\mathbf{x}) \neq 0$ *for all* $\mathbf{x} \in A$, *then* $\lim\limits_{\mathbf{x} \to \mathbf{x}_0} 1/f(\mathbf{x}) = 1/b$, *where* $1/f: A \to \mathbb{R}$ *is defined by* $\mathbf{x} \mapsto 1/f(\mathbf{x})$.

(v) *If* $f(\mathbf{x}) = (f_1(\mathbf{x}), \ldots, f_m(\mathbf{x}))$ *where* $f_i: A \to \mathbb{R}, i = 1, \ldots, m$, *are the component functions of* f, *then* $\lim\limits_{\mathbf{x} \to \mathbf{x}_0} f(\mathbf{x}) = \mathbf{b} = (b_1, \ldots, b_m)$ *if and only if* $\lim\limits_{\mathbf{x} \to \mathbf{x}_0} f_i(\mathbf{x}) = b_i$ *for each* $i = 1, \ldots, m$.

These results ought to be intuitively clear. For instance, rule (ii) says that if $f(\mathbf{x})$ is close to $\mathbf{b}_1$ and $g(\mathbf{x})$ is close to $\mathbf{b}_2$ when $\mathbf{x}$ is close to $\mathbf{x}_0$, then $f(\mathbf{x}) + g(\mathbf{x})$ is close to $\mathbf{b}_1 + \mathbf{b}_2$ when $\mathbf{x}$ is close to $\mathbf{x}_0$. The following example illustrates how this works.

EXAMPLE 6 Let $f: \mathbb{R}^2 \to \mathbb{R}$, $(x, y) \mapsto x^2 + y^2 + 2$. Compute $\displaystyle \lim_{(x,y)\to(0,1)} f(x, y)$.

SOLUTION Here f is the sum of the three functions $(x, y) \mapsto x^2$, $(x, y) \mapsto y^2$, and $(x, y) \mapsto 2$. The limit of a sum is the sum of the limits, and the limit of a product is the product of the limits (Theorem 3). Hence, using the fact that $\displaystyle \lim_{(x,y)\to(x_0,y_0)} x = x_0$ (Example 4), we obtain

$$\lim_{(x,y)\to(x_0,y_0)} x^2 = \left(\lim_{(x,y)\to(x_0,y_0)} x \right) \left(\lim_{(x,y)\to(x_0,y_0)} x \right) = x_0^2$$

and, using the same reasoning, $\displaystyle \lim_{(x,y)\to(x_0,y_0)} y^2 = y_0^2$. Consequently,

$$\lim_{(x,y)\to(0,1)} f(x, y) = 0^2 + 1^2 + 2 = 3. \quad \blacktriangle$$

In single-variable calculus we learned that the idea of a continuous function is based on the intuitive notion of a function whose graph is an unbroken curve, that is, a curve which has no *jumps*, or the kind of curve that would be traced by a particle in motion or by a moving pencil point that is not lifted from the paper.

To perform a detailed analysis of functions, we need concepts more precise than this rather vague notion. An example may clarify these ideas. Consider the specific function $f: \mathbb{R} \to \mathbb{R}$ defined by $f(x) = -1$ if $x \le 0$ and $f(x) = 1$ if $x > 0$. The graph of f is shown in Figure 2.2.12(a). [The little open circle denotes the fact that the point $(0, 1)$ does *not* lie on the graph of f]. Clearly, the graph of f is broken at $x = 0$. Consider also the function $g: x \mapsto x^2$. This function is pictured in Figure 2.2.12(b). The graph of g is not broken at any point.

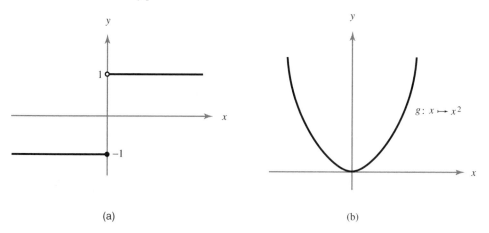

(a)   (b)

Figure 2.2.12 The function f in (a) is not continuous, because its value jumps as x crosses 0, while the function g in (b) is continuous.

If one examines examples of functions like f, whose graphs are broken at some point x_0, and functions like g, whose graphs are not broken, one sees that the principal difference between them is that for a function like g, the values of $g(x)$ get closer to $g(x_0)$ as x gets closer and closer to x_0. The same idea works for functions of several variables. But the notion of closer and closer does not suffice as a mathematical definition; thus we shall formulate these concepts precisely in terms of limits.

Since the condition $\lim_{\mathbf{x} \to \mathbf{x}_0} f(\mathbf{x}) = f(\mathbf{x}_0)$ means that $f(\mathbf{x})$ is close to $f(\mathbf{x}_0)$ when $\mathbf{x}$ is close to $\mathbf{x}_0$, we see that this limit condition does indeed correspond to the requirement that the graph of f be unbroken (see Figures 2.2.10, 2.2.13, where we illustrate the case $f: \mathbb{R} \to \mathbb{R}$). The case of several variables is easiest to visualize if we deal with real-valued functions, say $f: \mathbb{R}^2 \to \mathbb{R}$. In this case we can visualize f by drawing its graph, which consists of all points (x, y, z) in $\mathbb{R}^3$ with $z = f(x, y)$. The continuity of f thus means that its graph has no "breaks" in it (see Figure 2.2.14).

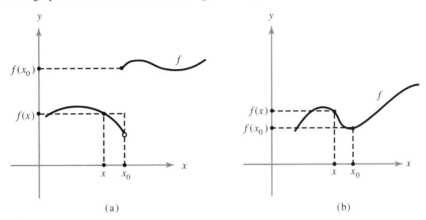

(a)   (b)

Figure 2.2.13 (a) Discontinuous function for which $\lim_{x \to x_0} f(x)$ does not exist; (b) continuous function for which this limit exists and equals $f(x_0)$.

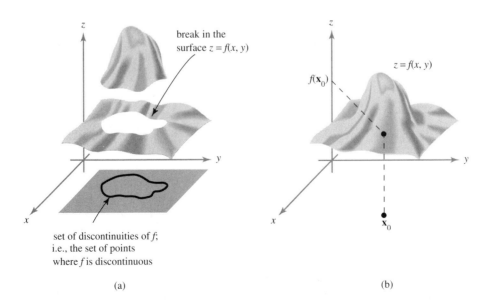

(a)   (b)

Figure 2.2.14 (a) A discontinuous function of two variables. (b) A continuous function.

DEFINITION: CONTINUITY Let $f\colon A \subset \mathbb{R}^n \to \mathbb{R}^m$ be a given function with domain A. Let $\mathbf{x}_0 \in A$. We say f is **continuous** at $\mathbf{x}_0$ if and only if

$$\lim_{\mathbf{x}\to\mathbf{x}_0} f(\mathbf{x}) = f(\mathbf{x}_0).$$

If we just say that f is **continuous**, we shall mean that f is continuous at each point $\mathbf{x}_0$ of A. If f is not continuous at $\mathbf{x}_0$, we say f is **discontinuous** at $\mathbf{x}_0$. If f is discontinuous at some point in its domain, we say f is **discontinuous**.

EXAMPLE 7   Any polynomial $p(x) = a_0 + a_1 x + \cdots + a_n x^n$ is continuous from $\mathbb{R}$ to $\mathbb{R}$. Indeed, from Theorem 3 and Example 4,

$$\lim_{x\to x_0} (a_0 + a_1 x + \cdots + a_n x^n) = \lim_{x\to x_0} a_0 + \lim_{x\to x_0} a_1 x + \cdots + \lim_{x\to x_0} a_n x^n$$
$$= a_0 + a_1 x_0 + \cdots + a_n x_0^n,$$

since the limit of a product is the product of the limits, which gives

$$\lim_{x\to x_0} x^n = \left(\lim_{x\to x_0} x\right)^n = x_0^n. \quad \blacktriangle$$

EXAMPLE 8   Let $f\colon \mathbb{R}^2 \to \mathbb{R}$, $f(x, y) = xy$. Then f is continuous, since, by the limit theorems and Example 4,

$$\lim_{(x,y)\to(x_0,y_0)} xy = \left(\lim_{(x,y)\to(x_0,y_0)} x\right)\left(\lim_{(x,y)\to(x_0,y_0)} y\right) = x_0 y_0. \quad \blacktriangle$$

One can see by the same method that any polynomial $p(x, y)$ [for example, $p(x, y) = 3x^2 - 6xy^2 + y^3$] in x and y is continuous.

EXAMPLE 9   The function $f\colon \mathbb{R}^2 \to \mathbb{R}$ defined by

$$f(x, y) = \begin{cases} 1 & \text{if } x \le 0 \text{ or } y \le 0 \\ 0 & \text{otherwise} \end{cases}$$

is not continuous at $(0, 0)$ or at any point on the positive x axis or positive y axis. Indeed, if $(x_0, y_0) = \mathbf{u}$ is such a point (i.e., $x_0 = 0$ and $y_0 \ge 0$, or $y_0 = 0$ and $x_0 \ge 0$) and $\delta \ge 0$, there are points $(x, y) \in D_\delta(\mathbf{u})$, a neighborhood of $\mathbf{u}$, with $f(x, y) = 1$ and other points $(x, y) \in D_\delta(\mathbf{u})$ with $f(x, y) = 0$. Thus it is *not* true that $f(x, y) \to f(x_0, y_0) = 1$ as $(x, y) \to (x_0, y_0)$. $\quad \blacktriangle$

To prove that specific functions are continuous, we can avail ourselves of the limit theorems (see Theorem 3 and Example 7). If we transcribe those results in terms of continuity, we are led to the following:

THEOREM 4: PROPERTIES OF CONTINUOUS FUNCTIONS

Let $f: A \subset \mathbb{R}^n \to \mathbb{R}^m$, $g: A \subset \mathbb{R}^n \to \mathbb{R}^m$, and let c be a real number.

(i)   *If f is continuous at $\mathbf{x}_0$, so is cf, where $(cf)(\mathbf{x}) = c[f(\mathbf{x})]$.*

(ii)   *If f and g are continuous at $\mathbf{x}_0$, so is $f + g$, where $(f + g)(\mathbf{x}) = f(\mathbf{x}) + g(\mathbf{x})$.*

(iii)   *If f and g are continuous at $\mathbf{x}_0$ and $m = 1$, then the product function fg defined by $(fg)(\mathbf{x}) = f(\mathbf{x})g(\mathbf{x})$ is continuous at $\mathbf{x}_0$.*

(iv)   *If $f: A \subset \mathbb{R}^n \to \mathbb{R}$ is continuous at $\mathbf{x}_0$ and nowhere zero on A, then the quotient $1/f$ is continuous at $\mathbf{x}_0$ where $(1/f)(\mathbf{x}) = 1/f(\mathbf{x})$.*

(v)   *If $f: A \subset \mathbb{R}^n \to \mathbb{R}^m$ and $f(\mathbf{x}) = (f_1(\mathbf{x}), \ldots, f_m(\mathbf{x}))$, then f is continuous at $\mathbf{x}_0$ if and only if each of the real-valued functions $f_1, \ldots, f_m$ is continuous at $\mathbf{x}_0$.*

A variant of (iv) is often used: If $f(\mathbf{x}_0) \neq 0$ and f is continuous, then $f(\mathbf{x}) \neq 0$ in a neighborhood of $\mathbf{x}_0$ and so $1/f$ is defined in that neighborhood, and $1/f$ is continuous at $\mathbf{x}_0$.

EXAMPLE 10   Let $f: \mathbb{R}^2 \to \mathbb{R}^2$, $(x, y) \mapsto (x^2 y, (y + x^3)/(1 + x^2))$. Show that f is continuous.

SOLUTION   To see this, it is sufficient, by property (v) above, to show that each component is continuous. Therefore we first show that $(x, y) \mapsto x^2 y$ is continuous. Now $(x, y) \mapsto x$ is continuous [see Example 4 and Theorem 4(v)], and so, by (iii), $(x, y) \mapsto x^2$ is continuous. Since $(x, y) \mapsto y$ is continuous, by (iii), the map $(x, y) \mapsto x^2 y$ is continuous. Since $1 + x^2$ is continuous and nonzero, by property (iv) we know that $1/(1 + x^2)$ is continuous; hence $(y + x^3)/(1 + x^2)$ is a product of continuous functions, and by (iii) is continuous.   ▲

Similar reasoning applies to examples like the function $\mathbf{c}: \mathbb{R} \to \mathbb{R}^3$ given by $\mathbf{c}(t) = (t^2, 1, t^3/(1 + t^2))$ to show they are continuous as well.

Next we discuss *composition*, another basic operation that can be performed on functions. If g maps A to B and f maps B to C, the **composition of g with f**, or of f on g, denoted by $f \circ g$, maps A to C by sending $\mathbf{x} \mapsto f(g(\mathbf{x}))$ (see Figure 2.2.15). For example, $\sin(x^2)$ is the composition of $x \mapsto x^2$ with $y \mapsto \sin y$.

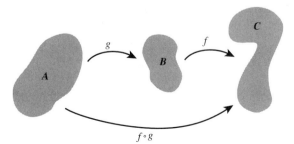

Figure 2.2.15   The composition of f on g.

THEOREM 5: CONTINUITY OF COMPOSITIONS
Let $g: A \subset \mathbb{R}^n \to \mathbb{R}^m$ and let $f: B \subset \mathbb{R}^m \to \mathbb{R}^p$. Suppose $g(A) \subset B$, so that $f \circ g$ is defined on A. If g is continuous at $\mathbf{x}_0 \in A$ and f is continuous at $\mathbf{y}_0 = g(\mathbf{x}_0)$, then $f \circ g$ is continuous at $\mathbf{x}_0$.

The intuition behind this is easy; the formal proof in Section 2.7 follows a similar pattern. Intuitively, we must show that as $\mathbf{x}$ gets close to $\mathbf{x}_0$, $f(g(\mathbf{x}))$ gets close to $f(g(\mathbf{x}_0))$. But as $\mathbf{x}$ gets close to $\mathbf{x}_0$, $g(\mathbf{x})$ gets close to $g(\mathbf{x}_0)$ (by continuity of g at $\mathbf{x}_0$); and as $g(\mathbf{x})$ gets close to $g(\mathbf{x}_0)$, $f(g(\mathbf{x}))$ gets close to $f(g(\mathbf{x}_0))$ [by continuity of f at $g(\mathbf{x}_0)$].

EXAMPLE 11   Let $f(x, y, z) = (x^2 + y^2 + z^2)^{30} + \sin z^3$. Show that f is continuous.

SOLUTION   Here we can write f as a sum of the two functions $(x^2 + y^2 + z^2)^{30}$ and $\sin z^3$, so it suffices to show that each is continuous. The first is the composite of $(x, y, z) \mapsto (x^2 + y^2 + z^2)$ with $u \mapsto u^{30}$, and the second is the composite of $(x, y, z) \mapsto z^3$ with $u \mapsto \sin u$, and so we have continuity by Theorem 5.   ▲

Supplement to Section 2.2:   LIMITS IN TERMS OF ε's AND δ's

We now state a theorem (proved in Section 2.7) giving a useful formulation of the notion of limit in terms of epsilons and deltas that is often taken as the *definition* of limit. It is another way of making precise the intuitive statement that "$f(\mathbf{x})$ is close to $\mathbf{b}$ when $\mathbf{x}$ is close to $\mathbf{x}_0$." To help understand this formulation, the reader should consider it with respect to each of the examples already presented.

THEOREM 6   *Let $f: A \subset \mathbb{R}^n \to \mathbb{R}^m$ and let $\mathbf{x}_0$ be in A or be a boundary point of A. Then $\lim\limits_{\mathbf{x} \to \mathbf{x}_0} f(\mathbf{x}) = \mathbf{b}$ if and only if for every number $\varepsilon > 0$ there is a $\delta > 0$ such that for any $\mathbf{x} \in A$ satisfying $0 < \|\mathbf{x} - \mathbf{x}_0\| < \delta$, we have $\|f(\mathbf{x}) - \mathbf{b}\| < \varepsilon$ (see Figure 2.2.16).*

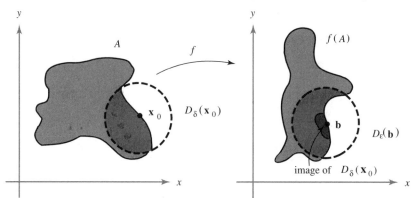

Figure 2.2.16   The geometry of the ε-δ definition of limit.

HISTORICAL NOTE

Although calculus was created by Newton and Leibniz in the 1600s and was developed extensively in the 1700s by the Bernoulli's, Euler, Lagrange, and many others, it was only in the 1800s with work of Dirichlet, Cauchy, Weierstrass, and others that its foundation in terms of limits was put on a firm foundation.

To illustrate the methodology of the epsilon-delta technique in Theorem 6, we consider the following examples.

EXAMPLE 12 Show that $\displaystyle\lim_{(x,y)\to(0,0)} x = 0$ using the ε-δ method.

SOLUTION Note that if $\delta > 0$, $\|(x, y) - (0, 0)\| = \sqrt{x^2 + y^2} < \delta$ implies $|x - 0| = |x| = \sqrt{x^2} \le \sqrt{x^2 + y^2} < \delta$. Thus, if $\|(x, y) - (0, 0)\| < \delta$, then $|x - 0|$ is also less than δ. Given $\varepsilon > 0$, we are required to find a $\delta > 0$ (generally depending on ε) with the property that $0 < \|(x, y) - (0, 0)\| < \delta$ implies $|x - 0| < \varepsilon$. What are we to pick as our δ? From the above calculation, we see that if we choose $\delta = \varepsilon$, then $\|(x, y) - (0, 0)\| < \delta$ implies $|x - 0| < \varepsilon$. This shows that $\displaystyle\lim_{(x,y)\to(0,0)} x = 0$. We could have also chosen $\delta = \varepsilon/2$ or $\varepsilon/3$, but it suffices to find just one δ satisfying the requirements of the definition of a limit. ▲

EXAMPLE 13 Consider the function

$$f(x, y) = \frac{\sin(x^2 + y^2)}{x^2 + y^2}.$$

Even though f is not defined at $(0, 0)$, determine whether $f(x, y)$ approaches some number as (x, y) approaches $(0, 0)$.

SOLUTION From elementary calculus we know that

$$\lim_{\alpha \to 0} \frac{\sin \alpha}{\alpha} = 1.$$

Thus it is reasonable to guess that

$$\lim_{v \to (0,0)} f(v) = \lim_{v \to (0,0)} \frac{\sin \|v\|^2}{\|v\|^2} = 1.$$

Indeed, since $\lim_{\alpha \to 0}(\sin \alpha)/\alpha = 1$, given $\varepsilon > 0$ we find a $\delta > 0$, with $1 > \delta > 0$, such that $0 < |\alpha| < \delta$ implies that $|(\sin \alpha)/\alpha - 1| < \varepsilon$. If $0 < \|v\| < \delta$, then $0 < \|v\|^2 < \delta^2 < \delta$, and therefore

$$|f(\mathbf{v}) - 1| = \left| \frac{\sin \|\mathbf{v}\|^2}{\|\mathbf{v}\|^2} - 1 \right| < \varepsilon.$$

Thus, $\displaystyle\lim_{\mathbf{v} \to (0,0)} f(\mathbf{v}) = 1$. If we plot $[\sin(x^2 + y^2)]/(x^2 + y^2)$ on a computer, we get a graph that is indeed well behaved near $(0, 0)$ (Figure 2.2.17).   ▲

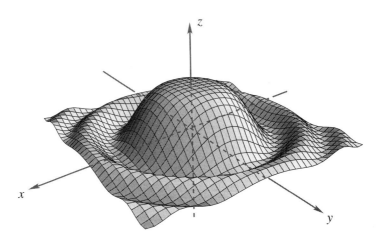

Figure 2.2.17   Graph of $f(x, y) = [\sin(x^2 + y^2)]/(x^2 + y^2)$.

EXAMPLE 14   Show that

$$\lim_{(x,y) \to (0,0)} \frac{x^2}{\sqrt{x^2 + y^2}} = 0.$$

SOLUTION   We must show that $x^2/\sqrt{x^2 + y^2}$ is small when (x, y) is close to the origin. To do this we use the following inequality:

$$0 \leq \frac{x^2}{\sqrt{x^2 + y^2}} \leq \frac{x^2 + y^2}{\sqrt{x^2 + y^2}} \qquad (\text{since } y^2 \geq 0)$$
$$= \sqrt{x^2 + y^2}.$$

Given $\varepsilon > 0$, choose $\delta = \varepsilon$. Then $\|(x, y) - (0, 0)\| = \|(x, y)\| = \sqrt{x^2 + y^2}$, and so $\|(x, y) - (0, 0)\| < \delta$ implies that

$$\left| \frac{x^2}{\sqrt{x^2 + y^2}} - 0 \right| = \frac{x^2}{\sqrt{x^2 + y^2}} \leq \sqrt{x^2 + y^2} = \|(x, y) - (0, 0)\| < \delta = \varepsilon.$$

Thus the conditions of Theorem 6 have been fulfilled and the limit is verified.   ▲

EXAMPLE 15   (a)   Does $\displaystyle\lim_{(x,y)\to(0,0)} x^2/(x^2 + y^2)$ exist? [See Figure 2.2.18(a).]

(b)   Prove that [see Figure 2.2.18(b)]

$$\lim_{(x,y)\to(0,0)} \frac{2x^2y}{x^2 + y^2} = 0.$$

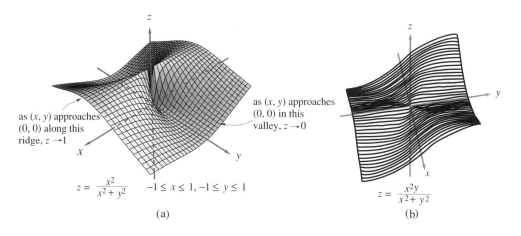

as (x, y) approaches $(0, 0)$ along this ridge, $z \to 1$

as (x, y) approaches $(0, 0)$ in this valley, $z \to 0$

$z = \dfrac{x^2}{x^2 + y^2}$ $-1 \le x \le 1, -1 \le y \le 1$

$z = \dfrac{x^2y}{x^2 + y^2}$

(a)   (b)

Figure 2.2.18   (a) The function $z = x^2/(x^2 + y^2)$ has no limit at $(0, 0)$. (b) The function $z = (2x^2y)/(x^2 + y^2)$ has limit 0 at $(0,0)$.

SOLUTION   (a)   If the limit exists, $x^2/(x^2 + y^2)$ should approach a definite value, say a, as (x, y) gets near $(0, 0)$. In particular, if (x, y) approaches zero along any given path, then $x^2/(x^2 + y^2)$ should approach the limiting value a. If (x, y) approaches $(0, 0)$ along the line $y = 0$, the limiting value is clearly 1 (just set $y = 0$ in the above expression to get $x^2/x^2 = 1$). If (x, y) approaches $(0, 0)$ along the line $x = 0$, the limiting value is

$$\lim_{(x,y)\to(0,0)} \frac{0^2}{0^2 + y^2} = 0 \ne 1.$$

Hence, $\displaystyle\lim_{(x,y)\to(0,0)} x^2/(x^2 + y^2)$ does not exist.

(b)   Note that

$$\left| \frac{2x^2y}{x^2 + y^2} \right| \le \left| \frac{2x^2y}{x^2} \right| = 2|y|.$$

Thus, given $\varepsilon > 0$, choose $\delta = \varepsilon/2$; then $0 < \|(x, y) - (0, 0)\| = \sqrt{x^2 + y^2} < \delta$ implies $|y| < \delta$, and thus

$$\left| \frac{2x^2y}{x^2 + y^2} - 0 \right| < 2\delta = \varepsilon. \quad \blacktriangle$$

Using the ε-δ notation, we are led to the following reformulation of the definition of continuity.

THEOREM 7   *Let $f: A \subset \mathbb{R}^n \to \mathbb{R}^m$ be given. Then f is continuous at $\mathbf{x}_0 \in A$ if and only if for every number $\varepsilon > 0$ there is a number $\delta > 0$ such that*

$$\mathbf{x} \in A \quad and \quad \|\mathbf{x} - \mathbf{x}_0\| < \delta \qquad implies \qquad \|f(\mathbf{x}) - f(\mathbf{x}_0)\| < \varepsilon.$$

The proof is almost immediate. Notice that in Theorem 6 we insisted that $0 < \|\mathbf{x} - \mathbf{x}_0\|$, that is, $\mathbf{x} \neq \mathbf{x}_0$. That is *not* imposed here; indeed, the conclusion of Theorem 7 is certainly valid when $\mathbf{x} = \mathbf{x}_0$, and so there is no need to exclude this case. Here we do care about the value of f at $\mathbf{x}_0$; we want f at nearby points to be close to *this* value.

EXERCISES

In the following exercises the reader may assume that the exponential, sine, and cosine functions are continuous and may freely use techniques from one-variable calculus.

1. Show that the following subsets of the plane are open:
   (a) $A = \{(x, y) \mid -1 < x < 1, -1 < y < 1\}$
   (b) $B = \{(x, y) \mid y > 0\}$
   (c) $C = \{(x, y) \mid 2 < x^2 + y^2 < 4\}$
   (d) $A \cup B \cup C$, where the sets A, B, and C are as defined in parts (a), (b), and (c)
   (e) $D = \{(x, y) \mid x \neq 0 \text{ and } y \neq 0\}$

2. (a) Prove that for $\mathbf{x} \in \mathbb{R}^n$ and $s < t$, $D_s(\mathbf{x}) \subset D_t(\mathbf{x})$.
   (b) Prove that if U and V are neighborhoods of $\mathbf{x} \in \mathbb{R}^n$, then so are $U \cap V$ and $U \cup V$.
   (c) Prove that the boundary points of an open interval $(a, b) \subset \mathbb{R}$ are the points a and b.

3. (Uses material in the supplement to this section.) Use the ε-δ formulation of limits to prove that $x^2 \to 4$ as $x \to 2$. Give another proof using Theorem 3.

4. Compute the following limits:

   (a) $\displaystyle\lim_{(x,y)\to(0,1)} x^3 y$
   (b) $\displaystyle\lim_{(x,y)\to(0,1)} e^x y$
   (c) $\displaystyle\lim_{x\to0} \frac{\sin^2 x}{x}$
    (d) $\displaystyle\lim_{x\to0} \frac{\sin^2 x}{x^2}$

5. Compute the following limits:

   (a) $\displaystyle\lim_{x\to3}(x^2 - 3x + 5)$
   (b) $\displaystyle\lim_{x\to0} \sin x$
   (c) $\displaystyle\lim_{h\to0} \frac{(x+h)^2 - x^2}{h}$
    (d) $\displaystyle\lim_{h\to0} \frac{e^h - 1}{h}$
   (e) $\displaystyle\lim_{x\to0} \frac{\cos x - 1}{x^2}$

6. Compute the following limits, if they exist:

(a) $\displaystyle\lim_{(x,y)\to(0,0)} (x^2 + y^2 + 3)$

(b) $\displaystyle\lim_{(x,y)\to(0,0)} \frac{xy}{x^2 + y^2 + 2}$

(c) $\displaystyle\lim_{(x,y)\to(0,0)} \frac{e^{xy}}{x + 1}$

(d) $\displaystyle\lim_{(x,y)\to(0,0)} \frac{\cos x - 1 - (x^2/2)}{x^4 + y^4}$

(e) $\displaystyle\lim_{(x,y)\to(0,0)} \frac{(x - y)^2}{x^2 + y^2}$

7. Compute $\displaystyle\lim_{\mathbf{x}\to\mathbf{x}_0} f(\mathbf{x})$, if it exists, for the following cases:

(a) $f\colon \mathbb{R} \to \mathbb{R}, x \mapsto |x|, x_0 = 1$

(b) $f\colon \mathbb{R}^n \to \mathbb{R}, \mathbf{x} \mapsto \|\mathbf{x}\|$, arbitrary $\mathbf{x}_0$

(c) $f\colon \mathbb{R} \to \mathbb{R}^2, x \mapsto (x^2, e^x), x_0 = 1$

(d) $f\colon \mathbb{R}^2\backslash\{(0,0)\} \to \mathbb{R}^2, (x, y) \mapsto (\sin(x - y), e^{x(y+1)} - x - 1)/\|(x, y)\|, \mathbf{x}_0 = (0, 0)$.

8. Let $A \subset \mathbb{R}^2$ be the open unit disk $D_1(0, 0)$ with the point $\mathbf{x}_0 = (1, 0)$ added, and let $f\colon A \to \mathbb{R}, \mathbf{x} \mapsto f(\mathbf{x})$ be the constant function $f(\mathbf{x}) = 1$. Show that $\displaystyle\lim_{\mathbf{x}\to\mathbf{x}_0} f(\mathbf{x}) = 1$.

9. Show that $\displaystyle\lim_{(x,y)\to(0,0)} (\sin xy)/xy = 1$.

10. Show that the map $f\colon \mathbb{R} \to \mathbb{R}, x \mapsto x^2 e^x/(2 - \sin x)$ is continuous.

11. Let $f(x, y, z) = (x^2 + 3y^2)/(x + 1)$. Compute

$$\lim_{(x,y,z)\to(0,0,0)} f(x, y, z).$$

12. Show that $f\colon \mathbb{R} \to \mathbb{R}, x \mapsto (1 - x)^8 + \cos(1 + x^3)$ is continuous.

13. If $f\colon \mathbb{R}^n \to \mathbb{R}$ and $g\colon \mathbb{R}^n \to \mathbb{R}$ are continuous, show that the functions

$$f^2 g\colon \mathbb{R}^n \to \mathbb{R}, \mathbf{x} \mapsto [f(\mathbf{x})]^2 g(\mathbf{x})$$

and

$$f^2 + g\colon \mathbb{R}^n \to \mathbb{R}, x \mapsto [f(\mathbf{x})]^2 + g(\mathbf{x})$$

are continuous.

14. Prove that $f\colon \mathbb{R}^2 \to \mathbb{R}, (x, y) \mapsto ye^x + \sin x + (xy)^4$ is continuous.

15. (a) Can $[\sin(x + y)]/(x + y)$ be made continuous by suitably defining it at (0,0)?

(b) Can $xy/(x^2 + y^2)$ be made continuous by suitably defining it at $(0, 0)$?

16. (a) Use l'Hôpital's rule to calculate $\displaystyle\lim_{x\to0} \frac{\sin 2x - 2x}{x^3}$.

(b) Does $\displaystyle\lim_{(x,y)\to(0,0)} \frac{\sin 2x - 2x + y}{x^3 + y}$ exist?

17. Suppose $\mathbf{x}$ and $\mathbf{y}$ are in $\mathbb{R}^n$ and $\mathbf{x} \neq \mathbf{y}$. Show that there is a continuous function $f \colon \mathbb{R}^n \to \mathbb{R}$ with $f(\mathbf{x}) = 1$, $f(\mathbf{y}) = 0$, and $0 \leq f(\mathbf{z}) \leq 1$ for every $\mathbf{z}$ in $\mathbb{R}^n$.

18. Let $f \colon A \subset \mathbb{R}^n \to \mathbb{R}$ be given and let $\mathbf{x}_0$ be a boundary point of A. We say that $\lim\limits_{\mathbf{x} \to \mathbf{x}_0} f(\mathbf{x}) = \infty$ if for every $N > 0$ there is a $\delta > 0$ such that $0 < \|\mathbf{x} - \mathbf{x}_0\| < \delta$ and $\mathbf{x} \in A$ implies $f(\mathbf{x}) > N$.
  - (a) Prove that $\lim\limits_{x \to 1} (x - 1)^{-2} = \infty$.
  - (b) Prove that $\lim\limits_{x \to 0} 1/|x| = \infty$. Is it true that $\lim\limits_{x \to 0} 1/x = \infty$?
  - (c) Prove that $\lim\limits_{(x,y) \to (0,0)} 1/(x^2 + y^2) = \infty$.

19. Let $b \in \mathbb{R}$ and $f \colon \mathbb{R} \backslash [b] \to \mathbb{R}$ be a function. We write $\lim\limits_{x \to b-} f(x) = L$ and say that L is the **_left-hand limit_** of f at b, if for every $\varepsilon > 0$, there is a $\delta > 0$ such that $x < b$ and $0 < |x - b| < \delta$ implies $|f(x) - L| < \varepsilon$.
  - (a) Formulate a definition of **_right-hand limit_**, or $\lim\limits_{x \to b+} f(x)$.
  - (b) Find $\lim\limits_{x \to 0-} 1/(1 + e^{1/x})$ and $\lim\limits_{x \to 0+} 1/(1 + e^{1/x})$.
  - (c) Sketch the graph of $1/(1 + e^{1/x})$.

20. Show that f is continuous at $\mathbf{x}_0$ if and only if

$$\lim_{\mathbf{x} \to \mathbf{x}_0} \|f(\mathbf{x}) - f(\mathbf{x}_0)\| = 0.$$

21. Let $f \colon A \subset \mathbb{R}^n \to \mathbb{R}^m$ satisfy $\|f(\mathbf{x}) - f(\mathbf{y})\| \leq K \|\mathbf{x} - \mathbf{y}\|^\alpha$ for all $\mathbf{x}$ and $\mathbf{y}$ in A for positive constants K and α. Show that f is continuous. (Such functions are called **_Hölder-continuous_** or, if $\alpha = 1$, **_Lipschitz-continuous._**)

22. Show that $f \colon \mathbb{R}^n \to \mathbb{R}^m$ is continuous at all points if and only if the inverse image of every open set is open.

23.
  - (a) Find a number $\delta > 0$ such that if $|a| < \delta$, then $|a^3 + 3a^2 + a| < 1/100$.
  - (b) Find a number $\delta > 0$ such that if $x^2 + y^2 < \delta^2$, then

$$|x^2 + y^2 + 3xy + 180xy^5| < 1/10,000.$$

2.3 DIFFERENTIATION

In Section 2.1 we considered a few methods for graphing functions. By these methods alone it may be impossible to compute enough information to grasp even the general features of a complicated function. From elementary calculus we know that the idea of the derivative can greatly aid us in this task; for example, it enables us to locate maxima and minima and to compute rates of change. The derivative also has many applications beyond this, as the student surely has discovered in elementary calculus.

Intuitively, we know from our work in Section 2.2 that a continuous function is one that has no "breaks" in its graph. A differentiable function from $\mathbb{R}^2$ to $\mathbb{R}$ ought to be such that not only are there no "breaks" in its graph, but there is a well-defined plane tangent to the graph at each point. Thus, there must not be any sharp folds, corners, or peaks in the graph (see Figure 2.3.1). In other words, the graph must be *smooth*.

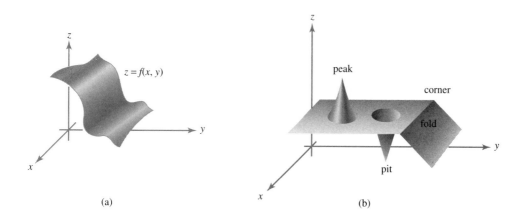

Figure 2.3.1 A smooth graph (a) and a nonsmooth one (b).

To make these ideas precise, we need a sound definition of what we mean by the phrase "$f(x_1, \ldots, x_n)$ is differentiable at $\mathbf{x} = (x_1, \ldots, x_n)$." Actually, this definition is not quite as simple as one might think. Toward this end, however, let us introduce the notion of the *partial derivative*. This notion relies only on our knowledge of one-variable calculus. (A quick review of the definition of the derivative in a one-variable calculus text might be advisable at this point.)

DEFINITION: PARTIAL DERIVATIVES Let $U \subset \mathbb{R}^n$ be an open set and $f: U \subset \mathbb{R}^n \to \mathbb{R}$ be a real-valued function. Then $\partial f/\partial x_1, \ldots, \partial f/\partial x_n$, the *partial derivatives* of f with respect to the first, second, $\ldots$, nth variable, are the real-valued functions of n variables, which, at the point $(x_1, \ldots, x_n) = \mathbf{x}$, are defined by

$$\frac{\partial f}{\partial x_j}(x_1, \ldots, x_n) = \lim_{h \to 0} \frac{f(x_1, x_2, \ldots, x_j + h, \ldots, x_n) - f(x_1, \ldots, x_n)}{h}$$

$$= \lim_{h \to 0} \frac{f(\mathbf{x} + h\mathbf{e}_j) - f(\mathbf{x})}{h}$$

if the limits exist, where $1 \le j \le n$ and $\mathbf{e}_j$ is the jth standard basis vector defined by $\mathbf{e}_j = (0, \ldots, 1, \ldots, 0)$, with 1 in the jth slot (see Section 1.5). The domain of the function $\partial f/\partial x_j$ is the set of $\mathbf{x} \in \mathbb{R}^n$ for which the limit exists.

In other words, $\partial f/\partial x_j$ is just the derivative of f with respect to the variable x_j, with the other variables held fixed. If $f: \mathbb{R}^3 \to \mathbb{R}$, we shall often use the notation $\partial f/\partial x, \partial f/\partial y, \partial f/\partial z$ in place of $\partial f/\partial x_1, \partial f/\partial x_2, \partial f/\partial x_3$. If $f: U \subset \mathbb{R}^n \to \mathbb{R}^m$, then we can write

$$f(x_1, \ldots, x_n) = (f_1(x_1, \ldots, x_n), \ldots, f_m(x_1, \ldots, x_n)),$$

so that we can speak of the partial derivatives of each component; for example, $\partial f_m / \partial x_n$ is the partial derivative of the mth component with respect to x_n, the nth variable.

EXAMPLE 1   If $f(x, y) = x^2 y + y^3$, find $\partial f / \partial x$ and $\partial f / \partial y$.

SOLUTION   To find $\partial f / \partial x$ we hold y constant (think of it as some number, say 1) and differentiate only with respect to x; this yields

$$\frac{\partial f}{\partial x} = \frac{d(x^2 y + y^3)}{dx} = 2xy.$$

Similarly, to find $\partial f / \partial y$ we hold x constant and differentiate only with respect to y:

$$\frac{\partial f}{\partial y} = \frac{d(x^2 y + y^3)}{dy} = x^2 + 3y^2. \quad \blacktriangle$$

To indicate that a partial derivative is to be evaluated at a particular point, for example, at (x_0, y_0), we write

$$\frac{\partial f}{\partial x}(x_0, y_0) \quad \text{or} \quad \left. \frac{\partial f}{\partial x} \right|_{x=x_0, y=y_0} \quad \text{or} \quad \left. \frac{\partial f}{\partial x} \right|_{(x_0, y_0)}$$

When we write $z = f(x, y)$ for the dependent variable, we sometimes write $\partial z / \partial x$ for $\partial f / \partial x$. Strictly speaking, this is an abuse of notation, but it is common practice to use these two notations interchangeably.

EXAMPLE 2   If $z = \cos xy + x \cos y = f(x, y)$, find the two partial derivatives $(\partial z / \partial x)(x_0, y_0)$ and $(\partial z / \partial y)(x_0, y_0)$.

SOLUTION   First we fix y_0 and differentiate with respect to x, giving

$$\frac{\partial z}{\partial x}(x_0, y_0) = \left. \frac{d(\cos x y_0 + x \cos y_0)}{dx} \right|_{x=x_0}$$

$$= (-y_0 \sin x y_0 + \cos y_0)|_{x=x_0}$$

$$= -y_0 \sin x_0 y_0 + \cos y_0.$$

Similarly, we fix x_0 and differentiate with respect to y to obtain

$$\frac{\partial z}{\partial y}(x_0, y_0) = \left. \frac{d(\cos x_0 y + x_0 \cos y)}{dy} \right|_{y=y_0}$$

$$= (-x_0 \sin x_0 y - x_0 \sin y)|_{y=y_0}$$

$$= -x_0 \sin x_0 y_0 - x_0 \sin y_0. \quad \blacktriangle$$

EXAMPLE 3 Find $\partial f/\partial x$ if $f(x, y) = xy/\sqrt{x^2 + y^2}$.

SOLUTION By the quotient rule,

$$\frac{\partial f}{\partial x} = \frac{y\sqrt{x^2 + y^2} - xy(x/\sqrt{x^2 + y^2})}{x^2 + y^2} = \frac{y(x^2 + y^2) - x^2 y}{(x^2 + y^2)^{3/2}} = \frac{y^3}{(x^2 + y^2)^{3/2}}. \quad \blacktriangle$$

A definition of differentiability which requires only the existence of partial derivatives turns out to be insufficient. Many standard results, such as the chain rule for functions of several variables would not follow, as Example 4 shows. Below, we shall see how to rectify this situation.

EXAMPLE 4 Let $f(x, y) = x^{1/3} y^{1/3}$. By definition,

$$\frac{\partial f}{\partial x}(0, 0) = \underset{h \to 0}{\text{limit}} \ \frac{f(h, 0) - f(0, 0)}{h} = \underset{h \to 0}{\text{limit}} \ \frac{0 - 0}{h} = 0,$$

and, similarly, $(\partial f/\partial y)(0, 0) = 0$ (these are not indeterminate forms!). It is necessary to use the original definition of partial derivatives, because the functions $x^{1/3}$ and $y^{1/3}$ are not themselves differentiable at 0. Suppose we restrict f to the line $y = x$ to get $f(x, x) = x^{2/3}$ (see Figure 2.3.2). We can view the substitution $y = x$ as the composition $f \circ g$ of the function $g \colon \mathbb{R} \to \mathbb{R}^2$, defined by $g(x) = (x, x)$, and $f \colon \mathbb{R}^2 \to \mathbb{R}$, defined by $f(x, y) = x^{1/3} y^{1/3}$.

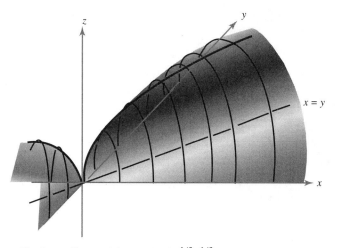

Figure 2.3.2 The "upper" part of the graph of $x^{1/3} y^{1/3}$.

Thus the composite $f \circ g$ is given by $(f \circ g)(x) = x^{2/3}$. Each component of g is differentiable in x, and f has partial derivatives at $(0, 0)$, but $f \circ g$ *is not differentiable* at $x = 0$, in the sense of one-variable calculus. In other words, *the composition of f with g is not differentiable* in contrast to the calculus of functions of one variable, where the composition of differentiable functions *is* differentiable. Below, we shall give a definition of differentiability which has the pleasant consequence that the composition of differentiable functions *is* differentiable.

There is another reason for being dissatisfied with the mere existence of partial derivatives of $f(x, y) = x^{1/3}y^{1/3}$; there is no plane tangent, in any reasonable sense, to the graph at $(0, 0)$. The xy plane is tangent to the graph along the x and y axes because f has slope zero at $(0, 0)$ along these axes; that is, $\partial f/\partial x = 0$ and $\partial f/\partial y = 0$ at $(0, 0)$. Thus, if there is a tangent plane, it must be the xy plane. However, as is evident from Figure 2.3.2, the xy plane is not tangent to the graph in other directions, since the graph has a severe crinkle and so the xy plane cannot be said to be tangent to the graph of f.   ▲

To "motivate" our definition of differentiability, let us compute what the equation of the plane tangent to the graph of $f: \mathbb{R}^2 \to \mathbb{R}$, $(x, y) \mapsto f(x, y)$ at (x_0, y_0) ought to be if f is smooth enough. In $\mathbb{R}^3$, a nonvertical plane has an equation of the form

$$z = ax + by + c.$$

If it is to be the plane tangent to the graph of f, the slopes along the x and y axes must be equal to $\partial f/\partial x$ and $\partial f/\partial y$, the rates of change of f with respect to x and y. Thus, $a = \partial f/\partial x, b = \partial f/\partial y$ [evaluated at (x_0, y_0)]. Finally, we may determine the constant c from the fact that $z = f(x_0, y_0)$ when $x = x_0, y = y_0$. Thus we get

$$z = f(x_0, y_0) + \left[\frac{\partial f}{\partial x}(x_0, y_0)\right](x - x_0) + \left[\frac{\partial f}{\partial y}(x_0, y_0)\right](y - y_0), \qquad (1)$$

which should be the equation of the plane tangent to the graph of f at (x_0, y_0), if f is "smooth enough" (see Figure 2.3.3).

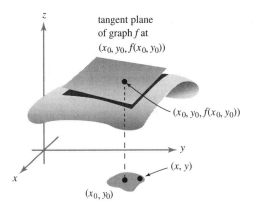

tangent plane
of graph f at
$(x_0, y_0, f(x_0, y_0))$

$(x_0, y_0, f(x_0, y_0))$

z

y

x

(x, y)

(x_0, y_0)

Figure 2.3.3   For points (x, y) near (x_0, y_0) the graph of the tangent plane is close to the graph of f.

Our definition of differentiability will mean in effect that the plane given by equation (1) is a "good" approximation of f near (x_0, y_0). To get an idea of what one might mean by a good approximation, let us return for a moment to one-variable calculus. If f is differentiable at a point x_0, then we know that

$$\underset{\Delta x \to 0}{\text{limit}} \frac{f(x_0 + \Delta x) - f(x_0)}{\Delta x} = f'(x_0).$$

Let $x = x_0 + \Delta x$ and rewrite this as

$$\underset{x \to x_0}{\text{limit}} \frac{f(x) - f(x_0)}{x - x_0} = f'(x_0).$$

Using the trivial limit $\underset{x \to x_0}{\text{limit}} f'(x_0) = f'(x_0)$, we can rewrite the above equation as

$$\underset{x \to x_0}{\text{limit}} \frac{f(x) - f(x_0)}{x - x_0} = \underset{x \to x_0}{\text{limit}} f'(x_0)$$

that is,

$$\underset{x \to x_0}{\text{limit}} \left[\frac{f(x) - f(x_0)}{x - x_0} - f'(x_0) \right] = 0$$

that is,

$$\underset{x \to x_0}{\text{limit}} \frac{f(x) - f(x_0) - f'(x_0)(x - x_0)}{x - x_0} = 0.$$

Thus, the tangent line l through $(x_0, f(x_0))$ with slope $f'(x_0)$ is close to f in the sense that the difference between $f(x)$ and $l(x) = f(x_0) + f'(x_0)(x - x_0)$ goes to zero *even* when divided by $x - x_0$ as x goes to x_0. This is the notion of a "good approximation" that we will adapt to functions of several variables, with the tangent line replaced by the tangent plane (see equation (1), above).

DEFINITION: DIFFERENTIABLE: TWO VARIABLES

Let $f: \mathbb{R}^2 \to \mathbb{R}$. We say f is **differentiable** at (x_0, y_0), if $\partial f / \partial x$ and $\partial f / \partial y$ exist at (x_0, y_0) and if

$$\frac{f(x, y) - f(x_0, y_0) - \left[\dfrac{\partial f}{\partial x}(x_0, y_0) \right](x - x_0) - \left[\dfrac{\partial f}{\partial y}(x_0, y_0) \right](y - y_0)}{\|(x, y) - (x_0, y_0)\|} \to 0 \quad (2)$$

as $(x, y) \to (x_0, y_0)$. This equation expresses what we mean by saying that

$$f(x_0, y_0) + \left[\frac{\partial f}{\partial x}(x_0, y_0) \right](x - x_0) + \left[\frac{\partial f}{\partial y}(x_0, y_0) \right](y - y_0)$$

is a **good approximation** to the function f.

It is not always easy to use this definition to see whether f is differentiable, but it will be easy to use another criterion, given in Theorem 9, below.

We have used the informal notion of the plane tangent to the graph of a function to motivate our definition of differentiability. Now we are ready to adopt a formal definition of the tangent plane.

> **DEFINITION: TANGENT PLANE** Let $f: \mathbb{R}^2 \to \mathbb{R}$ be differentiable at $\mathbf{x}_0 = (x_0, y_0)$. The plane in $\mathbb{R}^3$ defined by equation (1),
>
> $$z = f(x_0, y_0) + \left[\frac{\partial f}{\partial x}(x_0, y_0)\right](x - x_0) + \left[\frac{\partial f}{\partial y}(x_0, y_0)\right](y - y_0),$$
>
> is called the **tangent plane** of the graph of f at the point (x_0, y_0).

EXAMPLE 5   Compute the plane tangent to the graph of $z = x^2 + y^4 + e^{xy}$ at the point $(1, 0, 2)$.

SOLUTION   Here we use formula (1), with $x_0 = 1$, $y_0 = 0$, and $z_0 = f(x_0, y_0) = 2$. The partial derivatives are

$$\frac{\partial z}{\partial x} = 2x + ye^{xy} \qquad \text{and} \qquad \frac{\partial z}{\partial y} = 4y^3 + xe^{xy}.$$

At $(1, 0, 2)$, they are 2 and 1, respectively. Thus, by formula (1), the tangent plane is

$$z = 2(x - 1) + 1(y - 0) + 2, \qquad \text{that is,} \qquad z = 2x + y. \quad \blacktriangle$$

Let us write $\mathbf{D}f(x_0, y_0)$ for the row matrix

$$\left[\frac{\partial f}{\partial x}(x_0, y_0) \quad \frac{\partial f}{\partial y}(x_0, y_0)\right],$$

so that the definition of differentiability asserts that

$$f(x_0, y_0) + \mathbf{D}f(x_0, y_0)\begin{bmatrix} x - x_0 \\ y - y_0 \end{bmatrix}$$

$$= f(x_0, y_0) + \left[\frac{\partial f}{\partial x}(x_0, y_0)\right](x - x_0) + \left[\frac{\partial f}{\partial y}(x_0, y_0)\right](y - y_0) \qquad (3)$$

is our good approximation to f near (x_0, y_0). As above, "good" is taken in the sense that expression (3) differs from $f(x, y)$ by something small times $\sqrt{(x - x_0)^2 + (y - y_0)^2}$. We say that expression (3) is the **best linear approximation** to f near (x_0, y_0).

Now we are ready to give a definition of differentiability for maps f of $\mathbb{R}^n$ to $\mathbb{R}^m$, using the above discussion as motivation. The derivative $\mathbf{D}f(\mathbf{x}_0)$ of $f = (f_1, \ldots, f_m)$ at a point $\mathbf{x}_0$ is a matrix $\mathbf{T}$ whose elements are $t_{ij} = \partial f_i / \partial x_j$ evaluated at $\mathbf{x}_0$.[*]

---

[*]It turns out that we need only postulate the existence of *some* matrix giving the best linear approximation near $\mathbf{x}_0 \in \mathbb{R}^n$, because in fact this matrix is *necessarily* the matrix whose ijth entry is $\partial f_i / \partial x_j$ (see Section 2.7).

> **DEFINITION: DIFFERENTIABLE, n VARIABLES, m FUNCTIONS**
>
> Let U be an open set in $\mathbb{R}^n$ and let $f: U \subset \mathbb{R}^n \to \mathbb{R}^m$ be a given function. We say that f is **differentiable** at $\mathbf{x}_0 \in U$ if the partial derivatives of f exist at $\mathbf{x}_0$ and if
>
> $$\lim_{\mathbf{x} \to \mathbf{x}_0} \frac{\| f(\mathbf{x}) - f(\mathbf{x}_0) - \mathbf{T}(\mathbf{x} - \mathbf{x}_0)\|}{\|\mathbf{x} - \mathbf{x}_0\|} = 0, \tag{4}$$
>
> where $\mathbf{T} = \mathbf{D}f(\mathbf{x}_0)$ is the $m \times n$ matrix with matrix elements $\partial f_i / \partial x_j$ evaluated at $\mathbf{x}_0$ and $\mathbf{T}(\mathbf{x} - \mathbf{x}_0)$ means the product of $\mathbf{T}$ with $\mathbf{x} - \mathbf{x}_0$ (regarded as a column matrix). We call $\mathbf{T}$ the **derivative** of f at $\mathbf{x}_0$.

We shall always denote the derivative $\mathbf{T}$ of f at $\mathbf{x}_0$ by $\mathbf{D}f(\mathbf{x}_0)$, although in some books it is denoted $df(\mathbf{x}_0)$ and referred to as the **differential** of f. In the case where $m = 1$, the matrix $\mathbf{T}$ is just the row matrix

$$\left[\frac{\partial f}{\partial x_1}(\mathbf{x}_0) \quad \cdots \quad \frac{\partial f}{\partial x_n}(\mathbf{x}_0) \right].$$

(Sometimes, when there is danger of confusion, we separate the entries by commas.) Furthermore, setting $n = 2$ and putting the result back into equation (4), we see that conditions (2) and (4) do agree. Thus, if we let $\mathbf{h} = \mathbf{x} - \mathbf{x}_0$, a real-valued function f of n variables is differentiable at a point $\mathbf{x}_0$ if

$$\lim_{\mathbf{h} \to 0} \frac{1}{\|\mathbf{h}\|} \left| f(\mathbf{x}_0 + \mathbf{h}) - f(\mathbf{x}_0) - \sum_{j=1}^{n} \frac{\partial f}{\partial x_j}(\mathbf{x}_0) h_j \right| = 0,$$

since

$$\mathbf{T}\mathbf{h} = \sum_{j=1}^{n} h_j \frac{\partial f}{\partial x_j}(\mathbf{x}_0).$$

For the general case of f mapping a subset of $\mathbb{R}^n$ to $\mathbb{R}^m$, the derivative is the $m \times n$ matrix given by

$$\mathbf{D}f(\mathbf{x}_0) = \begin{bmatrix} \dfrac{\partial f_1}{\partial x_1} & \cdots & \dfrac{\partial f_1}{\partial x_n} \\ \vdots & & \vdots \\ \dfrac{\partial f_m}{\partial x_1} & \cdots & \dfrac{\partial f_m}{\partial x_n} \end{bmatrix},$$

where $\partial f_i / \partial x_j$ is evaluated at $\mathbf{x}_0$. $\mathbf{D}f(\mathbf{x}_0)$ is, appropriately, called the **matrix of partial derivatives of f at $\mathbf{x}_0$.**

EXAMPLE 6 Calculate the matrices of partial derivatives for the functions (a) $f(x, y) = (e^{x+y} + y, y^2 x)$, (b) $f(x, y) = (x^2 + \cos y, y e^x)$, and (c) $f(x, y, z) = (z e^x, -y e^z)$.

SOLUTION

(a)   Here $f: \mathbb{R}^2 \to \mathbb{R}^2$ is defined by $f_1(x, y) = e^{x+y} + y$ and $f_2(x, y) = y^2 x$. Hence $\mathbf{D}f(x, y)$ is the 2×2 matrix

$$\mathbf{D}f(x, y) = \begin{bmatrix} e^{x+y} & e^{x+y} + 1 \\ y^2 & 2xy \end{bmatrix}.$$

(b)   We have

$$\mathbf{D}f(x, y) = \begin{bmatrix} 2x & -\sin y \\ ye^x & e^x \end{bmatrix}.$$

(c)   Here

$$\mathbf{D}f(x, y, z) = \begin{bmatrix} ze^x & 0 & e^x \\ 0 & -e^z & -ye^z \end{bmatrix}. \quad \blacktriangle$$

DEFINITION: GRADIENT   Consider the special case $f: U \subset \mathbb{R}^n \to \mathbb{R}$. Here $\mathbf{D}f(\mathbf{x})$ is a $1 \times n$ matrix:

$$\mathbf{D}f(\mathbf{x}) = \begin{bmatrix} \dfrac{\partial f}{\partial x_1} & \cdots & \dfrac{\partial f}{\partial x_n} \end{bmatrix}.$$

We can form the corresponding vector $(\partial f/\partial x_1, \ldots, \partial f/\partial x_n)$, called the **gradient** of f and denoted by ∇f or grad f.

From the definition, we see that for $f: \mathbb{R}^3 \to \mathbb{R}$,

$$\nabla f = \frac{\partial f}{\partial x}\mathbf{i} + \frac{\partial f}{\partial y}\mathbf{j} + \frac{\partial f}{\partial z}\mathbf{k},$$

while for $f: \mathbb{R}^2 \to \mathbb{R}$,

$$\nabla f = \frac{\partial f}{\partial x}\mathbf{i} + \frac{\partial f}{\partial y}\mathbf{j}.$$

The geometric significance of the gradient will be discussed in Section 2.6. In terms of inner products, we can write the derivative of f as

$$\mathbf{D}f(\mathbf{x})(\mathbf{h}) = \nabla f(\mathbf{x}) \cdot \mathbf{h}.$$

EXAMPLE 7   Let $f: \mathbb{R}^3 \to \mathbb{R}$, $f(x, y, z) = xe^y$. Then

$$\nabla f = \left(\frac{\partial f}{\partial x}, \frac{\partial f}{\partial y}, \frac{\partial f}{\partial z} \right) = (e^y, xe^y, 0). \quad \blacktriangle$$

EXAMPLE 8   If $f: \mathbb{R}^2 \to \mathbb{R}$ is given by $(x, y) \mapsto e^{xy} + \sin xy$, then

$$\nabla f(x, y) = (ye^{xy} + y \cos xy)\mathbf{i} + (xe^{xy} + x \cos xy)\mathbf{j}$$
$$= (e^{xy} + \cos xy)(y\mathbf{i} + x\mathbf{j}). \quad \blacktriangle$$

In one-variable calculus it is shown that if f is differentiable then f is continuous. We will state in Theorem 8 that this is also true for differentiable functions of several variables. As we know, there are plenty of functions of one variable that are continuous but not differentiable, such as $f(x) = |x|$. Before stating the result, let us give an example of a function of two variables whose partial derivatives exist at a point but which is not continuous at that point.

EXAMPLE 9   Let $f: \mathbb{R}^2 \to \mathbb{R}$ be defined by

$$f(x, y) = \begin{cases} 1 & \text{if } x = 0 \text{ or if } y = 0 \\ 0 & \text{otherwise.} \end{cases}$$

Since f is constant on the x and y axes, where it equals 1,

$$\frac{\partial f}{\partial x}(0, 0) = 0 \qquad \text{and} \qquad \frac{\partial f}{\partial y}(0, 0) = 0.$$

But f is not continuous at $(0, 0)$, because $\displaystyle\lim_{(x,y) \to (0,0)} f(x, y)$ does not exist. $\quad \blacktriangle$

THEOREM 8   *Let $f: U \subset \mathbb{R}^n \to \mathbb{R}^m$ be differentiable at $\mathbf{x}_0 \in U$. Then f is continuous at $\mathbf{x}_0$.*

Consult Section 2.7 for the proof.

As we have seen, it is usually easy to tell when the partial derivatives of a function exist using what we know from one-variable calculus. However, the definition of differentiability looks somewhat complicated, and the required approximation condition in equation (4) may seem difficult to verify. Fortunately there is a simple criterion, given in the following theorem, that tells us when a function is differentiable.

THEOREM 9   *Let $f: U \subset \mathbb{R}^n \to \mathbb{R}^m$. Suppose the partial derivatives $\partial f_i / \partial x_j$ of f all exist and are continuous in a neighborhood of a point $\mathbf{x} \in U$. Then f is differentiable at $\mathbf{x}$.*

We shall give the proof in Section 2.7. Notice the following hierarchy:

Each converse statement, obtained by reversing an implication, is invalid. [For a counterexample to the converse of the first implication, use $f(x) = x^2 \sin(1/x)$, $f(0) = 0$; for the second, see Example 1 in Section 2.7 or use Example 4 in this section.]

A function whose partial derivatives exist and are continuous is said to be of *class* C^1. Thus Theorem 9 says that *any* C^1 *function is differentiable.*

EXAMPLE 10   Let

$$f(x, y) = \frac{\cos x + e^{xy}}{x^2 + y^2}.$$

Show that f is differentiable at all points $(x, y) \neq (0, 0)$.

SOLUTION   Observe that the partial derivatives

$$\frac{\partial f}{\partial x} = \frac{(x^2 + y^2)(ye^{xy} - \sin x) - 2x(\cos x + e^{xy})}{(x^2 + y^2)^2}$$

$$\frac{\partial f}{\partial y} = \frac{(x^2 + y^2)xe^{xy} - 2y(\cos x + e^{xy})}{(x^2 + y^2)^2}$$

are continuous except when $x = 0$ and $y = 0$ (by the results in Section 2.2). Thus, f is differentiable by Theorem 9.   ▲

EXERCISES

1. Find $\partial f/\partial x$, $\partial f/\partial y$ if

   (a) $f(x, y) = xy$
   (b) $f(x, y) = e^{xy}$
   (c) $f(x, y) = x \cos x \cos y$
   (d) $f(x, y) = (x^2 + y^2) \log(x^2 + y^2)$

2. Evaluate the partial derivatives $\partial z/\partial x$, $\partial z/\partial y$ for the given function at the indicated points.

   (a) $z = \sqrt{a^2 - x^2 - y^2}$; $(0, 0)$, $(a/2, a/2)$
   (b) $z = \log \sqrt{1 + xy}$; $(1, 2)$, $(0, 0)$
   (c) $z = e^{ax} \cos(bx + y)$; $(2\pi/b, 0)$

3. In each case following, find the partial derivatives $\partial w/\partial x$, $\partial w/\partial y$.

   (a) $w = xe^{x^2 + y^2}$ (b) $w = \dfrac{x^2 + y^2}{x^2 - y^2}$
   (c) $w = e^{xy} \log(x^2 + y^2)$ (d) $w = x/y$
   (e) $w = \cos ye^{xy} \sin x$

4. Show that each of the following functions is differentiable at each point in its domain. Decide which of the functions are C^1.

   (a) $f(x, y) = \dfrac{2xy}{(x^2 + y^2)^2}$ (b) $f(x, y) = \dfrac{x}{y} + \dfrac{y}{x}$

(c) $f(r, \theta) = \frac{1}{2}r \sin 2\theta, r > 0$

(d) $f(x, y) = \dfrac{xy}{\sqrt{x^2 + y^2}}$

(e) $f(x, y) = \dfrac{x^2 y}{x^4 + y^2}$

5. Find the equation of the plane tangent to the surface $z = x^2 + y^3$ at $(3, 1, 10)$.

6. Using the respective functions in Exercise 1, compute the plane tangent to the graphs at the indicated points.

    (a) $(0, 0)$ (b) $(0, 1)$ (c) $(0, \pi)$ (d) $(0, 1)$

7. Compute the matrix of partial derivatives of the following functions:

    (a) $f: \mathbb{R}^2 \rightarrow \mathbb{R}^2, f(x, y) = (x, y)$
    (b) $f: \mathbb{R}^2 \rightarrow \mathbb{R}^3, f(x, y) = (xe^y + \cos y, x, x + e^y)$
    (c) $f: \mathbb{R}^3 \rightarrow \mathbb{R}^2, f(x, y, z) = (x + e^z + y, yx^2)$
    (d) $f: \mathbb{R}^2 \rightarrow \mathbb{R}^3, f(x, y) = (xye^{xy}, x \sin y, 5xy^2)$

8. Compute the matrix of partial derivatives of

    (a) $f(x, y) = (e^x, \sin xy)$ (b) $f(x, y, z) = (x - y, y + z)$
    (c) $f(x, y) = (x + y, x - y, xy)$ (d) $f(x, y, z) = (x + z, y - 5z, x - y)$

9. Where does the plane tangent to $z = e^{x-y}$ at $(1, 1, 1)$ meet the z axis?

10. Why should the graphs of $f(x, y) = x^2 + y^2$ and $g(x, y) = -x^2 - y^2 + xy^3$ be called "tangent" at $(0, 0)$?

11. Let $f(x, y) = e^{xy}$. Show that $x(\partial f/\partial x) = y(\partial f/\partial y)$.

12. Use expression (1) in this section to approximate a suitable function $f(x, y)$ and thereby estimate the following:

    (a) $(0.99e^{0.02})^8$
    (b) $(0.99)^3 + (2.01)^3 - 6(0.99)(2.01)$
    (c) $\sqrt{(4.01)^2 + (3.98)^2 + (2.02)^2}$

13. Compute the gradients of the following functions:

    (a) $f(x, y, z) = x \exp(-x^2 - y^2 - z^2)$ (Note that $\exp u = e^u$.)

    (b) $f(x, y, z) = \dfrac{xyz}{x^2 + y^2 + z^2}$ (c) $f(x, y, z) = z^2 e^x \cos y$

14. Compute the tangent plane at $(1, 0, 1)$ for each of the functions in Exercise 13. [The solution to part (c) only is in the Study Guide to this text.]

15. Find the equation of the tangent plane to $z = x^2 + 2y^3$ at $(1, 1, 3)$.

16. Calculate $\nabla h(1, 1, 1)$ if $h(x, y, z) = (x + z)e^{x-y}$.

17. Let $f(x, y, z) = x^2 + y^2 - z^2$. Calculate $\nabla f(0, 0, 1)$.

18. Evaluate the gradient of $f(x, y, z) = \log(x^2 + y^2 + z^2)$ at $(1, 0, 1)$.

19. Describe all Hölder-continuous functions with $\alpha > 1$ (see Exercise 21, Section 2.2). [HINT: What is the derivative of such a function?]

20. Suppose $f: \mathbb{R}^n \to \mathbb{R}^m$ is a linear map. What is the derivative of f?

2.4 INTRODUCTION TO PATHS

In this section we introduce some of the basic geometry and computational methods for paths in the plane and space. This will be an important ingredient for the chain rule treated in the next section. We will return to paths with additional topics in Chapter 4.

One often thinks of a curve as a line drawn on paper, such as a straight line, a circle, or a sine curve. It is useful to think of a curve C mathematically as the set of values of a function that maps an interval of real numbers into the plane or space. We shall call such a map a ***path***. We usually denote a path by **c**. The image C of the path then corresponds to the curve we see on paper (see Figure 2.4.1). Often we write t for the independent variable and imagine it to be *time*, so that $\mathbf{c}(t)$ is the position at time t of a moving particle, which ***traces out*** a curve as t varies. We also say **c** ***parametrizes*** C. Strictly speaking, we should distinguish between $\mathbf{c}(t)$ as a *point* in space and as a *vector* based at the origin.

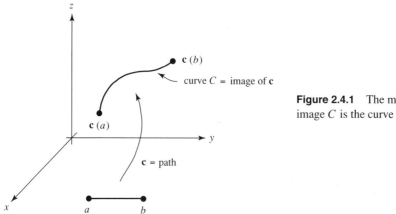

Figure 2.4.1 The map **c** is the path; its image C is the curve we "see."

EXAMPLE 1   The straight line L in $\mathbb{R}^3$ through the point (x_0, y_0, z_0) in the direction of vector **v** is the image of the path

$$\mathbf{c}(t) = (x_0, y_0, z_0) + t\mathbf{v}$$

for $t \in \mathbb{R}$ (see Figure 2.4.2). Thus, our notion of curve includes straight lines as special cases. ▲

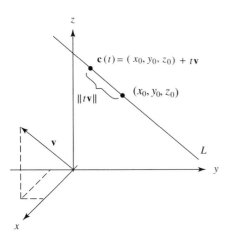

Figure 2.4.2 L is the straight line in space through (x_0, y_0, z_0) and in direction $\mathbf{v}$; its equation is $\mathbf{c}(t) = (x_0, y_0, z_0) + t\mathbf{v}$.

EXAMPLE 2 The unit circle $C \colon x^2 + y^2 = 1$ in the plane is the image of the path

$$\mathbf{c} \colon \mathbb{R} \to \mathbb{R}^2, \qquad \mathbf{c}(t) = (\cos t, \sin t), \qquad 0 \leq t \leq 2\pi,$$

(see Figure 2.4.3). The unit circle is also the image of the path $\tilde{\mathbf{c}}(t) = (\cos 2t, \sin 2t)$, $0 \leq t \leq \pi$. Thus, different paths may parametrize the same curve. ▲

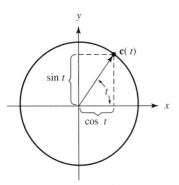

Figure 2.4.3 $\mathbf{c}(t) = (\cos t, \sin t)$ is a path whose image C is the unit circle.

PATHS AND CURVES A *path* in $\mathbb{R}^n$ is a map $\mathbf{c} \colon [a, b] \to \mathbb{R}^n$; it is a *path in the plane* if $n = 2$ and a *path in space* if $n = 3$. The collection C of points $\mathbf{c}(t)$ as t varies in $[a, b]$ is called a *curve*, and $\mathbf{c}(a)$ and $\mathbf{c}(b)$ are its *endpoints*. The path $\mathbf{c}$ is said to *parametrize* the curve C. We also say $\mathbf{c}(t)$ *traces out* C as t varies.

If $\mathbf{c}$ is a path in $\mathbb{R}^3$, we can write $\mathbf{c}(t) = (x(t), y(t), z(t))$ and we call $x(t)$, $y(t)$, and $z(t)$ the *component functions* of $\mathbf{c}$. We form component functions similarly in $\mathbb{R}^2$ or, generally, in $\mathbb{R}^n$.

EXAMPLE 3 The path $\mathbf{c}(t) = (t, t^2)$ traces out a parabolic arc. This curve coincides with the graph $f(x) = x^2$ (see Figure 2.4.4). ▲

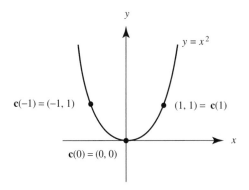

Figure 2.4.4   The image of $\mathbf{c}(t) = (t, t^2)$ is the parabola $y = x^2$.

EXAMPLE 4   A wheel of radius R rolls to the right along a straight line at speed v. Use vector methods to find the path $\mathbf{c}(t)$ of the point on the wheel that initially lies at a distance r below the center.

SOLUTION   We place the wheel in the xy plane with its center initially at $(0, R)$, so that the position of the center at time t is given by the path $\mathbf{C}(t) = (vt, R)$. (Refer to Figure 2.4.5.)

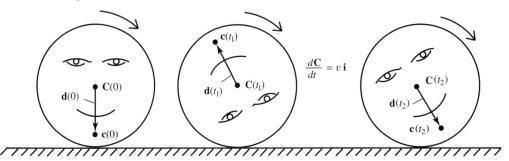

Figure 2.4.5   The vector $\mathbf{d}(t)$ points from the wheel's center, $\mathbf{C}(t)$, to the position $\mathbf{c}(t)$ of a point on the wheel and rotates in the clockwise direction while the wheel moves to the right.

The position of the point $\mathbf{c}(t)$ *relative to the center* is given by the vector $\mathbf{d}(t) = \mathbf{c}(t) - \mathbf{C}(t)$ that has the initial value $-r\mathbf{j}$ and rotates in the *clockwise* direction. The rate of rotation is such that the wheel makes a full rotation after the center has moved a distance $2\pi R$ (equal to the circumference of the wheel). This takes a time $2\pi R/v$, so the angular velocity $d\theta/dt$ of the wheel is v/R. Since the rotation is clockwise, the vector function $\mathbf{d}(t)$ is of the form

$$\mathbf{d}(t) = r\left(\cos\left[-\frac{v}{r}t + \theta\right]\mathbf{i} + \sin\left[-\frac{v}{r}t + \theta\right]\mathbf{j}\right)$$

for some initial angle θ. Since $\mathbf{d}(0) = -r\mathbf{j}$, we have $\cos\theta = 0$ and $\sin\theta = -1$, so

$\theta = -\pi/2$, and hence

$$\mathbf{d}(t) = r\left(\cos\left[-\frac{v}{r}t - \frac{\pi}{2}\right]\mathbf{i} + \sin\left[-\frac{v}{r}t - \frac{\pi}{2}\right]\mathbf{j}\right).$$

Using $\cos(\varphi - \pi/2) = \sin\varphi$ and $\sin(\varphi - \pi/2) = -\cos\varphi$, we get

$$\mathbf{d}(t) = r\left(-\sin\frac{vt}{r}\mathbf{i} - \cos\frac{vt}{r}\mathbf{j}\right).$$

Finally, the path $\mathbf{c}(t)$ is given by adding the components of the vector function $\mathbf{d}(t)$ to the coordinates of the path $\mathbf{C}(t)$; the result is

$$\mathbf{c}(t) = \left(vt - r\sin\frac{vt}{r}, R - r\cos\frac{vt}{r}\right).$$

In the special case $v = R = r = 1$, we get $\mathbf{c}(t) = (t - \sin t, 1 - \cos t)$. The image curve C of this path $\mathbf{c}$ is shown in Figure 2.4.6; it is called a *cycloid*. ▲

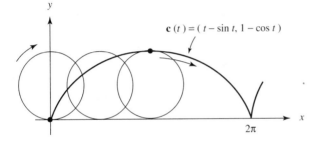

Figure 2.4.6 The curve traced by a point moving on the rim of a rolling circle is called a cycloid.

The preceding example considered the path of a point not necessarily on the rim of a wheel rolling along a straight line. When the wheel rolls on a circle, the resulting curve is called an *epicycle*. If the wheel is outside the circle and the point is on the rim, the curve is called an *epicycloid*, and when the wheel is inside the circle it is a *hypocycloid*. An example of the later is shown in Figure 2.4.7.

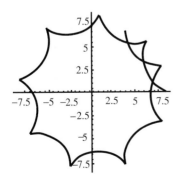

Figure 2.4.7 An example of a hypocycloid.

HISTORICAL NOTE

Since ancient times the circle and sphere have been considered to be the perfect forms in geometry. For the Greeks they were the symbols of the ultimate symmetry of the divine. What forms could be better suited to describe the immutable and eternal motion of the planets than circular motion? From the Earth, these motions appear to be quite complicated. The motion of the Sun and the Moon can be roughly described as circular with constant speed. The orbits of the other planets seem much more complicated, because as the planets go through one revolution, they reverse direction for a time, then reverse again to go forward all with changing speed. In the third century B.C., Apolonius of Perga suggested that the celestial orbits could be explained using epicycles. This idea came to fruition in the work of Ptolemy, a second century Greek astronomer. The resulting Ptolemaic system became the most important astronomical theory of the next 1200 years. It was ultimately replaced by the Copernicus-Kepler-Newton theory, which we will discuss later.

The French mathematician Blaise Pascal also studied the cycloid in 1649 as a way of distracting himself at a time when he was suffering from a painful toothache. When the pain disappeared, he took it as a sign that God was not displeased with his thoughts. Pascal's results stimulated other mathematicians to investigate this curve, and subsequently numerous remarkable properties were found. One of these was discovered by the Dutchman Christian Huygens, who used it in the construction of a "perfect" pendulum clock.

If we think of $\mathbf{c}(t)$ as the curve traced out by a particle and t as time, it is reasonable to define the velocity vector as follows.

DEFINITION: VELOCITY VECTOR   If $\mathbf{c}$ is a path and it is differentiable, we say $\mathbf{c}$ is a *differentiable path*. The *velocity* of $\mathbf{c}$ at time t is defined by

$$\mathbf{c}'(t) = \lim_{h \to 0} \frac{\mathbf{c}(t+h) - \mathbf{c}(t)}{h}.$$

We normally draw the vector $\mathbf{c}'(t)$ with its tail at the point $\mathbf{c}(t)$. The *speed* of the path $\mathbf{c}(t)$ is $s = \|\mathbf{c}'(t)\|$, the length of the velocity vector. If $\mathbf{c}(t) = (x(t), y(t))$ in $\mathbb{R}^2$, then

$$\mathbf{c}'(t) = (x'(t), y'(t)) = x'(t)\mathbf{i} + y'(t)\mathbf{j}$$

and if $\mathbf{c}(t) = (x(t), y(t), z(t))$ in $\mathbb{R}^3$, then

$$\mathbf{c}'(t) = (x'(t), y'(t), z'(t)) = x'(t)\mathbf{i} + y'(t)\mathbf{j} + z'(t)\mathbf{k}.$$

Here, $x'(t)$ is the one-variable derivative dx/dt. If we accept limits of vectors interpreted componentwise, the formulas for the velocity vector follow from the definition of the derivative. However, the limit can be interpreted in the sense of vectors as well. In Figure 2.4.8, we see that $[\mathbf{c}(t + h) - \mathbf{c}(t)]/h$ approaches the tangent to the path as $h \to 0$.

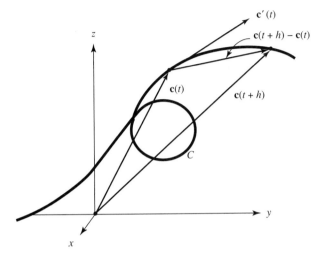

Figure 2.4.8 The vector $\mathbf{c}'(t)$ is tangent to the path $\mathbf{c}(t)$.

TANGENT VECTOR   The velocity $\mathbf{c}'(t)$ is a vector *tangent* to the path $\mathbf{c}(t)$ at time t. If C is a curve traced out by $\mathbf{c}$ and if $\mathbf{c}'(t)$ is not equal to $\mathbf{0}$, then $\mathbf{c}'(t)$ is a vector tangent to the curve C at the point $\mathbf{c}(t)$.

If we think of the derivative $\mathbf{Dc}(t)$ as a matrix, it will be a column vector with the entries $x'(t)$, $y'(t)$, and $z'(t)$. Thus, the derivative here is consistent with our earlier notion.

EXAMPLE 5   Compute the tangent vector to the path $\mathbf{c}(t) = (t, t^2, e^t)$ at $t = 0$.

SOLUTION   Here $\mathbf{c}'(t) = (1, 2t, e^t)$, and so at $t = 0$ we obtain the tangent vector $(1, 0, 1)$. ▲

EXAMPLE 6   Describe the path $\mathbf{c}(t) = (\cos t, \sin t, t)$. Find the velocity vector at the point on the image curve where $t = \pi/2$.

SOLUTION   For a given t, the point $(\cos t, \sin t, 0)$ lies on the circle $x^2 + y^2 = 1$ in the xy plane. Therefore the point $(\cos t, \sin t, t)$ lies t units above the point $(\cos t, \sin t, 0)$ if t is positive and $-t$ units below $(\cos t, \sin t, 0)$ if t is negative. As t increases, $(\cos t, \sin t, t)$ wraps around the cylinder $x^2 + y^2 = 1$ with the z-coordinate increasing. The curve this traces out is called a **helix**, which is depicted in Figure 2.4.9. At $t = \pi/2$, $\mathbf{c}'(\pi/2) = (-\sin \pi/2, \cos \pi/2, 1) = (-1, 0, 1) = -\mathbf{i} + \mathbf{k}$. ▲

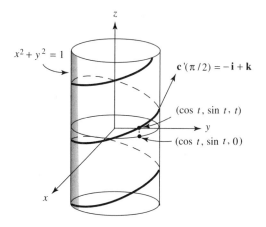

Figure 2.4.9   The helix $\mathbf{c}(t) = (\cos t, \sin t, t)$ wraps around the cylinder $x^2 + y^2 = 1$.

EXAMPLE 7   The cycloidal path of a particle on the edge of a wheel of radius R with velocity v is given by $\mathbf{c}(t) = (vt - R \sin (vt/R), R - R \cos (vt/R))$. (See Example 4.) Find the velocity $\mathbf{c}'(t)$ of the particle as a function of t. When is the velocity zero? Is the velocity vector ever vertical?

SOLUTION

$$
\mathbf{c}'(t) = \left(\frac{d}{dt} \left(vt - R \sin \frac{vt}{r} \right), \frac{d}{dt} \left(R - R \cos \frac{vt}{r} \right) \right)
$$
$$
= \left(v - v \cos \frac{vt}{R}, v \sin \frac{vt}{R} \right).
$$

In vector notation, $\mathbf{c}'(t) = (v - v \cos (vt/R))\mathbf{i} + (v \sin (vt/R))\mathbf{j}$. The component in the direction of $\mathbf{i}$ is $v(1 - \cos (vt/R))$, which is zero whenever vt/R is an integer multiple of 2π. For such values of t, $\sin (vt/R)$ is zero as well, so the only times at which the velocity is zero are when $t = 2\pi nR/v$ for some integer n. At such times, $\mathbf{c}(t) = (2\pi nR, 0)$, so the moving point is touching the ground. These moments occur at time intervals of $2\pi R/v$ (more frequently for small wheels, as well as for rapidly rolling ones).

The velocity vector is never vertical, since the horizontal component vanishes only when the vertical one does as well.   ▲

Figure 2.4.10 shows some velocity vectors superimposed on the cycloidal path of Figure 2.4.6.

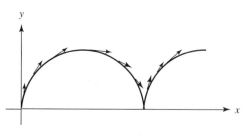

Figure 2.4.10   Velocity vectors for the curve traced out by a point on the rim of a rolling wheel.

The tangent line to a path at a point is the line through the point in the direction of the tangent vector. Using the point-direction form of the equation of a line, we obtain the parametric equation for the tangent line.

TANGENT LINE TO A PATH   If $\mathbf{c}(t)$ is a path, and if $\mathbf{c}(t_0) \neq \mathbf{0}$, the equation of its *tangent line* at the point $\mathbf{c}(t_0)$ is

$$l(t) = \mathbf{c}(t_0) + (t - t_0)\mathbf{c}'(t_0).$$

If C is the curve traced out by $\mathbf{c}$, then the line traced out by $\mathbf{l}$ is the tangent line to the curve C at $\mathbf{c}(t_0)$.

For convenience we have written the equation so l goes through $\mathbf{c}(t_0)$ at $t = t_0$ (rather than $t = 0$). See Figure 2.4.11.

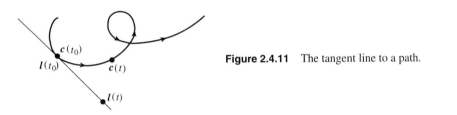

Figure 2.4.11   The tangent line to a path.

EXAMPLE 8   A path in $\mathbb{R}^3$ goes through the point $(3, 6, 5)$ at $t = 0$ with tangent vector $\mathbf{i} - \mathbf{j}$. Find the equation of the tangent line.

SOLUTION   The equation of the tangent line is

$$l(t) = (3, 6, 5) + t(\mathbf{i} - \mathbf{j}) = (3, 6, 5) + t(1, -1, 0) = (3 + t, 6 - t, 5).$$

In (x, y, z) coordinates, the tangent line is $x = 3 + t, y = 6 - t, z = 5$.   ▲

Physically, we can interpret motion along the tangent line as the path that a particle on a curve would follow if it were set free at a certain moment.

EXAMPLE 9   Suppose that a particle follows the path $\mathbf{c}(t) = (e^t, e^{-t}, \cos t)$ until it flies off on a tangent at $t = 1$. Where is it at $t = 3$?

SOLUTION   The velocity vector is $(e^t, e^{-t}, -\sin t)$, which at $t = 1$ is the vector $(e, -1/e, \sin 1)$. The particle is at $(e, 1/e, \cos 1)$ at $t = 1$. The tangent line is $l(t) =$

$(e, 1/e, \cos 1) + (t - 1)(e, -1/e, -\sin 1)$. At $t = 3$, the position on this line is

$$l(3) = \left(e, \frac{1}{e}, \cos 1\right) + 2\left(e, -\frac{1}{e}, -\sin 1\right) = \left(3e, -\frac{1}{e}, \cos 1 - 2\sin 1\right). \quad \blacktriangle$$

EXERCISES

Sketch the curves in Exercises 1 to 4.

1. $x = \sin t, y = 4\cos t, 0 \le t \le 2\pi$

2. $x = 2\sin t, y = 4\cos t, 0 \le t \le 2\pi$

3. $\mathbf{c}(t) = (2t - 1, t + 2, t)$

4. $\mathbf{c}(t) = (-t, 2t, 1/t), 1 \le t \le 3$

In Exercises 5 to 8, determine the velocity vector of the given path.

5. $\mathbf{c}(t) = 6t\mathbf{i} + 3t^2\mathbf{j} + t^3\mathbf{k}$

6. $\mathbf{c}(t) = (\sin 3t)\mathbf{i} + (\cos 3t)\mathbf{j} + 2t^{3/2}\mathbf{k}$

7. $\mathbf{r}(t) = (\cos^2 t, 3t - t^3, t)$

8. $\mathbf{r}(t) = (4e^t, 6t^4, \cos t)$

In Exercises 9 to 12, compute the tangent vector to the given path.

9. $\mathbf{c}(t) = (e^t, \cos t)$

10. $\mathbf{c}(t) = (3t^2, t^3)$

11. $\mathbf{c}(t) = (t \sin t, 4t)$

12. $\mathbf{c}(t) = (t^2, e^2)$

13. When is the velocity vector of a point on the rim of a rolling wheel *horizontal*? What is the speed at this point?

14. If the position of a particle in space is $(6t, 3t^2, t^3)$ at time t, what is its velocity vector at $t = 0$?

In Exercises 15 and 16, determine the equation of the tangent line to the given path at the specified value of t.

15. $(\sin 3t, \cos 3t, 2t^{5/2}); t = 1$

16. $(\cos^2 t, 3t - t^3, t); t = 0$

In Exercises 17 to 20, suppose that a particle following the given path $\mathbf{c}(t)$ flies off on a tangent at $t = t_0$. Compute the position of the particle at the given time t_1.

17. $\mathbf{c}(t) = (t^2, t^3 - 4t, 0), t_0 = 2, t_1 = 3$

18. $\mathbf{c}(t) = (e^t, e^{-t}, \cos t), t_0 = 1, t_1 = 2$

19. $\mathbf{c}(t) = (4e^t, 6t^4, \cos t), t_0 = 0, t_1 = 1$

20. $\mathbf{c}(t) = (\sin e^t, t, 4 - t^3), t_0 = 1, t_1 = 2$

2.5 PROPERTIES OF THE DERIVATIVE

In elementary calculus we learn how to differentiate sums, products, quotients, and composite functions. We now generalize these ideas to functions of several variables, paying particular attention to the differentiation of composite functions. The rule for differentiating composites, called the *chain rule*, takes on a more profound form for functions of several variables than for those of one variable.

If f is a real-valued function of one variable, written as $z = f(y)$, and y is a function of x, written $y = g(x)$, then z becomes a function of x through substitution, namely, $z = f(g(x))$, and we have the familiar chain rule:

$$\frac{dz}{dx} = \frac{dz}{dy}\frac{dy}{dx} = f'(g(x))g'(x).$$

If f is a real-valued function of three variables u, v, and w, written in the form $z = f(u, v, w)$, and the variables u, v, w are each functions of $x, u = g(x), v = h(x)$, and $w = k(x)$, then by substituting $g(x), h(x)$, and $k(x)$ for u, v, and w, we obtain z as a function of x: $z = f(g(x), h(x), k(x))$. The chain rule in this case reads:

$$\frac{dz}{dx} = \frac{\partial z}{\partial u}\frac{du}{dx} + \frac{\partial z}{\partial v}\frac{dv}{dx} + \frac{\partial z}{\partial w}\frac{dw}{dx}.$$

One of the objects of this section is to explain such formulas in detail.

We begin with the differentiation rules for sums, products, and quotients.

THEOREM 10: SUMS, PRODUCTS, QUOTIENTS

(i) **Constant Multiple Rule.** Let $f: U \subset \mathbb{R}^n \to \mathbb{R}^m$ be differentiable at $\mathbf{x}_0$ and let c be a real number. Then $h(\mathbf{x}) = cf(\mathbf{x})$ is differentiable at $\mathbf{x}_0$ and

$$\mathbf{D}h(\mathbf{x}_0) = c\mathbf{D}f(\mathbf{x}_0) \quad \text{(equality of matrices)}.$$

(ii) **Sum Rule.** Let $f: U \subset \mathbb{R}^n \to \mathbb{R}^m$ and $g: U \subset \mathbb{R}^n \to \mathbb{R}^m$ be differentiable at

$\mathbf{x}_0$. Then $h(\mathbf{x}) = f(\mathbf{x}) + g(\mathbf{x})$ is differentiable at $\mathbf{x}_0$ and

$$\mathbf{D}h(\mathbf{x}_0) = \mathbf{D}f(\mathbf{x}_0) + \mathbf{D}g(\mathbf{x}_0) \quad (\text{sum of matrices}).$$

(iii) **Product Rule**. Let $f: U \subset \mathbb{R}^n \to \mathbb{R}$ and $g: U \subset \mathbb{R}^n \to \mathbb{R}$ be differentiable at $\mathbf{x}_0$ and let $h(\mathbf{x}) = g(\mathbf{x}) f(\mathbf{x})$. Then $h: U \subset \mathbb{R}^n \to \mathbb{R}$ is differentiable at $\mathbf{x}_0$ and

$$\mathbf{D}h(\mathbf{x}_0) = g(\mathbf{x}_0)\mathbf{D}f(\mathbf{x}_0) + f(\mathbf{x}_0)\mathbf{D}g(\mathbf{x}_0).$$

(*Note that each side of this equation is a* $1 \times n$ *matrix; a more general product rule is presented in Exercise* 25 *at the end of this section.*)

(iv) **Quotient Rule**. *With the same hypotheses as in rule* (iii), *let* $h(\mathbf{x}) = f(\mathbf{x})/g(\mathbf{x})$ *and suppose* g *is never zero on* U. *Then* h *is differentiable at* $\mathbf{x}_0$ *and*

$$\mathbf{D}h(\mathbf{x}_0) = \frac{g(\mathbf{x}_0)\mathbf{D}f(\mathbf{x}_0) - f(\mathbf{x}_0)\mathbf{D}g(\mathbf{x}_0)}{[g(\mathbf{x}_0)]^2}.$$

PROOF   The proofs of rules (i) through (iv) proceed almost exactly as in the one-variable case with a slight difference in notation. We shall prove rules (i) and (ii), leaving the proofs of rules (iii) and (iv) as Exercise 25.

(i)   To show that $\mathbf{D}h(\mathbf{x}_0) = c\mathbf{D}f(\mathbf{x}_0)$, we must show that

$$\lim_{\mathbf{x} \to \mathbf{x}_0} \frac{\|h(\mathbf{x}) - h(\mathbf{x}_0) - c\mathbf{D}f(\mathbf{x}_0)(\mathbf{x} - \mathbf{x}_0)\|}{\|\mathbf{x} - \mathbf{x}_0\|} = 0,$$

that is, that

$$\lim_{\mathbf{x} \to \mathbf{x}_0} \frac{\|cf(\mathbf{x}) - cf(\mathbf{x}_0) - c\mathbf{D}f(\mathbf{x}_0)(\mathbf{x} - \mathbf{x}_0)\|}{\|\mathbf{x} - \mathbf{x}_0\|} = 0,$$

[see equation (4) of Section 2.3]. This is certainly true, since f is differentiable and the constant c can be factored out [see Theorem 3(i), Section 2.2].

(ii)   By the triangle inequality, we may write

$$\frac{\|h(\mathbf{x}) - h(\mathbf{x}_0) - [\mathbf{D}f(\mathbf{x}_0) + \mathbf{D}g(\mathbf{x}_0)](\mathbf{x} - \mathbf{x}_0)\|}{\|\mathbf{x} - \mathbf{x}_0\|}$$

$$= \frac{\|f(\mathbf{x}) - f(\mathbf{x}_0) - [\mathbf{D}f(\mathbf{x}_0)](\mathbf{x} - \mathbf{x}_0) + g(\mathbf{x}) - g(\mathbf{x}_0) - [\mathbf{D}g(\mathbf{x}_0)](\mathbf{x} - \mathbf{x}_0)\|}{\|\mathbf{x} - \mathbf{x}_0\|}$$

$$\leq \frac{\|f(\mathbf{x}) - f(\mathbf{x}_0) - [\mathbf{D}f(\mathbf{x}_0)](\mathbf{x} - \mathbf{x}_0)\|}{\|\mathbf{x} - \mathbf{x}_0\|} + \frac{\|g(\mathbf{x}) - g(\mathbf{x}_0) - [\mathbf{D}g(\mathbf{x}_0)](\mathbf{x} - \mathbf{x}_0)\|}{\|\mathbf{x} - \mathbf{x}_0\|},$$

and each term approaches 0 as $\mathbf{x} \to \mathbf{x}_0$. Hence rule (ii) holds.   ■

EXAMPLE 1   Verify the formula for $\mathbf{D}h$ in rule (iv) of Theorem 10 with $f(x, y, z) = x^2 + y^2 + z^2$ and $g(x, y, z) = x^2 + 1$.

SOLUTION   Here

$$h(x, y, z) = \frac{x^2 + y^2 + z^2}{x^2 + 1},$$

so that by direct differentiation

$$\mathbf{D}h(x, y, z) = \left[\frac{\partial h}{\partial x}, \frac{\partial h}{\partial y}, \frac{\partial h}{\partial z} \right] = \left[\frac{(x^2 + 1)2x - (x^2 + y^2 + z^2)2x}{(x^2 + 1)^2}, \frac{2y}{x^2 + 1}, \frac{2z}{x^2 + 1} \right]$$

$$= \left[\frac{2x(1 - y^2 - z^2)}{(x^2 + 1)^2}, \frac{2y}{x^2 + 1}, \frac{2z}{x^2 + 1} \right].$$

By rule (iv), we get

$$\mathbf{D}h = \frac{g\mathbf{D}f - f\mathbf{D}g}{g^2} = \frac{(x^2 + 1)[2x, 2y, 2z] - (x^2 + y^2 + z^2)[2x, 0, 0]}{(x^2 + 1)^2},$$

which is the same as what we obtained directly.   ▲

As we mentioned above, it is in the differentiation of composite functions that we meet apparently substantial alterations of the formula from one-variable calculus. However, if we use the $\mathbf{D}$ notation, that is, matrix notation for derivatives, the chain rule for functions of several variables looks similar to the one-variable rule.

THEOREM 11: CHAIN RULE *Let $U \subset \mathbb{R}^n$ and $V \subset \mathbb{R}^m$ be open. Let $g: U \subset \mathbb{R}^n \to \mathbb{R}^m$ and $f: V \subset \mathbb{R}^m \to \mathbb{R}^p$ be given functions such that g maps U into V, so that $f \circ g$ is defined. Suppose g is differentiable at $\mathbf{x}_0$ and f is differentiable at $\mathbf{y}_0 = g(\mathbf{x}_0)$. Then $f \circ g$ is differentiable at $\mathbf{x}_0$ and*

$$\mathbf{D}(f \circ g)(\mathbf{x}_0) = \mathbf{D}f(\mathbf{y}_0)\mathbf{D}g(\mathbf{x}_0). \tag{1}$$

The right-hand side is the matrix product of $\mathbf{D}f(\mathbf{y}_0)$ with $\mathbf{D}g(\mathbf{x}_0)$.

We shall now give a proof of the chain rule *under the additional assumption that the partial derivatives of f are continuous*, building up to the general case by developing two special cases that are themselves important. (The complete proof of Theorem 11 without the additional assumption of continuity is given in Section 2.7.)

FIRST SPECIAL CASE OF THE CHAIN RULE Suppose $\mathbf{c}: \mathbb{R} \to \mathbb{R}^3$ is a differentiable path and $f: \mathbb{R}^3 \to \mathbb{R}$. Let $h(t) = f(\mathbf{c}(t)) = f(x(t), y(t), z(t))$, where $\mathbf{c}(t) = (x(t), y(t), z(t))$. Then

$$\frac{dh}{dt} = \frac{\partial f}{\partial x}\frac{dx}{dt} + \frac{\partial f}{\partial y}\frac{dy}{dt} + \frac{\partial f}{\partial z}\frac{dz}{dt}. \tag{2}$$

That is,

$$\frac{dh}{dt} = \nabla f(\mathbf{c}(t)) \cdot \mathbf{c}'(t),$$

where $\mathbf{c}'(t) = (x'(t), y'(t), z'(t))$.

This is the special case of Theorem 11 in which we take $\mathbf{c} = g$ and f to be real-valued, and $m = 3$. Notice that

$$\nabla f(\mathbf{c}(t)) \cdot \mathbf{c}'(t) = \mathbf{D}f(\mathbf{c}(t))\mathbf{D}\mathbf{c}(t),$$

where the product on the left-hand side is the dot product of vectors while the product on the right-hand side is matrix multiplication, and where we regard $\mathbf{D}f(\mathbf{c}(t))$ as a *row* matrix and $\mathbf{D}\mathbf{c}(t)$ as *a column* matrix. The vectors $\nabla f(\mathbf{c}(t))$ and $\mathbf{c}'(t)$ have the same components as their matrix equivalents; the notational change indicates the switch from matrices to vectors.

PROOF OF EQUATION (2) By definition,

$$\frac{dh}{dt}(t_0) = \lim_{t \to t_0} \frac{h(t) - h(t_0)}{t - t_0}.$$

Adding and subtracting two terms, we write

$$\frac{h(t) - h(t_0)}{t - t_0} = \frac{f(x(t), y(t), z(t)) - f(x(t_0), y(t_0), z(t_0))}{t - t_0}$$

$$= \frac{f(x(t), y(t), z(t)) - f(x(t_0), y(t), z(t))}{t - t_0}$$

$$+ \frac{f(x(t_0), y(t), z(t)) - f(x(t_0), y(t_0), z(t))}{t - t_0}$$

$$+ \frac{f(x(t_0), y(t_0), z(t)) - f(x(t_0), y(t_0), z(t_0))}{t - t_0}.$$

Now we invoke the *mean value theorem* from one-variable calculus, which states: *If $g: [a, b] \to \mathbb{R}$ is continuous and is differentiable on the open interval (a, b), then there is a point c in (a, b) such that $g(b) - g(a) = g'(c)(b - a)$.* Applying this to f as a function of x, we can assert that for some c between x and x_0,

$$f(x, y, z) - f(x_0, y, z) = \left[\frac{\partial f}{\partial x}(c, y, z)\right](x - x_0).$$

In this way, we find that

$$\frac{h(t) - h(t_0)}{t - t_0} = \left[\frac{\partial f}{\partial x}(c, y(t), z(t))\right]\frac{x(t) - x(t_0)}{t - t_0} + \left[\frac{\partial f}{\partial y}(x(t_0), d, z(t))\right]\frac{y(t) - y(t_0)}{t - t_0}$$

$$+ \left[\frac{\partial f}{\partial z}(x(t_0), y(t_0), e)\right]\frac{z(t) - z(t_0)}{t - t_0},$$

where c, d, and e lie between $x(t)$ and $x(t_0)$, between $y(t)$ and $y(t_0)$, and between $z(t)$ and $z(t_0)$, respectively. Taking the limit $t \to t_0$, using the continuity of the partials $\partial f/\partial x, \partial f/\partial y, \partial f/\partial z$, and the fact that c, d, and e converge to $x(t_0), y(t_0)$, and $z(t_0)$, respectively, we obtain formula (2). ∎

SECOND SPECIAL CASE OF THE CHAIN RULE Let $f: \mathbb{R}^3 \to \mathbb{R}$ and let $g: \mathbb{R}^3 \to \mathbb{R}^3$. Write $g(x, y, z) = (u(x, y, z), v(x, y, z), w(x, y, z))$ and define $h: \mathbb{R}^3 \to \mathbb{R}$ by setting $h(x, y, z) = f(u(x, y, z), v(x, y, z), w(x, y, z))$. Then

$$
\begin{bmatrix} \dfrac{\partial h}{\partial x} & \dfrac{\partial h}{\partial y} & \dfrac{\partial h}{\partial z} \end{bmatrix} = \begin{bmatrix} \dfrac{\partial f}{\partial u} & \dfrac{\partial f}{\partial v} & \dfrac{\partial f}{\partial w} \end{bmatrix} \begin{bmatrix} \dfrac{\partial u}{\partial x} & \dfrac{\partial u}{\partial y} & \dfrac{\partial u}{\partial z} \\[2mm] \dfrac{\partial v}{\partial x} & \dfrac{\partial v}{\partial y} & \dfrac{\partial v}{\partial z} \\[2mm] \dfrac{\partial w}{\partial x} & \dfrac{\partial w}{\partial y} & \dfrac{\partial w}{\partial z} \end{bmatrix}. \tag{3}
$$

In this special case we have taken $n = m = 3$ and $p = 1$ for concreteness, and $U = \mathbb{R}^3$ and $V = \mathbb{R}^3$ for simplicity, and have written out the matrix product $[\mathbf{D}f(\mathbf{y}_0)][\mathbf{D}g(\mathbf{x}_0)]$ explicitly (with the arguments $\mathbf{x}_0$ and $\mathbf{y}_0$ suppressed in the matrices).

PROOF OF THE SECOND SPECIAL CASE OF THE CHAIN RULE
By definition, $\partial h/\partial x$ is obtained by differentiating h with respect to x, holding y and z fixed. But then $(u(x, y, z), v(x, y, z), w(x, y, z))$ may be regarded as a vector function of the single variable x. The first special case applies to this situation and, after the variables are renamed, gives

$$
\frac{\partial h}{\partial x} = \frac{\partial f}{\partial u}\frac{\partial u}{\partial x} + \frac{\partial f}{\partial v}\frac{\partial v}{\partial x} + \frac{\partial f}{\partial w}\frac{\partial w}{\partial x}. \tag{3$'$}
$$

Similarly,

$$
\frac{\partial h}{\partial y} = \frac{\partial f}{\partial u}\frac{\partial u}{\partial y} + \frac{\partial f}{\partial v}\frac{\partial v}{\partial y} + \frac{\partial f}{\partial w}\frac{\partial w}{\partial y} \tag{3$''$}
$$

and

$$
\frac{\partial h}{\partial z} = \frac{\partial f}{\partial u}\frac{\partial u}{\partial z} + \frac{\partial f}{\partial v}\frac{\partial v}{\partial z} + \frac{\partial f}{\partial w}\frac{\partial w}{\partial z}. \tag{3$'''$}
$$

These equations are exactly what would be obtained by multiplying out the matrices in equation (3). ∎

PROOF OF THEOREM 11 The general case in equation (1) may be proved in two steps. First, equation (2) is generalized to m variables; that is, for $f(x_1, \ldots, x_m)$ and $\mathbf{c}(t) = (x_1(t), \ldots, x_m(t))$, one has

$$
\frac{dh}{dt} = \sum_{i=1}^{m} \frac{\partial f}{\partial x_i}\frac{dx_i}{dt},
$$

where $h(t) = f(x_1(t), \ldots, x_m(t))$. Second, the result obtained in the first step is used to obtain the formula

$$
\frac{\partial h_j}{\partial x_i} = \sum_{k=1}^{m} \frac{\partial f_j}{\partial y_k}\frac{\partial y_k}{\partial x_i},
$$

where $f = (f_1, \ldots, f_p)$ is a vector function of arguments $y_1, \ldots, y_m$; $g(x_1, \ldots, x_n) = (y_1(x_1, \ldots, x_n), \ldots, y_m(x_1, \ldots, x_n))$; and $h_j(x_1, \ldots, x_n) = f_j(y_1(x_1, \ldots, x_n), \ldots, y_m(x_1, \ldots, x_n))$. (Using the letter y for both functions and arguments is an abuse of notation, but it can help one remember the formula.) This formula is equivalent to formula (1) after the matrices are multiplied out. ■

The pattern of the chain rule will become clear once the student has worked some additional examples. For instance,

$$\frac{\partial}{\partial x} f(u(x, y), v(x, y), w(x, y), z(x, y)) = \frac{\partial f}{\partial u} \frac{\partial u}{\partial x} + \frac{\partial f}{\partial v} \frac{\partial v}{\partial x} + \frac{\partial f}{\partial w} \frac{\partial w}{\partial x} + \frac{\partial f}{\partial z} \frac{\partial z}{\partial x},$$

with a similar formula for $\partial f / \partial y$.

The chain rule can help us understand the relationship between the geometry of a mapping $f \colon \mathbb{R}^2 \to \mathbb{R}^2$ and the geometry of curves in $\mathbb{R}^2$. (Similar statements may be made about $\mathbb{R}^3$ or, generally, $\mathbb{R}^n$.) If $\mathbf{c}(t)$ is a path in the plane, then as we saw in Section 2.4, $\mathbf{c}'(t)$ represents the tangent (or velocity) vector of the path $\mathbf{c}(t)$, and this tangent (or velocity) vector is thought of as beginning at $\mathbf{c}(t)$. Now let $\mathbf{p}(t) = f(\mathbf{c}(t))$, where $f \colon \mathbb{R}^2 \to \mathbb{R}^2$. The path $\mathbf{p}$ represents the image of the path $\mathbf{c}(t)$ under the mapping f. The tangent vector to $\mathbf{p}$ is given by the chain rule:

matrix
multiplication

$$\mathbf{p}'(t) = \mathbf{D}f(\mathbf{c}(t)) \, \mathbf{c}'(t).$$

matrix   column
vector

In other words, *the derivative matrix of f maps the tangent (or velocity) vector of a path $\mathbf{c}$ to the tangent (or velocity) vector of the corresponding image path $\mathbf{p}$* (see Figure 2.5.1). Thus, points are mapped by f, while tangent vectors to curves are mapped by the derivative of f, evaluated at the base point of the tangent vector in the domain.

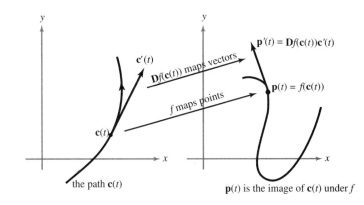

Figure 2.5.1   Tangent vectors are mapped by the derivative matrix.

EXAMPLE 2 Verify the chain rule in the form of formula (3′) for

$$f(u, v, w) = u^2 + v^2 - w,$$

where

$$u(x, y, z) = x^2 y, \quad v(x, y, z) = y^2, \quad w(x, y, z) = e^{-xz}.$$

SOLUTION Here

$$h(x, y, z) = f(u(x, y, z), v(x, y, z), w(x, y, z))$$
$$= (x^2 y)^2 + y^4 - e^{-xz} = x^4 y^2 + y^4 - e^{-xz}.$$

Thus, differentiating directly,

$$\frac{\partial h}{\partial x} = 4x^3 y^2 + ze^{-xz}.$$

On the other hand, using the chain rule,

$$\frac{\partial h}{\partial x} = \frac{\partial f}{\partial u}\frac{\partial u}{\partial x} + \frac{\partial f}{\partial v}\frac{\partial v}{\partial x} + \frac{\partial f}{\partial w}\frac{\partial w}{\partial x} = 2u(2xy) + 2v \cdot 0 + (-1)(ze^{-xz}) = (2x^2 y)(2xy) + ze^{-xz},$$

which is the same as the above. ▲

EXAMPLE 3 Given $g(x, y) = (x^2 + 1, y^2)$ and $f(u, v) = (u + v, u, v^2)$, compute the derivative of $f \circ g$ at $(1, 1)$ using the chain rule.

SOLUTION The matrices of partial derivatives are

$$\mathbf{D}f(u, v) = \begin{bmatrix} \dfrac{\partial f_1}{\partial u} & \dfrac{\partial f_1}{\partial v} \\[2mm] \dfrac{\partial f_2}{\partial u} & \dfrac{\partial f_2}{\partial v} \\[2mm] \dfrac{\partial f_3}{\partial u} & \dfrac{\partial f_3}{\partial v} \end{bmatrix} = \begin{bmatrix} 1 & 1 \\ 1 & 0 \\ 0 & 2v \end{bmatrix} \quad \text{and} \quad \mathbf{D}g(x, y) = \begin{bmatrix} 2x & 2 \\ 0 & 2y \end{bmatrix}.$$

When $(x, y) = (1, 1)$, $g(x, y) = (u, v) = (2, 1)$. Hence

$$\mathbf{D}(f \circ g)(1, 1) = \mathbf{D}f(2, 1)\mathbf{D}g(1, 1) = \begin{bmatrix} 1 & 1 \\ 1 & 0 \\ 0 & 2 \end{bmatrix}\begin{bmatrix} 2 & 0 \\ 0 & 2 \end{bmatrix} = \begin{bmatrix} 2 & 2 \\ 2 & 0 \\ 0 & 4 \end{bmatrix}$$

is the required derivative. ▲

EXAMPLE 4 Let $f(x, y)$ be given and make the substitution $x = r \cos\theta$, $y = r \sin\theta$ (polar coordinates). Write a formula for $\partial f / \partial \theta$.

SOLUTION   By the chain rule,

$$\frac{\partial f}{\partial \theta} = \frac{\partial f}{\partial x}\frac{\partial x}{\partial \theta} + \frac{\partial f}{\partial y}\frac{\partial y}{\partial \theta}$$

that is,

$$\frac{\partial f}{\partial \theta} = -r\sin\theta \frac{\partial f}{\partial x} + r\cos\theta \frac{\partial f}{\partial y}. \quad \blacktriangle$$

EXAMPLE 5   Let $f(x, y) = (\cos y + x^2, e^{x+y})$ and $g(u, v) = (e^{u^2}, u - \sin v)$.
(a) Write a formula for $f \circ g$. (b) Calculate $\mathbf{D}(f \circ g)(0, 0)$ using the chain rule.

SOLUTION   (a)   We have

$$(f \circ g)(u, v) = f(e^{u^2}, u - \sin v)$$
$$= (\cos(u - \sin v) + e^{2u^2}, e^{u^2 + u - \sin v}).$$

(b)   By the chain rule,

$$\mathbf{D}(f \circ g)(0, 0) = [\mathbf{D}f(g(0, 0))][\mathbf{D}g(0, 0)] = [\mathbf{D}f(1, 0)][\mathbf{D}g(0, 0)].$$

Now

$$\mathbf{D}g(0, 0) = \begin{bmatrix} 2ue^{u^2} & 0 \\ 1 & -\cos v \end{bmatrix}_{(u,v)=(0,0)} = \begin{bmatrix} 0 & 0 \\ 1 & -1 \end{bmatrix}$$

and

$$\mathbf{D}f(1, 0) = \begin{bmatrix} 2x & -\sin y \\ e^{x+y} & e^{x+y} \end{bmatrix}_{(x,y)=(1,0)} = \begin{bmatrix} 2 & 0 \\ e & e \end{bmatrix}.$$

[Remember that $\mathbf{D}f$ is evaluated at $g(0, 0)$, not at $(0, 0)$!] Thus

$$\mathbf{D}(f \circ g)(0, 0) = \begin{bmatrix} 2 & 0 \\ e & e \end{bmatrix}\begin{bmatrix} 0 & 0 \\ 1 & -1 \end{bmatrix} = \begin{bmatrix} 0 & 0 \\ e & -e \end{bmatrix}. \quad \blacktriangle$$

EXAMPLE 6   Let $f: U \subset \mathbb{R}^n \to \mathbb{R}^m$ be differentiable, with $f = (f_1, \ldots, f_m)$, and
let $g(\mathbf{x}) = \sin[f(\mathbf{x}) \cdot f(\mathbf{x})]$. Compute $\mathbf{D}g(\mathbf{x})$.

SOLUTION   By the chain rule, $\mathbf{D}g(\mathbf{x}) = \cos[f(\mathbf{x}) \cdot f(\mathbf{x})]\mathbf{D}h(\mathbf{x})$, where $h(\mathbf{x}) = [f(\mathbf{x}) \cdot f(\mathbf{x})] = f_1^2(\mathbf{x}) + \cdots + f_m^2(\mathbf{x})$. Then

$$\mathbf{D}h(x) = \begin{bmatrix} \dfrac{\partial h}{\partial x_1} & \cdots & \dfrac{\partial h}{\partial x_n} \end{bmatrix}$$
$$= \begin{bmatrix} 2f_1\dfrac{\partial f_1}{\partial x_1} + \cdots + 2f_m\dfrac{\partial f_m}{\partial x_1} & \cdots & 2f_1\dfrac{\partial f_1}{\partial x_n} + \cdots + 2f_m\dfrac{\partial f_m}{\partial x_n} \end{bmatrix},$$

which can be written $2f(\mathbf{x})\mathbf{D}f(\mathbf{x})$, where we regard f as a row matrix,

$$f = [f_1 \quad \cdots \quad f_m] \qquad \text{and} \qquad \mathbf{D}f = \begin{bmatrix} \dfrac{\partial f_1}{\partial x_1} & \cdots & \dfrac{\partial f_1}{\partial x_n} \\ \vdots & & \vdots \\ \dfrac{\partial f_m}{\partial x_1} & \cdots & \dfrac{\partial f_m}{\partial x_n} \end{bmatrix}.$$

Thus $\mathbf{D}g(\mathbf{x}) = 2[\cos(f(\mathbf{x}) \cdot f(\mathbf{x}))]f(\mathbf{x})\mathbf{D}f(\mathbf{x})$. ▲

EXERCISES

1. If $f: U \subset \mathbb{R}^n \to \mathbb{R}$ is differentiable, prove that $\mathbf{x} \mapsto f^2(\mathbf{x}) + 2f(\mathbf{x})$ is differentiable as well, and compute its derivative in terms of $\mathbf{D}f(\mathbf{x})$.

2. Prove that the following functions are differentiable, and find their derivatives at an arbitrary point:
(a) $f: \mathbb{R}^2 \to \mathbb{R}, (x, y) \mapsto 2$
(b) $f: \mathbb{R}^2 \to \mathbb{R}, (x, y) \mapsto x + y$
(c) $f: \mathbb{R}^2 \to \mathbb{R}, (x, y) \mapsto 2 + x + y$
(d) $f: \mathbb{R}^2 \to \mathbb{R}, (x, y) \mapsto x^2 + y^2$
(e) $f: \mathbb{R}^2 \to \mathbb{R}, (x, y) \mapsto e^{xy}$
(f) $f: U \to \mathbb{R}, (x, y) \mapsto \sqrt{1 - x^2 - y^2}, U = \{(x, y) \mid x^2 + y^2 < 1\}$
(g) $f: \mathbb{R}^2 \to \mathbb{R}, (x, y) \mapsto x^4 - y^4$

3. Write out the chain rule for each of the following functions and justify your answer in each case using Theorem 11.
(a) $\partial h/\partial x$ where $h(x, y) = f(x, u(x, y))$
(b) dh/dx where $h(x) = f(x, u(x), v(x))$
(c) $\partial h/\partial x$ where $h(x, y, z) = f(u(x, y, z), v(x, y), w(x))$

4. Verify the chain rule for $\partial h/\partial x$ where $h(x, y) = f(u(x, y), v(x, y))$ and

$$f(u, v) = \frac{u^2 + v^2}{u^2 - v^2} \qquad u(x, y) = e^{-x-y}, \qquad v(x, y) = e^{xy}.$$

5. Verify the first special case of the chain rule for the composition $f \circ \mathbf{c}$ in each of the cases:
(a) $f(x, y) = xy, \mathbf{c}(t) = (e^t, \cos t)$
(b) $f(x, y) = e^{xy}, \mathbf{c}(t) = (3t^2, t^3)$
(c) $f(x, y) = (x^2 + y^2)\log\sqrt{x^2 + y^2}, \mathbf{c}(t) = (e^t, e^{-t})$
(d) $f(x, y) = x\exp(x^2 + y^2), \mathbf{c}(t) = (t, -t)$

6. What is the velocity vector for each path $\mathbf{c}(t)$ in Exercise 5? [The solution to part (b) only is in the Study Guide to this text.]

7. Let $f: \mathbb{R}^3 \to \mathbb{R}$ and $g: \mathbb{R}^3 \to \mathbb{R}$ be differentiable. Prove that

$$\nabla(fg) = f\nabla g + g\nabla f.$$

8. Let $f: \mathbb{R}^3 \to \mathbb{R}$ be differentiable. Making the substitution

$$x = \rho\cos\theta\sin\phi, \quad y = \rho\sin\theta\sin\phi, \quad z = \rho\cos\phi$$

(spherical coordinates) into $f(x, y, z)$, compute $\partial f/\partial\rho$, $\partial f/\partial\theta$, and $\partial f/\partial\phi$.

9. Let $f(u, v) = (\tan(u - 1) - e^v, u^2 - v^2)$ and $g(x, y) = (e^{x-y}, x - y)$. Calculate $f \circ g$ and $\mathbf{D}(f \circ g)(1, 1)$.

10. Let $f(u, v, w) = (e^{u-w}, \cos(v + u) + \sin(u + v + w))$ and $g(x, y) = (e^x, \cos(y - x), e^{-y})$. Calculate $f \circ g$ and $\mathbf{D}(f \circ g)(0, 0)$.

11. Find $(\partial/\partial s)(f \circ T)(1, 0)$, where $f(u, v) = \cos u \sin v$ and $T(s, t) = (\cos(t^2 s), \log \sqrt{1 + s^2})$.

12. Suppose that the temperature at the point (x, y, z) in space is $T(x, y, z) = x^2 + y^2 + z^2$. Let a particle follow the right-circular helix $\boldsymbol{\sigma}(t) = (\cos t, \sin t, t)$ and let $T(t)$ be its temperature at time t.
  (a) What is $T'(t)$?
  (b) Find an approximate value for the temperature at $t = (\pi/2) + 0.01$.

13. Suppose that a duck is swimming in the circle $x = \cos t$, $y = \sin t$ and that the water temperature is given by the formula $T = x^2 e^y - xy^3$. Find dT/dt, the rate of change in temperature the duck might feel: (a) by the chain rule; (b) by expressing T in terms of t and differentiating.

14. Let $f: \mathbb{R}^n \to \mathbb{R}^m$ be a linear mapping so that (by Exercise 20, Section 2.3) $\mathbf{D}f(\mathbf{x})$ is the matrix of f. Check the validity of the chain rule directly for linear mappings.

15. Let $f: \mathbb{R}^2 \to \mathbb{R}^2$; $(x, y) \mapsto (e^{x+y}, e^{x-y})$. Let $\mathbf{c}(t)$ be a path with $\mathbf{c}(0) = (0, 0)$ and $\mathbf{c}'(0) = (1, 1)$. What is the tangent vector to the image of $\mathbf{c}(t)$ under f at $t = 0$?

16. Let $f(x, y) = 1/\sqrt{x^2 + y^2}$. Compute $\nabla f(x, y)$.

17. (a) Let $y(x)$ be defined implicitly by $G(x, y(x)) = 0$, where G is a given function of two variables. Prove that if $y(x)$ and G are differentiable, then

$$\frac{dy}{dx} = -\frac{\partial G/\partial x}{\partial G/\partial y} \quad \text{if} \quad \frac{\partial G}{\partial y} \neq 0.$$

  (b) Obtain a formula analogous to that in part (a) if y_1, y_2 are defined implicitly by

$$G_1(x, y_1(x), y_2(x)) = 0,$$
$$G_2(x, y_1(x), y_2(x)) = 0.$$

  (c) Let y be defined implicitly by

$$x^2 + y^3 + e^y = 0.$$

Compute dy/dx in terms of x and y.

18. Thermodynamics texts* use the relationship

$$\left(\frac{\partial y}{\partial x}\right)\left(\frac{\partial z}{\partial y}\right)\left(\frac{\partial x}{\partial z}\right) = -1.$$

---

*See S. M. Binder, "Mathematical Methods in Elementary Thermodynamics," *J. Chem. Educ.* 43 (1966): 85–92. A proper understanding of partial differentiation can be of significant use in applications; for example, see M. Feinberg, "Constitutive Equation for Ideal Gas Mixtures and Ideal Solutions as Consequences of Simple Postulates," *Chem. Eng. Sci.* 32 (1977): 75–78.

Explain the meaning of this equation and prove that it is true. [HINT: Start with a relationship $F(x, y, z) = 0$ that defines $x = f(y, z)$, $y = g(x, z)$, and $z = h(x, y)$ and differentiate implicitly.]

19. Dieterici's equation of state for a gas is

$$P(V - b)e^{a/RVT} = RT,$$

where a, b, and R are constants. Regard volume V as a function of temperature T and pressure P and prove that

$$\frac{\partial V}{\partial T} = \left(R + \frac{a}{TV} \right) \bigg/ \left(\frac{RT}{V - b} - \frac{a}{V^2} \right).$$

20. This exercise gives another example of the fact that the chain rule is not applicable if f is not differentiable. Consider the function

$$f(x, y) = \begin{cases} \dfrac{xy^2}{x^2 + y^2} & (x, y) \neq (0, 0) \\ 0 & (x, y) = (0, 0). \end{cases}$$

Show that
    (a) $\partial f/\partial x$ and $\partial f/\partial y$ exist at $(0, 0)$.
    (b) If $\mathbf{g}(t) = (at, bt)$ for constants a and b, then $f \circ \mathbf{g}$ is differentiable and $(f \circ \mathbf{g})'(0) = ab^2/(a^2 + b^2)$, but $\nabla f(0, 0) \cdot \mathbf{g}'(0) = 0$.

21. Prove that if $f: U \subset \mathbb{R}^n \to \mathbb{R}$ is differentiable at $\mathbf{x}_0 \in U$, there is a neighborhood V of $\mathbf{0} \in \mathbb{R}^n$ and a function $R_1: V \to \mathbb{R}$ such that for all $\mathbf{h} \in V$, $\mathbf{x}_0 + \mathbf{h} \in U$,

$$f(\mathbf{x}_0 + \mathbf{h}) = f(\mathbf{x}_0) + [\mathbf{D}f(\mathbf{x}_0)]\mathbf{h} + R_1(\mathbf{h})$$

and

$$\frac{R_1(\mathbf{h})}{\|\mathbf{h}\|} \to 0 \quad \text{as} \quad \mathbf{h} \to \mathbf{0}.$$

22. Suppose $\mathbf{x}_0 \in \mathbb{R}^n$ and $0 \leq r_1 < r_2$. Show that there is a C^1 function $f: \mathbb{R}^n \to \mathbb{R}$ such that $f(\mathbf{x}) = 0$ for $\|\mathbf{x} - \mathbf{x}_0\| \geq r_2$; $0 < f(\mathbf{x}) < 1$ for $r_1 < \|\mathbf{x} - \mathbf{x}_0\| < r_2$; and $f(\mathbf{x}) = 1$ for $\|\mathbf{x} - \mathbf{x}_0\| \leq r_1$. [HINT: Apply a cubic polynomial with $g(r_1^2) = 1$ and $g(r_2^2) = g'(r_2^2) = g'(r_1^2) = 0$ to $\|\mathbf{x} - \mathbf{x}_0\|^2$ when $r_1 < \|\mathbf{x} - \mathbf{x}_0\| < r_2$.]

23. Find a C^1 mapping $f: \mathbb{R}^3 \to \mathbb{R}^3$ that takes the vector $\mathbf{i} + \mathbf{j} + \mathbf{k}$ emanating from the origin to $\mathbf{i} - \mathbf{j}$ emanating from $(1, 1, 0)$ and takes $\mathbf{k}$ emanating from $(1, 1, 0)$ to $\mathbf{k} - \mathbf{i}$ emanating from the origin.

24. What is wrong with the following argument? Suppose $w = f(x, y, z)$ and $z = g(x, y)$. By the chain rule,

$$\frac{\partial w}{\partial x} = \frac{\partial w}{\partial x}\frac{\partial x}{\partial x} + \frac{\partial w}{\partial y}\frac{\partial y}{\partial x} + \frac{\partial w}{\partial z}\frac{\partial z}{\partial x} = \frac{\partial w}{\partial x} + \frac{\partial w}{\partial z}\frac{\partial z}{\partial x}.$$

Hence $0 = (\partial w/\partial z)(\partial z/\partial x)$, and so $\partial w/\partial z = 0$ or $\partial z/\partial x = 0$, which is, in general, absurd.

25. Prove rules (iii) and (iv) of Theorem 10. [HINT: Use the same addition and subtraction tricks as in the one-variable case and Theorem 8.]

26. Show that $h: \mathbb{R}^n \to \mathbb{R}^m$ is differentiable if and only if each of the m components $h_i: \mathbb{R}^n \to \mathbb{R}$ is differentiable. [HINT: Use the coordinate projection function and the chain rule for one implication and consider

$$\left[\frac{\|h(\mathbf{x}) - h(\mathbf{x}_0) - \mathbf{D}h(\mathbf{x}_0)(\mathbf{x} - \mathbf{x}_0)\|}{\|\mathbf{x} - \mathbf{x}_0\|} \right]^2 = \frac{\sum_{i=1}^{m} [h_i(\mathbf{x}) - h_i(\mathbf{x}_0)\mathbf{D}h_i(\mathbf{x}_0)(\mathbf{x} - \mathbf{x}_0)]^2}{\|\mathbf{x} - \mathbf{x}_0\|^2}$$

to obtain the other.]

27. Use the chain rule to show that

$$\frac{\partial}{\partial x} \int_0^x f(x, y) \, dy = f(x, x) + \int_0^x \frac{\partial f}{\partial x}(x, y) \, dy.$$

28. For what integers $p > 0$ is

$$f(x) = \begin{cases} x^p \sin(1/x) & x \neq 0 \\ 0 & x = 0 \end{cases}$$

differentiable? For what p is the derivative continuous?

29. Suppose $f: \mathbb{R}^n \to \mathbb{R}$ and $g: \mathbb{R}^n \to \mathbb{R}^m$ are differentiable. Show that the product function $h(\mathbf{x}) = f(\mathbf{x})g(\mathbf{x})$ from $\mathbb{R}^n$ to $\mathbb{R}^m$ is differentiable and that if $\mathbf{x}_0$ and $\mathbf{y}$ are in $\mathbb{R}^n$, then $[\mathbf{D}h(\mathbf{x}_0)]\mathbf{y} = f(\mathbf{x}_0)\{[\mathbf{D}g(\mathbf{x}_0)]\mathbf{y}\} + \{[\mathbf{D}f(\mathbf{x}_0)]\mathbf{y}\}g(\mathbf{x}_0)$.

2.6 GRADIENTS AND DIRECTIONAL DERIVATIVES

In Section 2.1 we studied the graphs of real-valued functions. Now we take up this study again, using the methods of calculus. Specifically, gradients will be used to obtain a formula for the plane tangent to a level surface. Let us recall how the gradient is defined.

DEFINITION: THE GRADIENT If $f: U \subset \mathbb{R}^3 \to \mathbb{R}$ is differentiable, the *gradient* of f at (x, y, z) is the vector in space given by

$$\nabla f = \left(\frac{\partial f}{\partial x}, \frac{\partial f}{\partial y}, \frac{\partial f}{\partial z} \right).$$

This vector is also denoted $\nabla f(x, y, z)$. Thus ∇f is just the matrix of the derivative $\mathbf{D}f$, written as a vector.

EXAMPLE 1 Let $f(x, y, z) = \sqrt{x^2 + y^2 + z^2} = r$, the distance from $\mathbf{0}$ to (x, y, z). Then

$$\nabla f(x, y, z) = \left(\frac{\partial f}{\partial x}, \frac{\partial f}{\partial y}, \frac{\partial f}{\partial z} \right)$$

$$= \left(\frac{x}{\sqrt{x^2 + y^2 + z^2}}, \frac{y}{\sqrt{x^2 + y^2 + z^2}}, \frac{z}{\sqrt{x^2 + y^2 + z^2}} \right) = \frac{\mathbf{r}}{r},$$

where $\mathbf{r}$ is the point (x, y, z). Thus ∇f is the unit vector in the direction of (x, y, z). ▲

EXAMPLE 2 If $f(x, y, z) = xy + z$, then

$$\nabla f(x, y, z) = \left(\frac{\partial f}{\partial x}, \frac{\partial f}{\partial y}, \frac{\partial f}{\partial z} \right) = (y, x, 1). \quad ▲$$

Suppose $f: \mathbb{R}^3 \to \mathbb{R}$ is a real-valued function. Let $\mathbf{v}$ and $\mathbf{x} \in \mathbb{R}^3$ be fixed vectors and consider the function from $\mathbb{R}$ to $\mathbb{R}$ defined by $t \mapsto f(\mathbf{x} + t\mathbf{v})$. The set of points of the form $\mathbf{x} + t\mathbf{v}, t \in \mathbb{R}$, is the line L through the point $\mathbf{x}$ parallel to the vector $\mathbf{v}$ (see Figure 2.6.1).

Figure 2.6.1 The equation of L is $\mathbf{l}(t) = \mathbf{x} + t\mathbf{v}$.

The function $t \mapsto f(\mathbf{x} + t\mathbf{v})$ represents the function f restricted to the line L. For example, if a bird flies along this line with velocity $\mathbf{v}$ so that $\mathbf{x} + t\mathbf{v}$ is its position at time t, and if f represents the temperature as a function of position then $f(\mathbf{x} + t\mathbf{v})$ is the temperature at time t. We may ask: How fast are the values of f changing along the line L at the point $\mathbf{x}$? Since the rate of change of a function is given by a derivative, we could say that the answer to this question is the value of the derivative of this function of t at $t = 0$ (when $t = 0$, $\mathbf{x} + t\mathbf{v}$ reduces to $\mathbf{x}$). This would be the derivative of f at the point $\mathbf{x}$ in the direction of L, that is, of $\mathbf{v}$. We can formalize this concept as follows.

DEFINITION: DIRECTIONAL DERIVATIVES   If $f: \mathbb{R}^3 \to \mathbb{R}$, the *directional derivative* of f at $\mathbf{x}$ along the vector $\mathbf{v}$ is given by

$$\frac{d}{dt} f(\mathbf{x} + t\mathbf{v}) \bigg|_{t=0}$$

if this exists.

From the definition, we can see that the directional derivative can also be defined by the formula

$$\lim_{h \to 0} \frac{f(\mathbf{x} + h\mathbf{v}) - f(\mathbf{x})}{h}.$$

THEOREM 12   *If $f: \mathbb{R}^3 \to \mathbb{R}$ is differentiable, then all directional derivatives exist. The directional derivative at $\mathbf{x}$ in the direction $\mathbf{v}$ is given by*

$$\mathbf{D}f(\mathbf{x})\mathbf{v} = \operatorname{grad} f(\mathbf{x}) \cdot \mathbf{v} = \nabla f(\mathbf{x}) \cdot \mathbf{v} = \left[\frac{\partial f}{\partial x}(\mathbf{x})\right] v_1 + \left[\frac{\partial f}{\partial y}(\mathbf{x})\right] v_2 + \left[\frac{\partial f}{\partial z}(\mathbf{x})\right] v_3,$$

where $\mathbf{v} = (v_1, v_2, v_3)$.

PROOF   Let $\mathbf{c}(t) = \mathbf{x} + t\mathbf{v}$, so that $f(\mathbf{x} + t\mathbf{v}) = f(\mathbf{c}(t))$. By the first special case of the chain rule, $(d/dt) f(\mathbf{c}(t)) = \nabla f(\mathbf{c}(t)) \cdot \mathbf{c}'(t)$. However, $\mathbf{c}(0) = \mathbf{x}$ and $\mathbf{c}'(0) = \mathbf{v}$, and so

$$\frac{d}{dt} f(\mathbf{x} + t\mathbf{v}) \bigg|_{t=0} = \nabla f(\mathbf{x}) \cdot \mathbf{v},$$

as we were required to prove.   ■

Notice that one does not have to use straight lines when computing the rate of change of f in a specific direction $\mathbf{v}$. Indeed, for a general path $\mathbf{c}(t)$ with $\mathbf{c}(0) = \mathbf{x}$ and $\mathbf{c}'(0) = \mathbf{v}$, we have from the chain rule,

$$\frac{d}{dt} f(\mathbf{c}(t)) \bigg|_{t=0} = \nabla f(\mathbf{c}(t)) \cdot \mathbf{c}'(t) \bigg|_{t=0} = \nabla f(\mathbf{x}) \cdot \mathbf{v}.$$

In the definition of a directional derivative, one often chooses $\mathbf{v}$ to be a *unit* vector. In this case we are moving in the direction $\mathbf{v}$ with unit speed and we refer to $\nabla f(\mathbf{x}) \cdot \mathbf{v}$ as the ***directional derivative of f in the direction $\mathbf{v}$***.

EXAMPLE 3 Let $f(x, y, z) = x^2 e^{-yz}$. Compute the rate of change of f in the direction of the unit vector

$$\mathbf{v} = \left(\frac{1}{\sqrt{3}}, \frac{1}{\sqrt{3}}, \frac{1}{\sqrt{3}} \right) \text{ at } (1, 0, 0).$$

SOLUTION The required rate of change is, using Theorem 12,

$$\nabla f \cdot \mathbf{v} = (2xe^{-yz}, -x^2 ze^{-yz}, -x^2 ye^{-yz}) \cdot \left(\frac{1}{\sqrt{3}}, \frac{1}{\sqrt{3}}, \frac{1}{\sqrt{3}} \right),$$

which at $(1, 0, 0)$ becomes

$$(2, 0, 0) \cdot \left(\frac{1}{\sqrt{3}}, \frac{1}{\sqrt{3}}, \frac{1}{\sqrt{3}} \right) = \frac{2}{\sqrt{3}}. \quad \blacktriangle$$

From Theorem 12 we can also obtain the geometrical significance of the gradient:

THEOREM 13 *Assume $\nabla f(\mathbf{x}) \neq \mathbf{0}$. Then $\nabla f(\mathbf{x})$ points in the direction along which f is increasing the fastest.*

PROOF If $\mathbf{n}$ is a unit vector, the rate of change of f in direction $\mathbf{n}$ is given by $\nabla f(\mathbf{x}) \cdot \mathbf{n} = \|\nabla f(\mathbf{x})\| \cos \theta$, where θ is the angle between $\mathbf{n}$ and $\nabla f(\mathbf{x})$. This is maximum when $\theta = 0$; that is, when $\mathbf{n}$ and ∇f are parallel. [If $\nabla f(\mathbf{x}) = \mathbf{0}$ this rate of change is 0 for any $\mathbf{n}$.] ∎

In other words, if one wishes to move in a direction in which f will increase most quickly, one should proceed in the direction $\nabla f(\mathbf{x})$. Analogously, if one wishes to move in a direction in which f decreases the fastest, one should proceed in the direction $-\nabla f(\mathbf{x})$.

EXAMPLE 4 In what direction from $(0, 1)$ does $f(x, y) = x^2 - y^2$ increase the fastest?

SOLUTION The gradient is

$$\nabla f = 2x\mathbf{i} - 2y\mathbf{j},$$

and so at $(0, 1)$ this is

$$\nabla f|_{(0,1)} = -2\mathbf{j}.$$

By Theorem 13, f increases fastest in the direction $-\mathbf{j}$. (Can you see why this answer is consistent with Figure 2.1.9?) ▲

Now we find the relationship between the gradient of a function f and its level surfaces. The gradient points in the direction in which the values of f change most rapidly, whereas a level surface lies in the directions in which they do not change at all. If f is reasonably well behaved, the gradient and the level surface will be perpendicular.

THEOREM 14: THE GRADIENT IS NORMAL TO LEVEL SURFACES *Let $f \colon \mathbb{R}^3 \to \mathbb{R}$ be a C^1 map and let (x_0, y_0, z_0) lie on the level surface S defined by $f(x, y, z) = k$, for k a constant. Then $\nabla f(x_0, y_0, z_0)$ is normal to the level surface in the following sense: If $\mathbf{v}$ is the tangent vector at $t = 0$ of a path $\mathbf{c}(t)$ in S with $\mathbf{c}(0) = (x_0, y_0, z_0)$, then $(\nabla f) \cdot \mathbf{v} = 0$ (see Figure 2.6.2).*

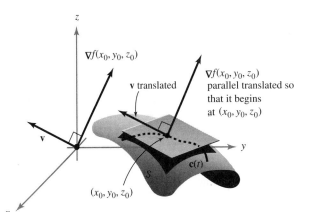

Figure 2.6.2 Geometric significance of the gradient: ∇f is orthogonal to the surface S on which f is constant.

PROOF   Let $\mathbf{c}(t)$ lie in S; then $f(\mathbf{c}(t)) = k$. Let $\mathbf{v}$ be as in the hypothesis; then $\mathbf{v} = \mathbf{c}'(0)$. Hence, the fact that $f(\mathbf{c}(t))$ is constant in t, and the chain rule give

$$0 = \left.\frac{d}{dt} f(\mathbf{c}(t)) \right|_{t=0} = \nabla f(\mathbf{c}(0)) \cdot \mathbf{v}. \qquad \blacksquare$$

If we study the conclusion of Theorem 14, we see that it is reasonable to *define* the plane tangent to S as the orthogonal plane to the gradient.

DEFINITION: TANGENT PLANES TO LEVEL SURFACES   Let S be the surface consisting of those (x, y, z) such that $f(x, y, z) = k$, for k a constant. The ***tangent plane*** of S at a point (x_0, y_0, z_0) of S is defined by the equation

$$\nabla f(x_0, y_0, z_0) \cdot (x - x_0, y - y_0, z - z_0) = 0 \qquad (1)$$

if $\nabla f(x_0, y_0, z_0) \neq \mathbf{0}$. That is, the tangent plane is the set of points (x, y, z) that satisfy equation (1).

This extends the definition we gave earlier for the tangent plane of the graph of a function (see Exercise 11 at the end of this section).

EXAMPLE 5   Compute the equation of the plane tangent to the surface defined by $3xy + z^2 = 4$ at $(1, 1, 1)$.

SOLUTION   Here $f(x, y, z) = 3xy + z^2$ and $\nabla f = (3y, 3x, 2z)$, which at $(1, 1, 1)$ is the vector $(3, 3, 2)$. Thus the tangent plane is

$$(3, 3, 2) \cdot (x - 1, y - 1, z - 1) = 0;$$

that is,

$$3x + 3y + 2z = 8. \quad \blacktriangle$$

In Theorem 14 and the above definition we could just as well have worked in two dimensions as in three. Thus if we have $f: \mathbb{R}^2 \to \mathbb{R}$ and consider a level curve

$$C = \{(x, y) \mid f(x, y) = k\},$$

then $\nabla f(x_0, y_0)$ is perpendicular to C for any point (x_0, y_0) on C. Likewise, the tangent line to C at (x_0, y_0) has the equation

$$\nabla f(x_0, y_0) \cdot (x - x_0, y - y_0) = 0 \tag{2}$$

if $\nabla f(x_0, y_0) \neq 0$; that is, the tangent line is the set of points (x, y) that satisfy equation (2) (see Figure 2.6.3).

Figure 2.6.3   In the plane, the gradient ∇f is orthogonal to the curve f = constant.

We often speak of ∇f as a **gradient vector field**. Note that ∇f assigns a vector to each point in the domain of f. In Figure 2.6.4 we describe the gradient ∇f not by drawing its graph, which, if $f: \mathbb{R}^3 \to \mathbb{R}$, would be a subset of $\mathbb{R}^6$, that is, the set of tuples $(\mathbf{x}, \nabla f(\mathbf{x}))$, but by representing $\nabla f(P)$, for each point P, as a vector emanating from the point P rather than from the origin. Like a graph, this pictorial method of depicting ∇f contains the point P and the value $\nabla f(P)$ in the same picture.

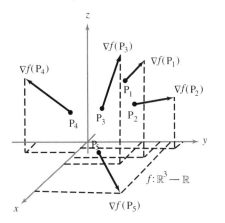

Figure 2.6.4   The gradient ∇f of a function $f \colon \mathbb{R}^3 \to \mathbb{R}$ is a vector field on $\mathbb{R}^3$; at each point P_i, $\nabla f(P_i)$ is a vector emanating from P_i.

The gradient vector field has important geometric significance. It shows the direction in which f is increasing the fastest and the direction that is orthogonal to the level surfaces (or curves in the plane) of f. That it does both of these at once is quite plausible. To see this, imagine a hill as shown in Figure 2.6.5(a). Let h be the height function, a function of two variables. If we draw level curves of h, these are just level contours of the hill. We could imagine them as level paths on the hill [see Figure 2.6.5(b)]. One thing should be obvious to anyone who has gone for a hike: To get to the top of the hill the fastest, we should walk perpendicular to level contours. This is consistent with Theorems 13 and 14, which state that the direction of fastest increase (the gradient) is orthogonal to the level curves.

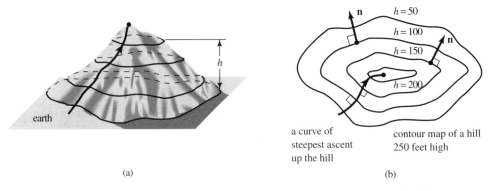

Figure 2.6.5   A physical illustration of the two facts (a) ∇f is the direction of fastest increase of f and (b) ∇f is orthogonal to the level curves.

EXAMPLE 6   The gravitational force on a unit mass m at (x, y, z) produced by a mass M at the origin in $\mathbb{R}^3$ is, according to Newton's law of gravitation, given by

$$\mathbf{F} = -\frac{GmM}{r^2}\mathbf{n},$$

where G is a constant; $r = \|\mathbf{r}\| = \sqrt{x^2 + y^2 + z^2}$, which is the distance of (x, y, z) from the origin; and $\mathbf{n} = \mathbf{r}/r$, the unit vector in the direction of $\mathbf{r} = x\mathbf{i} + y\mathbf{j} + z\mathbf{k}$, which is the position vector from the origin to (x, y, z).

Note that $\mathbf{F} = \nabla(GmM/r) = -\nabla V$, that is, $\mathbf{F}$ is the negative of the gradient of the gravitational potential $V = -GmM/r$. This can be verified as in Example 1. Notice that $\mathbf{F}$ is directed inward toward the origin. Also, the level surfaces of V are spheres. $\mathbf{F}$ is normal to these spheres, which confirms the result of Theorem 14.   ▲

EXAMPLE 7   Find a unit vector normal to the surface S given by $z = x^2 y^2 + y + 1$ at the point $(0, 0, 1)$.

SOLUTION   Let $f(x, y, z) = x^2 y^2 + y + 1 - z$, and consider the level surface defined by $f(x, y, z) = 0$. Since this is the set of points (x, y, z) with $z = x^2 y^2 + y + 1$, we see that this is the surface S. The gradient is given by

$$\nabla f(x, y, z) = \frac{\partial f}{\partial x}\mathbf{i} + \frac{\partial f}{\partial y}\mathbf{j} + \frac{\partial f}{\partial z}\mathbf{k} = 2xy^2\mathbf{i} + (2x^2 y + 1)\mathbf{j} - \mathbf{k},$$

and so

$$\nabla f(0, 0, 1) = \mathbf{j} - \mathbf{k}.$$

This vector is perpendicular to S at $(0, 0, 1)$, and so to find a unit normal $\mathbf{n}$ we divide this vector by its length to obtain

$$\mathbf{n} = \frac{\nabla f(0, 0, 1)}{\|\nabla f(0, 0, 1)\|} = \frac{1}{\sqrt{2}}(\mathbf{j} - \mathbf{k}).   \quad ▲$$

EXAMPLE 8   Consider two conductors, one charged positively and the other negatively. Between them, an electric potential is set up. This potential is a function $\phi: \mathbb{R}^3 \to \mathbb{R}$ (an example of a *scalar* field). The electric field is given by $\mathbf{E} = -\nabla\phi$. From Theorem 14 we know that $\mathbf{E}$ is perpendicular to level surfaces of ϕ. These level surfaces are called ***equipotential surfaces***, since the potential is constant on them (see Figure 2.6.6).   ▲

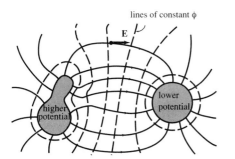

lines of constant ϕ

Figure 2.6.6   Equipotential surfaces (the dotted lines) are orthogonal to the electric force field $\mathbf{E}$.

EXERCISES

1. Show that the directional derivatives of $f(x, y, z) = z^2 x + y^3$ at $(1, 1, 2)$ in the direction $(1/\sqrt{5})\mathbf{i} + (2/\sqrt{5})\mathbf{j}$ is $2\sqrt{5}$.

2. Compute the directional derivatives of the following functions at the indicated points in the given directions:

(a) $f(x, y) = x + 2xy - 3y^2$, $(x_0, y_0) = (1, 2)$, $\mathbf{v} = \frac{3}{5}\mathbf{i} + \frac{4}{5}\mathbf{j}$

(b) $f(x, y) = \log \sqrt{x^2 + y^2}$, $(x_0, y_0) = (1, 0)$, $\mathbf{v} = (1/\sqrt{5})(2\mathbf{i} + \mathbf{j})$

(c) $f(x, y) = e^x \cos(\pi y)$, $(x_0, y_0) = (0, -1)$, $\mathbf{v} = -(1/\sqrt{5})\mathbf{i} + (2/\sqrt{5})\mathbf{j}$

(d) $f(x, y) = xy^2 + x^3 y$, $(x_0, y_0) = (4, -2)$, $\mathbf{v} = (1/\sqrt{10})\mathbf{i} + (3/\sqrt{10})\mathbf{j}$

3. Compute the directional derivatives of the following functions along unit vectors at the indicated points in directions *parallel* to the given vector:

(a) $f(x, y) = x^y$, $(x_0, y_0) = (e, e)$, $\mathbf{d} = 5\mathbf{i} + 12\mathbf{j}$

(b) $f(x, y, z) = e^x + yz$, $(x_0, y_0, z_0) = (1, 1, 1)$, $\mathbf{d} = (1, -1, 1)$

(c) $f(x, y, z) = xyz$, $(x_0, y_0, z_0) = (1, 0, 1)$, $\mathbf{d} = (1, 0, -1)$

4. Find the planes tangent to the following surfaces at the indicated points:

(a) $x^2 + 2y^2 + 3xz = 10$, $(1, 2, \frac{1}{3})$

(b) $y^2 - x^2 = 3$, $(1, 2, 8)$

(c) $xyz = 1$, $(1, 1, 1)$

5. Find the equation for the plane tangent to each surface $z = f(x, y)$ at the indicated point:

(a) $z = x^3 + y^3 - 6xy$, $(1, 2, -3)$

(b) $z = (\cos x)(\cos y)$, $(0, \pi/2, 0)$

(c) $z = (\cos x)(\sin y)$, $(0, \pi/2, 1)$

6. Compute the gradient ∇f for each of the following functions:

(a) $f(x, y, z) = 1/\sqrt{x^2 + y^2 + z^2}$

(b) $f(x, y, z) = xy + yz + xz$

(c) $f(x, y, z) = \dfrac{1}{x^2 + y^2 + z^2}$

7. For the functions in Exercise 6, what is the direction of fastest increase at $(1, 1, 1)$? [The solution to part (c) only is in the Study Guide to this text.]

8. Show that a unit normal to the surface $x^3 y^3 + y - z + 2 = 0$ at $(0, 0, 2)$ is given by $\mathbf{n} = (1/\sqrt{2})(\mathbf{j} - \mathbf{k})$.

9. Find a unit normal to the surface $\cos(xy) = e^z - 2$ at $(1, \pi, 0)$.

10. Verify Theorems 13 and 14 for $f(x, y, z) = x^2 + y^2 + z^2$.

11. Show that the definition following Theorem 14 yields, as a special case, the formula for the plane tangent to the graph of $f(x, y)$ by regarding the graph as a level surface of $F(x, y, z) = f(x, y) - z$ (see Section 2.3).

12. Let $f(x, y) = -(1 - x^2 - y^2)^{1/2}$ for (x, y) such that $x^2 + y^2 < 1$. Show that the plane tangent to the graph of f at $(x_0, y_0, f(x_0, y_0))$ is orthogonal to the vector with components $(x_0, y_0, f(x_0, y_0))$.

Interpret this geometrically.

13. For the following functions $f\colon \mathbb{R}^3 \to \mathbb{R}$ and $\mathbf{g}\colon \mathbb{R} \to \mathbb{R}^3$, find ∇f and $\mathbf{g}'$ and evaluate $(f \circ \mathbf{g})'(1)$.

(a) $f(x, y, z) = xz + yz + xy, \mathbf{g}(t) = (e^t, \cos t, \sin t)$

(b) $f(x, y, z) = e^{xyz}, \mathbf{g}(t) = (6t, 3t^2, t^3)$

(c) $f(x, y, z) = (x^2 + y^2 + z^2) \log \sqrt{x^2 + y^2 + z^2}, \mathbf{g}(t) = (e^t, e^{-t}, t)$

14. Compute the directional derivative of f in the given directions $\mathbf{v}$ at the given points P.

(a) $f(x, y, z) = xy^2 + y^2 z^3 + z^3 x, \mathrm{P} = (4, -2, -1), \mathbf{v} = 1/\sqrt{14}(\mathbf{i} + 3\mathbf{j} + 2\mathbf{k})$

(b) $f(x, y, z) = x^{yz}, \mathrm{P} = (e, e, 0), \mathbf{v} = \frac{12}{13}\mathbf{i} + \frac{3}{13}\mathbf{j} + \frac{4}{13}\mathbf{k}$

15. Let $\mathbf{r} = x\mathbf{i} + y\mathbf{j} + z\mathbf{k}$ and $r = \|\mathbf{r}\|$. Prove that

$$\nabla\left(\frac{1}{r}\right) = -\frac{\mathbf{r}}{r^3}.$$

16. Captain Ralph is in trouble near the sunny side of Mercury. The temperature of the ship's hull when he is at location (x, y, z) will be given by $T(x, y, z) = e^{-x^2 - 2y^2 - 3z^2}$, where x, y, and z are measured in meters. He is currently at $(1, 1, 1)$.

(a) In what direction should he proceed in order to decrease the temperature most rapidly?

(b) If the ship travels at e^8 meters per second, how fast will the temperature decrease if he proceeds in that direction?

(c) Unfortunately, the metal of the hull will crack if cooled at a rate greater than $\sqrt{14e^2}$ degrees per second. Describe the set of possible directions in which he may proceed to bring the temperature down at no more than that rate.

17. A function $f\colon \mathbb{R}^2 \to \mathbb{R}$ is said to be independent of the second variable if there is a function $g\colon \mathbb{R} \to \mathbb{R}$ such that $f(x, y) = g(x)$ for all x in $\mathbb{R}$. In this case, calculate ∇f in terms of g'.

18. Let f and g be functions from $\mathbb{R}^3$ to $\mathbb{R}$. Suppose f is differentiable and $\nabla f(\mathbf{x}) = g(\mathbf{x})\mathbf{x}$. Show that spheres centered at the origin are contained in the level sets for f; that is, f is constant on such spheres.

19. A function $f\colon \mathbb{R}^n \to \mathbb{R}$ is called an *even* function if $f(\mathbf{x}) = f(-\mathbf{x})$ for every $\mathbf{x}$ in $\mathbb{R}^n$. If f is differentiable and even, find $\mathbf{D}f$ at the origin.

20. Suppose that a mountain has the shape of an elliptic paraboloid $z = c - ax^2 - by^2$, where a, b, and c are positive constants, x and y are the east-west and north-south map coordinates, and z is the altitude above sea level (x, y, z are all measured in meters). At the point $(1, 1)$, in what direction is the altitude increasing most rapidly? If a marble were released at $(1, 1)$, in what direction would it begin to roll?

21. An engineer wishes to build a railroad up the mountain of Exercise 20. Straight up the mountain is much too steep for the power of the engines. At the point $(1, 1)$, in what directions may the track be laid so that it will be climbing with a 3% grade—that is, an angle whose tangent is 0.03? (There are two possibilities.) Make a sketch of the situation indicating the two possible directions for a 3% grade at $(1, 1)$.

22. In electrostatics, the force $\mathbf{P}$ of attraction between two particles of opposite charge is given by $\mathbf{P} = k(\mathbf{r}/\|\mathbf{r}\|^3)$ (*Coulomb's law*), where k is a constant and $\mathbf{r} = x\mathbf{i} + y\mathbf{j} + z\mathbf{k}$. Show that $\mathbf{P}$ is the gradient of $f = -k/\|\mathbf{r}\|$.

23. The electrostatic potential V due to two infinite parallel filaments with linear charge densities λ and $-\lambda$ is $V = (\lambda/2\pi\varepsilon_0)\ln(r_2/r_1)$, where $r_1^2 = (x - x_0)^2 + y^2$ and $r_2^2 = (x + x_0)^2 + y^2$. We think of the filaments as being in the z direction, passing through the xy plane at $(-x_0, 0)$ and $(x_0, 0)$. Find $\nabla V(x, y)$.

24. For each of the following, find the maximum and minimum values attained by the function f along the path $(\mathbf{c})(t)$:

    (a) $f(x, y) = xy$; $\mathbf{c}(t) = (\cos t, \sin t)$; $0 \le t \le 2\pi$.
    (b) $f(x, y) = x^2 + y^2$; $\mathbf{c}(t) = (\cos t, 2\sin t)$; $0 \le t \le 2\pi$.

25. Suppose that a particle is ejected from the surface $x^2 + y^2 - z^2 = -1$ at the point $(1, 1, \sqrt{3})$ along the normal directed toward the xy plane to the surface at time $t = 0$ with a speed of 10 units per second. When and where does it cross the xy plane?

26. Let $f: \mathbb{R}^3 \to \mathbb{R}$ and regard $\mathbf{D}f(x, y, z)$ as a linear map of $\mathbb{R}^3$ to $\mathbb{R}$. Show that the kernel (that is, the set of vectors mapped to zero) of $\mathbf{D}f$ is the plane in $\mathbb{R}^3$ orthogonal to ∇f.

---

*2.7 SOME TECHNICAL DIFFERENTIATION THEOREMS

In this section we examine the mathematical foundations of differential calculus in further detail and supply some of the proofs omitted from Sections 2.2, 2.3, and 2.5. We shall begin by supplying the proofs of the limit theorems presented in Section 2.2 (the theorem numbering corresponds to that earlier in this chapter). Let us recall the definition of a limit.

DEFINITION OF LIMIT Let $f: A \subset \mathbb{R}^n \to \mathbb{R}^m$ where A is open. Let $\mathbf{x}_0$ be in A or be a boundary point of A, and let N be a neighborhood of $\mathbf{b} \in \mathbb{R}^m$. We say f is *eventually* in N as $\mathbf{x}$ *approaches* $\mathbf{x}_0$ if there exists a neighborhood U of $\mathbf{x}_0$ such that $\mathbf{x} \ne \mathbf{x}_0$, $\mathbf{x} \in U$, and $\mathbf{x} \in A$ implies $f(\mathbf{x}) \in N$. We say $f(\mathbf{x})$ *approaches* $\mathbf{b}$ as $\mathbf{x}$ approaches $\mathbf{x}_0$, or, in symbols,

$$\underset{\mathbf{x} \to \mathbf{x}_0}{\text{limit}}\ f(\mathbf{x}) = \mathbf{b} \quad \text{or} \quad f(\mathbf{x}) \to \mathbf{b} \quad \text{as} \quad \mathbf{x} \to \mathbf{x}_0,$$

when, given *any* neighborhood N of $\mathbf{b}$, f is eventually in N as $\mathbf{x}$ approaches $\mathbf{x}_0$. If, as $\mathbf{x}$ approaches $\mathbf{x}_0$, the values $f(\mathbf{x})$ do not get close to any particular number, we say that $\underset{\mathbf{x} \to \mathbf{x}_0}{\text{limit}} f(\mathbf{x})$ *does not exist.*

Let us first establish that this definition is equivalent to the ε-δ formulation of limits.

THEOREM 6 *Let $f: A \subset \mathbb{R}^n \to \mathbb{R}^m$ and let $\mathbf{x}_0$ be in A or be a boundary point of A. Then $\underset{\mathbf{x} \to \mathbf{x}_0}{\text{limit}} f(\mathbf{x}) = \mathbf{b}$ if and only if for every number $\varepsilon > 0$ there is a $\delta > 0$ such that for any $\mathbf{x} \in A$ satisfying $0 < \|\mathbf{x} - \mathbf{x}_0\| < \delta$, we have $\|f(\mathbf{x}) - \mathbf{b}\| < \varepsilon$.*

PROOF First let us assume that $\underset{\mathbf{x} \to \mathbf{x}_0}{\text{limit}} f(\mathbf{x}) = \mathbf{b}$. Let $\varepsilon > 0$ be given, and consider the ε neighborhood $N = D_\varepsilon(\mathbf{b})$, the ball or disk of radius ε with center $\mathbf{b}$. By the definition of a limit, f is eventually in $D_\varepsilon(\mathbf{b})$ as $\mathbf{x}$ approaches $\mathbf{x}_0$, which means there is a neighborhood U of $\mathbf{x}_0$ such that $f(\mathbf{x}) \in D_\varepsilon(\mathbf{b})$ if $\mathbf{x} \in U$, $\mathbf{x} \in A$, and $\mathbf{x} \ne \mathbf{x}_0$. Now since U is open and $\mathbf{x}_0 \in U$, there is a $\delta > 0$ such that $D_\delta(\mathbf{x}_0) \subset U$. Consequently, $0 < \|\mathbf{x} - \mathbf{x}_0\| < \delta$ and $\mathbf{x} \in A$ implies $\mathbf{x} \in D_\delta(\mathbf{x}_0) \subset U$.

Thus $f(\mathbf{x}) \in D_\varepsilon(\mathbf{b})$, which means that $\|f(\mathbf{x}) - \mathbf{b}\| < \varepsilon$. This is the ε-δ assertion we wanted to prove.

We now prove the converse. Assume that for every $\varepsilon > 0$ there is a $\delta > 0$ such that $0 < \|\mathbf{x} - \mathbf{x}_0\| < \delta$ and $\mathbf{x} \in A$ implies $\|f(\mathbf{x}) - \mathbf{b}\| < \varepsilon$. Let N be a neighborhood of $\mathbf{b}$. We have to show that f is eventually in N as $\mathbf{x} \to \mathbf{x}_0$; that is, we must find an open set $U \subset \mathbb{R}^n$ such that $\mathbf{x} \in U, \mathbf{x} \in A$, and $\mathbf{x} \neq \mathbf{x}_0$ implies $f(\mathbf{x}) \in N$. Now since N is open, there is an $\varepsilon > 0$ such that $D_\varepsilon(\mathbf{b}) \subset N$. If we choose $U = D_\delta(\mathbf{x})$ (according to our assumption), then $\mathbf{x} \in U, \mathbf{x} \in A$, and $\mathbf{x} \neq \mathbf{x}_0$ means $\|f(\mathbf{x}) - \mathbf{b}\| < \varepsilon$, that is $f(\mathbf{x}) \in D_\varepsilon(\mathbf{b}) \subset N$. ∎

THEOREM 2: UNIQUENESS OF LIMITS *If* $\lim\limits_{\mathbf{x} \to \mathbf{x}_0} f(\mathbf{x}) = \mathbf{b}_1$ *and* $\lim\limits_{\mathbf{x} \to \mathbf{x}_0} f(\mathbf{x}) = \mathbf{b}_2$, *then* $\mathbf{b}_1 = \mathbf{b}_2$.

PROOF It is convenient to use the ε-δ formulation of Theorem 6. Suppose $f(\mathbf{x}) \to \mathbf{b}_1$ and $f(\mathbf{x}) \to \mathbf{b}_2$ as $\mathbf{x} \to \mathbf{x}_0$. Given $\varepsilon > 0$, we can, by assumption, find $\delta_1 > 0$ such that if $\mathbf{x} \in A$ and $0 < \|\mathbf{x} - \mathbf{x}_0\| < \delta_1$, then $\|f(\mathbf{x}) - \mathbf{b}_1\| < \varepsilon$, and similarly, we can find $\delta_2 > 0$ such that $0 < \|\mathbf{x} - \mathbf{x}_0\| < \delta_2$ implies $\|f(\mathbf{x}) - \mathbf{b}_2\| < \varepsilon$. Let δ be the smaller of δ_1 and δ_2. Choose $\mathbf{x}$ such that $0 < \|\mathbf{x} - \mathbf{x}_0\| < \delta$ and $\mathbf{x} \in A$. Such $\mathbf{x}$'s exist, because $\mathbf{x}_0$ is in A or is a boundary point of A. Thus, using the triangle inequality,

$$\|\mathbf{b}_1 - \mathbf{b}_2\| = \|(\mathbf{b}_1 - f(\mathbf{x})) + (f(\mathbf{x}) - \mathbf{b}_2)\|$$
$$\leq \|\mathbf{b}_1 - f(\mathbf{x})\| + \|f(\mathbf{x}) - \mathbf{b}_2\| < \varepsilon + \varepsilon = 2\varepsilon.$$

Thus for *every* $\varepsilon > 0$, $\|\mathbf{b}_1 - \mathbf{b}_2\| < 2\varepsilon$. Hence $\mathbf{b}_1 = \mathbf{b}_2$, for if $\mathbf{b}_1 \neq \mathbf{b}_2$ we could let $\varepsilon = \|\mathbf{b}_1 - \mathbf{b}_2\|/2 > 0$ and we would have $\|\mathbf{b}_1 - \mathbf{b}_2\| < \|\mathbf{b}_1 - \mathbf{b}_2\|$, an impossibility. ∎

THEOREM 3: PROPERTIES OF LIMITS *Let* $f: A \subset \mathbb{R}^n \to \mathbb{R}^m$, $g: A \subset \mathbb{R}^n \to \mathbb{R}^m$, $\mathbf{x}_0$ *be in* A *or be a boundary point of* A, $\mathbf{b} \in \mathbb{R}^m$, *and* $c \in \mathbb{R}$; *the following assertions then hold*:

(i) *If* $\lim\limits_{\mathbf{x} \to \mathbf{x}_0} f(\mathbf{x}) = \mathbf{b}$, *then* $\lim\limits_{\mathbf{x} \to \mathbf{x}_0} cf(\mathbf{x}) = c\mathbf{b}$, *where* $cf: A \to \mathbb{R}^m$ *is defined by* $\mathbf{x} \mapsto c(f(\mathbf{x}))$.

(ii) *If* $\lim\limits_{\mathbf{x} \to \mathbf{x}_0} f(\mathbf{x}) = \mathbf{b}_1$ *and* $\lim\limits_{\mathbf{x} \to \mathbf{x}_0} g(\mathbf{x}) = \mathbf{b}_2$, *then* $\lim\limits_{\mathbf{x} \to \mathbf{x}_0} (f + g)(\mathbf{x}) = \mathbf{b}_1 + \mathbf{b}_2$, *where* $(f + g): A \to \mathbb{R}^m$ *is defined by* $\mathbf{x} \mapsto f(\mathbf{x}) + g(\mathbf{x})$.

(iii) *If* $m = 1$, $\lim\limits_{\mathbf{x} \to \mathbf{x}_0} f(\mathbf{x}) = b_1$, *and* $\lim\limits_{\mathbf{x} \to \mathbf{x}_0} g(\mathbf{x}) = b_2$, *then* $\lim\limits_{\mathbf{x} \to \mathbf{x}_0} (fg)(\mathbf{x}) = b_1 b_2$, *where* $(fg): A \to \mathbb{R}$ *is defined by* $\mathbf{x} \mapsto f(\mathbf{x})g(\mathbf{x})$.

(iv) *If* $m = 1$, $\lim\limits_{\mathbf{x} \to \mathbf{x}_0} f(\mathbf{x}) = b \neq 0$, *and* $f(\mathbf{x}) \neq 0$ *for all* $\mathbf{x} \in A$, *then* $\lim\limits_{\mathbf{x} \to \mathbf{x}_0} 1/f = 1/b$, *where* $1/f: A \to \mathbb{R}$ *is defined by* $\mathbf{x} \mapsto 1/f(\mathbf{x})$.

(v) *If* $f(\mathbf{x}) = (f_1(\mathbf{x}), \ldots, f_m(\mathbf{x}))$, *where* $f_i: A \to \mathbb{R}, i = 1, \ldots, m$, *are the component functions of* f, *then* $\lim\limits_{\mathbf{x} \to \mathbf{x}_0} f(\mathbf{x}) = \mathbf{b} = (b_1, \ldots, b_m)$ *if and only if* $\lim\limits_{\mathbf{x} \to \mathbf{x}_0} f_i(\mathbf{x}) = b_i$ *for each* $i = 1, \ldots, m$.

PROOF We shall illustrate the technique of proof by proving assertions (i) and (ii). The proofs of the other assertions are only a bit more complicated and may be supplied by the reader. In each case, the ε-δ formulation of Theorem 6 is probably the most convenient approach.

To prove rule (i), let $\varepsilon > 0$ be given; we must produce a number $\delta > 0$ such that the inequality $\|cf(\mathbf{x}) - c\mathbf{b}\| < \varepsilon$ holds if $0 < \|\mathbf{x} - \mathbf{x}_0\| < \delta$. If $c = 0$, any δ will do, so we can suppose $c \neq 0$. Let $\varepsilon' = \varepsilon/|c|$; from the definition of limit, there is a δ with the property that $0 < \|\mathbf{x} - \mathbf{x}_0\| < \delta$ implies

$\|f(\mathbf{x}) - \mathbf{b}\| < \varepsilon' = \varepsilon/|c|$. Thus $0 < \|\mathbf{x} - \mathbf{x}_0\| < \delta$ implies $\|cf(\mathbf{x}) - c\mathbf{b}\| = |c| \|f(\mathbf{x}) - \mathbf{b}\| < \varepsilon$, which proves rule (i).

To prove rule (ii), let $\varepsilon > 0$ be given again. Choose $\delta_1 > 0$ such that $0 < \|\mathbf{x} - \mathbf{x}_0\| < \delta_1$ implies $\|f(\mathbf{x}) - \mathbf{b}_1\| < \varepsilon/2$. Similarly, choose $\delta_2 > 0$ such that $0 < \|\mathbf{x} - \mathbf{x}_0\| < \delta_2$ implies $\|g(\mathbf{x}) - \mathbf{b}_2\| < \varepsilon/2$. Let δ be the lesser of δ_1 and δ_2. Then $0 < \|\mathbf{x} - \mathbf{x}_0\| < \delta$ implies

$$\|f(\mathbf{x}) + g(\mathbf{x}) - \mathbf{b}_1 - \mathbf{b}_2\| \leq \|f(\mathbf{x}) - \mathbf{b}_1\| + \|g(\mathbf{x}) - \mathbf{b}_2\| < \frac{\varepsilon}{2} + \frac{\varepsilon}{2} = \varepsilon.$$

Thus we have proved that $(f + g)(\mathbf{x}) \to \mathbf{b}_1 + \mathbf{b}_2$ as $\mathbf{x} \to \mathbf{x}_0$.   ■

Let us recall the definition of a continuous function.

DEFINITION   Let $f : A \subset \mathbb{R}^n \to \mathbb{R}^m$ be a given function with domain A. Let $\mathbf{x}_0 \in A$. We say f is *continuous at* $\mathbf{x}_0$ if and only if

$$\lim_{\mathbf{x} \to \mathbf{x}_0} f(\mathbf{x}) = f(\mathbf{x}_0).$$

If we just say that f is *continuous*, we shall mean that f is continuous at each point $\mathbf{x}_0$ of A.

From Theorem 6, we get the ε-δ criterion for continuity (Theorem 7): f is continuous at $\mathbf{x}_0 \in A$ if and only if for every number $\varepsilon > 0$ there is a number $\delta > 0$ such that

$$\mathbf{x} \in A \quad \text{and} \quad \|\mathbf{x} - \mathbf{x}_0\| < \delta \quad \text{implies} \quad \|f(\mathbf{x}) - f(\mathbf{x}_0)\| < \varepsilon.$$

One of the properties of continuous functions stated without proof in Section 2.2 was the following:

THEOREM 5: CONTINUITY OF COMPOSITIONS   *Let $f : A \subset \mathbb{R}^n \to \mathbb{R}^m$ and let $g : B \subset \mathbb{R}^m \to \mathbb{R}^p$. Suppose $f(A) \subset B$ so that $g \circ f$ is defined on A. If f is continuous at $\mathbf{x}_0 \in A$ and g is continuous at $\mathbf{y}_0 = f(\mathbf{x}_0)$, then $g \circ f$ is continuous at $\mathbf{x}_0$.*

PROOF   We use the ε-δ criterion for continuity. Thus, given $\varepsilon > 0$, we must find $\delta > 0$ such that for $\mathbf{x} \in A$,

$$\|\mathbf{x} - \mathbf{x}_0\| < \delta \quad \text{implies} \quad \|(g \circ f)(\mathbf{x}) - (g \circ f)(\mathbf{x}_0)\| < \varepsilon.$$

Since g is continuous at $f(\mathbf{x}_0) = \mathbf{y}_0 \in B$, there is a $\gamma > 0$ such that for $\mathbf{y} \in B$,

$$\|\mathbf{y} - \mathbf{y}_0\| < \gamma \quad \text{implies} \quad \|g(\mathbf{y}) - g(\mathbf{y}_0)\| < \varepsilon.$$

Since f is continuous at $\mathbf{x}_0 \in A$, there is, for this γ, a $\delta > 0$ such that $\mathbf{x} \in A$, and

$$\|\mathbf{x} - \mathbf{x}_0\| < \delta \quad \text{implies} \quad \|f(\mathbf{x}) - f(\mathbf{x}_0)\| < \gamma,$$

which in turn implies

$$\|g(f(\mathbf{x})) - g(f(\mathbf{x}_0))\| < \varepsilon,$$

which is the desired conclusion.   ■

The exposition in Section 2.3 was simplified by assuming, as part of the definition of $\mathbf{D}f(\mathbf{x}_0)$, that the partial derivatives of f existed. Our next objective is to show that *this assumption can be omitted*. Let us begin by *redefining* "differentiable." Theorem 14 will show that the new definition is equivalent to the old one.

DEFINITION   Let U be an open set in $\mathbb{R}^n$ and let $f : U \subset \mathbb{R}^n \to \mathbb{R}^m$ be a given function. We say that f is *differentiable* at $\mathbf{x}_0 \in U$ if and only if there exists an $m \times n$ matrix $\mathbf{T}$ such that

$$\lim_{\mathbf{x} \to \mathbf{x}_0} \frac{\|f(\mathbf{x}) - f(\mathbf{x}_0) - \mathbf{T}(\mathbf{x} - \mathbf{x}_0)\|}{\|\mathbf{x} - \mathbf{x}_0\|} = 0. \tag{1}$$

We call $\mathbf{T}$ the *derivative* of f at $\mathbf{x}_0$ and denote it by $\mathbf{D}f(\mathbf{x}_0)$. In matrix notation, $\mathbf{T}(\mathbf{x} - \mathbf{x}_0)$ stands for

$$
\begin{bmatrix}
t_{11} & t_{12} & \cdots & t_{1n} \\
t_{21} & t_{22} & \cdots & t_{2n} \\
\vdots & \vdots & & \vdots \\
t_{m1} & t_{m2} & \cdots & t_{mn}
\end{bmatrix}
\begin{bmatrix}
x_1 - x_{01} \\
\vdots \\
x_n - x_{0n}
\end{bmatrix},
$$

where $\mathbf{x} = (x_1, \ldots, x_n)$ and $\mathbf{x}_0 = (x_{01}, \ldots, x_{0n})$. Sometimes we write $\mathbf{T}(\mathbf{y})$ as $\mathbf{T} \cdot \mathbf{y}$ or just $\mathbf{Ty}$.

Condition (1) can be rewritten as

$$
\underset{\mathbf{h} \to \mathbf{0}}{\text{limit}} \frac{\|f(\mathbf{x}_0 + \mathbf{h}) - f(\mathbf{x}_0) - \mathbf{Th}\|}{\|\mathbf{h}\|} = 0 \tag{2}
$$

by letting $\mathbf{h} = \mathbf{x} - \mathbf{x}_0$. Written in terms of ε-δ notation, equation (2) says that for every $\varepsilon > 0$ there is a $\delta > 0$ such that $0 < \|\mathbf{h}\| < \delta$ implies

$$
\frac{\|f(\mathbf{x}_0 + \mathbf{h}) - f(\mathbf{x}_0) - \mathbf{Th}\|}{\|\mathbf{h}\|} < \varepsilon,
$$

or, in other words,

$$
\|f(\mathbf{x}_0 + \mathbf{h}) - f(\mathbf{x}_0) - \mathbf{Th}\| < \varepsilon \|\mathbf{h}\|.
$$

Notice that because U is open, as long as δ is small enough, $\|\mathbf{h}\| < \delta$ implies $\mathbf{x}_0 + \mathbf{h} \in U$.

Our task is to show that the matrix $\mathbf{T}$ is necessarily the matrix of partial derivatives, and hence that this abstract definition agrees with the definition of differentiability given in Section 2.3.

THEOREM 15 *Suppose $f: U \subset \mathbb{R}^n \to \mathbb{R}^m$ is differentiable at $\mathbf{x}_0 \in \mathbb{R}^n$. Then all the partial derivatives of f exist at the point $\mathbf{x}_0$ and the $m \times n$ matrix $\mathbf{T}$ is given by*

$$
[t_{ij}] = \left[\frac{\partial f_i}{\partial x_j} \right],
$$

that is,

$$
\mathbf{T} = \mathbf{D}f(\mathbf{x}_0) =
\begin{bmatrix}
\dfrac{\partial f_1}{\partial x_1} & \cdots & \dfrac{\partial f_1}{\partial x_n} \\
\vdots & & \vdots \\
\dfrac{\partial f_m}{\partial x_1} & \cdots & \dfrac{\partial f_m}{\partial x_n}
\end{bmatrix},
$$

where $\partial f_i / \partial x_j$ is evaluated at $\mathbf{x}_0$. In particular, this implies that $\mathbf{T}$ is uniquely determined; there is no other matrix satisfying condition (1).

PROOF   By Theorem 3(v), condition (2) is the same as

$$\underset{\mathbf{h}\to\mathbf{0}}{\text{limit}}\ \frac{|f_i(\mathbf{x}_0+\mathbf{h})-f_i(\mathbf{x}_0)-(\mathbf{Th})_i|}{\|\mathbf{h}\|}=0, \qquad 1\le i\le m.$$

Here $(\mathbf{Th})_i$ stands for the ith component of the column vector $\mathbf{Th}$. Now let $\mathbf{h}=a\mathbf{e}_j=(0,\ldots,a,\ldots,0)$, which has the number a in the jth slot and zeros elsewhere. We get

$$\underset{a\to 0}{\text{limit}}\ \frac{|f_i(\mathbf{x}_0+a\mathbf{e}_j)-f_i(\mathbf{x}_0)-a(\mathbf{Te}_j)_i|}{|a|}=0,$$

or, in other words,

$$\underset{a\to 0}{\text{limit}}\ \left|\frac{f_i(\mathbf{x}_0+a\mathbf{e}_j)-f_i(\mathbf{x}_0)}{a}-(\mathbf{Te}_j)_i\right|=0,$$

so that

$$\underset{a\to 0}{\text{limit}}\ \frac{f_i(\mathbf{x}_0+a\mathbf{e}_j)-f_i(\mathbf{x}_0)}{a}=(\mathbf{Te}_j)_i.$$

But this limit is nothing more than the partial derivative $\partial f_i/\partial x_j$ evaluated at the point $\mathbf{x}_0$. Thus we have proved that $\partial f_i/\partial x_j$ exists and equals $(\mathbf{Te}_j)_i$. But $(\mathbf{Te}_j)_i=t_{ij}$ (see Section 1.5), and so the theorem follows. ∎

Our next task is to show that differentiability implies continuity.

THEOREM 8   *Let $f\colon U\subset\mathbb{R}^n\to\mathbb{R}^m$ be differentiable at $\mathbf{x}_0$. Then f is continuous at $\mathbf{x}_0$, and furthermore, $\|f(\mathbf{x})-f(\mathbf{x}_0)\|<M_1\|\mathbf{x}-\mathbf{x}_0\|$ for some constant M_1 and $\mathbf{x}$ near $\mathbf{x}_0$, $\mathbf{x}\ne\mathbf{x}_0$.*

PROOF   We shall use the result of Exercise 2 at the end of this section, namely, $\|\mathbf{D}f(\mathbf{x}_0)\cdot\mathbf{h}\|\le M\|\mathbf{h}\|$ where M is the square root of the sum of the squares of the matrix elements in $\mathbf{D}f(\mathbf{x}_0)$.

Choose $\varepsilon=1$. Then by the definition of the derivative (see formula (2)) there is a $\delta_1>0$ such that $0<\|\mathbf{h}\|<\delta_1$ implies

$$\|f(\mathbf{x}_0+\mathbf{h})-f(\mathbf{x}_0)-\mathbf{D}f(\mathbf{x}_0)\cdot\mathbf{h}\|<\varepsilon\|\mathbf{h}\|=\|\mathbf{h}\|.$$

If $\|\mathbf{h}\|<\delta_1$, then using the triangle inequality,

$$\begin{aligned}
\|f(\mathbf{x}_0+\mathbf{h})-f(\mathbf{x}_0)\| &=\|f(\mathbf{x}_0+\mathbf{h})-f(\mathbf{x}_0)-\mathbf{D}f(\mathbf{x}_0)\cdot\mathbf{h}+\mathbf{D}f(\mathbf{x}_0)\cdot\mathbf{h}\|\\
&\le\|f(\mathbf{x}_0+\mathbf{h})-f(\mathbf{x}_0)-\mathbf{D}f(\mathbf{x}_0)\cdot\mathbf{h}\|+\|\mathbf{D}f(\mathbf{x}_0)\cdot\mathbf{h}\|\\
&<\|\mathbf{h}\|+M\|\mathbf{h}\|=(1+M)\|\mathbf{h}\|.
\end{aligned}$$

Setting $\mathbf{x}=\mathbf{x}_0+\mathbf{h}$ and $M_1=1+M$ proves the second assertion of the theorem.

Now let ε' be any positive number, and let δ be the smaller of δ_1 and $\varepsilon'/(1+M)$. Then $\|\mathbf{h}\|<\delta$ implies

$$\|f(\mathbf{x}_0+\mathbf{h})-f(\mathbf{x}_0)\|<(1+M)\frac{\varepsilon'}{1+M}=\varepsilon',$$

which proves that (see Exercise 15)

$$\lim_{\mathbf{x}\to\mathbf{x}_0}f(\mathbf{x})=\underset{\mathbf{h}\to\mathbf{0}}{\text{limit}}\,f(\mathbf{x}_0+\mathbf{h})=f(\mathbf{x}_0)$$

or that f is continuous at $\mathbf{x}_0$. ∎

We asserted in Section 2.3 that an important criterion for differentiability is that the partial derivatives exist and are continuous. We now prove this.

THEOREM 9 *Let $f \colon U \subset \mathbb{R}^n \to \mathbb{R}^m$. Suppose the partial derivatives $\partial f_i / \partial x_j$ of f all exist and are continuous in a neighborhood of a point $\mathbf{x} \in U$. Then f is differentiable at $\mathbf{x}$.*

PROOF (In this proof we are going to use the mean value theorem from one-variable calculus. See Section 2.5 for the statement.)

We shall only consider the case $m = 1$, that is, $f \colon U \subset \mathbb{R}^n \to \mathbb{R}$, leaving the general case to the reader (see the proof of Theorem 15, above). We have to show that

$$\lim_{\mathbf{h} \to 0} \frac{\left| f(\mathbf{x} + \mathbf{h}) - f(\mathbf{x}) - \sum_{i=1}^{n} \left[\dfrac{\partial f}{\partial x_i}(\mathbf{x}) \right] h_i \right|}{\|\mathbf{h}\|} = 0.$$

Write

$$
\begin{aligned}
f(x_1 + h_1, &\ldots, x_n + h_n) - f(x_1, \ldots, x_n) \\
&= f(x_1 + h_1, \ldots, x_n + h_n) - f(x_1, x_2 + h_2, \ldots, x_n + h_n) \\
&\quad + f(x_1, x_2 + h_2, \ldots, x_n + h_n) - f(x_1, x_2, x_3 + h_3, \ldots, x_n + h) + \cdots \\
&\quad + f(x_1, \ldots, x_{n-1} + h_{n-1}, x_n + h_n) - f(x_1, \ldots, x_{n-1}, x_n + h_n) \\
&\quad + f(x_1, \ldots, x_{n-1}, x_n + h_n) - f(x_1, \ldots, x_n).
\end{aligned}
$$

(This is called a ***telescoping sum***, since each term cancels with the succeeding or preceding one, except the first and last.) By the mean value theorem, this expression may be written as

$$f(\mathbf{x} + \mathbf{h}) - f(\mathbf{x}) = \left[\frac{\partial f}{\partial x_1}(\mathbf{y}_1) \right] h_1 + \left[\frac{\partial f}{\partial x_2}(\mathbf{y}_2) \right] h_2 + \cdots + \left[\frac{\partial f}{\partial x_n}(\mathbf{y}_n) \right] h_n,$$

where $\mathbf{y}_1 = (c_1, x_2 + h_2, \ldots, x_n + h_n)$ with c_1 lying between x_1 and $x_1 + h_1$; $\mathbf{y}_2 = (x_1, c_2, x_3 + h_3, \ldots, x_n + h_n)$ with c_2 lying between x_2 and $x_2 + h_2$; and $\mathbf{y}_n = (x_1, \ldots, x_{n-1}, c_n)$ where c_n lies between x_n and $x_n + h_n$. Thus we can write

$$
\begin{aligned}
&\left| f(\mathbf{x} + \mathbf{h}) - f(\mathbf{x}) - \sum_{i=1}^{n} \left[\frac{\partial f}{\partial x_i}(\mathbf{x}) \right] h_i \right| \\
&\quad = \left| \left(\frac{\partial f}{\partial x_1}(\mathbf{y}_1) - \frac{\partial f}{\partial x_1}(\mathbf{x}) \right) h_1 + \cdots + \left(\frac{\partial f}{\partial x_n}(\mathbf{y}_n) - \frac{\partial f}{\partial x_n}(\mathbf{x}) \right) h_n \right|.
\end{aligned}
$$

By the triangle inequality, this expression is less than or equal to

$$
\begin{aligned}
&\left| \frac{\partial f}{\partial x_1}(\mathbf{y}_1) - \frac{\partial f}{\partial x_1}(\mathbf{x}) \right| |h_1| + \cdots + \left| \frac{\partial f}{\partial x_n}(\mathbf{y}_n) - \frac{\partial f}{\partial x_n}(\mathbf{x}) \right| |h_n| \\
&\quad \leq \left\{ \left| \frac{\partial f}{\partial x_1}(\mathbf{y}_1) - \frac{\partial f}{\partial x_1}(\mathbf{x}) \right| + \cdots + \left| \frac{\partial f}{\partial x_n}(\mathbf{y}_n) - \frac{\partial f}{\partial x_n}(\mathbf{x}) \right| \right\} \|\mathbf{h}\|,
\end{aligned}
$$

since $|h_i| \leq \|\mathbf{h}\|$. Thus we have proved that

$$\frac{\left| f(\mathbf{x} + \mathbf{h}) - f(\mathbf{x}) - \sum_{i=1}^{n} \left[\dfrac{\partial f}{\partial x_i}(\mathbf{x}) \right] h_i \right|}{\|\mathbf{h}\|} \leq \left| \frac{\partial f}{\partial x_1}(\mathbf{y}_1) - \frac{\partial f}{\partial x_1}(\mathbf{x}) \right| + \cdots + \left| \frac{\partial f}{\partial x_n}(\mathbf{y}_n) - \frac{\partial f}{\partial x_n}(\mathbf{x}) \right|.$$

But since the partial derivatives are continuous by assumption, the right side approaches 0 as $\mathbf{h} \to \mathbf{0}$, so that the left side approaches 0 as well. ∎

THEOREM 11: CHAIN RULE *Let $U \subset \mathbb{R}^n$ and $V \subset \mathbb{R}^m$ be open. Let $g: U \subset \mathbb{R}^n \to \mathbb{R}^m$ and $f: V \subset \mathbb{R}^m \to \mathbb{R}^p$ be given functions such that g maps U into V, so that $f \circ g$ is defined. Suppose g is differentiable at $\mathbf{x}_0$ and f is differentiable at $\mathbf{y}_0 = g(\mathbf{x}_0)$. Then $f \circ g$ is differentiable at $\mathbf{x}_0$ and*

$$\mathbf{D}(f \circ g)(\mathbf{x}_0) = \mathbf{D}f(\mathbf{y}_0)\mathbf{D}g(\mathbf{x}_0).$$

PROOF   According to the definition of the derivative, we must verify that

$$\underset{\mathbf{x} \to \mathbf{x}_0}{\text{limit}} \frac{\| f(g(\mathbf{x})) - f(g(\mathbf{x}_0)) - \mathbf{D}f(\mathbf{y}_0)\mathbf{D}g(\mathbf{x}_0) \cdot (\mathbf{x} - \mathbf{x}_0) \|}{\| \mathbf{x} - \mathbf{x}_0 \|} = 0.$$

First rewrite the numerator and apply the triangle inequality as follows:

$$\| f(g(\mathbf{x})) - f(g(\mathbf{x}_0)) - \mathbf{D}f(\mathbf{y}_0) \cdot (g(\mathbf{x}) - g(\mathbf{x}_0)) + \mathbf{D}f(\mathbf{y}_0) \cdot [g(\mathbf{x}) - g(\mathbf{x}_0) - \mathbf{D}g(\mathbf{x}_0) \cdot (\mathbf{x} - \mathbf{x}_0)] \|$$

$$\leq \| f(g(\mathbf{x})) - f(g(\mathbf{x}_0)) - \mathbf{D}f(\mathbf{y}_0) \cdot (g(\mathbf{x}) - g(\mathbf{x}_0)) \|$$

$$+ \| \mathbf{D}f(\mathbf{y}_0) \cdot [g(\mathbf{x}) - g(\mathbf{x}_0) - \mathbf{D}g(\mathbf{x}_0) \cdot (\mathbf{x} - \mathbf{x}_0)] \|. \tag{3}$$

As in the proof of Theorem 8, $\| \mathbf{D}f(\mathbf{y}_0) \cdot \mathbf{h} \| \leq M \| \mathbf{h} \|$ for some constant M. Thus the right-hand side of inequality (3) is less than or equal to

$$\| f(g(\mathbf{x})) - f(g(\mathbf{x}_0)) - \mathbf{D}f(\mathbf{y}_0) \cdot (g(\mathbf{x}) - g(\mathbf{x}_0)) \| + M \| g(\mathbf{x}) - g(\mathbf{x}_0) - \mathbf{D}g(\mathbf{x}_0) \cdot (\mathbf{x} - \mathbf{x}_0) \|. \tag{4}$$

Since g is differentiable at $\mathbf{x}_0$, given $\varepsilon > 0$, there is a $\delta_1 > 0$ such that $0 < \| \mathbf{x} - \mathbf{x}_0 \| < \delta_1$ implies

$$\frac{\| g(\mathbf{x}) - g(\mathbf{x}_0) - \mathbf{D}g(\mathbf{x}_0) \cdot (\mathbf{x} - \mathbf{x}_0) \|}{\| \mathbf{x} - \mathbf{x}_0 \|} < \frac{\varepsilon}{2M}.$$

This makes the second term in expression (4) less than $\varepsilon \| \mathbf{x} - \mathbf{x}_0 \| / 2$.

Let us turn to the first term in expression (4). By Theorem 8, $\| g(\mathbf{x}) - g(\mathbf{x}_0) \| < M_1 \| \mathbf{x} - \mathbf{x}_0 \|$ for a constant M_1 if $\mathbf{x}$ is near $\mathbf{x}_0$, say $0 < \| \mathbf{x} - \mathbf{x}_0 \| < \delta_2$. Now choose δ_3 such that $0 < \| \mathbf{y} - \mathbf{y}_0 \| < \delta_3$ implies

$$\| f(\mathbf{y}) - f(\mathbf{y}_0) - \mathbf{D}f(\mathbf{y}_0) \cdot (\mathbf{y} - \mathbf{y}_0) \| < \frac{\varepsilon \| \mathbf{y} - \mathbf{y}_0 \|}{2M_1}.$$

Since $\mathbf{y} = g(\mathbf{x})$ and $\mathbf{y}_0 = g(\mathbf{x}_0)$, $\| \mathbf{y} - \mathbf{y}_0 \| < \delta_3$ if $\| \mathbf{x} - \mathbf{x}_0 \| < \delta_3 / M_1$ and $\| \mathbf{x} - \mathbf{x}_0 \| < \delta_2$, and so

$$\| f(g(\mathbf{x})) - f(g(\mathbf{x}_0)) - \mathbf{D}f(\mathbf{y}_0) \cdot (g(\mathbf{x}) - g(\mathbf{x}_0)) \| \leq \frac{\varepsilon \| g(\mathbf{x}) - g(\mathbf{x}_0) \|}{2M_1} < \frac{\varepsilon \| \mathbf{x} - \mathbf{x}_0 \|}{2}.$$

Thus if $\delta = \min\,(\delta_1, \delta_2, \delta_3/M_1)$, expression (4) is less than

$$\frac{\varepsilon\,\|\mathbf{x} - \mathbf{x}_0\|}{2} + \frac{\varepsilon\,\|\mathbf{x} - \mathbf{x}_0\|}{2} = \varepsilon\,\|\mathbf{x} - \mathbf{x}_0\|,$$

and so

$$\frac{\|f(g(\mathbf{x})) - f(g(\mathbf{x}_0)) - \mathbf{D}f(\mathbf{y}_0)\,\mathbf{D}g(\mathbf{x}_0)(\mathbf{x} - \mathbf{x}_0)\|}{\|\mathbf{x} - \mathbf{x}_0\|} < \varepsilon$$

for $0 < \|\mathbf{x} - \mathbf{x}_0\| < \delta$. This proves the theorem. ∎

The student has already met with a number of examples illustrating the above theorems. Let us consider one more of a more technical nature.

EXAMPLE 1   Let

$$f(x, y) = \begin{cases} \dfrac{xy}{\sqrt{x^2 + y^2}} & (x, y) \neq (0, 0) \\ 0 & (x, y) = (0, 0). \end{cases}$$

Is f differentiable at $(0, 0)$? (See Figure 2.7.1.)

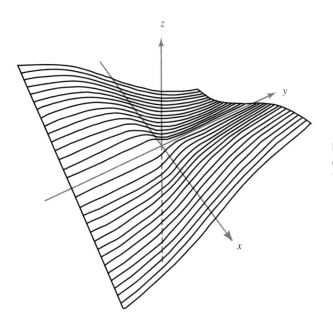

Figure 2.7.1   This function is not differentiable at $(0, 0)$, because it is "crinkled."

SOLUTION   We note that

$$\frac{\partial f}{\partial x}(0, 0) = \underset{x \to 0}{\text{limit}}\ \frac{f(x, 0) - f(0, 0)}{x} = \underset{x \to 0}{\text{limit}}\ \frac{(x \cdot 0)/\sqrt{x^2 + 0} - 0}{x} = \underset{x \to 0}{\text{limit}}\ \frac{0 - 0}{x} = 0$$

and similarly, $(\partial f/\partial y)(0, 0) = 0$. Thus the partial derivatives exist at $(0, 0)$. Also, if $(x, y) \neq (0, 0)$, then

$$\frac{\partial f}{\partial x} = \frac{y\sqrt{x^2 + y^2} - 2x(xy)/2\sqrt{x^2 + y^2}}{x^2 + y^2} = \frac{y}{\sqrt{x^2 + y^2}} - \frac{x^2 y}{(x^2 + y^2)^{3/2}},$$

which does not have a limit as $(x, y) \to (0, 0)$. Different limits are obtained for different paths of approach, as can be seen by letting $x = My$. Thus the partial derivatives are *not continuous* at $(0, 0)$, and so we cannot use Theorem 9.

We might now try to show that f is not differentiable (f is continuous, however). If $\mathbf{D}f(0, 0)$ existed, by Theorem 15 it would have to be the zero matrix, since $\partial f / \partial x$ and $\partial f / \partial y$ are zero at $(0, 0)$. Thus, by the definition of differentiability, for any $\varepsilon > 0$ there would be a $\delta > 0$ such that $0 < \|(h_1, h_2)\| < \delta$ implies

$$\frac{|f(h_1, h_2) - f(0, 0)|}{\|(h_1, h_2)\|} < \varepsilon$$

that is, $|f(h_1, h_2)| < \varepsilon \|(h_1, h_2)\|$, or $|h_1 h_2| < \varepsilon(h_1^2 + h_2^2)$. But if we choose $h_1 = h_2$, this reads $\frac{1}{2} < \varepsilon$, which is untrue if we choose $\varepsilon \leq \frac{1}{2}$. Hence, f is *not differentiable* at $(0, 0)$.   ▲

EXERCISES

1. Let $f(x, y, z) = (e^x, \cos y, \sin z)$. Compute $\mathbf{D}f$. In general, when will $\mathbf{D}f$ be a diagonal matrix?

2. (a) Let $\mathbf{A}: \mathbb{R}^n \to \mathbb{R}^m$ be a linear transformation with matrix $[a_{ij}]$ so that $\mathbf{A}\mathbf{x}$ has components $y_i = \sum a_{ij} x_j$. Let $M = (\sum a_{ij}^2)^{1/2}$. Use the Cauchy-Schwarz inequality to prove that $\|\mathbf{A}\mathbf{x}\| \leq M \|\mathbf{x}\|$.

(b) Use the inequality derived in part (a) to show that a linear transformation $\mathbf{T}: \mathbb{R}^n \to \mathbb{R}^m$ with matrix $[t_{ij}]$ is continuous.

(c) Let $\mathbf{A}: \mathbb{R}^n \to \mathbb{R}^m$ be a linear transformation. If

$$\lim_{\mathbf{x} \to \mathbf{0}} \frac{\mathbf{A}\mathbf{x}}{\|\mathbf{x}\|} = \mathbf{0},$$

show that $\mathbf{A} = \mathbf{0}$.

3. Let $f: A \to B$ and $g: B \to C$ be maps between open subsets of Euclidean space, and let $\mathbf{x}_0$ be in A or be a boundary point of A and $\mathbf{y}_0$ be in B or be a boundary point of B.

(a) If $\lim_{\mathbf{x} \to \mathbf{x}_0} f(\mathbf{x}) = \mathbf{y}_0$ and $\lim_{\mathbf{y} \to \mathbf{y}_0} g(\mathbf{y}) = \mathbf{w}$, show that $\lim_{\mathbf{x} \to \mathbf{x}_0} g(f(\mathbf{x}))$ *need not equal* $\mathbf{w}$.

(b) If $\mathbf{y}_0 \in B$, and $\mathbf{w} = g(\mathbf{y}_0)$, show that $\lim_{\mathbf{x} \to \mathbf{x}_0} g(f(\mathbf{x})) = \mathbf{w}$.

4. A function $f: A \subset \mathbb{R}^n \to \mathbb{R}^m$ is said to be *uniformly continuous* if for every $\varepsilon > 0$ there is a $\delta > 0$ such that for all $\mathbf{p}$ and $\mathbf{q} \in A$, $\|\mathbf{p} - \mathbf{q}\| < \delta$ implies $\|f(\mathbf{p}) - f(\mathbf{q})\| < \varepsilon$. (Note that a uniformly continuous function is continuous; try to describe explicitly the extra property that a uniformly continuous function has.)

(a) Prove that a linear map $\mathbf{T}: \mathbb{R}^n \to \mathbb{R}^m$ is uniformly continuous. [HINT: Use Exercise 2.]

(b) Prove that $x \mapsto 1/x^2$ on $(0, 1]$ is continuous, but not uniformly continuous.

5. Let $\mathbf{A} = (a_{ij})$ be a *symmetric* $n \times n$ matrix (that is, $a_{ij} = a_{ji}$) and define $f(\mathbf{x}) = \mathbf{x} \cdot \mathbf{A}\mathbf{x}$, so $f: \mathbb{R}^n \to \mathbb{R}$. Show that $\nabla f(\mathbf{x})$ is the vector $2\mathbf{A}\mathbf{x}$.

6. The following function is graphed in Figure 2.7.2:

$$f(x, y) = \begin{cases} \dfrac{2xy^2}{x^2 + y^4} & (x, y) \neq (0, 0) \\[2mm] 0 & (x, y) = (0, 0). \end{cases}$$

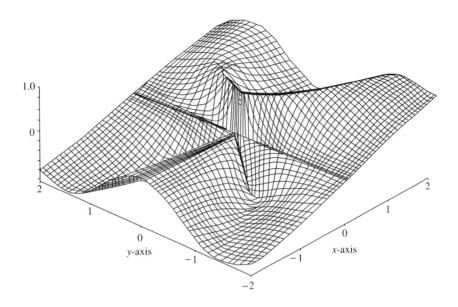

Figure 2.7.2   Graph of $z = 2xy^2/(x^2 + y^4)$.

Show that $\partial f/\partial x$ and $\partial f/\partial y$ exist everywhere; in fact all directional derivatives exist. But show that f is not continuous at $(0, 0)$. Is f differentiable?

7.   Let $f(x, y) = g(x) + h(y)$, and suppose g is differentiable at x_0 and h is differentiable at y_0. Prove from the definition that f is differentiable at (x_0, y_0).

8.   Use the Cauchy-Schwarz inequality to prove that for any vector $\mathbf{v} \in \mathbb{R}^n$,

$$\lim_{\mathbf{x} \to \mathbf{x}_0} \mathbf{v} \cdot \mathbf{x} = \mathbf{v} \cdot \mathbf{x}_0.$$

9.   Prove that if $\displaystyle\lim_{\mathbf{x} \to \mathbf{x}_0} f(\mathbf{x}) = b$ for $f: A \subset \mathbb{R}^n \to \mathbb{R}$, then

$$\lim_{\mathbf{x} \to \mathbf{x}_0} [f(\mathbf{x})]^2 = b^2 \quad \text{and} \quad \lim_{\mathbf{x} \to \mathbf{x}_0} \sqrt{|f(\mathbf{x})|} = \sqrt{|b|}.$$

[You may use Exercise 3(b).]

10.   Show that in Theorem 9 with $m = 1$ it is enough to assume that $n - 1$ partial derivatives are continuous and merely that the other one exists. Does this agree with what you expect when $n = 1$?

11.   Define $f: \mathbb{R}^2 \to \mathbb{R}$ by

$$f(x, y) = \begin{cases} \dfrac{xy}{(x^2 + y^2)^{1/2}} & (x, y) \neq (0, 0) \\ 0 & (x, y) = (0, 0). \end{cases}$$

Show that f is continuous.

12. (a) Does $\displaystyle\lim_{(x,\,y)\to(0,\,0)} \frac{x}{x^2+y^2}$ exist?

   (b) Does $\displaystyle\lim_{(x,\,y)\to(0,\,0)} \frac{x^3}{x^2+y^2}$ exist?

13. Find $\displaystyle\lim_{(x,\,y)\to(0,\,0)} \frac{xy^2}{\sqrt{x^2+y^2}}$.

14. Prove that $s\colon \mathbb{R}^2 \to \mathbb{R}$, $(x, y) \mapsto x + y$ is continuous.

15. Using the definition of continuity, prove that f is continuous at $\mathbf{x}$ if and only if

$$\lim_{\mathbf{h}\to\mathbf{0}} f(\mathbf{x}+\mathbf{h}) = f(\mathbf{x}).$$

16. (a) A sequence $\mathbf{x}_n$ of points in $\mathbb{R}^m$ is said to **converge to** $\mathbf{x}$, written $\mathbf{x}_n \to \mathbf{x}$ as $n \to \infty$, if for any $\varepsilon > 0$ there is an N such that $n \geq N$ implies $\|\mathbf{x} - \mathbf{x}_n\| < \varepsilon$. Show that $\mathbf{y}$ is a boundary point of an open set A if and only if $\mathbf{y}$ is not in A and there is a sequence of distinct points of A converging to $\mathbf{y}$.

   (b) Let $f\colon A \subset \mathbb{R}^n \to \mathbb{R}^m$ and $\mathbf{y}$ be in A or be a boundary point of A. Prove that $\displaystyle\lim_{\mathbf{x}\to\mathbf{y}} f(\mathbf{x}) = \mathbf{b}$ if and only if $f(\mathbf{x}_n) \to \mathbf{b}$ for every sequence $\mathbf{x}_n$ of points in A with $\mathbf{x}_n \to \mathbf{y}$.

   (c) If $U \subset \mathbb{R}^m$ is open, show that $f\colon U \to \mathbb{R}^p$ is continuous if and only if $\mathbf{x}_n \to \mathbf{x} \in U$ implies $f(\mathbf{x}_n) \to f(\mathbf{x})$.

17. If $f(\mathbf{x}) = g(\mathbf{x})$ for all $\mathbf{x} \neq \mathbf{a}$ and if $\displaystyle\lim_{\mathbf{x}\to\mathbf{a}} f(\mathbf{x}) = \mathbf{b}$, then show that $\displaystyle\lim_{\mathbf{x}\to\mathbf{a}} g(\mathbf{x}) = \mathbf{b}$ as well.

18. Let $A \subset \mathbb{R}^n$ and let $\mathbf{x}_0$ be a boundary point of A. Let $f\colon A \to \mathbb{R}$ and $g\colon A \to \mathbb{R}$ be functions defined on A such that $\displaystyle\lim_{\mathbf{x}\to\mathbf{x}_0} f(\mathbf{x})$ and $\displaystyle\lim_{\mathbf{x}\to\mathbf{x}_0} g(\mathbf{x})$ exist, and assume that for all $\mathbf{x}$ in some deleted neighborhood of $\mathbf{x}_0$, $f(\mathbf{x}) \leq g(\mathbf{x})$. (A deleted neighborhood of $\mathbf{x}_0$ is any neighborhood of $\mathbf{x}_0$, less $\mathbf{x}_0$ itself.)

   (a) Prove that $\displaystyle\lim_{\mathbf{x}\to\mathbf{x}_0} f(\mathbf{x}) \leq \lim_{\mathbf{x}\to\mathbf{x}_0} g(\mathbf{x})$. [HINT: Consider the function $\phi(\mathbf{x}) = g(\mathbf{x}) - f(\mathbf{x})$; prove that $\displaystyle\lim_{\mathbf{x}\to\mathbf{x}_0} \phi(\mathbf{x}) \geq 0$, and then use the fact that the limit of the sum of two functions is the sum of their limits.]

   (b) If $f(\mathbf{x}) < g(\mathbf{x})$, do we necessarily have strict inequality of the limits?

19. Given $f\colon A \subset \mathbb{R}^n \to \mathbb{R}^m$, we say that "$f$ is $o(\mathbf{x})$ as $\mathbf{x} \to \mathbf{0}$" if $\displaystyle\lim_{\mathbf{x}\to\mathbf{0}} f(\mathbf{x})/\|\mathbf{x}\| = \mathbf{0}$.

   (a) If f_1 and f_2 are $o(\mathbf{x})$ as $\mathbf{x} \to \mathbf{0}$, prove that $f_1 + f_2$ is also $o(\mathbf{x})$ as $\mathbf{x} \to \mathbf{0}$.

   (b) Let $g\colon A \to \mathbb{R}$ be a function with the property that there is a number $c > 0$ such that $|g(\mathbf{x})| \leq c$ for all $\mathbf{x}$ in A (the function g is said to be **bounded**). If f is $o(\mathbf{x})$ as $\mathbf{x} \to \mathbf{0}$, prove that gf is also $o(\mathbf{x})$ as $\mathbf{x} \to \mathbf{0}$ [where $(gf)(\mathbf{x}) = g(\mathbf{x})f(\mathbf{x})$].

   (c) Show that $f(x) = x^2$ is $o(x)$ as $x \to 0$. Is $g(x) = x$ also $o(x)$ as $x \to 0$?

REVIEW EXERCISES FOR CHAPTER 2

1. Describe the graphs of:

   (a) $f(x, y) = 3x^2 + y^2$
   (b) $f(x, y) = xy + 3x$

2. Describe some appropriate level surfaces and sections of the graphs of:

    (a) $f(x, y, z) = 2x^2 + y^2 + z^2$

    (b) $f(x, y, z) = x^2$

    (c) $f(x, y, z) = xyz$

3. Compute the derivative $\mathbf{D}f(\mathbf{x})$ of each of the following functions:

    (a) $f(x, y) = (x^2 y, e^{-xy})$

    (b) $f(x) = (x, x)$

    (c) $f(x, y, z) = e^x + e^y + e^z$

    (d) $f(x, y, z) = (x, y, z)$

4. Suppose $f(x, y) = f(y, x)$ for all (x, y). Prove that

$$(\partial f/\partial x)(a, b) = (\partial f/\partial y)(b, a).$$

5. Let $f(x, y) = (1 - x^2 - y^2)^{1/2}$. Show that the plane tangent to the graph of f at $(x_0, y_0, f(x_0, y_0))$ is orthogonal to the vector $(x_0, y_0, f(x_0, y_0))$. Interpret geometrically.

6. Let $F(u, v)$ and $u = h(x, y, z)$, $v = k(x, y, z)$ be given (differentiable) real valued functions and let $f(x, y, z)$ be defined by $f(x, y, z) = F(h(x, y, z), k(x, y, z))$. Write a formula for the gradient of f in terms of the partial derivatives of F, h, and k.

7. Find an equation for the tangent plane of the graph of f at the point $(x_0, y_0, f(x_0, y_0))$ for:

    (a) $f: \mathbb{R}^2 \to \mathbb{R}, (x, y) \mapsto x - y + 2,$ $(x_0, y_0) = (1, 1)$

    (b) $f: \mathbb{R}^2 \to \mathbb{R}, (x, y) \mapsto x^2 + 4y^2,$ $(x_0, y_0) = (2, -1)$

    (c) $f: \mathbb{R}^2 \to \mathbb{R}, (x, y) \mapsto xy,$ $(x_0, y_0) = (-1, -1)$

    (d) $f(x, y) = \log(x + y) + x \cos y + \arctan(x + y),$ $(x_0, y_0) = (1, 0)$

    (e) $f(x, y) = \sqrt{x^2 + y^2},$ $(x_0, y_0) = (1, 1)$

    (f) $f(x, y) = xy,$ $(x_0, y_0) = (2, 1)$

8. Compute an equation for the tangent planes of the following surfaces at the indicated points.

    (a) $x^2 + y^2 + z^2 = 3,$ $(1, 1, 1)$

    (b) $x^3 - 2y^3 + z^3 = 0,$ $(1, 1, 1)$

    (c) $(\cos x)(\cos y)e^z = 0,$ $(\pi/2, 1, 0)$

    (d) $e^{xyz} = 1,$ $(1, 1, 0)$

9. Draw some level curves for the following functions:

    (a) $f(x, y) = 1/xy$

    (b) $f(x, y) = x^2 - xy - y^2$

10. Consider a temperature function $T(x, y) = x \sin y$. Plot a few level curves. Compute ∇T and explain its meaning.

11. Find the following limits if they exist:

    (a) $\displaystyle \lim_{(x, y) \to (0, 0)} \frac{\cos xy - 1}{x}$ (b) $\displaystyle \lim_{(x, y) \to (0, 0)} \sqrt{|(x + y)/(x - y)|}, x \neq y$

12. Compute the first partial derivatives and gradients of the following functions:

    (a) $f(x, y, z) = xe^z + y \cos x$

    (b) $f(x, y, z) = (x + y + z)^{10}$

    (c) $f(x, y, z) = (x^2 + y)/z$

13. Compute $\frac{\partial}{\partial x}[x \exp(1 + x^2 + y^2)]$

14. Let $y(x)$ be a differentiable function defined implicitly by $F(x, y(x)) = 0$. From Exercise 17(a), Section 2.5, we know that

$$\frac{dy}{dx} = -\frac{\partial F/\partial x}{\partial F/\partial y}.$$

Consider the surface $z = F(x, y)$, and suppose F is increasing as a function of x and as a function of y; that is, $\partial F/\partial x > 0$ and $\partial F/\partial y > 0$. By considering the graph and the plane $z = 0$, show that for z fixed at $z = 0$, y should *decrease* as x increases and x should *decrease* as y increases. Does this agree with the minus sign in the formula for dy/dx?

15. (a) Consider the graph of a function $f(x, y)$ [Figure 2.R.1(a)]. Let (x_0, y_0) lie on a level curve C, so $\nabla f(x_0, y_0)$ is perpendicular to this curve. Show that the tangent plane of the graph is the plane that (i) contains the line perpendicular to $\nabla f(x_0, y_0)$ and lying in the horizontal plane $z = f(x_0, y_0)$, and (ii) has slope $\|\nabla f(x_0, y_0)\|$ relative to the xy plane. (By the *slope* of a plane P relative to the xy plane we mean the tangent of the angle θ, $0 \leq \theta \leq \pi$, between the upward-pointing normal $\mathbf{p}$ to P and the unit vector $\mathbf{k}$.)

(b) Use this method to show that the tangent plane of the graph of $f(x, y) = (x + \cos y)x^2$ at $(1, 0, 2)$ is as sketched in Figure 2.R.1(b).

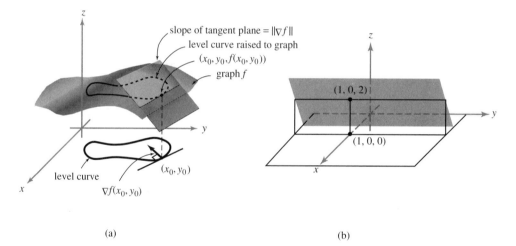

(a)   (b)

Figure 2.R.1   (a) Relationship between the gradient of a function and the tangent plane to the graph [Exercise 15(a)]. A specific instance of the plane in (b) for Exercise 15(b).

16. Find the plane tangent to the surface $z = x^2 + y^2$ at the point $(1, -2, 5)$. Explain the geometric significance, for this surface, of the gradient of $f(x, y) = x^2 + y^2$ (see Exercise 15).

17. In which direction is the directional derivative of $f(x, y) = (x^2 - y^2)/(x^2 + y^2)$ at $(1, 1)$ equal to zero?

18. Find the directional derivative of the given function at the given point and in the direction of the given vector.

(a) $f(x, y, z) = e^x \cos(yz)$, $p_0 = (0, 0, 0)$, $\mathbf{v} = (2, 1, -2)$
(b) $f(x, y, z) = xy + yz + zx$, $p_0 = (1, 1, 2)$, $\mathbf{v} = (10, -1, 2)$

19. Find the tangent plane and normal to the hyperboloid $x^2 + y^2 - z^2 = 18$ at $(3, 5, -4)$.

20. Let $(x(t), y(t))$ be a path in the plane, $0 \le t \le 1$, and let $f(x, y)$ be a C^1 function of two variables. Assume that $(dx/dt)f_x + (dy/dt)f_y \le 0$. Show that $f(x(1), y(1)) \le f(x(0), y(0))$.

21. A bug finds itself in a toxic environment. The toxicity level is given by $T(x, y) = 2x^2 - 4y^2$. The bug is at $(-1, 2)$. In what direction should it move to lower the toxicity the fastest?

22. Find the direction in which the function $w = x^2 + xy$ increases most rapidly at the point $(-1, 1)$. What is the magnitude of ∇w in this direction? Interpret this magnitude geometrically.

23. Let f be defined on an open set S in $\mathbb{R}^n$. We say that f is **homogeneous of degree** p over S if $f(\lambda \mathbf{x}) = \lambda^p f(\mathbf{x})$ for every real λ and for every $\mathbf{x}$ in S for which $\lambda \mathbf{x} \in S$.
 (a) If such a function is differentiable at $\mathbf{x}$, show that $\mathbf{x} \cdot \nabla f(\mathbf{x}) = pf(\mathbf{x})$. This is known as **Euler's theorem** for homogeneous functions. [HINT: For fixed $\mathbf{x}$, define $g(\lambda) = f(\lambda \mathbf{x})$ and compute $g'(1)$.]
 (b) Find p and check Euler's theorem for the function $f(x, y, z) = x - 2y - \sqrt{xz}$, $xz > 0$.

24. If $z = [f(x - y)]/y$ (where f is differentiable and $y \ne 0$), show that $z + y(\partial z/\partial x) + y(\partial z/\partial y) = 0$.

25. Given $z = f((x + y)/(x - y))$ for f a C^1 function, show that

$$x\frac{\partial z}{\partial x} + y\frac{\partial z}{\partial y} = 0.$$

26. Let f have partial derivatives $\partial f(\mathbf{x})/\partial x_i$, where $i = 1, 2, \ldots, n$, at each point $\mathbf{x}$ of an open set U in $\mathbb{R}^n$. If f has a local maximum or a local minimum at the point $\mathbf{x}_0$ in U, show that $\partial f(\mathbf{x}_0)/\partial x_i = 0$ for each i.

27. Consider the functions defined in $\mathbb{R}^2$ by the following formulas:
 (i) $f(x, y) = xy/(x^2 + y^2)$ if $(x, y) \ne (0, 0)$, $f(0, 0) = 0$
 (ii) $f(x, y) = x^2 y^2/(x^2 + y^4)$ if $(x, y) \ne (0, 0)$, $f(0, 0) = 0$

 (a) In each case show that the partial derivatives $\partial f(x, y)/\partial x$ and $\partial f(x, y)/\partial y$ exist for every (x, y) in $\mathbb{R}^2$, and evaluate these derivatives explicitly in terms of x and y.
 (b) Explain why the functions described in (i) and (ii) are or are not differentiable at $(0, 0)$.

28. Compute the gradient vector $\nabla f(x, y)$ at all points (x, y) in $\mathbb{R}^2$ for each of the following functions:
 (a) $f(x, y) = x^2 y^2 \log(x^2 + y^2)$ if $(x, y) \ne (0, 0)$, $f(0, 0) = 0$
 (b) $f(x, y) = xy \sin[1/(x^2 + y^2)]$ if $(x, y) \ne (0, 0)$, $f(0, 0) = 0$

29. Find the directional derivatives of the following functions at the point $(1, 1)$ in the direction $(\mathbf{i} + \mathbf{j})/\sqrt{2}$:
 (a) $f(x, y) = x \tan^{-1}(x/y)$
 (b) $f(x, y) = \cos(\sqrt{x^2 + y^2})$
 (c) $f(x, y) = \exp(-x^2 - y^2)$

30. (a) Let $\mathbf{u} = \mathbf{i} - 2\mathbf{j} + 2\mathbf{k}$ and $\mathbf{v} = 2\mathbf{i} + \mathbf{j} - 3\mathbf{k}$. Find: $\|\mathbf{u}\|$, $\mathbf{u} \cdot \mathbf{v}$, $\mathbf{u} \times \mathbf{v}$, and a vector in the same direction as $\mathbf{u}$ but of unit length.
 (b) Find the rate of change of $e^{xy} \sin(xyz)$ in the direction $\mathbf{u}$ at $(0, 1, 1)$.

31. Let $h(x, y) = 2e^{-x^2} + e^{-3y^2}$ denote the height on a mountain at position (x, y). In what direction from $(1, 0)$ should one begin walking in order to climb the fastest?

32. Compute an equation for the plane tangent to the graph of

$$f(x, y) = \frac{e^x}{x^2 + y^2}$$

at $x = 1, y = 2$.

33. (a) Give a careful statement of the general form of the chain rule.
(b) Let $f(x, y) = x^2 + y$ and $\mathbf{h}(u) = (\sin 3u, \cos 8u)$. Let $g(u) = f(\mathbf{h}(u))$. Compute dg/du at $u = 0$ both directly *and* by using the chain rule.

34. (a) Sketch the level curves of $f(x, y) = -x^2 - 9y^2$ for $c = 0, -1, -10$.
(b) On your sketch, draw in ∇f at $(1, 1)$. Discuss.

35. At time $t = 0$, a particle is ejected from the surface $x^2 + 2y^2 + 3z^2 = 6$ at the point $(1, 1, 1)$ in a direction normal to the surface at a speed of 10 units per second. At what time does it cross the sphere $x^2 + y^2 + z^2 = 103$?

36. At what point(s) on the surface in Exercise 35 is the normal vector parallel to the line $x = y = z$?

37. Compute $\partial z/\partial x$ and $\partial z/\partial y$ if

$$z = \frac{u^2 + v^2}{u^2 - v^2}, \qquad u = e^{x-y}, \qquad v = e^{xy}$$

(a) by substitution and direct calculation and (b) by the chain rule.

38. Compute the partial derivatives as in Exercise 37 if $z = uv, u = x + y$, and $v = x - y$.

39. What is wrong with the following argument? Suppose that $w = f(x, y)$ and $y = x^2$. By the chain rule,

$$\frac{\partial w}{\partial x} = \frac{\partial w}{\partial x}\frac{\partial x}{\partial x} + \frac{\partial w}{\partial y}\frac{\partial y}{\partial x} = \frac{\partial w}{\partial x} + 2x\frac{\partial w}{\partial y}.$$

Hence $0 = 2x(\partial w/\partial y)$, and so $\partial w/\partial y = 0$. Choose an explicit example to really see this is incorrect.

40. A boat is sailing northeast at 20 km/h. Assuming that the temperature drops at a rate of 0.2°C/km in the northerly direction and 0.3°C/km in the easterly direction, what is the time rate of change of temperature as observed on the boat?

41. Use the chain rule to find a formula for $(d/dt)\exp[f(t)g(t)]$.

42. Use the chain rule to find a formula for $(d/dt)(f(t)^{g(t)})$.

43. Verify the chain rule for the function $f(x, y, z) = [\ln(1 + x^2 + 2z^2)]/(1 + y^2)$ and the path $\mathbf{c}(t) = (t, 1 - t^2, \cos t)$.

44. Verify the chain rule for the function $f(x, y) = x^2/(2 + \cos y)$ and the path $x = e^t, y = e^{-t}$.

45.   Suppose that $u(x, t)$ satisfies the differential equation $u_t + uu_x = 0$ and that x, as a function $x = f(t)$ of t, satisfies $dx/dt = u(x, t)$. Prove that $u(f(t), t)$ is constant in t.

46.   The displacement at time t and horizontal position on a line x of a certain violin string is given by $u = \sin(x - 6t) + \sin(x + 6t)$. Calculate the velocity of the string at $x = 1$ when $t = \frac{1}{3}$.

47.   The *ideal gas law* $PV = nRT$ involves a constant R, the number n of moles of the gas, the volume V, the Kelvin temperature T, and the pressure P.
   (a)   Show that each of n, P, T, V is a function of the remaining variables, and determine explicitly the defining equations.
   (b)   Calculate $\partial V/\partial T$, $\partial T/\partial P$, $\partial P/\partial V$ and show that their product equals -1.

48.   The *potential temperature* θ is defined in terms of temperature T and pressure p by

$$\theta = T\left(\frac{1000}{p}\right)^{0.286}.$$

The temperature and pressure may be thought of as functions of position (x, y, z) in the atmosphere and also of time t.
   (a)   Find formulas for $\partial\theta/\partial x$, $\partial\theta/\partial y$, $\partial\theta/\partial z$, $\partial\theta/\partial t$ in terms of partial derivatives of T and p.
   (b)   The condition $\partial\theta/\partial z < 0$ is regarded as unstable atmosphere, for it leads to large vertical excursions of air parcels from a single upward or downward impetus. Meteorologists use the formula

$$\frac{\partial\theta}{\partial z} = \frac{\theta}{T}\left(\frac{\partial T}{\partial z} + \frac{g}{C_p}\right)$$

where $g = 32.2$ and C_p is a positive constant. How does the temperature change in the upward direction for an unstable atmosphere?

49.   The specific volume V, pressure P, and temperature T of a van der Waals gas are related by $P = RT/(V - \beta) - \alpha/V^2$, where α, β, and R are constants.
   (a)   Explain why any two of V, P, and T can be considered independent variables which determine the third variable.
   (b)   Find $\partial T/\partial P$, $\partial P/\partial V$, $\partial V/\partial T$. Identify which variables are constant, and interpret each partial derivative physically.
   (c)   Verify that $(\partial T/\partial P)(\partial P/\partial V)(\partial V/\partial T) = -1$ (not $+1$!).

50.   The height h of the Hawaiian volcano Mauna Loa is (roughly) described by the function $h(x, y) = 2.59 - 0.00024y^2 - 0.00065x^2$, where h is the height above sea level in miles and x and y measure east-west and north-south distances in miles from the top of the mountain. At $(x, y) = (-2, -4)$:
   (a)   How fast is the height increasing in the direction $(1, 1)$ (that is, northeastward)? Express your answer in miles of height per mile of horizontal distance traveled.
   (b)   In what direction is the steepest upward path?

51.   (a)   In what direction is the directional derivative of $f(x, y) = (x^2 - y^2)/(x^2 + y^2)$ at $(1, 1)$ equal to zero?
   (b)   How about at an arbitrary point (x_0, y_0) in the first quadrant?
   (c)   Describe the level curves of f. In particular, discuss them in terms of the result of (b).

52.   (a)   Show that the curve $x^2 - y^2 = c$, for *any* value of c, satisfies the differential equation $dy/dx = x/y$.
   (b)   Draw in a few of the curves $x^2 - y^2 = c$, say for $c = \pm 1$. At several points (x, y) along each of these curves, draw a short segment of slope x/y; check that these segments appear to be tangent to the curve. What happens when $y = 0$? What happens when $c = 0$?

53. Suppose that f is a differentiable function of one variable and that a function $u = g(x, y)$ is defined by

$$u = g(x, y) = xyf\left(\frac{x+y}{xy}\right).$$

Show that u satisfies a (partial) differential equation of the form

$$x^2\frac{\partial u}{\partial x} - y^2\frac{\partial u}{\partial y} = G(x, y)u$$

and find the function $G(x, y)$.

54. (a)   Let F be a function of one variable and f a function of two variables. Show that the gradient vector of $g(x, y) = F(f(x, y))$ is parallel to the gradient vector of $f(x, y)$.

(b)   Let $f(x, y)$ and $g(x, y)$ be functions such that $\nabla f = \lambda\nabla g$ for some function $\lambda(x, y)$. What is the relation between the level curves of f and g? Explain why there might be a function F such that $g(x, y) = F(f(x, y))$.

3 HIGHER-ORDER DERIVATIVES; MAXIMA AND MINIMA

> ... namely, because the shape of the whole universe is most perfect, and, in fact designed by the wisest creator, nothing in all of the world will occur in which no maximum or minimum rule is somehow shining forth.
>
> Leonhard Euler

In one-variable calculus, to test a function $f(x)$ for a local maximum or minimum we look for critical points x_0, that is, points x_0 for which $f'(x_0) = 0$, and at each such point we check the sign of the second derivative $f''(x_0)$. If $f''(x_0) < 0$, $f(x_0)$ is a local maximum of f; if $f''(x_0) > 0$, $f(x_0)$ is a local minimum of f; if $f''(x_0) = 0$, the test fails.

This chapter extends these methods to real-valued functions of several variables. We begin in Section 3.1 with a discussion of iterated and higher-order partial derivatives and in Section 3.2 we discuss the multivariable form of Taylor's theorem; this is used in Section 3.3 to derive tests for maxima, minima, and saddle points. As with functions of one variable, such methods help one to visualize the shape of a graph.

In Section 3.4 we study the problem of maximizing a real-valued function subject to supplementary conditions, also referred to as constraints. For example, we might wish to maximize $f(x, y, z)$ among those (x, y, z) constrained to lie on the unit sphere, $x^2 + y^2 + z^2 = 1$. Section 3.5 gives a technical theorem (the implicit function theorem) useful for studying constraints. It will also be useful later in our study of surfaces.

In Section 3.6 we describe a few applications to geometry, to economics, and to equilibrium points of physical systems and their stability.

3.1 ITERATED PARTIAL DERIVATIVES

The preceding chapter developed considerable information concerning the derivative of a map and investigated the geometry associated with the derivative of real-valued functions by making use of the gradient. In this section, we proceed to study higher-order derivatives, with the goal of proving the equality of the "mixed second partial derivatives" of a function. We begin by defining the necessary terms.

Let $f: \mathbb{R}^3 \to \mathbb{R}$ be of class C^1. Recall that this means that $\partial f/\partial x$, $\partial f/\partial y$, and $\partial f/\partial z$ exist and are continuous. If these derivatives, in turn, have continuous partial derivatives, we say that f is of *class* C^2, or is *twice continuously differentiable*. Likewise, if we say f is of class C^3, we mean f has continuous iterated partial derivatives of third order, and so on. A few examples of how second-order derivatives are written follow:

$$\frac{\partial^2 f}{\partial x^2} = \frac{\partial}{\partial x}\left(\frac{\partial f}{\partial x}\right), \qquad \frac{\partial^2 f}{\partial x\,\partial y} = \frac{\partial}{\partial x}\left(\frac{\partial f}{\partial y}\right), \qquad \frac{\partial^2 f}{\partial z\,\partial y} = \frac{\partial}{\partial z}\left(\frac{\partial f}{\partial y}\right), \qquad \text{etc.}$$

The process can, of course, be repeated for third-order derivatives, and so on. If f is a function of only x and y and $\partial f/\partial x$, $\partial f/\partial y$ are continuously differentiable, then by taking second partial derivatives we get the four functions

$$\frac{\partial^2 f}{\partial x^2}, \qquad \frac{\partial^2 f}{\partial y^2}, \qquad \frac{\partial^2 f}{\partial x\,\partial y}, \qquad \text{and} \qquad \frac{\partial^2 f}{\partial y\,\partial x}.$$

All of these are called *iterated partial derivatives*, while $\partial^2 f/\partial x\,\partial y$ and $\partial^2 f/\partial y\,\partial x$ are called *mixed partial derivatives*.

EXAMPLE 1   Find all second partial derivatives of $f(x, y) = xy + (x + 2y)^2$.

SOLUTION

$$\frac{\partial f}{\partial x} = y + 2(x + 2y), \qquad \frac{\partial f}{\partial y} = x + 4(x + 2y)$$

$$\frac{\partial^2 f}{\partial x^2} = 2, \qquad \frac{\partial^2 f}{\partial y^2} = 8$$

$$\frac{\partial^2 f}{\partial x\,\partial y} = 5, \qquad \frac{\partial^2 f}{\partial y\,\partial x} = 5. \quad \blacktriangle$$

EXAMPLE 2   Find all second partial derivatives of $f(x, y) = \sin x \sin^2 y$.

SOLUTION

$$\frac{\partial f}{\partial x} = \cos x \sin^2 y, \qquad \frac{\partial f}{\partial y} = 2 \sin x \sin y \cos y = \sin x \sin 2y;$$

$$\frac{\partial^2 f}{\partial x^2} = -\sin x \sin^2 y \qquad \frac{\partial^2 f}{\partial y^2} = 2\sin x \cos 2y;$$

$$\frac{\partial^2 f}{\partial x \, \partial y} = \cos x \sin 2y, \qquad \frac{\partial^2 f}{\partial y \, \partial x} = 2\cos x \sin y \cos y = \cos x \sin 2y. \quad \blacktriangle$$

EXAMPLE 3   Let $f(x, y, z) = e^{xy} + z \cos x$. Then

$$\frac{\partial f}{\partial x} = ye^{xy} - z\sin x, \qquad \frac{\partial f}{\partial y} = xe^{xy}, \qquad \frac{\partial f}{\partial z} = \cos x,$$

$$\frac{\partial^2 f}{\partial z \, \partial x} = -\sin x, \qquad \frac{\partial^2 f}{\partial x \, \partial z} = -\sin x, \qquad \text{etc.} \quad \blacktriangle$$

In all these examples note that the pairs of mixed partial derivatives such as $\partial^2 f/\partial x \, \partial y$ and $\partial^2 f/\partial y \, \partial x$, or $\partial^2 f/\partial z \, \partial x$ and $\partial^2 f/\partial x \, \partial z$, are equal. It is a basic and perhaps surprising fact that this is always the case for C^2 functions. We shall prove this in the next theorem for functions $f(x, y)$ of two variables, but the proof can be readily extended to functions of n variables.

> **THEOREM 1: EQUALITY OF MIXED PARTIALS**   *If $f(x, y)$ is of class C^2 (is twice continuously differentiable), then the mixed partial derivatives are equal; that is,*
>
> $$\frac{\partial^2 f}{\partial x \, \partial y} = \frac{\partial^2 f}{\partial y \, \partial x}.$$

PROOF   Consider the following expression (see Figure 3.1.1):

$$S(\Delta x, \Delta y) = f(x_0 + \Delta x, y_0 + \Delta y) - f(x_0 + \Delta x, y_0) - f(x_0, y_0 + \Delta y) + f(x_0, y_0).$$

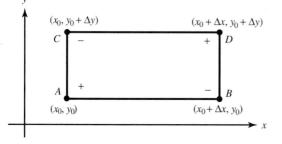

Figure 3.1.1   The algebra behind the equality of mixed partials, writing the difference of differences in two ways.

Holding y_0 and Δy fixed, define

$$g(x) = f(x, y_0 + \Delta y) - f(x, y_0),$$

so that $S(\Delta x, \Delta y) = g(x_0 + \Delta x) - g(x_0)$, *which expresses S as a difference of differences.*
By the mean value theorem for functions of one variable, $g(x_0 + \Delta x) - g(x_0)$ equals
$g'(\bar{x})\Delta x$ for some $\bar{x}$ between x_0 and $x_0 + \Delta x$. Hence,

$$S(\Delta x, \Delta y) = \left[\frac{\partial f}{\partial x}(\bar{x}, y_0 + \Delta y) - \frac{\partial f}{\partial x}(\bar{x}, y_0) \right]\Delta x.$$

Applying the mean value theorem again, there is a $\bar{y}$ between y_0 and $y_0 + \Delta y$ such that

$$S(\Delta x, \Delta y) = \frac{\partial^2 f}{\partial y\, \partial x}(\bar{x}, \bar{y})\, \Delta x\, \Delta y.$$

Since $\partial^2 f/\partial y\, \partial x$ is continuous, it follows that

$$\frac{\partial^2 f}{\partial y\, \partial x}(x_0, y_0) = \underset{(\Delta x, \Delta y) \to (0,0)}{\text{limit}} \frac{1}{\Delta x \Delta y}[S(\Delta x, \Delta y)].$$

Noting that S is symmetric in Δx and Δy, one shows in a similar way that $\partial^2 f/\partial x\, \partial y$ is given by the *same limit formula*, which proves the result. ∎

This theorem was first proved by Leonhard Euler in 1734 in connection with his studies of hydrodynamics.

In Exercise 7 we ask the reader to deduce from Theorem 1 that for a C^3 function of x, y, and z,

$$\frac{\partial^3 f}{\partial x\, \partial y\, \partial z} = \frac{\partial^3 f}{\partial z\, \partial y\, \partial x} = \frac{\partial^3 f}{\partial y\, \partial z\, \partial x}, \quad \text{etc.}$$

In other words, we can compute iterated partial derivatives in any order we please.

EXAMPLE 4   Verify the equality of the mixed second partial derivatives for the function

$$f(x, y) = xe^y + yx^2.$$

SOLUTION   Here

$$\frac{\partial f}{\partial x} = e^y + 2xy, \qquad \frac{\partial f}{\partial y} = xe^y + x^2,$$

$$\frac{\partial^2 f}{\partial y\, \partial x} = e^y + 2x, \qquad \frac{\partial^2 f}{\partial x\, \partial y} = e^y + 2x,$$

and so we have

$$\frac{\partial^2 f}{\partial y\, \partial x} = \frac{\partial^2 f}{\partial x\, \partial y}. \quad \blacktriangle$$

Sometimes the notation f_x, f_y, f_z is used for the partial derivatives: $f_x = \partial f/\partial x$, and so on. With this notation, we write $f_{xy} = (f_x)_y$, and so equality of the mixed partials is denoted by $f_{xy} = f_{yx}$. Notice that $f_{xy} = \partial^2 f/\partial y\, \partial x$, so the order of x and y is reversed in the two notations; fortunately, the equality of mixed partials makes this potential ambiguity irrelevant. The following example illustrates this subscript notation.

EXAMPLE 5 Let $z = f(x, y) = e^x \sin xy$ and write $x = g(s, t)$, $y = h(s, t)$ for functions g and h. Let $k(s, t) = f(g(s, t), h(s, t))$. Calculate k_{st}.

SOLUTION By the chain rule,

$$k_s = f_x g_s + f_y h_s = (e^x \sin xy + y e^x \cos xy)g_s + (x e^x \cos xy)h_s.$$

Differentiating in t using the product rules gives

$$k_{st} = (f_x)_t g_s + f_x(g_s)_t + (f_y)_t h_s + f_y(h_s)_t$$

Applying the chain rule again to $(f_x)_t$ and $(f_y)_t$ gives

$$(f_x)_t = f_{xx}g_t + f_{xy}h_t \qquad \text{and} \qquad (f_y)_t = f_{yx}g_t + f_{yy}h_t$$

and so k_{st} becomes

$$k_{st} = (f_{xx}g_t + f_{xy}h_t)g_s + f_x g_{st} + (f_{yx}g_t + f_{yy}h_t)h_s + f_y h_{st}$$

$$= f_{xx}g_t g_s + f_{xy}(h_t g_s + h_s g_t) + f_{yy}h_t h_s + f_x g_{st} + f_y h_{st}.$$

Notice that this last formula is symmetric in (s, t), verifying the equality $k_{st} = k_{ts}$. Computing f_{xx}, f_{xy}, and f_{yy}, we get

$$k_{st} = (e^x \sin xy + 2y e^x \cos xy - y^2 e^x \sin xy)g_t g_s$$

$$+ (x e^x \cos xy + e^x \cos xy - xy e^x \sin xy)(h_t g_s + h_s g_t)$$

$$- (x^2 e^x \sin xy)h_t h_s + (e^x \sin xy + y e^x \cos xy)g_{st} + (x e^x \cos xy)h_{st},$$

in which it is understood that $x = g(s, t)$ and $y = h(s, t)$. $\quad \blacktriangle$

EXAMPLE 6 The partial differential equation $u_t + u_{xxx} + u u_x = 0$, called the ***Korteweg–de Vries equation*** (or KdV equation, for short), describes the motion of water waves in a shallow channel.

(a)   Show that for any positive constant c, the function

$$u(x, t) = 3c \operatorname{sech}^2\left[\tfrac{1}{2}(x - ct)\sqrt{c}\right]$$

is a solution of the Korteweg–de Vries equation. (This solution represents a traveling "hump" of water in the channel and is called a **soliton**.)*

(b)   How do the shape and speed of the soliton depend on c?

SOLUTION   (a) We compute u_t, u_x, u_{xx}, and u_{xxx} using the chain rule and the differentiation formula $(d/dx) \operatorname{sech} x = -\operatorname{sech} x \tanh x$ from one-variable calculus. Letting $\alpha = (x - ct)\sqrt{c}/2$,

$$u_t = 6c \operatorname{sech} \alpha \frac{\partial}{\partial t} \operatorname{sech} \alpha = -6c \operatorname{sech}^2 \alpha \tanh \alpha \frac{\partial \alpha}{\partial t}$$

$$= 3c^{5/2} \operatorname{sech}^2 \alpha \tanh \alpha = c^{3/2} u \tanh \alpha.$$

Also,

$$u_x = -6c \operatorname{sech}^2 \alpha \, \tanh^2 \alpha \frac{\partial \alpha}{\partial x}$$

$$= -3c^{3/2} \operatorname{sech}^2 \alpha \tanh \alpha = -\sqrt{c} \, u \tanh \alpha,$$

and so $u_t = -cu_x$ and

$$u_{xx} = -\sqrt{c}\left[u_x \tanh \alpha + u \, (\operatorname{sech}^2 \alpha) \frac{\sqrt{c}}{2} \right] = -\sqrt{c} \, (\tanh \alpha) u_x - \frac{u^2}{6}$$

$$= c(\tanh^2 \alpha) u - \frac{u^2}{6} = c(1 - \operatorname{sech}^2 \alpha) \, u - \frac{u^2}{6}$$

$$= cu - \frac{u^2}{3} - \frac{u^2}{6} = cu - \frac{u^2}{2}.$$

Thus,

$$u_{xxx} = cu_x - uu_x; \text{ that is }, u_{xxx} + uu_x = cu_x.$$

Hence,

$$u_t + u_{xxx} + uu_x = u_t + cu_x = 0.$$

(b) The speed of the soliton is c, since $u(x + ct, t) = u(x, 0)$. The soliton is higher and thinner when c is larger. Its shape at time $t = 10$ is shown in Figure 3.1.2.   ▲

---

*Solitons were first observed by J. Scott Russell around 1840 in barge canals near Edinburgh. He reported his results in *Trans. R. Soc. Edinburgh* **14** (1840): 47–109.

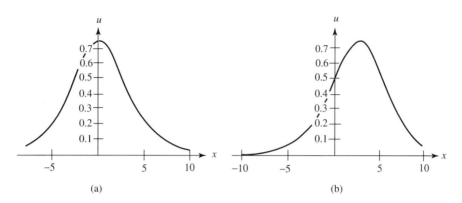

Figure 3.1.2 The graph of $u(x, t) = 3 \ \text{sech}^2 \ (\sqrt{c}(x - ct)/2)$ for $c = \frac{1}{4}$ at times (a) $t = 0$ and (b) $t = 10$.

HISTORICAL NOTE: SOME PARTIAL DIFFERENTIAL EQUATIONS

> Philosophy [nature] is written in that great book which ever is before our eyes—I mean the universe—but we cannot understand it if we do not first learn the language and grasp the symbols in which it is written. The book is written in mathematical language, and the symbols are triangles, circles and other geometrical figures, without whose help it is impossible to comprehend a single word of it; without which one wanders in vain through a dark labyrinth.
>
> GALILEO

This quotation illustrates the belief, popular in the time of Galileo, that much of nature could be described using mathematics. In the latter part of the seventeenth century this thinking was dramatically reinforced when Newton used his law of gravitation to derive Kepler's three laws of celestial motion (see Section 4.1), to explain the tides, and to show that the earth was flattened at the poles. The impact of this philosophy on mathematics was substantial, and many mathematicians sought to "mathematize" nature. The extent to which mathematics pervades the physical sciences today (and, to an increasing amount, economics and the social and life sciences) is testament to the success of these endeavors. Correspondingly, the attempts to mathematize nature have often led to new mathematical discoveries.

Many of the laws of nature were described in terms of either ordinary differential equations (equations involving the derivatives of functions of one variable alone, such as the laws of planetary motion) or partial differential equations, that is, equations involving partial derivatives of functions. To give the reader some historical perspective and offer motivation for studying partial derivatives, we present a brief description of three of the most famous partial differential equations: the heat equation, the potential equation (or Laplace's equation), and the wave equation. (All three of these will be discussed in greater depth in Section 8.5.)

THE HEAT EQUATION.   In the early part of the nineteenth century the French mathematician Joseph Fourier (1768–1830) took up the study of heat. Heat flow had obvious applications to both industrial and scientific problems: a better understanding of it would, for example, make possible more efficient smelting of metals and would enable scientists to determine the temperature of a body given the temperature at its boundary, and to approximate the temperature of the earth's interior.

Let a homogeneous body $\mathcal{B} \subset \mathbb{R}^3$ (Figure 3.1.3) be represented by some region in 3-space. Let $T(x, y, z, t)$ denote the temperature of the body at the point (x, y, z) at time t. Fourier proved, on the basis of physical principles (described in Section 8.5) that T must satisfy the partial differential equation called the **heat equation**,

$$k\left(\frac{\partial^2 T}{\partial x^2} + \frac{\partial^2 T}{\partial y^2} + \frac{\partial^2 T}{\partial z^2}\right) = \frac{\partial T}{\partial t}, \tag{1}$$

where k is a constant whose value depends on the conductivity of the material comprising the body.

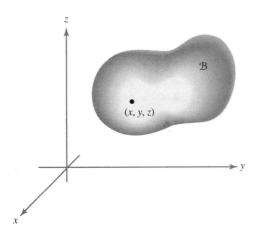

Figure 3.1.3   A homogeneous body in space.

Fourier used this equation to solve problems in heat conduction. In fact, his investigations into the solutions of equation (1) led him to the discovery of *Fourier series*.

THE POTENTIAL EQUATION.   Consider the gravitational potential V (often called Newton's potential) of a mass m at a point (x, y, z) caused by a point mass M situated at the origin. This potential is given by $V = -GmM/r$, where $r = \sqrt{x^2 + y^2 + z^2}$. The potential V satisfies the equation

$$\frac{\partial^2 V}{\partial x^2} + \frac{\partial^2 V}{\partial y^2} + \frac{\partial^2 V}{\partial z^2} = 0 \tag{2}$$

everywhere except at the origin, as we will check in the next chapter (see also Exercise 19). This equation is known as *Laplace's equation*. Pierre Simon de Laplace (1749–1827) had worked on the gravitational attraction of nonpoint masses and was the first to consider equation (2) with regard to gravitational attraction. He gave arguments (later shown to be incorrect) that equation (2) held for any body and any point whether inside or outside that body. However, Laplace was not the first person to write down equation (2). The potential equation appeared for the first time in one of Euler's major papers in 1752, "Principles of the Motions of Fluids," in which he derived the potential equation with regard to the motion of (incompressible) fluids. Euler remarked that he had no idea how to solve equation (2). Poisson later showed that if (x, y, z) lies inside an attracting body, then V satisfies the equation

$$\frac{\partial^2 V}{\partial x^2} + \frac{\partial^2 V}{\partial y^2} + \frac{\partial^2 V}{\partial z^2} = -4\pi\rho \tag{3}$$

where ρ is the mass density of the attracting body. Equation (3) is now called *Poisson's equation*. Poisson was also the first to point out the importance of this equation for problems involving electric fields. Notice that if the temperature T is constant in time, then the heat equation reduces to Laplace's equation.

Laplace's and Poisson's equations are fundamental to many fields besides fluid mechanics, gravitational fields, and electrostatic fields. For example, they are useful for studying soap films and liquid crystals; see *The Parsimonious Universe*: *Shape and Form in the Natural World* by S. Hildebrandt and A. Tromba, Springer-Verlag, New York/Berlin, 1995.

THE WAVE EQUATION. The linear wave equation in space has the form

$$\frac{\partial^2 f}{\partial x^2} + \frac{\partial^2 f}{\partial y^2} + \frac{\partial^2 f}{\partial z^2} = c^2 \frac{\partial^2 f}{\partial t^2}. \tag{4}$$

The one-dimensional wave equation

$$\frac{\partial^2 f}{\partial x^2} = c^2 \frac{\partial^2 f}{\partial t^2} \tag{4'}$$

was derived in about 1727 by Johann II Bernoulli and several years later by Jean Le Rond d'Alembert in the study of how to determine the motion of a vibrating string (such as a violin string). Equation (4) became useful in the study of both vibrating bodies and elasticity. As we shall see when we consider Maxwell's equations for electromagnetism in Section 8.5, this equation also arises in the study of the propagation of electromagnetic radiation and sound waves.

EXERCISES

1. Compute the second partial derivatives $\partial^2 f/\partial x^2$, $\partial^2 f/\partial x\,\partial y$, $\partial^2 f/\partial y\,\partial x$, $\partial^2 f/\partial y^2$ for each of the following functions. Verify Theorem 1 in each case:

(a) $f(x, y) = 2xy/(x^2 + y^2)^2$, $(x, y) \neq (0, 0)$
(b) $f(x, y, z) = e^z + (1/x) + xe^{-y}$, $x \neq 0$
(c) $f(x, y) = \cos(xy^2)$
(d) $f(x, y) = e^{-xy^2} + y^3 x^4$
(e) $f(x, y) = 1/(\cos^2 x + e^{-y})$

2. Let

$$f(x, y) = \begin{cases} xy(x^2 - y^2)/(x^2 + y^2), & (x, y) \neq (0, 0) \\ 0, & (x, y) = (0, 0) \end{cases}$$

(see Figure 3.1.4).

   (a)   If $(x, y) \neq (0, 0)$, calculate $\partial f/\partial x$ and $\partial f/\partial y$.

   (b)   Show that $(\partial f/\partial x)(0, 0) = 0 = (\partial f/\partial y)(0, 0)$.

   (c)   Show that $(\partial^2 f/\partial x\, \partial y)(0, 0) = 1$, $(\partial^2 f/\partial y\, \partial x)(0, 0) = -1$.

   (d)   What went wrong? Why are the mixed partials not equal?

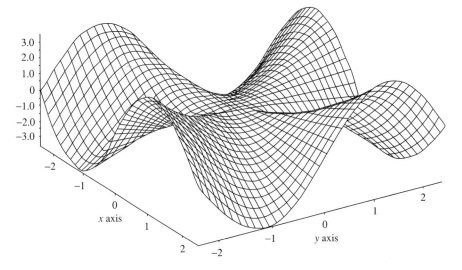

Figure 3.1.4   Graph of the function in Exercise 2.

3. Find $\partial^2 z/\partial x^2$, $\partial^2 z/\partial x\, \partial y$, $\partial^2 z/\partial y\, \partial x$, and $\partial^2 z/\partial y^2$ for

   (a) $z = 3x^2 + 2y^2$

   (b) $z = (2x^2 + 7x^2 y)/3xy$, $(x, y) \neq (0, 0)$

4. Find all the second partial derivatives of

   (a) $z = \sin(x^2 - 3xy)$

   (b) $z = x^2 y^2 e^{2xy}$

5. Find f_{xy}, f_{yz}, f_{zx}, and f_{xyz} for

$$f(x, y, z) = x^2 y + xy^2 + yz^2.$$

6. Let $z = x^4 y^3 - x^8 + y^4$.

   (a)   Compute $\partial^3 z/\partial y\, \partial x\, \partial x$, $\partial^3 z/\partial x\, \partial y\, \partial x$, and $\partial^3 z/\partial x\, \partial x\, \partial y$ (also denoted $\partial^3 z/\partial x^2 \partial y$).

   (b)   Compute $\partial^3 z/\partial x\, \partial y\, \partial y$, $\partial^3 z/\partial y\, \partial x\, \partial y$, and $\partial^3 z/\partial y\, \partial y\, \partial x$ (also denoted $\partial^3 z/\partial y^2 \partial x$).

7. Use Theorem 1 to show that if $f(x, y, z)$ is of class C^3, then

$$\frac{\partial^3 f}{\partial x\, \partial y\, \partial z} = \frac{\partial^3 f}{\partial y\, \partial z\, \partial x}.$$

8. Verify that

$$\frac{\partial^3 f}{\partial x \, \partial y \, \partial z} = \frac{\partial^3 f}{\partial z \, \partial y \, \partial x}$$

for $f(x, y, z) = ze^{xy} + yz^3x^2$.

9. Verify that $f_{xzw} = f_{zwx}$ for $f(x, y, z, w) = e^{xyz} \sin(xw)$.

10. If $f(x, y, z, w)$ is of class C^3, show that $f_{xzw} = f_{zwx}$.

11. Evaluate all first and second partial derivatives of the following functions:
    (a) $f(x, y) = x \arctan(x/y)$
    (b) $f(x, y) = \cos\sqrt{x^2 + y^2}$
    (c) $f(x, y) = \exp(-x^2 - y^2)$

12. Let $w = f(x, y)$ be a function of two variables and let $x = u + v$, $y = u - v$. Show that

$$\frac{\partial^2 w}{\partial u \, \partial v} = \frac{\partial^2 w}{\partial x^2} - \frac{\partial^2 w}{\partial y^2}.$$

13. Let $f \colon \mathbb{R}^2 \to \mathbb{R}$ be a C^2 function and let $\mathbf{c}(t)$ be a C^2 curve in $\mathbb{R}^2$. Write a formula for the second derivative $(d^2/dt^2)((f \circ \mathbf{c})(t))$ using the chain rule twice.

14. Let $f(x, y, z) = e^{xz} \tan(yz)$ and let $x = g(s, t)$, $y = h(s, t)$, $z = k(s, t)$, and define the function $m(s, t) = f(g(s, t), h(s, t), k(s, t))$. Find a formula for m_{st} using the chain rule and verify that your answer is symmetric in s and t.

15. A function $u = f(x, y)$ with continuous second partial derivatives satisfying Laplace's equation

$$\frac{\partial^2 u}{\partial x^2} + \frac{\partial^2 u}{\partial y^2} = 0$$

is called a ***harmonic function***. Show that the function $u(x, y) = x^3 - 3xy^2$ is harmonic.

16. Which of the following functions are harmonic? (See Exercise 15.)

    (a) $f(x, y) = x^2 - y^2$ (b) $f(x, y) = x^2 + y^2$
    (c) $f(x, y) = xy$ (d) $f(x, y) = y^3 + 3x^2y$
    (e) $f(x, y) = \sin x \cosh y$ (f) $f(x, y) = e^x \sin y$

17. Let f and g be differentiable functions of one variable. Set $\phi = f(x - t) + g(x + t)$.
    (a) Prove that ϕ satisfies the wave equation: $\partial^2 \phi / \partial t^2 = \partial^2 \phi / \partial x^2$.
    (b) Sketch the graph of ϕ against t and x if $f(x) = x^2$ and $g(x) = 0$.

18. (a) Show that function $g(x, t) = 2 + e^{-t} \sin x$ satisfies the heat equation: $g_t = g_{xx}$. [Here $g(x, t)$ represents the temperature in a metal rod at position x and time t.]
    (b) Sketch the graph of g for $t \geq 0$. (HINT: Look at sections by the planes $t = 0$, $t = 1$, and $t = 2$.)

(c) What happens to $g(x, t)$ as $t \to \infty$? Interpret this limit in terms of the behavior of heat in the rod.

19. Show that Newton's potential $V = -GmM/r$ satisfies Laplace's equation

$$\frac{\partial^2 V}{\partial x^2} + \frac{\partial^2 V}{\partial y^2} + \frac{\partial^2 V}{\partial z^2} = 0 \qquad \text{for} \qquad (x, y, z) \neq (0, 0, 0).$$

3.2 TAYLOR'S THEOREM

Taylor's theorem provides "higher-order" approximations to a function generalizing the linear approximation based on the first derivative of the function. We use Taylor's theorem in several variables to derive a test for different types of extrema, finally obtaining a test much like the second-derivative test that is studied in one-variable calculus. There are other important applications of this theorem as well.

For smooth functions of one variable $f \colon \mathbb{R} \to \mathbb{R}$, Taylor's theorem asserts that

$$f(x) = f(a) + f'(a)(x-a) + \frac{f''(a)}{2!}(x-a)^2 + \cdots + \frac{f^{(k)}(a)}{k!}(x-a)^k + R_k(x, a), \quad (1)$$

where

$$R_k(x, a) = \int_a^x \frac{(x-t)^k}{k!} f^{(k+1)}(t)\, dt \qquad (1')$$

is the remainder. For x near a, this error $R_k(x, a)$ is small "to order k." This means that

$$\frac{R_k(x, a)}{(x-a)^k} \to 0 \qquad \text{as} \qquad x \to a. \qquad (2)$$

In other words, $R_k(x, a)$ is small compared with the (already small) quantity $(x-a)^k$. This result is proved as follows. We begin with the fundamental theorem of calculus:

$$f(x) - f(a) = \int_a^x f'(t)\, dt.$$

Then the right-hand side is integrated by parts using the formula

$$\int u\, dv = uv - \int v\, du,$$

with $u = f'(t)$ and $v = x - t$. The result is

$$\int_a^x f'(t)\, dt = -\int_a^x u\, dv = -\left(uv \big|_a^x - \int_a^x v\, du \right)$$

$$= f'(a)(x-a) + \int_a^x (x-t) f''(t)\, dt.$$

Thus we have proved the first-order form of Taylor's theorem (the theorem with $k = 1$):

$$f(x) = f(a) + f'(a)(x-a) + \int_a^x (x-t) f''(t)\, dt.$$

Note that the first two terms on the right-hand side equal the first two terms in the Taylor series of f. If we integrate by parts again with $u = f''(t)$ and $v = (x - t)^2/2$, we get

$$\int_a^x (x - t) f''(t)\, dt = -\int_a^x u\, dv = -uv\big|_a^x + \int_a^x v\, du$$

$$= \frac{f''(a)}{2}(x - a)^2 + \int_a^x \frac{(x - t)^2}{2} f'''(t)\, dt;$$

so, substituting this into the first-order form of Taylor's theorem gives

$$f(x) = f(a) + f'(a)(x - a) + \frac{f''(a)}{2}(x - a)^2 + \int_a^x \frac{(x - t)^2}{2} f'''(t)\, dt.$$

Repeating this argument proves Taylor's theorem for one variable. Another common way to write this equation is to let $a = x_0$ and $h = x - a$, so that

$$f(x_0 + h) = f(x_0) + f'(x_0)h + \frac{1}{2} f''(x_0)h^2 + \frac{1}{2} \int_{x_0}^{x_0 + h} (x_0 + h - t)^2 f'''(t)\, dt.$$

Our goal in this section is to prove an analogous theorem valid for functions of several variables. We already know a first-order version, that is, when $k = 1$. Indeed, if $f \colon \mathbb{R}^n \to \mathbb{R}$ is differentiable at $\mathbf{x}_0$ and we define

$$R_1(\mathbf{x}, \mathbf{x}_0) = f(\mathbf{x}) - f(\mathbf{x}_0) - [\mathbf{D}f(\mathbf{x}_0)](\mathbf{x} - \mathbf{x}_0),$$

so that

$$f(\mathbf{x}) = f(\mathbf{x}_0) + [\mathbf{D}f(\mathbf{x}_0)](\mathbf{x} - \mathbf{x}_0) + R_1(\mathbf{x}, \mathbf{x}_0),$$

then by the definition of differentiability,

$$\frac{|R_1(\mathbf{x}, \mathbf{x}_0)|}{\|\mathbf{x} - \mathbf{x}_0\|} \to 0 \qquad \text{as} \qquad \mathbf{x} \to \mathbf{x}_0;$$

that is, $R_1(\mathbf{x}, \mathbf{x}_0)$ vanishes to first order at $\mathbf{x}_0$. Writing $\mathbf{h} = \mathbf{x} - \mathbf{x}_0$ and $R_1(\mathbf{x}, \mathbf{x}_0) = R_1(\mathbf{h}, \mathbf{x}_0)$ (an admitted abuse of notation!) we have:

THEOREM 2: FIRST-ORDER TAYLOR FORMULA

Let $f \colon U \subset \mathbb{R}^n \to \mathbb{R}$ be differentiable at $\mathbf{x}_0 \in U$. Then

$$f(\mathbf{x}_0 + \mathbf{h}) = f(\mathbf{x}_0) + \sum_{i=1}^{n} h_i \frac{\partial f}{\partial x_i}(\mathbf{x}_0) + R_1(\mathbf{h}, \mathbf{x}_0),$$

where $R_1(\mathbf{h}, \mathbf{x}_0)/\|\mathbf{h}\| \to 0$ as $\mathbf{h} \to \mathbf{0}$ in $\mathbb{R}^n$.

The second-order version is as follows:

THEOREM 3: SECOND-ORDER TAYLOR FORMULA

Let $f: U \subset \mathbb{R}^n \to \mathbb{R}$ have continuous partial derivatives of third order.* Then we may write

$$f(\mathbf{x}_0 + \mathbf{h}) = f(\mathbf{x}_0) + \sum_{i=1}^{n} h_i \frac{\partial f}{\partial x_i}(\mathbf{x}_0) + \frac{1}{2} \sum_{i,j=1}^{n} h_i h_j \frac{\partial^2 f}{\partial x_i \, \partial x_j}(\mathbf{x}_0) + R_2(\mathbf{h}, \mathbf{x}_0),$$

where $R_2(\mathbf{h}, \mathbf{x}_0)/\|\mathbf{h}\|^2 \to 0$ as $\mathbf{h} \to \mathbf{0}$ and the second sum is over all i's and j's between 1 and n (so there are n^2 terms).

In the course of the proof, we shall obtain a useful explicit formula for the remainder R_2 (See equation (5′), below). This formula is a generalization of formula (1′).

PROOF OF THEOREM 3 If $g(t) = f(\mathbf{x}_0 + t\mathbf{h})$ with $\mathbf{x}_0$ and $\mathbf{h}$ fixed, g is a C^3 function of t. Now apply the single-variable Taylor theorem (1) to g, with $k = 2$ to obtain

$$g(1) = g(0) + g'(0) + \frac{g''(0)}{2!} + R_2$$

where

$$R_2 = \int_0^1 \frac{(t-1)^2}{2!} g'''(t)\, dt.$$

$$(3)$$

By the chain rule,

$$g'(t) = \sum_{i=1}^{n} \frac{\partial f}{\partial x_i}(\mathbf{x}_0 + t\mathbf{h})\mathbf{h}_i$$

$$g''(t) = \sum_{i,j=1}^{n} \frac{\partial^2 f}{\partial x_i \, \partial x_j}(\mathbf{x}_0 + t\mathbf{h})\mathbf{h}_i \mathbf{h}_j$$

and

$$g'''(t) = \sum_{i,j,k=1}^{n} \frac{\partial^3 f}{\partial x_i \, \partial x_j \, \partial x_k}(\mathbf{x}_0 + t\mathbf{h})h_i h_j h_k.$$

---

*For the statement of the theorem as given here, f actually needs only to be of class C^2, but for a convenient form of the remainder we assume f is of class C^3.

Writing $R_2 = R_2(\mathbf{h}, \mathbf{x}_0)$ we have thus proved:

$$
\left.
\begin{aligned}
f(\mathbf{x}_0 + \mathbf{h}) &= f(\mathbf{x}_0) + \sum_{i=1}^{n} h_i \frac{\partial f}{\partial x_i}(\mathbf{x}_0) + \frac{1}{2} \sum_{i,j=1}^{n} h_i h_j \frac{\partial^2 f}{\partial x_i \, \partial x_j}(\mathbf{x}_0) + R_2(\mathbf{h}, \mathbf{x}_0) \\
&\text{where} \\
R_2(\mathbf{h}, \mathbf{x}_0) &= \sum_{i,j,k=1}^{n} \int_0^1 \frac{(t-1)^2}{2} \frac{\partial^3 f}{\partial x_i \, \partial x_j \, \partial x_k}(\mathbf{x} + t\mathbf{h}) h_i h_j h_k \, dt.
\end{aligned}
\right\} \quad (4)
$$

The integrand is a continuous function of t and is therefore bounded on a small neighborhood of $\mathbf{x}_0$ (since it has to be close to its value at $\mathbf{x}_0$). Also note that $|h_i| \leq \|\mathbf{h}\|$, for $\|\mathbf{h}\|$ small, and so

$$
|R_2(\mathbf{h}, \mathbf{x}_0)| \leq \|\mathbf{h}\|^3 M.
$$

In particular,

$$
\frac{|R_2(\mathbf{h}, \mathbf{x}_0)|}{\|\mathbf{h}\|^2} \leq \|\mathbf{h}\| M \to 0 \qquad \text{as} \qquad \mathbf{h} \to \mathbf{0},
$$

as required by the theorem.

The proof of Theorem 2 follows analogously from the Taylor formula (1) with $k = 1$. A similar argument for R_1 shows that $|R_1(\mathbf{h}, \mathbf{x}_0)|/\|\mathbf{h}\| \to 0$ as $\mathbf{h} \to \mathbf{0}$, although this also follows directly from the definition of differentiability. ∎

EXPLICIT FORMS OF THE REMAINDER   In Theorem 2,

$$
R_1(\mathbf{h}, \mathbf{x}_0) = \sum_{i,j=1}^{n} \int_0^1 (1-t) \frac{\partial^2 f}{\partial x_i \, \partial x_j}(\mathbf{x}_0 + t\mathbf{h}) h_i h_j \, dt = \sum_{i,j=1}^{n} \frac{1}{2} \frac{\partial^2 f}{\partial x_i \, \partial x_j}(\mathbf{c}_{ij}) h_i h_j \quad (5)
$$

where $\mathbf{c}_{ij}$ lies somewhere on the line joining $\mathbf{x}_0$ to $\mathbf{x}_0 + \mathbf{h}$.

In Theorem 3,

$$
\begin{aligned}
R_2(\mathbf{h}, \mathbf{x}_0) &= \sum_{i,j,k=1}^{n} \int_0^1 \frac{(t-1)^2}{2} \frac{\partial^3 f}{\partial x_i \, \partial x_j \, \partial x_k}(\mathbf{x}_0 + t\mathbf{h}) h_i h_j h_k \, dt \\
&= \sum_{i,j,k=1}^{n} \frac{1}{3!} \frac{\partial^3 f}{\partial x_i \, \partial x_j \, \partial x_k}(\mathbf{c}_{ijk}) h_i h_j h_k
\end{aligned} \quad (5')
$$

where $\mathbf{c}_{ijk}$ lies somewhere on the line joining $\mathbf{x}_0$ to $\mathbf{x}_0 + \mathbf{h}$.

The formulas involving $\mathbf{c}_{ij}$ and $\mathbf{c}_{ijk}$ (called *Lagrange's form of the remainder*) are obtained from the *second mean value theorem for integrals*. This states that

$$
\int_a^b h(t) g(t) \, dt = h(c) \int_a^b g(t) \, dt,
$$

provided h and g are continuous and $g \geq 0$ on $[a, b]$; here c is some number between a and b.* This is applied in formula (5) for the explicit form of the remainder with $h(t) = (\partial^2 f / \partial x_i \partial x_j)(x_0 + t\mathbf{h})$ and $g(t) = 1 - t$.

The third-order Taylor formula is

$$f(\mathbf{x}_0 + \mathbf{h}) = f(\mathbf{x}_0) + \sum_{i=1}^{n} h_i \frac{\partial f}{\partial x_i}(\mathbf{x}_0) + \frac{1}{2} \sum_{i,j=1}^{n} h_i h_j \frac{\partial^2 f}{\partial x_i \, \partial x_j}(\mathbf{x}_0)$$

$$+ \frac{1}{3!} \sum_{i,j,k=1}^{n} h_i h_j h_k \frac{\partial^3 f}{\partial x_i \, \partial x_j \, \partial x_k}(\mathbf{x}_0) + R_3(\mathbf{h}, \mathbf{x}_0),$$

where $R_3(\mathbf{h}, \mathbf{x}_0) / \|\mathbf{h}\|^3 \to 0$ as $\mathbf{h} \to \mathbf{0}$, and so on. The general formula can be proved by induction, using the method of proof already given.

EXAMPLE 1 Compute the second-order Taylor formula for $f(x, y) = \sin(x + 2y)$, about the point $\mathbf{x}_0 = (0, 0)$.

SOLUTION Notice that

$$f(0, 0) = 0,$$

$$\frac{\partial f}{\partial x}(0, 0) = \cos(0 + 2 \cdot 0) = 1, \qquad \frac{\partial f}{\partial y}(0, 0) = 2 \cos(0 + 2 \cdot 0) = 2,$$

$$\frac{\partial^2 f}{\partial x^2}(0, 0) = 0, \qquad \frac{\partial^2 f}{\partial y^2}(0, 0) = 0, \qquad \frac{\partial^2 f}{\partial x \, \partial y}(0, 0) = 0.$$

Thus,

$$f(\mathbf{h}) = f(h_1, h_2) = h_1 + 2h_2 + R_2(\mathbf{h}, \mathbf{0}),$$

where

$$\frac{R_2(\mathbf{h}, \mathbf{0})}{\|\mathbf{h}\|^2} \to 0 \qquad \text{as} \qquad \mathbf{h} \to \mathbf{0}. \quad \blacktriangle$$

EXAMPLE 2 Compute the second-order Taylor formula for $f(x, y) = e^x \cos y$ about the point $x_0 = 0, \ y_0 = 0$.

Proof If $g = 0$ the result is clear, so we can suppose $g \neq 0$; thus we can assume $\int_a^b g(t)\, dt > 0$. Let M and m be the maximum and minimum values of h, achieved at t_M and t_m, respectively. Since $g(t) \geq 0$,

$$m \int_a^b g(t)\, dt \leq \int_a^b h(t) g(t)\, dt \leq M \int_a^b g(t)\, dt.$$

Thus $\left(\int_a^b h(t) g(t)\, dt \right) \Big/ \left(\int_a^b g(t)\, dt \right)$ lies between $m = h(t_m)$ and $M = h(t_M)$ and therefore, by the intermediate value theorem, equals $h(c)$ for some intermediate c. ∎

SOLUTION   Here

$$f(0, 0) = 1, \qquad \frac{\partial f}{\partial x}(0, 0) = 1, \qquad \frac{\partial f}{\partial y}(0, 0) = 0,$$

$$\frac{\partial^2 f}{\partial x^2}(0, 0) = 1, \qquad \frac{\partial^2 f}{\partial y^2}(0, 0) = -1, \qquad \frac{\partial^2 f}{\partial x \, \partial y}(0, 0) = 0,$$

and so

$$f(\mathbf{h}) = f(h_1, h_2) = 1 + h_1 + \tfrac{1}{2}h_1^2 - \tfrac{1}{2}h_2^2 + R_2(\mathbf{h}, \mathbf{0}),$$

where

$$\frac{R_2(\mathbf{h}, \mathbf{0})}{\|\mathbf{h}\|^2} \to 0 \qquad \text{as} \qquad \mathbf{h} \to \mathbf{0}. \quad \blacktriangle$$

In the case of functions of one variable, one can expand $f(x)$ in an infinite power series, called the **Taylor series**:

$$f(x_0 + h) = f(x_0) + f'(x_0)h + \frac{f''(x_0)h^2}{2} + \cdots + \frac{f^{(k)}(x_0)h^k}{k!} + \cdots,$$

provided one can show that $R_k(h, x_0) \to 0$ as $k \to \infty$. Similarly, for functions of several variables the above terms are replaced by the corresponding ones involving partial derivatives, as we have seen in Theorem 3. Again, one can represent such a function by its Taylor series provided one can show that $R_k \to 0$ as $k \to \infty$. This point is examined further in Exercise 7.

EXAMPLE 3   Find the first- and second-order Taylor approximations to $f(x, y) = \sin(xy)$ at the point $(x_0, y_0) = (1, \pi/2)$.

SOLUTION   Here

$$f(x_0, y_0) = \sin(x_0 y_0) = \sin(\pi/2) = 1$$

$$f_x(x_0, y_0) = y_0 \cos(x_0 y_0) = \frac{\pi}{2} \cos(\pi/2) = 0$$

$$f_y(x_0, y_0) = x_0 \cos(x_0 y_0) = \cos(\pi/2) = 0$$

$$f_{xx}(x_0, y_0) = -y_0^2 \sin(x_0 y_0) = -\frac{\pi^2}{4} \sin(\pi/2) = -\pi^2/4$$

$$f_{xy}(x_0, y_0) = \cos(x_0 y_0) - x_0 y_0 \sin(x_0 y_0) = -\frac{\pi}{2} \sin(\pi/2) = -\pi/2$$

$$f_{yy}(x_0, y_0) = -x_0^2 \sin(x_0 y_0) = -\sin(\pi/2) = -1.$$

Thus, the linear (first order) approximation is

$$l(x, y) = f(x_0, y_0) + f_x(x_0, y_0)(x - x_0)) + f_y(x_0, y_0)(y - y_0)$$

$$= 1 + 0 + 0 = 1$$

and the second-order (or quadratic) approximation is

$$g(x, y) = 1 + 0 + 0 + \frac{1}{2}\left(-\frac{\pi^2}{4}\right)(x-1)^2 + \left(-\frac{\pi}{2}\right)(x-1)\left(y - \frac{\pi}{2}\right)$$
$$+ \frac{1}{2}(-1)\left(y - \frac{\pi}{2}\right)^2$$
$$= 1 - \frac{\pi^2}{8}(x-1)^2 - \frac{\pi}{2}(x-1)\left(y - \frac{\pi}{2}\right) - \frac{1}{2}\left(y - \frac{\pi}{2}\right)^2.$$

See Fig. 3.2.1. ▲

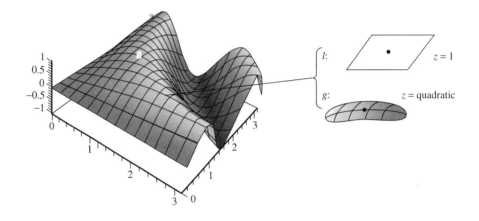

Figure 3.2.1 The linear and quadratic approximations to $z = \sin(xy)$ near $(1, \pi/2)$.

EXAMPLE 4 Find linear and quadratic approximations to $(3.98 - 1)^2/(5.97 - 3)^2$. Compare with the exact value.

SOLUTION Let $f(x, y) = (x-1)^2/(y-3)^2$. The desired expression is close to $f(4, 6) = 1$. To find the approximations, we differentiate:

$$f_x = \frac{2(x-1)}{(y-3)^2}, \qquad f_y = \frac{-2(x-1)^2}{(y-3)^3}$$

$$f_{xy} = f_{yx} = \frac{-4(x-1)}{(y-3)^3}, \qquad f_{xx} = \frac{2}{(y-3)^2}, \qquad f_{yy} = \frac{6(x-1)^2}{(y-3)^4}.$$

At the point of approximation, we have

$$f_x(4, 6) = \frac{2}{3}, \qquad f_y = -\frac{2}{3}, \qquad f_{xy} = f_{yx} = -\frac{4}{9}, \qquad f_{xx} = \frac{2}{9}, \qquad f_{yy} = \frac{2}{3}.$$

The linear approximation is then

$$1 + \frac{2}{3}(-0.02) - \frac{2}{3}(-0.03) = 1.00666.$$

The quadratic approximation is

$$1 + \frac{2}{3}(-0.02) - \frac{2}{3}(-0.03) + \frac{2}{9}\frac{(-0.02)^2}{2} - \frac{4}{9}(-0.02)(-0.03) + \frac{2}{3}\frac{(-0.03)^2}{2}$$

$$= 1.00674.$$

The "exact" value is 1.00675. ▲

EXERCISES

In each of Exercises 1 to 6, determine the second-order Taylor formula for the given function about the given point (x_0, y_0).

1. $f(x, y) = (x + y)^2$, $x_0 = 0$, $y_0 = 0$

2. $f(x, y) = 1/(x^2 + y^2 + 1)$, $x_0 = 0$, $y_0 = 0$

3. $f(x, y) = e^{x+y}$, $x_0 = 0$, $y_0 = 0$

4. $f(x, y) = e^{-x^2-y^2} \cos(xy)$, $x_0 = 0$, $y_0 = 0$

5. $f(x, y) = \sin(xy) + \cos(xy)$, $x_0 = 0$, $y_0 = 0$

6. $f(x, y) = e^{(x-1)^2} \cos y$, $x_0 = 1$, $y_0 = 0$

7. A function $f: \mathbb{R} \to \mathbb{R}$ is called an ***analytic*** function provided

$$f(x + h) = f(x) + f'(x)h + \cdots + \frac{f^{(k)}(x)}{k!}h^k + \cdots$$

[i.e., the series on the right-hand side converges and equals $f(x + h)$].
   (a) Suppose f satisfies the following condition: On any closed interval $[a, b]$ there is a constant M such that for all $k = 1, 2, 3, \ldots, |f^{(k)}(x)| \le M^k$ for all $x \in [a, b]$. Prove that f is analytic.

   (b) Let $f(x) = \begin{cases} e^{-1/x} & x > 0 \\ 0 & x \le 0. \end{cases}$

Show that f is a C^∞ function, but f is not analytic.
   (c) Give a definition of analytic functions from $\mathbb{R}^n$ to $\mathbb{R}$. Generalize the proof of part (a) to this class of functions.
   (d) Develop $f(x, y) = e^{x+y}$ in a power series about $x_0 = 0$, $y_0 = 0$.

3.3 EXTREMA OF REAL-VALUED FUNCTIONS

HISTORICAL NOTE

Throughout history, people have searched for laws to describe the phenomena of the physical world. However, no general principle encompassing all phenomena was proposed until 1744, when the French scientist Pierre Louis Moreau de Maupertuis put forth his grand scheme of the universe.

The "metaphysical principle" of Maupertuis is the assumption that nature always operates with the greatest possible economy. In short, physical laws are a consequence of a principle of "economy of means"; nature always acts in such a way as to minimize some quantity which Maupertuis called the "action." For example, physical systems often try to "rearrange themselves" to have a minimum energy, such as a ball rolling from a mountain peak to a valley, or the primordial irregular Earth assuming a more nearly spherical shape. As another example, the spherical shape of soap bubbles is connected with the fact that spheres are the surfaces of least area containing a fixed volume.

Minimization principles provide a philosophical cornerstone of many aspects of physics, engineering, and mathematics. Thus, describing many physical phenomena amounts to finding the maxima and minima of some scalar function. The Swiss mathematician Leonhard Euler provided much of the mathematical groundwork for the theory of maxima and minima of scalar quantities.[*]

Among the most basic geometric features of the graph of a function are its extreme points, at which the function attains its greatest and least values. In this section, we shall derive a method for determining these points. In fact, the method locates local extrema as well. These are points at which the function attains a maximum or minimum value relative only to nearby points. Let us begin by defining our terms.

DEFINITION   If $f: U \subset \mathbb{R}^n \to \mathbb{R}$ is a given scalar function, a point $\mathbf{x}_0 \in U$ is called a **local minimum** of f if there is a neighborhood V of $\mathbf{x}_0$ such that for all points $\mathbf{x}$ in V, $f(\mathbf{x}) \geq f(\mathbf{x}_0)$. (See Figure 3.3.1.) Similarly, $\mathbf{x}_0 \in U$ is a **local maximum** if there is a neighborhood V of $\mathbf{x}_0$ such that $f(\mathbf{x}) \leq f(\mathbf{x}_0)$ for all $\mathbf{x} \in V$. The point $\mathbf{x}_0 \in U$ is said to be a **local**, or **relative**, **extremum** if it is either a local minimum or a local maximum. A point $\mathbf{x}_0$ is a **critical point** of f if $\mathbf{D}f(\mathbf{x}_0) = \mathbf{0}$. A critical point that is not a local extremum is called a **saddle point**.[†]

---

[*]For more information and history, consult *The Parsimonious Universe: Shape and Form in the Natural World* by S. Hildebrandt and A. J. Tromba, Springer-Verlag, New York/Berlin, 1995.

[†]The term "saddle point" is sometimes not used this generally; we shall discuss saddle points further in the subsequent development.

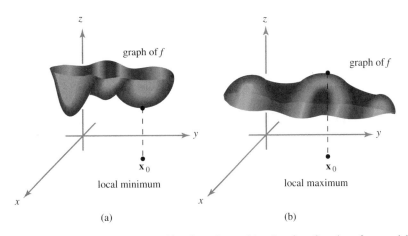

Figure 3.3.1   Local minimum (a) and local maximum (b) points for a function of two variables.

The location of extrema is based on the following fact, which should be familiar from one-variable calculus (the case $n = 1$): Every extremum is a critical point.

THEOREM 4: FIRST DERIVATIVE TEST   *If $U \subset \mathbb{R}^n$ is open, the function $f\colon U \subset \mathbb{R}^n \to \mathbb{R}$ is differentiable, and $\mathbf{x}_0 \in U$ is a local extremum, then $\mathbf{D}f(\mathbf{x}_0) = \mathbf{0}$; that is, $\mathbf{x}_0$ is a critical point of f.*

PROOF   Suppose that f achieves a local maximum at $\mathbf{x}_0$. Then for any $\mathbf{h} \in \mathbb{R}^n$, the function $g(t) = f(\mathbf{x}_0 + t\mathbf{h})$ has a local maximum at $t = 0$. Thus from one-variable calculus $g'(0) = 0$. [*Proof*   Since $g(0)$ is a local maximum, $g(t) \leq g(0)$ for small $t > 0$, so $g(t) - g(0) \leq 0$, and hence $g'(0) = \underset{t \to 0^+}{\text{limit}}(g(t) - g(0))/t \leq 0$, where $\underset{t \to 0^+}{\text{limit}}$ means the limit as $t \to 0$, $t > 0$. For small $t < 0$, we similarly have $g'(0) = \underset{t \to 0^-}{\text{limit}}(g(t) - g(0))/t \geq 0$, so $g'(0) = 0$.] On the other hand, by the chain rule,

$$g'(0) = [\mathbf{D}f(\mathbf{x}_0)]\mathbf{h}.$$

Thus $[\mathbf{D}f(\mathbf{x}_0)]\mathbf{h} = 0$ for every $\mathbf{h}$, and so $\mathbf{D}f(\mathbf{x}_0) = \mathbf{0}$. The case in which f achieves a local minimum at $\mathbf{x}_0$ is entirely analogous.   ■

If we remember that $\mathbf{D}f(\mathbf{x}_0) = \mathbf{0}$ means that all the components of $\mathbf{D}f(\mathbf{x}_0)$ are zero, we can rephrase the result of Theorem 4: If $\mathbf{x}_0$ is a local extremum, then

$$\frac{\partial f}{\partial x_i}(\mathbf{x}_0) = 0, \qquad i = 1, \dots, n;$$

that is, each partial derivative is zero at $\mathbf{x}_0$. In other words, $\nabla f(\mathbf{x}_0) = \mathbf{0}$, where ∇f is the gradient of f.

If we seek to find the extrema or local extrema of a function, then Theorem 4 states that we should look among the critical points. Sometimes these can be tested by inspection, but usually we use tests (to be developed below) analogous to the second-derivative test in one-variable calculus.

EXAMPLE 1   Find the maxima and minima of the function $f\colon \mathbb{R}^2 \to \mathbb{R}$, defined by $f(x, y) = x^2 + y^2$. (Ignore the fact that this example can be done by inspection.)

SOLUTION   We must identify the two critical points of f by solving the two equations $\partial f / \partial x = 0$ and $\partial f / \partial y = 0$, for x and y. But

$$\frac{\partial f}{\partial x} = 2x \qquad \text{and} \qquad \frac{\partial f}{\partial y} = 2y,$$

and so the only critical point is the origin $(0, 0)$, where the value of the function is zero. Since $f(x, y) \geq 0$, this point is a relative minimum—in fact, an absolute, or global, minimum—of f.   ▲

EXAMPLE 2   Consider the function $f\colon \mathbb{R}^2 \to \mathbb{R}$, $(x, y) \mapsto x^2 - y^2$. Ignoring for the moment that this function has a saddle and no extrema, apply the method of Theorem 4 for the location of extrema.

SOLUTION   As in Example 1, we find that f has only one critical point, at the origin, and the value of f there is zero. Examining values of f directly for points near the origin, we see that $f(x, 0) \geq f(0, 0)$ and $f(0, y) \leq f(0, 0)$. Since x or y can be taken arbitrarily small, the origin cannot be either a relative minimum or a relative maximum (so it is a saddle point). Therefore this function can have no relative extrema (see Figure 3.3.2).   ▲

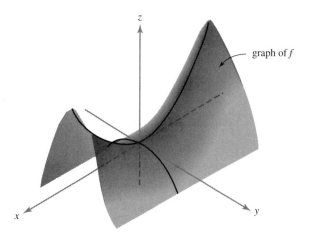

graph of f

Figure 3.3.2   A function of two variables with a saddle point.

EXAMPLE 3 Find all the critical points of $z = x^2y + y^2x$.

SOLUTION Differentiating, we obtain

$$\frac{\partial z}{\partial x} = 2xy + y^2, \qquad \frac{\partial z}{\partial y} = 2xy + x^2.$$

Equating the partial derivatives to zero yields

$$2xy + y^2 = 0, \qquad 2xy + x^2 = 0.$$

Subtracting, we obtain $x^2 = y^2$. Thus $x = \pm y$. Substituting $x = +y$ in the first equation above, we find that

$$2y^2 + y^2 = 3y^2 = 0,$$

so that $y = 0$ and thus $x = 0$. If $x = -y$, then

$$-2y^2 + y^2 = -y^2 = 0,$$

and so $y = 0$ and therefore $x = 0$. Hence the only critical point is $(0, 0)$. For $x = y$, $z = 2x^3$, which is both positive and negative for x near zero. Thus $(0, 0)$ is not a relative extremum. ▲

EXAMPLE 4 Refer to Figure 3.3.3, a computer-drawn graph of the function $z = 2(x^2 + y^2)\, e^{-x^2-y^2}$. Where are the critical points? (Note from the figure that there is a local (in fact, global) minimum at the volcano's center $(0, 0)$, where $z = 0$, and that there are maximum points all around the crater's rim (the circle $x^2 + y^2 = 1$).)

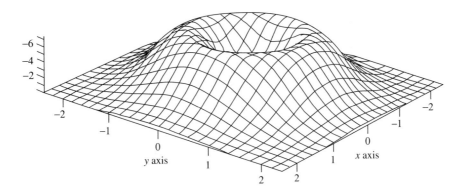

Figure 3.3.3 The volcano: $z = 2(x^2 + y^2)\exp(-x^2 - y^2)$.

SOLUTION   Since $z = 2(x^2 + y^2)e^{-x^2-y^2}$, we have

$$\frac{\partial z}{\partial x} = 4x(e^{-x^2-y^2}) + 2(x^2 + y^2)e^{-x^2-y^2}(-2x)$$

$$= e^{-x^2-y^2}[4x - 4x(x^2 + y^2)]$$

$$= 4x(e^{-x^2-y^2})(1 - x^2 - y^2)$$

and

$$\frac{\partial z}{\partial y} = 4y(e^{-x^2-y^2})(1 - x^2 - y^2).$$

These both vanish when $x = y = 0$ or when $x^2 + y^2 = 1$.   ▲

The remainder of this section is devoted to deriving a criterion, depending on the second derivative, for a critical point to be a relative extremum. In the special case $n = 1$, our criterion will reduce to the familiar condition $f''(x_0) > 0$ for a minimum and $f''(x_0) < 0$ for a maximum. But in the general context, the second derivative is a fairly complicated mathematical object. To state our criterion, we will introduce a version of the second derivative called the Hessian, which in turn is related to quadratic functions. **Quadratic functions** are functions $g: \mathbb{R}^n \to \mathbb{R}$ that have the form

$$g(h_1, \ldots, h_n) = \sum_{i,j=1}^{n} a_{ij} h_i h_j$$

for a $n \times n$ matrix $[a_{ij}]$. In terms of matrix multiplication, we can write

$$g(h_1, \ldots, h_n) = [h_1 \cdots h_n] \begin{bmatrix} a_{11} & a_{12} & \cdots & a_{1n} \\ \vdots & \vdots & & \vdots \\ a_{n1} & a_{n2} & \cdots & a_{nn} \end{bmatrix} \begin{bmatrix} h_1 \\ \vdots \\ h_n \end{bmatrix}.$$

For example, if $n = 3$,

$$g(h_1, h_2, h_3) = h_1^2 - 2h_1 h_2 + h_3^2 = [h_1 \, h_2 \, h_3] \begin{bmatrix} 1 & -1 & 0 \\ -1 & 0 & 0 \\ 0 & 0 & 1 \end{bmatrix} \begin{bmatrix} h_1 \\ h_2 \\ h_3 \end{bmatrix}$$

is a quadratic function.

We can, if we wish, assume that $[a_{ij}]$ is symmetric; in fact, g is unchanged if we replace $[a_{ij}]$ by the symmetric matrix $[b_{ij}]$, where $b_{ij} = \frac{1}{2}(a_{ij} + a_{ji})$, since $h_i h_j = h_j h_i$ and the sum is over all i and j. The *quadratic* nature of g is reflected in the identity

$$g(\lambda h_1, \ldots, \lambda h_n) = \lambda^2 g(h_1, \ldots, h_n),$$

which follows from the definition.

Now we are ready to define Hessian functions (named after Ludwig Otto Hesse, who introduced them in 1844).

DEFINITION   Suppose that $f: U \subset \mathbb{R}^n \to \mathbb{R}$ has second-order partial derivatives $(\partial^2 f / \partial x_i \, \partial x_j)(\mathbf{x}_0)$, for $i, j = 1, \ldots, n$, at a point $\mathbf{x}_0 \in U$. The **Hessian of f at $\mathbf{x}_0$** is the quadratic function defined by

$$H f(\mathbf{x}_0)(\mathbf{h}) = \frac{1}{2} \sum_{i,j=1}^{n} \frac{\partial^2 f}{\partial x_i \, \partial x_j}(\mathbf{x}_0) h_i h_j.$$

This function is usually used at critical points $\mathbf{x}_0 \in U$. In this case, $\mathbf{D}f(\mathbf{x}_0) = \mathbf{0}$, and thus the Taylor formula (see Theorem 3, Section 3.2) may be written in the form

$$f(\mathbf{x}_0 + \mathbf{h}) = f(\mathbf{x}_0) + H f(\mathbf{x}_0)(\mathbf{h}) + R_2(\mathbf{h}, \mathbf{x}_0).$$

Thus at a critical point the Hessian equals the first nonconstant term in the Taylor series of f.

A quadratic function $g: \mathbb{R}^n \to \mathbb{R}$ is called **positive-definite** if $g(\mathbf{h}) \geq 0$ for all $\mathbf{h} \in \mathbb{R}^n$ and $g(\mathbf{h}) = 0$ only for $\mathbf{h} = \mathbf{0}$. Similarly, g is **negative-definite** if $g(\mathbf{h}) \leq 0$ and $g(\mathbf{h}) = 0$ for $\mathbf{h} = \mathbf{0}$ only.

Note that if $n = 1$, $H f(x_0)(h) = \frac{1}{2} f''(x_0) h^2$, which is positive-definite if and only if $f''(x_0) > 0$.

THEOREM 5: SECOND DERIVATIVE TEST FOR RELATIVE MAXIMA AND MINIMA   *If $f: U \subset \mathbb{R}^n \to \mathbb{R}$ is of class C^3, $\mathbf{x}_0 \in U$ is a critical point of f, and the Hessian $H f(\mathbf{x}_0)$ is positive-definite, then $\mathbf{x}_0$ is a relative minimum of f. Similarly, if $H f(\mathbf{x}_0)$ is negative-definite, then $\mathbf{x}_0$ is a relative maximum.*

Actually, we shall prove that the extrema given by this criterion are *strict*. A relative maximum $\mathbf{x}_0$ is said to be *strict* if $f(\mathbf{x}) < f(\mathbf{x}_0)$ for nearby $\mathbf{x} \neq \mathbf{x}_0$. A strict relative minimum is defined similarly. Also, the theorem is valid even if f is only C^2 but we have assumed C^3 for simplicity.

The proof of Theorem 5 requires Taylor's theorem and the following result from linear algebra.

LEMMA 1   *If $B = [b_{ij}]$ is an $n \times n$ real matrix, and if the associated quadratic function*

$$H: \mathbb{R}^n \to \mathbb{R}, (h_1, \ldots, h_n) \mapsto \frac{1}{2} \sum_{i,j=1}^{n} b_{ij} h_i h_j$$

is positive-definite, then there is a constant $M > 0$ such that for all $\mathbf{h} \in \mathbb{R}^n$,

$$H(\mathbf{h}) \geq M \|\mathbf{h}\|^2.$$

PROOF   For $\|\mathbf{h}\| = 1$, set $g(\mathbf{h}) = H(\mathbf{h})$. Then g is a continuous function of $\mathbf{h}$ for $\|\mathbf{h}\| = 1$ and so achieves a minimum value, say M.* Since H is quadratic, we have

$$H(\mathbf{h}) = H\left(\frac{\mathbf{h}}{\|\mathbf{h}\|}\|\mathbf{h}\|\right) = H\left(\frac{\mathbf{h}}{\|\mathbf{h}\|}\right)\|\mathbf{h}\|^2 = g\left(\frac{\mathbf{h}}{\|\mathbf{h}\|}\right)\|\mathbf{h}\|^2 \geq M\|\mathbf{h}\|^2$$

for any $\mathbf{h} \neq \mathbf{0}$. (The result is obviously valid if $\mathbf{h} = \mathbf{0}$). ■

Note that the quadratic function associated with the symmetric matrix $\frac{1}{2}(\partial^2 f/\partial x_i\,\partial x_j)$ is exactly the Hessian.

PROOF OF THEOREM 5   Recall that if $f: U \subset \mathbb{R}^n \to \mathbb{R}$ is of class C^3 and $\mathbf{x}_0 \in U$ is a critical point, Taylor's theorem may be expressed in the form

$$f(\mathbf{x}_0 + \mathbf{h}) - f(\mathbf{x}_0) = Hf(\mathbf{x}_0)(\mathbf{h}) + R_2(\mathbf{h}, \mathbf{x}_0),$$

where

$$\frac{R_2(\mathbf{h}, \mathbf{x}_0)}{\|\mathbf{h}\|^2} \to 0$$

as $\mathbf{h} \to \mathbf{0}$.

Since $Hf(\mathbf{x}_0)$ is positive-definite, lemma 1 assures us of a constant $M > 0$ such that for all $\mathbf{h} \in \mathbb{R}^n$

$$Hf(\mathbf{x}_0)(\mathbf{h}) \geq M\|\mathbf{h}\|^2.$$

Since $R_2(\mathbf{h}, \mathbf{x}_0)/\|\mathbf{h}\|^2 \to 0$ as $\mathbf{h} \to \mathbf{0}$, there is a $\delta > 0$ such that for $0 < \|\mathbf{h}\| < \delta$

$$|R_2(\mathbf{h}, \mathbf{x}_0)| < M\|\mathbf{h}\|^2.$$

Thus $0 < Hf(\mathbf{x}_0)(\mathbf{h}) + R_2(\mathbf{h}, \mathbf{x}_0) = f(\mathbf{x}_0 + \mathbf{h}) - f(\mathbf{x}_0)$ for $0 < \|\mathbf{h}\| < \delta$, so that $\mathbf{x}_0$ is a relative minimum, in fact, a strict relative minimum.

The proof in the negative-definite case is similar, or else follows by applying the above to $-f$, and is left as an exercise. ■

EXAMPLE 5   Consider again the function $f: \mathbb{R}^2 \to \mathbb{R}$, $(x, y) \mapsto x^2 + y^2$. Then $(0, 0)$ is a critical point, and f is already in the form of Taylor's theorem:

$$f((0, 0) + (h_1, h_2)) = f(0, 0) + (h_1^2 + h_2^2) + 0.$$

We can see directly that the Hessian at $(0, 0)$ is

$$Hf(\mathbf{0})(\mathbf{h}) = h_1^2 + h_2^2,$$

---

*Here we are using, without proof, a theorem analogous to a theorem in calculus that states that every continuous function on an interval $[a, b]$ achieves a maximum and a minimum; see Theorem 7 below.

which is clearly positive-definite. Thus $(0, 0)$ is a relative minimum. This simple case can, of course, be done without calculus. Indeed, it is clear that $f(x, y) > 0$ for all $(x, y) \neq (0, 0)$.   ▲

For functions of two variables $f(x, y)$, the Hessian may be written as follows:

$$Hf(x, y)(\mathbf{h}) = \tfrac{1}{2}[h_1, h_2] \begin{bmatrix} \dfrac{\partial^2 f}{\partial x^2} & \dfrac{\partial^2 f}{\partial y\, \partial x} \\[2ex] \dfrac{\partial^2 f}{\partial x\, \partial y} & \dfrac{\partial^2 f}{\partial y^2} \end{bmatrix} \begin{bmatrix} h_1 \\ h_2 \end{bmatrix}.$$

Now we shall give a useful criterion for when a quadratic function defined by such a 2×2 matrix is positive-definite. This will be useful in conjunction with Theorem 5 .

LEMMA 2 Let

$$B = \begin{bmatrix} a & b \\ b & c \end{bmatrix} \qquad \text{and} \qquad H(\mathbf{h}) = \tfrac{1}{2}[h_1, h_2] B \begin{bmatrix} h_1 \\ h_2 \end{bmatrix}.$$

Then $H(\mathbf{h})$ is positive-definite if and only if $a > 0$ and $\det B = ac - b^2 > 0$.

PROOF We have

$$H(\mathbf{h}) = \tfrac{1}{2}[h_1, h_2] \begin{bmatrix} ah_1 + bh_2 \\ bh_1 + ch_2 \end{bmatrix} = \tfrac{1}{2}(ah_1^2 + 2bh_1 h_2 + ch_2^2).$$

Let us complete the square, writing

$$H(\mathbf{h}) = \tfrac{1}{2} a \left(h_1 + \frac{b}{a} h_2 \right)^2 + \frac{1}{2} \left(c - \frac{b^2}{a} \right) h_2^2.$$

Suppose H is positive-definite. Setting $h_2 = 0$, we see that $a > 0$. Setting $h_1 = -(b/a)h_2$, we get $c - b^2/a > 0$ or $ac - b^2 > 0$. Conversely, if $a > 0$ and $c - b^2/a > 0$, $H(\mathbf{h})$ is a sum of squares, so that $H(\mathbf{h}) \geq 0$. If $H(\mathbf{h}) = 0$, then each square must be zero. This implies that both h_1 and h_2 must be zero, so that $H(\mathbf{h})$ is positive definite. ∎

Similarly, $H(\mathbf{h})$ is negative-definite if and only if $a < 0$ and $ac - b^2 > 0$.

We note that an alternative formulation is that $H(\mathbf{h})$ is positive (respectively, negative) definite if $a + c = \text{trace } B > 0$ (respectively, < 0) and $\det B > 0$. There are similar criteria for an $n \times n$ symmetric matrix B. Consider the n square submatrices along the diagonal (see Figure 3.3.4). B is positive-definite (that is, the quadratic function associated with B is positive-definite) if and only if the determinants of these diagonal submatrices are all greater than zero. For negative-definite B, the signs should be alternately < 0 and > 0.

We shall not prove this general case here.*   In case the determinants of the diagonal submatrices are all nonzero but the matrix is not positive- or negative-definite, the critical point is of **saddle type**; in this case one can show that the point is neither a maximum nor a minimum.

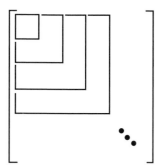

Figure 3.3.4   "Diagonal" submatrices are used in the criterion for positive definiteness; they must all have determinant >0.

Lemma 2 and Theorem 5 imply the following result:

THEOREM 6: MAXIMUM-MINIMUM TEST FOR FUNCTIONS OF TWO VARIABLES   *Let $f(x, y)$ be of class C^3 on an open set U in $\mathbb{R}^2$. A point (x_0, y_0) is a (strict) local minimum of f provided the following three conditions hold:*

(i) $\dfrac{\partial f}{\partial x}(x_0, y_0) = \dfrac{\partial f}{\partial y}(x_0, y_0) = 0$

(ii) $\dfrac{\partial^2 f}{\partial x^2}(x_0, y_0) > 0$

(iii) $D = \left(\dfrac{\partial^2 f}{\partial x^2}\right)\left(\dfrac{\partial^2 f}{\partial y^2}\right) - \left(\dfrac{\partial^2 f}{\partial x\, \partial y}\right)^2 > 0$ at (x_0, y_0)

*(D is called the **discriminant** of the Hessian.) If in (ii) we have <0 instead of >0 and condition (iii) is unchanged, then we have a (strict) local maximum.*

EXAMPLE 6   Classify the critical points of the function $f: \mathbb{R}^2 \to \mathbb{R}, (x, y) \mapsto x^2 - 2xy + 2y^2$.

SOLUTION   As in Example 5, we find that $f(0, 0) = 0$, the origin is the only critical point, and the Hessian is

$$Hf(\mathbf{0})(\mathbf{h}) = h_1^2 - 2h_1 h_2 + 2h_2^2 = (h_1 - h_2)^2 + h_2^2,$$

---

*This is proved in, for example, K. Hoffman and R. Kunze, *Linear Algebra*, Prentice-Hall, Englewood Cliffs, N.J., 1961, pp. 249–251. For students with the background in linear algebra, it should be noted that B is positive-definite when all of its eigenvalues (which are real, since B is symmetric) are positive.

which is clearly positive-definite. Thus f has a relative minimum at $(0, 0)$. Alternatively, we can apply Theorem 6. At $(0, 0)$, $\partial^2 f / \partial x^2 = 2$, $\partial^2 f / \partial y^2 = 4$, and $\partial^2 f / \partial x\, \partial y = -2$. Conditions (i), (ii), and (iii) hold, so f has a relative minimum at $(0, 0)$ ▲

If $D < 0$ in Theorem 5, then we have a saddle point. In fact, one can prove that $f(x, y)$ is larger than $f(x_0, y_0)$ as we move away from (x_0, y_0) in some direction and smaller in the orthogonal direction (see Exercise 26). The general appearance is thus similar to that shown in Figure 3.3.2. The appearance of the graph near (x_0, y_0) in the case $D = 0$ must be examined by further analysis.

We summarize the procedure for dealing with functions of two variables: After all critical points have been found and their associated Hessians computed, some of these Hessians may be positive-definite, indicating relative minima; some may be negative-definite, indicating relative maxima; and some may be neither positive nor negative-definite, indicating saddle points. The shape of the graph at a saddle point where $D < 0$ is like that in Figure 3.3.2. Critical points for which $D \neq 0$ are called ***nondegenerate critical points***. Such points are maxima, minima, or saddle points. The remaining critical points, where $D = 0$, may be tested directly, with level sets and sections or by some other method. Such critical points are said to be ***degenerate***; the methods developed in this chapter fail to provide a picture of the behavior of a function near such points, so we examine them case by case.

EXAMPLE 7   Locate the relative maxima, minima, and saddle points of the function

$$f(x, y) = \log (x^2 + y^2 + 1).$$

SOLUTION   We must first locate the critical points of this function; therefore, according to Theorem 3, we calculate

$$\nabla f(x, y) = \frac{2x}{x^2 + y^2 + 1}\, \mathbf{i} + \frac{2y}{x^2 + y^2 + 1}\, \mathbf{j}.$$

Thus $\nabla f(x, y) = \mathbf{0}$ if and only if $(x, y) = (0, 0)$, and so the only critical point of f is $(0, 0)$. Now we must determine whether this is a maximum, a minimum, or a saddle point. The second partial derivatives are

$$\frac{\partial^2 f}{\partial x^2} = \frac{2(x^2 + y^2 + 1) - (2x)(2x)}{(x^2 + y^2 + 1)^2},$$

$$\frac{\partial^2 f}{\partial y^2} = \frac{2(x^2 + y^2 + 1) - (2y)(2y)}{(x^2 + y^2 + 1)^2},$$

and

$$\frac{\partial^2 f}{\partial x\, \partial y} = \frac{-2x(2y)}{(x^2 + y^2 + 1)^2}.$$

Therefore

$$\frac{\partial^2 f}{\partial x^2}(0, 0) = 2 = \frac{\partial^2 f}{\partial y^2}(0, 0) \qquad \text{and} \qquad \frac{\partial^2 f}{\partial x \, \partial y}(0, 0) = 0,$$

which yields

$$D = 2 \cdot 2 = 4 > 0.$$

Since $(\partial^2 f / \partial x^2)(0, 0) > 0$, we concluded by Theorem 6 that $(0, 0)$ is a local minimum. (Can you show this just from the fact that $\log t$ is an increasing function of $t > 0$?) ▲

EXAMPLE 8   The graph of the function $g(x, y) = 1/xy$ is a surface S in $\mathbb{R}^3$. Find the points on S that are closest to the origin $(0, 0, 0)$. See Figure 3.3.5.

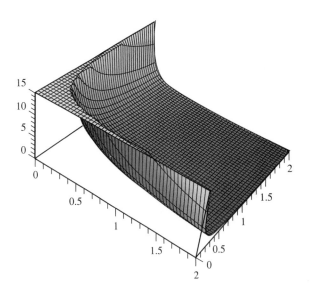

Figure 3.3.5   The surface $z = 1/xy$ in the first quadrant.

SOLUTION   Each point on S is of the form $(x, y, 1/xy)$. The distance from this point to the origin is

$$d(x, y) = \sqrt{x^2 + y^2 + \frac{1}{x^2 y^2}}.$$

It is easier to work with the square of d, so let $f(x, y) = x^2 + y^2 + (1/x^2 y^2)$, which will have the same minimum point. Notice that $f(x, y)$ becomes very large as x and y get large and larger as (x, y) approaches the x or y axis where f is not defined, so f must attain a minimum at some critical point. The critical points are determined by:

$$\frac{\partial f}{\partial x} = 2x - \frac{2}{x^3 y^2} = 0,$$

$$\frac{\partial f}{\partial y} = 2y - \frac{2}{y^3 x^2} = 0,$$

that is, $x^4 y^2 - 1 = 0$, and $x^2 y^4 - 1 = 0$. From the first equation we get $y^2 = 1/x^4$, and, substituting this into the second equation, we obtain

$$\frac{x^2}{x^8} = 1 = \frac{1}{x^6}.$$

Thus $x = \pm 1$ and $y = \pm 1$, and it therefore follows that f has four critical points, namely, $(1, 1)$, $(1, -1)$, $(-1, 1)$, and $(-1, -1)$. Note that f has the value 3 for all these points, so they are all minima. Therefore, the points on the surface closest to the point $(0, 0, 0)$ are $(1, 1, 1)$, $(1, -1, -1)$, $(-1, 1, -1)$, and $(-1, -1, 1)$ and the minimum distance is $\sqrt{3}$. ▲

EXAMPLE 9 Analyze the behavior of $z = x^5 y + x y^5 + xy$ at its critical points.

SOLUTION The first partial derivatives are

$$\frac{\partial z}{\partial x} = 5x^4 y + y^5 + y = y(5x^4 + y^4 + 1)$$

and

$$\frac{\partial z}{\partial y} = x(5y^4 + x^4 + 1).$$

The terms $5x^4 + y^4 + 1$ and $5y^4 + x^4 + 1$ are always greater than or equal to 1, and so it follows that the only critical point is $(0, 0)$.

The second partial derivatives are

$$\frac{\partial^2 z}{\partial x^2} = 20x^3 y, \qquad \frac{\partial^2 z}{\partial y^2} = 20xy^3,$$

and

$$\frac{\partial^2 z}{\partial x \, \partial y} = 5x^4 + 5y^4 + 1.$$

Thus at $(0, 0)$, $D = -1$, and so $(0, 0)$ is a nondegenerate saddle point and the graph of z near $(0, 0)$ looks like Figure 3.3.2. ▲

We end this section with a discussion of the theory of *absolute*, or *global*, maxima and minima of functions of several variables. Unfortunately, the location of absolute maxima and minima for functions on $\mathbb{R}^n$ is, in general, a more difficult problem than for functions of one variable.

DEFINITION   Suppose $f: A \to \mathbb{R}$ is a function on a set A in $\mathbb{R}^2$ or $\mathbb{R}^3$. A point $\mathbf{x}_0 \in A$ is said to be an ***absolute maximum*** (or ***absolute minimum***) point of f if $f(\mathbf{x}) \leq f(\mathbf{x}_0)$ [or $f(\mathbf{x}) \geq f(\mathbf{x}_0)$] for all $\mathbf{x} \in A$.

In one-variable calculus one learns that every continuous function on a closed interval I assumes its absolute maximum (or minimum) value at some point $\mathbf{x}_0$ in I. A generalization of this theoretical fact also holds in $\mathbb{R}^n$.

DEFINITION   A set $D \subset \mathbb{R}^n$ is said to be ***bounded*** if there is a number $M > 0$ such that $\|\mathbf{x}\| < M$ for all $\mathbf{x} \in D$. A set is ***closed*** if it contains all its boundary points.

Thus a set is bounded if it can be strictly contained in some (large) ball. The appropriate generalization of the one-variable theorem on maxima and minima is the following result, stated without proof.

THEOREM 7: MAXIMUM-MINIMUM THEOREM   *Let D be closed and bounded in $\mathbb{R}^n$ and let $f: D \to \mathbb{R}$ be continuous. Then f assumes its absolute maximum and minimum values at some points $\mathbf{x}_0$ and $\mathbf{x}_1$ of D.*

Simply stated, $\mathbf{x}_0$ and $\mathbf{x}_1$ are points where f assumes its largest and smallest values. As in one-variable calculus, these points need not be uniquely determined.

Suppose now that $D = U \cup \partial U$, where U is open and ∂U is its boundary. If $D \subset \mathbb{R}^2$, we suppose that ∂U is a piecewise smooth curve (i.e., D is a region bounded by a collection of smooth curves (such as a square) (Figure 3.3.6). Assume further that D is closed and bounded.

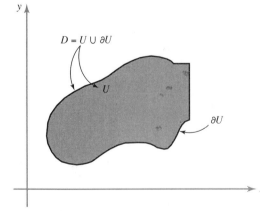

Figure 3.3.6 $D = U \cup \partial U$, a region whose boundary is a piecewise smooth curve.

If $\mathbf{x}_0$ and $\mathbf{x}_1$ are in U, we know by Theorem 4 that they are critical points of f. If they are in ∂U, and ∂U is a smooth curve (i.e., the image of a smooth path $\mathbf{c}$ with $\mathbf{c}' \neq 0$), then they are maximum or minimum points of f viewed as a function on ∂U.

This gives us a method of finding the absolute maximum and minimum values of f on a region D. See Figure 3.3.7.

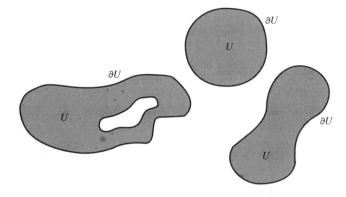

Figure 3.3.7 Regions whose boundaries are smooth closed curves.

THEOREM 8: FINDING THE ABSOLUTE MAXIMUM AND MINIMUM OF f ON A REGION $D = U \cup \partial U$

   Let f be a differentiable function of two variables defined on a closed and bounded region D in $\mathbb{R}^2$, which is bounded by a smooth closed curve. To find the absolute maximum and minimum of f on D:

 (i) *Locate all critical points for f in U.*

 (ii) *Find the maximum and minimum of f viewed as a function only on ∂U.*

(iii) *Compute the value of f at all of these critical points.*

(iv) *Compare all these values and select the largest and the smallest.*

If D is a region bounded by a collection of smooth curves (such as a square), then one follows a similar procedure, but including in step (iii) the points where the curves meet (such as the corners of the square).

All the steps except step (ii) should now be familiar to the student. To carry out step (ii) in the plane, one way is to find a smooth parametrization of ∂U; that is, we find a path $\mathbf{c}: I \to \partial U$, where I is some interval, which is onto ∂U. Second, we consider the function of one variable $t \mapsto f(\mathbf{c}(t))$, where $t \in I$, and locate the maximum and minimum points $t_0, t_1 \in I$ (remember to check the endpoints!). Then $\mathbf{c}(t_0), \mathbf{c}(t_1)$ will be maximum and minimum *points* for f as a function on ∂U. Another method for dealing with step (ii) is the Lagrange multiplier method, to be presented in the next section.

EXAMPLE 10   Find the maximum and minimum values of the function $f(x, y) = x^2 + y^2 - x - y + 1$ in the disk D defined by $x^2 + y^2 \le 1$.

SOLUTION   (i) To find the critical points we set $\partial f/\partial x = \partial f/\partial y = 0$. Thus, $2x - 1 = 0$, $2y - 1 = 0$, and hence $(x, y) = (\frac{1}{2}, \frac{1}{2})$ is the only critical point in the open disk $U = \{(x, y) \mid x^2 + y^2 < 1\}$.

(ii) The boundary ∂U can be parametrized by $\mathbf{c}(t) = (\sin t, \cos t)$, $0 \le t \le 2\pi$. Thus

$$f(\mathbf{c}(t)) = \sin^2 t + \cos^2 t - \sin t - \cos t + 1 = 2 - \sin t - \cos t = g(t).$$

To find the maximum and minimum of f on ∂U it suffices to locate the maximum and minimum of g. Now $g'(t) = 0$ only when

$$\sin t = \cos t, \qquad \text{that is,} \qquad t = \frac{\pi}{4}, \frac{5\pi}{4}.$$

Thus the candidates for the maximum and minimum for f on ∂U are the points $\mathbf{c}(\frac{\pi}{4})$, $\mathbf{c}(\frac{5}{4}\pi)$ and the endpoints $\mathbf{c}(0) = \mathbf{c}(2\pi)$.

(iii) The values of f at the critical points are: $f(\frac{1}{2}, \frac{1}{2}) = \frac{1}{2}$ from step (i) and, from step (ii),

$$f\left(\mathbf{c}\left(\frac{\pi}{4}\right)\right) = f\left(\frac{\sqrt{2}}{2}, \frac{\sqrt{2}}{2}\right) = 2 - \sqrt{2},$$

$$f\left(\mathbf{c}\left(\frac{5\pi}{4}\right)\right) = f\left(-\frac{\sqrt{2}}{2}, -\frac{\sqrt{2}}{2}\right) = 2 + \sqrt{2},$$

and

$$f(\mathbf{c}(0)) = f(\mathbf{c}(2\pi)) = f(0, 1) = 1.$$

(iv) Comparing all the values $\frac{1}{2}$, $2 - \sqrt{2}$, $2 + \sqrt{2}$, 1, it is clear that the absolute minimum occurs at $(\frac{1}{2}, \frac{1}{2})$ and the absolute maximum occurs at $(-\sqrt{2}/2, -\sqrt{2}/2)$.   ▲

In Section 3.4 we shall consider a generalization of Theorem 8 to regions D in $\mathbb{R}^n$.

EXERCISES

In Exercises 1 to 16, find the critical points of the given function and then determine whether they are local maxima, local minima, or saddle points.

1. $f(x, y) = x^2 - y^2 + xy$

2. $f(x, y) = x^2 + y^2 - xy$

3. $f(x, y) = x^2 + y^2 + 2xy$

4. $f(x, y) = x^2 + y^2 + 3xy$

5. $f(x, y) = e^{1+x^2-y^2}$

6. $f(x, y) = x^2 - 3xy + 5x - 2y + 6y^2 + 8$

7. $f(x, y) = 3x^2 + 2xy + 2x + y^2 + y + 4$

8. $f(x, y) = \sin(x^2 + y^2)$ [consider only the critical point $(0, 0)$]

9. $f(x, y) = \cos(x^2 + y^2)$ [consider only the critical point $(0, 0)$, $(\sqrt{\pi/2}, \sqrt{\pi/2})$, and $(0, \sqrt{\pi})$]

10. $f(x, y) = y + x \sin y$

11. $f(x, y) = e^x \cos y$

12. $f(x, y) = (x - y)(xy - 1)$

13. $f(x, y) = xy + \dfrac{1}{x} + \dfrac{1}{y}$

14. $f(x, y) = \log(2 + \sin xy)$

15. $f(x, y) = x \sin y$

16. $f(x, y) = (x + y)(xy + 1)$

17. Find the local maxima and minima for $z = (x^2 + 3y^2)e^{1-x^2-y^2}$. (See Figure 3.3.8.)

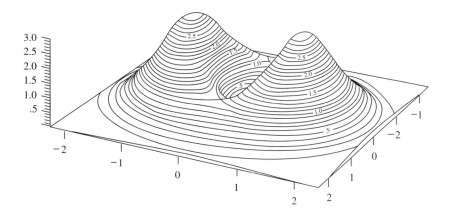

Figure 3.3.8 The graph of $z = (x^2 + 3y^2)e^{1-(x^2+y^2)}$.

18. Let $f(x, y) = x^2 + y^2 + kxy$. If you imagine the graph changing as k increases, at what values of k does the shape of the graph change qualitatively?

19. An examination of the function $f: \mathbb{R}^2 \to \mathbb{R}, (x, y) \mapsto (y - 3x^2)(y - x^2)$ will give an idea of the difficulty of finding conditions that guarantee that a critical point is a relative extremum when Theorem 6 fails. Show that

(a) the origin is a critical point of f;

(b) f has a relative minimum at $(0, 0)$ on every straight line through $(0, 0)$; that is, if $g(t) = (at, bt)$, then $f \circ g: \mathbb{R} \to \mathbb{R}$ has a relative minimum at 0, for every choice of a and b;

(c) the origin is not a relative minimum of f.

20. Let $f(x, y) = Ax^2 + E$ where A and E are constants. What are the critical points of f? Are they local maxima or local minima?

21. Let $f(x, y) = x^2 - 2xy + y^2$. Here $D = 0$. Can you say whether the critical points are local minima, local maxima, or saddle points?

22. Find the point in the plane $2x - y + 2z = 20$ nearest the origin.

23. Show that rectangular box of given volume has minimum surface area when the box is a cube.

24. Show that the rectangular parallelepiped with fixed surface area and maximum volume is a cube.

25. Write the number 120 as a sum of three numbers so that the sum of the products taken two at a time is a maximum.

26. Show that if (x_0, y_0) is a critical point of a quadratic function $f(x, y)$ and $D < 0$ then there are points (x, y) near (x_0, y_0) at which $f(x, y) > f(x_0, y_0)$ and, similarly, points for which $f(x, y) < f(x_0, y_0)$.

27. Determine the nature of the critical points of the function

$$f(x, y, z) = x^2 + y^2 + z^2 + xy.$$

28. Let n be an integer greater than 2 and set $f(x, y) = ax^n + cy^n$, where $ac \neq 0$. Determine the nature of the critical points of f.

29. Determine the nature of the critical points of $f(x, y) = x^3 + y^2 - 6xy + 6x + 3y$.

30. Find the absolute maximum and minimum values of the function $f(x, y) = (x^2 + y^2)^4$ on the disk $x^2 + y^2 \leq 1$. (You do not *have* to use calculus.)

31. Repeat Exercise 30 for the function $f(x, y) = x^2 + xy + y^2$.

32. A curve C in space is defined *implicitly* on the cylinder $x^2 + y^2 = 1$ by the additional equation $x^2 - xy + y^2 - z^2 = 1$. Find the point or points on C closest to the origin.

33. Find the absolute maximum and minimum values for $f(x, y) = \sin x + \cos y$ on the rectangle R defined by $0 \leq x \leq 2\pi, 0 \leq y \leq 2\pi$.

34. Find the absolute maximum and minimum values for the function $f(x, y) = xy$ on the rectangle R defined by $-1 \leq x \leq 1, -1 \leq y \leq 1$.

35. Determine the nature of the critical points of $f(x, y) = xy + 1/x + 8/y$.

In Exercises 36 through 40, D denotes the unit disk.

36. Let u be a function on D which is "strictly subharmonic"; that is, $\nabla^2 u = (\partial^2 u/\partial x^2) + (\partial^2 u/\partial y^2) > 0$. Show that u cannot have a maximum point in $D \backslash \partial D$ (the set of points in D but not in ∂D).

37. Let u be a harmonic function; that is, $\nabla^2 u = 0$. Show that if u achieves its maximum value in $D \backslash \partial D$ it also achieves it on ∂D. This is sometimes called the "weak maximum principle" for harmonic functions. [HINT: Consider $\nabla^2(u + \varepsilon e^x)$, $\varepsilon > 0$. You can use the following fact, which is proved in more advanced texts: Given a sequence $\{\mathbf{p}_n\}$, $n = 1, 2, \ldots$, of points in a closed bounded set A in $\mathbb{R}^2$ or $\mathbb{R}^3$, there exists a point $\mathbf{q}$ such that every neighborhood of $\mathbf{q}$ contains at least one member of $\{\mathbf{p}_n\}$.]

38. Define the notion of a strict superharmonic function u on D by mimicking Exercise 36. Show that u cannot have a minimum in $D \backslash \partial D$.

39. Let u be harmonic in D as in Exercise 37. Show that if u achieves its minimum value in $D \backslash \partial D$ it also achieves it on ∂D. This is sometimes called the "weak minimum principle" for harmonic functions.

40. Let $\phi \colon \partial D \to \mathbb{R}$ be continuous and let T be a solution on D to $\nabla^2 T = 0$, $T = \phi$ on ∂D.
    (a) Use Exercises 36 to 39 to show that such a solution, if it exists, must be unique.
    (b) Suppose that $T(x, y)$ represents a temperature function that is independent of time, with ϕ representing the temperature of a circular plate at its boundary. Can you give a physical interpretation of the principle stated in part (a)?

41. (a) Let f be a C^1 function on the real line $\mathbb{R}$. Suppose that f has exactly one critical point x_0 which is a strict local minimum for f. Show that x_0 is also an absolute minimum for f, that is, that $f(x) \geq f(x_0)$ for all x.
    (b) The next example shows that the conclusion of part (a) does not hold for functions of more than one variable. Let $f \colon \mathbb{R}^2 \to \mathbb{R}$ be defined by

$$f(x, y) = -y^4 - e^{-x^2} + 2y^2\sqrt{e^x + e^{-x^2}}.$$

(i) Show that $(0, 0)$ is the only critical point for f and that it is a local minimum.
(ii) Argue informally that f has no absolute minimum.

42. Suppose that a pentagon is composed of a rectangle topped by an isosceles triangle (see Figure 3.3.9). If the length of the perimeter is fixed, find the maximum possible area.

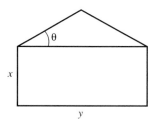

Figure 3.3.9 Maximize the area for fixed perimeter.

3.4 CONSTRAINED EXTREMA AND LAGRANGE MULTIPLIERS

Often we want to maximize a function subject to certain *constraints* or *side conditions*. For example, we might want to maximize $f(x, y)$ subject to the condition that $x^2 + y^2 = 1$; that is, that (x, y) lie on the unit circle. More generally we might need to maximize or minimize $f(x, y)$ subject to the side condition that (x, y) also satisfies an equation $g(x, y) = c$ where g is some function and c equals a constant [in the example above, $g(x, y) = x^2 + y^2$, and $c = 1$]. The set of such (x, y) is a level curve for g.

The purpose of this section is to develop some methods for handling this sort of problem. In Figure 3.4.1 we picture a graph of a function $f(x, y)$. In this picture, the maximum of f occurs at $(0, 0)$. However, suppose we are not interested in this maximum but only the maximum of $f(x, y)$ when (x, y) belongs to the unit circle; that is, when $x^2 + y^2 = 1$. The cylinder over $x^2 + y^2 = 1$ intersects the graph of $z = f(x, y)$ in a curve that lies on this graph. The problem of maximizing or minimizing $f(x, y)$ subject to the constraint $x^2 + y^2 = 1$ amounts to finding the point on this curve where z is the greatest or the least.

Figure 3.4.1   The geometric meaning of maximizing f subject to the constraint $x^2 + y^2 = 1$.

In general, let $f: U \subset \mathbb{R}^n \to \mathbb{R}$ and $g: U \subset \mathbb{R}^n \to \mathbb{R}$ be given C^1 functions, and let S be the level set for g with value c [recall that this is the set of points $\mathbf{x} \in \mathbb{R}^n$ with $g(\mathbf{x}) = c$].

When f is restricted to S we again have the notion of local maxima or local minima of f (local extrema), and an absolute maximum (largest value) or absolute minimum (smallest value) must be a local extremum. The following method provides a necessary condition for a constrained extremum:

THEOREM 9: THE METHOD OF LAGRANGE MULTIPLIERS
Let $f: U \subset \mathbb{R}^n \to \mathbb{R}$ and $g: U \subset \mathbb{R}^n \to \mathbb{R}$ be given C^1 real valued functions. Let $\mathbf{x}_0 \in U$ and $g(\mathbf{x}_0) = c$, and let S be the level set for g with value c (recall that this is the set of points $\mathbf{x} \in \mathbb{R}^n$ satisfying $g(\mathbf{x}) = c$). Assume $\nabla g(\mathbf{x}_0) \neq \mathbf{0}$.

If $f|S$, which denotes "f restricted to S," has a local maximum or minimum on S at $\mathbf{x}_0$, then there is a real number λ such that

$$\nabla f(\mathbf{x}_0) = \lambda \nabla g(\mathbf{x}_0). \tag{1}$$

In general, a point $\mathbf{x}_0$ where equation (1) holds is said to be a ***critical point*** of $f|S$.

PROOF Actually, we have not developed enough techniques to give a thorough proof, but we can provide the essential points. (The additional technicalities needed are given in Section 3.5).

Recall that for $n = 3$ the tangent space or tangent plane of S at $\mathbf{x}_0$ is the space orthogonal to $\nabla g(\mathbf{x}_0)$ (see Section 2.6), and for arbitrary n we can give the same definition for the tangent space of S at $\mathbf{x}_0$. This definition can be motivated by considering tangents to paths $\mathbf{c}(t)$ that lie in S, as follows: if $\mathbf{c}(t)$ is a path in S and $\mathbf{c}(0) = \mathbf{x}_0$, then $\mathbf{c}'(0)$ is a tangent vector to S at $\mathbf{x}_0$; but

$$\frac{d}{dt} g(\mathbf{c}(t)) = \frac{d}{dt} c = 0,$$

and on the other hand, by the chain rule,

$$\left. \frac{d}{dt} g(\mathbf{c}(t)) \right|_{t=0} = \nabla g(\mathbf{x}_0) \cdot \mathbf{c}'(0),$$

so that $\nabla g(\mathbf{x}_0) \cdot \mathbf{c}'(0) = 0$; that is, $\mathbf{c}'(0)$ is orthogonal to $\nabla g(\mathbf{x}_0)$.

If $f|S$ has a maximum at $\mathbf{x}_0$, then $f(\mathbf{c}(t))$ has a maximum at $t = 0$. By one-variable calculus, $df(\mathbf{c}(t))/dt|_{t=0} = 0$. Hence, by the chain rule,

$$0 = \left. \frac{d}{dt} f(\mathbf{c}(t)) \right|_{t=0} = \nabla f(\mathbf{x}_0) \cdot \mathbf{c}'(0).$$

Thus $\nabla f(\mathbf{x}_0)$ is perpendicular to the tangent of every curve in S and so is perpendicular to the whole tangent space to S at $\mathbf{x}_0$. Since the space perpendicular to this tangent space is a line, $\nabla f(\mathbf{x}_0)$ and $\nabla g(\mathbf{x}_0)$ are parallel. Since $\nabla g(\mathbf{x}_0) \neq \mathbf{0}$, it follows that $\nabla f(\mathbf{x}_0)$ is a multiple of $\nabla g(\mathbf{x}_0)$, which is the conclusion of the theorem. ∎

Let us extract some geometry from this proof.

COROLLARY *If f, when constrained to a surface S, has a maximum or minimum at $\mathbf{x}_0$, then $\nabla f(\mathbf{x}_0)$ is perpendicular to S at $\mathbf{x}_0$ (see Figure 3.4.2).*

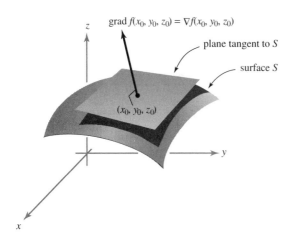

Figure 3.4.2   The geometry of constrained extrema.

These results tell us that in order to find the constrained extrema of f we must look among those points $\mathbf{x}_0$ satisfying the conclusions of the theorem or of the corollary. We shall give several illustrations of how to use each.

When the method of Theorem 9 is used, we look for a point $\mathbf{x}_0$ and a constant λ, called a ***Lagrange multiplier***, such that $\nabla f(\mathbf{x}_0) = \lambda \nabla g(\mathbf{x}_0)$. This method is more analytic in nature than the geometric method of the corollary.

Equation (1) says that the partial derivatives of f are proportional to those of g. Finding such points $\mathbf{x}_0$ at which this occurs means solving the simultaneous equations

$$\left.\begin{aligned}
\frac{\partial f}{\partial x_1}(x_1, \ldots, x_n) &= \lambda \frac{\partial g}{\partial x_1}(x_1, \ldots, x_n) \\[2mm]
\frac{\partial f}{\partial x_2}(x_1, \ldots, x_n) &= \lambda \frac{\partial g}{\partial x_2}(x_1, \ldots, x_n) \\[1mm]
&\;\;\vdots \\[1mm]
\frac{\partial f}{\partial x_n}(x_1, \ldots, x_n) &= \lambda \frac{\partial g}{\partial x_n}(x_1, \ldots, x_n) \\[2mm]
g(x_1, \ldots, x_n) &= c
\end{aligned}\right\} \tag{2}$$

for $x_1, \ldots, x_n$ and λ.

Another way of looking at these equations is as follows: Think of λ as an additional variable and form the auxiliary function

$$h(x_1, \ldots, x_n, \lambda) = f(x_1, \ldots, x_n) - \lambda[g(x_1, \ldots, x_n) - c].$$

The Lagrange multiplier theorem says that to find the extreme points of $f|S$, we should examine the critical points of h. These are found by solving the equations

$$
\left.
\begin{aligned}
0 &= \frac{\partial h}{\partial x_1} = \frac{\partial f}{\partial x_1} - \lambda \frac{\partial g}{\partial x_1} \\
&\;\;\vdots \\
0 &= \frac{\partial h}{\partial x_n} = \frac{\partial f}{\partial x_n} - \lambda \frac{\partial g}{\partial x_n} \\
0 &= \frac{\partial h}{\partial \lambda} = g(x_1, \ldots, x_n) - c
\end{aligned}
\right\}
\tag{3}
$$

which are the same as equations (2) above.

Second-derivative tests for maxima and minima analogous to those in Section 3.3 will be given in Theorem 10 below. However, in many problems it is possible to distinguish between maxima and minima by geometric means. Since this is often simpler, we consider examples of the latter type first.

EXAMPLE 1   Let $S \subset \mathbb{R}^2$ be the line through $(-1, 0)$ inclined at $45°$, and let $f\colon \mathbb{R}^2 \to \mathbb{R}, (x, y) \mapsto x^2 + y^2$. Find the extrema of $f|S$.

SOLUTION   Here $S = \{(x, y) \mid y - x - 1 = 0\}$, and therefore we set $g(x, y) = y - x - 1$ and $c = 0$. We have $\nabla g(x, y) = -\mathbf{i} + \mathbf{j} \neq \mathbf{0}$. The relative extrema of $f|S$ must be found among the points at which ∇f is orthogonal to S, that is, inclined at $-45°$. But $\nabla f(x, y) = (2x, 2y)$, which has the desired slope only when $x = -y$, or when (x, y) lies on the line L through the origin inclined at $-45°$. This can occur in the set S only for the single point at which L and S intersect (see Figure 3.4.3). Reference to the level curves of f indicates that this point, $(-\frac{1}{2}, \frac{1}{2})$, is a relative minimum of $f|S$ (but not of f).   ▲

Figure 3.4.3   The geometry associated with finding the extrema of $f(x, y) = x^2 + y^2$ restricted to $S = \{(x, y) \mid y - x - 1 = 0\}$.

EXAMPLE 2   Let $f\colon \mathbb{R}^2 \to \mathbb{R}, (x, y) \mapsto x^2 - y^2$, and let S be the circle of radius 1 around the origin. Find the extrema of $f|S$.

SOLUTION   The set S is the level curve for g with value 1, where $g\colon \mathbb{R}^2 \to \mathbb{R}$, $(x, y) \mapsto$ $x^2 + y^2$. Since both of these functions have been studied in previous examples, we know their level curves; these are shown in Figure 3.4.4. In two dimensions, the condition that $\nabla f = \lambda \nabla g$ at $\mathbf{x}_0$, i.e., that ∇f and ∇g are parallel at $\mathbf{x}_0$ is the same as the level curves being tangent at $\mathbf{x}_0$ (why?). Thus the extreme points of $f|S$ are $(0, \pm 1)$ and $(\pm 1, 0)$. Evaluating f, we find $(0, \pm 1)$ are minima and $(\pm 1, 0)$ are maxima.

Let us also do this problem analytically by the method of Lagrange multipliers. Clearly,

$$\nabla f(x, y) = \left(\frac{\partial f}{\partial x}, \frac{\partial f}{\partial y} \right) = (2x, -2y) \quad \text{and} \quad \nabla g(x, y) = (2x, 2y).$$

Note that $\nabla g(x, y) \neq \mathbf{0}$ if $x^2 + y^2 = 1$. Thus, according to the Lagrange multiplier theorem, we must find a λ such that

$$(2x, -2y) = \lambda(2x, 2y) \quad \text{and} \quad (x, y) \in S, \quad \text{i.e., } x^2 + y^2 = 1.$$

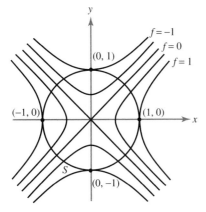

Figure 3.4.4   The geometry associated with the problem of finding the extrema of $x^2 - y^2$ on $S = \{(x, y) \mid x^2 + y^2 = 1\}$.

These conditions yield three equations, which can be solved for the three unknowns x, y, and λ. From $2x = \lambda 2x$ we conclude that either $x = 0$ or $\lambda = 1$. If $x = 0$, then $y = \pm 1$ and $-2y = \lambda 2y$ implies $\lambda = -1$. If $\lambda = 1$, then $y = 0$ and $x = \pm 1$. Thus we get the points $(0, \pm 1)$ and $(\pm 1, 0)$, as before. As we have mentioned, this method only locates potential extrema; whether they are maxima, minima, or neither must be determined by other means, such as geometric arguments or the second-derivative test given below.*   ▲

EXAMPLE 3   Maximize the function $f(x, y, z) = x + z$ subject to the constraint $x^2 + y^2 + z^2 = 1$.

---

*In these examples, $\nabla g(\mathbf{x}_0) \neq \mathbf{0}$ on the surface S, as required by the Lagrange multiplier theorem. If $\nabla g(\mathbf{x}_0)$ were zero for some $\mathbf{x}_0$ on S, then it would have to be included among the possible extrema.

SOLUTION By Theorem 7 we know that f restricted to the sphere $x^2 + y^2 + z^2 = 1$ has a maximum (and also a minimum). To find the maximum we again use the Lagrange multiplier theorem. We seek λ and (x, y, z) such that

$$1 = 2x\lambda, \qquad 0 = 2y\lambda \qquad \text{and} \qquad 1 = 2z\lambda,$$

and

$$x^2 + y^2 + z^2 = 1.$$

From the first or the third equation, we see that $\lambda \neq 0$. Thus, from the second equation, we get $y = 0$. From the first and third equations, $x = z$, and so from the fourth, $x = \pm 1/\sqrt{2} = z$. Hence our points are $(1/\sqrt{2}, 0, 1/\sqrt{2})$ and $(-1/\sqrt{2}, 0, -1/\sqrt{2})$. Comparing the values of f at these points, we can see that the first point yields the maximum of f (restricted to the constraint) and the second the minimum. ▲

EXAMPLE 4 Assume that among all rectangular boxes with fixed surface area of 10 square meters there is a box of largest possible volume. Find its dimensions.

SOLUTION If x, y, and z are the lengths of the sides, $x \geq 0, y \geq 0, z \geq 0$, respectively, and the volume is $f(x, y, z) = xyz$. The constraint is $2(xy + xz + yz) = 10$; that is, $xy + xz + yz = 5$. Thus, the Lagrange multiplier conditions are

$$yz = \lambda(y + z)$$
$$xz = \lambda(x + z)$$
$$xy = \lambda(y + x)$$
$$xy + xz + yz = 5.$$

First of all, $x \neq 0$, since $x = 0$ implies $yz = 5$ and $0 = \lambda z$, so that $\lambda = 0$ and we get the contradictory equation $yz = 0$. Similarly, $y \neq 0, z \neq 0, x + y \neq 0$. Elimination of λ from the first two equations gives $yz/(y+z) = xz/(x+z)$, which gives $x = y$; similarly, $y = z$. Substituting these values into the last equation, we obtain $3x^2 = 5$, or $x = \sqrt{5/3}$. Thus we get the solution $x = y = z = \sqrt{5/3}$, and $xyz = (5/3)^{3/2}$. This (cubical) shape must therefore maximize the volume, assuming there is a box of maximum volume. ▲

We should note that the solution to Example 4 does *not* demonstrate that the cube is the rectangular box of largest volume with a given fixed surface area; it proves that the cube is the only possible candidate for a maximum. We shall sketch a proof that it really is the maximum below. The distinction between showing that there is *only one possible solution* to a problem and that, in fact, *a solution exists* is a subtle one that many (even great) mathematicians have overlooked.

For example, Queen Dido (ca. 900 B.C.) realized that among all planar regions with fixed circumference, the disc is the region of maximum area. It is not difficult to prove this fact under the assumption that there is a region of maximum area; however, proving that such a region of maximum area exists is quite another (difficult) matter. A complete proof was not given until the second half of the nineteenth century by the German mathematician Weierstrass.

Let us consider a nonmathematical parallel to this situation. Put yourself in the place of Lord Peter Wimsey, Dorothy Sayers' famous detective:

> "Undoubtedly," said Wimsey, "but if you think that this identification is going to make life one grand, sweet song for you, you are mistaken.... Since we have devoted a great deal of time and thought to the case on the assumption that it was murder, it's a convenience to know that the assumption is correct."

Wimsey has found the body of a dead man, and after some time has located ten suspects. He is sure that no one else other than one of the suspects could be the murderer. By collecting all the evidence and checking alibis, he then reduces the number of suspects one by one, until, finally, only the butler remains; hence he is the murderer! But wait, Peter is a very cautious man. By checking everything once again, he discovers that the man died by suicide; so there is no murder. You see the point: It does not suffice to find a clear and uniquely determined suspect in a criminal case where murder is suspected; you must prove that a murder actually took place.

The same goes for our cube; the fact that it is the only possible candidate for a maximum does not prove that it is maximum. (For more information see S. Hildebrandt and A. Tromba, *The Parsimonious Universe*: *Shape and Form in the Natural World*, Springer-Verlag, New York/Berlin, 1995.)

The key to showing that $f(x, y, z) = xyz$ really has a maximum lies in the fact that f is a continuous function which is defined on the unbounded surface $S: xy + xz + yz = 5$, and not on a bounded set which includes its boundary. We have already seen problems of this sort for functions of one and two variables.

The way to show that $f(x, y, z) = xyz \geq 0$ does indeed have a maximum on $xy + yz + xz = 5$ is to show that if either x, y, or z tend to ∞, then $f(x, y, z) \to 0$. We may then conclude that the maximum of f on S must exist by appealing to Theorem 7 (the student should supply the details). So, suppose (x, y, z) lies in S and $x \to \infty$, then $y \to 0$ and $z \to 0$ (why?). Multiplying the equation defining S by z we obtain the equation $xyz + xz^2 + yz^2 = 5z \to 0$ as $x \to \infty$. Since $x, y, z \geq 0$, $xyz = f(x, y, z) \to 0$. Similarly, $f(x, y, z) \to 0$ if either y or z tend to ∞. Thus a box of maximum volume must exist.

Some general guidelines may be useful for maximum and minimum problems with constraints. First of all, if the surface S is bounded (as an ellipsoid is, for example), then f must have a maximum and a minimum on S. (See Theorem 7 in the preceding section.) In particular, if f has only two points satisfying the conditions of the Lagrange multiplier theorem or its corollary, then one must be a maximum and one must be a minimum. Evaluating f at each point will tell the maximum from the minimum. However, if there are more than two such points, some can be saddle points. Also, if S is not bounded (for example, if it is a hyperboloid), then f need not have any maxima nor minima.

If a surface S is defined by a number of constraints, namely,

$$\left.\begin{array}{c} g_1(x_1, \ldots, x_n) = c_1 \\ g_2(x_1, \ldots, x_n) = c_2 \\ \vdots \\ g_k(x_1, \ldots, x_n) = c_k \end{array}\right\}, \tag{4}$$

then the Lagrange multiplier theorem may be generalized as follows: *If f has a maximum or a minimum at $\mathbf{x}_0$ on S, there must exist constants $\lambda_1, \ldots, \lambda_k$ such that*[*]

$$\nabla f(\mathbf{x}_0) = \lambda_1 \nabla g_1(\mathbf{x}_0) + \cdots + \lambda_k \nabla g_k(\mathbf{x}_0). \tag{5}$$

This case may be proved by generalizing the method used to prove the Lagrange multiplier theorem. Let us give an example of how this more general formulation is used.

EXAMPLE 5   Find the extreme points of $f(x, y, z) = x + y + z$ subject to the two conditions $x^2 + y^2 = 2$ and $x + z = 1$.

SOLUTION   Here there are two constraints:

$$g_1(x, y, z) = x^2 + y^2 - 2 = 0, \qquad \text{and} \qquad g_2(x, y, z) = x + z - 1 = 0.$$

Thus we must find x, y, z, λ_1, and λ_2 such that

$$\nabla f(x, y, z) = \lambda_1 \nabla g_1(x, y, z) + \lambda_2 \nabla g_2(x, y, z),$$
$$g_1(x, y, z) = 0, \qquad \text{and} \qquad g_2(x, y, z) = 0.$$

Computing the gradients and equating components, we get

$$1 = \lambda_1 \cdot 2x + \lambda_2 \cdot 1,$$
$$1 = \lambda_1 \cdot 2y + \lambda_2 \cdot 0,$$
$$1 = \lambda_1 \cdot 0 + \lambda_2 \cdot 1,$$
$$x^2 + y^2 = 2, \qquad \text{and} \qquad x + z = 1.$$

These are five equations for x, y, z, λ_1, and λ_2. From the third equation, $\lambda_2 = 1$, and so $2x\lambda_1 = 0, 2y\lambda_1 = 1$. Since the second implies $\lambda_1 \neq 0$, we have $x = 0$. Thus $y = \pm\sqrt{2}$ and $z = 1$. Hence the desired extrema are $(0, \pm\sqrt{2}, 1)$. By inspection, $(0, \sqrt{2}, 1)$ gives a relative maximum, and $(0, -\sqrt{2}, 1)$ a relative minimum.

The condition $x^2 + y^2 = 2$ implies that x and y must be bounded. The condition $x + z = 1$ implies that z is also bounded. If follows that the constraint set S is closed and

---

[*]As with the hypothesis $\nabla g(\mathbf{x}_0) \neq \mathbf{0}$ in the Lagrange multiplier theorem, here one must assume that the vectors $\nabla g_1(\mathbf{x}_0), \ldots, \nabla g_k(\mathbf{x}_0)$ are linearly independent.

bounded. By Theorem 7 it follows that f has a maximum and minimum on S that must therefore occur at $(0, \sqrt{2}, 1)$ and $(0, -\sqrt{2}, 1)$, respectively. ▲

The method of Lagrange multipliers provides us with another tool to locate the absolute maxima and minima of differentiable functions on bounded regions in $\mathbb{R}^2$ (see Theorem 8).

EXAMPLE 6 Find the absolute maximum of $f(x, y) = xy$ on the unit disk D, where D is the set of points (x, y) with $x^2 + y^2 \leq 1$.

SOLUTION First we find all the critical points of f in U, the set of points (x, y) with $x^2 + y^2 < 1$. Since

$$\frac{\partial f}{\partial x} = y \quad \text{and} \quad \frac{\partial f}{\partial y} = x,$$

$(0, 0)$ is the only critical point of f in U. Now consider f on the unit circle, the level curve $g(x, y) = 1$, where $g(x, y) = x^2 + y^2$. To locate the maximum and minimum of f on C, we write down the Lagrange multiplier equations: $\nabla f(x, y) = (y, x) = \lambda \nabla g(x, y) = \lambda(2x, 2y)$ and $x^2 + y^2 = 1$. Rewriting these in component form, we get

$$y = 2\lambda x,$$
$$x = 2\lambda y,$$
$$x^2 + y^2 = 1.$$

Thus,

$$y = 4\lambda^2 y$$

or $\lambda = \pm 1/2$ and $y = \pm x$ which means that $x^2 + x^2 = 2x^2 = 1$ or $x = \pm 1/\sqrt{2}$, $y = \pm 1/\sqrt{2}$. On C we compute four candidates for the absolute maximum and minimum, namely,

$$\left(-\frac{1}{\sqrt{2}}, -\frac{1}{\sqrt{2}}\right), \left(-\frac{1}{\sqrt{2}}, \frac{1}{\sqrt{2}}\right), \left(\frac{1}{\sqrt{2}}, \frac{1}{\sqrt{2}}\right), \left(\frac{1}{\sqrt{2}}, -\frac{1}{\sqrt{2}}\right).$$

The value of f at both $(-1/\sqrt{2}, -1/\sqrt{2})$ and $(1/\sqrt{2}, 1/\sqrt{2})$ is $1/2$. The value of f at $(-1/\sqrt{2}, 1/\sqrt{2})$ and $(1/\sqrt{2}, -1/\sqrt{2})$ is $-1/2$, and the value of f at $(0, 0)$ is 0. Therefore the absolute maximum of f is $1/2$ and the absolute minimum is $-1/2$, both occurring on C. At $(0, 0)$, $\partial^2 f/\partial x^2 = 0$, $\partial^2 f/\partial y^2 = 0$ and $\partial^2 f/\partial x \, \partial y = 1$, so the discriminant is -1 and thus $(0, 0)$ is a saddle point. ▲

EXAMPLE 7 Find the absolute maximum and minimum of $f(x, y) = \frac{1}{2}x^2 + \frac{1}{2}y^2$ in the elliptical region D defined by $\frac{1}{2}x^2 + y^2 \leq 1$.

SOLUTION We first locate the critical points of f in U, the set of points (x, y) with $\frac{1}{2}x^2 + y^2 < 1$. Since

$$\frac{\partial f}{\partial x} = x, \qquad \frac{\partial f}{\partial y} = y,$$

the only critical point is the origin $(0, 0)$.

We now find the maximum and minimum of f on C, the boundary of U, which is the level curve $g(x, y) = 1$, where $g(x, y) = \frac{1}{2}x^2 + y^2$. The Lagrange multiplier equations are

$$\nabla f(x, y) = (x, y) = \lambda \nabla g(x, y) = \lambda(x, 2y)$$

and $(x^2/2) + y^2 = 1$. In other words,

$$x = \lambda x$$
$$y = 2\lambda y$$
$$\frac{x^2}{2} + y^2 = 1.$$

If $x = 0$, then $y = \pm 1$ and $\lambda = \frac{1}{2}$. If $y = 0$, then $x = \pm\sqrt{2}$ and $\lambda = 1$. If $x \neq 0$ and $y \neq 0$ we get both $\lambda = 1$ and $\frac{1}{2}$, which is impossible. Thus the candidates for the maxima and minima of f on C are $(0, \pm 1)$, $(\pm\sqrt{2}, 0)$ and for f inside D, the candidate is $(0, 0)$. The value of f at $(0, \pm 1)$ is $\frac{1}{2}$, at $(\pm\sqrt{2}, 0)$ it is 1 and at $(0, 0)$ it is 0. Thus the absolute minimum of f occurs at $(0, 0)$ and is 0. The absolute maximum of f on D is thus 1 and occurs at the points $(\pm\sqrt{2}, 0)$. ▲

Using the nature of a level set we can now generalize Theorem 8 to $\mathbb{R}^n$.

DEFINITION Let U be an open region in $\mathbb{R}^n$ with boundary ∂U. We say that ∂U is *smooth* if ∂U is the level set of a smooth function g whose gradient ∇g never vanishes (i.e., $\nabla g \neq 0$). Then we have the following theorem.

THEOREM 8′: FINDING THE ABSOLUTE MAXIMUM AND MINIMUM OF f ON $D = U \cup \partial U$ IN $\mathbb{R}^n$ Let f be a differentiable function on a closed and bounded region $D = U \cup \partial U$, U open in $\mathbb{R}^n$, with ∂U smooth.

To find the absolute maximum and minimum of f on D:

(i) Locate all critical points of f in U.

(ii) Use the method of Lagrange multiplier to locate all the critical points of $f|\partial U$.

(iii) Compare the value of f at all of these critical points.

(iv) Compute all of these values and select the largest and the smallest.

EXAMPLE 8   Find the absolute maximum and minimum of $f(x, y, z) = x + y + z$ on $D = \{(x, y, z) \mid x^2 + y^2 + z^2 \leq 1\}$.

SOLUTION $D = U \cup \partial U$, where

$$U = \{(x, y, z) \mid x^2 + y^2 + z^2 < 1\}$$

and

$$\partial U = \{(x, y, z) \mid x^2 + y^2 + z^2 = 1\}.$$

Now $\nabla f = (1, 1, 1)$, and so f has no critical points in U. Therefore the maximum and minimum values of f must occur on ∂U.

Let $g(x, y, z) = x^2 + y^2 + z^2$. Then ∂U is the level set $g(x, y, z) = 1$. By the method of Lagrange multipliers, the maximum and minimum must occur at a critical point of $f \mid \partial U$, that is, at a point $\mathbf{x}_0$ where $\nabla f(\mathbf{x}_0) = \lambda \nabla g(\mathbf{x}_0)$ for some scalar λ. Thus,

$$(1, 1, 1) = \lambda(2x, 2y, 2z); \qquad \text{that is,} \qquad x = \frac{\lambda}{2}, \qquad y = \frac{\lambda}{2}, \qquad z = \frac{\lambda}{2}.$$

Since $x^2 + y^2 + z^2 = 1$, we obtain $\lambda = \pm 2/\sqrt{3}$ and so $\mathbf{x}_0 = \pm(1/\sqrt{3}, 1/\sqrt{3}, 1/\sqrt{3})$. Clearly, $-(1/\sqrt{3}, 1/\sqrt{3}, 1/\sqrt{3})$ is the point where f assumes its absolute minimum (namely, $-\sqrt{3}$) and $(1/\sqrt{3}, 1/\sqrt{3}, 1\sqrt{3})$, the point where f assumes its maximum value $\sqrt{3}$.   ▲

Supplement to Section 3.4   A Second-Derivative Test
for Constrained Extrema

In Section 3.3, we developed a second-derivative test for extrema of functions of several variables by looking at the second-degree term in the Taylor series of f. If the Hessian matrix of second partial derivatives is either positive-definite or negative-definite at a critical point of f, this point is a relative minimum or maximum, respectively.

As above, however, we are not interested in all values of f but only in those obtained by restricting f to a set S that is the level set of another function g. The situation is complicated, first, because the constrained extrema of f need not come at critical points of f and, second, because the variable is allowed to move only in the set S. Nevertheless, a second-derivative test can be given in terms of what is called a *bordered Hessian*. We will show how this arises for the case of a function $f(x, y)$ of two variables subject to the constraint $g(x, y) = c$.

According to the remarks following the Lagrange multiplier theorem, the constrained extrema of f are found by looking at the critical points of the auxiliary function

$$h(x, y, \lambda) = f(x, y) - \lambda(g(x, y) - c).$$

Suppose (x_0, y_0, λ) is such a point and let $\mathbf{v}_0 = (x_0, y_0)$. That is,

$$\left.\frac{\partial f}{\partial x}\right|_{\mathbf{v}_0} = \lambda \left.\frac{\partial g}{\partial x}\right|_{\mathbf{v}_0}, \qquad \left.\frac{\partial f}{\partial y}\right|_{\mathbf{v}_0} = \lambda \left.\frac{\partial g}{\partial y}\right|_{\mathbf{v}_0}, \qquad \text{and} \qquad g(x_0, y_0) = c.$$

In a sense this is a one-variable problem. If the function g is at all reasonable, then the set S defined by $g(x, y) = c$ is a curve and we are interested in how f varies as we move along this curve. If we can solve the equation $g(x, y) = c$ for one variable in terms of the other, then we can make this explicit and use the one-variable second-derivative test. If $\partial g/\partial y|_{\mathbf{v}_0} \neq 0$, then the curve S is not vertical at $\mathbf{v}_0$ and it is reasonable that we can solve for y as a function of x in a neighborhood of x_0. We will, in fact, prove this in Section 3.5. (If $\partial g/\partial x|_{\mathbf{v}_0} \neq 0$, we can solve for x as a function of y.)

Suppose S is the graph of $y = \phi(x)$. Then $f|S$ can be written as a function of one variable, $f(x, y) = f(x, \phi(x))$. The chain rule gives

$$
\begin{rcases}
\dfrac{df}{dx} = \dfrac{\partial f}{\partial x} + \dfrac{\partial f}{\partial y}\dfrac{d\phi}{dx} \\[4mm]
\text{and} \\[2mm]
\dfrac{d^2 f}{dx^2} = \dfrac{\partial^2 f}{\partial x^2} + 2\dfrac{\partial^2 f}{\partial x\,\partial y}\dfrac{d\phi}{dx} + \dfrac{\partial^2 f}{\partial y^2}\left(\dfrac{d\phi}{dx}\right)^2 + \dfrac{\partial f}{\partial y}\dfrac{d^2\phi}{dx^2}.
\end{rcases} \tag{6}
$$

The relation $g(x, \phi(x)) = c$ can be used to find $d\phi/dx$ and $d^2\phi/dx^2$. Differentiating both sides of $g(x, \phi(x)) = c$ with respect to x gives

$$
\frac{\partial g}{\partial x} + \frac{\partial g}{\partial y}\frac{d\phi}{dx} = 0
$$

and

$$
\frac{\partial^2 g}{\partial x^2} + 2\frac{\partial^2 g}{\partial x\,\partial y}\frac{d\phi}{dx} + \frac{\partial^2 g}{\partial y^2}\left(\frac{d\phi}{dx}\right)^2 + \frac{\partial g}{\partial y}\frac{d^2\phi}{dx^2} = 0,
$$

so that

$$
\begin{rcases}
\dfrac{d\phi}{dx} = -\dfrac{\partial g/\partial x}{\partial g/\partial y} \\[4mm]
\text{and} \\[2mm]
\dfrac{d^2\phi}{dx^2} = -\dfrac{1}{\partial g/\partial y}\left[\dfrac{\partial^2 g}{\partial x^2} - 2\dfrac{\partial^2 g}{\partial x\,\partial y}\dfrac{\partial g/\partial x}{\partial g/\partial y} + \dfrac{\partial^2 g}{\partial y^2}\left(\dfrac{\partial g/\partial x}{\partial g/\partial y}\right)^2\right].
\end{rcases} \tag{7}
$$

Substituting equation (7) into equation (6) gives

$$
\begin{rcases}
\dfrac{df}{dx} = \dfrac{\partial f}{\partial x} - \dfrac{\partial f/\partial y}{\partial g/\partial y}\dfrac{\partial g}{\partial x} \\[4mm]
\text{and} \\[2mm]
\dfrac{d^2 f}{dx^2} = \dfrac{1}{(\partial g/\partial y)^2}\left\{\left[\dfrac{\partial^2 f}{\partial x^2} - \dfrac{\partial f/\partial y}{\partial g/\partial y}\dfrac{\partial^2 g}{\partial x^2}\right]\left(\dfrac{\partial g}{\partial y}\right)^2 - 2\left[\dfrac{\partial^2 f}{\partial x\,\partial y} - \dfrac{\partial f/\partial y}{\partial g/\partial y}\dfrac{\partial^2 g}{\partial x\,\partial y}\right]\dfrac{\partial g}{\partial x}\dfrac{\partial g}{\partial y} \right. \\[4mm]
\qquad\qquad \left. + \left[\dfrac{\partial^2 f}{\partial y^2} - \dfrac{\partial f/\partial y}{\partial g/\partial y}\dfrac{\partial^2 g}{\partial y^2}\right]\left(\dfrac{\partial g}{\partial x}\right)^2\right\}.
\end{rcases} \tag{8}
$$

At $\mathbf{v}_0$, we know that $\partial f/\partial y = \lambda \partial g/\partial y$ and $\partial f/\partial x = \lambda \partial g/\partial x$, and so equation (8) becomes

$$
\left.\frac{df}{dx}\right|_{x_0} = \left.\frac{\partial f}{\partial x}\right|_{x_0} - \lambda\left.\frac{\partial g}{\partial x}\right|_{x_0} = 0
$$

and

$$\frac{d^2 f}{dx^2}\bigg|_{x_0} = \frac{1}{(\partial g/\partial y)^2}\left[\frac{\partial^2 h}{\partial x^2}\left(\frac{\partial g}{\partial y}\right)^2 - 2\frac{\partial^2 h}{\partial x\,\partial y}\frac{\partial g}{\partial x}\frac{\partial g}{\partial y} + \frac{\partial^2 h}{\partial y^2}\left(\frac{\partial g}{\partial x}\right)^2\right]$$

$$= -\frac{1}{(\partial g/\partial y)^2}\begin{vmatrix} 0 & -\dfrac{\partial g}{\partial x} & -\dfrac{\partial g}{\partial y} \\[2mm] -\dfrac{\partial g}{\partial x} & \dfrac{\partial^2 h}{\partial x^2} & \dfrac{\partial^2 h}{\partial x\,\partial y} \\[2mm] -\dfrac{\partial g}{\partial y} & \dfrac{\partial^2 h}{\partial x\,\partial y} & \dfrac{\partial^2 h}{\partial y^2} \end{vmatrix}$$

where the quantities are evaluated at x_0 and h is the auxiliary function introduced above. This 3×3 determinant is called a ***bordered Hessian***, and its sign is opposite that of $d^2 f/dx^2$. Therefore, if it is negative, we must be at a local minimum. If it is positive, we are at a local maximum; and if it is zero, the test is inconclusive. This reasoning leads to the following test (see Exercise 26 for an alternative approach).

THEOREM 10   *Let $f: U \subset \mathbb{R}^2 \to \mathbb{R}$ and $g: U \subset \mathbb{R}^2 \to \mathbb{R}$ be smooth (at least C^2) functions. Let $\mathbf{v}_0 \in U$, $g(\mathbf{v}_0) = c$, and S be the level curve for g with value c. Assume that $\nabla g(\mathbf{v}_0) \neq \mathbf{0}$ and that there is a real number λ such that $\nabla f(\mathbf{v}_0) = \lambda \nabla g(\mathbf{v}_0)$. Form the auxiliary function $h = f - \lambda g$ and the bordered Hessian determinant*

$$|\bar{H}| = \begin{vmatrix} 0 & -\dfrac{\partial g}{\partial x} & -\dfrac{\partial g}{\partial y} \\[2mm] -\dfrac{\partial g}{\partial x} & \dfrac{\partial^2 h}{\partial x^2} & \dfrac{\partial^2 h}{\partial x\,\partial y} \\[2mm] -\dfrac{\partial g}{\partial y} & \dfrac{\partial^2 h}{\partial x\,\partial y} & \dfrac{\partial^2 h}{\partial y^2} \end{vmatrix} \quad \textit{evaluated at } \mathbf{v}_0.$$

(i)   *If $|\bar{H}| > 0$, then $\mathbf{v}_0$ is a local maximum point for $f\,|\,S$.*

(ii)   *If $|\bar{H}| < 0$, then $\mathbf{v}_0$ is a local minimum point for $f\,|\,S$.*

(iii)   *If $|\bar{H}| = 0$, the test is inconclusive and $\mathbf{v}_0$ may be a minimum, a maximum, or neither.*

EXAMPLE 9   Find extreme points of $f(x, y) = (x - y)^n$ subject to the constraint $x^2 + y^2 = 1$, where $n \geq 1$.

SOLUTION   We set the first derivatives of the auxiliary function h defined by $h(x, y, \lambda) = (x - y)^n - \lambda(x^2 + y^2 - 1)$ equal to 0:

$$n(x - y)^{n-1} - 2\lambda x = 0$$

$$-n(x - y)^{n-1} - 2\lambda y = 0$$

$$-(x^2 + y^2 - 1) = 0.$$

From the first two equations we see that $\lambda(x + y) = 0$. If $\lambda = 0$, then $x = y = \pm\sqrt{2}/2$. If $\lambda \neq 0$, then $x = -y$. The four critical points are represented in Figure 3.4.5 and the corresponding values of $f(x, y)$ are listed below:

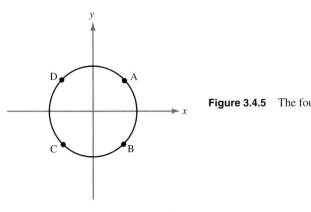

Figure 3.4.5 The four critical points in Example 9.

(A) $x = \sqrt{2}/2$ $\quad y = \sqrt{2}/2$ $\quad \lambda = 0$ $\qquad\qquad f(x, y) = 0$

(B) $x = \sqrt{2}/2$ $\quad y = -\sqrt{2}/2$ $\quad \lambda = n(\sqrt{2})^{n-2}$ $\qquad f(x, y) = (\sqrt{2})^n$

(C) $x = -\sqrt{2}/2$ $\quad y = -\sqrt{2}/2$ $\quad \lambda = 0$ $\qquad\qquad f(x, y) = 0$

(D) $x = -\sqrt{2}/2$ $\quad y = \sqrt{2}/2$ $\quad \lambda = (-1)^{n-2} n(\sqrt{2})^{n-2}$ $\quad f(x, y) = -(\sqrt{2})^n.$

By inspection, we see that if n is even, then A and C are minimum points and B and D are maxima. If n is odd, then B is a maximum point, D is a minimum, and A and C are neither. Let us see whether Theorem 10 is consistent with these observations.

The bordered Hessian determinant is

$$|\bar{H}| = \begin{vmatrix} 0 & -2x & -2y \\ -2x & n(n-1)(x-y)^{n-2} - 2\lambda & -n(n-1)(x-y)^{n-2} \\ -2y & -n(n-1)(x-y)^{n-2} & n(n-1)(x-y)^{n-2} - 2\lambda \end{vmatrix}$$

$$= -4n(n-1)(x-y)^{n-2}(x+y)^2 + 8\lambda(x^2 - y^2).$$

If $n = 1$ or if $n \geq 3$, $|\bar{H}| = 0$ at A, B, C, and D. If $n = 2$, then $|\bar{H}| = 0$ at B and D and -16 at A and C. Thus the second-derivative test picks up the minima at A and C but is inconclusive in testing the maxima at B and D for $n = 2$. It is also inconclusive for all other values of n. ▲

Just as in the unconstrained case, there is also a second-derivative test for functions of more than two variables. If we are to find extreme points for $f(x_1, \ldots, x_n)$ subject to a single constraint $g(x_1, \ldots, x_n) = c$, we first form the bordered Hessian for the auxiliary function $h(x_1, \ldots, x_n) = f(x_1, \ldots, x_n) - \lambda(g(x_1, \ldots, x_n) - c)$ as follows:

$$\begin{vmatrix} 0 & -\partial g/\partial x_1 & -\partial g/\partial x_2 & \cdots & -\partial g/\partial x_n \\ -\partial g/\partial x_1 & \partial^2 h/\partial x_1^2 & \partial^2 h/\partial x_1 \partial x_2 & \cdots & \partial^2 h/\partial x_1 \partial x_n \\ -\partial g/\partial x_2 & \partial^2 h/\partial x_1 \partial x_2 & \partial^2 h/\partial x_2^2 & \cdots & \partial^2 h/\partial x_2 \partial x_n \\ \vdots & \vdots & \vdots & & \vdots \\ -\partial g/\partial x_n & \partial^2 h/\partial x_1 \, \partial x_n & \partial^2 h/\partial x_2 \partial x_n & \cdots & \partial^2 h/\partial x_n^2 \end{vmatrix}$$

Second, we examine the determinants of the diagonal submatrices of order ≥ 3 at the critical points of h. If they are all negative, that is, if

$$
\begin{vmatrix}
0 & -\dfrac{\partial g}{\partial x_1} & -\dfrac{\partial g}{\partial x_2} \\[2mm]
-\dfrac{\partial g}{\partial x_1} & \dfrac{\partial^2 h}{\partial x_1^2} & \dfrac{\partial^2 h}{\partial x_1\,\partial x_2} \\[2mm]
-\dfrac{\partial g}{\partial x_2} & \dfrac{\partial^2 h}{\partial x_1\,\partial x_2} & \dfrac{\partial^2 h}{\partial x_2^2}
\end{vmatrix} < 0,
\qquad
\begin{vmatrix}
0 & -\dfrac{\partial g}{\partial x_1} & -\dfrac{\partial g}{\partial x_2} & -\dfrac{\partial g}{\partial x_3} \\[2mm]
-\dfrac{\partial g}{\partial x_1} & \dfrac{\partial^2 h}{\partial x_1^2} & \dfrac{\partial^2 h}{\partial x_1\,\partial x_2} & \dfrac{\partial^2 h}{\partial x_1\,\partial x_3} \\[2mm]
-\dfrac{\partial g}{\partial x_2} & \dfrac{\partial^2 h}{\partial x_1\,\partial x_2} & \dfrac{\partial^2 h}{\partial x_2^2} & \dfrac{\partial^2 h}{\partial x_2\,\partial x_3} \\[2mm]
-\dfrac{\partial g}{\partial x_3} & \dfrac{\partial^2 h}{\partial x_1\,\partial x_3} & \dfrac{\partial^2 h}{\partial x_2\,\partial x_3} & \dfrac{\partial^2 h}{\partial x_3^2}
\end{vmatrix} < 0, \ldots,
$$

then we are at a local minimum of $f\,|S$. If they start out with a positive 3×3 subdeterminant and alternate in sign (that is, >0, <0, >0, $<0, \ldots$), then we are at a local maximum. If they are all nonzero and do not fit one of these patterns, then the point is neither a maximum nor a minimum (it is said to be of the saddle type).[*]

EXAMPLE 10   Study the local extreme points of $f(x, y, z) = xyz$ on the surface of the unit sphere $x^2 + y^2 + z^2 = 1$ using the second-derivative test.

SOLUTION   Setting the partial derivatives of the auxiliary function $h(x, y, z, \lambda) = xyz - \lambda(x^2 + y^2 + z^2 - 1)$ equal to zero gives

$$yz = 2\lambda x$$

$$xz = 2\lambda y$$

$$xy = 2\lambda z$$

$$x^2 + y^2 + z^2 = 1.$$

Thus, $3xyz = 2\lambda(x^2 + y^2 + z^2) = 2\lambda$. If $\lambda = 0$, the solutions are $(x, y, z, \lambda) = (\pm 1, 0, 0, 0)$, $(0, \pm 1, 0, 0)$, and $(0, 0, \pm 1, 0)$. If $\lambda \neq 0$, then we have $2\lambda = 3xyz = 6\lambda z^2$ and so $z^2 = \frac{1}{3}$. Similarly, $x^2 = y^2 = \frac{1}{3}$. Thus the solutions are given by $\lambda = \frac{3}{2}xyz = \pm\sqrt{3}/6$. The critical points of h and the corresponding values of f are given in Table 3.1. From it, we see that points E, F, G, and K are minima. Points D, H, I, and J are maxima. To see whether this is in accord with the second-derivative test, we need to consider two determinants. First we look at the following:

$$
|\bar{H}_2| =
\begin{vmatrix}
0 & -\partial g/\partial x & -\partial g/\partial y \\
-\partial g/\partial x & \partial^2 h/\partial x^2 & \partial^2/\partial x\,\partial y \\
-\partial g/\partial y & \partial^2 h/\partial x\,\partial y & \partial^2 h/\partial y^2
\end{vmatrix}
=
\begin{vmatrix}
0 & -2x & -2y \\
-2x & -2\lambda & z \\
-2y & z & -2\lambda
\end{vmatrix}
$$

$$= 8\lambda x^2 + 8\lambda y^2 + 8xyz = 8\lambda(x^2 + y^2 + 2z^2).$$

[*]For a detailed discussion, see C. Caratheodory, *Calculus of Variations and Partial Differential Equations*, Holden-Day, San Francisco, 1965; Y. Murata, *Mathematics for Stability and Optimization of Economic Systems*, Academic Press, New York, 1977, pp. 263–271; or D. Spring, *Am. Math. Mon.* 92 (1985): 631–643. See also Exercise 26.

Table 3.1. The critical points A, B, ..., J, K of h and corresponding values of f

	x	y	z	λ	$f(x, y, z)$
A	1	0	0	0	0
B	0	1	0	0	0
C	0	0	1	0	0
D	$\sqrt{3}/3$	$\sqrt{3}/3$	$\sqrt{3}/3$	$\sqrt{3}/6$	$\sqrt{3}/9$
E	$-\sqrt{3}/3$	$\sqrt{3}/3$	$\sqrt{3}/3$	$-\sqrt{3}/6$	$-\sqrt{3}/9$
F	$\sqrt{3}/3$	$-\sqrt{3}/3$	$\sqrt{3}/3$	$-\sqrt{3}/6$	$-\sqrt{3}/9$
G	$\sqrt{3}/3$	$\sqrt{3}/3$	$-\sqrt{3}/3$	$-\sqrt{3}/6$	$-\sqrt{3}/9$
H	$\sqrt{3}/3$	$-\sqrt{3}/3$	$-\sqrt{3}/3$	$\sqrt{3}/6$	$\sqrt{3}/9$
I	$-\sqrt{3}/3$	$\sqrt{3}/3$	$-\sqrt{3}/3$	$\sqrt{3}/6$	$\sqrt{3}/9$
J	$-\sqrt{3}/3$	$-\sqrt{3}/3$	$\sqrt{3}/3$	$\sqrt{3}/6$	$\sqrt{3}/9$
K	$-\sqrt{3}/3$	$-\sqrt{3}/3$	$-\sqrt{3}/3$	$-\sqrt{3}/6$	$-\sqrt{3}/9$

Observe that sign $(|\bar{H}_2|) = $ sign $\lambda = $ sign (xyz), where sign $\alpha = +1$ if $\alpha > 0$ and -1 if $\alpha < 0$. Second, we consider

$$|\bar{H}_3| = \begin{vmatrix} 0 & -\partial g/\partial x & -\partial g/\partial y & -\partial g/\partial z \\ -\partial g/\partial x & \partial^2 h/\partial x^2 & \partial^2 h/\partial x\,\partial y & \partial^2 h/\partial x\,\partial z \\ -\partial g/\partial y & \partial^2 h/\partial x\,\partial y & \partial^2 h/\partial y^2 & \partial^2 h/\partial y\,\partial z \\ -\partial g/\partial z & \partial^2 h/\partial x\,\partial z & \partial^2 h/\partial y\,\partial z & \partial^2 h/\partial z^2 \end{vmatrix} = \begin{vmatrix} 0 & -2x & -2y & -2z \\ -2x & -2\lambda & z & y \\ -2y & z & -2\lambda & x \\ -2z & y & x & -2\lambda \end{vmatrix},$$

which works out to be $+4$ at points $\pm$ A, $\pm$ B, and $\pm$ C and $-\frac{16}{3}$ at the other eight points. At E, F, G, and K, we have $|\bar{H}_2| < 0$ *and* $|\bar{H}_3| < 0$, and so the test indicates these are local minima. At D, H, I, and J we have $|\bar{H}_2| > 0$ and $|\bar{H}_3| < 0$, and so the test says these are local maxima. Finally, the second-derivative test shows that $\pm$ A, $\pm$ B, and $\pm$ C are saddle points. ▲

EXERCISES

In Exercises 1 to 5 find the extrema of f subject to the stated constraints.

1. $f(x, y, z) = x - y + z, x^2 + y^2 + z^2 = 2$

2. $f(x, y) = x - y, x^2 - y^2 = 2$

3. $f(x, y) = x, x^2 + 2y^2 = 3$

4. $f(x, y, z) = x + y + z, x^2 - y^2 = 1, 2x + z = 1$

5. $f(x, y) = 3x + 2y, 2x^2 + 3y^2 = 3$

Find the relative extrema of $f|S$ in Exercises 6 to 9.

6. $f: \mathbb{R}^2 \to \mathbb{R}, (x, y) \mapsto x^2 + y^2, S = \{(x, 2) \mid x \in \mathbb{R}\}$

7. $f: \mathbb{R}^2 \to \mathbb{R}, (x, y) \mapsto x^2 + y^2, S = \{(x, y) \mid y \geq 2\}$

8. $f: \mathbb{R}^2 \to \mathbb{R}, (x, y) \mapsto x^2 - y^2, S = \{(x, \cos x) \mid x \in \mathbb{R}\}$

9. $f: \mathbb{R}^3 \to \mathbb{R}, (x, y, z) \mapsto x^2 + y^2 + z^2, S = \{(x, y, z) \mid z \geq 2 + x^2 + y^2\}$

10. Use the method of Lagrange multipliers to find the absolute maximum and minimum values of $f(x, y) = x^2 + y^2 - x - y + 1$ on the unit disk (see Example 10 of Section 3.3).

11. Consider the function $f(x, y) = x^2 + xy + y^2$ on the unit disk $D = \{(x, y) \mid x^2 + y^2 \leq 1\}$. Use the method of Lagrange multipliers to locate the maximum and minimum points for f on the unit circle. Use this to determine the absolute maximum and minimum values for f on D.

12. A rectangular box with no top is to have a surface area of 16 m^2. Find the dimensions that maximize its volume.

13. Design a cylindrical can (with a lid) to contain 1 liter of water, using the minimum amount of metal.

14. Show that solutions of equations (4) and (5) are in one-to-one correspondence with the critical points of $h(x_1, \ldots, x_n, \lambda_1, \ldots, \lambda_k) = f(x_1, \ldots, x_n) - \lambda_1[g_1(x_1, \ldots, x_n) - c_1] - \cdots - \lambda_k[g_k(x_1, \ldots, x_n) - c_k]$.

15. Find the absolute maximum and minimum for the function $f(x, y, z) = x + y - z$ on the ball $B = \{(x, y, z) \mid x^2 + y^2 + z^2 \leq 1\}$.

16. Repeat Exercise 15 for $f(x, y, z) = x + yz$.

17. A rectangular mirror with area A square feet is to have trim along the edges. If the trim along the horizontal edges costs p cents per foot and that for the vertical edges costs q cents per foot, find the dimensions which will minimize the total cost.

18. An irrigation canal in Arizona has concrete sides and bottom with trapezoidal cross section of area $A = y(x + y \tan \theta)$ and wetted perimeter $P = x + 2y/\cos \theta$, where $x =$ bottom width, $y =$ water depth, $\theta =$ side inclination, measured from vertical. The best design for a fixed inclination θ is found by solving $P = \text{minimum}$ subject to the condition $A = \text{constant}$. Show that $y^2 = (A \cos \theta)/(2 - \sin \theta)$.

19. Apply the second-derivative test to study the nature of the extrema in Exercises 1 and 5.

20. A light ray travels from point A to point B crossing a boundary between two media (see Figure 3.4.6). In the first medium its speed is v_1, and in the second it is v_2. Show that the trip is made in minimum time when *Snell's law* holds:

$$\frac{\sin \theta_1}{\sin \theta_2} = \frac{v_1}{v_2}.$$

Figure 3.4.6 Snell's law of refraction.

21. A parcel delivery service requires that the dimensions of a rectangular box be such that the length plus twice the width plus twice the height be no more than 108 inches ($l + 2w + 2h \leq 108$). What is the volume of the largest-volume box the company will deliver?

22. Let P be a point on a surface S in $\mathbb{R}^3$ defined by the equation $f(x, y, z) = 1$, where f is of class C^1. Suppose that P is a point where the distance from the origin to S is maximized. Show that the vector emanating from the origin and ending at P is perpendicular to S.

23. Let A be a nonzero symmetric 3×3 matrix. Thus its entries satisfy $a_{ij} = a_{ji}$. Consider the function $f(\mathbf{x}) = \frac{1}{2}(A\mathbf{x}) \cdot \mathbf{x}$.
    (a) What is ∇f?
    (b) Consider the restriction of f to the unit sphere $S = \{(x, y, z) \mid x^2 + y^2 + z^2 = 1\}$ in $\mathbb{R}^3$. By Theorem 7 we know that f must have a maximum and a minimum on S. Show that there must be an $\mathbf{x} \in S$ and a $\lambda \neq 0$ such that $A\mathbf{x} = \lambda\mathbf{x}$. (The vector $\mathbf{x}$ is called an *eigenvector*, while the scalar λ is called an *eigenvalue*.)
    (c) What are the maxima and minima for f on $B = \{(x, y, z) \mid x^2 + y^2 + z^2 \leq 1\}$?

24. Suppose that A in the function f defined in Exercise 23 is not necessarily symmetric.
    (a) What is ∇f?
    (b) Can one conclude the existence of an eigenvector and eigenvalues as in Exercise 23?

25. (a) Find the critical points of $x + y^2$ subject to the constraint $2x^2 + y^2 = 1$.
    (b) Use the bordered Hessian to classify the critical points.

26. Show that the bordered Hessian of $f(x_1, \ldots, x_n)$ subject to the single constraint $g(x_1, \ldots, x_n) = c$ is the Hessian of the function $f(x_1, \ldots, x_n) - \lambda g(x_1, \ldots, x_n)$ of the $n + 1$ variables $\lambda, x_1, \ldots, x_n$ (evaluated at the critical point). Can you use this observation to give another proof of the constrained second-derivative test using the unconstrained one? [HINT: If λ_0 denotes the value of λ determined by the Lagrange multiplier theorem, consider the function

$$F(x_1, \ldots, x_n, \lambda) = f(x_1, \ldots, x_n) - \lambda g(x_1, \ldots, x_n) \pm (\lambda - \lambda_0)^2.]$$

*3.5   THE IMPLICIT FUNCTION THEOREM

This section begins with a statement and proof of a special version of the implicit function theorem. This version is appropriate for a study of surfaces and, in particular, allows us to complete the proof of the Lagrange multiplier theorem in Section 3.4. Upon proving this theorem, we state, without proof, the general implicit (and attendent inverse) function theorem. This discussion will be useful later when we study the change of variables theorem in Chapter 6. However, the topics covered in this section are not essential for an understanding of the principal results and applications of the rest of this book.

In one-variable calculus we learn the importance of the inversion process. For example, $x = \ln y$ is the inverse of $y = e^x$, and $x = \sin^{-1} y$ is the inverse of $y = \sin x$. The inversion process is also important for functions of several variables; for example, the switch between Cartesian and polar coordinates in the plane involves inverting two functions of two variables.

Recall from one-variable calculus that if $y = f(x)$ is a C^1 function and $f'(x_0) \neq 0$, then locally near x_0 we can solve for x to give the inverse function: $x = f^{-1}(y)$. We learn that $(f^{-1})'(y) = 1/f'(x)$; that is, $dx/dy = 1/(dy/dx)$. That $y = f(x)$ can be inverted is plausible because $f'(x_0) \neq 0$ means that the slope of $y = f(x)$ is nonzero, so that the graph is rising or falling near x_0. Thus, if we reflect the graph across the line $y = x$ it is still a graph *near* (x_0, y_0) where $y_0 = f(x_0)$. In Figure 3.5.1 we can invert $y = f(x)$ in the shaded box, so in this range, $x = f^{-1}(y)$ is defined.

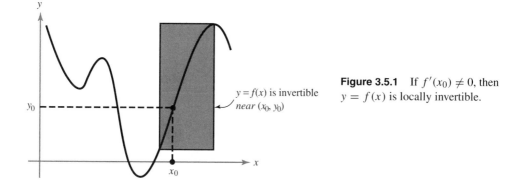

$y = f(x)$ is invertible
near (x_0, y_0)

Figure 3.5.1   If $f'(x_0) \neq 0$, then $y = f(x)$ is locally invertible.

THEOREM 11:SPECIAL IMPLICIT FUNCTION THEOREM   *Suppose that $F: \mathbb{R}^{n+1} \to \mathbb{R}$ has continuous partial derivatives. Denoting points in $\mathbb{R}^{n+1}$ by $(\mathbf{x}, z)$, where $\mathbf{x} \in \mathbb{R}^n$ and $z \in \mathbb{R}$, assume that $(\mathbf{x}_0, z_0)$ satisfies*

$$F(\mathbf{x}_0, z_0) = 0 \quad and \quad \frac{\partial F}{\partial z}(\mathbf{x}_0, z_0) \neq 0.$$

Then there is a ball U containing $\mathbf{x}_0$ in $\mathbb{R}^n$ and a neighborhood V of z_0 in $\mathbb{R}$ such that there is a unique function $z = g(\mathbf{x})$ defined for $\mathbf{x}$ in U and z in V that satisfies $F(\mathbf{x}, g(\mathbf{x})) = 0$. Moreover, if $\mathbf{x}$ in U and z in

V satisfy $F(\mathbf{x}, z) = 0$, then $z = g(\mathbf{x})$. Finally, $z = g(\mathbf{x})$ is continuously differentiable, with the derivative given by

$$\mathbf{D}g(\mathbf{x}) = -\frac{1}{\dfrac{\partial F}{\partial z}(\mathbf{x}, z)} \, \mathbf{D_x}F(\mathbf{x}, z)\Bigg|_{z=g(\mathbf{x})}$$

where $\mathbf{D_x}F$ denotes the (partial) derivative of F with respect to the variable $\mathbf{x}$, that is, we have $\mathbf{D_x}F = [\partial F/\partial x_1, \ldots, \partial F/\partial x_n]$; in other words

$$\frac{\partial g}{\partial x_i} = -\frac{\partial F/\partial x_i}{\partial F/\partial z}, \qquad i = 1, \ldots, n. \tag{1}$$

PROOF   We shall prove the case $n = 2$, so that $F \colon \mathbb{R}^3 \to \mathbb{R}$. The case for all n is similar, but the notation must be modified. We write $\mathbf{x} = (x, y)$ and $\mathbf{x}_0 = (x_0, y_0)$. Since $(\partial F/\partial z)(x_0, y_0, z_0) \neq 0$, it is either positive or negative. Suppose for definiteness that it is positive. By continuity, we can find numbers $a > 0$ and $b > 0$ such that if $|\mathbf{x} - \mathbf{x}_0| < a$ and $|z - z_0| < a$, then $(\partial F/\partial x)(\mathbf{x}, z) > b$. We can also assume that the other partial derivatives are bounded by a number M in this region, that is, $|(\partial F/\partial y)(\mathbf{x}, z)| \leq M$ and $|(\partial F/\partial z)(\mathbf{x}, z)| \leq M$. This also follows from continuity. We can now write

$$F(\mathbf{x}, z) = F(\mathbf{x}, z) - F(\mathbf{x}_0, z_0)$$
$$= [F(\mathbf{x}, z) - F(\mathbf{x}_0, z)] + [F(\mathbf{x}_0, z) - F(\mathbf{x}_0, z_0)]. \tag{2}$$

Consider the function

$$h(t) = F(t\mathbf{x} + (1 - t)\mathbf{x}_0, z)$$

for fixed $\mathbf{x}$ and z. By the mean value theorem, there is a number θ between 0 and 1 such that

$$h(1) - h(0) = h'(\theta)(1 - 0) = h'(\theta),$$

that is,

$$F(\mathbf{x}, z) - F(\mathbf{x}_0, z) = [\mathbf{D_x}F(\theta\mathbf{x} + (1 - \theta)\mathbf{x}_0, z)](\mathbf{x} - \mathbf{x}_0).$$

Substitution of this formula into equation (2) along with a similar formula for the second term of the equation gives

$$F(\mathbf{x}, z) = [\mathbf{D_x}F(\theta\mathbf{x} + (1 - \theta)\mathbf{x}_0, z)](\mathbf{x} - \mathbf{x}_0) + \left[\frac{\partial F}{\partial z}(\mathbf{x}_0, \phi z + (1 - \phi)z_0)\right](z - z_0), \tag{3}$$

where ϕ is between 0 and 1. Let a_0 satisfy $0 < a_0 < a$ and choose $\delta > 0$ such that $\delta < a_0$ and $\delta < ba_0/2M$. Then, if $|\mathbf{x} - \mathbf{x}_0| < \delta$, both $|x - x_0|$ and $|y - y_0|$ are less than δ, so that the absolute value of each of the two terms in

$$[\mathbf{D_x}F(\theta\mathbf{x} + (1 - \theta)\mathbf{x}_0, z)](\mathbf{x} - \mathbf{x}_0)$$
$$= \left[\frac{\partial F}{\partial x}(\theta\mathbf{x} + (1 - \theta)\mathbf{x}_0, z)\right](x - x_0) + \left[\frac{\partial F}{\partial y}(\theta\mathbf{x} + (1 - \theta)\mathbf{x}_0, z)\right](y - y_0)$$

is less than $M\delta < M(ba_0/2M) = ba_0/2$. Thus, $|\mathbf{x} - \mathbf{x}_0| < \delta$ implies

$$|[\mathbf{D_x}F(\theta\mathbf{x} + (1 - \theta)\mathbf{x}_0, z)](\mathbf{x} - \mathbf{x}_0)| < ba_0.$$

Therefore, from equation (3) and the choice of b, $|\mathbf{x} - \mathbf{x}_0| < \delta$ implies that

$$F(\mathbf{x}, z_0 + a_0) > 0 \qquad \text{and} \qquad F(\mathbf{x}, z_0 - a_0) < 0.$$

[The inequalities are reversed if $(\partial F/\partial z)(\mathbf{x}_0, z_0) < 0$.] Thus, by the intermediate value theorem applied to $F(\mathbf{x}, z)$ as a function of z, for each $\mathbf{x}$ there is a z between $z_0 - a_0$ and $z_0 + a_0$ such that $F(\mathbf{x}, z) = 0$. This z is unique, since, by elementary calculus, a function with a positive derivative is increasing and thus can have no more than one zero.

Let U be the open ball of radius δ and center $\mathbf{x}_0$ in $\mathbb{R}^n$ and let V be the open interval on $\mathbb{R}$ from $z_0 - a_0$ to $z_0 + a_0$. We have proved that if $\mathbf{x}$ is confined to U, there is a unique z in V such that $F(\mathbf{x}, z) = 0$. This defines the function $z = g(\mathbf{x}) = g(x, y)$ required by the theorem. We leave it to the reader to prove from this construction that $z = g(x, y)$ is a continuous function.

It remains to establish the continuous differentiability of $z = g(\mathbf{x})$. From equation (3), and since $F(\mathbf{x}, z) = 0$ and $z_0 = g(\mathbf{x}_0)$, we have

$$g(\mathbf{x}) - g(\mathbf{x}_0) = -\frac{[\mathbf{D_x}F(\theta\mathbf{x} + (1 - \theta)\mathbf{x}_0, z)](\mathbf{x} - \mathbf{x}_0)}{\dfrac{\partial F}{\partial z}(\mathbf{x}_0, \phi z + (1 - \phi)z_0)}.$$

If we let $\mathbf{x} = (x_0 + h, y_0)$, then this equation becomes

$$\frac{g(x_0 + h, y_0) - g(x_0, y_0)}{h} = -\frac{\dfrac{\partial F}{\partial x}(\theta\mathbf{x} + (1 - \theta)\mathbf{x}_0, z)}{\dfrac{\partial F}{\partial z}(\mathbf{x}_0, \phi z + (1 - \phi)z_0)}.$$

As $h \to 0$, it follows that $x \to x_0$ and $z \to z_0$, and so we get

$$\frac{\partial g}{\partial x}(x_0, y_0) = \underset{h \to 0}{\text{limit}}\frac{g(x_0 + h, y_0) - g(x_0, y_0)}{h} = -\frac{\dfrac{\partial F}{\partial x}(x_0, z)}{\dfrac{\partial F}{\partial z}(x_0, z)}.$$

The formula

$$\frac{\partial g}{\partial y}(x_0, y_0) = -\frac{\dfrac{\partial F}{\partial y}(x_0, z)}{\dfrac{\partial F}{\partial z}(x_0, z)}$$

is proved in the same way. This derivation holds at any point (x, y) in U by the same argument, and so we have proved formula (1). Since the right-hand side of formula (1) is continuous, we have proved the theorem.　■

Once it is known that $z = g(\mathbf{x})$ exists and is differentiable, formula (1) may be checked by implicit differentiation; that is, the chain rule applied to $F(\mathbf{x}, g(\mathbf{x})) = 0$ gives

$$\mathbf{D_x}F(\mathbf{x}, g(\mathbf{x})) + \left[\frac{\partial F}{\partial z}(\mathbf{x}, g(\mathbf{x}))\right][\mathbf{D}g(\mathbf{x})] = 0,$$

which is equivalent to formula (1).

EXAMPLE 1   In the special implicit function theorem, it is important to recognize the necessity of taking sufficiently small neighborhoods U and V. For example, consider the equation

$$x^2 + z^2 - 1 = 0,$$

that is, $F(x, z) = x^2 + z^2 - 1$, with $n = 1$. Here $(\partial F/\partial z)(x, z) = 2z$, and so the special implicit function theorem applies to a point (x_0, z_0) satisfying $x_0^2 + z_0^2 - 1 = 0$ and $z_0 \neq 0$. Thus, near such points, z is a unique function of x. This function is $z = -\sqrt{1 - x^2}$ if $z_0 > 0$ and $z = -\sqrt{1 - x^2}$ if $z_0 < 0$. Note that z is defined only for $|x| < 1$ (U must not be too big) and z is unique only if it is near z_0 (V must not be too big). These facts and the nonexistence of $\partial z/\partial x$ at $z_0 = 0$ are, of course, clear from the fact that $x^2 + z^2 = 1$ defines a circle in the xz plane (Figure 3.5.2).   ▲

Figure 3.5.2   It is necessary to take small neighborhoods in the implicit function theorem.

Let us apply Theorem 11 to the study of surfaces. We are concerned with the level set of a function $g: U \subset \mathbb{R}^n \to \mathbb{R}$, that is, with the surface S consisting of the set of $\mathbf{x}$ satisfying $g(\mathbf{x}) = c_0$, where $c_0 = g(\mathbf{x}_0)$ and where $\mathbf{x}_0$ is given. Let us take $n = 3$ for concreteness. Thus we are dealing with the level surface of a function $g(x, y, z)$ through a given point (x_0, y_0, z_0). As in the Lagrange multiplier theorem, let us assume that $\nabla g(x_0, y_0, z_0) \neq \mathbf{0}$. This means that at least one of the partial derivatives of g is nonzero. For definiteness, suppose that $(\partial g/\partial z)(x_0, y_0, z_0) \neq 0$. By applying Theorem 11 to the function $(x, y, z) \mapsto g(x, y, z) - c_0$, we know there is a unique function $z = k(x, y)$ satisfying $g(x, y, k(x, y)) = c_0$ for (x, y) near (x_0, y_0) and z near z_0. Thus, near z_0 the surface S is the graph of the function k. Since k is continuously differentiable, this surface has a tangent plane at (x_0, y_0, z_0) defined by

$$z = z_0 + \left[\frac{\partial k}{\partial x}(x_0, y_0) \right](x - x_0) + \left[\frac{\partial k}{\partial y}(x_0, y_0) \right](y - y_0). \tag{4}$$

But by formula (1),

$$\frac{\partial k}{\partial x}(x_0, y_0) = -\frac{\dfrac{\partial g}{\partial x}(x_0, y_0, z_0)}{\dfrac{\partial g}{\partial z}(x_0, y_0, z_0)} \quad \text{and} \quad \frac{\partial k}{\partial y}(x_0, y_0) = -\frac{\dfrac{\partial g}{\partial y}(x_0, y_0, z_0)}{\dfrac{\partial g}{\partial z}(x_0, y_0, z_0)}.$$

Substituting these two equations into equation (4) gives this equivalent description:

$$0 = (z - z_0)\frac{\partial g}{\partial z}(x_0, y_0, z_0) + (x - x_0)\frac{\partial g}{\partial x}(x_0, y_0, z_0) + (y - y_0)\frac{\partial g}{\partial y}(x_0, y_0, z_0);$$

that is,

$$(x - x_0, y - y_0, z - z_0) \cdot \nabla g(x_0, y_0, z_0) = 0.$$

Thus the tangent plane to the level surface of g is the orthogonal complement to $\nabla g(x_0, y_0, z_0)$ through the point (x_0, y_0, z_0). This agrees with our characterization of tangent planes to level sets from Chapter 2.

We are now ready to complete the proof of the Lagrange multiplier theorem. To do this we must show that every vector tangent to S at (x_0, y_0, z_0) is tangent to a curve in S. By Theorem 11, we need only show this for a graph of the form $z = k(x, y)$. However, if $\mathbf{v} = (x - x_0, y - y_0, z - z_0)$ is tangent to the graph [that is, if it satisfies equation (4)], then $\mathbf{v}$ is tangent to the path in S given by

$$\mathbf{c}(t) = (x_0 + t(x - x_0), y_0 + t(y - y_0), k(x_0 + t(x - x_0), y_0 + t(y - y_0)))$$

at $t = 0$. This can be checked by using the chain rule. (See Figure 3.5.3.)

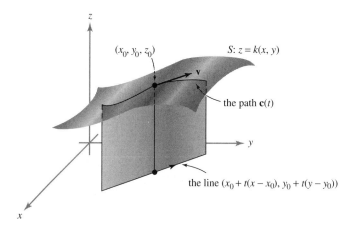

Figure 3.5.3   The construction of a path $\mathbf{c}(t)$ in the surface S whose tangent vector is $\mathbf{v}$.

E X A M P L E   2   Near what points may the surface

$$x^3 + 3y^2 + 8xz^2 - 3z^3y = 1$$

be represented as a graph of a differentiable function $z = k(x, y)$?

S O L U T I O N   Here we take $F(x, y, z) = x^3 + 3y^2 + 8xz^2 - 3z^3y - 1$ and attempt to solve $F(x, y, z) = 0$ for z as a function of (x, y). By Theorem 11, this may be done near a point (x_0, y_0, z_0) if

$(\partial F/\partial z)(x_0, y_0, z_0) \neq 0$, that is, if

$$z_0(16x_0 - 9z_0 y_0) \neq 0,$$

which means, in turn,

$$z_0 \neq 0 \quad \text{and} \quad 16x_0 \neq 9z_0 y_0. \quad \blacktriangle$$

Next we shall state, without proof, the *general implicit function theorem.* * Instead of attempting to solve one equation for one variable, we attempt to solve m equations for m variables $z_1, \ldots, z_m$:

$$\begin{aligned}
F_1(x_1, \ldots, x_n, z_1, \ldots, z_m) &= 0 \\
F_2(x_1, \ldots, x_n, z_1, \ldots, z_m) &= 0 \\
\vdots \qquad\qquad \vdots \qquad\qquad \vdots \\
F_m(x_1, \ldots, x_n, z_1, \ldots, z_m) &= 0.
\end{aligned} \tag{5}$$

In Theorem 11 we had the condition $\partial F/\partial z \neq 0$. The condition appropriate to the general implicit function theorem is that $\Delta \neq 0$,[†] where Δ is the determinant of the $m \times m$ matrix

$$\begin{bmatrix}
\dfrac{\partial F_1}{\partial z_1} & \cdots & \dfrac{\partial F_1}{\partial z_m} \\
\vdots & & \vdots \\
\dfrac{\partial F_m}{\partial z_1} & \cdots & \dfrac{\partial F_m}{\partial z_m}
\end{bmatrix}$$

evaluated at the point $(\mathbf{x}_0, \mathbf{z}_0)$; in the neighborhood of such a point we can uniquely solve for $\mathbf{z}$ in terms of $\mathbf{x}$.

THEOREM 12: GENERAL IMPLICIT FUNCTION THEOREM   *If $\Delta \neq 0$, then near the point $(\mathbf{x}_0, \mathbf{z}_0)$, equation (5) defines unique (smooth) functions*

$$z_i = k_i(x_1, \ldots, x_n) \qquad (i = 1, \ldots, m).$$

Their derivatives may be computed by implicit differentiation.

---

[*]For three different proofs of the general case, consult:

(a) E. Goursat, *A Course in Mathematical Analysis*, I, Dover, New York, 1959, p. 45. (This proof derives the general theorem by successive application of Theorem 11.)

(b) T. M. Apostol, *Mathematical Analysis*, 2d ed., Addison-Wesley, Reading, Mass., 1974.

(c) J. E. Marsden and M. Hoffman, *Elementary Classical Analysis*, 2d ed., Freeman, New York, 1993.

Of these sources, the last two use more sophisticated ideas that are usually not covered until a junior-level course in analysis. The first, however, is easily understood by the reader who has some knowledge of linear algebra.

[†]For students who have had linear algebra: The condition $\Delta \neq 0$ has a simple interpretation in the case that *F is linear*; namely, $\Delta \neq 0$ is equivalent to the rank of F being equal to m, which in turn is equivalent to the fact that the solution space of $F = 0$ is m-dimensional.

EXAMPLE 3   Show that near the point $(x, y, u, v) = (1, 1, 1, 1)$ we can solve

$$xu + yvu^2 = 2$$
$$xu^3 + y^2v^4 = 2$$

uniquely for u and v as functions of x and y. Compute $\partial u/\partial x$ at the point $(1, 1)$.

SOLUTION   To check solvability, we form the equations

$$F_1(x, y, u, v) = xu + yvu^2 - 2$$
$$F_2(x, y, u, v) = xu^3 + y^2v^4 - 2$$

and the determinant

$$
\Delta =
\begin{vmatrix}
\dfrac{\partial F_1}{\partial u} & \dfrac{\partial F_1}{\partial v} \\[2mm]
\dfrac{\partial F_2}{\partial u} & \dfrac{\partial F_2}{\partial v}
\end{vmatrix}
\qquad \text{at} \qquad (1, 1, 1, 1)
$$

$$
=
\begin{vmatrix}
x + 2yuv & yu^2 \\[2mm]
3u^2x & 4y^2v^3
\end{vmatrix}
\qquad \text{at} \qquad (1, 1, 1, 1)
$$

$$
=
\begin{vmatrix}
3 & 1 \\
3 & 4
\end{vmatrix}
= 9.
$$

Since $\Delta \neq 0$, solvability is assured by the general implicit function theorem. To find $\partial u/\partial x$, we implicitly differentiate the given equations in x using the chain rule:

$$x\frac{\partial u}{\partial x} + u + y\frac{\partial v}{\partial x}u^2 + 2yvu\frac{\partial u}{\partial x} = 0$$
$$3xu^2\frac{\partial u}{\partial x} + u^3 + 4y^2v^3\frac{\partial v}{\partial x} = 0.$$

Setting $(x, y, u, v) = (1, 1, 1, 1)$ gives

$$3\frac{\partial u}{\partial x} + \frac{\partial v}{\partial x} = -1$$
$$3\frac{\partial u}{\partial x} + 4\frac{\partial v}{\partial x} = -1.$$

Solving for $\partial u/\partial x$ by multiplying the first equation by 4 and subtracting gives $\partial u/\partial x = -\frac{1}{3}$.   ▲

A special case of the general implicit function theorem is the *inverse function theorem*. Here we attempt to solve the n equations

$$
\left.
\begin{aligned}
f_1(x_1, \ldots, x_n) &= y_1 \\
&\cdots \\
f_n(x_1, \ldots, x_n) &= y_n
\end{aligned}
\right\}
\qquad (6)
$$

for $x_1, \ldots, x_n$ as functions of $y_1, \ldots, y_n$; that is, we are trying to invert the equations of system (6). This is analogous to forming the inverses of functions like $\sin x = y$ and $e^x = y$, with which the reader should be familiar from elementary calculus. Now, however, we are concerned with functions of several variables. The question of solvability is answered by the general implicit function theorem applied to the functions $y_i - f_i(x_1, \ldots, x_n)$ with the unknowns $x_1, \ldots, x_n$ (called $z_1, \ldots, z_n$ above). The condition for solvability in a neighborhood of a point $\mathbf{x}_0$ is $\Delta \neq 0$, where Δ is the determinant of the matrix $\mathbf{D}f(\mathbf{x}_0)$, and $f = (f_1, \ldots, f_n)$. The quantity Δ is denoted by $\partial(f_1, \ldots, f_n)/\partial(x_1, \ldots, x_n)$, or $\partial(y_1, \ldots, y_n)/\partial(x_1, \ldots, x_n)$ or $J(f)(\mathbf{x}_0)$ and is called the **Jacobian determinant** of f. Explicitly,

$$\frac{\partial(f_1, \ldots, f_n)}{\partial(x_1, \ldots, x_n)}\bigg|_{\mathbf{x}=\mathbf{x}_0} = J(f)(\mathbf{x}_0) = \begin{vmatrix} \dfrac{\partial f_1}{\partial x_1}(\mathbf{x}_0) & \cdots & \dfrac{\partial f_1}{\partial x_n}(\mathbf{x}_0) \\ \vdots & & \vdots \\ \dfrac{\partial f_n}{\partial x_1}(\mathbf{x}_0) & \cdots & \dfrac{\partial f_n}{\partial x_n}(\mathbf{x}_0) \end{vmatrix}. \tag{7}$$

The reader should note that in the case when f is linear, for example $f(x) = Ax$, where A is an $n \times n$ matrix, the condition $\Delta \neq 0$ is equivalent to the fact that the determinant of A, $\det A \neq 0$, and from Section 1.5 we know that A, and therefore f, has an inverse.

The Jacobian determinant will play an important role in our work on integration (see Chapter 5). The following theorem summarizes this discussion:

THEOREM 13: INVERSE FUNCTION THEOREM *Let $U \subset \mathbb{R}^n$ be open and let $f_1: U \to \mathbb{R}, \ldots, f_n: U \to \mathbb{R}$ have continuous partial derivatives. Consider the equations (6) near a given solution $\mathbf{x}_0, \mathbf{y}_0$. If $[\partial(f_1, \ldots, f_n)]/[\partial(x_1, \ldots, x_n)] = J(f)(\mathbf{x}_0)$ [defined by equation (7)] is nonzero, then (6) can be solved uniquely as $\mathbf{x} = g(\mathbf{y})$ for $\mathbf{x}$ near $\mathbf{x}_0$ and $\mathbf{y}$ near $\mathbf{y}_0$. Moreover, the function g has continuous partial derivatives.*

EXAMPLE 4 Consider the equations

$$\frac{x^4 + y^4}{x} = u, \qquad \sin x + \cos y = v.$$

Near which points (x, y) can we solve for x, y in terms of u, v?

SOLUTION Here the functions are $u = f_1(x, y) = (x^4 + y^4)/x$ and $v = f_2(x, y) = \sin x + \cos y$. We want to know the points near which we can solve for x, y as functions of u and v. According to the inverse function theorem, we must first compute $\partial(f_1, f_2)/\partial(x, y)$. We take the domain of $f = (f_1, f_2)$ to be $U = \{(x, y) \in \mathbb{R}^2 \mid x \neq 0\}$. Now

$$\frac{\partial(f_1, f_2)}{\partial(x, y)} = \begin{vmatrix} \dfrac{\partial f_1}{\partial x} & \dfrac{\partial f_1}{\partial y} \\ \dfrac{\partial f_2}{\partial x} & \dfrac{\partial f_2}{\partial y} \end{vmatrix} = \begin{vmatrix} \dfrac{3x^4 - y^4}{x^2} & \dfrac{4y^3}{x} \\ \cos x & -\sin y \end{vmatrix} = \frac{\sin y}{x^2}(y^4 - 3x^4) - \frac{4y^3}{x}\cos x.$$

Therefore, at points where this does not vanish we can solve for x, y in terms of u and v. In other words, we can solve for x, y near those x, y for which $x \neq 0$ and $(\sin y)(y^4 - 3x^4) \neq 4xy^3 \cos x$. Such conditions generally cannot be solved explicitly. For example, if $x_0 = \pi/2$, $y_0 = \pi/2$, we can solve for x, y near (x_0, y_0) because there, $\partial(f_1, f_2)/\partial(x, y) \neq 0$.   ▲

EXERCISES

1. Let $F(x, y) = 0$ define a curve in the xy plane through the point (x_0, y_0), where F is C^1. Assume that $(\partial F/\partial y)(x_0, y_0) \neq 0$. Show that this curve can be locally represented by the graph of a function $y = g(x)$. Show that (i) the line orthogonal to $\nabla F(x_0, y_0)$ agrees with (ii) the tangent line to the graph of $y = g(x)$.

2. Show that $xy + z + 3xz^5 = 4$ is solvable for z as a function of (x, y) near $(1, 0, 1)$. Compute $\partial z/\partial x$ and $\partial z/\partial y$ at $(1, 0)$.

3. (a)   Check directly (i.e., without using Theorem 11) where we can solve the equation $F(x, y) = y^2 + y + 3x + 1 = 0$ for y in terms of x.
(b)   Check that your answer in part (a) agrees with the answer you expect from the implicit function theorem. Compute dy/dx.

4. Repeat Exercise 3 with $F(x, y) = xy^2 - 2y + x^2 + 2 = 0$.

5. Show that $x^3 z^2 - z^3 yx = 0$ is solvable for z as a function of (x, y) near $(1, 1, 1)$, but not near the origin. Compute $\partial z/\partial x$ and $\partial z/\partial y$ at $(1, 1)$.

6. Discuss the solvability in the system

$$3x + 2y + z^2 + u + v^2 = 0$$
$$4x + 3y + z + u^2 + v + w + 2 = 0$$
$$x + z + w + u^2 + 2 = 0$$

for u, v, w in terms of x, y, z near $x = y = z = 0$, $u = v = 0$, and $w = -2$.

7. Discuss the solvability of

$$y + x + uv = 0$$
$$uxy + v = 0$$

for u, v in terms of x, y near $x = y = u = v = 0$ and check directly.

8. Investigate whether or not the system

$$u(x, y, z) = x + xyz$$
$$v(x, y, z) = y + xy$$
$$w(x, y, z) = z + 2x + 3z^2$$

can be solved for x, y, z in terms of u, v, w near $(x, y, z) = (0, 0, 0)$.

9. Consider $f(x, y) = ((x^2 - y^2)/(x^2 + y^2), xy/(x^2 + y^2))$. Does this map of $\mathbb{R}^2 \backslash (0, 0)$ to $\mathbb{R}^2$ have a local inverse near $(x, y) = (0, 1)$?

10. (a) Define $x\colon \mathbb{R}^2 \to \mathbb{R}$ by $x(r, \theta) = r \cos \theta$ and define $y\colon \mathbb{R}^2 \to \mathbb{R}$ by $y(r, \theta) = r \sin \theta$. Show that

$$\frac{\partial(x, y)}{\partial(r, \theta)}\bigg|_{(r_0, \theta_0)} = r_0.$$

(b) When can we form a smooth inverse function $(r(x, y), \theta(x, y))$? Check directly and with the inverse function theorem.

(c) Consider the following transformations for spherical coordinates (see Section 1.4):

$$x(\rho, \phi, \theta) = \rho \sin \phi \cos \theta$$
$$x(\rho, \phi, \theta) = \rho \sin \phi \sin \theta$$
$$z(\rho, \phi, \theta) = \rho \cos \phi.$$

Show that

$$\frac{\partial(x, y, z)}{\partial(\rho, \phi, \theta)} = \rho^2 \sin \phi.$$

(d) When can we solve for (ρ, ϕ, θ) in terms of (x, y, z)?

11. Let (x_0, y_0, z_0) be a point of the locus defined by $z^2 + xy - a = 0, z^2 + x^2 - y^2 - b = 0$, where a and b are constants.

(a) Under what conditions may the part of the locus near (x_0, y_0, z_0) be represented in the form $x = f(z), y = g(z)$?

(b) Compute $f'(z)$ and $g'(z)$.

12. Is it possible to solve

$$xy^2 + xzu + yv^2 = 3$$
$$u^3yz + 2xv - u^2v^2 = 2$$

for $u(x, y, z), v(x, y, z)$ near $(x, y, z) = (1, 1, 1), (u, v) = (1, 1)$? Compute $\partial v/\partial y$ at $(x, y, z) = (1, 1, 1)$.

13. The problem of factoring a polynomial $x^n + a_{n-1}x^{n-1} + \cdots + a_0$ into linear factors is, in a sense, an "inverse function" problem. The coefficients a_i may be thought of as functions of the n roots r_j. We would like to find the roots as functions of the coefficients in some region. With $n = 3$, apply the inverse function theorem to this problem and state what it tells you about the possibility of doing this.

3.6 SOME APPLICATIONS

In this section we give some applications of the mathematical methods that we have developed in the preceding sections. The applications are made to mechanics, geometry, and economics. The student should consult the instructor concerning which of these examples should be studied.

Let $\mathbf{F}$ denote a force field defined on a certain domain U of $\mathbb{R}^3$. Thus $\mathbf{F}: U \to \mathbb{R}^3$ is a given vector field. Let us agree that a particle (with mass m) is to move along a path $\mathbf{c}(t)$ in such a way that Newton's law holds; mass $\times$ acceleration = force; that is, the path $\mathbf{c}(t)$ is to satisfy the equation

$$m\mathbf{c}''(t) = \mathbf{F}(\mathbf{c}(t)). \tag{1}$$

If $\mathbf{F}$ is a potential field with potential V, that is, if $\mathbf{F} = -\nabla V$, then

$$\tfrac{1}{2}m\|\mathbf{c}'(t)\|^2 + V(\mathbf{c}(t)) = \text{ constant}. \tag{2}$$

(The first term is called the ***kinetic energy***.) Indeed, by differentiating the left side of (2) using the chain rule

$$\frac{d}{dt}\left[\frac{1}{2}m\|\mathbf{c}'(t)\|^2 + V(\mathbf{c}(t))\right] = m\mathbf{c}'(t) \cdot \mathbf{c}''(t) + \nabla V(\mathbf{c}(t)) \cdot \mathbf{c}'(t)$$

$$= [m\mathbf{c}''(t) + \nabla V(\mathbf{c}(t))] \cdot \mathbf{c}'(t) = 0,$$

since $m\mathbf{c}''(t) = -\nabla V(\mathbf{c}(t))$. This proves formula (2).

DEFINITION   A point $\mathbf{x}_0 \in U$ is called a ***position of equilibrium*** if the force at that point is zero: $\mathbf{F}(\mathbf{x}_0) = \mathbf{0}$. A point $\mathbf{x}_0$ that is a position of equilibrium is said to be ***stable*** if for every $\rho > 0$ and $\varepsilon > 0$, we can choose numbers $\rho_0 > 0$ and $\varepsilon_0 > 0$ such that a point situated anywhere at a distance less than ρ_0 from $\mathbf{x}_0$, after initially receiving kinetic energy in an amount less than ε_0, will forever remain a distance from $\mathbf{x}_0$ less then ρ and possess kinetic energy less than ε (see Figure 3.6.1).

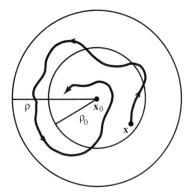

Figure 3.6.1   Motion near a stable point $\mathbf{x}_0$.

Thus if we have a position of equilibrium, stability at $\mathbf{x}_0$ means that a slowly moving particle near $\mathbf{x}_0$ will always remain near $\mathbf{x}_0$ and keep moving slowly. If we have an unstable equilibrium point $\mathbf{x}_0$, then $\mathbf{c}(t) = \mathbf{x}_0$ solves the equation $m\mathbf{c}''(t) = \mathbf{F}(\mathbf{c}(t))$, but nearby solutions may move away from $\mathbf{x}_0$ as time progresses. For example, a pencil balancing on its tip illustrates an unstable configuration, whereas a ball hanging on a spring illustrates a stable equilibrium.

THEOREM 14

(i) *Critical points of a potential are the positions of equilibrium.*

(ii) *In a potential field, a point $\mathbf{x}_0$ at which the potential takes a strict local minimum is a position of stable equilibrium. (Recall that a function f is said to have a* strict *local minimum at the point $\mathbf{x}_0$ if there exists a neighborhood U of $\mathbf{x}_0$ such that $f(\mathbf{x}) > f(\mathbf{x}_0)$ for all $\mathbf{x}$ in U other than $\mathbf{x}_0$.)*

PROOF The first assertion is quite obvious from the definition $\mathbf{F} = -\nabla V$; equilibrium points $\mathbf{x}_0$ are exactly critical points of V, at which $\nabla V(\mathbf{x}_0) = \mathbf{0}$.

To prove assertion (ii), we shall make use of the law of conservation of energy, that is, equation (2). We have

$$\tfrac{1}{2}m\|\mathbf{c}'(t)\|^2 + V(\mathbf{c}(t)) = \tfrac{1}{2}m\|\mathbf{c}'(0)\|^2 + V(\mathbf{c}(0)).$$

We shall argue slightly informally to amplify and illuminate the central ideas involved. Let us choose a small neighborhood of $\mathbf{x}_0$ and start our particle with a small kinetic energy. As t increases, the particle moves away from $\mathbf{x}_0$ on a path $\mathbf{c}(t)$ and $V(\mathbf{c}(t))$ increases [since $V(\mathbf{x}_0) = V(\mathbf{c}(0))$ is a strict minimum], so that the kinetic energy must decrease. If the initial kinetic energy is sufficiently small, then in order for the particle to escape from our neighborhood of $\mathbf{x}_0$, outside of which V has increased by a definite amount, the kinetic energy would have to become negative (which is impossible). Thus the particle cannot escape the neighborhood. ■

EXAMPLE 1 Find the points that are positions of equilibrium, and determine whether or not they are stable, if the force field $\mathbf{F} = F_x\mathbf{i} + F_y\mathbf{j} + F_z\mathbf{k}$ is given by $F_x = -k^2x$, $F_y = -k^2y$, $F_z = -k^2z$ $(k \neq 0)$.*

SOLUTION The force field $\mathbf{F}$ is a potential field with potential given by the function $V = \tfrac{1}{2}k^2(x^2 + y^2 + z^2)$. The only critical point of V is the origin. The Hessian of V at the origin is $\tfrac{1}{2}k^2(h_1^2 + h_2^2 + h_3^2)$, which is positive-definite. It follows that the origin is a strict minimum of V. Thus, by (i) and (ii) of Theorem 14, we have shown that the origin is a position of stable equilibrium. ▲

Let a point in a potential field V be constrained to remain on the level surface S given by the equation $\phi(x, y, z) = 0$, with $\nabla\phi \neq \mathbf{0}$. If in formula (1) we replace $\mathbf{F}$ by the

---

*The force field in this example is that governing the motion of a three-dimensional harmonic oscillator.

component of $\mathbf{F}$ parallel to S, we ensure that the particle will remain on S.[†] By analogy with Theorem 14, we have:

THEOREM 15

(i) *If at a point* P *on the surface* S *the potential* $V\,|\,S$ *has an extreme value, then the point* P *is a position of equilibrium on the surface.*

(ii) *If a point* $P \in S$ *is a strict local minimum of the potential* $V\,|\,S$, *then the point* P *is a position of stable equilibrium.*

The proof of this theorem will be omitted. It is similar to the proof of Theorem 14, with the additional fact that the equation of motion uses only the component of $\mathbf{F}$ along the surface.[*]

EXAMPLE 2   Let $\mathbf{F}$ be the gravitational field near the surface of the earth; that is, let $\mathbf{F} = (F_x, F_y, F_z)$, where $F_x = 0$, $F_y = 0$, and $F_z = -mg$, where g is the acceleration due to gravity. What are the positions of equilibrium, if a particle with mass m is constrained to the sphere $\phi(x, y, z) = x^2 + y^2 + z^2 - r^2 = 0\,(r > 0)$? Which of these are stable?

SOLUTION   Notice that $\mathbf{F}$ is a potential field with $V = mgz$. Using the method of Lagrange multipliers introduced in Section 3.4 to locate the possible extrema, we have the equations

$$\nabla V = \lambda \nabla \phi$$

$$\phi = 0$$

or, in terms of components,

$$0 = 2\lambda x$$

$$0 = 2\lambda y$$

$$mg = 2\lambda z$$

$$x^2 + y^2 + z^2 - r^2 = 0.$$

The solution of these simultaneous equations is $x = 0$, $y = 0$, $z = \pm r$, $\lambda = \pm mg/2r$. By Theorem 15 it follows that the points $P_1 = (0, 0, -r)$ and $P_2 = (0, 0, r)$ are positions

---

[†] If $\phi(x, y, z) = x^2 + y^2 + z^2 - r^2$, the particle is constrained to move on a sphere; for instance, it may be whirling on a string. The part subtracted from $\mathbf{F}$ to make it parallel to S is normal to S and is called the **centripetal force.**

[*] These ideas can be applied to quite a number of interesting physical situations, such as molecular vibrations. The stability of such systems is an important question. For further information consult the physics literature (e.g., H. Goldstein, *Classical Mechanics,* Addison-Wesley, Reading, Mass., 1950, Chapter 10) and the mathematics literature (e.g., M. Hirsch and S. Smale, *Differential Equations, Dynamical Systems and Linear Algebra,* Academic Press, New York, 1974).

of equilibrium. By observation of the potential function $V = mgz$ and by Theorem 15, part (ii), it follows that P_1 is a strict minimum and hence a stable point, whereas P_2 is not. This conclusion should be physically obvious.   ▲

Next we turn to a geometric application.

EXAMPLE 3   Suppose we have a curve defined by the equation

$$\phi(x, y) = Ax^2 + 2Bxy + Cy^2 - 1 = 0.$$

Find the maximum and minimum distance of the curve to the origin. (These are the lengths of the **semimajor** and the **semiminor** axis of this quadric.)

SOLUTION   The problem is equivalent to finding the extreme values of $f(x, y) = x^2 + y^2$ subject to the constraining condition $\phi(x, y) = 0$. Using the Lagrange multiplier method, we have the following equations:

$$2x + \lambda(2Ax + 2By) = 0 \tag{1}$$

$$2y + \lambda(2Bx + 2Cy) = 0 \tag{2}$$

$$Ax^2 + 2Bxy + Cy^2 = 1. \tag{3}$$

Adding x times equation (1) to y times equation (2), we obtain

$$2(x^2 + y^2) + 2\lambda(Ax^2 + 2Bxy + Cy^2) = 0.$$

By equation (3), it follows that $x^2 + y^2 + \lambda = 0$. Let $t = -1/\lambda = 1/(x^2 + y^2)$ [$\lambda = 0$ is impossible, since $(0, 0)$ is not on the curve $\phi(x, y) = 0$]. Then equations (1) and (2) can be written as follows:

$$2(A - t)x + 2By = 0$$
$$2Bx + 2(C - t)y = 0 \tag{4}$$

If these two equations are to have a nontrivial solution [remember that $(x, y) = (0, 0)$ is not on our curve and so is not a solution], it follows from a theorem of linear algebra that their determinant vanishes:*

$$\begin{vmatrix} A - t & B \\ B & C - t \end{vmatrix} = 0.$$

Since this equation is quadratic in t, there are two solutions, which we shall call t_1 and t_2. Since $-\lambda = x^2 + y^2$, we have $\sqrt{x^2 + y^2} = \sqrt{-\lambda}$. Now $\sqrt{x^2 + y^2}$ is the distance from the point (x, y) to the origin. Therefore, if (x_1, y_1) and (x_2, y_2) denote the

---

*The matrix of coefficients of the equations cannot have an inverse, because this would imply that the solution is zero. Recall that a matrix that does not have an inverse has determinant zero.

nontrivial solutions to equation (4) corresponding to t_1 and t_2, we get $\sqrt{x_2^2 + y_2^2} = 1/\sqrt{t_2}$ and $\sqrt{x_1^2 + y_1^2} = 1/\sqrt{t_1}$. Consequently, if $t_1 > t_2$, the lengths of the semiminor and semimajor axes are $1/\sqrt{t_1}$ and $1/\sqrt{t_2}$, respectively. If the curve is an ellipse, both t_1 and t_2 are real and positive. What happens with a hyperbola or a parabola? ▲

Finally, we discuss an application to economics.

EXAMPLE 4   Suppose that the output of a manufacturing firm is a quantity Q of a certain product, where Q is a function $f(K, L)$, where K is the amount of capital equipment (or investment) and L is the amount of labor used. If the price of labor is p, the price of capital is q, and the firm can spend no more than B dollars, how can we find the amount of capital and labor to maximize the output Q?

SOLUTION   We would expect that if the amount of capital or labor is increased, then the output Q should also increase; that is,

$$\frac{\partial Q}{\partial K} \geq 0 \quad \text{and} \quad \frac{\partial Q}{\partial L} \geq 0.$$

We also expect that as more labor is added to a given amount of capital equipment, we get less additional output for our effort; that is,

$$\frac{\partial^2 Q}{\partial L^2} < 0.$$

Similarly,

$$\frac{\partial^2 Q}{\partial K^2} < 0.$$

With these assumptions on Q, it is reasonable to expect the level curves of output (called *isoquants*) $Q(K, L) = c$ to look something like the curves sketched in Figure 3.6.2, with $c_1 < c_2 < c_3$.

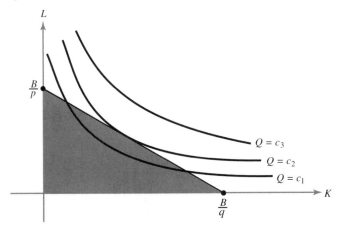

Figure 3.6.2   What is the largest value of Q in the shaded triangle?

We can interpret the convexity of the isoquants as follows: As one moves to the right along a given isoquant, it takes more and more capital to replace a unit of labor and still produce the same output. The budget constraint means that we must stay inside the triangle bounded by the axes and the line $pL + qK = B$. Geometrically, it is clear that we produce the most by spending all our money in such a way as to pick the isoquant that just touches, but does not cross, the budget line.

Since the maximum point lies on the boundary of our domain, we apply the method of Lagrange multipliers to find the maximum. To maximize $Q = f(K, L)$ subject to the constraint $pL + qK = B$, we look for critical points of the auxiliary function,

$$h(K, L, \lambda) = f(K, L) - \lambda(pL + qK - B).$$

Thus we want

$$\frac{\partial Q}{\partial K} = \lambda q, \qquad \frac{\partial Q}{\partial L} = \lambda p, \qquad \text{and} \qquad pL + qK = B.$$

These are the conditions we must meet in order to maximize output. (The reader is asked to work out a specific case in Exercise 11.) ▲

In the preceding example, λ represents something interesting. Let $k = qK$ and $l = pL$, so that k is the dollar value of the capital used and l is the dollar value of the labor used. Then the first two equations become

$$\frac{\partial Q}{\partial k} = \frac{1}{q}\frac{\partial Q}{\partial K} = \lambda = \frac{1}{p}\frac{\partial Q}{\partial L} = \frac{\partial Q}{\partial l}.$$

Thus at the optimum production point the marginal change in output per dollar's worth of additional capital investment is equal to the marginal change of output per dollar's worth of additional labor, and λ is this common value. At the optimum point, the exchange of a dollar's worth of capital for a dollar's worth of labor does not change the output. Away from the optimum point the marginal outputs are different, and one exchange or the other will increase the output.

EXERCISES

1. Let a particle move in a potential field in $\mathbb{R}^2$ given by $V(x, y) = 3x^2 + 2xy + 2x + y^2 + y + 4$. Find the stable equilibrium points, if any.

2. Let a particle move in a potential field in $\mathbb{R}^2$ given by $V(x, y) = x^2 - 2xy + y^2 + y^3 + x^4$. Is the point $(0, 0)$ a position of stable equilibrium?

3. Let a particle move in a potential field in $\mathbb{R}^2$ given by $V(x, y) = x^2 + 4xy - y^2 - 8x - 6y$. Find all the equilibrium points. Which, if any, are stable?

4. Let a particle be constrained to move on the circle $x^2 + y^2 = 25$ subject to gravitational forces (as in Example 2) as well as an additional potential $V(x, y) = x^2 + 24xy + 8y^2$. Find the stable equilibrium points, if any.

5. Let a particle be constrained to move on the sphere $x^2 + y^2 + z^2 = 1$, subject to gravitational forces (as in Example 2) as well as an additional potential $V(x, y, z) = x + y$. Find the stable equilibrium points, if any.

6. Attempt to formulate a definition and a theorem saying that if a potential has a maximum at x_0, then x_0 is a position of unstable equilibrium. Watch out for pitfalls in your argument.

7. Try to find the extrema of $xy + yz$ among points satisfying $xz = 1$.

8. Answer the question posed in the last line of Example 3.

9. Find the point on the curve $(\cos t, \sin t, \sin(t/2))$ which is farthest from the origin.

10. A company's production function is $Q(x, y) = xy$. The cost of production is $C(x, y) = 2x + 3y$. If this company can spend $C(x, y) = 10$, what is the maximum quantity that can be produced?

11. Carry out the analysis of Example 4 for the production function $Q(K, L) = AK^{\alpha}L^{1-\alpha}$, where A and α are positive constants and $0 < \alpha < 1$. This is called a **Cobb-Douglas production function** and is sometimes used as a simple model for the national economy. Q is then the aggregate output of the economy for a given input of capital and labor.

12. A firm uses wool and cotton fiber to produce cloth. The amount of cloth produced is given by $Q(x, y) = xy - x - y + 1$, where x is the number of pounds of wool, y the number of pounds of cotton, $x > 1$, and $y > 1$. If wool costs p dollars per pound, and cotton q dollars per pound and the firm can spend B dollars on material, what should the ratio of cotton and wool be to produce the most cloth?

REVIEW EXERCISES FOR CHAPTER 3

1. Analyze the behavior of the following functions at the indicated points. (Your answer may depend on the constant C.)

  (a) $z = x^2 - y^2 + 3xy$, $(x, y) = (0, 0)$
  (b) $z = x^2 - y^2 + Cxy$, $(x, y) = (0, 0)$

2. Find and classify the extreme values (if any) of the functions on $\mathbb{R}^2$ defined by the following expressions:

  (a) $y^2 - x^3$ (b) $(x - 1)^2 + (x - y)^2$ (c) $x^2 + xy^2 + y^4$

3. (a) Find the minimum distance from the origin in $\mathbb{R}^3$ to the surface $z = \sqrt{x^2 - 1}$.
  (b) Repeat part (a) for the surface $z = 6xy + 7$.

4. Find the first few terms in the Taylor expansion of $f(x, y) = e^{xy} \cos x$ about $x = 0$, $y = 0$.

5. Prove that

$$z = \frac{3x^4 - 4x^3 - 12x^2 + 18}{12(1 + 4y^2)}$$

has one local maximum, one local minimum, and one saddle point. (The graph is shown in Figure 3.R.1.)

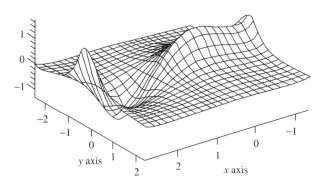

Figure 3.R.1   Graph of $z = (3x^4 - 4x^3 - 12x^2 + 18)/12(1 + 4y^2)$.

6.   Find the maxima, minima, and saddles of the function $z = (2 + \cos \pi x)(\sin \pi y)$, which is graphed in Figure 3.R.2.

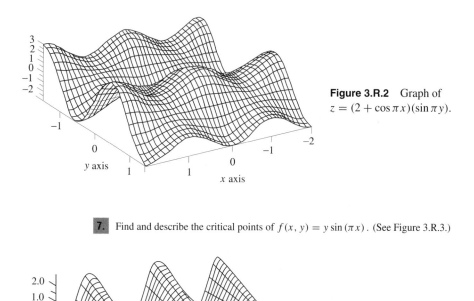

Figure 3.R.2   Graph of $z = (2 + \cos \pi x)(\sin \pi y)$.

7.   Find and describe the critical points of $f(x, y) = y \sin (\pi x)$. (See Figure 3.R.3.)

Figure 3.R.3   Graph of $z = y \sin(\pi x)$.

8.   A graph of the function $z = \sin(\pi x)/(1 + y^2)$ is shown in Figure 3.R.4. Verify that this function has alternating maxima and minima on the x axis, with no other critical points.

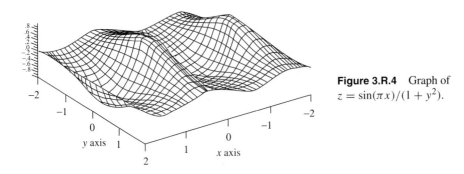

Figure 3.R.4 Graph of $z = \sin(\pi x)/(1 + y^2)$.

In Exercise 9 to 14 find the extrema of the given functions subject to the given constraints.

9. $f(x, y) = x^2 - 2xy + 2y^2; x^2 + y^2 = 1$

10. $f(x, y) = xy - y^2; x^2 + y^2 = 1$

11. $f(x, y) = \cos(x^2 - y^2); x^2 + y^2 = 1$

12. $f(x, y) = \dfrac{x^2 - y^2}{x^2 + y^2}; x + y = 1$

13. $z = xy$, subject to the condition $x + y = 1$.

14. $z = \cos^2 x + \cos^2 y$ subject to the condition $x + y = \pi/4$.

15. Find the points on the surface $z^2 - xy = 1$ nearest to the origin.

16. Use the implicit function theorem to compute dy/dx for

(a) $x/y = 10$ (b) $x^3 - \sin y + y^4 = 4$ (c) $e^{x+y^2} + y^3 = 0$

17. Find the shortest distance from the point $(0, b)$ to the parabola $x^2 - 4y = 0$. Solve this problem using the Lagrange multiplier method and also without using Lagrange's method.

18. Solve the following geometric problems by Lagrange's method.
(a)   Find the shortest distance from the point (a_1, a_2, a_3) in $\mathbb{R}^3$ to the plane whose equation is given by $b_1x_1 + b_2x_2 + b_3x_3 + b_0 = 0$, where $(b_1, b_2, b_3) \neq (0, 0, 0)$.
(b)   Find the point on the line of intersection of the two planes $a_1x_1 + a_2x_2 + a_3x_3 = 0$ and $b_1x_1 + b_2x_2 + b_3x_3 + b_0 = 0$ that is nearest to the origin.
(c)   Show that the volume of the largest rectangular parallelepiped that can be inscribed in the ellipsoid

$$\frac{x^2}{a^2} + \frac{y^2}{b^2} + \frac{z^2}{c^2} = 1$$

is $8abc/3\sqrt{3}$.

19.   A particle moves in a potential $V(x, y) = x^3 - y^2 + x^2 + 3xy$. Determine whether $(0, 0)$ is a stable equilibrium point.

20.   Study the nature of the function $f(x, y) = x^3 - 3xy^2$ near $(0, 0)$. Show that the point $(0, 0)$ is a degenerate critical point, that is, $D = 0$. This surface is called a "monkey saddle."

21.   Find and sketch the maximum of $f(x, y) = xy$ on the curve $(x + 1)^2 + y^2 = 1$.

22.   Find the maximum and minimum of $f(x, y) = xy - y + x - 1$ on the set $x^2 + y^2 \leq 2$.

23.   The Baraboo, Wisconsin, plant of International Widget Co., Inc., uses aluminium, iron, and magnesium to produce high-quality widgets. The quantity of widgets that may be produced using x tons of aluminum, y tons of iron, and z tons of magnesium is $Q(x, y, z) = xyz$. The cost of raw materials is aluminum, \$6 per ton; iron, \$4 per ton; and magnesium, \$8 per ton. How many tons each of aluminum, iron, and magnesium should be used to manufacture 1000 widgets at the lowest possible cost? (HINT: Find an extreme value for what function subject to what constraint?)

24.   Let $f: \mathbb{R} \to \mathbb{R}$ be of class C^1 and let

$$u = f(x)$$
$$v = -y + xf(x).$$

If $f'(x_0) \neq 0$, show that this transformation of $\mathbb{R}^2$ to $\mathbb{R}^2$ is invertible near (x_0, y_0) and its inverse is given by

$$x = f^{-1}(u)$$
$$y = -v + uf^{-1}(u).$$

25.   Show that the equations

$$x^2 - y^2 - u^3 + v^2 + 4 = 0$$
$$2xy + y^2 - 2u^2 + 3v^4 + 8 = 0$$

determine functions $u(x, y)$ and $v(x, y)$ defined for (x, y) near $x = 2$ and $y = -1$ such that $u(2, -1) = 2$ and $v(2, -1) = 1$. Compute $\partial u / \partial x$ at $(2, -1)$.

26.   Show that there exist positive numbers p and q such that there are unique functions u and v from the interval $(-1 - p, -1 + p)$ into the interval $(1 - q, 1 + q)$ satisfying

$$xe^{u(x)} + u(x)e^{v(x)} = 0 = xe^{v(x)} + v(x)e^{u(x)}$$

for all x in the interval $(-1 - p, -1 + p)$ with $u(-1) = 1 = v(-1)$.

27.   To work this exercise the reader should be familiar with the technique of diagonalizing a 2×2 matrix. Let $a(x)$, $b(x)$, and $c(x)$ be three continuous functions defined on $U \cup \partial U$, where U is an open set and ∂U denotes its set of boundary points (see Section 2.2). Use the notation of Lemma 2 in Section 3.3, and assume that for each $x \in U \cup \partial U$ the quadratic form defined by the matrix

$$\begin{bmatrix} a & b \\ b & c \end{bmatrix}$$

is positive-definite. For a C^2 function v on $U \cup \partial U$ we define a differential operator L by $Lv = a(\partial^2 v/\partial x^2) + 2b(\partial^2 v/\partial x \partial y) + c(\partial^2 v/\partial y^2)$. With this positive-definite condition, such an operator is said to be *elliptic*. A function v is said to be *strictly subharmonic relative* to L if $Lv > 0$. Show that a strictly subharmonic function cannot have a maximum point in U.

28. A function v is said to be in the *kernel* of the operator L described in Exercise 27 if $Lv = 0$ on $U \cup \partial U$. Arguing as in Exercise 37 of Section 3.3, show that if v achieves its maximum on U it also achieves it on ∂U. This is called the weak maximum principle for elliptic operators.

29. Let L be an elliptic differential operator as in Exercises 27 and 28.
(a) Define the notion of a strict superharmonic function.
(b) Show that such functions cannot achieve a minimum on U.
(c) If v is as in Exercise 28, show that if v achieves its minimum on U it also achieves it on ∂U.

*The following **method of least squares** should be applied to Exercises 30 to 35.*

It often happens that the theory behind an experiment indicates that the experimental data should lie approximately along a straight line of the form $y = mx + b$. The actual results, of course, never match the theory exactly. We are then faced with the problem of finding the straight line that best fits some set of experimental data $(x_1, y_1), \ldots, (x_n, y_n)$ as in Figure 3.R.5. If we guess at a straight line $y = mx + b$ to fit the data, each point will deviate vertically from the line by an amount $d_i = y_i - (mx_i + b)$.

Figure 3.R.5 The method of least squares tries to find a straight line that best approximates a set of data.

We would like to choose m and b in such a way as to make the total effect of these deviations as small as possible. However, since some are negative and some positive, we could get a lot of cancellations and still have a pretty bad fit. This leads us to suspect that a better measure of the total error might be the sum of the *squares* of these deviations. Thus we are led to the problem of finding the m and b that minimize the function

$$s = f(m, b) = d_1^2 + d_2^2 + \cdots + d_n^2 = \sum_{i=1}^{n} (y_i - mx_i - b)^2,$$

where $x_1, \ldots, x_n$ and $y_1, \ldots, y_n$ are given data.

30. For each set of three data points, plot the points, write down the function $f(m, b)$ above, find m and b to give the best straight-line fit according to the method of least squares, and plot the straight line.

(a) $(x_1, y_1) = (1, 1)$
$(x_2, y_2) = (2, 3)$
$(x_3, y_3) = (4, 3)$

(b) $(x_1, y_1) = (0, 0)$
$(x_2, y_2) = (1, 2)$
$(x_3, y_3) = (2, 3)$

31. Show that if only two data points (x_1, y_1) and (x_2, y_2) are given, this method produces the line through (x_1, y_1) and (x_2, y_2).

32. Show that the equations for a critical point, $\partial s/\partial b = 0$ and $\partial s/\partial m = 0$, are equivalent to

$$m\left(\sum x_i\right) + nb = \left(\sum y_i\right) \quad \text{and} \quad m\left(\sum x_i^2\right) + b\left(\sum x_i\right) = \left(\sum x_i y_i\right),$$

where all the sums run from $i = 1$ to $i = n$.

33. If $y = mx + b$ is the best-fitting straight line to the data points $(x_1, y_1), \dots, (x_n, y_n)$ according to the least squares method, show that

$$\sum_{i=1}^{n}(y_i - mx_i - b) = 0;$$

that is, the positive and negative deviations cancel (see Exercise 32).

34. Use the second-derivative test to show that the critical point of f is a minimum.

35. Use the method of least squares to find the straight line that best fits the points $(0, 1)$, $(1, 3)$, $(2, 2)$, $(3, 4)$, and $(4, 5)$. Plot the points and line.*

---

*The method of least squares may be varied and generalized in a number of ways. The basic idea can be applied to equations of more complicated curves than the straight line. For example, this might be done to find the parabola that best fits a given set of data points. These ideas also formed part of the basis for the development of the science of cybernetics by Norbert Wiener. Another version of the data is the following problem of least squares approximation: Given a function f defined and integrable on an interval $[a, b]$, find a polynomial P of degree $\leq n$ such that the mean square error

$$\int_a^b |f(x) - P(x)|^2 \, dx$$

is as small as possible.

4 VECTOR-VALUED FUNCTIONS

... who by vigor of mind almost divine, the motions and figures of
the planets, the paths of comets, and the tides of the seas first
demonstrated.

Newton's Epitaph

Chapters 2 and 3 focused on *real*-valued functions. This chapter is largely concerned with *vector*-valued functions. We begin in the first section with a continuation of our study of paths, adding applications of Newton's second law. Then we study arc length of paths. Following this, we introduce the divergence and curl of a vector field which, in addition to the gradient, are basic operations in vector *differential* calculus. The basic geometry and calculus of the divergence and curl are studied. The associated *integral* calculus will be given in Chapter 8.

4.1 ACCELERATION AND NEWTON'S SECOND LAW

In Section 2.4 we studied the basic geometry of paths, learning how to sketch curves (the images of paths) and compute tangent lines. We also learned to think of, as the name suggests, a path as the trajectory of a particle and to regard the derivative of the path as its velocity vector.

In this section we continue our study of paths, including additional topics, especially acceleration and Newton's second law.

As we know, a path in $\mathbb{R}^n$ is a map $\mathbf{c}: \mathbb{R} \to \mathbb{R}^n$. If the path is differentiable, its derivative at each time t is an $n \times 1$ matrix. Specifically, if $x_1(t), \dots, x_n(t)$ are the component functions of $\mathbf{c}$, the derivative matrix is

$$\mathbf{c}'(t) = \begin{bmatrix} dx_1/dt \\ dx_2/dt \\ \vdots \\ dx_n/dt \end{bmatrix},$$

which can also be written in vector form as

$$(dx_1/dt, \ldots, dx_n/dt) \qquad \text{or as} \qquad (x_1'(t), \ldots, x_n'(t)).$$

Recall from Section 2.4 that $\mathbf{c}'(t)$ is the *tangent vector* to the path at the point $\mathbf{c}(t)$. Also recall that if $\mathbf{c}$ represents the path of a moving particle, then its *velocity vector* is

$$\mathbf{v} = \mathbf{c}'(t)$$

and its *speed* is $s = \|\mathbf{v}\|$.

The differentiation of paths is facilitated by the following rules.

DIFFERENTIATION RULES Let $\mathbf{b}(t)$ and $\mathbf{c}(t)$ be differentiable paths in $\mathbb{R}^3$ and $p(t)$ and $q(t)$ be differentiable scalar functions:

$$\textit{Sum Rule:} \qquad \frac{d}{dt}[\mathbf{b}(t) + \mathbf{c}(t)] = \mathbf{b}'(t) + \mathbf{c}'(t)$$

$$\textit{Scalar Multiplication Rule:} \qquad \frac{d}{dt}[p(t)\mathbf{c}(t)] = p'(t)\mathbf{c}(t) + p(t)\mathbf{c}'(t)$$

$$\textit{Dot Product Rule:} \qquad \frac{d}{dt}[\mathbf{b}(t) \cdot \mathbf{c}(t)] = \mathbf{b}'(t) \cdot \mathbf{c}(t) + \mathbf{b}(t) \cdot \mathbf{c}'(t)$$

$$\textit{Cross Product Rule:} \qquad \frac{d}{dt}[\mathbf{b}(t) \times \mathbf{c}(t)] = \mathbf{b}'(t) \times \mathbf{c}(t) + \mathbf{b}(t) \times \mathbf{c}'(t)$$

$$\textit{Chain Rule:} \qquad \frac{d}{dt}[\mathbf{c}(q(t))] = q'(t)\mathbf{c}'(q(t))$$

These rules follow by applying the usual differentiation rules to the components.

EXAMPLE 1 Show that if $\mathbf{c}(t)$ is a vector function such that $\|\mathbf{c}(t)\|$ is constant, then $\mathbf{c}'(t)$ is perpendicular to $\mathbf{c}(t)$ for all t.

SOLUTION Since $\|\mathbf{c}(t)\|$ is constant, so is its square $\|\mathbf{c}(t)\|^2 = \mathbf{c}(t) \cdot \mathbf{c}(t)$. The derivative of this constant is zero, so by the dot product rule,

$$0 = \frac{d}{dt}[\mathbf{c}(t) \cdot \mathbf{c}(t)] = \mathbf{c}'(t) \cdot \mathbf{c}(t) + \mathbf{c}(t) \cdot \mathbf{c}'(t) = 2\mathbf{c}(t) \cdot \mathbf{c}'(t);$$

thus, $\mathbf{c}(t) \cdot \mathbf{c}'(t) = 0$; that is, $\mathbf{c}'(t)$ is perpendicular to $\mathbf{c}(t)$. ▲

For a path describing uniform rectilinear motion, the velocity vector is constant. In general, the velocity vector is a vector function $\mathbf{v} = \mathbf{c}'(t)$ that depends on t. The derivative $\mathbf{a} = d\mathbf{v}/dt = \mathbf{c}''(t)$ is called the ***acceleration*** of the curve. If the curve is $(x(t), y(t), z(t))$, then the acceleration at time t, $\mathbf{a}(t)$ is given by

$$\mathbf{a}(t) = x''(t)\mathbf{i} + y''(t)\mathbf{j} + z''(t)\mathbf{k}.$$

EXAMPLE 2   A particle moves in such a way that its acceleration is constantly equal to $-\mathbf{k}$. If the position when $t = 0$ is $(0, 0, 1)$ and the velocity at $t = 0$ is $\mathbf{i} + \mathbf{j}$, when and where does the particle fall below the plane $z = 0$? Describe the path traveled by the particle (assume $t \geq 0$).

SOLUTION   Let $(x(t), y(t), z(t))$ be the path traced out by the particle, so that the velocity vector is $\mathbf{c}'(t) = x'(t)\mathbf{i} + y'(t)\mathbf{j} + z'(t)\mathbf{k}$. The acceleration $\mathbf{c}''(t)$ is $-\mathbf{k}$, so $x''(t) = 0$, $y''(t) = 0$, and $z''(t) = -1$. It follows that $x'(t)$ and $y'(t)$ are constant functions, and $z'(t)$ is a linear function with slope -1. Since $\mathbf{c}'(0) = \mathbf{i} + \mathbf{j}$, we get $\mathbf{c}'(t) = \mathbf{i} + \mathbf{j} - t\mathbf{k}$. Integrating again and using the initial position $(0, 0, 1)$, we find that $(x(t), y(t), z(t)) = (t, t, 1 - \frac{1}{2}t^2)$. The particle drops below the plane $z = 0$ when $1 - \frac{1}{2}t^2 = 0$; that is, $t = \sqrt{2}$. At that instant, the position is $(\sqrt{2}, \sqrt{2}, 0)$. The path traveled by the particle is a parabola in the plane $y = x$ (see Figure 4.1.1) because in this plane the equation is described by $z = 1 - \frac{1}{2}x^2$. ▲

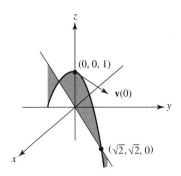

Figure 4.1.1   The path of the parabola with initial position $(0, 0, 1)$, initial velocity $\mathbf{i} + \mathbf{j}$, and constant acceleration $-\mathbf{k}$ is a parabola in the plane $y = x$.

The image of a C^1 path is not necessarily "very smooth"; indeed, it may have sharp bends or changes of direction. For instance, the cycloid $\mathbf{c}(t) = (t - \sin t, 1 - \cos t)$ has cusps at all points where $\mathbf{c}$ touches the x axis (that is, where $t = 2\pi n, n = 0, \pm 1, \ldots$). Another example is the *hypocycloid of four cusps*, $\mathbf{c} : [0, 2\pi] \to \mathbb{R}^2, t \mapsto (\cos^3 t, \sin^3 t)$, which has cusps at four points (Figure 4.1.2). At all such points, however, $\mathbf{c}'(t) = \mathbf{0}$, and the tangent line is not well defined. Evidently the direction of $\mathbf{c}'(t)$ may change abruptly at points where it slows to rest.

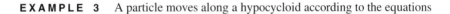

Figure 4.1.2 The image of the smooth path $\mathbf{c}(t) = (\cos^3 t, \ \sin^3 t)$, a hypocycloid, does not "look smooth."

EXAMPLE 3 A particle moves along a hypocycloid according to the equations

$$x = \cos^3 t, \qquad y = \sin^3 t, \qquad a \le t \le b.$$

What are the velocity and speed of the particle?

SOLUTION The velocity vector of the particle is

$$\mathbf{v} = \frac{dx}{dt}\mathbf{i} + \frac{dy}{dt}\mathbf{j} = -(3 \ \sin t \ \cos^2 t)\mathbf{i} + (3 \ \cos t \ \sin^2 t)\mathbf{j}$$

and its speed is

$$s = \|\mathbf{v}\| = (9 \ \sin^2 t \ \cos^4 t + 9 \ \cos^2 t \ \sin^4 t)^{1/2} = 3 \ |\sin t| \ |\cos t|. \quad \blacktriangle$$

If a particle of mass m moves in $\mathbb{R}^3$, the force $\mathbf{F}$ acting on it at the point $\mathbf{c}(t)$ is related to the acceleration $\mathbf{a}(t)$ by *Newton's second law*:

$$\mathbf{F}(\mathbf{c}(t)) = m\mathbf{a}(t).$$

In particular, if no forces act on a particle, then $\mathbf{a}(t) = \mathbf{0}$, so $\mathbf{c}'(t)$ is constant and the particle follows a straight line.

ACCELERATION AND NEWTON'S SECOND LAW   The *accelera-tion* of a path $\mathbf{c}(t)$ is

$$\mathbf{a}(t) = \mathbf{c}''(t).$$

If $\mathbf{F}$ is the force acting and m is the mass of the particle, then

$$\mathbf{F} = m\mathbf{a}.$$

In the problem of determining the path $\mathbf{c}(t)$ of a particle under the influence of a given force field $\mathbf{F}$, Newton's law becomes a differential equation (i.e., an equation involving derivatives) for $\mathbf{c}(t)$.

For example, the motion of a planet moving along a path $\mathbf{r}(t)$ around the sun (considered to be located at the origin in $\mathbb{R}^3$) obeys the law

$$m\mathbf{r}'' = -\frac{GmM}{r^3}\mathbf{r}$$

where M is the mass of the sun, m that of the planet, $r = \|\mathbf{r}\|$, and G is the gravitational constant. The relation used in determining the force, $\mathbf{F} = -GmM\mathbf{r}/r^3$, is called *Newton's law of gravitation* (see Figure 4.1.3). We shall not make a general study of such equations in this book, but content ourselves with the special case of circular orbits.

Figure 4.1.3   A mass M attracts a mass m with a force $\mathbf{F}$ given by Newton's law of gravitation: $\mathbf{F} = -GmM\mathbf{r}/r^3$.

Consider a particle of mass m moving at constant speed s in a circular path of radius r_0. Supposing that it moves in the xy plane, we can suppress the third component and write

$$\mathbf{r}(t) = \left(r_0 \cos\frac{st}{r_0}, r_0 \sin\frac{st}{r_0}\right),$$

since this is a circle of radius r_0 and $\|\mathbf{r}'(t)\| = s$. Then

$$\mathbf{a}(t) = \mathbf{r}''(t) = \left(-\frac{s^2}{r_0} \cos \frac{st}{r_0}, \; -\frac{s^2}{r_0} \sin \frac{st}{r_0} \right) = -\frac{s^2}{r_0^2} \mathbf{r}(t).$$

Thus, the acceleration is in a direction opposite to $\mathbf{r}(t)$; that is, it is directed toward the center of the circle (see Figure 4.1.4). This acceleration multiplied by the mass of the particle is called the **centripetal force**. Even though the speed is constant, the direction of the velocity is continuously changing and therefore the acceleration which is a rate of change in either speed or direction or both is nonzero.

Figure 4.1.4 The position, velocity, and acceleration of a particle in circular motion.

Newton's law helps us discover a relationship between the radius of the orbit of a revolving body and the period, that is, the time it takes for one complete revolution. Consider a satellite of mass m moving with a speed s around a central body with mass M in a *circular* orbit of radius r_0 (distance from the *center* of the spherical central body). By Newton's second law $\mathbf{F} = m\mathbf{a}$, we get

$$-\frac{s^2 m}{r_0^2} \mathbf{r}(t) = -\frac{GmM}{r_0^3} \mathbf{r}(t).$$

The lengths of the vectors on both sides of this equation must be equal. Hence

$$s^2 = \frac{GM}{r_0}.$$

If T denotes the period, then $s = 2\pi r_0/T$; substituting this value for s in the above equation and solving for T, we obtain the following:

KEPLER'S LAW

$$T^2 = r_0^3 \frac{(2\pi)^2}{GM}.$$

Thus, the *square of the period is proportional to the cube of the radius.*

We have defined two basic concepts associated with a path; its velocity and its acceleration. Both involve *differential* calculus. The basic concept of the length of a path, which involves *integral* calculus, will be taken up in the next section.

HISTORICAL NOTE

The law of planetary motion in the preceding box is one of the three that Kepler observed before Newton formulated his general laws of motion. It enables one to compute the period of a satellite about the Earth or a planet about the Sun when the radius of its orbit is given, and vice versa. Kepler discovered and used formulas like this not only for circular orbits, but more generally for elliptical orbits. Newton was able to derive Kepler's three celestial laws from his own law of gravitation. The neat mathematical order of the universe that these laws provided had a great impact on eighteenth-century thought.

Newton never wrote down his laws as differential equations. This was first done by Euler around 1730. Newton made most of his deductions (at least in published form) by geometric methods.

EXAMPLE 4   Suppose that a satellite is to be in a circular orbit about the earth such that it stays fixed in the sky over one point on the equator. What is the radius of such a *geosynchronous* orbit? (The mass of the Earth is 5.98×10^{24} kilograms and $G = 6.67 \times 10^{-11}$ in the meter-kilogram-second system of units.)

SOLUTION   The period of the satellite should be 1 day, so $T = 60 \times 60 \times 24 = 86,400$. From the formula $T^2 = r_0^3 (2\pi)^2 / GM$ we get $r_0^3 = T^2 GM / (2\pi)^2$ and so

$$r_0^3 = \frac{T^2 GM}{(2\pi)^2} = \frac{(86,400)^2 \times (6.67 \times 10^{-11}) \times (5.98 \times 10^{24})}{(2\pi)^2} \approx 7.54 \times 10^{22} \, \text{m}^3.$$

Thus $r_0 = 4.23 \times 10^7 \, \text{m} = 42,300 \, \text{km} \approx 26,200 \, \text{mi}$.   ▲

EXERCISES

In Exercises 1 to 4, find the velocity and acceleration vectors and the equation of the tangent line for each of the following curves, at the given value of t.

1. $\mathbf{r}(t) = (\cos t)\mathbf{i} + (\sin 2t)\mathbf{j}, t = 0$

2. $\mathbf{c}(t) = (t \sin t, t \cos t, \sqrt{3}t), t = 0$

3. $\mathbf{r}(t) = \sqrt{2}t\mathbf{i} + e^t\mathbf{j} + e^{-t}\mathbf{k}, t = 0$

4. $\mathbf{c}(t) = t\mathbf{i} + t\mathbf{j} + \frac{2}{3}t^{3/2}\mathbf{k}, t = 9$

In Exercises 5 to 8, let $\mathbf{c}_1(t) = e^t\mathbf{i} + (\sin t)\mathbf{j} + t^3\mathbf{k}$ and $\mathbf{c}_2(t) = e^{-t}\mathbf{i} + (\cos t)\mathbf{j} - 2t^3\mathbf{k}$. Find each of the stated derivatives in two different ways to verify the rules in the box preceding Example 1.

5. $\dfrac{d}{dt}[\mathbf{c}_1(t) + \mathbf{c}_2(t)]$

6. $\dfrac{d}{dt}[\mathbf{c}_1(t) \cdot \mathbf{c}_2(t)]$

7. $\dfrac{d}{dt}[\mathbf{c}_1(t) \times \mathbf{c}_2(t)]$

8. $\dfrac{d}{dt}\{\mathbf{c}_1(t) \cdot [2\mathbf{c}_2(t) + \mathbf{c}_1(t)]\}$

9. If $\mathbf{r}(t) = 6t\mathbf{i} + 3t^2\mathbf{j} + t^3\mathbf{k}$, what force acts on a particle of mass m moving along $\mathbf{r}$ at $t = 0$?

10. Let a particle mass of 1 gram (g) follow the path in Exercise 1, with units in seconds and centimeters. What force acts on it at $t = 0$? (Give the units in your answer.)

11. A body of mass 2 kg moves in a circular path on a circle of radius 3 meters, making one revolution every 5 seconds. Find the centripetal force acting on the body.

12. Find the centripetal force acting on a body of mass 4 kg, moving on a circle of radius 10 m with a frequency of 2 revolutions per second.

13. Show that if the acceleration of an object is always perpendicular to the velocity, then the speed of the object is constant. (HINT: See Example 1.)

14. Show that, at a local maximum or minimum of $\|\mathbf{r}(t)\|$, $\mathbf{r}'(t)$ is perpendicular to $\mathbf{r}(t)$.

15. A satellite is in a circular orbit 500 mi above the surface of the Earth. What is the period of the orbit? (You may take the radius of the Earth to be 4000 mi.)

16. What is the acceleration of the satellite in Exercise 15? The centripetal force?

17. Find the path $\mathbf{c}$ such that $\mathbf{c}(0) = (0, -5, 1)$ and $\mathbf{c}'(t) = (t, e^t, t^2)$.

18. Let $\mathbf{c}$ be a path in $\mathbb{R}^3$ with zero acceleration. Prove that $\mathbf{c}$ is a straight line or a point.

19. Find paths $\mathbf{c}(t)$ that represent the following curves or trajectories.

(a) $\{(x, y) \mid y = e^x\}$ (b) $\{(x, y) \mid 4x^2 + y^2 = 1\}$

(c) A straight line in $\mathbb{R}^3$ passing through   (d) $\{(x, y) \mid 9x^2 + 16y^2 = 4\}$
the origin and the point (a, b, c)

20. Let $\mathbf{c}(t)$ be a path, $\mathbf{v}(t)$ its velocity, and $\mathbf{a}(t)$ the acceleration. Suppose $\mathbf{F}$ is a C^1 mapping of $\mathbb{R}^3$ to $\mathbb{R}^3$, $m > 0$, and $\mathbf{F}(\mathbf{c}(t)) = m\mathbf{a}(t)$ (Newton's second law). Prove that

$$\frac{d}{dt}[m\mathbf{c}(t) \times \mathbf{v}(t)] = \mathbf{c}(t) \times \mathbf{F}(\mathbf{c}(t))$$

(i.e., "rate of change of angular momentum = torque"). What can you conclude if $\mathbf{F}(\mathbf{c}(t))$ is parallel to $\mathbf{c}(t)$? Is this the case in planetary motion?

21. Continue the investigations in Exercise 20 to prove Kepler's law that a planet moving under the influence of gravity about the sun does so in a fixed plane.

4.2 ARC LENGTH

What is the length of a path $\mathbf{c}(t)$? Since the speed $\|\mathbf{c}'(t)\|$ is the rate of change of distance traveled with respect to time, the distance traveled by a point moving along the curve should be the integral of speed with respect to the time over the interval $[t_0, t_1]$ of travel time; that is, the length of the path, also called its *arc length*, is

$$l = \int_{t_0}^{t_1} \|\mathbf{c}'(t)\| \, dt.$$

There is the question as to whether or not this formula actually corresponds to the true arc length. For example, suppose we take a curve in space and glue a string tightly to it, cutting the string so it exactly fits the curve. If we then remove the string, straighten it out and measure it with a straight edge, we surely should obtain the length of the curve. That our formula for arc length agrees with such a process is justified in the supplement at the end of this section.

EXAMPLE 1   The arc length of the path $\mathbf{c}(t) = (r \cos t, r \sin t)$, for $0 \le t \le 2\pi$, is

$$l = \int_0^{2\pi} \sqrt{(-r \sin t)^2 + (r \cos t)^2} \, dt = 2\pi r,$$

the circumference of a circle of radius r. If we had allowed $0 \le t \le 4\pi$, we would have obtained $4\pi r$, because the path traverses the same circle *twice* (Figure 4.2.1). ▲

Figure 4.2.1   The arc length of a circle traversed twice is $4\pi r$.

ARC LENGTH   The length of the path $\mathbf{c}(t) = (x(t), y(t), z(t))$ for $t_0 \le t \le t_1$, is

$$l = \int_{t_0}^{t_1} \sqrt{[x'(t)]^2 + [y'(t)]^2 + [z'(t)]^2} \, dt.$$

For planar curves one omits the $z'(t)$ term, as in Example 1. Here is an example in $\mathbb{R}^3$.

EXAMPLE 2   Find the arc length of $(\cos t, \sin t, t^2), 0 \le t \le \pi$.

SOLUTION   The path $\mathbf{c}(t) = (\cos t, \sin t, t^2)$ has the velocity vector given by $\mathbf{v} = (-\sin t, \cos t, 2t)$. Since

$$\|\mathbf{v}\| = \sqrt{\sin^2 t + \cos^2 t + 4t^2} = \sqrt{1 + 4t^2} = 2\sqrt{t^2 + \left(\frac{1}{2}\right)^2},$$

the arc length is

$$l = \int_0^\pi 2\sqrt{t^2 + \left(\frac{1}{2}\right)^2}\, dt.$$

This integral may be evaluated using the following formula from the table of integrals:

$$\int \sqrt{x^2 + a^2}\, dx = \frac{1}{2}\left[x\sqrt{x^2 + a^2} + a^2 \log\left(x + \sqrt{x^2 + a^2}\right)\right] + C.$$

Thus,

$$l = 2 \cdot \frac{1}{2}\left[t\sqrt{t^2 + \left(\frac{1}{2}\right)^2} + \left(\frac{1}{2}\right)^2 \log\left(t + \sqrt{t^2 + \left(\frac{1}{2}\right)^2}\right)\right]\Bigg|_{t=0}^{\pi}$$

$$= \pi\sqrt{\pi^2 + \frac{1}{4}} + \frac{1}{4}\log\left(\pi + \sqrt{\pi^2 + \frac{1}{4}}\right) - \frac{1}{4}\log\left(\sqrt{\frac{1}{4}}\right)$$

$$= \frac{\pi}{2}\sqrt{1 + 4\pi^2} + \frac{1}{4}\log\left(2\pi + \sqrt{1 + 4\pi^2}\right) \approx 10.63.$$

As a check on our answer, we may note that the path $\mathbf{c}$ connects the points $(1, 0, 0)$ and $(-1, 0, \pi^2)$. The distance between these points is $\sqrt{4 + \pi^2} \approx 10.06$ which is less than 10.63, so the answer is reasonable.   ▲

If a curve is made up of a finite number of pieces each of which is smooth, we compute the arc length by adding the lengths of the component pieces. Such curves are called *piecewise smooth*.

EXAMPLE 3   A billiard ball on a square table follows the path $\mathbf{c}\colon [-1, 1] \to \mathbb{R}^3$ defined by $\mathbf{c}(t) = (x(t), y(t), z(t)) = (|t|, |t - \frac{1}{2}|, 0)$. Find the distance traveled by the ball.

SOLUTION   This path is not smooth, because $x(t) = |t|$ is not differentiable at 0, nor is $y(t) = |t - \frac{1}{2}|$ differentiable at $\frac{1}{2}$. However, if we divide the interval $[-1, 1]$ into the pieces $[-1, 0]$, $[0, \frac{1}{2}]$, and $[\frac{1}{2}, 1]$, we see that $x(t)$ and $y(t)$ have continuous derivatives on each of the intervals $[-1, 0]$, $[0, \frac{1}{2}]$, and $[\frac{1}{2}, 1]$. (See Figure 4.2.2.)

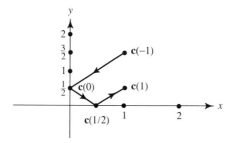

Figure 4.2.2   A piecewise smooth path.

On $[-1, 0]$, $x(t) = -t$, $y(t) = -t + \frac{1}{2}$, and $z(t) = 0$, so $\|\mathbf{c}'(t)\| = \sqrt{2}$. Hence, the arc length of $\mathbf{c}$ between -1 and 0 is $\int_{-1}^{0} \sqrt{2}\, dt = \sqrt{2}$. Similarly, on $[0, \frac{1}{2}]$, $x(t) = t$, $y(t) = -t + \frac{1}{2}$, $z(t) = 0$, and again $\|\mathbf{c}'(t)\| = \sqrt{2}$, so that the arc length of $\mathbf{c}$ between 0 and $\frac{1}{2}$ is $\frac{1}{2}\sqrt{2}$. Finally, on $[\frac{1}{2}, 1]$ we have $x(t) = t$, $y(t) = t - \frac{1}{2}$, $z(t) = 0$, and the arc length of $\mathbf{c}$ between $\frac{1}{2}$ and 1 is $\frac{1}{2}\sqrt{2}$. Thus the total arc length of $\mathbf{c}$ is $2\sqrt{2}$. Of course, one can also compute the answer as the sum of the distances from $\mathbf{c}(-1)$ to $\mathbf{c}(0)$ to $\mathbf{c}(\frac{1}{2})$ to $\mathbf{c}(1)$.   ▲

We have given a definition of arc length which was motivated by the concept of total distance traversed by the moving particle. However there are situations where a more exacting ("rigorous") treatment is needed. For example, some curves, like a mountain range silhouette or a coastline are quite complex and their "length" perhaps depends on the detail with which they are measured and could even turn out to be infinite! Computers can generate complex curves of this sort, and it is a serious problem to decide what their length might be. See Figure 4.2.3 for an example. You may also refer to B. Mandelbrot, *The Fractal Geometry of Nature*, Freeman, New York, 1982.

Figure 4.2.3   A complicated computer-generated curve. What is its length?

E X A M P L E   4   Consider the point with position function

$$\mathbf{c}(t) = (t - \sin t, 1 - \cos t)$$

tracing out the cycloid discussed in Section 2.4. Find the velocity, the speed, and the length of one arc.

SOLUTION The velocity vector is $\mathbf{c}'(t) = (1 - \cos t, \ \sin t)$, so the speed of the point $\mathbf{c}(t)$ is

$$\|\mathbf{c}'(t)\| = \sqrt{(1 - \cos t)^2 + \sin^2 t} = \sqrt{2 - 2 \cos t}.$$

Hence, $\mathbf{c}(t)$ moves at variable speed although the circle rolls at constant speed. Furthermore, the speed of $\mathbf{c}(t)$ is zero when t is an integral multiple of 2π. At these values of t, the y coordinate of the point $\mathbf{c}(t)$ is zero and so the point lies on the x axis. The arc length of one cycle is

$$
\begin{aligned}
l &= \int_0^{2\pi} \sqrt{2 - 2 \cos t} \, dt = 2 \int_0^{2\pi} \sqrt{\frac{1 - \cos t}{2}} \, dt \\
&= 2 \int_0^{2\pi} \sin \frac{t}{2} \, dt \ \left(\text{since } 1 - \cos t = 2 \sin^2 \frac{t}{2} \text{ and } \sin \frac{t}{2} \geq 0 \text{ on } [0, 2\pi] \right) \\
&= 4 \left(- \cos \frac{t}{2} \right) \Big|_0^{2\pi} = 8. \quad \blacktriangle
\end{aligned}
$$

The arc-length formula suggests that one introduce the following notation, which will be useful in Chapter 7 in our discussion of line integrals.

ARC-LENGTH DIFFERENTIAL An *infinitesimal displacement* of a particle following a path $\mathbf{c}(t) = x(t)\mathbf{i} + y(t)\mathbf{j} + z(t)\mathbf{k}$ is

$$d\mathbf{s} = dx\mathbf{i} + dy\mathbf{j} + dz\mathbf{k} = \left(\frac{dx}{dt}\mathbf{i} + \frac{dy}{dt}\mathbf{j} + \frac{dz}{dt}\mathbf{k} \right) dt$$

and its length

$$ds = \sqrt{dx^2 + dy^2 + dz^2} = \sqrt{\left(\frac{dx}{dt} \right)^2 + \left(\frac{dy}{dt} \right)^2 + \left(\frac{dz}{dt} \right)^2} \, dt$$

is the *differential of arc length*. See Figure 4.2.4.

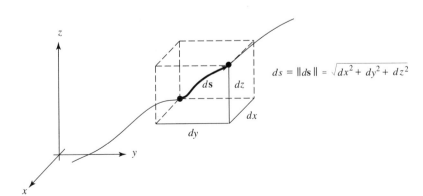

Figure 4.2.4 Differential of arc length.

These formulas help us remember the arc-length formula as

$$\text{arc length} = \int_{t_0}^{t_1} ds.$$

As we have done before with such geometric concepts as length and angle, we can extend the notion of arc length to paths in n-dimensional space.

ARC LENGTH IN $\mathbb{R}^n$ Let $\mathbf{c}$: $[t_0, t_1] \to \mathbb{R}^n$ be a path. Its *length* is defined to be

$$l = \int_{t_0}^{t_1} \|\mathbf{c}'(t)\| \, dt.$$

The integrand is the square root of the sum of the squares of the coordinate functions of $\mathbf{c}'(t)$: If

$$\mathbf{c}(t) = (x_1(t), x_2(t), \ldots, x_n(t)),$$

then

$$l = \int_{t_0}^{t_1} \sqrt{(x_1'(t))^2 + (x_2'(t))^2 + \cdots + (x_n'(t))^2} \, dt.$$

EXAMPLE 5   Find the length of the path $\mathbf{c}(t) = (\cos t, \sin t, \cos 2t, \sin 2t)$ in $\mathbb{R}^4$, defined on the interval from 0 to π.

SOLUTION   We have $\mathbf{c}'(t) = (-\sin t, \cos t, -2 \sin 2t, 2 \cos 2t)$, and so

$$\|\mathbf{c}'(t)\| = \sqrt{\sin^2 t + \cos^2 t + 4 \sin^2 2t + 4 \cos^2 2t} = \sqrt{1 + 4} = \sqrt{5},$$

a constant, so the length of the path is

$$\int_0^{\pi} \sqrt{5} \, dt = \sqrt{5}\pi. \quad \blacktriangle$$

It is common to introduce the *arc-length function* $s(t)$ associated to a path $\mathbf{c}(t)$ given by

$$s(t) = \int_a^t \|\mathbf{c}'(u)\| \, du,$$

so that (by the fundamental theorem of calculus)

$$s'(t) = \|\mathbf{c}'(t)\|$$

and

$$\int_a^b s'(t) \, dt = s(b) - s(a) = s(b).$$

Supplement to Section 4.2

The following discussion assumes an acquaintance with the definite integral defined in terms of Riemann sums. If your background in this is weak, the material may be postponed until after Chapter 5.

In $\mathbb{R}^3$ there is another way to justify the arc-length formula based on polygonal approximations. We partition the interval $[a, b]$ into N subintervals of equal length:

$$a = t_0 < t_1 < \cdots < t_N = b;$$
$$t_{i+1} - t_i = \frac{b - a}{N} \quad \text{for} \quad 0 \le i \le N - 1.$$

We then consider the polygonal line obtained by joining the successive pairs of points $\mathbf{c}(t_i)$, $\mathbf{c}(t_{i+1})$ for $0 \le i \le N - 1$. This yields a polygonal approximation to $\mathbf{c}$ as in Figure 4.2.5. By the formula for distance in $\mathbb{R}^3$, it follows that the line segment from $\mathbf{c}(t_i)$ to $\mathbf{c}(t_{i+1})$ has length

$$\|\mathbf{c}(t_{i+1}) - \mathbf{c}(t_i)\| = \sqrt{[x(t_{i+1}) - x(t_i)]^2 + [y(t_{i+1}) - y(t_i)]^2 + [z(t_{i+1}) - z(t_i)]^2},$$

where $\mathbf{c}(t) = (x(t), y(t), z(t))$. Applying the mean value theorem to $x(t)$, $y(t)$, and $z(t)$ on $[t_i, t_{i+1}]$, we obtain three points t_i^*, t_i^{**}, and t_i^{***} such that

$$x(t_{i+1}) - x(t_i) = x'(t_i^*)(t_{i+1} - t_i),$$
$$y(t_{i+1}) - y(t_i) = y'(t_i^{**})(t_{i+1} - t_i),$$

and

$$z(t_{i+1}) - z(t_i) = z'(t_i^{***})(t_{i+1} - t_i).$$

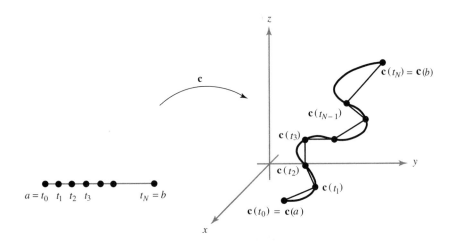

Figure 4.2.5   A path $\mathbf{c}$ may be approximated by a polygonal path obtained by joining each $\mathbf{c}(t_i)$ to $\mathbf{c}(t_{i+1})$ by a straight line.

Thus the line segment from $\mathbf{c}(t_i)$ to $\mathbf{c}(t_{i+1})$ has length

$$\sqrt{[x'(t_i^*)]^2 + [y'(t_i^{**})]^2 + [z'(t_i^{***})]^2}\,(t_{i+1} - t_i).$$

Therefore the length of our approximating polygonal line is

$$S_N = \sum_{i=0}^{N-1} \sqrt{[x'(t_i^*)]^2 + [y'(t_i^{**})]^2 + [z'(t_i^{***})]^2}\,(t_{i+1} - t_i).$$

As $N \to \infty$, this polygonal line approximates the image of $\mathbf{c}$ more closely. Therefore, we define the arc length of $\mathbf{c}$ as the limit, if it exists, of the sequence S_N as $N \to \infty$. Since the derivatives x', y', and z' are all assumed to be continuous on $[a, b]$, we can conclude that, in fact, the limit does exist and is given by

$$\lim_{N \to \infty} S_N = \int_a^b \sqrt{[x'(t)]^2 + [y'(t)]^2 + [z'(t)]^2}\, dt.$$

(The theory of integration relates the integral to sums by the formula

$$\int_a^b f(t)\, dt = \lim_{N \to \infty} \sum_{i=0}^{N-1} f(t_i^*)(t_{i+1} - t_i),$$

where $t_0, \ldots, t_N$ is a partition of $[a, b]$, $t_i^* \in [t_i, t_{i+1}]$ is arbitrary, and f is a continuous function. Here we have *possibly different points* t_i^*, t_i^{**}, and t_i^{***}, and so this formula must be extended slightly.)

EXERCISES

Find the arc length of the given curve on the specified interval in Exercises 1 to 6.*

1. $(2 \cos t, 2 \sin t, t); 0 \le t \le 2\pi$

2. $(1, 3t^2, t^3); 0 \le t \le 1$

3. $(\sin 3t, \cos 3t, 2t^{3/2}); 0 \le t \le 1$

4. $\left(t + 1, \dfrac{2\sqrt{2}}{3}t^{3/2} + 7, \dfrac{1}{2}t^2\right); 1 \le t \le 2$

5. $(t, t, t^2); 1 \le t \le 2$

6. $(t, t \sin t, t \cos t); 0 \le t \le \pi$

---

*Several of these problems make use of the formula

$$\int \sqrt{x^2 + a^2}\, dx = \tfrac{1}{2}\left[x\sqrt{x^2 + a^2} + a^2 \log\left(x + \sqrt{x^2 + a^2}\right)\right] + C$$

from the table of integrals in the back of the book.

7. Find the length of the path $\mathbf{c}(t)$, defined by $\mathbf{c}(t) = (2\cos t, 2\sin t, t)$, if $0 \le t \le 2\pi$ and $\mathbf{c}(t) = (2, t - 2\pi, t)$, if $2\pi \le t \le 4\pi$.

8. Let $\mathbf{c}$ be the path $\mathbf{c}(t) = (t, t\sin t, t\cos t)$. Find the arc length of $\mathbf{c}$ between $(0, 0, 0)$ and $(2\pi, 0, -2\pi)$.

9. Let $\mathbf{c}$ be the path $\mathbf{c}(t) = (2t, t^2, \log t)$, defined for $t > 0$. Find the arc length of $\mathbf{c}$ between the points $(2, 1, 0)$ and $(4, 4, \log 2)$.

10. The arc-length function $s(t)$ for a given path $\mathbf{c}(t)$, defined by $s(t) = \int_a^t \|\mathbf{c}'(\tau)\| \, d\tau$, represents the distance a particle traversing the trajectory of $\mathbf{c}$ will have traveled by time t if it starts out at time a; that is, it gives the length of $\mathbf{c}$ between $\mathbf{c}(a)$ and $\mathbf{c}(t)$. Find the arc-length functions for the curves $\boldsymbol{\alpha}(t) = (\cosh t, \sinh t, t)$ and $\boldsymbol{\beta}(t) = (\cos t, \sin t, t)$, with $a = 0$.

11. Let $\mathbf{c}(t)$ be a given path, $a \le t \le b$. Let $s = \alpha(t)$ be a new variable, where α is a strictly increasing C^1 function given on $[a, b]$. For each s in $[\alpha(a), \alpha(b)]$ there is a unique t with $\alpha(t) = s$. Define the function $\mathbf{d}: [\alpha(a), \alpha(b)] \to \mathbb{R}^3$ by $\mathbf{d}(s) = \mathbf{c}(t)$.
  (a) Argue that the image curves of $\mathbf{c}$ and $\mathbf{d}$ are the same.
  (b) Show that $\mathbf{c}$ and $\mathbf{d}$ have the same arc length.
  (c) Let $s = \alpha(t) = \int_a^t \|\mathbf{c}'(\tau)\| \, d\tau$. Define $\mathbf{d}$ as above by $\mathbf{d}(s) = \mathbf{c}(t)$. Show that

$$\left\| \frac{d}{ds}\mathbf{d}(s) \right\| = 1.$$

The path $s \mapsto \mathbf{d}(s)$ is said to be a ***reparametrization*** of $\mathbf{c}$, according to arc length (see also Exercise 13).

Exercises 12 to 17 develop some of the classical differential geometry of curves.

12. Let $\mathbf{c}: [a, b] \to \mathbb{R}^3$ be an infinitely differentiable path (derivatives of all orders exist). Assume $\mathbf{c}'(t) \ne \mathbf{0}$ for any t. The vector $\mathbf{c}'(t)/\|\mathbf{c}'(t)\| = \mathbf{T}(t)$ is tangent to $\mathbf{c}$ at $\mathbf{c}(t)$, and, since $\|\mathbf{T}(t)\| = 1$, $\mathbf{T}$ is called the ***unit tangent*** to $\mathbf{c}$.
  (a) Show that $\mathbf{T}'(t) \cdot \mathbf{T}(t) = 0$. [HINT: Differentiate $\mathbf{T}(t) \cdot \mathbf{T}(t) = 1$.]
  (b) Write down a formula for $\mathbf{T}'(t)$ in terms of $\mathbf{c}$.

13. (a) A path $\mathbf{c}(s)$ is said to be ***parametrized by arc length*** or, what is the same thing, to have ***unit speed*** if $\|\mathbf{c}'(s)\| = 1$. For a path parametrized by arc length on $[a, b]$, show that $l(\mathbf{c}) = b - a$.
  (b) The ***curvature*** at a point $\mathbf{c}(s)$ on a path is defined by $k = \|\mathbf{T}'(s)\|$ when the path is parametrized by arc length. Show that $k = \|\mathbf{c}''(s)\|$.
  (c) If $\mathbf{c}$ is given in terms of some other parameter t and $\mathbf{c}'(t)$ is never $\mathbf{0}$, show that $k = \|\mathbf{c}'(t) \times \mathbf{c}''(t)\|/\|\mathbf{c}'(t)\|^3$.
  (d) Calculate the curvature of the helix $\mathbf{c}(t) = (1/\sqrt{2})(\cos t, \sin t, t)$. (This $\mathbf{c}$ is a scalar multiple of the right-circular helix.)

14. If $\mathbf{T}'(t) \ne \mathbf{0}$, it follows from Exercise 12 that $\mathbf{N}(t) = \mathbf{T}'(t)/\|\mathbf{T}'(t)\|$ is normal (i.e., perpendicular) to $\mathbf{T}(t)$; $\mathbf{N}$ is called the ***principal normal vector***. Let a third unit vector that is perpendicular to both $\mathbf{T}$ and $\mathbf{N}$ be defined by $\mathbf{B} = \mathbf{T} \times \mathbf{N}$; $\mathbf{B}$ is called the ***binormal vector***. Together, $\mathbf{T}$, $\mathbf{N}$, and $\mathbf{B}$ form a right-handed system of mutually orthogonal vectors that may be thought of as moving along the path (Figure 4.2.6). Show that

  (a) $\dfrac{d\mathbf{B}}{dt} \cdot \mathbf{B} = 0.$ (b) $\dfrac{d\mathbf{B}}{dt} \cdot \mathbf{T} = 0.$
  (c) $d\mathbf{B}/dt$ is a scalar multiple of $\mathbf{N}$.

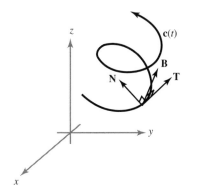

Figure 4.2.6   The tangent **T**, principal normal **N**, and binormal **B**.

15. If $\mathbf{c}(s)$ is parametrized by arc length, we use the result of Exercise 14(c) to define a scalar-valued function τ, called the ***torsion***, by

$$\frac{d\mathbf{B}}{ds} = -\tau\mathbf{N}.$$

(a)   Show that $\tau = [\mathbf{c}'(s) \times \mathbf{c}''(s)] \cdot \mathbf{c}'''(s)/\|\mathbf{c}''(s)\|^2$.
(b)   Show that if $\mathbf{c}$ is given in terms of some other parameter t,

$$\tau = \frac{[\mathbf{c}'(t) \times \mathbf{c}''(t)] \cdot \mathbf{c}'''(t)}{\|\mathbf{c}'(t) \times \mathbf{c}''(t)\|^2}.$$

Compare with Exercise 13(c).
(c)   Compute the torsion of the helix $\mathbf{c}(t) = (1/\sqrt{2})(\cos t,\ \sin t, t)$.

16.   Show that if a path lies in a plane then the torsion is zero. Do this by demonstrating that **B** is constant and is a normal vector to the plane in which $\mathbf{c}$ lies. (If the torsion is not zero, it gives a measure of how fast the curve is twisting out of the plane of **T** and **N**.)

17.   (a)   Use the results of Exercises 14 and 15 to prove the following ***Frenet formulas*** for a unit-speed curve:

$$\frac{d\mathbf{T}}{ds} = k\mathbf{N}; \quad \frac{d\mathbf{N}}{ds} = -k\mathbf{T} + \tau\mathbf{B}; \quad \frac{d\mathbf{B}}{ds} = -\tau\mathbf{N}.$$

(b)   Reexpress the results of part (a) as

$$\frac{d}{ds}\begin{pmatrix} \mathbf{T} \\ \mathbf{N} \\ \mathbf{B} \end{pmatrix} = \boldsymbol{\omega} \times \begin{pmatrix} \mathbf{T} \\ \mathbf{N} \\ \mathbf{B} \end{pmatrix}$$

for a suitable vector $\boldsymbol{\omega}$.

18.   In special relativity, the ***proper time*** of a path $\mathbf{c}: [a, b] \to \mathbb{R}^4$ with $\mathbf{c}(\lambda) = (x(\lambda), y(\lambda), z(\lambda), t(\lambda))$ is defined to be the quantity

$$\int_a^b \sqrt{-[x'(\lambda)]^2 - [y'(\lambda)]^2 - [z'(\lambda)]^2 + c^2[t'(\lambda)]^2}\, d\lambda,$$

where c is the velocity of light, a constant. In Figure 4.2.7, show that, using self-explanatory notation, the "twin paradox inequality" holds:

proper time (AB) + proper time (BC) < proper time (AC).

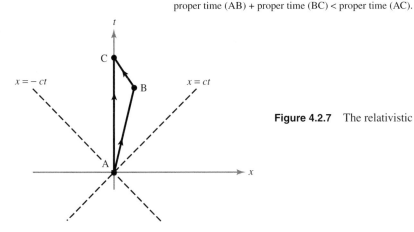

Figure 4.2.7   The relativistic triangle inequality.

4.3 VECTOR FIELDS

In Chapter 2 we introduced a particular kind of vector field, the gradient. In this section we study *general* vector fields, discussing their geometric and physical significance.

VECTOR FIELDS   A *vector field* in $\mathbb{R}^n$ is a map $\mathbf{F}: A \subset \mathbb{R}^n \to \mathbb{R}^n$ that assigns to each point $\mathbf{x}$ in its domain A a vector $\mathbf{F}(\mathbf{x})$. If $n = 2$, $\mathbf{F}$ is called a *vector field in the plane*, and if $n = 3$, $\mathbf{F}$ is a *vector field in space*.

Picture $\mathbf{F}$ as attaching an *arrow* to each point (Figure 4.3.1).   By contrast, a map $f: A \subset \mathbb{R}^n \to \mathbb{R}$ that assigns a *number* to each point is a *scalar field*. A vector field $\mathbf{F}(x, y, z)$ on $\mathbb{R}^3$ has three *component scalar fields* F_1, F_2, and F_3, so that

$$\mathbf{F}(x, y, z) = (F_1(x, y, z), F_2(x, y, z), F_3(x, y, z)).$$

Similarly, a vector field on $\mathbb{R}^n$ has n components $F_1, \ldots, F_n$. If each component is a C^k function, we say the vector field $\mathbf{F}$ is of *class C^k*. Vector fields will be assumed to be at least of class C^1 unless otherwise noted.

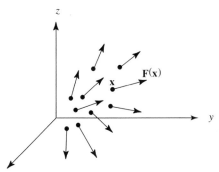

Figure 4.3.1   A vector field $\mathbf{F}$ assigns a vector $\mathbf{F}(\mathbf{x})$ to each point $\mathbf{x}$ of its domain.

In many applications, $\mathbf{F}(\mathbf{x})$ represents a physical vector quantity (force, velocity, etc.) associated with the position $\mathbf{x}$, as in the following examples.

EXAMPLE 1   The flow of water through a pipe is said to be **steady** if, at each point inside the pipe, the velocity of the fluid passing through that point does not change with time. (Note that this is quite different from saying that the water in the pipe is not moving.) Attaching to each point the fluid velocity at that point, we obtain the **velocity field** $\mathbf{V}$ of the fluid (see Figure 4.3.2). Notice that the length of the arrows (the speed), as well as the direction of flow, may change from point to point.   ▲

Figure 4.3.2   A vector field describing the velocity of flow in a pipe.

EXAMPLE 2   A piece of material is heated on one side and cooled on another. The temperature at each point within the body is described at a given moment by a scalar field $T(x, y, z)$. The flow of heat may be marked by a field of arrows indicating the direction and magnitude of the flow (Figure 4.3.3). This **energy** or **heat flux vector field** is given by $\mathbf{J} = -k\nabla T$, where $k > 0$ is a constant called the **conductivity** and ∇T is the gradient of the real-valued function T. Level sets of T are called **isotherms**. Note that the heat flows from hot regions toward cold ones, since $-\nabla T$ points in the direction of decreasing T.   ▲

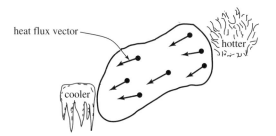
heat flux vector
hotter
cooler

Figure 4.3.3   A vector field describing the direction and magnitude of heat flow.

EXAMPLE 3   The force of attraction of the earth on a mass m can be described by a vector field called the **gravitational force field**. Place the origin of a coordinate system at the center of the Earth (assumed spherical). According to Newton's law of gravity, this field is given by

$$\mathbf{F} = -\frac{mMG}{r^3}\mathbf{r},$$

where $\mathbf{r}(x, y, z) = (x, y, z)$, and $r = \|\mathbf{r}\|$ (see Figure 4.3.4). The domain of this vector field consists of those $\mathbf{r}$ for which $\|\mathbf{r}\|$ is greater than the radius of the Earth. As we saw in Example 6, Section 2.6, $\mathbf{F}$ is a gradient field, $\mathbf{F} = -\nabla V$, where $V = -(mMG)/r$. Note again that **F** points in the direction of decreasing V. Writing **F** out in terms of components, we see that

$$\mathbf{F}(x, y, z) = \left(\frac{-mMG}{r^3}x, \; \frac{-mMG}{r^3}y, \; \frac{-mMG}{r^3}z \right). \quad \blacktriangle$$

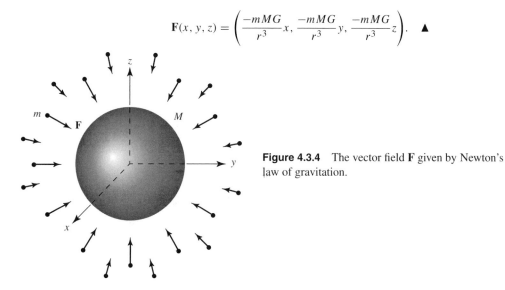

Figure 4.3.4 The vector field **F** given by Newton's law of gravitation.

EXAMPLE 4   Some forms of rotary motion (such as the motion of particles on a turntable) can be described by the vector field

$$\mathbf{V}(x, y) = -y\mathbf{i} + x\mathbf{j}.$$

See Figure 4.3.5, in which we have shown instead of **V** the shorter vector field $\frac{1}{4}\mathbf{V}$ so that the arrows do not overlap. This is a common convention in drawing pictures of vector fields.   ▲

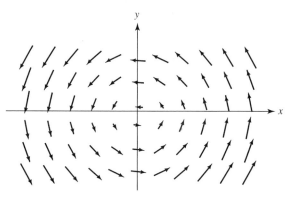

Figure 4.3.5 A rotary vector field.

EXAMPLE 5   In the plane, $\mathbb{R}^2$, the function **V** defined by

$$\mathbf{V}(x, y) = \frac{y\mathbf{i}}{x^2 + y^2} - \frac{x\mathbf{j}}{x^2 + y^2} = \left(\frac{y}{x^2 + y^2}, \; -\frac{x}{x^2 + y^2} \right)$$

is a vector field on $\mathbb{R}^2$, except at the origin, where it is not defined. This vector field is a good approximation to the planar part of the velocity of water flowing toward a hole in the bottom of a tub (Figure 4.3.6). Notice that the velocity becomes *larger* as you approach the hole.   ▲

Figure 4.3.6   The vector field describing circular flow in a tub.

E X A M P L E   6   According to *Coulomb's law*, the force acting on a charge e at position **r** due to a charge Q at the origin is

$$\mathbf{F} = \frac{\varepsilon Q e}{r^3}\mathbf{r} = -\nabla V,$$

where $V = \varepsilon Q e / r$ and ε is a constant that depends on the units used. For $Qe > 0$ (like charges) the force is repulsive [Figure 4.3.7(a)], and for $Qe < 0$ (unlike charges) the force is attractive [Figure 4.3.7(b)]. Since the potential V is constant on the level surfaces of V, they are called *equipotential surfaces*. Note that the force field is orthogonal to the equipotential surfaces (the force field is radial and the equipotential surfaces are concentric spheres).   ▲

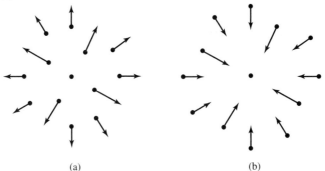

(a)   (b)

Figure 4.3.7   The vector fields associated with (a) like charges ($Qe > 0$) and (b) unlike charges ($Qe < 0$).

The next example shows that not every vector field is a gradient.

EXAMPLE 7 Show that the vector field $\mathbf{V}$ on $\mathbb{R}^2$ defined by $\mathbf{V}(x, y) = y\mathbf{i} - x\mathbf{j}$ is not a gradient vector field; that is, there is no C^1 function f such that

$$\mathbf{V}(x, y) = \nabla f(x, y) = \frac{\partial f}{\partial x}\mathbf{i} + \frac{\partial f}{\partial y}\mathbf{j}.$$

SOLUTION Suppose that such an f exists. Then $\partial f/\partial x = y$ and $\partial f/\partial y = -x$. Since these are C^1 functions, f itself must have continuous first- *and* second-order partial derivatives. But, $\partial^2 f/\partial x\,\partial y = -1$ and $\partial^2 f/\partial y\,\partial x = 1$ which violates the equality of mixed partials. Thus $\mathbf{V}$ cannot be a gradient vector field. ▲

An important concept related to vector fields is that of a flow line.

FLOW LINES If $\mathbf{F}$ is a vector field, a *flow line* for $\mathbf{F}$ is a path $\mathbf{c}(t)$ such that

$$\mathbf{c}'(t) = \mathbf{F}(\mathbf{c}(t)).$$

That is, $\mathbf{F}$ yields the velocity field of the path $\mathbf{c}(t)$.

In the context of Example 1, a flow line is the path followed by a small particle suspended in the fluid (Figure 4.3.8). Flow lines are also appropriately called *streamlines* or *integral curves*.

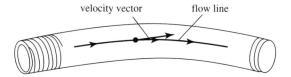

velocity vector   flow line

Figure 4.3.8 The velocity vector of a fluid is tangent to a flow line.

Geometrically, a flow line for a given vector filed $\mathbf{F}$ is a curve that threads its way through the domain of the vector field in such a way that the tangent vector of the curve coincides with the vector field, as in Figure 4.3.9.

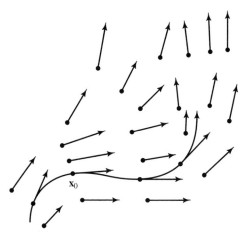

Figure 4.3.9 A flow line threading its way through a vector field in the plane.

A flow line may be viewed as a solution of a system of differential equations. Indeed, we can write the definition $\mathbf{c}'(t) = \mathbf{F}(\mathbf{c}(t))$ as

$$x'(t) = P(x(t), y(t), z(t)),$$
$$y'(t) = Q(x(t), y(t), z(t)),$$
$$z'(t) = R(x(t), y(t), z(t)),$$

where $\mathbf{c}(t) = x(t)\mathbf{i} + y(t)\mathbf{j} + z(t)\mathbf{k}$ and where

$$\mathbf{F} = P\mathbf{i} + Q\mathbf{j} + R\mathbf{k}.$$

One learns about such systems in courses on differential equations, but we are not presuming such a course has been taken.

EXAMPLE 8   Show that the path $\mathbf{c}(t) = (\cos t, \sin t)$ is a flow line of $\mathbf{F}(x, y) = -y\mathbf{i} + x\mathbf{j}$. Can you find others?

SOLUTION   We must verify that $\mathbf{c}'(t) = \mathbf{F}(\mathbf{c}(t))$. The left side is $(-\sin t)\mathbf{i} + (\cos t)\mathbf{j}$ while the right side is $\mathbf{F}(\cos t, \sin t) = (-\sin t)\mathbf{i} + (\cos t)\mathbf{j}$, so we have a flow line. As suggested by Figure 4.3.5, the other flow lines are also circles. They have the form

$$\mathbf{c}(t) = (r\,\cos(t - t_0), r\,\sin(t - t_0))$$

for constants r and t_0.   ▲

In many cases explicit formulas for flow lines are not available, so one must resort to numerical methods. Figure 4.3.10 shows some output from a program that computes flow lines numerically and plots them on the screen.

Figure 4.3.10   Computer-generated integral curves of the vector field $\mathbf{F}(x, y) = (\sin y)\mathbf{i} + (x^2 - y)\mathbf{j}$, using "MacMath" by Hubbard and West, Springer-Verlag, New York.

Supplement to Section 4.3   Flows of Vector Fields

It is convenient to give the unique solution through a given point at time 0 a special notation:

$$\phi(\mathbf{x}, t) = \left\{ \begin{matrix} \text{the position of the point on the flow line} \\ \text{through } \mathbf{x} \text{ after time } t \text{ has elapsed.} \end{matrix} \right\}.$$

With $\mathbf{x}$ as the initial condition, follow along the flow line for a time period t until the new position $\phi(\mathbf{x}, t)$ is reached (see Figure 4.3.11). Alternatively, $\phi(\mathbf{x}, t)$ is defined by:

$$\left. \begin{matrix} \dfrac{\partial}{\partial t}\phi(\mathbf{x}, t) = \mathbf{F}(\phi(\mathbf{x}, t)) \\ \phi(\mathbf{x}, 0) = \mathbf{x} \end{matrix} \right\}. \tag{1}$$

We call the mapping ϕ, which is regarded as the function of the variables $\mathbf{x}$ and t, the *flow* of $\mathbf{F}$.

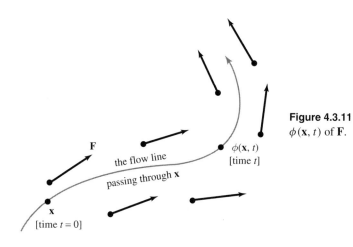

Figure 4.3.11   The definition of the flow $\phi(\mathbf{x}, t)$ of $\mathbf{F}$.

Let $\mathbf{D_x}$ denote differentiation with respect to $\mathbf{x}$, holding t fixed. It is proved in courses on differential equations that ϕ is a differentiable function of $\mathbf{x}$. By differentiating equation (1) with respect to $\mathbf{x}$, we get

$$\mathbf{D_x}\frac{\partial}{\partial t}\phi(\mathbf{x}, t) = \mathbf{D_x}[\mathbf{F}(\phi(\mathbf{x}, t))].$$

The equality of mixed partial derivatives may be used on the left-hand side of this equation and the chain rule applied to the right-hand side, yielding

$$\frac{\partial}{\partial t}\mathbf{D_x}\phi(\mathbf{x}, t) = \mathbf{DF}(\phi(\mathbf{x}, t))\mathbf{D_x}\phi(\mathbf{x}, t), \tag{2}$$

where $\mathbf{DF}(\phi(\mathbf{x}, t))$ denotes the derivative of $\mathbf{F}$ evaluated at $\phi(\mathbf{x}, t)$. Equation (2), a linear differential equation for $\mathbf{D_x}\phi(\mathbf{x}, t)$, is called the *equation of first variation.* It will be useful in our discussion of divergence and curl in the next section. Both $\mathbf{D_x}\mathbf{F}(\phi)$ and $\mathbf{D_x}\phi$ are 3×3 matrices since $\mathbf{F}$ and ϕ take values in $\mathbb{R}^3$ and are differentiated with respect to $\mathbf{x} \in \mathbb{R}^3$; for vector fields in the plane, they would be 2×2 matrices.

EXERCISES

In Exercises 1 to 8, sketch the given vector field or a small multiple of it.

1. $\mathbf{F}(x, y) = (2, 2)$

2. $\mathbf{F}(x, y) = (4, 0)$

3. $\mathbf{F}(x, y) = (x, y)$

4. $\mathbf{F}(x, y) = (-x, y)$

5. $\mathbf{F}(x, y) = (2y, x)$

6. $\mathbf{F}(x, y) = (y, -2x)$

7. $\mathbf{F}(x, y) = \left(\dfrac{x}{\sqrt{x^2 + y^2}}, \dfrac{y}{\sqrt{x^2 + y^2}} \right)$

8. $\mathbf{F}(x, y) = \left(\dfrac{y}{\sqrt{x^2 + y^2}}, \dfrac{x}{\sqrt{x^2 + y^2}} \right)$

In Exercises 9 to 12, sketch a few flow lines of the given vector field.

9. $\mathbf{F}(x, y) = (y, -x)$

10. $\mathbf{F}(x, y) = (x, -y)$

11. $\mathbf{F}(x, y) = (x, x^2)$

12. $\mathbf{F}(x, y, z) = (y, -x, 0)$

In Exercises 13 to 16, show that the given curve $\mathbf{c}(t)$ is a flow line of the given velocity vector field $\mathbf{F}(x, y, z)$.

13. $\mathbf{c}(t) = (e^{2t}, \log |t|, 1/t), t \neq 0; \mathbf{F}(x, y, z) = (2x, z, -z^2)$

14. $\mathbf{c}(t) = (t^2, 2t - 1, \sqrt{t}), t > 0; \mathbf{F}(x, y, z) = (y + 1, 2, 1/2z)$

15. $\mathbf{c}(t) = (\sin t, \cos t, e^t); \mathbf{F}(x, y, z) = (y, -x, z)$

16. $\mathbf{c}(t) = \left(\dfrac{1}{t^3}, e^t, \dfrac{1}{t} \right); \mathbf{F}(x, y, z) = (-3z^4, y, -z^2)$

17. Let a particle of mass m move on a path $\mathbf{r}(t)$ according to Newton's law in a force field $\mathbf{F} = -\nabla V$ on $\mathbb{R}^3$, where V is a given potential energy function.
   (a) Prove that the energy $E = \frac{1}{2}m\|\mathbf{r}'(t)\|^2 + V(\mathbf{r}(t))$ is constant in time by calculating dE/dt.
   (b) If the particle moves on an equipotential surface, show that its speed is constant.

18. Let $\mathbf{c}(t)$ be a flow line of a gradient field $\mathbf{F} = -\nabla V$. Prove that $V(\mathbf{c}(t))$ is a decreasing function of t.

19. Suppose that the isotherms in a region are all concentric spheres centered at the origin. Prove that the energy flux vector field points either toward or away from the origin.

20. Sketch the gradient field $-\nabla V$ for $V(x, y) = (x + y)/(x^2 + y^2)$ and the equipotential surface $V = 1$.

21. If $f(\mathbf{x}, t)$ is a real-valued function of $\mathbf{x}$ and t, define the ***material derivative*** of f relative to a vector field $\mathbf{F}$ as

$$\frac{Df}{Dt} = \frac{\partial f}{\partial t} + \nabla f(\mathbf{x}) \cdot \mathbf{F}.$$

Show that Df/Dt is the t derivative of $f(\phi(\mathbf{x}, t), t)$ (i.e., the t derivative of f transported by the flow of $\mathbf{F}$).

22. (a) Assuming uniqueness of flow lines through a given point at a given time, prove the following property of the flow $\phi(\mathbf{x}, t)$ of a vector field $\mathbf{F}$:

$$\phi(\mathbf{x}, t + s) = \phi(\phi(\mathbf{x}, s), t).$$

(b) What is the corresponding property for $\mathbf{D_x}\phi$?

4.4 DIVERGENCE AND CURL

For each of the divergence and curl operations, we use the ***del operator***

$$\nabla = \mathbf{i}\frac{\partial}{\partial x} + \mathbf{j}\frac{\partial}{\partial y} + \mathbf{k}\frac{\partial}{\partial z}.$$

For functions of one variable, taking a derivative can be thought of as an operation or process; that is, given a function $y = f(x)$, its derivative is the result of *operating* on y by the derivative *operator d/dx*. Similarly, we can write the gradient as

$$\nabla f = \left(\mathbf{i}\frac{\partial}{\partial x} + \mathbf{j}\frac{\partial}{\partial y}\right)f = \mathbf{i}\frac{\partial f}{\partial x} + \mathbf{j}\frac{\partial f}{\partial y}$$

for functions of two variables, and

$$\nabla f = \left(\mathbf{i}\frac{\partial}{\partial x} + \mathbf{j}\frac{\partial}{\partial y} + \mathbf{k}\frac{\partial}{\partial z}\right)f = \mathbf{i}\frac{\partial f}{\partial x} + \mathbf{j}\frac{\partial f}{\partial y} + \mathbf{k}\frac{\partial f}{\partial z}$$

for three variables. In operational terms, the gradient of f is obtained by taking the ∇ operator and applying it to f.

We define the divergence of a vector field $\mathbf{F}$ by taking the *dot product* of ∇ with $\mathbf{F}$.

DIVERGENCE   If $\mathbf{F} = F_1\mathbf{i} + F_2\mathbf{j} + F_3\mathbf{k}$, the ***divergence*** of $\mathbf{F}$ is the scalar field

$$\text{div } \mathbf{F} = \nabla \cdot \mathbf{F} = \frac{\partial F_1}{\partial x} + \frac{\partial F_2}{\partial y} + \frac{\partial F_3}{\partial z}.$$

Similarly, if $\mathbf{F} = (F_1, \ldots, F_n)$ is a vector field on $\mathbb{R}^n$, its divergence is

$$\text{div } \mathbf{F} = \sum_{i=1}^{n} \frac{\partial F_i}{\partial x_i}.$$

EXAMPLE 1   Compute the divergence of

$$\mathbf{F} = x^2 y\mathbf{i} + z\mathbf{j} + xyz\mathbf{k}.$$

SOLUTION

$$\text{div } \mathbf{F} = \frac{\partial}{\partial x}(x^2 y) + \frac{\partial}{\partial y}(z) + \frac{\partial}{\partial z}(xyz) = 2xy + 0 + xy = 3xy. \quad \blacktriangle$$

The divergence has an important physical interpretation. If we imagine $\mathbf{F}$ to be the velocity field of a gas (or a fluid), then div $\mathbf{F}$ *represents the rate of expansion per unit volume under the flow of the gas (or fluid)*. If div $\mathbf{F} < 0$, it *means the gas is compressing*. For a vector field $\mathbf{F}(x, y) = F_1\mathbf{i} + F_2\mathbf{j}$ on the plane, *the divergence*

$$\nabla \cdot \mathbf{F} = \frac{\partial F_1}{\partial x} + \frac{\partial F_2}{\partial y}$$

measures the rate of expansion of area. We will justify this interpretation in Chapter 8, but we will now present some examples.

EXAMPLE 2   Consider the vector field in the plane given by $\mathbf{V}(x, y) = x\mathbf{i}$. Relate the sign of the divergence of $\mathbf{V}$ with the rate of change of areas under the flow.

SOLUTION   We think of $\mathbf{V}$ as the velocity field of a fluid in the plane. The vector field $\mathbf{V}$ points to the right for $x > 0$ and to the left if $x < 0$ as we see in Figure 4.4.1. The length of $\mathbf{V}$ gets shorter toward the origin. As the fluid moves, it expands (the area of the shaded rectangle increases), so we expect div $\mathbf{V} > 0$. Indeed, div $\mathbf{V} = 1$. $\quad \blacktriangle$

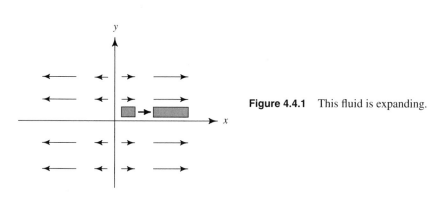

Figure 4.4.1 This fluid is expanding.

EXAMPLE 3 The flow lines of the vector field $\mathbf{F} = x\mathbf{i} + y\mathbf{j}$ are straight lines directed away from the origin (Figure 4.4.2).

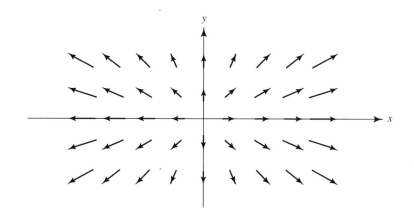

Figure 4.4.2 The vector field $\mathbf{F}(x, y) = x\mathbf{i} + y\mathbf{j}$.

If these flow lines are those of a fluid, the fluid is expanding as it moves out from the origin, so div $\mathbf{F}$ should be positive. In fact,

$$\nabla \cdot \mathbf{F} = \frac{\partial}{\partial x}x + \frac{\partial}{\partial y}y = 2 > 0. \quad \blacktriangle$$

EXAMPLE 4 Consider the vector field $\mathbf{F} = -x\mathbf{i} - y\mathbf{j}$. Here the flow lines point toward the origin instead of away from it (see Figure 4.4.3). Therefore, the fluid is compressing, so we expect (div $\mathbf{F}$) < 0. Calculating, we see that

$$\nabla \cdot \mathbf{F} = \frac{\partial}{\partial x}(-x) + \frac{\partial}{\partial y}(-y) = -1 - 1 = -2 < 0. \quad \blacktriangle$$

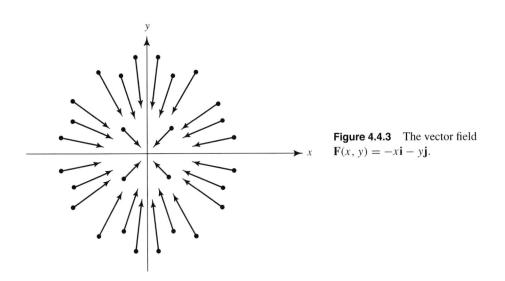

Figure 4.4.3 The vector field
$\mathbf{F}(x, y) = -x\mathbf{i} - y\mathbf{j}$.

EXAMPLE 5   As we saw in the last section, the flow lines of $\mathbf{F} = -y\mathbf{i} + x\mathbf{j}$ are concentric circles about the origin, moving counterclockwise (see Figure 4.4.4). From this figure, it appears that the fluid is neither compressing nor expanding. This is confirmed by calculating

$$\nabla \cdot \mathbf{F} = \frac{\partial}{\partial x}(-y) + \frac{\partial}{\partial y}(x) = 0 + 0 = 0. \quad \blacktriangle$$

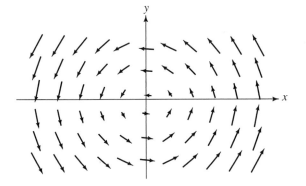

Figure 4.4.4 The vector field
$\mathbf{F}(x, y) = -y\mathbf{i} + x\mathbf{j}$ has zero
divergence.

EXAMPLE 6   Some flow lines of $\mathbf{F} = x\mathbf{i} - y\mathbf{j}$ are shown in Figure 4.4.5. Here our intuition about expansion or compression is less clear. However, it is true that the shaded regions shown have the same area and we calculate that

$$\nabla \cdot \mathbf{F} = \frac{\partial}{\partial x}x + \frac{\partial}{\partial y}(-y) = 1 + (-1) = 0. \quad \blacktriangle$$

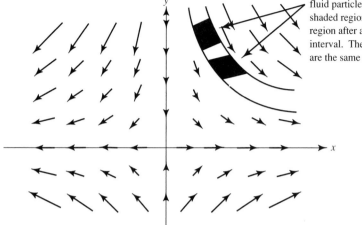

fluid particles move from shaded region to shaded region after a fixed time interval. The two areas are the same

Figure 4.4.5 The vector field $\mathbf{F}(x, y) = x\mathbf{i} - y\mathbf{j}$.

To calculate the curl, the second basic operation performed on vector fields, we take the *cross product* of ∇ with $\mathbf{F}$.

CURL OF A VECTOR FIELD If $\mathbf{F} = F_1\mathbf{i} + F_2\mathbf{j} + F_3\mathbf{k}$, the ***curl*** of $\mathbf{F}$ is the *vector field*

$$\text{curl } \mathbf{F} = \nabla \times \mathbf{F} = \begin{vmatrix} \mathbf{i} & \mathbf{j} & \mathbf{k} \\ \dfrac{\partial}{\partial x} & \dfrac{\partial}{\partial y} & \dfrac{\partial}{\partial z} \\ F_1 & F_2 & F_3 \end{vmatrix}$$

$$= \left(\frac{\partial F_3}{\partial y} - \frac{\partial F_2}{\partial z} \right)\mathbf{i} + \left(\frac{\partial F_1}{\partial z} - \frac{\partial F_3}{\partial x} \right)\mathbf{j} + \left(\frac{\partial F_2}{\partial x} - \frac{\partial F_1}{\partial y} \right)\mathbf{k}.$$

We sometimes write curl $\mathbf{F}$ for $\nabla \times \mathbf{F}$.

EXAMPLE 7 Let $\mathbf{F}(x, y, z) = x\mathbf{i} + xy\mathbf{j} + \mathbf{k}$. Find $\nabla \times \mathbf{F}$.

SOLUTION

$$\nabla \times \mathbf{F} = \begin{vmatrix} \mathbf{i} & \mathbf{j} & \mathbf{k} \\ \dfrac{\partial}{\partial x} & \dfrac{\partial}{\partial y} & \dfrac{\partial}{\partial z} \\ x & xy & 1 \end{vmatrix} = (0 - 0)\mathbf{i} - (0 - 0)\mathbf{j} - (y - 0)\mathbf{k}.$$

Thus $\nabla \times \mathbf{F} = y\mathbf{k}$. ▲

EXAMPLE 8   Find the curl of $xy\mathbf{i} - \sin z\mathbf{j} + \mathbf{k}$.

SOLUTION   Letting $\mathbf{F} = xy\mathbf{i} - \sin z\mathbf{j} + \mathbf{k}$,

$$\nabla \times \mathbf{F} = \begin{vmatrix} \mathbf{i} & \mathbf{j} & \mathbf{k} \\ \dfrac{\partial}{\partial x} & \dfrac{\partial}{\partial y} & \dfrac{\partial}{\partial z} \\ xy & -\sin z & 1 \end{vmatrix}$$

$$= \begin{vmatrix} \dfrac{\partial}{\partial y} & \dfrac{\partial}{\partial z} \\ -\sin z & 1 \end{vmatrix} \mathbf{i} - \begin{vmatrix} \dfrac{\partial}{\partial x} & \dfrac{\partial}{\partial z} \\ xy & 1 \end{vmatrix} \mathbf{j} + \begin{vmatrix} \dfrac{\partial}{\partial x} & \dfrac{\partial}{\partial y} \\ xy & -\sin z \end{vmatrix} \mathbf{k}$$

$$= \cos z\mathbf{i} - x\mathbf{k}. \quad \blacktriangle$$

Unlike the divergence, which can be defined in $\mathbb{R}^n$ for any n, we only define the curl in three-dimensional space.

The physical significance of the curl will be discussed in Chapter 8, when we study Stokes' theorem. However, we can now consider a specific situation, in which the curl is associated with rotations.

EXAMPLE 9   Consider a solid rigid body B rotating about an axis L. The rotational motion of the body can be described by a vector $\mathbf{w}$ along the axis of rotation, the direction being chosen so that the body rotates about $\mathbf{w}$ as in Figure 4.4.6. We call $\mathbf{w}$ the **angular velocity vector**. The length $\omega = \|\mathbf{w}\|$ is taken to be the angular speed of the body B, that is, the speed of any point in B divided by its distance from the axis L of rotation. The motion of points in the rotating body is described by the vector field $\mathbf{v}$ whose value at each point is the velocity at that point. To find $\mathbf{v}$, let Q be any point in B and let α be the distance from Q to L.

Figure 4.4.6   The velocity $\mathbf{v}$ and angular velocity $\mathbf{w}$ of a rotating body are related by $\mathbf{v} = \mathbf{w} \times \mathbf{r}$.

Figure 4.4.6 shows that $\alpha = \|\mathbf{r}\| \sin\theta$, where $\mathbf{r}$ is the vector whose initial point is the origin and whose terminal point is Q and θ is the angle between $\mathbf{r}$ and the axis L of rotation. The tangential velocity $\mathbf{v}$ of Q is directed counterclockwise along the tangent to a circle parallel to the xy plane with radius α and has magnitude

$$\|\mathbf{v}\| = \omega\alpha = \omega\|\mathbf{r}\| \sin \theta = \|\mathbf{w}\|\|\mathbf{r}\| \sin \theta.$$

The direction and magnitude of $\mathbf{v}$ imply that $\mathbf{v} = \mathbf{w} \times \mathbf{r}$. Selecting a coordinate system in which L is the z axis, we can write $\mathbf{w} = \omega\mathbf{k}$ and $\mathbf{r} = x\mathbf{i} + y\mathbf{j} + z\mathbf{k}$. Thus,

$$\mathbf{v} = \mathbf{w} \times \mathbf{r} = -\omega y\mathbf{i} + \omega x\mathbf{j}$$

and so

$$\text{curl } \mathbf{v} = \begin{vmatrix} \mathbf{i} & \mathbf{j} & \mathbf{k} \\ \dfrac{\partial}{\partial x} & \dfrac{\partial}{\partial y} & \dfrac{\partial}{\partial z} \\ -\omega y & \omega x & 0 \end{vmatrix} = 2\,\omega\mathbf{k} = 2\mathbf{w}.$$

Hence, for the rotation of a rigid body, the curl of the velocity vector field is a vector field whose value is the same at each point. It is directed along the axis of rotation with magnitude *twice* the angular speed. ▲

If a vector field represents the flow of a *fluid*, then the value of $\nabla \times \mathbf{F}$ at a point is twice the angular velocity vector of a *rigid* body that rotates as the fluid does near that point. In particular, $\nabla \times \mathbf{F} = \mathbf{0}$ at a point P means that the fluid is free from rigid rotations at P; that is, it has no whirlpools. Another justification of this idea depends on Stokes' theorem from Chapter 8. However, we can say informally that curl $\mathbf{F} = \mathbf{0}$ means that if a *small* rigid paddle wheel is placed in the fluid, it will move with the fluid, but will not rotate around its axis. Such a vector field is called ***irrotational***. For example, it has been determined from experiments that fluid draining from a tub is usually irrotational except right at the center, even though the fluid is "rotating" around the drain (see Figure 4.4.7). In Example 10 below, the flow lines of the vector field $\mathbf{V}$ are circles about the origin, yet the flow is irrotational. Thus the reader should be warned of the possible confusion the word "irrotational" can cause.

Figure 4.4.7 Looking at a paddle wheel from above a moving fluid. The velocity field $\mathbf{V}(x, y, z) = (y\mathbf{i} - x\mathbf{j})/(x^2 + y^2)$ is irrotational; the paddle wheel does not rotate around its axis $\mathbf{w}$.

EXAMPLE 10   Verify that the vector field

$$\mathbf{V}(x, y, z) = \frac{y\mathbf{i} - x\mathbf{j}}{x^2 + y^2}$$

is irrotational when $(x, y) \neq (0, 0)$, (i.e., except where $\mathbf{V}$ is not defined).

SOLUTION   The curl is

$$\nabla \times \mathbf{V} = \begin{vmatrix} \mathbf{i} & \mathbf{j} & \mathbf{k} \\ \dfrac{\partial}{\partial x} & \dfrac{\partial}{\partial y} & \dfrac{\partial}{\partial z} \\ \dfrac{y}{x^2 + y^2} & \dfrac{-x}{x^2 + y^2} & 0 \end{vmatrix}$$

$$= 0\mathbf{i} + 0\mathbf{j} + \left[\frac{\partial}{\partial x}\left(\frac{-x}{x^2 + y^2}\right) - \frac{\partial}{\partial y}\left(\frac{y}{x^2 + y^2}\right)\right]\mathbf{k}$$

$$= \left[\frac{-(x^2 + y^2) + 2x^2}{(x^2 + y^2)^2} + \frac{-(x^2 + y^2) + 2y^2}{(x^2 + y^2)^2}\right]\mathbf{k} = \mathbf{0}. \quad \blacktriangle$$

The following identity is a basic relation between the gradient and curl, which should be compared with the fact that for any vector $\mathbf{v}$, we have $\mathbf{v} \times \mathbf{v} = \mathbf{0}$.

THEOREM 1: CURL OF A GRADIENT   *For any C^2 function f,*

$$\nabla \times (\nabla f) = \mathbf{0}.$$

That is, the curl of any gradient is the zero vector.

PROOF   Since $\nabla f = (\partial f/\partial x, \partial f/\partial y, \partial f/\partial z)$ we have, by definition,

$$\nabla \times \nabla f = \begin{vmatrix} \mathbf{i} & \mathbf{j} & \mathbf{k} \\ \dfrac{\partial}{\partial x} & \dfrac{\partial}{\partial y} & \dfrac{\partial}{\partial z} \\ \dfrac{\partial f}{\partial x} & \dfrac{\partial f}{\partial y} & \dfrac{\partial f}{\partial z} \end{vmatrix}$$

$$= \left(\frac{\partial^2 f}{\partial y\, \partial z} - \frac{\partial^2 f}{\partial z\, \partial y}\right)\mathbf{i} + \left(\frac{\partial^2 f}{\partial z\, \partial x} - \frac{\partial^2 f}{\partial x\, \partial z}\right)\mathbf{j} + \left(\frac{\partial^2 f}{\partial x\, \partial y} - \frac{\partial^2 f}{\partial y\, \partial x}\right)\mathbf{k}.$$

Each component is zero because of the equality of mixed partial derivatives.   ■

EXAMPLE 11   Let $\mathbf{V}(x, y, z) = y\mathbf{i} - x\mathbf{j}$. Show that $\mathbf{V}$ is not a gradient field.

SOLUTION If $\mathbf{V}$ were a gradient field, then it would satisfy curl $\mathbf{V} = \mathbf{0}$ by Theorem 1. But

$$\text{curl } \mathbf{V} = \begin{vmatrix} \mathbf{i} & \mathbf{j} & \mathbf{k} \\ \dfrac{\partial}{\partial x} & \dfrac{\partial}{\partial y} & \dfrac{\partial}{\partial z} \\ y & -x & 0 \end{vmatrix} = -2\mathbf{k} \neq \mathbf{0}. \quad \blacktriangle$$

There is an operation on vector fields in the plane that is closely related to the curl. If $\mathbf{F} = P(x, y)\mathbf{i} + Q(x, y)\mathbf{j}$ is a vector field in the plane, it can also be regarded as a vector field in space for which the $\mathbf{k}$-component is zero and the other two components are independent of z. The curl of $\mathbf{F}$ then reduces to

$$\nabla \times \mathbf{F} = \left(\frac{\partial Q}{\partial x} - \frac{\partial P}{\partial y} \right)\mathbf{k}$$

and always points in the $\mathbf{k}$ direction. The function

$$\frac{\partial Q}{\partial x} - \frac{\partial P}{\partial y}$$

of x and y is called the ***scalar curl*** of $\mathbf{F}$.

EXAMPLE 12 Find the scalar curl of $\mathbf{V}(x, y) = -y^2\mathbf{i} + x\mathbf{j}$.

SOLUTION The curl is

$$\nabla \times \mathbf{V} = \begin{vmatrix} \mathbf{i} & \mathbf{j} & \mathbf{k} \\ \dfrac{\partial}{\partial x} & \dfrac{\partial}{\partial y} & \dfrac{\partial}{\partial z} \\ -y^2 & x & 0 \end{vmatrix} = (1 + 2y)\,\mathbf{k},$$

so the scalar curl, which is the coefficient of $\mathbf{k}$, is $1 + 2y$. $\quad \blacktriangle$

A basic relation between the divergence and curl operations is given next.

THEOREM 2: DIVERGENCE OF A CURL *For any C^2 vector field* $\mathbf{F}$,

$$\text{div curl } \mathbf{F} = \nabla \cdot (\nabla \times \mathbf{F}) = 0.$$

That is, the divergence of any curl is zero.

As with the curl of a gradient, the proof rests on the equality of the mixed partial derivatives. The student should write out the details.

E X A M P L E   13   Show that the vector field $\mathbf{V}(x, y) = x\mathbf{i} + y\mathbf{j} + z\mathbf{k}$ cannot be the curl of some vector field $\mathbf{F}$; that is, there is no $\mathbf{F}$ with $\mathbf{V} = \text{curl } \mathbf{F}$.

S O L U T I O N   If this were so, then div $\mathbf{V}$ would be zero by Theorem 2. But

$$\text{div } \mathbf{V} = \frac{\partial x}{\partial x} + \frac{\partial y}{\partial y} + \frac{\partial z}{\partial z} = 3 \neq 0,$$

so $\mathbf{V}$ cannot be curl $\mathbf{F}$ for any $\mathbf{F}$.   ▲

The **Laplace operator** ∇^2, which operates on functions f, is defined as the divergence of the gradient:

$$\nabla^2 f = \nabla \cdot (\nabla f) = \frac{\partial^2 f}{\partial x^2} + \frac{\partial^2 f}{\partial y^2} + \frac{\partial^2 f}{\partial z^2}.$$

This operator plays an important role in many physical laws, as we have mentioned in Section 3.1.

E X A M P L E   14   Show that $\nabla^2 f = 0$ for

$$f(x, y, z) = \frac{1}{\sqrt{x^2 + y^2 + z^2}} = \frac{1}{r} \quad \text{and} \quad (x, y, z) \neq (0, 0, 0)$$

where $\mathbf{r} = x\mathbf{i} + y\mathbf{j} + z\mathbf{k}$ and $r = \|\mathbf{r}\|$.

S O L U T I O N   The first derivatives are

$$\frac{\partial f}{\partial x} = \frac{-x}{(x^2 + y^2 + z^2)^{3/2}}, \quad \frac{\partial f}{\partial y} = \frac{-y}{(x^2 + y^2 + z^2)^{3/2}},$$

$$\frac{\partial f}{\partial z} = \frac{-z}{(x^2 + y^2 + z^2)^{3/2}}.$$

Computing the second derivatives, we find that

$$\frac{\partial^2 f}{\partial x^2} = \frac{3x^2}{(x^2 + y^2 + z^2)^{5/2}} - \frac{1}{(x^2 + y^2 + z^2)^{3/2}},$$

$$\frac{\partial^2 f}{\partial y^2} = \frac{3y^2}{(x^2 + y^2 + z^2)^{5/2}} - \frac{1}{(x^2 + y^2 + z^2)^{3/2}},$$

$$\frac{\partial^2 f}{\partial z^2} = \frac{3z^2}{(x^2 + y^2 + z^2)^{5/2}} - \frac{1}{(x^2 + y^2 + z^2)^{3/2}}.$$

Thus,

$$\frac{\partial^2 f}{\partial x^2} + \frac{\partial^2 f}{\partial y^2} + \frac{\partial^2 f}{\partial z^2} = \frac{3(x^2 + y^2 + z^2)}{(x^2 + y^2 + z^2)^{5/2}} - \frac{3}{(x^2 + y^2 + z^2)^{3/2}}$$

$$= \frac{3}{(x^2 + y^2 + z^2)^{3/2}} - \frac{3}{(x^2 + y^2 + z^2)^{3/2}} = 0. \quad ▲$$

We now have these basic operations on hand: gradient, divergence, curl, and the Laplace operator. The following table lists some basic general formulas that are useful when computing with vector fields.

BASIC IDENTITIES OF VECTOR ANALYSIS

1. $\nabla(f + g) = \nabla f + \nabla g$

2. $\nabla(cf) = c\nabla f$, for a constant c

3. $\nabla(fg) = f\nabla g + g\nabla f$

4. $\nabla(f/g) = (g\nabla f - f\nabla g)/g^2$, at points $\mathbf{x}$ where $g(\mathbf{x}) \neq 0$

5. $\text{div}\,(\mathbf{F} + \mathbf{G}) = \text{div}\,\mathbf{F} + \text{div}\,\mathbf{G}$

6. $\text{curl}\,(\mathbf{F} + \mathbf{G}) = \text{curl}\,\mathbf{F} + \text{curl}\,\mathbf{G}$

7. $\text{div}\,(f\mathbf{F}) = f\,\text{div}\,\mathbf{F} + \mathbf{F} \cdot \nabla f$

8. $\text{div}\,(\mathbf{F} \times \mathbf{G}) = \mathbf{G} \cdot \text{curl}\,\mathbf{F} - \mathbf{F} \cdot \text{curl}\,\mathbf{G}$

9. $\text{div}\,\text{curl}\,\mathbf{F} = 0$

10. $\text{curl}\,(f\mathbf{F}) = f\,\text{curl}\,\mathbf{F} + \nabla f \times \mathbf{F}$

11. $\text{curl}\,\nabla f = \mathbf{0}$

12. $\nabla^2(fg) = f\nabla^2 g + g\nabla^2 f + 2(\nabla f \cdot \nabla g)$

13. $\text{div}\,(\nabla f \times \nabla g) = 0$

14. $\text{div}\,(f\nabla g - g\nabla f) = f\nabla^2 g - g\nabla^2 f$

EXAMPLE 15 Prove identity 7 in the preceding box.

SOLUTION The vector field $f\mathbf{F}$ has components fF_i, for $i = 1, 2, 3$, and so

$$\text{div}\,(f\mathbf{F}) = \frac{\partial}{\partial x}(fF_1) + \frac{\partial}{\partial y}(fF_2) + \frac{\partial}{\partial z}(fF_3).$$

However, $(\partial/\partial x)(fF_1) = f\,\partial F_1/\partial x + F_1\,\partial f/\partial x$ by the product rule, with similar expressions for the other terms. Therefore

$$\text{div}\,(f\mathbf{F}) = f\left(\frac{\partial F_1}{\partial x} + \frac{\partial F_2}{\partial y} + \frac{\partial F_3}{\partial z}\right) + F_1\frac{\partial f}{\partial x} + F_2\frac{\partial f}{\partial y} + F_3\frac{\partial f}{\partial z}$$

$$= f(\nabla \cdot \mathbf{F}) + \mathbf{F} \cdot \nabla f. \quad \blacktriangle$$

Let us use these identities to redo Example 14.

EXAMPLE 16   Show that for $\mathbf{r} \neq \mathbf{0}$, $\nabla^2(1/r) = 0$.

SOLUTION   As in the case of the gravitational potential, $\nabla(1/r) = -\mathbf{r}/r^3$. In general $\nabla(r^n) = nr^{n-2}\mathbf{r}$ (see Exercise 30). By the identity $\nabla \cdot (f\mathbf{F}) = f\nabla \cdot \mathbf{F} + \nabla f \cdot \mathbf{F}$, we get

$$\nabla \cdot \left(\frac{\mathbf{r}}{r^3}\right) = \frac{1}{r^3}\nabla \cdot \mathbf{r} + \mathbf{r} \cdot \nabla\left(\frac{1}{r^3}\right)$$

$$= \frac{3}{r^3} + \mathbf{r} \cdot \left(\frac{-3\mathbf{r}}{r^5}\right) = 0. \quad \blacktriangle$$

Supplement to Section 4.4   The Geometry of the Divergence

We now study the geometric meaning of the divergence in more detail. This discussion depends on the concept of the flow $\phi(\mathbf{x}, t)$ of a vector field $\mathbf{F}$ given at the end of Section 4.3. See Exercises 33 to 35 for the corresponding discussion of curl.

Fix a point $\mathbf{x}$ and consider the three standard basis vectors $\mathbf{i}, \mathbf{j}, \mathbf{k}$ emanating from $\mathbf{x}$. Let $\varepsilon > 0$ be small and consider the basis vectors $\mathbf{v}_1 = \varepsilon\mathbf{i}$, $\mathbf{v}_2 = \varepsilon\mathbf{j}$, $\mathbf{v}_3 = \varepsilon\mathbf{k}$, also emanating from $\mathbf{x}$. These vectors span a parallelepiped $P(0)$. As time increases or decreases, the flow $\phi(\mathbf{x}, t)$ carries $P(0)$ into some object. For *fixed* time, ϕ is a differentiable function of $\mathbf{x}$ (that is, ϕ is a differentiable map of $\mathbb{R}^3$ to $\mathbb{R}^3$). When ε is small, the image of $P(0)$ under ϕ can be approximated by its image under the derivative of ϕ with respect to $\mathbf{x}$. Recall that if $\mathbf{v}$ is a short vector based at a point P_1 and ending at P_2, so $\mathbf{v} = P_2 - P_1$, then $\phi(P_2, t) - \phi(P_1, t) \approx \mathbf{D_x}\phi(\mathbf{x}, t) \cdot \mathbf{v}$. Thus for fixed time and small positive ε, $P(0)$ is approximately carried into a parallelepiped spanned by the vectors $\mathbf{v}_1(t), \mathbf{v}_2(t), \mathbf{v}_3(t)$ given by

$$\left.\begin{array}{l} \mathbf{v}_1(t) = \mathbf{D_x}\phi(\mathbf{x}, t) \cdot \mathbf{v}_1 \\ \mathbf{v}_2(t) = \mathbf{D_x}\phi(\mathbf{x}, t) \cdot \mathbf{v}_2 \\ \mathbf{v}_3(t) = \mathbf{D_x}\phi(\mathbf{x}, t) \cdot \mathbf{v}_3 \end{array}\right\} \tag{1}$$

Since $\phi(\mathbf{x}, 0) = \mathbf{x}$ for all $\mathbf{x}$, it follows that $\mathbf{v}_1(0) = \mathbf{v}_1$, $\mathbf{v}_2(0) = \mathbf{v}_2$, and $\mathbf{v}_3(0) = \mathbf{v}_3$. The vectors $\mathbf{v}_1(t)$, $\mathbf{v}_2(t), \mathbf{v}_3(t)$ span a parallelepiped $P(t)$ that moves in time (see Figure 4.4.8).

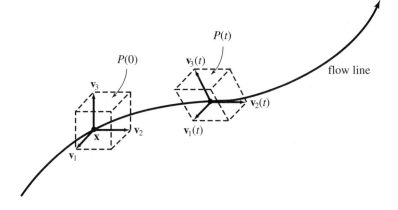

Figure 4.4.8   The moving basis $\mathbf{v}_1(t), \mathbf{v}_2(t), \mathbf{v}_3(t)$ and the associated parallelepiped.

Let the volume of $P(t)$ be denoted by $\mathcal{V}(t)$. The main geometric meaning of divergence is given by the following theorem.

THEOREM 3

$$\text{div } \mathbf{F}(\mathbf{x}) = \frac{1}{\mathcal{V}(0)} \frac{d}{dt} \mathcal{V}(t) \bigg|_{t=0}.$$

PROOF By equation (2) of the previous section,

$$\frac{d}{dt} \mathbf{v}_i(t) = \mathbf{DF}(\phi(\mathbf{x}, t)) \cdot (\mathbf{D_x}\phi(\mathbf{x}, t) \cdot \mathbf{v}_i) \tag{2}$$

for $i = 1, 2, 3$.

Since $\phi(\mathbf{x}, 0) = \mathbf{x}$, $\mathbf{D_x}\phi(\mathbf{x}, 0)$ is the identity matrix, so that evaluation at $t = 0$ gives

$$\frac{d}{dt} \mathbf{v}_i(t) \bigg|_{t=0} = \mathbf{DF}(\mathbf{x}) \cdot \mathbf{v}_i.$$

The volume $\mathcal{V}(t)$ is given by the triple product:

$$\mathcal{V}(t) = \mathbf{v}_1(t) \cdot [\mathbf{v}_2(t) \times \mathbf{v}_3(t)].$$

Using the differentiation rules of Section 4.1 and the identities $\mathbf{v}_1 \cdot [\mathbf{v}_2 \times \mathbf{v}_3] = \mathbf{v}_2 \cdot [\mathbf{v}_3 \times \mathbf{v}_1] = \mathbf{v}_3 \cdot [\mathbf{v}_1 \times \mathbf{v}_2]$, we get

$$\frac{d\mathcal{V}}{dt} = \frac{d\mathbf{v}_1}{dt} \cdot [\mathbf{v}_2(t) \times \mathbf{v}_3(t)] + \mathbf{v}_1(t) \cdot \left[\frac{d\mathbf{v}_2}{dt} \times \mathbf{v}_3(t)\right] + \mathbf{v}_1(t) \cdot \left[\mathbf{v}_2(t) \times \frac{d\mathbf{v}_3}{dt}\right]$$

$$= \frac{d\mathbf{v}_1}{dt} \cdot [\mathbf{v}_2(t) \times \mathbf{v}_3(t)] + \frac{d\mathbf{v}_2}{dt} \cdot [\mathbf{v}_3(t) \times \mathbf{v}_1(t)] + \frac{d\mathbf{v}_3}{dt} \cdot [\mathbf{v}_1(t) \times \mathbf{v}_2(t)].$$

At $t = 0$, substitution from formula (1) and the fact that $\mathbf{v}_1 \times \mathbf{v}_2 = \varepsilon \mathbf{v}_3$, $\mathbf{v}_3 \times \mathbf{v}_1 = \varepsilon \mathbf{v}_2$, and $\mathbf{v}_2 \times \mathbf{v}_3 = \varepsilon \mathbf{v}_1$ gives

$$\frac{d\mathcal{V}}{dt} \bigg|_{t=0} = \varepsilon^3 [\mathbf{DF}(\mathbf{x})\mathbf{i}] \cdot \mathbf{i} + \varepsilon^3 [\mathbf{DF}(\mathbf{x})\mathbf{j}] \cdot \mathbf{j} + \varepsilon^3 [\mathbf{DF}(\mathbf{x})\mathbf{k}] \cdot \mathbf{k} \tag{3}$$

But $\mathcal{V}(0) = \varepsilon^3$, $\mathbf{F} = F_1 \mathbf{i} + F_2 \mathbf{j} + F_3 \mathbf{k}$, $[\mathbf{DF}(\mathbf{x})\mathbf{i}] \cdot \mathbf{i} = \partial F_1 / \partial x$, and, similarly, the second and third terms of equation (2) are $\varepsilon^3 (\partial F_2 / \partial y)$ and $\varepsilon^3 (\partial F_3 / \partial z)$. Substituting these into equation (3) and dividing by ε^3 proves the theorem. ∎

The reader who is familiar with a little more linear algebra can prove this generalization of Theorem 3:* *Let $\mathbf{v}_1$, $\mathbf{v}_2$, and $\mathbf{v}_3$ be any three noncoplanar vectors emanating from $\mathbf{x}$ that flow according to the formula*

$$\mathbf{v}_i(t) = \mathbf{D_x}\phi(\mathbf{x}, t) \cdot \mathbf{v}_i, \qquad i = 1, 2, 3.$$

---

*The reader will need to know how to write the matrix of a linear transformation with respect to a given basis and be familiar with the fact that the trace of a matrix is independent of the basis.

The vectors $\mathbf{v}_1(t)$, $\mathbf{v}_2(t)$, $\mathbf{v}_3(t)$ span a parallelepiped $P(t)$ with volume $\mathcal{V}(t)$. Then

$$\frac{1}{\mathcal{V}(0)} \frac{d\mathcal{V}}{dt}\bigg|_{t=0} = \text{div } \mathbf{F}(\mathbf{x}). \tag{4}$$

In other words, the divergence of $\mathbf{F}$ at $\mathbf{x}$ is the rate at which the volumes change, per unit volume. "Rate" refers to the rate of change with respect to time as the volumes are transported by the flow.

EXERCISES

Find the divergence of the vector fields in Exercises 1 to 4.

1. $\mathbf{V}(x, y, z) = e^{xy}\mathbf{i} - e^{xy}\mathbf{j} + e^{yz}\mathbf{k}$

2. $\mathbf{V}(x, y, z) = yz\mathbf{i} + xz\mathbf{j} + xy\mathbf{k}$

3. $\mathbf{V}(x, y, z) = x\mathbf{i} + (y + \cos x)\mathbf{j} + (z + e^{xy})\mathbf{k}$

4. $\mathbf{V}(x, y, z) = x^2\mathbf{i} + (x + y)^2\mathbf{j} + (x + y + z)^2\mathbf{k}$

5. Figure 4.4.9 shows some flow lines and moving regions for a fluid moving in the plane field velocity field $\mathbf{V}$. Where is div $\mathbf{V} > 0$ and also where is div $\mathbf{V} < 0$?

Figure 4.4.9 The flow lines of a fluid moving in the plane.

6. Let $V(x, y, z) = x\mathbf{i}$ be the velocity field of a fluid in space. Relate the sign of the divergence with the rate of change of volume under the flow.

7. Sketch a few flow lines for $\mathbf{F}(x, y) = y\mathbf{i}$. Calculate $\nabla \cdot \mathbf{F}$ and explain why your answer is consistent with your sketch.

8. Sketch a few flow lines for $\mathbf{F}(x, y) = -3x\mathbf{i} - y\mathbf{j}$. Calculate $\nabla \cdot \mathbf{F}$ and explain why your answer is consistent with your sketch.

Calculate the divergence of the vector fields in Exercises 9 to 12.

9. $\mathbf{F}(x, y) = x^3\mathbf{i} - x\sin(xy)\mathbf{j}$

10. $\mathbf{F}(x, y) = y\mathbf{i} - x\mathbf{j}$

11. $\mathbf{F}(x, y) = \sin(xy)\mathbf{i} - \cos(x^2 y)\mathbf{j}$

12. $\mathbf{F}(x, y) = xe^y\mathbf{i} - [y/(x + y)]\mathbf{j}$

Compute the curl, $\nabla \times \mathbf{F}$, of the vector fields in Exercises 13 to 16.

13. $\mathbf{F}(x, y, z) = x\mathbf{i} + y\mathbf{j} + z\mathbf{k}$

14. $\mathbf{F}(x, y, z) = yz\mathbf{i} + xz\mathbf{j} + xy\mathbf{k}$

15. $\mathbf{F}(x, y, z) = (x^2 + y^2 + z^2)(3\mathbf{i} + 4\mathbf{j} + 5\mathbf{k})$

16. $\mathbf{F}(x, y, z) = \dfrac{yz\mathbf{i} - xz\mathbf{j} + xy\mathbf{k}}{x^2 + y^2 + z^2}$

Calculate the scalar curl of each of the vector fields in Exercises 17 to 20.

17. $\mathbf{F}(x, y) = \sin x\mathbf{i} + \cos x\mathbf{j}$

18. $\mathbf{F}(x, y) = y\mathbf{i} - x\mathbf{j}$

19. $\mathbf{F}(x, y) = xy\mathbf{i} + (x^2 - y^2)\mathbf{j}$

20. $\mathbf{F}(x, y) = x\mathbf{i} + y\mathbf{j}$

Verify that $\nabla \times (\nabla f) = \mathbf{0}$ for the functions in Exercises 21 to 24.

21. $f(x, y, z) = \sqrt{x^2 + y^2 + z^2}$

22. $f(x, y, z) = xy + yz + xz$

23. $f(x, y, z) = 1/(x^2 + y^2 + z^2)$

24. $f(x, y, z) = x^2 y^2 + y^2 z^2$

25. Show that $\mathbf{F} = y(\cos x)\mathbf{i} + x(\sin y)\mathbf{j}$ is *not* a gradient vector field.

26. Show that $\mathbf{F} = (x^2 + y^2)\mathbf{i} - 2xy\mathbf{j}$ is *not* a gradient field.

27. Prove identity 10 in the table of vector identities.

28. Suppose that $\nabla \cdot \mathbf{F} = 0$ and $\nabla \cdot \mathbf{G} = 0$. Which of the following necessarily have zero divergence?

(a) $\mathbf{F} + \mathbf{G}$ 　　　　　　　　　　　　　　　(b) $\mathbf{F} \times \mathbf{G}$

29. Let $\mathbf{F} = 2xz^2\mathbf{i} + \mathbf{j} + y^3zx\mathbf{k}$ and $f = x^2y$. Compute the following quantities:

(a) ∇f 　　　　(b) $\nabla \times \mathbf{F}$ 　　　　(c) $\mathbf{F} \times \nabla f$ 　　　　(d) $\mathbf{F} \cdot (\nabla f)$

30. Let $\mathbf{r}(x, y, z) = (x, y, z)$ and $r = \sqrt{x^2 + y^2 + z^2} = \|\mathbf{r}\|$. Prove the following identities.
(a) $\nabla(1/r) = -\mathbf{r}/r^3, r \neq 0$; and, in general, $\nabla(r^n) = nr^{n-2}\mathbf{r}$ and $\nabla(\log r) = \mathbf{r}/r^2$.
(b) $\nabla^2(1/r) = 0, r \neq 0$; and, in general, $\nabla^2 r^n = n(n + 1)r^{n-2}$.
(c) $\nabla \cdot (\mathbf{r}/r^3) = 0$; and, in general, $\nabla \cdot (r^n\mathbf{r}) = (n + 3)r^n$.
(d) $\nabla \times \mathbf{r} = \mathbf{0}$; and, in general, $\nabla \times (r^n\mathbf{r}) = \mathbf{0}$.

31. Does $\nabla \times \mathbf{F}$ have to be perpendicular to $\mathbf{F}$?

32. Let $\mathbf{F}(x, y, z) = 3x^2y\mathbf{i} + (x^3 + y^3)\mathbf{j}$.
(a) Verify that curl $\mathbf{F} = \mathbf{0}$.
(b) Find a function f such that $\mathbf{F} = \nabla f$. (Techniques for constructing f in general are given in Chapter 8. The one in this problem should be sought by trial and error.)

33. Let $\mathbf{v}$ and $\mathbf{w}$ be two vectors emanating from the origin and let them be moved by the derivative of the flow as in the supplement to Section 4.4:

$$\mathbf{v}(t) = \mathbf{D_x}\phi(\mathbf{0}, t)\mathbf{v}, \qquad \mathbf{w}(t) = \mathbf{D_x}\phi(\mathbf{0}, t)\mathbf{w},$$

so that at time $t = 0$ and at the origin $\mathbf{0}$ in $\mathbb{R}^3$

$$\frac{d\mathbf{v}}{dt}\bigg|_{t=0} = \mathbf{D_x}\mathbf{F}(\mathbf{0}) \cdot \mathbf{v} \qquad \text{and} \qquad \frac{d\mathbf{w}}{dt}\bigg|_{t=0} = \mathbf{D_x}\mathbf{F}(\mathbf{0}) \cdot \mathbf{w}. \tag{5}$$

Show that

$$\frac{d}{dt}\mathbf{v} \cdot \mathbf{w}\bigg|_{t=0} = [\mathbf{D_x}\mathbf{F}(\mathbf{0}) \cdot \mathbf{v}] \cdot \mathbf{w} + \mathbf{v} \cdot [\mathbf{D_x}\mathbf{F}(\mathbf{0}) \cdot \mathbf{w}]$$
$$= \{(\mathbf{D_x}\mathbf{F}(\mathbf{0}) + [\mathbf{D_x}\mathbf{F}(\mathbf{0})]^T)\mathbf{v}\} \cdot \mathbf{w}$$

34. Any matrix A can be written (uniquely) as the sum of a symmetric matrix (a matrix S is *symmetric* if $S^T = S$) and an antisymmetric matrix ($W^T = -W$) as follows:

$$A = \tfrac{1}{2}(A + A^T) + \tfrac{1}{2}(A - A^T) = S + W.$$

In particular, for $A = \mathbf{D_x}\mathbf{F}(\mathbf{0})$,

$$S = \tfrac{1}{2}\{\mathbf{D_x}\mathbf{F}(\mathbf{0}) + [\mathbf{D_x}\mathbf{F}(\mathbf{0})]^T\}$$

and

$$W = \tfrac{1}{2}\{\mathbf{D_x}\mathbf{F}(\mathbf{0}) - [\mathbf{D_x}\mathbf{F}(\mathbf{0})]^T\}.$$

We call S the *deformation matrix* and W the *rotation matrix*. Show that the entries of W are determined by

$$w_{12} = -\tfrac{1}{2}(\text{curl } \mathbf{F})_3, \qquad w_{23} = -\tfrac{1}{2}(\text{curl } \mathbf{F})_1, \qquad \text{and} \qquad w_{31} = -\tfrac{1}{2}(\text{curl } \mathbf{F})_2.$$

35. Let $\mathbf{w} = \tfrac{1}{2}(\nabla \times \mathbf{F})(\mathbf{0})$. Assume that axes are chosen so that $\mathbf{w}$ is parallel to the z axis and points in the direction of $\mathbf{k}$. Let $\mathbf{v} = \mathbf{w} \times \mathbf{r}$, where $\mathbf{r} = x\mathbf{i} + y\mathbf{j} + z\mathbf{k}$, so that $\mathbf{v}$ is the velocity field of a rotation about the axis $\mathbf{w}$ with angular velocity $\omega = \|\mathbf{w}\|$ and with curl $\mathbf{v} = 2\mathbf{w}$. Since $\mathbf{r}$ is a function of (x, y, z), $\mathbf{v}$ is also a function of (x, y, z). Show that the derivative of $\mathbf{v}$ at the origin is given by

$$\mathbf{Dv}(\mathbf{0}) = W = \begin{bmatrix} 0 & -\omega & 0 \\ \omega & 0 & 0 \\ 0 & 0 & 0 \end{bmatrix}.$$

Interpret the result, as in the supplement to this section.

36. Let

$$\mathbf{V}(x, y, z) = -y\mathbf{i} + x\mathbf{j}, \qquad \mathbf{W}(x, y, z) = \frac{\mathbf{V}(x, y, z)}{(x^2 + y^2)^{1/2}},$$

and

$$\mathbf{Y}(x, y, z) = \frac{\mathbf{V}(x, y, z)}{(x^2 + y^2)}.$$

(a) Compute the divergence and curl of $\mathbf{V}$, $\mathbf{W}$, and $\mathbf{Y}$.
(b) Find the flow lines of $\mathbf{V}$, $\mathbf{W}$, and $\mathbf{Y}$.
(c) How will a small paddle wheel behave in the flow of each of $\mathbf{V}$, $\mathbf{W}$, and $\mathbf{Y}$?

37. Let $\phi(\mathbf{x}, t)$ be the flow of a vector field $\mathbf{F}$. Let $\mathbf{x}$ and t be fixed. For small vectors $\mathbf{v}_1 = \varepsilon\mathbf{i}$, $\mathbf{v}_2 = \varepsilon\mathbf{j}$ and $\mathbf{v}_3 = \varepsilon\mathbf{k}$ emanating from $\mathbf{x}$, let $P(0)$ be the parallelepiped spanned by $\mathbf{v}_1, \mathbf{v}_2, \mathbf{v}_3$. Argue that for small positive ε, $P(0)$ is carried by the flow to an approximate parallelepiped spanned by $\mathbf{v}_1(t), \mathbf{v}_2(t), \mathbf{v}_3(t)$, given by formula (1).

38. Show that the real and imaginary parts of each of the following complex functions form the components of an irrotational and incompressible vector field in the plane; here $i = \sqrt{-1}$.

(a) $(x - iy)^2$ (b) $(x - iy)^3$ (c) $e^{x-iy} = e^x(\cos y - i \sin y)$

REVIEW EXERCISES FOR CHAPTER 4

For Exercises 1 to 4, at the indicated point, compute the velocity vector, the acceleration vector, the speed, and the equation of the tangent line.

1. $\mathbf{c}(t) = (t^3 + 1, e^{-t}, \cos(\pi t/2)); t = 1$

2. $\mathbf{c}(t) = (t^2 - 1, \cos(t^2), t^4); t = \sqrt{\pi}$

3. $\mathbf{c}(t) = (e^t, \sin t, \cos t); t = 0$

4. $\mathbf{c}(t) = \dfrac{t^2}{1+t^2}\mathbf{i} + t\mathbf{j} + \mathbf{k};\, t = 2$

5. Calculate the tangent and acceleration vectors for the helix $\mathbf{c}(t) = (\cos t, \sin t, t)$ at $t = \pi/4$.

6. Calculate the tangent and acceleration vector for the cycloid $\mathbf{c}(t) = (t - \sin t, 1 - \cos t)$ at $t = \pi/4$ and sketch.

7. Let a particle of mass m move on the path $\mathbf{c}(t) = (t^2, \sin t, \cos t)$. Compute the force acting on the particle at $t = 0$.

8. (a) Let $\mathbf{c}(t)$ be a path with $\|\mathbf{c}(t)\| = $ constant; that is, the curve lies on a sphere. Show that $\mathbf{c}'(t)$ is orthogonal to $\mathbf{c}(t)$.

(b) Let $\mathbf{c}$ be a path whose speed is never zero. Show that $\mathbf{c}$ has constant speed if and only if the acceleration vector $\mathbf{c}''$ is always perpendicular to the velocity vector $\mathbf{c}'$.

9. Express as an integral the arc length of the curve $x^2 = y^3 = z^5$ between $x = 1$ and $x = 4$, using a suitable parametrization.

10. Find the arc length of $\mathbf{c}(t) = t\mathbf{i} + (\log t)\mathbf{j} + 2\sqrt{2}t\mathbf{k}$ for $1 \le t \le 2$.

11. A particle is constrained to move around the unit circle in the xy plane according to the formula $(x, y, z) = (\cos(t^2), \sin(t^2), 0), t \ge 0$.

(a) What are the velocity vector and speed of the particle as functions of t?

(b) At what point on the circle should the particle be released to hit a target at $(2, 0, 0)$? (Be careful about which direction the particle is moving around the circle.)

(c) At what time t should the release take place? (Use the smallest $t > 0$ that will work.)

(d) What are the velocity and speed at the time of release?

(e) At what time is the target hit?

12. A particle of mass m moves under the influence of a force $\mathbf{F} = -k\mathbf{r}$, where k is a constant and $\mathbf{r}(t)$ is the position of the particle at time t.

(a) Write down differential equations for the components of $\mathbf{r}(t)$.

(b) Solve the equations in (a) subject to the initial conditions $\mathbf{r}(0) = \mathbf{0}, \mathbf{r}'(0) = 2\mathbf{j} + \mathbf{k}$.

13. Write in parametric form the curve described by the equations $x - 1 = 2y + 1 = 3z + 2$.

14. Write in parametric form the curve $x = y^3 = z^2 + 1$.

15. Show that $\mathbf{c}(t) = (1/(1-t), 0, e^t/(1-t))$ is a flow line of the vector field defined by $\mathbf{F}(x, y, z) = (x^2, 0, z(1+x))$.

16. Let $\mathbf{F}(x, y) = f(x^2 + y^2)[-y\mathbf{i} + x\mathbf{j}]$ for a function f of one variable. What equation must $g(t)$ satisfy for

$$\mathbf{c}(t) = [\cos g(t)]\mathbf{i} + [\sin g(t)]\mathbf{j}$$

to be a flow line for $\mathbf{F}$?

Compute $\nabla \cdot \mathbf{F}$ and $\nabla \times \mathbf{F}$ for the vector fields in Exercises 17 to 20.

17. $\mathbf{F} = 2x\mathbf{i} + 3y\mathbf{j} + 4z\mathbf{k}$

18. $\mathbf{F} = x^2\mathbf{i} + y^2\mathbf{j} + z^2\mathbf{k}$

19. $\mathbf{F} = (x + y)\mathbf{i} + (y + z)\mathbf{j} + (z + x)\mathbf{k}$

20. $\mathbf{F} = x\mathbf{i} + 3xy\mathbf{j} + z\mathbf{k}$

Compute the divergence and curl of the vector fields in Exercises 21 and 22 at the points indicated.

21. $\mathbf{F}(x, y, z) = y\mathbf{i} + z\mathbf{j} + x\mathbf{k}$; $(1, 1, 1)$

22. $\mathbf{F}(x, y, z) = (x + y)^3\mathbf{i} + (\sin xy)\mathbf{j} + (\cos xyz)\mathbf{k}$; $(2, 0, 1)$

Calculate the gradients of the functions in Exercises 23 to 26, and verify that $\nabla \times \nabla f = \mathbf{0}$.

23. $f(x, y) = e^{xy} + \cos(xy)$

24. $f(x, y) = \dfrac{x^2 - y^2}{x^2 + y^2}$

25. $f(x, y) = e^{x^2} - \cos(xy^2)$

26. $f(x, y) = \tan^{-1}(x^2 + y^2)$

27. (a) Let $f(x, y, z) = xyz^2$; compute ∇f.
  (b) Let $\mathbf{F}(x, y, z) = xy\mathbf{i} + yz\mathbf{j} + zy\mathbf{k}$; compute $\nabla \times \mathbf{F}$.
  (c) Compute $\nabla \times (f\mathbf{F})$ using identity 10 of the table of vector identities. Compare with a direct computation.

28. (a) Let $\mathbf{F} = 2xye^z\mathbf{i} + e^z x^2\mathbf{j} + (x^2 ye^z + z^2)\mathbf{k}$. Compute $\nabla \cdot \mathbf{F}$ and $\nabla \times \mathbf{F}$.
  (b) Find a function $f(x, y, z)$ such that $\mathbf{F} = \nabla f$.

29. Let $\mathbf{F}(x, y) = f(x^2 + y^2)[-y\mathbf{i} + x\mathbf{j}]$, as in Exercise 16. Calculate div $\mathbf{F}$ and curl $\mathbf{F}$ and discuss your answers in view of the results of Exercise 16.

30. (a) Write in parametric form the curve that is the intersection of the surfaces $x^2 + y^2 + z^2 = 3$ and $y = 1$.
  (b) Find the equation of the line tangent to this curve at $(1,1,1)$.
  (c) Write an integral expression for the arc length of this curve. What is the value of this integral?

31. In meteorology, the **negative pressure gradient** $\mathbf{G}$ is a vector quantity that points from regions of high pressure to regions of low pressure, normal to the lines of constant pressure (*isobars*).
  (a) In an xy coordinate system,

$$\mathbf{G} = -\frac{\partial P}{\partial x}\mathbf{i} - \frac{\partial P}{\partial y}\mathbf{j}.$$

Write a formula for the magnitude of the negative pressure gradient.

(b)   If the horizontal pressure gradient provided the only horizontal force acting on the air, the wind would blow directly across the isobars in the direction of **G**, and for a given air mass, with acceleration proportional to the magnitude of **G**. Explain, using Newton's second law.

(c)   Because of the rotation of the Earth, the wind does not blow in the direction that (b) would suggest. Instead, it obeys **Buys–Ballot's law** which states: "If in the Northern Hemisphere, you stand with your back to the wind, the high pressure is on your right and the low pressure is on your left." Draw a figure and introduce xy coordinates so that **G** points in the proper direction.

(d)   State and graphically illustrate Buys–Ballot's law for the southern Hemisphere, in which the orientation of high and low pressure is reversed.

32.   A sphere of mass m, radius a, and uniform density has potential u and gravitational force **F**, at a distance r from the center $(0,0,0)$, given by

$$u = \frac{3m}{2a} - \frac{mr^2}{2a^3}, \qquad \mathbf{F} = -\frac{m}{a^3}\mathbf{r} \qquad (r \le a);$$

$$u = \frac{m}{r}, \qquad \mathbf{F} = -\frac{m}{r^3}\mathbf{r} \qquad (r > a).$$

Here, $r = \|\mathbf{r}\|, \mathbf{r} = x\mathbf{i} + y\mathbf{j} + z\mathbf{k}$.

(a)   Verify that $\mathbf{F} = \nabla u$ on the inside and outside of the sphere.

(b)   Check that u satisfies Poisson's equation: $\partial^2 u/\partial x^2 + \partial^2 u/\partial y^2 + \partial^2 u/\partial z^2 = \text{constant inside}$ the sphere.

(c)   Show that u satisfies Laplace's equation: $\partial^2 u/\partial x^2 + \partial^2 u/\partial y^2 + \partial^2 u/\partial z^2 = 0$ outside the sphere.

33.   A circular helix that lies on the cylinder $x^2 + y^2 = R^2$ with pitch ρ may be described parametrically by

$$x = R\cos\theta, \qquad y = R\sin\theta, \qquad z = \rho\theta, \qquad \theta \ge 0.$$

A particle slides under the action of gravity (which acts parallel to the z axis) without friction along the helix. If the particle starts out at the height $z_0 > 0$, then when it reaches the height z along the helix, its speed is given by

$$\frac{ds}{dt} = \sqrt{(z_0 - z)2g},$$

where s is arc length along the helix, g is the constant of gravity, t is time, and $0 \le z \le z_0$.

(a)   Find the length of the part of the helix between the planes $z = z_0$ and $z = z_1, 0 \le z_1 < z_0$.

(b)   Compute the time T_0 it takes the particle to reach the plane $z = 0$.

34.   A sphere of radius 10 centimeters (cm) with center at $(0, 0, 0)$ rotates about the z axis with angular velocity 4 in such a direction that the rotation looks counterclockwise from the positive z axis.

(a)   Find the rotation vector $\boldsymbol{\omega}$ (see Example 9, in Section 4.4).

(b)   Find the velocity $\mathbf{v} = \boldsymbol{\omega} \times \mathbf{r}$ when $\mathbf{r} = 5\sqrt{2}(\mathbf{i} - \mathbf{j})$ is on the "equator."

(c)   Find the velocity of the point $(0, 5\sqrt{3}, 5)$ on the sphere.

35.   Find the velocity of the students in a classroom located at a latitude 49°N due to the rotation of the Earth. (Ignore the motion of the Earth about the Sun, the Sun in the Galaxy, etc.; the radius of the Earth is 3960 miles).

5 DOUBLE AND TRIPLE INTEGRALS

It is to Archimedes himself (c. 225 B.C.) that we owe the nearest approach to actual integration to be found among the Greeks. His first noteworthy advance in this direction was concerned with his proof that the area of a parabolic segment is four thirds of the triangle with the same base and vertex, or two thirds of the circumscribed parallelogram.

D. E. SMITH, *History of Mathematics*

In this chapter and the next we study the integration of real-valued functions of several variables; this chapter treats integrals of functions of two and three variables, or *double* and *triple integrals*. The double integral has a basic geometric interpretation as volume, and can be defined rigorously as a limit of approximating sums. We shall present several techniques for evaluating double and triple integrals and consider some applications.

5.1 INTRODUCTION

This section discusses some geometric aspects of the double integral, deferring a more rigorous discussion in terms of Riemann sums until Section 5.2.

Consider a continuous function of two variables $f: R \subset \mathbb{R}^2 \to \mathbb{R}$ whose domain R is a rectangle with sides parallel to the coordinate axes. The rectangle R can be described in terms of the two closed intervals $[a, b]$ and $[c, d]$, representing the sides of R along the x and y axes, respectively, as in Figure 5.1.1. In this case, we say that R is the **Cartesian product** of $[a, b]$ and $[c, d]$ and write $R = [a, b] \times [c, d]$.

Assume that $f(x, y) \geq 0$ on R, so that the graph of $z = f(x, y)$ is a surface lying above the rectangle R. This surface, the rectangle R, and the four planes $x = a, x = b, y = c$, and $y = d$ form the boundary of a region V in space (see Figure 5.1.1).

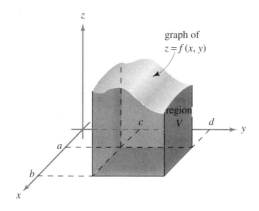

Figure 5.1.1   The region V in space is bounded by the graph of f, the rectangle R, and the four vertical sides indicated.

The problem of how to rigorously define the volume of V has to be faced, and we shall solve it in Section 5.2 by the classical method of exhaustion, or rather, in more modern terms, the method of Riemann sums. To gain an intuitive grasp of this method, we provisionally assume that the volume of a region has been defined.

DOUBLE INTEGRALS   The volume of the region above R and under the graph of f is called the *(double) integral* of f over R and is denoted by

$$\iint_R f(x, y)\, dA, \qquad \text{or} \qquad \iint_R f(x, y)\, dx\, dy.$$

EXAMPLE 1   (a)   If f is defined by $f(x, y) = k$, where k is a positive constant, then $\iint_R f(x, y)\, dA = k(b - a)(d - c)$, since the integral is equal to the volume of a rectangular box with base R and height k.

(b)   If $f(x, y) = 1 - x$ and $R = [0, 1] \times [0, 1]$, then $\iint_R f(x, y)\, dA = \frac{1}{2}$, since the integral is equal to the volume of the triangular solid shown in Figure 5.1.2.   ▲

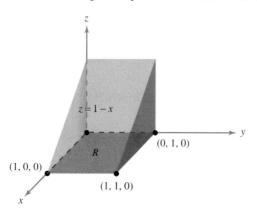

Figure 5.1.2   Volume under the graph $z = 1 - x$ and over $R = [0, 1] \times [0, 1]$.

EXAMPLE 2   Suppose $z = f(x, y) = x^2 + y^2$ and $R = [-1, 1] \times [0, 1]$. Then the integral $\iint_R (x^2 + y^2)\, dx\, dy$ is equal to the volume of the solid sketched in Figure 5.1.3. We shall compute this integral in Example 3.   ▲

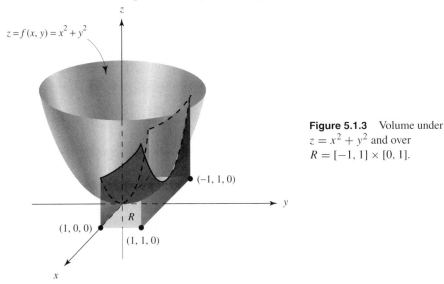

$z = f(x, y) = x^2 + y^2$

$(-1, 1, 0)$

$(1, 0, 0)$

R

$(1, 1, 0)$

Figure 5.1.3   Volume under $z = x^2 + y^2$ and over $R = [-1, 1] \times [0, 1]$.

These ideas are similar to the idea of a single integral $\int_a^b f(x)\, dx$, which represents the area under the graph of f if $f \geq 0$; see Figure 5.1.4.*

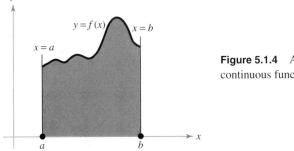

$y = f(x)$

$x = a$

$x = b$

a

b

Figure 5.1.4   Area under the graph of a nonnegative continuous function f from $x = a$ to $x = b$ is $\int_a^b f(x)\, dx$.

Single integrals $\int_a^b f(x)\, dx$ can be rigorously defined, without recourse to the area concept, as a limit of Riemann sums. Thus we can approximate $\int_a^b f(x)\, dx$ by choosing a partition $a = x_0 < x_1 < \cdots < x_n = b$ of $[a, b]$, selecting points $c_i \in [x_i, x_{i+1}]$, and forming the Riemann sum

$$\sum_{i=0}^{n-1} f(c_i)(x_{i+1} - x_i) \approx \int_a^b f(x)\, dx$$

---

*Readers not already familiar with this idea should review the appropriate sections of their introductory calculus text.

(see Figure 5.1.5). We examine the analogous process for double integrals in the next section.

Figure 5.1.5   The sum of the areas of the shaded rectangles is a Riemann sum approximating the area under f from $x = a$ to $x = b$.

There is a useful method for computing volumes known as *Cavalieri's principle.* Suppose we have a solid body and we let $A(x)$ denote its cross-sectional area in a plane P_x measured at a distance x from a reference plane (Figure 5.1.6).

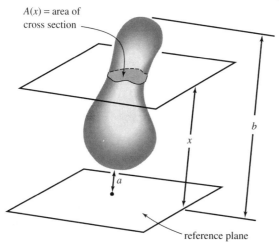

Figure 5.1.6   A solid body with cross-sectional area $A(x)$ at distance x from a reference plane.

According to Cavalieri's principle, the volume of the body is given by

$$\text{volume} = \int_a^b A(x)\,dx,$$

where a and b are the minimum and maximum distances from the reference plane. This can be made intuitively clear. If we partition $[a, b]$ into $a = x_0 < x_1 < \cdots < x_n = b$, then an approximating Riemann sum for the preceding integral is

$$\sum_{i=0}^{n-1} A(c_i)(x_{i+1} - x_i).$$

But this sum also approximates the volume of the body, since $A(x)\,\Delta x$ is the volume of a slab with cross-sectional area $A(x)$ and thickness Δx (Figure 5.1.7). Therefore it is reasonable to accept the above formula for the volume. A more careful justification of this method is given below.

Δx

$A(x)$

Figure 5.1.7 Volume of a slab with cross-sectional area $A(x)$ and thickness Δx equals $A(x)\Delta x$. The total volume of the body is $\int_a^b A(x)\,dx$.

THE SLICE METHOD—CAVALIERI'S PRINCIPLE Let S be a solid and for x satisfying $a \le x \le b$, P_x be a family of parallel planes such that:

1. S lies between P_a and P_b;

2. The area of the slice of S cut by P_x is $A(x)$.

Then the volume of S is equal to

$$\int_a^b A(x)\,dx.$$

HISTORICAL NOTE

Bonaventura Cavalieri (1598–1647) was a pupil of Galileo and a professor in Bologna. His investigations into area and volume were important building blocks of the foundations of calculus. Although his methods were criticized by his contemporaries, similar ideas had been used by Archimedes in antiquity, and were later taken up by the "fathers" of calculus, Newton and Leibniz.

We now use Cavalieri's principle to evaluate double integrals. Consider the solid region under a graph $z = f(x, y)$ defined on the region $[a, b] \times [c, d]$, where f is continuous and greater than zero. There are two natural cross-sectional area functions: one obtained by using cutting planes perpendicular to the x axis, and the other obtained by using cutting planes perpendicular to the y axis. The cross section determined by a cutting plane $x = x_0$, of the first sort, is the plane region under the graph of $z = f(x_0, y)$ from $y = c$ to $y = d$ (Figure 5.1.8).

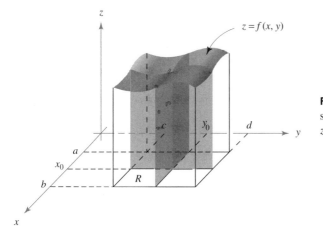

Figure 5.1.8   Two different cross sections sweeping out the volume under $z = f(x, y)$.

When we fix $x = x_0$, we have the function $y \mapsto f(x_0, y)$, which is continuous on $[c, d]$. The cross-sectional area $A(x_0)$ is, therefore, equal to the integral $\int_c^d f(x_0, y)\, dy$. Thus the cross-sectional area function A has domain $[a, b]$, and $A \colon x \mapsto \int_c^d f(x, y)\, dy$. By Cavalieri's principle, the volume V of the region under $z = f(x, y)$ must be equal to

$$V = \int_a^b A(x)\, dx = \int_a^b \left[\int_c^d f(x, y)\, dy \right] dx.$$

The integral $\int_a^b \left[\int_c^d f(x, y)\, dy \right] dx$ is known as an ***iterated integral***, because it is obtained by integrating with respect to y and then integrating the result with respect to x. Since $\iint_R f(x, y)\, dA$ is equal to the volume V, we get the following result.

DOUBLE AND ITERATED INTEGRALS

$$\iint_R f(x, y)\, dA = \int_a^b \left[\int_c^d f(x, y)\, dy \right] dx. \tag{1}$$

If we use cutting planes perpendicular to the y axis, we obtain

$$\iint_R f(x, y)\, dA = \int_c^d \left[\int_a^b f(x, y)\, dx \right] dy. \tag{2}$$

The expression on the right of formula (2) is the iterated integral obtained by integrating with respect to x and then integrating the result with respect to y.

Thus, if our intuition about volumes is correct, formulas (1) and (2) ought to be valid. This is in fact true when the concepts we are discussing are defined rigorously, and is known as Fubini's theorem. We give a proof of this theorem in the next section.

As the following examples illustrate, the notion of the iterated integral and equations (1) and (2) provide a powerful method for *computing* the double integral of a function of two variables.

EXAMPLE 3   Let $z = f(x, y) = x^2 + y^2$ and let $R = [-1, 1] \times [0, 1]$. Evaluate the integral $\iint_R (x^2 + y^2)\, dx\, dy$.

SOLUTION   By equation (2),

$$\iint_R (x^2 + y^2)\, dx\, dy = \int_0^1 \left[\int_{-1}^1 (x^2 + y^2)\, dx \right] dy.$$

To find $\int_{-1}^1 (x^2 + y^2)\, dx$, we treat y as a constant and integrate with respect to x. Since $x^3/3 + y^2 x$ is an antiderivative of $x^2 + y^2$ with respect to x, we can integrate, using the fundamental theorem of calculus, to obtain

$$\int_{-1}^1 (x^2 + y^2)\, dx = \left[\frac{x^3}{3} + y^2 x \right]_{x=-1}^1 = \frac{2}{3} + 2y^2.$$

Next we integrate $\frac{2}{3} + 2y^2$ with respect to y from 0 to 1, to obtain

$$\int_0^1 \left(\frac{2}{3} + 2y^2 \right) dy = \left[\frac{2}{3}y + \frac{2}{3}y^3 \right]_{y=0}^1 = \frac{4}{3}.$$

Hence the volume of the solid we saw in Figure 5.1.3 is 4/3.

For completeness, let us evaluate $\iint_R (x^2 + y^2)\, dx\, dy$ using equation (1)—that is, integrating with respect to y first and then with respect to x. We have

$$\iint_R (x^2 + y^2)\, dx\, dy = \int_{-1}^1 \left[\int_0^1 (x^2 + y^2)\, dy \right] dx.$$

Treating x as a constant in the y integration, we obtain

$$\int_0^1 (x^2 + y^2)\, dy = \left[x^2 y + \frac{y^3}{3} \right]_{y=0}^1 = x^2 + \frac{1}{3}.$$

Next we evaluate $\int_{-1}^1 \left(x^2 + \frac{1}{3} \right) dx$ to obtain

$$\int_{-1}^1 \left(x^2 + \frac{1}{3} \right) dx = \left[\frac{x^3}{3} + \frac{x}{3} \right]_{x=-1}^1 = \frac{4}{3},$$

which agrees with our previous answer.   ▲

EXAMPLE 4 Compute the double integral $\iint_S \cos x \sin y \, dx \, dy$, where S is the square $[0, \pi/2] \times [0, \pi/2]$ (see Figure 5.1.9).

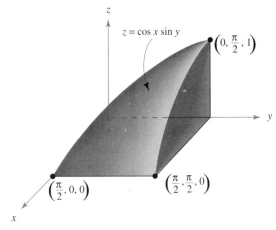

Figure 5.1.9 Volume under $z = \cos x \sin y$ and over the rectangle $[0, \pi/2] \times [0, \pi/2]$.

SOLUTION By equation (2),

$$\iint_S \cos x \sin y \, dx \, dy = \int_0^{\pi/2} \left[\int_0^{\pi/2} \cos x \sin y \, dx \right] dy = \int_0^{\pi/2} \sin y \left[\int_0^{\pi/2} \cos x \, dx \right] dy$$

$$= \int_0^{\pi/2} \sin y \, dy = 1. \quad \blacktriangle$$

In the next section, we shall use Riemann sums to rigorously define the double integral for a wide class of functions of two variables without recourse to the notion of volume. Although we shall drop the requirement that $f(x, y) \geq 0$, equations (1) and (2) will remain valid. Therefore, the iterated integral will again provide the key to computing the double integral. In Section 5.3 we shall treat double integrals over regions more general than rectangles.

Finally, we remark that it is common to delete the brackets in iterated integrals such as equations (1) and (2) above and write

$$\int_a^b \int_c^d f(x, y) \, dy \, dx \qquad \text{in place of} \qquad \int_a^b \left[\int_c^d f(x, y) \, dy \right] dx$$

and

$$\int_c^d \int_a^b f(x, y) \, dx \, dy \qquad \text{in place of} \qquad \int_c^d \left[\int_a^b f(x, y) \, dx \right] dy.$$

EXERCISES

1. Evaluate the following iterated integrals:
(a) $\int_{-1}^1 \int_0^1 (x^4 y + y^2) \, dy \, dx$
(c) $\int_0^1 \int_0^1 (xye^{x+y}) \, dy \, dx$
 (b) $\int_0^{\pi/2} \int_0^1 (y \cos x + 2) \, dy \, dx$
(d) $\int_{-1}^0 \int_1^2 (-x \log y) \, dy \, dx$

2. Evaluate the integrals in Exercise 1 by integrating with respect to x and then with respect to y. [The solution to part (b) only is in the Study Guide to this text.]

3. Use Cavalieri's principle to show that the volumes of two cylinders with the same base and height are equal (see Figure 5.1.10).

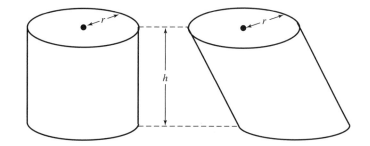

Figure 5.1.10   Two cylinders with same base and height have the same volume.

4. Using Cavalieri's principle, compute the volume of the structure shown in Figure 5.1.11; each cross section is a rectangle of length 5 and width 3.

Figure 5.1.11   Compute this volume.

5. A lumberjack cuts out a wedge-shaped piece W of a cylindrical tree of radius r obtained by making two saw cuts to the tree's center, one horizontally and one at an angle θ. Compute the volume of the wedge W using Cavalieri's principle. (See Figure 5.1.12.)

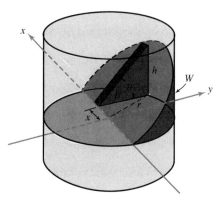

Figure 5.1.12   Find the volume of W.

6. (a) Show that the volume of the solid of revolution shown in Figure 5.1.13(a) is

$$\pi \int_a^b [f(x)]^2 \, dx.$$

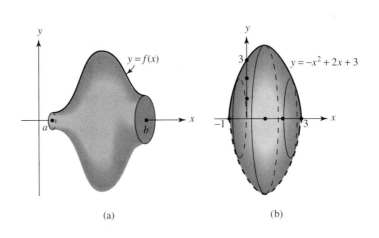

Figure 5.1.13   The solid of revolution (a) has volume $\pi \int_a^b [f(x)]^2 \, dx$. Part (b) shows the region between the graph of $y = -x^2 + 2x + 3$ and the x axis rotated about the x axis.

(b) Show that the volume of the region obtained by rotating the region under the graph of the parabola $y = -x^2 + 2x + 3$, $-1 \leq x \leq 3$, about the x axis is $512\pi/15$ [see Figure 5.1.13(b)].

Evaluate the double integrals in Exercises 7 to 9, where R is the rectangle $[0, 2] \times [-1, 0]$.

7. $\iint_R (x^2 y^2 + x) \, dy \, dx$

8. $\iint_R \left(|y| \cos \frac{1}{4} \pi x \right) dy \, dx$

9. $\iint_R \left(-x e^x \sin \frac{1}{2} \pi y \right) dy \, dx$

10. Find the volume bounded by the graph of $f(x, y) = 1 + 2x + 3y$, the rectangle $[1, 2] \times [0, 1]$, and the four vertical sides of the rectangle R, as in Figure 5.1.1.

11. Repeat Exercise 10 for the function $f(x, y) = x^4 + y^2$ and the rectangle $[-1, 1] \times [-3, -2]$.

5.2   THE DOUBLE INTEGRAL OVER A RECTANGLE

We are ready to give a rigorous definition of the double integral as the limit of a sequence of sums. This will then be used to *define* the volume of the region under the graph of a function $f(x, y)$. We shall not require that $f(x, y) \geq 0$; but if $f(x, y)$ assumes negative values, we shall interpret the integral as a signed volume, just as for the area under the graph of a function of one variable. In addition, we shall discuss some of the fundamental algebraic properties of the double integral and prove Fubini's theorem, which states that the double integral can be calculated as an iterated integral. To begin, let us establish some notation for partitions and sums.

Consider a closed rectangle $R \subset \mathbb{R}^2$; that is, R is a Cartesian product $R = [a, b] \times [c, d]$. By a ***regular partition*** of R of order n we mean the two ordered collections of $n+1$ equally spaced points $\{x_j\}_{j=0}^n$ and $\{y_k\}_{k=0}^n$, that is, points satisfying

$$a = x_0 < x_1 < \cdots < x_n = b, \qquad c = y_0 < y_1 < \cdots < y_n = d$$

and

$$x_{j+1} - x_j = \frac{b-a}{n}, \qquad y_{k+1} - y_k = \frac{d-c}{n}$$

(see Figure 5.2.1).

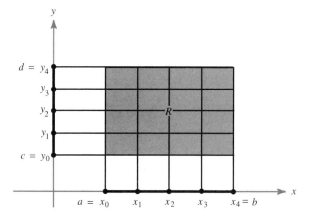

Figure 5.2.1 A regular partition of a rectangle R, with $n = 4$.

A function $f(x, y)$ is said to be ***bounded*** if there is a number $M > 0$ such that $-M \le f(x, y) \le M$ for all (x, y) in the domain of f. A continuous function on a *closed* rectangle is always bounded, but, for example, $f(x, y) = 1/x$ on $(0, 1] \times [0, 1]$ is not bounded, because $1/x$ becomes arbitrarily large for x near 0. The rectangle $(0, 1] \times [0, 1]$ is not closed since the endpoint 0 is missing in the first factor.

Let R_{jk} be the rectangle $[x_j, x_{j+1}] \times [y_k, y_{k+1}]$, and let $\mathbf{c}_{jk}$ be *any* point in R_{jk}. Suppose $f: R \to \mathbb{R}$ is a bounded real-valued function. Form the sum

$$S_n = \sum_{j,k=0}^{n-1} f(\mathbf{c}_{jk}) \, \Delta x \, \Delta y = \sum_{j,k=0}^{n-1} f(\mathbf{c}_{jk}) \, \Delta A, \tag{1}$$

where

$$\Delta x = x_{j+1} - x_j = \frac{b-a}{n}, \qquad \Delta y = y_{k+1} - y_k = \frac{d-c}{n},$$

and

$$\Delta A = \Delta x \, \Delta y.$$

This sum is taken over all j's and k's from 0 to $n - 1$, and so there are n^2 terms. A sum of this type is called a ***Riemann sum*** for f.

DEFINITION: DOUBLE INTEGRAL If the sequence $\{S_n\}$ converges to a limit S as $n \to \infty$ and the limit S is the same for any choice of points $\mathbf{c}_{jk}$ in the rectangles R_{jk}, then we say that f is ***integrable*** over R and we write

$$\iint_R f(x, y) \, dA, \qquad \iint_R f(x, y) \, dx \, dy, \qquad \text{or} \qquad \iint_R f \, dx \, dy$$

for the limit S.

Thus, we can rewrite integrability in the following way:

$$\lim_{n \to \infty} \sum_{j,k=0}^{n-1} f(\mathbf{c}_{jk}) \, \Delta x \, \Delta y = \iint_R f \, dx \, dy$$

for any choice of $\mathbf{c}_{jk} \in R_{jk}$.

The proof of the following basic theorem is not difficult, but because it requires some technicalities that are not essential to the main text, it is presented in Section 5.5.[*]

THEOREM 1 *Any continuous function defined on a closed rectangle R is integrable.*

If $f(x, y) \geq 0$, the existence of $\lim_{n \to \infty} S_n$ has a straightforward geometric meaning. Consider the graph of $z = f(x, y)$ as the top of a solid whose base is the rectangle R. If we take each $\mathbf{c}_{jk}$ to be a point where $f(x, y)$ has its minimum value[†] on R_{jk}, then $f(\mathbf{c}_{jk}) \, \Delta x \, \Delta y$ represents the volume of a rectangular box with base R_{jk}. The sum $\sum_{j,k=0}^{n-1} f(\mathbf{c}_{jk}) \, \Delta x \, \Delta y$ equals the volume of an inscribed solid, part of which is shown in Figure 5.2.2.

---

[*]For additional technical details concerning the theory of integration, see J. Marsden and H. Hoffman, *Elementary Classical Analysis*, 2nd ed., Freeman, New York, 1993.

[†]Such $\mathbf{c}_{jk}$ exist by virtue of the continuity of f on R, but we shall not prove this fact.

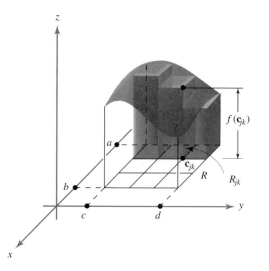

Figure 5.2.2 The sum of inscribed boxes approximates the volume under the graph of $z = f(x, y)$.

Similarly, if $\mathbf{c}_{jk}$ is a point where $f(x, y)$ has its maximum on R_{jk}, then the sum $\sum_{j,k=0}^{n-1} f(\mathbf{c}_{jk}) \, \Delta x \, \Delta y$ is equal to the volume of a circumscribed solid (see Figure 5.2.3).

Figure 5.2.3 The volume of circumscribed boxes also approximates the volume under $z = f(x, y)$.

Therefore, if $\text{limit}_{n \to \infty} S_n$ exists and is independent of $\mathbf{c}_{jk} \in R_{jk}$, it follows that the volumes of the inscribed and circumscribed solids approach the same limit as $n \to \infty$. It is therefore reasonable to call this limit the exact volume of the solid under the graph of f. Thus the method of Riemann sums supports the concepts introduced on an intuitive basis in Section 5.1.

There is a theorem guaranteeing the existence of the integral of certain discontinuous functions as well. We shall need this result in the next section in order to discuss the integrals of functions over regions more general than rectangles. We shall be specifically interested in functions whose discontinuities lie on curves in the xy plane. Figure 5.2.4 shows two functions defined on a rectangle R whose discontinuities lie along curves. In other words, f is continuous at each point that is in R but not necessarily on the curve.

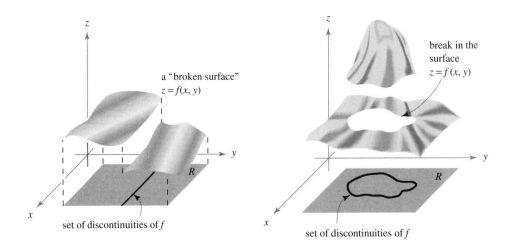

Figure 5.2.4   What the graphs of discontinuous functions of two variables might look like.

Useful curves are graphs of functions $y = \phi(x)$, $a \leq x \leq b$, or $x = \psi(y)$, $c \leq y \leq d$, or finite unions of such graphs. Some examples are shown in Figure 5.2.5.

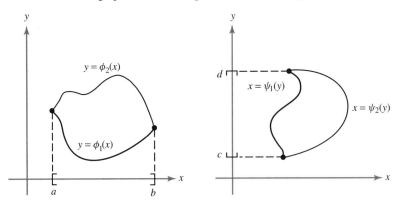

Figure 5.2.5   Curves in the plane represented as graphs.

The next theorem provides an important criterion for determining whether a function is integrable. The proof is discussed in Section 5.5.

THEOREM 2: INTEGRABILITY OF BOUNDED FUNCTIONS

Let $f: R \to \mathbb{R}$ be a bounded real-valued function on the rectangle R, and suppose that the set of points where f is discontinuous lies on a finite union of graphs of continuous functions. Then f is integrable over R.

Using Theorem 2 and the remarks preceding it, we see that the functions sketched in Figure 5.2.4 are integrable over R, since these functions are bounded and continuous except on graphs of continuous functions.

From the definition of the integral as a limit of sums and the limit theorems, we can deduce some fundamental properties of the integral $\iint_R f(x, y)\, dA$; these properties are essentially the same as for the integral of a real-valued function of a single variable.

Let f and g be integrable functions on the rectangle R, and let c be a constant. Then $f + g$ and cf are integrable, and

(i) (*Linearity*)

$$\iint_R [f(x, y) + g(x, y)]\, dA = \iint_R f(x, y)\, dA + \iint_R g(x, y)\, dA.$$

(ii) (*Homogeneity*)

$$\iint_R cf(x, y)\, dA = c \iint_R f(x, y)\, dA.$$

(iii) (*Monotonicity*)   If $f(x, y) \geq g(x, y)$, then

$$\iint_R f(x, y)\, dA \geq \iint_R g(x, y)\, dA.$$

(iv) (*Additivity*)   If $R_i, i = 1, \ldots, m$, are pairwise disjoint rectangles such that f is bounded and integrable over each R_i and if $Q = R_1 \cup R_2 \cup \cdots \cup R_m$ is a rectangle, then $f \colon Q \to \mathbb{R}$ is integrable over Q and

$$\iint_Q f(x, y)\, dA = \sum_{i=1}^{m} \iint_{R_i} f(x, y)\, dA.$$

Properties (i) and (ii) are a consequence of the definition of the integral as a limit of a sum and the following facts for convergent sequences $\{S_n\}$ and $\{T_n\}$, which are proved as we did for the limit theorems in Chapter 2:

$$\lim_{n \to \infty} (T_n + S_n) = \lim_{n \to \infty} T_n + \lim_{n \to \infty} S_n$$

$$\lim_{n \to \infty} (cS_n) = c \lim_{n \to \infty} S_n.$$

To demonstrate monotonicity, we first observe that if $h(x, y) \geq 0$ and $\{S_n\}$ is a sequence of Riemann sums that converges to $\iint_R h(x, y)\, dA$, then $S_n \geq 0$ for all n, so that $\iint_R h(x, y)\, dA = \lim_{n \to \infty} S_n \geq 0$. If $f(x, y) \geq g(x, y)$ for all $(x, y) \in R$, then $(f - g)(x, y) \geq 0$ for all (x, y) and, using properties (i) and (ii), we have

$$\iint_R f(x, y)\, dA - \iint_R g(x, y)\, dA = \iint_R [f(x, y) - g(x, y)]\, dA \geq 0.$$

This proves property (iii). The proof of property (iv) is more technical and a special case is proved in Section 5.5. It should be intuitively obvious.

Another important result is the inequality

$$\left| \iint_R f \, dA \right| \le \iint_R |f| \, dA. \tag{2}$$

To see why formula (2) is true, note that, by the definition of absolute value,

$$-|f| \le f \le |f|;$$

therefore, from the monotonicity and homogeneity of integration (with $c = -1$),

$$-\iint_R |f| \, dA \le \iint_R f \, dA \le \iint_R |f| \, dA,$$

which is equivalent to formula (2).

Although we have noted the integrability of a variety of functions, we have not yet established rigorously a general method of computing integrals. In the case of one variable we avoid computing $\int_a^b f(x) \, dx$ from its definition as a limit of a sum by using the *fundamental theorem of integral calculus.* This important theorem tells us that *if f is continuous, then*

$$\int_a^b f(x) \, dx = F(b) - F(a),$$

where F is an antiderivative of f; that is, $F' = f$.

This technique will not work as stated for functions $f(x, y)$ of two variables. However, as we indicated in Section 5.1, we can often reduce a double integral over a rectangle to iterated single integrals; the fundamental theorem then applies to each of these single integrals. Fubini's theorem, which was mentioned in the last section, establishes this reduction to iterated integrals rigorously, by using Riemann sums. As we saw in Section 5.1, the reduction,

$$\iint_R f(x, y) \, dA = \int_a^b \left[\int_c^d f(x, y) \, dy \right] dx = \int_c^d \left[\int_a^b f(x, y) \, dx \right] dy,$$

is a consequence of Cavalieri's principle, at least if $f(x, y) \ge 0$. In terms of Riemann sums, it corresponds to the following equality:

$$\sum_{j,k=0}^{n-1} f(\mathbf{c}_{jk}) \, \Delta x \, \Delta y = \sum_{j=0}^{n-1} \left(\sum_{k=0}^{n-1} f(\mathbf{c}_{jk}) \, \Delta y \right) \Delta x = \sum_{k=0}^{n-1} \left(\sum_{j=0}^{n-1} f(\mathbf{c}_{jk}) \, \Delta x \right) \Delta y,$$

which may be proved more generally as follows: *Let $[a_{jk}]$ be an $n \times n$ matrix, where $0 \le j \le n-1$ and $0 \le k \le n-1$. Let $\sum_{j,k=0}^{n-1} a_{jk}$ be the sum of the n^2 matrix entries. Then*

$$\sum_{j,k=0}^{n-1} a_{jk} = \sum_{j=0}^{n-1} \left(\sum_{k=0}^{n-1} a_{jk} \right) = \sum_{k=0}^{n-1} \left(\sum_{j=0}^{n-1} a_{jk} \right). \tag{3}$$

In the first equality, the right-hand side represents summing the matrix entries first by rows and then adding the results:

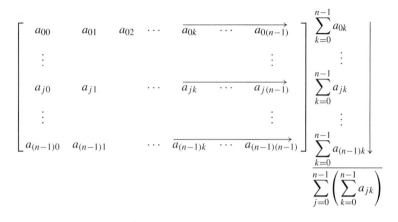

Clearly this is equal to $\sum_{j,k=0}^{n-1} a_{jk}$, that is, the sum of all the a_{jk}'s. Similarly, the sum $\sum_{k=0}^{n-1} \left(\sum_{j=0}^{n-1} a_{jk} \right)$ represents a summing of the matrix entries by columns. This establishes equation (3) and makes the reduction to iterated integrals quite plausible if we remember that integrals can be approximated by the corresponding Riemann sums. The actual proof of Fubini's theorem exploits this idea.

THEOREM 3: FUBINI'S THEOREM *Let f be a continuous function with a rectangular domain $R = [a, b] \times [c, d]$. Then*

$$\int_a^b \int_c^d f(x, y) \, dy \, dx = \int_c^d \int_a^b f(x, y) \, dx \, dy = \iint_R f(x, y) \, dA. \qquad (4)$$

PROOF We shall first show that

$$\int_a^b \int_c^d f(x, y) \, dy \, dx = \iint_R f(x, y) \, dA.$$

Let $c = y_0 < y_1 < \cdots < y_n = d$ be a partition of $[c, d]$ into n equal parts. Define

$$F(x) = \int_c^d f(x, y) \, dy.$$

Then

$$F(x) = \sum_{k=0}^{n-1} \int_{y_k}^{y_{k+1}} f(x, y) \, dy.$$

Using the integral version of the mean value theorem,* for each fixed x and for each k we have

$$\int_{y_k}^{y_{k+1}} f(x, y)\, dy = f(x, Y_k(x))(y_{k+1} - y_k)$$

(see Figure 5.2.6), where the point $Y_k(x)$ belongs to $[y_k, y_{k+1}]$ and may depend on x, k, and n.

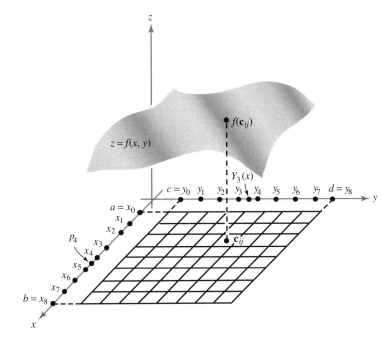

Figure 5.2.6   The notation needed in the proof of Fubini's theorem; $n = 8$.

We have thus shown that

$$F(x) = \sum_{k=0}^{n-1} f(x, Y_k(x))(y_{k+1} - y_k). \tag{5}$$

Now, by the definition of the integral in one variable as a limit of Riemann sums,

$$\int_a^b F(x)\, dx = \int_a^b \left[\int_c^d f(x, y)\, dy \right] dx = \lim_{n \to \infty} \sum_{j=0}^{n-1} F(p_j)(x_{j+1} - x_j),$$

---

*This states that if $g(x)$ is continuous on $[a, b]$, then $\int_a^b g(x)\, dx = g(c)(b - a)$ for some point $c \in [a, b]$. The more general second mean value theorem was proved in Section 3.2.

where $a = x_0 < x_1 < \cdots < x_n = b$ is a partition of the interval $[a, b]$ into n equal parts and p_j is any point in $[x_j, x_{j+1}]$. Setting $\mathbf{c}_{jk} = (p_j, Y_k(p_j)) \in R_{jk}$, we have [substituting p_j for x in equation (5)]

$$F(p_j) = \sum_{k=0}^{n-1} f(\mathbf{c}_{jk})(y_{k+1} - y_k).$$

Therefore,

$$\int_a^b \int_c^d f(x, y)\, dy\, dx = \int_a^b F(x)\, dx = \lim_{n \to \infty} \sum_{j=0}^{n-1} F(p_j)(x_{j+1} - x_j)$$

$$= \lim_{n \to \infty} \sum_{j=0}^{n-1} \sum_{k=0}^{n-1} f(\mathbf{c}_{jk})(y_{k+1} - y_k)(x_{j+1} - x_j) = \iint_R f(x, y)\, dA.$$

Thus we have proved that

$$\int_a^b \int_c^d f(x, y)\, dy\, dx = \iint_R f(x, y)\, dA.$$

By the same reasoning we can show that

$$\int_c^d \int_a^b f(x, y)\, dx\, dy = \iint_R f(x, y)\, dA.$$

These two conclusions are exactly what we wanted to prove. ∎

Fubini's theorem can be generalized to the case where f is not necessarily continuous. Although we shall not present a proof, we state here this more general version.

THEOREM 3′: FUBINI'S THEOREM *Let f be a bounded function with domain a rectangle $R = [a, b] \times [c, d]$, and suppose the discontinuities of f lie on a finite union of graphs of continuous functions. If the integral $\int_c^d f(x, y)\, dy$ exists for each $x \in [a, b]$, then*

$$\int_a^b \left[\int_c^d f(x, y)\, dy \right] dx$$

exists and

$$\int_a^b \int_c^d f(x, y)\, dy\, dx = \iint_R f(x, y)\, dA.$$

Similarly, if $\int_a^b f(x, y)\, dx$ exists for each $y \in [c, d]$, then

$$\int_c^d \left[\int_a^b f(x, y)\, dx \right] dy$$

exists and

$$\int_c^d \int_a^b f(x, y)\, dx\, dy = \iint_R f(x, y)\, dA.$$

Thus, if all these conditions hold simultaneously,

$$\int_a^b \int_c^d f(x, y)\, dy\, dx = \int_c^d \int_a^b f(x, y)\, dx\, dy = \iint_R f(x, y)\, dA.$$

The assumptions made for this version of Fubini's theorem are more complicated than those we made in Theorem 3. They are necessary because if f is not continuous everywhere, for example, there is no guarantee that $\int_c^d f(x, y)\, dy$ will exist for each x.

EXAMPLE 1   Compute $\iint_R (x^2 + y)\, dA$, where R is the square $[0, 1] \times [0, 1]$.

SOLUTION   By Fubini's theorem,

$$\iint_R (x^2 + y)\, dA = \int_0^1 \int_0^1 (x^2 + y)\, dx\, dy = \int_0^1 \left[\int_0^1 (x^2 + y)\, dx \right] dy.$$

By the fundamental theorem of calculus, the x integration may be performed:

$$\int_0^1 (x^2 + y)\, dx = \left[\frac{x^3}{3} + yx \right]_{x=0}^1 = \frac{1}{3} + y.$$

Thus

$$\iint_R (x^2 + y)\, dA = \int_0^1 \left[\frac{1}{3} + y \right] dy = \left[\frac{1}{3}y + \frac{y^2}{2} \right]_0^1 = \frac{5}{6}.$$

What we have done is hold y fixed, integrate with respect to x, and then evaluate the result between the given limits for the x variable. Next we integrated the remaining function (of y alone) with respect to y to obtain the final answer.   ▲

EXAMPLE 2   A consequence of Fubini's theorem is that interchanging the order of integration in the iterated integrals does not change the answer. Verify this for Example 1.

SOLUTION   We carry out the integration in the other order:

$$\int_0^1 \int_0^1 (x^2 + y)\, dy\, dx = \int_0^1 \left[x^2 y + \frac{y^2}{2} \right]_{y=0}^1 dx = \int_0^1 \left[x^2 + \frac{1}{2} \right] dx = \left[\frac{x^3}{3} + \frac{x}{2} \right]_0^1 = \frac{5}{6}.   ▲$$

We have seen that when $f(x, y) \geq 0$ on $R = [a, b] \times [c, d]$, the integral $\iint_R f(x, y)\, dA$ can be interpreted as a volume. If the function also takes on negative values, then the double integral can be thought of as the sum of all volumes lying between the surface $z = f(x, y)$ and the plane $z = 0$, bounded by the planes $x = a$, $x = b$, $y = c$, and $y = d$; here the volumes above $z = 0$ are counted as positive and those below as negative. However,

Fubini's theorem as stated remains valid in the case where $f(x, y)$ is negative or changes sign on R; that is, there is no restriction on the sign of f in the hypotheses of the theorem.

EXAMPLE 3 Let R be the rectangle $[-2, 1] \times [0, 1]$ and let f be defined by $f(x, y) = y(x^3 - 12x)$; $f(x, y)$ takes on both positive and negative values on R. Evaluate the integral $\iint_R f(x, y)\, dx\, dy = \iint_R y(x^3 - 12x)\, dx\, dy$.

SOLUTION By Fubini's theorem, we may write

$$\iint_R y(x^3 - 12x)\, dx\, dy = \int_0^1 \left[\int_{-2}^1 y(x^3 - 12x)\, dx\right] dy = \frac{57}{4} \int_0^1 y\, dy = \frac{57}{8}.$$

Alternatively, integrating first with respect to y, we find

$$\iint_R y(x^3 - 12x)\, dy\, dx = \int_{-2}^1 \left[\int_0^1 (x^3 - 12x)y\, dy\right] dx$$

$$= \frac{1}{2} \int_{-2}^1 (x^3 - 12x)\, dx = \frac{1}{2}\left[\frac{x^4}{4} - 6x^2\right]_{-2}^1 = \frac{57}{8}. \quad \blacktriangle$$

HISTORICAL NOTE

Although Theorem 3 on the equality of iterated integrals is named after the Italian mathematician Guido Fubini (1879–1943), who proved a very general result of this type in 1907, it was certainly known to Cauchy and his contemporaries that equality held for continuous functions. Cauchy was the first to show that equality need not hold when f is unbounded, and, somewhat later, examples were also found of bounded functions where equality does not hold.

EXERCISES

1. Evaluate each of the following integrals if $R = [0, 1] \times [0, 1]$.

   (a) $\iint_R (x^3 + y^2)\, dA$

   (b) $\iint_R ye^{xy}\, dA$

   (c) $\iint_R (xy)^2 \cos x^3\, dA$

   (d) $\iint_R \ln[(x + 1)(y + 1)]\, dA$

2. Evaluate each of the following integrals if $R = [0, 1] \times [0, 1]$.

   (a) $\iint_R (x^m y^n)\, dx\, dy$, where $m, n > 0$

   (b) $\iint_R (ax + by + c)\, dx\, dy$

   (c) $\iint_R \sin(x + y)\, dx\, dy$

   (d) $\iint_R (x^2 + 2xy + y\sqrt{x})\, dx\, dy$

3. Let f be continuous, $f \geq 0$, on the rectangle R. If $\iint_R f\, dA = 0$, prove that $f = 0$ on R.

4. Compute the volume of the solid bounded by the xz plane, the yz plane, the xy plane, the planes $x = 1$ and $y = 1$, and the surface $z = x^2 + y^4$.

5. Let f be continuous on $[a, b]$ and g continuous on $[c, d]$. Show that

$$\iint_R [f(x)g(y)] \, dx \, dy = \left[\int_a^b f(x) \, dx\right]\left[\int_c^d g(y) \, dy\right],$$

where $R = [a, b] \times [c, d]$.

6. Compute the volume of the solid bounded by the surface $z = \sin y$, the planes $x = 1$, $x = 0$, $y = 0$, and $y = \pi/2$, and the xy plane.

7. Compute the volume of the solid bounded by the graph $z = x^2 + y$, the rectangle $R = [0, 1] \times [1, 2]$, and the "vertical sides" of R.

8. Let f be continuous on $R = [a, b] \times [c, d]$; for $a < x < b, c < y < d$, define

$$F(x, y) = \int_a^x \int_c^y f(u, v) \, dv \, du.$$

Show that $\partial^2 F/\partial x \, \partial y = \partial^2 F/\partial y \, \partial x = f(x, y)$. Use this example to discuss the relationship between Fubini's theorem and the equality of mixed partial derivatives.

9. Let $f: [0, 1] \times [0, 1] \to \mathbb{R}$ be defined by

$$f(x, y) = \begin{cases} 1 & x \text{ rational} \\ 2y & x \text{ irrational.} \end{cases}$$

Show that the iterated integral $\int_0^1 \left[\int_0^1 f(x, y) \, dy\right] dx$ exists but f is not integrable.

10. Express $\iint_R \cosh xy \, dx \, dy$ as a convergent series, where $R = [0, 1] \times [0, 1]$.

11. Although Fubini's theorem holds for most functions met in practice, one must still exercise some caution. It certainly does not hold for all functions. For example, one could divide the unit square into infinitely many rectangles of the form $[1/(m+1), 1/m] \times [1/(n+1), 1/n]$, as in Figure 5.2.7. Define f in such a way that the volume under the graph of f over each rectangle takes on values according to the following table:

	$\frac{1}{32}$	$\frac{1}{16}$	$\frac{1}{8}$	$\frac{1}{4}$	$\frac{1}{2}$	-1
$\cdots$	$\frac{1}{16}$	$\frac{1}{8}$	$\frac{1}{4}$	$\frac{1}{2}$	-1	0
$\cdots$	$\frac{1}{8}$	$\frac{1}{4}$	$\frac{1}{2}$	-1	0	0
$\cdots$	$\frac{1}{4}$	$\frac{1}{2}$	-1	0	0	0
	$\vdots$	$\vdots$	$\vdots$	$\vdots$	$\vdots$	$\vdots$

Define f to be zero at $(0, 0)$. Each row adds to zero, so that adding rows and then columns gives a result of zero. On the other hand the columns add to

$$\cdots \quad -\frac{1}{32} \quad -\frac{1}{16} \quad -\frac{1}{8} \quad -\frac{1}{4} \quad -\frac{1}{2} \quad -1,$$

so that adding columns and then rows gives a result of -2. Why does Fubini's theorem not hold for this function?

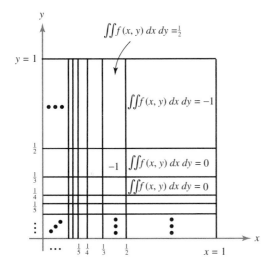

Figure 5.2.7 Construction of a function that does not satisfy Fubini's theorem (Exercise 11).

5.3 THE DOUBLE INTEGRAL OVER MORE GENERAL REGIONS

Our goal in this section is twofold: first, we wish to define the double integral of a function $f(x, y)$ over regions D more general than rectangles; and second, we want to develop a technique for evaluating this type of integral. To accomplish this, we shall define three special types of subsets of the xy plane, and then extend the notion of the double integral to them.

Suppose we are given two continuous real-valued functions $\phi_1 : [a, b] \to \mathbb{R}$, and $\phi_2 : [a, b] \to \mathbb{R}$ that satisfy $\phi_1(x) \leq \phi_2(x)$ for all $x \in [a, b]$. Let D be the set of all points (x, y) such that $x \in [a, b]$, and $\phi_1(x) \leq y \leq \phi_2(x)$.

This region D is said to be of *type 1*. Figure 5.3.1 shows various examples of regions of type 1. The curves and straight line segments that bound the region taken together constitute the *boundary* of D, denoted ∂D.

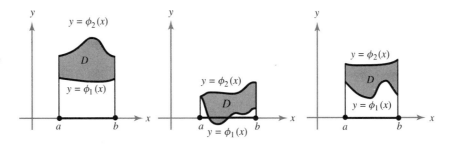

Figure 5.3.1 Some regions of type 1.

We say that a region D is of ***type 2*** if there are continuous functions ψ_1 and ψ_2 defined on $[c, d]$ such that D is the set of points (x, y) satisfying

$$y \in [c, d], \qquad \psi_1(y) \leq x \leq \psi_2(y),$$

where $\psi_1(y) \leq \psi_2(y)$ for all $y \in [c, d]$. Again, the curves that bound the region D constitute its boundary ∂D. Some examples of type 2 regions are shown in Figure 5.3.2.

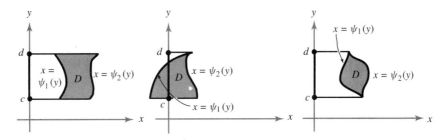

Figure 5.3.2   Some regions of type 2.

Finally, a region of ***type 3*** is one that is both type 1 and type 2; that is, the region can be described both as a region of type 1 and as a region of type 2. An example of a type 3 region is the unit disk (Figure 5.3.3).

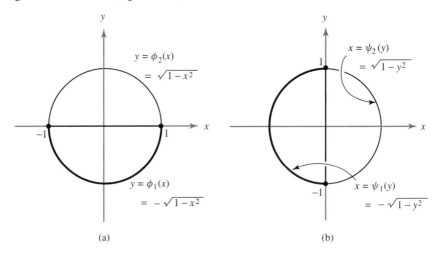

Figure 5.3.3   The unit disk, a region of type 3: (a) as a type 1 region, and (b) as a type 2 region.

Sometimes we shall refer to regions of types 1, 2, and 3 as ***elementary regions***. Note that the boundary ∂D of an elementary region is the type of set of discontinuities of a function allowed in Theorem 2.

DEFINITION: INTEGRAL OVER AN ELEMENTARY REGION
If D is an elementary region in the plane, choose a rectangle R that contains D. Given $f: D \to \mathbb{R}$, where f is continuous (and hence bounded), define $\iint_D f(x, y)\, dA$, the *integral of f over the set D* as follows: extend f to a function f^* defined on all of R by

$$f^*(x, y) \text{ if } = \begin{cases} f(x, y) & \text{if } (x, y) \in D \\ 0 & \text{if } (x, y) \notin D \text{ and } (x, y) \in R. \end{cases}$$

Note that f^* is bounded (since f is) and continuous except possibly on the boundary of D (see Figure 5.3.4). The boundary of D consists of graphs of continuous functions, and so f^* is integrable over R by Theorem 2, Section 5.2. Therefore we can define

$$\iint_D f(x, y)\, dA = \iint_R f^*(x, y)\, dA.$$

When $f(x, y) \geq 0$ on D, we can interpret the integral $\iint_D f(x, y)\, dA$ as the volume of the three-dimensional region between the graph of f and D, as is evident from Figure 5.3.4.

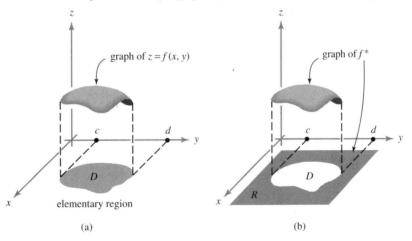

Figure 5.3.4 (a) Graph of $z = f(x, y)$ over an elementary region D. (b) Shaded region shows graph of $z = f^*(x, y)$ on some rectangle R containing D. From this picture we see that boundary points of D may be points of discontinuity of f^*, since the graph of $z = f^*(x, y)$ can be broken at these points.

We have defined $\iint_D f(x, y)\, dx\, dy$ by choosing a rectangle R that encloses D. It should be intuitively clear that the value of $\iint_D f(x, y)\, dx\, dy$ does not depend on the particular R we select; we shall demonstrate this fact at the end of this section.

If $R = [a, b] \times [c, d]$ is a rectangle containing D, we can use the results on iterated integrals in Section 5.2 to obtain

$$\iint_D f(x, y)\, dA = \iint_R f^*(x, y)\, dA = \int_a^b \int_c^d f^*(x, y)\, dy\, dx = \int_c^d \int_a^b f^*(x, y)\, dx\, dy,$$

where f^* equals f in D and zero outside D, as above. Assume that D is a region of type 1 determined by functions $\phi_1 \colon [a, b] \to \mathbb{R}$ and $\phi_2 \colon [a, b] \to \mathbb{R}$. Consider the iterated integral

$$\int_a^b \int_c^d f^*(x, y)\, dy\, dx$$

and, in particular, the inner integral $\int_c^d f^*(x, y)\, dy$ for some fixed x (Figure 5.3.5). Since by definition $f^*(x, y) = 0$ if $y < \phi_1(x)$ or $y > \phi_2(x)$, we obtain

$$\int_c^d f^*(x, y)\, dy = \int_{\phi_1(x)}^{\phi_2(x)} f^*(x, y)\, dy = \int_{\phi_1(x)}^{\phi_2(x)} f(x, y)\, dy.$$

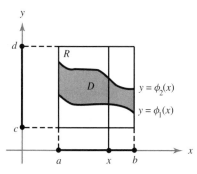

Figure 5.3.5   The region between two graphs—a type 1 region.

We summarize what we have obtained in the following.

THEOREM 4: REDUCTION TO ITERATED INTEGRALS *If D is a region of type 1, as shown in Figure 5.3.5, then*

$$\iint_D f(x, y)\, dA = \int_a^b \int_{\phi_1(x)}^{\phi_2(x)} f(x, y)\, dy\, dx. \tag{1}$$

In the case $f(x, y) = 1$ for all $(x, y) \in D$, $\iint_D f(x, y)\, dA$ is the area of D. We can check this for formula (1) as follows:

$$\int_a^b \int_{\phi_1(x)}^{\phi_2(x)} f(x, y)\, dy\, dx = \int_a^b [\phi_2(x) - \phi_1(x)]\, dx = A(D),$$

which is the formula for the area of D learned in elementary calculus.

EXAMPLE 1   Find $\iint_T (x^3 y + \cos x)\, dA$, where T is the triangle consisting of all points (x, y) such that $0 \le x \le \pi/2, 0 \le y \le x$.

SOLUTION Referring to Figure 5.3.6 and formula (1), we have

$$\iint_T (x^3 y + \cos x)\, dA = \int_0^{\pi/2} \int_0^x (x^3 y + \cos x)\, dy\, dx$$

$$= \int_0^{\pi/2} \left[\frac{x^3 y^2}{2} + y \cos x \right]_{y=0}^x dx = \int_0^{\pi/2} \left(\frac{x^5}{2} + x \cos x \right) dx$$

$$= \left[\frac{x^6}{12} \right]_0^{\pi/2} + \int_0^{\pi/2} (x \cos x)\, dx = \frac{\pi^6}{(12)(64)} + [x \sin x + \cos x]_0^{\pi/2}$$

$$= \frac{\pi^6}{768} + \frac{\pi}{2} - 1. \quad \blacktriangle$$

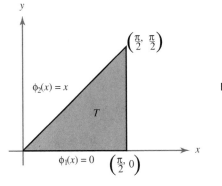

$\left(\frac{\pi}{2}, \frac{\pi}{2} \right)$

$\phi_2(x) = x$

T

$\phi_1(x) = 0$ $\left(\frac{\pi}{2}, 0 \right)$

Figure 5.3.6 A triangle T represented as a region of type 1.

In the next example we shall use formula (1) to find the volume of a solid whose base is a nonrectangular region D.

EXAMPLE 2 Find the volume of the tetrahedron bounded by the planes $y = 0, z = 0, x = 0$, and $y - x + z = 1$ (Figure 5.3.7).

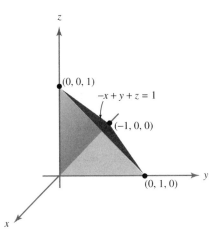

$(0, 0, 1)$

$-x + y + z = 1$

$(-1, 0, 0)$

$(0, 1, 0)$

Figure 5.3.7 A tetrahedron bounded by the planes $y = 0, z = 0, x = 0$, and $y - x + z = 1$.

SOLUTION   We first note that the given tetrahedron has a triangular base D whose points (x, y) satisfy $-1 \le x \le 0$ and $0 \le y \le 1 + x$; hence D is a region of type 1. In fact, D is of type 3; see Figure 5.3.8.

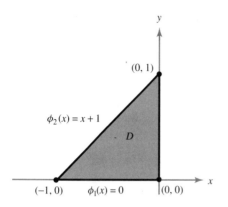

Figure 5.3.8   The base of the tetrahedron in Figure 5.3.7 represented as a region of type 1.

For any point (x, y) in D, the height of the surface z above (x, y) is $1 - y + x$. Thus, the volume we seek is given by the integral

$$\iint_D (1 - y + x)\, dA.$$

Using formula (1) with $\phi_1(x) = 0$ and $\phi_2(x) = x + 1$, we have

$$\iint_D (1 - y + x)\, dA = \int_{-1}^{0} \int_{0}^{1+x} (1 - y + x)\, dy\, dx = \int_{-1}^{0} \left[(1 + x)y - \frac{y^2}{2} \right]_{y=0}^{1+x} dx$$

$$= \int_{-1}^{0} \left[\frac{(1 + x)^2}{2} \right] dx = \left[\frac{(1 + x)^3}{6} \right]_{-1}^{0} = \frac{1}{6}. \quad \blacktriangle$$

EXAMPLE 3   Let D be a region of type 1. Describe its area $A(D)$ as a limit of Riemann sums.

SOLUTION   If we recall the definition, $A(D) = \iint_D dx\, dy$ is the integral over a containing rectangle R of the function $f = 1$. A Riemann sum S_n for this integral is obtained by dividing R into subrectangles and forming the sum $S_n = \sum_{j,k=0}^{n-1} f^*(\mathbf{c}_{jk})\Delta x \Delta y$ as in formula (1) of Section 5.2. Now $f^*(\mathbf{c}_{jk})$ is 1 or 0, depending on whether or not $\mathbf{c}_{jk}$ is in D. Consider those subrectangles R_{jk} that have nonvoid intersection with D, and choose $\mathbf{c}_{jk}$ in $D \cap R_{jk}$. Thus S_n is the sum of the areas of the subrectangles that meet D and $A(D)$ is the limit of these as $n \to \infty$. Thus $A(D)$ is the limit of the areas of the rectangles "circumscribing" D. The reader should draw a figure to accompany this discussion. $\blacktriangle$

The methods for treating regions of type 2 are entirely analogous. Specifically, we have the following.

THEOREM 4': ITERATED INTEGRALS FOR TYPE 2 REGIONS

Suppose that D is the set of points (x, y) such that $y \in [c, d]$ and $\psi_1(y) \leq x \leq \psi_2(y)$. If f is continuous on D, then

$$\iint_D f(x, y)\, dA = \int_c^d \left[\int_{\psi_1(y)}^{\psi_2(y)} f(x, y)\, dx \right] dy. \qquad (2)$$

To find the area of D we substitute $f = 1$ in formula (2); this yields

$$\iint_D dA = \int_c^d (\psi_2(y) - \psi_1(y))\, dy.$$

Again, this result for area agrees with the results of single-variable calculus for the area of a region between two curves.

Either the method for type 1 or the method for type 2 regions can be used for integrals over regions of type 3.

It follows from formulas (1) and (2) that $\iint_D f\, dA$ is independent of the choice of the rectangle R enclosing D used in the definition of $\iint_D f\, dA$, since, if we had picked another rectangle enclosing D we would have arrived at the same formula (1).

EXERCISES

1. Evaluate the following iterated integrals and draw the regions D determined by the limits. State whether the regions are of type 1, type 2, or both.

(a) $\int_0^1 \int_0^{x^2} dy\, dx$

(b) $\int_1^2 \int_{2x}^{3x+1} dy\, dx$

(c) $\int_0^1 \int_1^{e^x} (x + y)\, dy\, dx$

(d) $\int_0^1 \int_{x^3}^{x^2} y\, dy\, dx$

2. Repeat Exercise 1 for the following iterated integrals:

(a) $\int_{-3}^2 \int_0^{y^2} (x^2 + y)\, dx\, dy$

(b) $\int_{-1}^1 \int_{-2|x|}^{|x|} e^{x+y}\, dy\, dx$

(c) $\int_0^1 \int_0^{(1-x^2)^{1/2}} dy\, dx$

(d) $\int_0^{\pi/2} \int_0^{\cos x} y \sin x\, dy\, dx$

(e) $\int_0^1 \int_{y^2}^y (x^n + y^m)\, dx\, dy, \quad m, n > 0$

(f) $\int_{-1}^0 \int_0^{2(1-x^2)^{1/2}} x\, dy\, dx$

3. Use double integrals to compute the area of a circle of radius r.

4. Using double integrals, determine the area of an ellipse with semiaxes of length a and b.

5. What is the volume of a barn that has a rectangular base 20 ft by 40 ft, and vertical walls 30 ft high at the front (which we assume is on the 20-ft side of the barn) and 40 ft high at the rear? The barn has a flat roof. Use double integrals to compute the volume.

6. Let D be the region bounded by the positive x and y axes and the line $3x + 4y = 10$. Compute

$$\iint_D (x^2 + y^2)\, dA.$$

7. Let D be the region bounded by the y axis and the parabola $x = -4y^2 + 3$. Compute

$$\iint_D x^3 y\, dx\, dy.$$

8. Evaluate $\int_0^1 \int_0^{x^2} (x^2 + xy - y^2)\, dy\, dx$. Describe this iterated integral as an integral over a certain region D in the xy plane.

9. Let D be the region given as the set of (x, y) where $1 \le x^2 + y^2 \le 2$ and $y \ge 0$. Is D an elementary region? Evaluate $\iint_D f(x, y)\, dA$ where $f(x, y) = 1 + xy$.

10. Use the formula $A(D) = \iint_D dx\, dy$ to find the area enclosed by one period of the sine function $\sin x$, for $0 \le x \le 2\pi$, and the x axis.

11. Find the volume of the region inside the surface $z = x^2 + y^2$ and between $z = 0$ and $z = 10$.

12. Set up the integral required to calculate the volume of a cone of base radius r and height h.

13. Evaluate $\iint_D y\, dA$ where D is the set of points (x, y) such that $0 \le 2x/\pi \le y,\ y \le \sin x$.

14. From Exercise 5, Section 5.2, $\int_a^b \int_c^d f(x) g(y)\, dy\, dx = \left(\int_a^b f(x)\, dx \right) \left(\int_c^d g(y)\, dy \right)$. Is it true that $\iint_D f(x) g(y)\, dx\, dy = \left(\int_a^b f(x)\, dx \right) \left(\int_{\phi_1(a)}^{\phi_2(b)} g(y)\, dy \right)$ for regions of type 1?

15. Let D be a region given as the set of (x, y) with $-\phi(x) \le y \le \phi(x)$ and $a \le x \le b$, where ϕ is a nonnegative continuous function on the interval $[a, b]$. Let $f(x, y)$ be a function on D such that $f(x, y) = -f(x, -y)$ for all $(x, y) \in D$. Argue that $\iint_D f(x, y)\, dA = 0$.

16. Use the methods of this section to show that the area of the parallelogram D determined by vectors $\mathbf{a}$ and $\mathbf{b}$ is $|a_1 b_2 - a_2 b_1|$, where $\mathbf{a} = a_1 \mathbf{i} + a_2 \mathbf{j}$ and $\mathbf{b} = b_1 \mathbf{i} + b_2 \mathbf{j}$.

17. Describe the area $A(D)$ of a region as a limit of areas of inscribed rectangles, as in Example 3.

5.4 CHANGING THE ORDER OF INTEGRATION

Suppose that D is a region of type 3. Thus, being of types 1 and 2, it can be given as the set of points (x, y) such that

$$a \le x \le b, \qquad \phi_1(x) \le y \le \phi_2(x),$$

and also as the set of points (x, y) such that

$$c \le y \le d, \qquad \psi_1(y) \le x \le \psi_2(y).$$

Hence we have the formulas

$$\iint_D f(x, y) \, dA = \int_a^b \int_{\phi_1(x)}^{\phi_2(x)} f(x, y) \, dy \, dx = \int_c^d \int_{\psi_1(y)}^{\psi_2(y)} f(x, y) \, dx \, dy.$$

If we are required to compute one of the iterated integrals above, we may do so by evaluating the other iterated integral; this technique is called *changing the order of integration*. It is often useful to make such a change when evaluating iterated integrals, since one of the iterated integrals may be more difficult to compute than the other.

EXAMPLE 1 By changing the order of integration, evaluate

$$\int_0^a \int_0^{(a^2-x^2)^{1/2}} (a^2 - y^2)^{1/2} \, dy \, dx.$$

SOLUTION Note that x varies between 0 and a, and for fixed x, $0 \leq y \leq (a^2-x^2)^{1/2}$. Thus the iterated integral is equivalent to the double integral

$$\iint_D (a^2 - y^2)^{1/2} \, dy \, dx,$$

where D is the set of points (x, y) such that $0 \leq x \leq a$ and $0 \leq y \leq (a^2 - x^2)^{1/2}$. But this is the representation of one quarter (the positive quadrant portion) of the disk of radius a; hence D can also be described as the set of points (x, y) satisfying

$$0 \leq y \leq a, \qquad 0 \leq x \leq (a^2 - y^2)^{1/2}$$

(see Figure 5.4.1). Thus,

$$\int_0^a \int_0^{(a^2-x^2)^{1/2}} (a^2 - y^2)^{1/2} \, dy \, dx = \int_0^a \left[\int_0^{(a^2-y^2)^{1/2}} (a^2 - y^2)^{1/2} \, dx \right] dy$$

$$= \int_0^a [x(a^2 - y^2)^{1/2}]_{x=0}^{(a^2-y^2)^{1/2}} \, dy$$

$$= \int_0^a (a^2 - y^2) \, dy = \left[a^2 y - \frac{y^3}{3} \right]_0^a = \frac{2a^3}{3}. \quad \blacktriangle$$

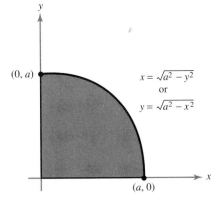

$$x = \sqrt{a^2 - y^2}$$
or
$$y = \sqrt{a^2 - x^2}$$

Figure 5.4.1 The positive-quadrant portion of a disk of radius a.

We could have evaluated the initial iterated integral directly, but, as the reader can easily verify, changing the order of integration makes the problem simpler. The next sample shows that it may even be "impossible" to evaluate an iterated integral and yet be possible to evaluate the iterated integral obtained by changing the order of integration.

EXAMPLE 2   Evaluate

$$\int_1^2 \int_0^{\log x} (x-1)\sqrt{1+e^{2y}}\, dy\, dx.$$

SOLUTION   First observe that we could not compute this integral in the order given by using the fundamental theorem. However, the integral is equal to $\iint_D (x-1)\sqrt{1+e^{2y}}\, dA$, where D is the set of (x, y) such that

$$1 \le x \le 2 \qquad \text{and} \qquad 0 \le y \le \log x.$$

The region D is of type 3 (see Figure 5.4.2) and can be described by

$$0 \le y \le \log 2 \qquad \text{and} \qquad e^y \le x \le 2.$$

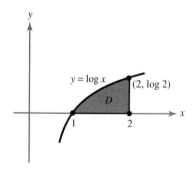

Figure 5.4.2 D is the region of integration for Example 2.

Thus the given iterated integral is equal to

$$\int_0^{\log 2} \int_{e^y}^2 (x-1)\sqrt{1+e^{2y}}\, dx\, dy = \int_0^{\log 2} \sqrt{1+e^{2y}} \left[\int_{e^y}^2 (x-1)\, dx \right] dy$$

$$= \int_0^{\log 2} \sqrt{1+e^{2y}} \left[\frac{x^2}{2} - x \right]_{e^y}^2 dy$$

$$= -\int_0^{\log 2} \left(\frac{e^{2y}}{2} - e^y \right)\sqrt{1+e^{2y}}\, dy$$

$$= -\frac{1}{2}\int_0^{\log 2} e^{2y}\sqrt{1+e^{2y}}\, dy + \int_0^{\log 2} e^y\sqrt{1+e^{2y}}\, dy. \qquad (1)$$

In the first integral in expression (1) we substitute $u = e^{2y}$, and in the second, $v = e^y$. Hence we obtain

$$-\frac{1}{4}\int_1^4 \sqrt{1+u}\, du + \int_1^2 \sqrt{1+v^2}\, dv. \tag{2}$$

Both integrals in expression (2) are easily found with techniques of calculus (or by consulting the table of integrals at the back of the book). For the first integral, we get

$$\frac{1}{4}\int_1^4 \sqrt{1+u}\, du = \left[\frac{1}{6}(1+u)^{3/2}\right]_1^4 = \frac{1}{6}[(1+4)^{3/2} - 2^{3/2}] = \frac{1}{6}[5^{3/2} - 2^{3/2}]. \tag{3}$$

The second integral is

$$\int_1^2 \sqrt{1+v^2}\, dv = \frac{1}{2}[v\sqrt{1+v^2} + \log(\sqrt{1+v^2}+v)]_1^2$$

$$= \frac{1}{2}[2\sqrt{5} + \log(\sqrt{5}+2)] - \frac{1}{2}[\sqrt{2} + \log(\sqrt{2}+1)] \tag{4}$$

(see formula 43 in the table of integrals). Finally, we subtract equation (3) from equation (4) to obtain the answer

$$\frac{1}{2}\left(2\sqrt{5} - \sqrt{2} + \log\frac{\sqrt{5}+2}{\sqrt{2}+1}\right) - \frac{1}{6}[5^{3/2} - 2^{3/2}]. \quad \blacktriangle$$

To conclude this section we mention an important analog of the mean value theorem of integral calculus.

THEOREM 5: MEAN VALUE THEOREM: DOUBLE INTEGRALS
Suppose $f: D \to \mathbb{R}$ is continuous and D is an elementary region. Then for some point (x_0, y_0) in D we have

$$\iint_D f(x, y)\, dA = f(x_0, y_0)A(D),$$

where $A(D)$ denotes the area of D.

PROOF We cannot prove this theorem with complete rigor, because it requires some concepts about continuous functions not proved in this course; but we can sketch the main ideas that underlie the proof.

Since f is continuous on D, it has a maximum value M and a minimum value m. Thus

$$m \leq f(x, y) \leq M \tag{5}$$

for all $(x, y) \in D$. Furthermore, $f(x_1, y_1) = m$ and $f(x_2, y_2) = M$ for some pairs

(x_1, y_1) and (x_2, y_2) in D. From inequality (5) it follows that

$$mA(D) = \iint_D m\, dA \leq \iint_D f(x, y)\, dA \leq \iint_D M\, dA = MA(D).$$

Therefore, dividing through by $A(D)$, we get

$$m \leq \frac{1}{A(D)} \iint_D f(x, y)\, dA \leq M. \tag{6}$$

Since a continuous function on D takes on every value between its maximum and minimum values (this is the intermediate value theorem proved in advanced calculus), and since the number $[1/A(D)] \iint_D f(x, y)\, dA$ is by inequality (6) between these values, there must be a point $(x_0, y_0) \in D$ with

$$f(x_0, y_0) = \frac{1}{A(D)} \iint_D f(x, y)\, dA,$$

which is precisely the conclusion of Theorem 5.   ∎

EXERCISES

1. In the following integrals, change the order of integration, sketch the corresponding regions, and evaluate the integral both ways.

(a) $\int_0^1 \int_x^1 xy\, dy\, dx$

(b) $\int_0^{\pi/2} \int_0^{\cos\theta} \cos\theta\, dr\, d\theta$

(c) $\int_0^1 \int_1^{2-y} (x + y)^2\, dx\, dy$

(d) $\int_a^b \int_a^y f(x, y)\, dx\, dy$ (express your answer in terms of antiderivatives).

2. Find

(a) $\int_{-1}^1 \int_{|y|}^1 (x + y)^2\, dx\, dy$

(b) $\int_{-3}^1 \int_{-\sqrt{(9-y^2)}}^{\sqrt{(9-y^2)}} x^2\, dx\, dy$

(c) $\int_0^4 \int_{y/2}^2 e^{x^2}\, dx\, dy$

(d) $\int_0^1 \int_{\tan^{-1} y}^{\pi/4} (\sec^5 x)\, dx\, dy$

3. If $f(x, y) = e^{\sin(x+y)}$ and $D = [-\pi, \pi] \times [-\pi, \pi]$, show that

$$\frac{1}{e} \leq \frac{1}{4\pi^2} \iint_D f(x, y)\, dA \leq e.$$

4. Show that

$$\frac{1}{2}(1 - \cos 1) \leq \iint_{[0,1]\times[0,1]} \frac{\sin x}{1 + (xy)^4}\, dx\, dy \leq 1.$$

5. If $D = [-1, 1] \times [-1, 2]$, show that

$$1 \leq \iint_D \frac{dx\, dy}{x^2 + y^2 + 1} \leq 6.$$

6. Using the mean value theorem, show that

$$\frac{1}{6} \leq \iint_D \frac{dA}{y - x + 3} \leq \frac{1}{4},$$

where D is the triangle with vertices $(0, 0)$, $(1, 1)$ and $(1, 0)$.

7. Compute the volume of an ellipsoid with semiaxes a, b, and c. (HINT: Use symmetry and first find the volume of one half of the ellipsoid.)

8. Compute $\iint_D f(x, y) \, dA$, where $f(x, y) = y^2 \sqrt{x}$ and D is the set of (x, y) where $x > 0$, $y > x^2$, and $y < 10 - x^2$.

9. Find the volume of the region determined by $x^2 + y^2 + z^2 \leq 10$, $z \geq 2$. Use the disk method from one-variable calculus and state how the method is related to Cavalieri's principle.

10. Evaluate $\iint_D e^{x-y} \, dx \, dy$, where D is the interior of the triangle with vertices $(0, 0)$, $(1, 3)$, and $(2, 2)$.

11. Evaluate $\iint_D y^3 (x^2 + y^2)^{-3/2} \, dx \, dy$, where D is the region determined by the conditions $\frac{1}{2} \leq y \leq 1$ and $x^2 + y^2 \leq 1$.

12. Given that the double integral $\iint_D f(x, y) \, dx \, dy$ of a positive continuous function f equals the iterated integral $\int_0^1 [\int_{x^2}^x f(x, y) \, dy] \, dx$, sketch the region D and interchange the order of integration.

13. Given that the double integral $\iint_D f(x, y) \, dx \, dy$ of a positive continuous function f equals the iterated integral $\int_0^1 \left[\int_y^{\sqrt{2-y^2}} f(x, y) \, dx \right] dy$, sketch the region D and interchange the order of integration.

14. Prove that $2 \int_a^b \int_x^b f(x) f(y) \, dy \, dx = \left(\int_a^b f(x) \, dx \right)^2$. [HINT: Notice that $\left(\int_a^b f(x) \, dx \right)^2 = \iint_{[a,b] \times [a,b]} f(x) f(y) \, dx \, dy$.]

15. Show that

$$\frac{d}{dx} \int_a^x \int_c^d f(x, y, z) \, dz \, dy = \int_c^d f(x, x, z) \, dz + \int_a^x \int_c^d f_x(x, y, z) \, dz \, dy.$$

*5.5 SOME TECHNICAL INTEGRATION THEOREMS

This section provides the main ideas of the proofs of the existence and additivity of the integral that were stated in Section 5.2. These proofs require more advanced concepts than those needed for the rest of this chapter. The notions of uniform continuity and the completeness of the real numbers, both of which are usually treated in a junior-level course in mathematical analysis or real-variable theory, are called upon here.

DEFINITION   Let $D \subset \mathbb{R}^n$ and $f: D \to \mathbb{R}$. Recall that f is said to be **continuous at $\mathbf{x}_0 \in D$** provided that for all $\varepsilon > 0$ there is a number $\delta > 0$ such that if $\mathbf{x} \in D$ and

$\|\mathbf{x} - \mathbf{x}_0\| < \delta$, then $\|f(\mathbf{x}) - f(\mathbf{x}_0)\| < \varepsilon$. We say f is **continuous on** D if it is continuous at each point of D.

The function f is said to be **uniformly continuous** on D if for every number $\varepsilon > 0$ there is a $\delta > 0$ such that whenever $\mathbf{x}, \mathbf{y} \in D$ and $\|\mathbf{x} - \mathbf{y}\| < \delta$, then $\|f(\mathbf{x}) - f(\mathbf{y})\| < \varepsilon$.

The main difference between continuity and uniform continuity is that for continuity, δ can depend on $\mathbf{x}_0$ as well as ε, whereas in uniform continuity δ depends only on ε. Thus any uniformly continuous function is also continuous. An example of a function that is continuous but not uniformly continuous is given in Exercise 2. The distinction between the notions of continuity and uniform continuity can be rephrased: For a function f that is continuous but not uniformly continuous, δ cannot be chosen independently of the point of the domain (the $\mathbf{x}_0$ in the definition). The definition of uniform continuity states explicitly that once you are given an $\varepsilon > 0$, a δ can be found independent of any point of D.

Recall from Section 3.3 that a set $D \subset \mathbb{R}^n$ is **bounded** *if there exists a number* $M > 0$ *such that* $\|\mathbf{x}\| \leq M$ *for all* $\mathbf{x} \in D$. *A set is* **closed** *if it contains all its boundary points.* Thus a set is bounded if it can be strictly contained in some (large) ball. The next theorem states that under some conditions a continuous function is actually uniformly continuous.

THEOREM 6: THE UNIFORM CONTINUITY PRINCIPLE *Every function that is continuous on a closed and bounded set D in $\mathbb{R}^n$ is uniformly continuous on D.*

The proof of this theorem will take us too far afield;* however, we can prove a special case of it, which is, in fact, sufficient for many situations relevant to this text.

PROOF OF A SPECIAL CASE OF THEOREM 6   Let us assume that $D = [a, b]$ is a closed interval on the line, that $f: D \to \mathbb{R}$ is continuous, that df/dx exists on the open interval (a, b), and that df/dx is bounded (that is, there is a constant $C > 0$ such that $|df(x)/dx| \leq C$ for all x in (a, b)). To show that these conditions imply f is uniformly continuous, we use the mean value theorem as follows: Let $\varepsilon > 0$ be given and let x and y lie in D. Then by the mean value theorem,

$$f(x) - f(y) = f'(c)(x - y)$$

for some c between x and y. By the assumed boundedness of the derivative,

$$|f(x) - f(y)| \leq C|x - y|.$$

Let $\delta = \varepsilon/C$. If $|x - y| < \delta$, then

$$|f(x) - f(y)| < C\frac{\varepsilon}{C} = \varepsilon.$$

---

*The proof can be found in texts on mathematical analysis. See, for example, J. Marsden and M. Hoffman, *Elementary Classical Analysis*, 2nd ed., Freeman, New York, 1993, or W. Rudin, *Principles of Mathematical Analysis*, 3rd ed., McGraw-Hill, New York, 1976.

Thus f is uniformly continuous. (Note that δ depends on neither x nor y, which is a crucial part of the definition.) ∎

This proof also works for regions in $\mathbb{R}^n$ that are convex; that is, for any two points $\mathbf{x}$, $\mathbf{y}$ in D, the line segment $\mathbf{c}(t) = t\mathbf{x} + (1 - t)\mathbf{y}$, $0 \leq t \leq 1$, joining them also lies in D. We assume f is differentiable (on an open set containing D) and that $\|\nabla f(\mathbf{x})\| \leq C$ for a constant C. Then the mean value theorem applied to the function $h(t) = f(\mathbf{c}(t))$ gives a point t_0 such that

$$h(1) - h(0) = [h'(t_0)][1 - 0]$$

or

$$f(\mathbf{x}) - f(\mathbf{y}) = h'(t_0) = \nabla f(\mathbf{c}(t_0)) \cdot \mathbf{c}'(t_0) = \nabla f(\mathbf{c}(t_0)) \cdot (\mathbf{x} - \mathbf{y})$$

by the chain rule. Thus by the Cauchy-Schwarz inequality,

$$|f(\mathbf{x}) - f(\mathbf{y})| \leq \|\nabla f(\mathbf{c}(t_0))\| \|\mathbf{x} - \mathbf{y}\| \leq C\|\mathbf{x} - \mathbf{y}\|.$$

Then, as above, given $\varepsilon > 0$, we can let $\delta = \varepsilon/C$.

We now move on to the notion of a *Cauchy sequence* of real numbers. In the definition of Riemann sums we obtained a sequence of numbers $\{S_n\}$, $n = 1, \ldots$. It would be nice if we could say that this sequence of numbers converges to S (or has a limit S), but how can we obtain such a limit? In the abstract setting, we know no more about S_n than that it is the Riemann sum of a (say, continuous) function, and, although this has not yet been proved, it should be enough information to ensure its convergence.

Thus we must determine a property for sequences that guarantees their convergence. We shall define a class of sequences called Cauchy sequences, and then take as an axiom of the real number system that all such sequences converge to a limit.* The determination in the nineteenth century that such an axiom was necessary for the foundations of calculus was a major breakthrough in the history of mathematics and paved the way for the modern rigorous approach to mathematical analysis. We shall say more about this shortly.

DEFINITION A sequence of real numbers $\{S_n\}$, $n = 1, \ldots$, is said to satisfy the *Cauchy criterion* if for every $\varepsilon > 0$ there exists an N such that for all $m, n \geq N$, we have $|S_n - S_m| < \varepsilon$.

If a sequence S_n converges to a limit S, then S_n is a Cauchy sequence. To see this, we use the definition: For every $\varepsilon > 0$ there is an N such that for all $n \geq N$, $|S_n - S| < \varepsilon$. Given $\varepsilon > 0$, choose N_1 such that for $n \geq N_1$, $|S_n - S| < \varepsilon/2$ (use the definition with $\varepsilon/2$ in place of ε). Then if $n, m \geq N_1$,

---

*Texts on mathematical analysis, such as those mentioned in the preceding footnote, sometimes use different axioms, such as the least upper bound property. In such a setting, our completeness axiom becomes a theorem.

$$|S_n - S_m| = |S_n - S + S - S_m| \leq |S_n - S| + |S - S_m| < \frac{\varepsilon}{2} + \frac{\varepsilon}{2} = \varepsilon,$$

which proves our contention. The completeness axiom asserts that the converse is true as well:

COMPLETENESS AXIOM OF THE REAL NUMBERS   Every Cauchy sequence $\{S_n\}$ converges to some limit S.

HISTORICAL NOTE

Augustin Louis Cauchy (1789–1857), one of the greatest mathematicians of all time, defined what we now call Cauchy sequences in his *Cours d'analyse*, published in 1821. This book was a basic work on the foundations of analysis, although it would be considered somewhat loosely written by the standards of our time. Cauchy knew that a convergent sequence was "Cauchy" and remarked that a Cauchy sequence converges. He did not have a proof, nor could he have had one, since such a proof depends on the rigorous development of the real number system that was achieved only in 1872 by the German mathematician Georg Cantor (1845–1918).

It is now clear what we must do to ensure that the Riemann sums $\{S_n\}$ of, say, a continuous function on a rectangle converge to some limit S, which would prove that continuous functions on rectangles are integrable; *we must show that $\{S_n\}$ is a Cauchy sequence*. In demonstrating this, we use the uniform continuity principle. The integrability of continuous functions will be a consequence of the following two lemmas.

LEMMA 1   *Let f be a continuous function on a rectangle R in the plane, and let $\{S_n\}$ be a sequence of Riemann sums for f. Then $\{S_n\}$ converges to some number S.*

PROOF   Given a rectangle $R \subset \mathbb{R}^2$, $R = [a, b] \times [c, d]$, we have the regular partition of R, $a = x_0 < x_1 < \cdots < x_n = b$, $c = y_0 < y_1 < \cdots < y_n = d$, discussed in Section 5.2. Recall that

$$\Delta x = x_{j+1} - x_j = \frac{b - a}{n}, \qquad \Delta y = y_{k+1} - y_k = \frac{d - c}{n},$$

and

$$S_n = \sum_{j,k=0}^{n-1} f(\mathbf{c}_{jk}) \, \Delta x \, \Delta y,$$

where $\mathbf{c}_{jk}$ is an arbitrarily chosen point in $R_{jk} = [x_j, x_{j+1}] \times [y_k, y_{k+1}]$. The sequence $\{S_n\}$ is determined only by the selection of the points $\mathbf{c}_{jk}$.

For the purpose of the proof we shall introduce a slightly more complicated but very precise notation: we set

$$\Delta x^n = \frac{b-a}{n} \quad \text{and} \quad \Delta y^n = \frac{d-c}{n}.$$

With this notation we have

$$S_n = \sum_{j,k}^{n-1} f(\mathbf{c}_{jk})\, \Delta x^n \Delta y^n. \tag{1}$$

To show that $\{S_n\}$ satisfies the Cauchy criterion, we must show that given $\varepsilon > 0$ there exists an N such that for all $n, m \geq N, |S_n - S_m| \leq \varepsilon$. By the uniform continuity principle, f is uniformly continuous on R. Thus given $\varepsilon > 0$ there exists a $\delta > 0$ such that when $\mathbf{x}, \mathbf{y} \in R, \|\mathbf{x} - \mathbf{y}\| < \delta$, then $|f(\mathbf{x}) - f(\mathbf{y})| < \varepsilon/[2\,\text{area}(R)]$ (the quantity $\varepsilon/[2\,\text{area}(R)]$ is used in place of ε in the definition). Let N be so large that for any $m \geq N$ the diameter (length of a diagonal) of any subrectangle R_{jk} in the mth regular partition of R is less than δ. Thus if $\mathbf{x}, \mathbf{y}$ are points in the same subrectangle, we will have the inequality $|f(\mathbf{x} - f(\mathbf{y}))| < \varepsilon/[2\,\text{area}(R)]$.

Fix $m, n \geq N$. We will show that $|S_n - S_m| < \varepsilon$. This shows that $\{S_n\}$ is a Cauchy sequence and hence converges. Consider the mnth $= (m$ times $n)$th regular partition of R. Then

$$S_{mn} = \sum_{r,t} f(\tilde{\mathbf{c}}_{rt})\, \Delta x^{mn} \Delta y^{mn},$$

where $\tilde{\mathbf{c}}_{rt}$ is a point in the rtth subrectangle. Note that each subrectangle of the mnth partition is a subrectangle of both the mth and the nth regular partitions (see Figure 5.5.1).

Figure 5.5.1   The shaded box shows a subrectangle in the mnth partition, and the darkly outlined box, a subrectangle in the mth partition.

Let us denote the subrectangles in the mnth subdivision by $\tilde{R}_{rt}$ and those in the nth subdivision by R_{jk}. Thus each $\tilde{R}_{rt} \subset R_{jk}$ for some jk, and hence we can rewrite formula (1) as

$$S_n = \sum_{j,k}^{n-1} \left(\sum_{\tilde{R}_{rt} \subset R_{jk}} f(\mathbf{c}_{jk}) \, \Delta x^{mn} \, \Delta y^{mn} \right). \tag{1'}$$

Here we are using the fact that

$$\sum_{\tilde{R}_{rt} \subset R_{jk}} f(\mathbf{c}_{jk}) \, \Delta x^{mn} \, \Delta y^{mn} = f(\mathbf{c}_{jk}) \, \Delta x^n \, \Delta y^n,$$

where the sum is taken over all subrectangles in the mnth subdivision contained in a *fixed* rectangle R_{jk} in the nth subdivision. We also have the identity

$$S_{mn} = \sum_{r,t}^{mn-1} f(\tilde{\mathbf{c}}_{rt}) \, \Delta x^{mn} \, \Delta y^{mn}. \tag{2}$$

This relation can also be rewritten as

$$S_{mn} = \sum_{j,k} \sum_{\tilde{R}_{rt} \subset R_{jk}} f(\tilde{\mathbf{c}}_{rt}) \, \Delta x^{mn} \, \Delta y^{mn}, \tag{2'}$$

where in equation (2′) we are first summing over those subrectangles in the mnth partition contained in a fixed R_{jk} and then summing over j, k. Subtracting equation (2′) from equation (1′), we get

$$|S_n - S_{mn}| = \left| \sum_{j,k} \sum_{\tilde{R}_{rt} \subset R_{jk}} \left[f(\mathbf{c}_{jk}) \, \Delta x^{mn} \, \Delta y^{mn} - f(\tilde{\mathbf{c}}_{rt}) \, \Delta x^{mn} \, \Delta y^{mn} \right] \right|$$

$$\leq \sum_{j,k} \sum_{\tilde{R}_{rt} \subset R_{jk}} |f(\mathbf{c}_{jk}) - f(\tilde{\mathbf{c}}_{rt})| \, \Delta x^{mn} \, \Delta y^{mn}.$$

By our choice of δ and N, $|f(\mathbf{c}_{jk}) - f(\tilde{\mathbf{c}}_{rt})| < \varepsilon / [2 \, \text{area}(R)]$, and consequently the above inequality becomes

$$|S_n - S_{mn}| \leq \sum_{j,k} \sum_{\tilde{R}_{rt} \subset R_{jk}} \frac{\varepsilon}{2 \, \text{area}(R)} \Delta x^{mn} \, \Delta y^{mn} = \frac{\varepsilon}{2}.$$

Thus $|S_n - S_{mn}| < \varepsilon/2$ and similarly one shows that $|S_n - S_{mn}| < \varepsilon/2$. Since

$$|S_n - S_m| = |S_n - S_{mn} + S_{mn} - S_m| \leq |S_n - S_{mn}| + |S_{mn} - S_m| < \varepsilon$$

for $m, n \geq N$, we have shown $\{S_n\}$ satisfies the Cauchy criterion and thus has a limit S. ∎

We have already remarked that each Riemann sum depends on the selection of a collection of points $\mathbf{c}_{jk}$. In order to show that a continuous function on a rectangle R is integrable, we must further demonstrate that the limit S we obtained in Lemma 1 is independent of the choices of the points c_{jk}.

LEMMA 2 *The limit S in Lemma 1 does not depend on the choice of points c_{jk}.*

PROOF Suppose we have two sequences of Riemann sums $\{S_n\}$ and $\{S_n^*\}$ obtained by selecting two different sets of points, say c_{jk} and c_{jk}^* in each nth partition. By Lemma 1 we know that $\{S_n\}$ converges to some number S and $\{S_n^*\}$ must also converge to some number, say S^*. We want to show that $S = S^*$ and shall do this by showing that given any $\varepsilon > 0$, $|S - S^*| < \varepsilon$, which implies that S must be equal to S^* (why?).

To start, we know that f is uniformly continuous on R. Consequently, given $\varepsilon > 0$, there exists a δ such that $|f(\mathbf{x}) - f(\mathbf{y})| < \varepsilon/[3\ \text{area}(R)]$ whenever $\|\mathbf{x} - \mathbf{y}\| < \delta$. We choose N so large that whenever $n \geq N$ the diameter of each subrectangle in the nth regular partition is less than δ. Since $\lim_{n\to\infty} S_n = S$ and $\lim_{n\to\infty} S_n^* = S^*$, we can assume that N has been chosen so large that $n \geq N$ implies that $|S_n - S| < \varepsilon/3$ and $|S_n^* - S^*| < \varepsilon/3$. Also, for $n \geq N$ we know by uniform continuity that if c_{jk} and c_{jk}^* are points in the same subrectangle R_{jk} of the nth partition, then $|f(c_{jk}) - f(c_{jk}^*)| < \varepsilon/[3\ \text{area}(R)]$. Thus

$$|S_n - S_n^*| = \left| \sum_{j,k} f(c_{jk})\, \Delta x^n \Delta y^n - \sum_{j,k} f(c_{jk}^*)\, \Delta x^n\, \Delta y^n \right|$$
$$\leq \sum_{j,k} |f(c_{jk}) - f(c_{jk}^*)|\, \Delta x^n \Delta y^n < \frac{\varepsilon}{3}.$$

We now write

$$|S - S^*| = |S - S_n + S_n - S_n^* + S_n^* - S^*| \leq |S - S_n| + |S_n - S_n^*| + |S_n^* - S| < \varepsilon$$

and so the lemma is proved. ■

Putting lemmas 1 and 2 together proves Theorem 1 of Section 5.2:

THEOREM 1 *Any continuous function defined on a rectangle R is integrable.*

HISTORICAL NOTE

Cauchy presented the first published proof of this theorem in his résumé of 1823, in which he points out the need to prove the existence of the integral as a limit of a sum. In this paper he first treats continuous functions (as we are doing now), but on an interval $[a, b]$. (The proof is essentially the same.) However, his proof was not rigorous, since it lacked the notion of uniform continuity, which was not available at that time.

The notion of a Riemann sum S_n for a function f certainly predates Bernhard Riemann (1826–1866). The sums are probably named after him because he developed a theoretical approach to the study of integration in a fundamental paper on trigonometric series in 1854. His approach, although later generalized by Darboux (1875) and Stieltjes (1894), was to last more than half a century until it was augmented by the theory Lebesgue presented to the mathematical world in 1902. This latter approach to integration theory is generally studied in graduate courses in mathematics.

The proof of Theorem 2 (Section 5.2) is left to the reader in Exercises 4 to 6 at the end of this section. The main ideas are essentially contained in the proof of Theorem 1.

Our next goal will be to present a proof of property (iv) of the integral from Section 5.2, namely, its additivity. However, because of some technical difficulties in establishing this result in its full generality, we shall prove it only in the case in which f is continuous.

THEOREM: ADDITIVITY OF THE INTEGRAL *Let R_1 and R_2 be two disjoint rectangles (rectangles whose intersection contains no rectangle) such that $Q = R_1 \cup R_2$ is again a rectangle, as in Figure 5.5.2. If f is a function that is continuous over Q and hence over each R_i, then*

$$\iint_Q f = \iint_{R_1} f + \iint_{R_2} f. \tag{3}$$

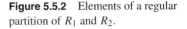
Figure 5.5.2 Elements of a regular partition of R_1 and R_2.

PROOF The proof depends on the ideas that have already been presented in the proof of Theorem 1.

The fact that f is integrable over Q, R_1, and R_2 follows from Theorem 1. Thus all three integrals in equation (3) exist, and it is necessary only to establish equality.

Without loss of generality we can assume that

$$R_1 = [a, b_1] \times [c, d] \quad \text{and} \quad R_2 = [b_1, b] \times [c, d]$$

(see Figure 5.5.2). Again we must develop some notation. Let

$$\Delta x_1^n = \frac{b_1 - a}{n}, \qquad \Delta x_2^n = \frac{b - b_1}{n}, \qquad \Delta x^n = \frac{b - a}{n}, \qquad \text{and} \qquad \Delta y^n = \frac{d - c}{n}.$$

Let

$$S_n^1 = \sum_{j,k} f\left(\mathbf{c}_{jk}^1\right) \Delta x_1^n \, \Delta y^n \tag{4}$$

$$S_n^2 = \sum_{j,k} f\left(\mathbf{c}_{jk}^2\right) \Delta x_2^n \, \Delta y^n \tag{5}$$

$$S_n = \sum_{j,k} f(\mathbf{c}_{jk}) \Delta x^n \, \Delta y^n \tag{6}$$

where $\mathbf{c}_{jk}^1, \mathbf{c}_{jk}^2$, and $\mathbf{c}_{jk}$ are points in the jkth subrectangle of the nth regular partition of R_1, R_2, and Q, respectively. Let $S^i = \lim_{n \to \infty} S_n^i$, where $i = 1, 2$, and $S = \lim_{n \to \infty} S_n$. It must be shown that $S = S^1 + S^2$, which we will accomplish by showing that for arbitrary $\varepsilon > 0, |S - S^1 - S^2| < \varepsilon$.

By the uniform continuity of f on Q we know that given $\varepsilon > 0$ there is a $\delta > 0$ such that whenever $\|\mathbf{x} - \mathbf{y}\| < \delta, |f(\mathbf{x}) - f(\mathbf{y})| < \varepsilon$. Let N be so big that for all $n \geq N$, $|S_n - S| < \varepsilon/3, |S_n^i - S^i| < \varepsilon/3, i = 1, 2$, and if $\mathbf{x}, \mathbf{y}$ are any two points in any subrectangle of the nth partition of either R_1, R_2, or Q, then $|f(\mathbf{x}) - f(\mathbf{y})| < \varepsilon/[3 \text{ area} (Q)]$. Let us consider the nth regular partition of R_1, R_2, and Q. These form a collection of subrectangles that we shall denote by $R_{jk}^1, R_{jk}^2, R_{jk}$, respectively [see Figures 5.5.2 and 5.5.3(a)].

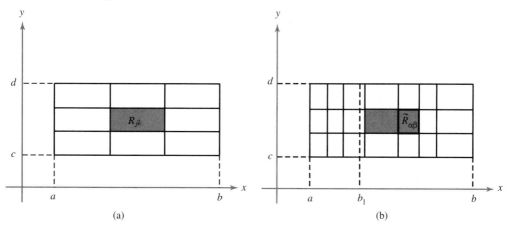

(a)   (b)

Figure 5.5.3   (a) A regular partition of Q. (b) The vertical and horizontal lines of this subdivision are obtained by taking the union of the vertical and horizontal of Figures 5.5.2 and 5.5.3(a).

If we superimpose the subdivision of Q on the nth subdivisions of R_1 and R_2, we get a new collection of rectangles, say $\tilde{R}_{\alpha\beta}$, $\beta = 1, \ldots, n$ and $\alpha = 1, \ldots, m, m > n$ [see Figure 5.5.3(b)].

Each $\tilde{R}_{\alpha\beta}$ is contained in some subrectangle R_{jk} of Q and in some subrectangle of the nth partition of either R_1 or R_2. Consider equalities (4), (5), and (6) above. These can be rewritten as

$$S_n^i = \sum_{j,k} \sum_{\tilde{R}_{\alpha\beta} \subset R_i} f(\mathbf{c}_{jk}^i) \text{ area } (\tilde{R}_{\alpha\beta}) = \sum_{\substack{\alpha,\beta \\ \tilde{R}_{\alpha\beta} \subset R_i}} f(\tilde{\mathbf{c}}_{\alpha\beta}) \text{ area } (\tilde{R}_{\alpha\beta}),$$

where $\tilde{\mathbf{c}}_{\alpha\beta} = \mathbf{c}_{jk}^i$ if $\tilde{R}_{\alpha\beta} \subset R_{jk}^i$, $i = 1, 2$, and

$$S_n = \sum_{j,k} \sum_{\tilde{R}_{\alpha\beta} \subset R_{jk}} f(\mathbf{c}_{jk}) \text{ area } (\tilde{R}_{\alpha\beta}) = \sum_{\alpha,\beta} f(\mathbf{c}_{\alpha\beta}^*) \text{ area } (\tilde{R}_{\alpha\beta}),$$

where $\mathbf{c}_{\alpha\beta}^* = \mathbf{c}_{jk}$ if $\tilde{R}_{\alpha\beta} \subset R_{jk}$.

For the reader encountering such index notation for the first time, we point out that

$$\sum_{\substack{\alpha,\beta \\ \tilde{R}_{\alpha\beta} \subset R_i}}$$

means that the summation is taken over those α's and β's such that the corresponding rectangle $\tilde{R}_{\alpha\beta}$ is contained in the rectangle R_i.

Now the sum for S_n can be split into two parts:

$$S_n = \sum_{\substack{\alpha,\beta \\ \tilde{R}_{\alpha\beta} \subset R_1}} f(\mathbf{c}_{\alpha\beta}^*) \text{ area } (\tilde{R}_{\alpha\beta}) + \sum_{\substack{\alpha,\beta \\ \tilde{R}_{\alpha\beta} \subset R_2}} f(\mathbf{c}_{\alpha\beta}^*) \text{ area } (\tilde{R}_{\alpha\beta}).$$

From these representations and the triangle inequality, it follows that

$$\left| S_n - S_n^1 - S_n^2 \right| \leq \left| \sum_{\substack{\alpha,\beta \\ \tilde{R}_{\alpha\beta} \subset R_1}} [f(\mathbf{c}_{\alpha\beta}^*) - f(\tilde{\mathbf{c}}_{\alpha\beta})] \text{ area } (\tilde{R}_{\alpha\beta}) \right|$$

$$+ \left| \sum_{\substack{\alpha,\beta \\ \tilde{R}_{\alpha\beta} \subset R_2}} [f(\mathbf{c}_{\alpha\beta}^*) - f(\tilde{\mathbf{c}}_{\alpha\beta})] \text{ area } (\tilde{R}_{\alpha\beta}) \right|$$

$$\leq \frac{\varepsilon}{3 \text{ area } Q} \sum_{\substack{\alpha,\beta \\ \tilde{R}_{\alpha\beta} \subset R_1}} \text{ area } (\tilde{R}_{\alpha\beta}) + \frac{\varepsilon}{3 \text{ area } Q} \sum_{\substack{\alpha,\beta \\ \tilde{R}_{\alpha\beta} \subset R_2}} \text{ area } (\tilde{R}_{\alpha\beta}) < \frac{\varepsilon}{3}.$$

In this step we used the uniform continuity of f. Thus $|S_n - S_n^1 - S_n^2| < \varepsilon/3$ for $n \geq N$. But $|S - S_n| < \varepsilon/3$, $|S_n^1 - S^1| < \varepsilon/3$, and $|S_n^2 - S^2| < \varepsilon/3$. As in Lemma 2 an application of the triangle inequality shows that $|S - S^1 - S^2| < \varepsilon$, which completes the proof. ∎

EXERCISES

1. Show that if a and b are two numbers such that for any $\varepsilon > 0$, $|a - b| < \varepsilon$, then $a = b$.

2. (a) Let f be the function on the half-open interval $(0, 1]$ defined by $f(x) = 1/x$. Show that f is continuous at every point of $(0, 1]$ but not uniformly continuous.
   (b) Generalize this example to $\mathbb{R}^2$.

3. Let R be the rectangle $[a, b] \times [c, d]$ and f be a bounded function that is integrable over R.
   (a) Show that f is integrable over $[(a + b)/2, b] \times [c, d]$.
   (b) Let N be any positive integer. Show that f is integrable over $[(a + b)/N, b] \times [c, d]$.

Exercises 4 to 6 are intended to give a proof of Theorem 2 of Section 5.2.

4. Let C be the graph of a continuous function $\phi: [a, b] \to \mathbb{R}$. Let $\varepsilon > 0$ be any positive number. Show that C can be placed in a finite union of boxes $B_i = [a_i, b_i] \times [c_i, d_i]$ such that C does not contain a boundary point of $\cup B_i$ and such that $\sum \text{area } (B_i) \leq \varepsilon$. (HINT: Use the uniform continuity principle presented in this section.)

5. Let R and B be rectangles and let $B \subset R$. Consider the nth regular partition of R and let b_n be the sum of the areas of all rectangles in the partition that have a nonempty intersection with B. Show that $\lim_{n \to \infty} b_n = \text{area } (B)$.

6. Let R be a rectangle and $C \subset R$ the graph of a continuous function ϕ. Suppose that $f: R \to \mathbb{R}$ is bounded and continuous except on C. Use Exercises 4 and 5 above and the techniques used in the proof of Theorem 1 of this section to show that f is integrable over R.

7. (a) Use the uniform continuity principle to show that if $\phi: [a, b] \to \mathbb{R}$ is a continuous function then ϕ is bounded.
   (b) Generalize part (a) to show that the continuous function $f: [a, b] \times [c, d] \to \mathbb{R}$ is bounded.
   (c) Generalize part (b) still further to show f is bounded where $f: D \to \mathbb{R}$ is a continuous function on a closed and bounded set $D \subset \mathbb{R}^n$.

5.6 THE TRIPLE INTEGRAL

Triple integrals are needed for many physical problems. For example, if the temperature inside an oven is not uniform, determining the average temperature involves "summing" the values of the temperature function at all points in the solid region enclosed by the oven walls and then dividing the answer by the total volume of the oven. Such a sum is expressed mathematically as a triple integral.

Our objective now is to define the triple integral of a function $f(x, y, z)$ over a box (rectangular parallelepiped) $B = [a, b] \times [c, d] \times [p, q]$. Proceeding as in double integrals, we partition the three sides of B into n equal parts and form the sum

$$S_n = \sum_{i=0}^{n-1} \sum_{j=0}^{n-1} \sum_{k=0}^{n-1} f(\mathbf{c}_{ijk}) \, \Delta V,$$

where $\mathbf{c}_{ijk}$ is a point in B_{ijk}, the ijkth rectangular parallelepiped (or box) in the partition of B, and ΔV is the volume of B_{ijk} (see Figure 5.6.1).

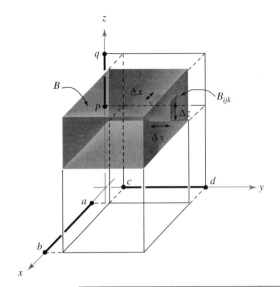

Figure 5.6.1   A partition of a box B into n^3 subboxes B_{ijk}.

> **DEFINITION: TRIPLE INTEGRALS**   Let f be a bounded function of three variables defined on B. If limit$_{n \to \infty} S_n = S$ exists and is independent of any choice of $\mathbf{c}_{ijk}$, we call f *integrable* and call S the *triple integral* (or simply the integral) of f over B and denote it by
>
> $$\iiint_B f \, dV, \qquad \iiint_B f(x, y, z) \, dV \qquad \text{or} \qquad \iiint_B f(x, y, z) \, dx \, dy \, dz.$$

As before, one can prove that continuous functions defined on B are integrable. Moreover, bounded functions whose discontinuities are confined to graphs of continuous functions [such as $x = \alpha(y, z)$, $y = \beta(x, z)$, or $z = \gamma(x, y)$] are integrable. The other basic properties (such as the fact that the integral of a sum is the sum of the integrals) for double integrals also hold for triple integrals. Especially important is the reduction to iterated integrals:

> **REDUCTION TO ITERATED INTEGRALS**   Let $f(x, y, z)$ be integrable on the box $B = [a, b] \times [c, d] \times [p, q]$. Then any iterated integral that exists is equal to the triple integral; that is,
>
> $$\iiint_B f(x, y, z) \, dx \, dy \, dz = \int_p^q \int_c^d \int_a^b f(x, y, z) \, dx \, dy \, dz$$
>
> $$= \int_p^q \int_a^b \int_c^d f(x, y, z) \, dy \, dx \, dz$$
>
> $$= \int_a^b \int_p^q \int_c^d f(x, y, z) \, dy \, dz \, dx,$$
>
> and so on. (There are six possible orders altogether.)

EXAMPLE 1 (a) Let B be the box $[0, 1] \times \left[-\frac{1}{2}, 0\right] \times \left[0, \frac{1}{3}\right]$. Evaluate

$$\iiint_B (x + 2y + 3z)^2 \, dx \, dy \, dz.$$

(b) Verify that we get the same answer if the integration is done in the order y first, then z, and then x.

SOLUTION (a) According to the principle of reduction to iterated integrals, this integral may be evaluated as

$$\int_0^{1/3} \int_{-1/2}^0 \int_0^1 (x + 2y + 3z)^2 \, dx \, dy \, dz$$

$$= \int_0^{1/3} \int_{-1/2}^0 \left[\frac{(x + 2y + 3z)^3}{3} \Big|_{x=0}^1 \right] dy \, dz$$

$$= \int_0^{1/3} \int_{-1/2}^0 \frac{1}{3} [(1 + 2y + 3z)^3 - (2y + 3z)^3] \, dy \, dz$$

$$= \int_0^{1/3} \frac{1}{24} [(1 + 2y + 3z)^4 - (2y + 3z)^4] \big|_{y=-1/2}^0 \, dz$$

$$= \int_0^{1/3} \frac{1}{24} [(3z + 1)^4 - 2(3z)^4 + (3z - 1)^4] \, dz$$

$$= \frac{1}{24 \cdot 15} [(3z + 1)^5 - 2(3z)^5 + (3z - 1)^5] \big|_{z=0}^{1/3}$$

$$= \frac{1}{24 \cdot 15} (2^5 - 2) = \frac{1}{12}.$$

(b)

$$\iiint_B (x + 2y + 3z)^2 \, dy \, dz \, dx$$

$$= \int_0^1 \int_0^{1/3} \int_{-1/2}^0 (x + 2y + 3z)^2 \, dy \, dz \, dx$$

$$= \int_0^1 \int_0^{1/3} \left[\frac{(x + 2y + 3z)^3}{6} \Big|_{y=-1/2}^0 \right] dz \, dx$$

$$= \int_0^1 \int_0^{1/3} \frac{1}{6} [(x + 3z)^3 - (x + 3z - 1)^3] \, dz \, dx$$

$$= \int_0^1 \frac{1}{6} \left\{ \left[\frac{(x + 3z)^4}{12} - \frac{(x + 3z - 1)^4}{12} \right] \Big|_{z=0}^{1/3} \right\} dx$$

$$= \int_0^1 \frac{1}{72} [(x + 1)^4 + (x - 1)^4 - 2x^4] \, dx$$

$$= \frac{1}{72} \frac{1}{5} [(x + 1)^5 + (x - 1)^5 - 2x^5] \big|_{x=0}^1 = \frac{1}{12}. \quad \blacktriangle$$

EXAMPLE 2   Integrate e^{x+y+z} over the box $[0, 1] \times [0, 1] \times [0, 1]$.

SOLUTION

$$
\begin{aligned}
\int_0^1 \int_0^1 \int_0^1 e^{x+y+z}\,dx\,dy\,dz &= \int_0^1 \int_0^1 (e^{x+y+z}|_{x=0}^1)\,dy\,dz \\
&= \int_0^1 \int_0^1 (e^{1+y+z} - e^{y+z})\,dy\,dz = \int_0^1 [e^{1+y+z} - e^{y+z}]_{y=0}^1\,dz \\
&= \int_0^1 [e^{2+z} - 2e^{1+z} + e^z]\,dz = [e^{2+z} - 2e^{1+z} + e^z]_0^1 \\
&= e^3 - 3e^2 + 3e - 1 = (e-1)^3. \quad \blacktriangle
\end{aligned}
$$

As in the two-variable case, we define the integral of a function f over a bounded region W by defining a new function f^*, equal to f on W and zero outside W, and then setting

$$
\iiint_W f(x, y, z)\,dx\,dy\,dz = \iiint_B f^*(x, y, z)\,dx\,dy\,dz,
$$

where B is any box containing the region W.

As before, we restrict our attention to particularly simple regions. An ***elementary region*** in three-dimensional space is one defined by restricting one of the variables to be between two functions of the remaining variables, the domains of these functions being an elementary region in the plane. For example, if D is an elementary region in the xy plane and if $\gamma_1(x, y)$ and $\gamma_2(x, y)$ are two functions with $\gamma_2(x, y) \geq \gamma_1(x, y)$, an elementary region consists of all (x, y, z) such that (x, y) lies in D and $\gamma_1(x, y) \leq z \leq \gamma_2(x, y)$. Figure 5.6.2 shows two elementary regions.

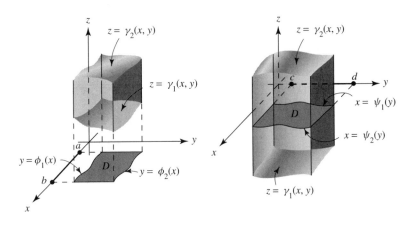

Figure 5.6.2   Two elementary regions in space.

EXAMPLE 3   Describe the unit ball $x^2 + y^2 + z^2 \leq 1$ as an elementary region.

SOLUTION   This can be done in several ways. One is:

$$-1 \leq x \leq 1,$$
$$-\sqrt{1 - x^2} \leq y \leq \sqrt{1 - x^2},$$
$$-\sqrt{1 - x^2 - y^2} \leq z \leq \sqrt{1 - x^2 - y^2}.$$

In doing this, we first write the top and bottom hemispheres as $z = \sqrt{1 - x^2 - y^2}$ and $z = -\sqrt{1 - x^2 - y^2}$, respectively, where x and y vary over the unit disk (that is, $-\sqrt{1 - x^2} \leq y \leq \sqrt{1 - x^2}$ and x varies between -1 and 1). (See Figure 5.6.3.) We can describe the region in other ways by interchanging the roles of x, y, and z in the defining inequalities.   ▲

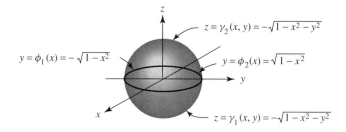

Figure 5.6.3   The unit ball as an elementary region in space.

As with integrals in the plane, any function of three variables that is continuous over an elementary region is integrable on that region. An argument like that for double integrals shows that a triple integral over an elementary region can be rewritten as an iterated integral in which the limits of integration are functions. The formulas for such iterated integrals are given in the following box.

TRIPLE INTEGRALS BY ITERATED INTEGRATION   Suppose that W is an elementary region described by bounding z between two functions of x and y. Then either

$$\iiint_W f(x, y, z)\, dx\, dy\, dz = \int_a^b \int_{\phi_1(x)}^{\phi_2(x)} \int_{\gamma_1(x,y)}^{\gamma_2(x,y)} f(x, y, z)\, dz\, dy\, dx$$

[see Figure 5.6.2 (left)] or

$$\iiint_W f(x, y, z)\, dx\, dy\, dz = \int_c^d \int_{\psi_1(y)}^{\psi_2(y)} \int_{\gamma_1(x,y)}^{\gamma_2(x,y)} f(x, y, z)\, dz\, dx\, dy$$

[see Figure 5.6.2 (right)].

If $f = 1$, we get the integral $\iiint_W dx\,dy\,dz$, which is the **volume** of the region W.

EXAMPLE 4   Verify the volume formula for the ball of radius 1:

$$\iiint_W dx\,dy\,dz = \frac{4}{3}\pi,$$

where W is the set of (x, y, z) with $x^2 + y^2 + z^2 \leq 1$.

SOLUTION   We use the description of the unit ball from Example 3. From the first formula in the preceding box, the integral is

$$\int_{-1}^{1} \int_{-\sqrt{1-x^2}}^{\sqrt{1-x^2}} \int_{-\sqrt{1-x^2-y^2}}^{\sqrt{1-x^2-y^2}} dz\,dy\,dx.$$

Holding y and x fixed and integrating with respect to z yields

$$\int_{-1}^{1} \int_{-\sqrt{1-x^2}}^{\sqrt{1-x^2}} \left[z \Big|_{-\sqrt{1-x^2-y^2}}^{\sqrt{1-x^2-y^2}} \right] dy\,dx = 2 \int_{-1}^{1} \left[\int_{-\sqrt{1-x^2}}^{\sqrt{1-x^2}} (1 - x^2 - y^2)^{1/2}\,dy \right] dx.$$

Since x is fixed in the y-integral, it can be expressed as $\int_{-a}^{a} (a^2 - y^2)^{1/2}\,dy$, where $a = (1 - x^2)^{1/2}$. This integral is the area of a semicircular region of radius a, so that

$$\int_{-a}^{a} (a^2 - y^2)^{1/2}\,dy = \frac{a^2}{2}\pi.$$

(We could also have used a trigonometric substitution or a table of integrals.) Thus,

$$\int_{-\sqrt{1-x^2}}^{\sqrt{1-x^2}} (1 - x^2 - y^2)^{1/2}\,dy = \frac{1 - x^2}{2}\pi,$$

and so

$$2 \int_{-1}^{1} \int_{-\sqrt{1-x^2}}^{\sqrt{1-x^2}} (1 - x^2 - y^2)^{1/2}\,dy\,dx = 2 \int_{-1}^{1} \pi \frac{1 - x^2}{2}\,dx$$

$$= \pi \int_{-1}^{1} (1 - x^2)\,dx = \pi \left(x - \frac{x^3}{3} \right) \Big|_{x=-1}^{1} = \frac{4}{3}\pi. \quad \blacktriangle$$

Other types of elementary regions are shown in Figure 5.6.4. For instance, in the second region, (y, z) lies in an elementary region in the yz plane and x lies between two graphs:

$$\rho_1(y, z) \leq x \leq \rho_2(y, z).$$

As shown in Figure 5.6.5, some elementary regions can be simultaneously described in all three ways. We shall call these regions **symmetric elementary regions.**

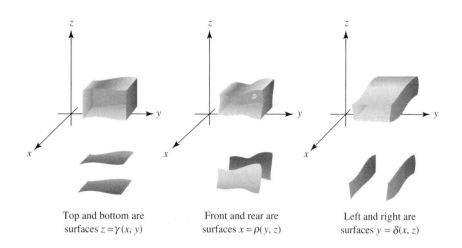

Top and bottom are
surfaces $z = \gamma(x, y)$

Front and rear are
surfaces $x = \rho(y, z)$

Left and right are
surfaces $y = \delta(x, z)$

Figure 5.6.4 Types of elementary regions in space.

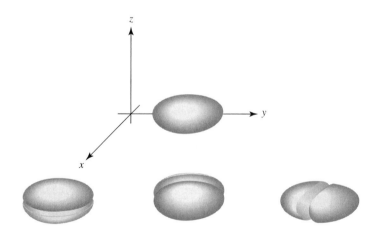

Figure 5.6.5 A symmetric elementary region can be described in three overall ways.

Corresponding to each description of a region as an elementary region is an integration formula. For instance, if W is expressed as the set of all (x, y, z) such that

$$c \leq y \leq d, \qquad \psi_1(y) \leq z \leq \psi_2(y), \qquad \rho_1(y, z) \leq x \leq \rho_2(y, z),$$

then

$$\iiint_W f(x, y, z) \, dx \, dy \, dz = \int_c^d \int_{\psi_1(y)}^{\psi_2(y)} \int_{\rho_1(y,z)}^{\rho_2(y,z)} f(x, y, z) \, dx \, dz \, dy.$$

EXAMPLE 5   Let W be the region bounded by the planes $x = 0$, $y = 0$, and $z = 2$, and the surface $z = x^2 + y^2$ and lying in the quadrant $x \geq 0$, $y \geq 0$. Compute $\iiint_W x \, dx \, dy \, dz$ and sketch the region.

SOLUTION   *Method 1.* The region W is sketched in Figure 5.6.6. As indicated in the figure, we may describe this region by the inequalities

$$0 \leq x \leq \sqrt{2}, \quad 0 \leq y \leq \sqrt{2 - x^2}, \quad x^2 + y^2 \leq z \leq 2.$$

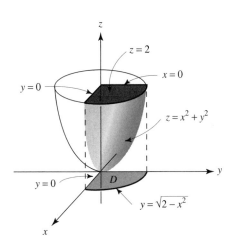

Figure 5.6.6 W is the region below the plane $z = 2$, above the paraboloid $z = x^2 + y^2$, and on the positive sides of the planes $x = 0$, $y = 0$.

Therefore,

$$
\begin{aligned}
\iiint_W x \, dx \, dy \, dz &= \int_0^{\sqrt{2}} \left[\int_0^{\sqrt{2 - x^2}} \left(\int_{x^2 + y^2}^{2} x \, dz \right) dy \right] dx \\
&= \int_0^{\sqrt{2}} \int_0^{\sqrt{2 - x^2}} x(2 - x^2 - y^2) \, dy \, dx \\
&= \int_0^{\sqrt{2}} x \left[(2 - x^2)^{3/2} - \frac{(2 - x^2)^{3/2}}{3} \right] dx \\
&= \int_0^{\sqrt{2}} \frac{2x}{3} (2 - x^2)^{3/2} \, dx = \frac{-2(2 - x^2)^{5/2}}{15} \Big|_0^{\sqrt{2}} \\
&= 2 \cdot \frac{2^{5/2}}{15} = \frac{8\sqrt{2}}{15}.
\end{aligned}
$$

Method 2. We can also place limits on x first and describe W by $0 \leq x \leq (z - y^2)^{1/2}$ and (y, z) in D, where D is the subset of the yz plane with $0 \leq z \leq 2$ and $0 \leq y \leq z^{1/2}$ (see Figure 5.6.7). Therefore,

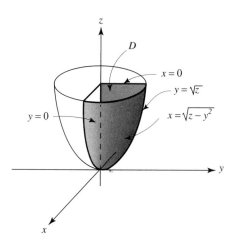

Figure 5.6.7 A different description of the region in Example 5.

$$\iiint_W x\,dx\,dy\,dz = \iint_D \left(\int_0^{(z-y^2)^{1/2}} x\,dx \right) dy\,dz$$

$$= \int_0^2 \left[\int_0^{z^{1/2}} \left(\int_0^{(z-y^2)^{1/2}} x\,dx \right) dy \right] dz$$

$$= \int_0^2 \int_0^{z^{1/2}} \left(\frac{z-y^2}{2} \right) dy\,dz$$

$$= \frac{1}{2} \int_0^2 \left(z^{3/2} - \frac{z^{3/2}}{3} \right) dz = \frac{1}{2} \int_0^2 \frac{2}{3} z^{3/2}\,dz$$

$$= \left[\frac{2}{15} z^{5/2} \right]_0^2 = \frac{2}{15} 2^{5/2} = \frac{8\sqrt{2}}{15},$$

which agrees with our other answer. ▲

EXAMPLE 6 Evaluate

$$\int_0^1 \int_0^x \int_{x^2+y^2}^2 dz\,dy\,dx.$$

Sketch the region W of integration and interpret.

SOLUTION

$$\int_0^1 \int_0^x \int_{x^2+y^2}^2 dz\,dy\,dx = \int_0^1 \int_0^x (2 - x^2 - y^2)\,dy\,dx$$

$$= \int_0^1 \left(2x - x^3 - \frac{x^3}{3} \right) dx = 1 - \frac{1}{4} - \frac{1}{12} = \frac{2}{3}.$$

This integral is the volume of the region sketched in Figure 5.6.8. ▲

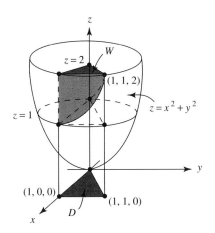

Figure 5.6.8   The region W lies between the paraboloid $z = x^2 + y^2$ and the plane $z = 2$ and above the region D.

EXERCISES

In Exercises 1 to 4, perform the indicated integration over the given box.

1. $\iiint_B x^2 \, dx \, dy \, dz$, $B = [0, 1] \times [0, 1] \times [0, 1]$

2. $\iiint_B e^{-xy} y \, dx \, dy \, dz$, $B = [0, 1] \times [0, 1] \times [0, 1]$

3. $\iiint_B (2x + 3y + z) \, dx \, dy \, dz$, $B = [0, 2] \times [-1, 1] \times [0, 1]$

4. $\iiint_B z e^{x+y} \, dx \, dy \, dz$, $B = [0, 1] \times [0, 1] \times [0, 1]$

In Exercises 5 to 8, describe the given region as an elementary region.

5. The region between the cone $z = \sqrt{x^2 + y^2}$ and the paraboloid $z = x^2 + y^2$.

6. The region cut out of the ball $x^2 + y^2 + z^2 \leq 4$ by the elliptic cylinder $2x^2 + z^2 = 1$, that is, the region inside the cylinder and the ball.

7. The region inside the sphere $x^2 + y^2 + z^2 = 1$ and above the plane $z = 0$.

8. The region bounded by the planes $x = 0$, $y = 0$, $z = 0$, $x + y = 4$, and $x = z - y - 1$.

Find the volume of the region in Exercises 9 to 12.

9. The region bounded by $z = x^2 + y^2$ and $z = 10 - x^2 - 2y^2$.

10. The solid bounded by $x^2 + 2y^2 = 2$, $z = 0$, and $x + y + 2z = 2$.

11. The solid bounded by $x = y$, $z = 0$, $y = 0$, $x = 1$, and $x + y + z = 0$.

12. The region common to the intersecting cylinders $x^2 + y^2 \leq a^2$ and $x^2 + z^2 \leq a^2$.

Evaluate the integrals in Exercises 13 to 21.

13. $\int_0^1 \int_1^2 \int_2^3 \cos\left[\pi(x+y+z)\right] dx\, dy\, dz$

14. $\int_0^1 \int_0^x \int_0^y (y+xz)\, dz\, dy\, dx$

15. $\iiint_W (x^2+y^2+z^2)\, dx\, dy\, dz$; W is the region bounded by $x+y+z = a$ (where $a > 0$), $x = 0$, $y = 0$, and $z = 0$.

16. $\iiint_W z\, dx\, dy\, dz$; W is the region bounded by the planes $x = 0$, $y = 0$, $z = 0$, $z = 1$, and the cylinder $x^2 + y^2 = 1$, with $x \geq 0$, $y \geq 0$.

17. $\iiint_W x^2 \cos z\, dx\, dy\, dz$; W is the region bounded by $z = 0$, $z = \pi$, $y = 0$, $y = 1$, $x = 0$, and $x + y = 1$.

18. $\int_0^2 \int_0^x \int_0^{x+y} dz\, dy\, dx$

19. $\iiint_W (1 - z^2)\, dx\, dy\, dz$; W is the pyramid with top vertex at $(0, 0, 1)$ and base vertices at $(0, 0, 0)$, $(1, 0, 0)$, $(0, 1, 0)$, and $(1, 1, 0)$.

20. $\iiint_W (x^2 + y^2)\, dx\, dy\, dz$; W is the same pyramid as in Exercise 19.

21. $\int_0^1 \int_0^{2x} \int_{x^2+y^2}^{x+y} dz\, dy\, dx$.

22. (a) Sketch the region for the integral

$$\int_0^1 \int_0^x \int_0^y f(x, y, z)\, dz\, dy\, dx.$$

(b) Write the integral with the integration order $dx\, dy\, dz$.

For the regions in Exercise 23 and 24, find the appropriate limits $\phi_1(x)$, $\phi_2(x)$, $\gamma_1(x, y)$, and $\gamma_2(x, y)$, and write the triple integral over the region W as an iterated integral in the form

$$\iiint_W f\, dV = \int_a^b \left\{ \int_{\phi_1(x)}^{\phi_2(x)} \left[\int_{\gamma_1(x,y)}^{\gamma_2(x,y)} f(x, y, z)\, dz \right] dy \right\} dx.$$

23. $W = \{(x, y, z) \mid \sqrt{x^2 + y^2} \leq z \leq 1\}$

24. $W = \{(x, y, z) \mid \frac{1}{2} \leq z \leq 1 \text{ and } x^2 + y^2 + z^2 \leq 1\}$

25. Show that the formula using triple integrals for the volume under the graph of a positive function $f(x, y)$, on an elementary region D in the plane, reduces to the double integral of f over D.

26. Let W be the region bounded by the planes $x = 0$, $y = 0$, $z = 0$, $x + y = 1$, and $z = x + y$.

(a) Find the volume of W.

(b) Evalute $\iiint_W x\, dx\, dy\, dz$.

(c) Evalute $\iiint_W y\, dx\, dy\, dz$.

27. Let f be continuous and let B_ε be the ball of radius ε centered at the point (x_0, y_0, z_0). Let vol (B_ε) be the volume of B_ε. Prove that

$$\lim_{\varepsilon \to 0} \frac{1}{\text{vol}(B_\varepsilon)} \iiint_{B_\varepsilon} f(x, y, z)\, dV = f(x_0, y_0, z_0).$$

REVIEW EXERCISES FOR CHAPTER 5

1. Evaluate each of the following integrals:

(a) $\int_0^3 \int_{-x^2+1}^{x^2+1} xy\, dy\, dx$

(b) $\int_0^1 \int_{\sqrt{x}}^1 (x + y)^2\, dy\, dx$

(c) $\int_0^1 \int_{e^x}^{e^{2x}} x \ln y\, dy\, dx$

(d) $\int_0^1 \int_1^2 \int_2^3 \cos[\pi(x + y + z)]\, dx\, dy\, dz$.

(e) $\int_0^1 \int_0^x \int_0^y (y + xz)\, dz\, dy\, dx$.

2. Reverse the order of integration of the integrals in Exercise 1 and evaluate. [The solution to part (b) only is in the Study Guide to this text.]

3. Evaluate the following integrals:

(a) $\int_0^1 \int_y^{y^3} e^{x/y}\, dx\, dy$

(b) $\int_0^1 \int_0^{(\arcsin y)/y} y \cos xy\, dx\, dy$

4. Change the order of integration and evaluate

$$\int_0^2 \int_{y/2}^1 (x + y)^2\, dx\, dy.$$

5. Show that evaluating $\iint_D dx\, dy$, where D is a region of type 1, reproduces the formula from one-variable calculus for the area between two curves.

6. Change the order of integration and evaluate

$$\int_0^1 \int_{y^{1/2}}^1 (x^2 + y^3x)\, dx\, dy.$$

7. Let D be the region in the xy plane inside the unit circle $x^2 + y^2 = 1$. Evaluate $\iint_D f(x, y)\, dx\, dy$ in each of the following cases:

(a) $f(x, y) = xy$

(b) $f(x, y) = x^2 y^2$

(c) $f(x, y) = x^3 y^3$

8. Find $\iint_D y[1 - \cos(\pi x/4)]\, dx\, dy$, where D is the region in Figure 5.R.1.

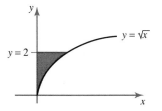

Figure 5.R.1   The region of integration for Exercise 8.

Evaluate the integrals in Exercises 9 to 16. Sketch and identify the type of the region (corresponding to the way the integral is written).

9. $\int_0^\pi \int_{\sin x}^{3\sin x} x(1+y)\,dy\,dx$

10. $\int_0^1 \int_{x-1}^{x\cos(\pi x/2)} (x^2 + xy + 1)\,dy\,dx$

11. $\int_{-1}^1 \int_{y^{2/3}}^{(2-y)^2} \left(\frac{3}{2}\sqrt{x} - 2y\right)\,dx\,dy$

12. $\int_0^2 \int_{-3(\sqrt{4-x^2})/2}^{3(\sqrt{4-x^2})/2} \left(\frac{5}{\sqrt{2+x}} + y^3\right)\,dy\,dx$

13. $\int_0^1 \int_0^{x^2} (x^2 + xy - y^2)\,dy\,dx$

14. $\int_2^4 \int_{y^2-1}^{y^3} 3\,dx\,dy$

15. $\int_0^1 \int_{x^2}^{x} (x+y)^2\,dy\,dx$

16. $\int_0^1 \int_0^{3y} e^{x+y}\,dx\,dy$

In Exercises 17 to 19, integrate the given function f over the given region D.

17. $f(x,y) = x - y$; D is the triangle with vertices $(0, 0)$, $(1, 0)$, and $(2, 1)$.

18. $f(x,y) = x^3 y + \cos x$; D is the triangle defined by $0 \le x \le \pi/2, 0 \le y \le x$.

19. $f(x,y) = x^2 + 2xy^2 + 2$; D is the region bounded by the graph of $y = -x^2 + x$, the x axis, and the lines $x = 0$ and $x = 2$.

In Exercises 20 and 21, sketch the region of integration, interchange the order, and evaluate.

20. $\int_1^4 \int_1^{\sqrt{x}} (x^2 + y^2)\,dy\,dx$

21. $\int_0^1 \int_{1-y}^1 (x + y^2)\,dx\,dy$

22. Show that

$$4e^5 \le \iint_{[1,3]\times[2,4]} e^{x^2+y^2}\,dA \le 4e^{25}.$$

23. Show that

$$4\pi \le \iint_D (x^2 + y^2 + 1)\,dx\,dy \le 20\pi,$$

where D is the disk of radius 2 centered at the origin.

24. Suppose W is a ***path-connected region***, i.e., given any two points of W there is a continuous path joining them. If f is a continuous function on W, use the intermediate value theorem to show that there is at least one point in W at which the value of f is equal to the average of f over W, i.e., the integral of f over D divided by the area of D. (Compare this with the mean value theorem for double integrals.) What happens if W is not connected?

25. Prove: $\int_0^x \left[\int_0^t F(u)\,du \right] dt = \int_0^x (x - u) F(u)\,du.$

Evaluate the integrals in Exercises 26 to 28.

26. $\int_0^1 \int_0^z \int_0^y xy^2 z^3 \, dx \, dy \, dz$

27. $\int_0^1 \int_0^y \int_0^{x/\sqrt{3}} \frac{x}{x^2 + z^2} \, dz \, dx \, dy$

28. $\int_1^2 \int_1^z \int_{1/y}^2 yz^2 \, dx \, dy \, dz$

29. Write the iterated integral $\int_0^1 \int_{1-x}^1 \int_x^1 f(x, y, z) \, dz \, dy \, dx$ as an integral over a region in $\mathbb{R}^3$ and then rewrite it in five other possible orders of integration.

6
THE CHANGE OF VARIABLES
FORMULA AND APPLICATIONS
OF INTEGRATION

If you are stuck in a calculus problem and don't know what else to do, try integrating by parts or changing variables.

JERRY KAZDAN

In this chapter we develop the change of variables theorem, which is of importance in evaluating integrals in polar, cylindrical, and spherical coordinates. The change of variables theorem is the multidimensional generalization of the substitution method learned in single-variable calculus. The chapter also includes some of the important physical applications of the integral.

6.1 THE GEOMETRY OF MAPS FROM $\mathbb{R}^2$ TO $\mathbb{R}^2$

In this section we shall be interested in maps from subsets of $\mathbb{R}^2$ to $\mathbb{R}^2$. The resulting geometric understanding will be useful in the next section, when we discuss the change of variables formula for multiple integrals.

Let D^* be a subset of $\mathbb{R}^2$; suppose we consider a continuously differentiable map $T: D^* \to \mathbb{R}^2$, so T takes points in D^* to points in $\mathbb{R}^2$. We denote the set of image points by D or by $T(D^*)$; hence, $D = T(D^*)$ is the set of all points $(x, y) \in \mathbb{R}^2$ such that

$$(x, y) = T(x^*, y^*) \qquad \text{for some} \qquad (x^*, y^*) \in D^*.$$

One way to understand the geometry of a map T is to see how it *deforms* or changes D^*. For example, Figure 6.1.1 illustrates a map T that takes a slightly twisted region into a disk.

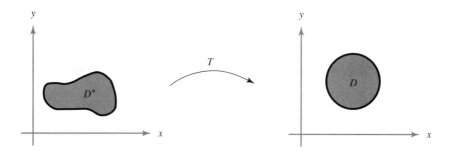

Figure 6.1.1   A function T from a region D^* to a disk D.

EXAMPLE 1   Let $D^* \subset \mathbb{R}^2$ be the rectangle $D^* = [0, 1] \times [0, 2\pi]$. Then all points in D^* are of the form (r, θ), where $0 \le r \le 1, 0 \le \theta \le 2\pi$. Let T be the polar coordinate "change of variables" defined by $T(r, \theta) = (r \cos \theta, r \sin \theta)$. Find the image set D.

SOLUTION   Let $(x, y) = (r \cos \theta, r \sin \theta)$. Since $x^2 + y^2 = r^2 \cos^2 \theta + r^2 \sin^2 \theta = r^2 \le 1$, the set of points $(x, y) \in \mathbb{R}^2$ such that $(x, y) \in D$ has the property that $x^2 + y^2 \le 1$, and so D is contained in the unit disk. In addition, any point (x, y) in the unit disk can be written as $(r \cos \theta, r \sin \theta)$ for some $0 \le r \le 1$ and $0 \le \theta \le 2\pi$. Thus D is the unit disk (see Figure 6.1.2).   ▲

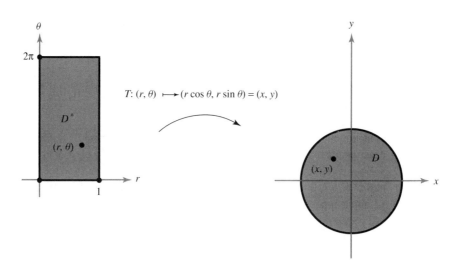

Figure 6.1.2 T gives a change of variables to polar coordinates. The unit circle is the image of a rectangle.

EXAMPLE 2 Let T be defined by $T(x, y) = ((x + y)/2, (x - y)/2)$ and let $D^* = [-1, 1] \times [-1, 1] \subset \mathbb{R}^2$ be a square with side of length 2 centered at the origin. Determine the image D obtained by applying T to D^*.

SOLUTION Let us first determine the effect of T on the line $\mathbf{c}_1(t) = (t, 1)$, where $-1 \leq t \leq 1$. We have $T(\mathbf{c}_1(t)) = ((t + 1)/2, (t - 1)/2)$. The map $t \mapsto T(\mathbf{c}_1(t))$ is a parametrization of the line $y = x - 1, 0 \leq x \leq 1$, since $(t - 1)/2 = (t + 1)/2 - 1$ (see Figure 6.1.3).

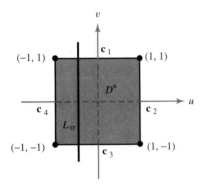

Figure 6.1.3 Domain for the transformation T of Example 2.

Let

$$\mathbf{c}_2(t) = (1, t), \qquad -1 \leq t \leq 1$$
$$\mathbf{c}_3(t) = (t, -1), \qquad -1 \leq t \leq 1$$
$$\mathbf{c}_4(t) = (-1, t), \qquad -1 \leq t \leq 1$$

be parametrizations of the other edges of the square D^*. Using the same argument as above, we see that $T \circ \mathbf{c}_2$ is a parametrization of the line $y = 1 - x, 0 \leq x \leq 1$; $T \circ \mathbf{c}_3$ is the line $y = x + 1, -1 \leq x \leq 0$; and $T \circ \mathbf{c}_4$ is the line $y = -x - 1, -1 \leq x \leq 0$. By this time it seems reasonable to guess that T "flips" the square D^* over and takes it to the square D whose vertices are $(1, 0), (0, 1), (-1, 0), (0, -1)$ (Figure 6.1.4). To prove that this is indeed the case, let $-1 \leq \alpha \leq 1$ and let L_α (Figure 6.1.3) be a fixed line parametrized by $\mathbf{c}(t) = (\alpha, t), -1 \leq t \leq 1$; then $T(\mathbf{c}(t)) = ((\alpha + t)/2, (\alpha - t)/2)$ is a parametrization of the line $y = -x + \alpha, (\alpha - 1)/2 \leq x \leq (\alpha + 1)/2$. This line begins, for $t = -1$, at the point $((\alpha - 1)/2, (1 + \alpha)/2)$ and ends at the point $((1 + \alpha)/2, (\alpha - 1)/2)$; as is easily checked, these points lie on the lines $T \circ \mathbf{c}_3$ and $T \circ \mathbf{c}_1$, respectively. Thus as α varies between -1 and 1, L_α sweeps out the square D^* while $T(L_\alpha)$ sweeps out the square D determined by the vertices $(-1, 0), (0, 1)(1, 0)$ and $(0, -1)$. ▲

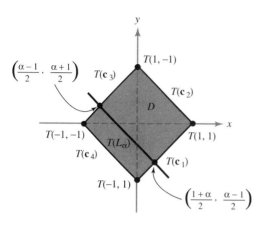

Figure 6.1.4   The effect of T on the region D^*.

The following theorem is a useful way to describe the image $T(D^*)$.

THEOREM 1   *Let A be a 2×2 matrix with $\det A \neq 0$ and let T be the linear mapping of $\mathbb{R}^2$ to $\mathbb{R}^2$ given by $T(x) = Ax$ (matrix multiplication). Then T transforms parallelograms into parallelograms and vertices into vertices. Moreover, if $T(D^*)$ is a parallelogram, D^* must be a parallelogram.*

The proof of Theorem 1 is left as Exercise 10 at the end of this section. This theorem simplifies the result of Example 2, because we need only find the vertices of $T(D^*)$ and then connect them by straight lines.

Although we cannot visualize the graph of a function $T\colon \mathbb{R}^2 \to \mathbb{R}^2$, it does help to consider how the function deforms subsets. However, simply looking at these deformations does not give us a complete picture of the behavior of T. We may characterize T further by using the notion of a one-to-one correspondence.

DEFINITION   The function T is *one-to-one* on D^* if for (u, v) and $(u', v') \in D^*$, $T(u, v) = T(u', v')$ implies that $u = u'$ and $v = v'$.

This statement means that *two different points of D^* are not sent into the same point of D by T*. For example, the function $T(x, y) = (x^2 + y^2, y^4)$ is not one-to-one, because $T(1, -1) = (2, 1) = T(1, 1)$ and yet $(1, -1) \neq (1, 1)$.

EXAMPLE 3   Consider the polar coordinate mapping function $T\colon \mathbb{R}^2 \to \mathbb{R}^2$ considered in Example 1, defined by $T(r, \theta) = (r \cos \theta, r \sin \theta)$. Show that T is not one-to-one if its domain is all of $\mathbb{R}^2$.

SOLUTION   If $\theta_1 \neq \theta_2$, then $T(0, \theta_1) = T(0, \theta_2)$ and so T cannot be one-to-one. This observation implies that if L is the side of the rectangle $D^* = [0, 1] \times [0, 2\pi]$ where

$0 \le \theta \le 2\pi$ and $r = 0$ (Figure 6.1.5), then T maps all of L into a single point, the center of the unit disk D. However, if we consider the set $S^* = (0, 1] \times [0, 2\pi)$, then $T: S^* \to S$ is one-to-one (see Exercise 1). Evidently, in determining whether a function is one-to-one, the domain chosen must be carefully considered. ▲

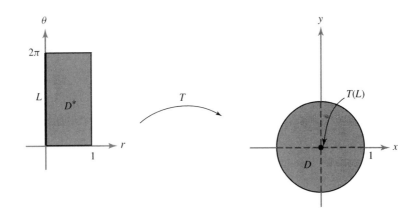

Figure 6.1.5 The polar-coordinate transformation T takes the line L to the point $(0, 0)$.

EXAMPLE 4   Show that the function $T: \mathbb{R}^2 \to \mathbb{R}^2$ of Example 2 is one-to-one.

SOLUTION   Suppose $T(x, y) = T(x', y')$; then

$$\left(\frac{x+y}{2}, \frac{x-y}{2} \right) = \left(\frac{x'+y'}{2}, \frac{x'-y'}{2} \right)$$

and we have

$$x + y = x' + y',$$
$$x - y = x' - y'.$$

Adding, we have

$$2x = 2x'.$$

Thus $x = x'$ and hence $y = y'$, which shows that T is one-to-one (with domain all of $\mathbb{R}^2$). Actually, since T is linear and $T(\mathbf{x}) = A\mathbf{x}$, where A is a 2×2 matrix, it would also suffice to show that $\det A \ne 0$ (see Exercise 8). ▲

In Examples 1 and 2 we have been determining the image $D = T(D^*)$ of a region D^* under a mapping T. What will be of interest to us in the next section is, in part, the inverse problem: namely, given D and a one-to-one mapping T of $\mathbb{R}^2$ to $\mathbb{R}^2$, find D^* such that $T(D^*) = D$.

Before we examine this question in more detail, we introduce the notion of "onto."

DEFINITION   The mapping T is **onto** D if for every point $(x, y) \in D$ there exists at least one point (u, v) in the domain of T such that $T(u, v) = (x, y)$.

Thus if T is onto, we can solve the equation $T(u, v) = (x, y)$ for (u, v), given $(x, y) \in D$. If T is, in addition, one-to-one, this solution is unique.

For *linear* mappings T of $\mathbb{R}^2$ to $\mathbb{R}^2$ (or $\mathbb{R}^n$ to $\mathbb{R}^n$) it turns out that one-to-one and onto are equivalent notions (see Exercises 8 and 9).

If we are given a region D and a mapping T, the determination of a region D^* such that $T(D^*) = D$ will be possible only when for every $(x, y) \in D$ there is a (u, v) in the domain of T such that $T(u, v) = (x, y)$ (that is, T must be onto D). The next example shows that this cannot always be done.

EXAMPLE 5   Let $T: \mathbb{R}^2 \to \mathbb{R}^2$ be given by $T(u, v) = (u, 0)$. Let D be the square, $D = [0, 1] \times [0, 1]$. Since T takes all of $\mathbb{R}^2$ to one axis, it is impossible to find a D^* such that $T(D^*) = D$.   ▲

EXAMPLE 6   Let T be defined as in Example 2 and let D be the square whose vertices are $(1, 0), (0, 1), (-1, 0), (0, -1)$. Find a D^* with $T(D^*) = D$.

SOLUTION   Since T is linear and $T(\mathbf{x}) = A\mathbf{x}$, where A is a 2×2 matrix satisfying $\det A \neq 0$, we know that $T: \mathbb{R}^2 \to \mathbb{R}^2$ is onto (see Exercises 8 and 9), and thus D^* can be found. By Theorem 1, D^* must be a parallelogram. In order to find D^*, it suffices to find the four points that are mapped onto vertices of D; then, by connecting these points, we will have found D^*. For the vertex $(1, 0)$ of D, we must solve $T(x, y) = (1, 0) = ((x + y)/2, (x - y)/2)$ so that $(x + y)/2 = 1, (x - y)/2 = 0$. Thus $(x, y) = (1, 1)$ is a vertex of D^*. Solving for the other vertices, we find that $D^* = [-1, 1] \times [-1, 1]$.   ▲

EXAMPLE 7   Let D be the region in the first quadrant lying between the arcs of the circles $x^2 + y^2 = a^2, x^2 + y^2 = b^2, 0 < a < b$ (see Figure 6.1.6). These circles have equations $r = a$ and $r = b$ in polar coordinates. Let T be the polar-coordinate transformation given by $T(r, \theta) = (r \cos \theta, r \sin \theta) = (x, y)$. Find D^* such that $T(D^*) = D$.

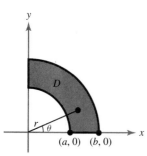

Figure 6.1.6   We seek a region D^* in the θr plane whose image under the polar-coordinate mapping is D.

SOLUTION   In the region $D, a^2 \leq x^2 + y^2 \leq b^2$; and since $r^2 = x^2 + y^2$, we see that $a \leq r \leq b$. Clearly for this region θ varies between $0 \leq \theta \leq \pi/2$. Thus if $D^* = [a, b] \times [0, \pi/2]$, we have $T(D^*) = D$ and T is one-to-one.   ▲

REMARK   The inverse function theorem discussed in Section 3.5 is relevant to the material here. It states that if the determinant of $\mathbf{D}T(u_0, v_0)$ [which is the matrix of partial derivatives of T evaluated at (u_0, v_0)] is not zero, then for (u, v) near (u_0, v_0) and (x, y) near $(x_0, y_0) = T(u_0, v_0)$, the equation $T(u, v) = (x, y)$ can be *uniquely solved* for (u, v) as functions of (x, y). In particular, by uniqueness, T is one-to-one near (u_0, v_0); also, T is onto a neighborhood of (x_0, y_0), since $T(u, v) = (x, y)$ is solvable for (u, v) if (x, y) is near (x_0, y_0). If D^* and D are elementary regions and $T: D^* \to D$ has the property that the determinant of $\mathbf{D}T(u, v)$ is not zero for any (u, v) in D^* and T maps the boundary of D^* in a one-to-one and onto manner to the boundary of D, then T is one-to-one and onto from D^* to D. (This proof is beyond the scope of this text.)

In summary, we have:

ONE-TO-ONE AND ONTO MAPPINGS   A mapping $T: D^* \to D$ is *one-to-one* when it maps distinct points to distinct points. It is *onto* when the image of D^* under T is all of D.

A *linear* transformation of $\mathbb{R}^n$ to $\mathbb{R}^n$ given by multiplication by a matrix A is one-to-one and onto when $\det A \neq 0$.

EXERCISES

1.   Let $S^* = (0, 1] \times [0, 2\pi)$ and define $T(r, \theta) = (r\cos\theta, r\sin\theta)$. Determine the image set S. Show that T is one-to-one on S^*.

2.   Define

$$T(x^*, y^*) = \left(\frac{x^* - y^*}{\sqrt{2}}, \frac{x^* + y^*}{\sqrt{2}} \right).$$

Show that T rotates the unit square, $D^* = [0, 1] \times [0, 1]$.

3.   Let $D^* = [0, 1] \times [0, 1]$ and define T on D^* by $T(u, v) = (-u^2 + 4u, v)$. Find the image D. Is T one-to-one?

4.   Let D^* be the parallelogram bounded by the lines $y = 3x - 4$, $y = 3x$, $y = \frac{1}{2}x$, and $y = \frac{1}{2}(x + 4)$. Let $D = [0, 1] \times [0, 1]$. Find a T such that D is the image of D^* under T.

5.   Let $D^* = [0, 1] \times [0, 1]$ and define T on D^* by $T(x^*, y^*) = (x^*y^*, x^*)$. Determine the image set D. Is T one-to-one? If not, can we eliminate some subset of D^* so that on the remainder T is one-to-one?

6. Let D^* be the parallelogram with vertices at $(-1, 3), (0, 0), (2, -1)$, and $(1, 2)$, and D be the rectangle $D = [0, 1] \times [0, 1]$. Find a T such that D is the image set of D^* under T.

7. Let $T: \mathbb{R}^3 \to \mathbb{R}^3$ be the spherical coordinate mapping defined by $(\rho, \phi, \theta) \mapsto (x, y, z)$, where

$$x = \rho \sin \phi \cos \theta, \quad y = \rho \sin \phi \sin \theta, \quad z = \rho \cos \phi.$$

Let D^* be the set of points (ρ, ϕ, θ) such that $\phi \in [0, \pi], \theta \in [0, 2\pi], \rho \in [0, 1]$. Find $D = T(D^*)$. Is T one-to-one? If not, can we eliminate some subset of D^* (as we did, in Exercise 1, to D^* in Example 1) so that, on the remainder T will be one-to-one?

In Exercises 8 and 9 let $T(\mathbf{x}) = A\mathbf{x}$, where A is a 2×2 matrix.

8. Show that T is one-to-one if and only if the determinant of A is not zero.

9. Show that $\det A \neq 0$ if and only if T is onto.

10. Suppose $T: \mathbb{R}^2 \to \mathbb{R}^2$ is linear; $T(\mathbf{x}) = A\mathbf{x}$, where A is a 2×2 matrix. Show that if $\det A \neq 0$, T takes parallelograms onto parallelograms. (HINT: The general parallelogram in $\mathbb{R}^2$ can be described by the set of points $\mathbf{q} = \mathbf{p} + \lambda\mathbf{v} + \mu\mathbf{w}$ for $\lambda, \mu \in [0, 1]$ where $\mathbf{p}, \mathbf{v}, \mathbf{w}$ are vectors in $\mathbb{R}^2$ with $\mathbf{v}$ not a scalar multiple of $\mathbf{w}$.)

11. Suppose $T: \mathbb{R}^2 \to \mathbb{R}^2$ is as in Exercise 10 and that $T(P^*) = P$ is a parallelogram. Show that P^* is a parallelogram.

6.2 THE CHANGE OF VARIABLES THEOREM

Given two regions D and D^* of type 1 or 2 in $\mathbb{R}^2$, a differentiable map T on D^* with image D, that is, $T(D^*) = D$, and any real-valued integrable function $f: D \to \mathbb{R}$, we would like to express $\iint_D f(x, y) \, dA$ as an integral over D^* of the composite function $f \circ T$. In this section we shall see how to do this.

Assume that the region D^* is of type 1, with the coordinate variables designated by (u, v). Furthermore, assume that D is a type 1 subset of the xy plane. The map T is given by two coordinate functions:

$$T(u, v) = (x(u, v), y(u, v)) \qquad \text{for} \qquad (u, v) \in D^*.$$

As a first guess one might conjecture that

$$\iint_D f(x, y) \, dx \, dy \overset{?}{=} \iint_{D^*} f(x(u, v), y(u, v)) \, du \, dv, \tag{1}$$

where $f \circ T(u, v) = f(x(u, v), y(u, v))$ is the composite function defined on D^*. However, if we consider the function $f: D \to \mathbb{R}^2$ where $f(x, y) = 1$, then equation (1) would imply

$$A(D) = \iint_D dx \, dy \overset{?}{=} \iint_{D^*} du \, dv = A(D^*). \tag{2}$$

But equation (2) will hold only for a few special cases and not for a general map T. For example, define T by $T(u, v) = (-u^2 + 4u, v)$. Restrict T to the unit square $D^* = [0, 1] \times [0, 1]$ in the uv plane (see Figure 6.2.1). Then, as in Exercise 3, Section 6.1, T takes D^* onto $D = [0, 3] \times [0, 1]$. Clearly $A(D) \neq A(D^*)$, and so formula (2) is *not valid*.

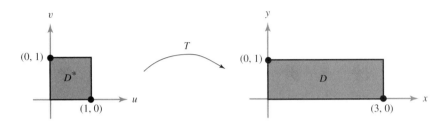

Figure 6.2.1   The map $T: (u, v) \mapsto (-u^2 + 4u, v)$ takes the square D^* onto the rectangle D.

What is needed is a measure of how a transformation $T: \mathbb{R}^2 \to \mathbb{R}^2$ distorts the area of a region. This is given by the *Jacobian determinant*, which is defined as follows.

DEFINITION: JACOBIAN DETERMINANT   Let $T: D^* \subset \mathbb{R}^2 \to \mathbb{R}^2$ be a C^1 transformation given by $x = x(u, v)$ and $y = y(u, v)$. The ***Jacobian determinant*** of T, written $\partial(x, y)/\partial(u, v)$, is the determinant of the derivative matrix $\mathbf{D}T(x, y)$ of T:

$$\frac{\partial(x, y)}{\partial(u, v)} = \begin{vmatrix} \dfrac{\partial x}{\partial u} & \dfrac{\partial x}{\partial v} \\[2mm] \dfrac{\partial y}{\partial u} & \dfrac{\partial y}{\partial v} \end{vmatrix}.$$

EXAMPLE 1   The function from $\mathbb{R}^2$ to $\mathbb{R}^2$ that transforms polar coordinates into Cartesian coordinates is given by

$$x = r \cos \theta, \qquad y = r \sin \theta$$

and its Jacobian determinant is

$$\frac{\partial(x, y)}{\partial(r, \theta)} = \begin{vmatrix} \cos \theta & -r \sin \theta \\ \sin \theta & r \cos \theta \end{vmatrix} = r(\cos^2 \theta + \sin^2 \theta) = r. \quad \blacktriangle$$

Under suitable restrictions on the function T, we will argue below that the area of $D = T(D^*)$ is obtained by integrating the absolute value of the Jacobian $\partial(x, y)/\partial(u, v)$

over D^*; that is, we have the equation

$$A(D) = \iint_D dx\,dy = \iint_{D^*} \left| \frac{\partial(x,\,y)}{\partial(u,\,v)} \right| du\,dv. \tag{3}$$

To illustrate: from Example 1 in Section 6.1 take $T\colon D^* \to D$, where $D = T(D^*)$ is the set of $(x,\,y)$ with $x^2 + y^2 \leq 1$ and $D^* = [0,\,1] \times [0,\,2\pi]$, and $T(r,\,\theta) = (r\cos\theta,\,r\sin\theta)$. By formula (3),

$$A(D) = \iint_{D^*} \left| \frac{\partial(x,\,y)}{\partial(r,\,\theta)} \right| dr\,d\theta = \iint_{D^*} r\,dr\,d\theta \tag{4}$$

(here r and θ play the role of u and v). From the above computation it follows that

$$\iint_{D^*} r\,dr\,d\theta = \int_0^{2\pi} \int_0^1 r\,dr\,d\theta = \int_0^{2\pi} \left[\frac{r^2}{2} \right]_0^1 d\theta = \frac{1}{2} \int_0^{2\pi} d\theta = \pi$$

is the area of the unit disk D, confirming formula (3) in this case. In fact, we may recall from first-year calculus that equation (4) is the correct formula for the area of a region in polar coordinates.

It is not easy to rigorously prove assertion (3). However, looked at in the proper way, it becomes quite plausible. Recall that $A(D) = \iint_D dx\,dy$ was obtained by dividing up D into little rectangles, summing their areas, and then taking the limit of this sum as the size of the subrectangles tended to zero. The problem is that T may map rectangles into regions whose area is not easy to compute. The solution is to approximate these images by simpler regions whose area we can compute. A useful tool for doing this is the derivative of T, which we know (from Chapter 2) gives the best linear approximation to T.

Consider a small rectangle D^* in the uv plane as shown in Figure 6.2.2. Let T' denote the derivative of T evaluated at $(u_0,\,v_0)$, so T' is a 2×2 matrix. From our work in Chapter 2

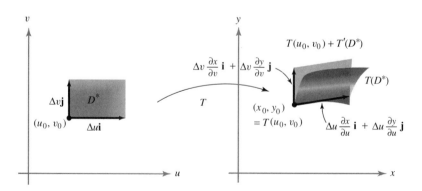

Figure 6.2.2   Effect of the transformation T on a small rectangle D^*.

we know that a good approximation to $T(u, v)$ is given by

$$T(u_0, v_0) + T'\begin{pmatrix} \Delta u \\ \Delta v \end{pmatrix},$$

where $\Delta u = u - u_0$ and $\Delta v = v - v_0$. This mapping takes D^* into a parallelogram with vertex at $T(u_0, v_0)$ and with adjacent sides given by the vectors

$$T'(\Delta u \mathbf{i}) = \begin{bmatrix} \dfrac{\partial x}{\partial u} & \dfrac{\partial x}{\partial v} \\ \dfrac{\partial y}{\partial u} & \dfrac{\partial y}{\partial v} \end{bmatrix} \begin{bmatrix} \Delta u \\ 0 \end{bmatrix} = \Delta u \begin{bmatrix} \dfrac{\partial x}{\partial u} \\ \dfrac{\partial y}{\partial u} \end{bmatrix} = \Delta u \mathbf{T}_u$$

$$T'(\Delta v \mathbf{j}) = \begin{bmatrix} \dfrac{\partial x}{\partial u} & \dfrac{\partial x}{\partial v} \\ \dfrac{\partial y}{\partial u} & \dfrac{\partial y}{\partial v} \end{bmatrix} \begin{bmatrix} 0 \\ \Delta v \end{bmatrix} = \Delta v \begin{bmatrix} \dfrac{\partial x}{\partial v} \\ \dfrac{\partial y}{\partial v} \end{bmatrix} = \Delta v \mathbf{T}_v,$$

where

$$\mathbf{T}_u = \frac{\partial x}{\partial u}\mathbf{i} + \frac{\partial y}{\partial u}\mathbf{j} \qquad \text{and} \qquad \mathbf{T}_v = \frac{\partial x}{\partial v}\mathbf{i} + \frac{\partial y}{\partial v}\mathbf{j}$$

are evaluated at (u_0, v_0).

Recall from Section 1.3 that the area of the parallelogram with sides equal to the vectors $a\mathbf{i} + b\mathbf{j}$ and $c\mathbf{i} + d\mathbf{j}$ is equal to the absolute value of the determinant

$$\begin{vmatrix} a & b \\ c & d \end{vmatrix} = \begin{vmatrix} a & c \\ b & d \end{vmatrix}.$$

Thus, the area of $T(D^*)$ is approximately equal to the *absolute value* of

$$\begin{vmatrix} \dfrac{\partial x}{\partial u}\Delta u & \dfrac{\partial x}{\partial v}\Delta v \\ \dfrac{\partial y}{\partial u}\Delta u & \dfrac{\partial y}{\partial v}\Delta v \end{vmatrix} = \begin{vmatrix} \dfrac{\partial x}{\partial u} & \dfrac{\partial x}{\partial v} \\ \dfrac{\partial y}{\partial u} & \dfrac{\partial y}{\partial v} \end{vmatrix} \Delta u\, \Delta v = \frac{\partial(x, y)}{\partial(u, v)}\Delta u\, \Delta v$$

evaluated at (u_0, v_0).

This fact and a partitioning argument should make formula (3) plausible. Indeed, if we partition D^* into small rectangles with sides of length Δu and Δv, the images of these rectangles are approximated by parallelograms with sides $\mathbf{T}_u \Delta u$ and $\mathbf{T}_v \Delta v$ and hence with area $|\partial(x, y)/\partial(u, v)|\, \Delta u\, \Delta v$. Thus the area of D^* is approximately $\sum \Delta u\, \Delta v$, where the sum is taken over all the rectangles R inside D^* (see Figure 6.2.3). Hence the area of $T(D^*)$ is approximately the sum $\sum |\partial(x, y)/\partial(u, v)|\Delta u\, \Delta v$. In the limit, this sum becomes

$$\iint_{D^*} \left| \frac{\partial(x, y)}{\partial(u, v)} \right| du\, dv.$$

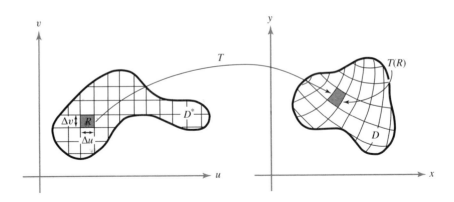

Figure 6.2.3   The area of the little rectangle R is $\Delta u \, \Delta v$. The area of $T(R)$ is approximately $|\partial(x, y)/\partial(u, v)| \Delta u \, \Delta v$.

Let us give another argument for the special case (4) of formula (3), that is, the case of polar coordinates. Consider a region D in the xy plane and a grid corresponding to a partition of the r and θ variables (Figure 6.2.4). The area of the shaded region shown is approximately $(\Delta r)(r_{jk} \, \Delta \theta)$, since the arc length of a segment of a circle of radius r subtending an angle ϕ is $r\phi$. The total area is then the limit of $\sum r_{jk} \, \Delta r \Delta \theta$; that is, $\iint_{D^*} r \, dr \, d\theta$. The key idea is thus that the jkth "polar rectangle" in the grid has area approximately equal to $r_{jk} \, \Delta r \, \Delta \theta$. (For n large, the jkth polar rectangle will look like a rectangle with sides of length $r_{jk} \, \Delta \theta$ and Δr). This should provide some insight into why we say the "area element $dx \, dy$" is transformed into the "area element $r \, dr \, d\theta$."

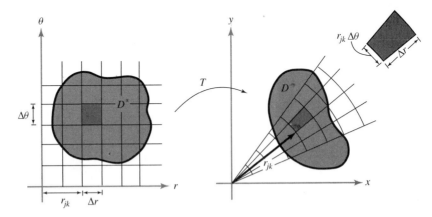

Figure 6.2.4 D^* is mapped to D under the polar-coordinate mapping T.

E X A M P L E   2   Let the elementary region D in the xy plane be bounded by the graph of a polar equation $r = f(\theta)$, where $\theta_0 \le \theta \le \theta_1$ and $f(\theta) \ge 0$ (see Figure 6.2.5). In the

$r\theta$ plane we consider the type 2 region D^* where $\theta_0 \le \theta \le \theta_1$ and $0 \le r \le f(\theta)$. Under the transformation $x = r\cos\theta$, $y = r\sin\theta$, the region D^* is carried onto the region D. Use (4) to calculate the area of D.

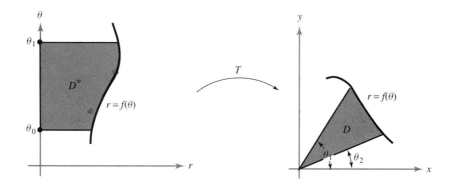

Figure 6.2.5 Effect on the region D^* of the polar-coordinate mapping.

SOLUTION

$$A(D) = \iint_D dx\,dy = \iint_{D^*} \left| \frac{\partial(x, y)}{\partial(r, \theta)} \right| dr\,d\theta$$

$$= \iint_{D^*} r\,dr\,d\theta = \int_{\theta_0}^{\theta_1} \left[\int_0^{f(\theta)} r\,dr \right] d\theta$$

$$= \int_{\theta_0}^{\theta_1} \left[\frac{r^2}{2} \right]_0^{f(\theta)} d\theta = \int_{\theta_0}^{\theta_1} \frac{[f(\theta)]^2}{2} d\theta.$$

This formula for $A(D)$ should be familiar from one-variable calculus. ▲

Before stating the change of variables formula, which is the culmination of this discussion, let us recall the corresponding theorem from one-variable calculus:

$$\int_a^b f(x(u)) \frac{dx}{du}\,du = \int_{x(a)}^{x(b)} f(x)\,dx, \tag{5}$$

where f is continuous and $u \mapsto x(u)$ is continuously differentiable on $[a, b]$.

PROOF Let F be an antiderivative of f; that is, $F' = f$, whose existence is guaranteed by the fundamental theorem of calculus. The right-hand side of equation (5) becomes

$$\int_{x(a)}^{x(b)} f(x)\,dx = F(x(b)) - F(x(a)).$$

To evaluate the left-hand side of equation (5), let $G(u) = F(x(u))$. By the chain rule, $G'(u) = F'(x(u))x'(u) = f(x(u))x'(u)$. Hence, again by the fundamental theorem,

$$\int_a^b f(x(u))x'(u)\,du = \int_a^b G'(u)\,du = G(b) - G(a) = F(x(b)) - F(x(a)),$$

as required. ∎

Suppose now that we have a C^1 *function* $u \mapsto x(u)$ that is one-to-one on $[a, b]$. Thus, we must have either $dx/du \geq 0$ on $[a, b]$ or $dx/du \leq 0$ on $[a, b]$. [If dx/du is positive and then negative, the function $x = x(u)$ rises and then falls, and thus is not one-to-one; a similar statement applies if dx/du is negative and then positive.] Let I^* denote the interval $[a, b]$, and let I denote the closed interval with endpoints $x(a)$ and $x(b)$. (Thus, $I = [x(a), x(b)]$ if $u \mapsto x(u)$ is increasing and $I = [x(b), x(a)]$ if $u \mapsto x(u)$ is decreasing.) With these conventions we can rewrite formula (5) as

$$\int_{I^*} f(x(u)) \left| \frac{dx}{du} \right| du = \int_I f(x)\,dx.$$

This formula generalizes to double integrals: I^* becomes D^*, I becomes D, and $|dx/du|$ is replaced by $|\partial(x, y)/\partial(u, v)|$. Let us state the result formally (the technical proof is omitted).

THEOREM 2: CHANGE OF VARIABLES: DOUBLE INTEGRALS

Let D and D^* be elementary regions in the plane and let $T: D^* \to D$ be of class C^1; suppose that T is one-to-one on D^*. Furthermore, suppose that $D = T(D^*)$. Then for any integrable function $f: D \to \mathbb{R}$, we have

$$\iint_D f(x, y)\,dx\,dy = \iint_{D^*} f(x(u, v), y(u, v)) \left| \frac{\partial(x, y)}{\partial(u, v)} \right| du\,dv. \qquad (6)$$

One of the purposes of the change of variables theorem is to supply a method by which some double integrals can be simplified. One might encounter an integral $\iint_D f\,dA$ for which either the integrand f or the region D is complicated and for which direct evaluation is difficult. Therefore, a mapping T is chosen so that the integral is easier to evaluate with the new integrand $f \circ T$ and with the new region D^* [defined by $T(D^*) = D$]. Unfortunately, the problem may actually become more complicated if T is not selected carefully.

EXAMPLE 3 Let P be the parallelogram bounded by $y = 2x$, $y = 2x - 2$, $y = x$, and $y = x+1$ (see Figure 6.2.6). Evaluate $\iint_P xy \, dx \, dy$ by making the change of variables

$$x = u - v, \qquad y = 2u - v,$$

that is, $T(u, v) = (u - v, 2u - v)$.

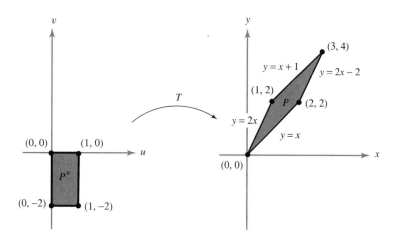

Figure 6.2.6 The effect of $T(u, v) = (u - v, 2u - v)$ on the rectangle P^*.

SOLUTION The transformation T is one-to-one (see Exercise 8, Section 6.1) and is designed so that it takes the *rectangle P^** bounded by $v = 0$, $v = -2$, $u = 0$, $u = 1$ onto P. The use of T simplifies the region of integration from P to P^*. Moreover,

$$\left| \frac{\partial(x, y)}{\partial(u, v)} \right| = \left| \det \begin{bmatrix} 1 & -1 \\ 2 & -1 \end{bmatrix} \right| = 1.$$

Therefore, by the change of variables formula,

$$\iint_P xy \, dx \, dy = \iint_{P^*} (u - v)(2u - v) \, du \, dv = \int_{-2}^0 \int_0^1 (2u^2 - 3vu + v^2) \, du \, dv$$

$$= \int_{-2}^0 \left[\frac{2}{3}u^3 - \frac{3u^2 v}{2} + v^2 u \right]_0^1 dv = \int_{-2}^0 \left[\frac{2}{3} - \frac{3}{2}v + v^2 \right] dv$$

$$= \left[\frac{2}{3}v - \frac{3}{4}v^2 + \frac{v^3}{3} \right]_{-2}^0 = -\left[\frac{2}{3}(-2) - 3 - \frac{8}{3} \right] = -\left[-\frac{12}{3} - 3 \right] = 7. \quad \blacktriangle$$

EXAMPLE 4 Evaluate $\iint_D \log(x^2 + y^2) \, dx \, dy$, where D is the region in the first quadrant lying between the arcs of the circles $x^2 + y^2 = a^2$ and $x^2 + y^2 = b^2$, where $0 < a < b$ (Figure 6.2.7).

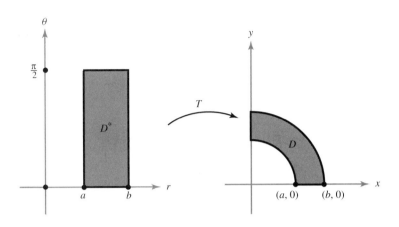

Figure 6.2.7 The polar-coordinate mapping takes a rectangle D^* onto part of an annulus D.

SOLUTION These circles have the simple equations $r = a$ and $r = b$ in polar coordinates. Moreover, $r^2 = x^2 + y^2$ appears in the integrand. Thus, a change to polar coordinates will simplify both the integrand and the region of integration. From Example 7, Section 6.1, the polar-coordinate transformation

$$x = r \cos \theta, \qquad y = r \sin \theta$$

sends the rectangle D^* given by $a \le r \le b, 0 \le \theta \le \frac{1}{2}\pi$ onto the region D. This transformation is one-to-one on D^* and so, by Theorem 2, we have

$$\iint_D \log (x^2 + y^2)\, dx\, dy = \iint_{D^*} \log r^2 \left| \frac{\partial(x, y)}{\partial(r, \theta)} \right| dr\, d\theta.$$

Now $|\partial(x, y)/\partial(r, \theta)| = r$, as we have seen before; hence the right-hand integral becomes

$$\int_a^b \int_0^{\pi/2} r \log r^2\, d\theta\, dr = \frac{\pi}{2} \int_a^b r \log r^2\, dr = \frac{\pi}{2} \int_a^b 2r \log r\, dr.$$

Applying integration by parts, or using the formula

$$\int x \log x\, dx = \frac{x^2}{2} \log x - \frac{x^2}{4}$$

from the table of integrals at the back of the book, we obtain the result

$$\frac{\pi}{2} \int_a^b 2r \log r\, dr = \frac{\pi}{2} \left[b^2 \log b - a^2 \log a - \frac{1}{2}(b^2 - a^2) \right]. \quad \blacktriangle$$

Suppose we consider the rectangle D^* defined by $0 \le \theta \le 2\pi, 0 \le r \le a$ in the $r\theta$ plane. The transformation T given by $T(r, \theta) = (r \cos \theta, r \sin \theta)$ takes D^* onto the disk D with equation $x^2 + y^2 \le a^2$ in the xy plane. This transformation represents the change from Cartesian coordinates to polar coordinates. However, T does not satisfy the requirements of the change of variables theorem, since it is not one-to-one on D^*: in particular, T sends all points with $r = 0$ to $(0, 0)$ (see Figure 6.2.8 and Example 3 of Section 6.1). Nevertheless, the change of variables theorem is valid in this case. Basically, the reason for this is that the set of points where T is not one-to-one lies on an edge of D^*, which is the graph of a smooth curve and therefore, for the purpose of integration, can be neglected. In summary, the formula

---

CHANGE OF VARIABLES—POLAR COORDINATES

$$\iint_D f(x, y) \, dx \, dy = \iint_{D^*} f(r \cos \theta, r \sin \theta) r \, dr \, d\theta \qquad (7)$$

---

is valid when T sends D^* onto D in a one-to-one fashion except possibly for points on the boundary of D^*.

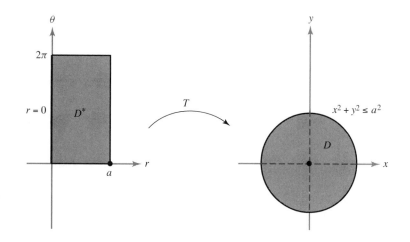

Figure 6.2.8 The image of the rectangle D^* under the polar coordinate transformation is the disk D.

Now we shall consider a more challenging example.

EXAMPLE 5 Evaluate $\iint_R \sqrt{x^2 + y^2} \, dx \, dy$, where $R = [0, 1] \times [0, 1]$.

SOLUTION This double integral is equal to the volume of the three-dimensional region shown in Figure 6.2.9.

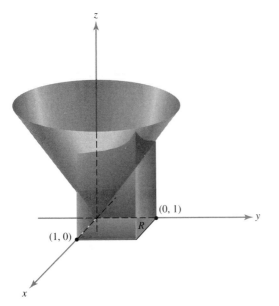

Figure 6.2.9   Volume of the region under $z = \sqrt{x^2 + y^2}$ and over $R = [0, 1] \times [0, 1]$.

As it stands, this integral is difficult to evaluate. Since the integrand is a simple function of $r^2 = x^2 + y^2$, we again try a change of variables to polar coordinates. This will result in a simplification of the integrand but, unfortunately, not in the domain of integration. However, the simplification is sufficient to enable us to evaluate the integral. To apply Theorem 2 with polar coordinates, refer to Figure 6.2.10.

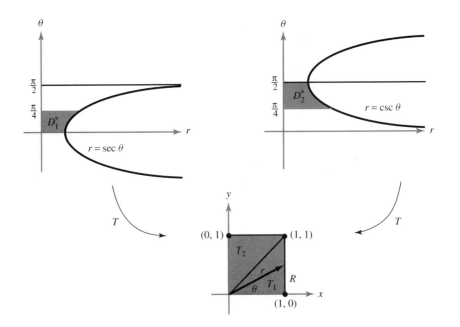

Figure 6.2.10   The polar-coordinate transformation takes D_1^* to the triangle T_1 and D_2^* to T_2.

The reader can verify that R is the image under $T(r, \theta) = (r \cos \theta, r \sin \theta)$ of the region $D^* = D_1^* \cup D_2^*$ where for D_1^* we have $0 \le \theta \le \frac{1}{4}\pi$ and $0 \le r \le \sec\theta$; for D_2^* we have $\frac{1}{4}\pi \le \theta \le \frac{1}{2}\pi, 0 \le r \le \csc\theta$. The transformation T sends D_1^* onto a triangle T_1 and D_2^* onto a triangle T_2. The transformation T is one-to-one except when $r = 0$, and so we can apply Theorem 2. From the symmetry of $z = \sqrt{x^2 + y^2}$ on R, we can see that

$$\iint_R \sqrt{x^2 + y^2}\, dx\, dy = 2 \iint_{T_1} \sqrt{x^2 + y^2}\, dx\, dy.$$

Applying formula (7), we obtain

$$\iint_{T_1} \sqrt{x^2 + y^2}\, dx\, dy = \iint_{D_1^*} \sqrt{r^2}\, r\, dr\, d\theta = \iint_{D_1^*} r^2\, dr\, d\theta.$$

Next we use iterated integration to obtain

$$\iint_{D_1^*} r^2\, dr\, d\theta = \int_0^{\pi/4} \left[\int_0^{\sec\theta} r^2\, dr \right] d\theta = \frac{1}{3} \int_0^{\pi/4} \sec^3\theta\, d\theta.$$

Consulting a table of integrals (see the back of the book) to find $\int \sec^3 x\, dx$, we have

$$\int_0^{\pi/4} \sec^3\theta\, d\theta = \left[\frac{\sec\theta \tan\theta}{2} \right]_0^{\pi/4} + \frac{1}{2}\int_0^{\pi/4} \sec\theta\, d\theta = \frac{\sqrt{2}}{2} + \frac{1}{2}\int_0^{\pi/4} \sec\theta\, d\theta.$$

Consulting the table again for $\int \sec x\, dx$, we find

$$\frac{1}{2}\int_0^{\pi/4} \sec\theta\, d\theta = \frac{1}{2}[\log|\sec\theta + \tan\theta|]_0^{\pi/4} = \frac{1}{2}\log(1 + \sqrt{2}).$$

Combining these results and recalling the factor $\frac{1}{3}$, we obtain

$$\iint_{D_1^*} r^2\, dr\, d\theta = \frac{1}{3}\left[\frac{\sqrt{2}}{2} + \frac{1}{2}\log(1 + \sqrt{2}) \right] = \frac{1}{6}[\sqrt{2} + \log(1 + \sqrt{2})].$$

Multiplying by 2, we obtain the answer

$$\iint_R \sqrt{x^2 + y^2}\, dx\, dy = \frac{1}{3}[\sqrt{2} + \log(1 + \sqrt{2})]. \quad \blacktriangle$$

There is also a change of variables formula for triple integrals. To state it, we first define the Jacobian of a transformation from $\mathbb{R}^3$ to $\mathbb{R}^3$—it is a simple extension of the two-variable case.

DEFINITION Let $T\colon W \subset \mathbb{R}^3 \to \mathbb{R}^3$ be a C^1 function defined by the equations $x = x(u, v, w), y = y(u, v, w), z = z(u, v, w)$. Then the **Jacobian** of T, which is denoted $\partial(x, y, z)/\partial(u, v, w)$, is the determinant

$$
\begin{vmatrix}
\dfrac{\partial x}{\partial u} & \dfrac{\partial x}{\partial v} & \dfrac{\partial x}{\partial w} \\[2ex]
\dfrac{\partial y}{\partial u} & \dfrac{\partial y}{\partial v} & \dfrac{\partial y}{\partial w} \\[2ex]
\dfrac{\partial z}{\partial u} & \dfrac{\partial z}{\partial v} & \dfrac{\partial z}{\partial w}
\end{vmatrix}.
$$

The absolute value of this determinant is equal to the volume of the parallelepiped determined by the three vectors

$$
\mathbf{T}_u = \frac{\partial x}{\partial u}\mathbf{i} + \frac{\partial y}{\partial u}\mathbf{j} + \frac{\partial z}{\partial u}\mathbf{k},
$$

$$
\mathbf{T}_v = \frac{\partial x}{\partial v}\mathbf{i} + \frac{\partial y}{\partial v}\mathbf{j} + \frac{\partial z}{\partial v}\mathbf{k},
$$

$$
\mathbf{T}_w = \frac{\partial x}{\partial w}\mathbf{i} + \frac{\partial y}{\partial w}\mathbf{j} + \frac{\partial z}{\partial w}\mathbf{k}.
$$

Just as in the two-variable case, the Jacobian measures how the transformation T distorts its domain. Hence for volume (triple) integrals, the change of variables formula takes the following form:

CHANGE OF VARIABLES FORMULA: TRIPLE INTEGRALS

$$
\iiint_D f(x, y, z)\, dx\, dy\, dz
$$

$$
= \iiint_{D^*} f(x(u, v, w), y(u, v, w), z(u, v, w)) \left| \frac{\partial(x, y, z)}{\partial(u, v, w)} \right| du\, dv\, dw, \quad (8)
$$

where D^* is an elementary region in uvw space corresponding to D in xyz space, under a mapping $T: (u, v, w) \mapsto (x(u, v, w), y(u, v, w), z(u, v, w))$, provided T is of class C^1 and one-to-one except possibly on a set that is the union of graphs of functions of two variables.

Let us apply formula (8) to cylindrical and spherical coordinates. First we compute the Jacobian for the map defining the change to cylindrical coordinates. Since

$$
x = r\cos\theta, \quad y = r\sin\theta, \quad z = z,
$$

we have

$$
\frac{\partial(x, y, z)}{\partial(r, \theta, z)} =
\begin{vmatrix}
\cos\theta & -r\sin\theta & 0 \\
\sin\theta & r\cos\theta & 0 \\
0 & 0 & 1
\end{vmatrix} = r.
$$

Thus, we obtain the formula

CHANGE OF VARIABLES—CYLINDRICAL COORDINATES

$$\iiint_D f(x, y, z)\, dx\, dy\, dz = \iiint_{D^*} f(r \cos\theta, r \sin\theta, z) r\, dr\, d\theta\, dz. \qquad (9)$$

Next we consider the spherical coordinate system. Recall that it is given by

$$x = \rho \sin\phi \cos\theta, \qquad y = \rho \sin\phi \sin\theta, \qquad z = \rho \cos\phi.$$

Therefore we have

$$\frac{\partial(x, y, z)}{\partial(\rho, \theta, \phi)} = \begin{vmatrix} \sin\phi \cos\theta & -\rho \sin\phi \sin\theta & \rho \cos\phi \cos\theta \\ \sin\phi \sin\theta & \rho \sin\phi \cos\theta & \rho \cos\phi \sin\theta \\ \cos\phi & 0 & -\rho \sin\phi \end{vmatrix}$$

Expanding along the last row, we get

$$\begin{aligned} \frac{\partial(x, y, z)}{\partial(\rho, \theta, \phi)} &= \cos\phi \begin{vmatrix} -\rho \sin\phi \sin\theta & \rho \cos\phi \cos\theta \\ \rho \sin\phi \cos\theta & \rho \cos\phi \sin\theta \end{vmatrix} \\ &\quad - \rho \sin\phi \begin{vmatrix} \sin\phi \cos\theta & -\rho \sin\phi \sin\theta \\ \sin\phi \sin\theta & \rho \sin\phi \cos\theta \end{vmatrix} \\ &= -\rho^2 \cos^2\phi \sin\phi \sin^2\theta - \rho^2 \cos^2\phi \sin\phi \cos^2\theta \\ &\quad - \rho^2 \sin^3\phi \cos^2\theta - \rho^2 \sin^3\phi \sin^2\theta \\ &= -\rho^2 \cos^2\phi \sin\phi - \rho^2 \sin^3\phi = -\rho^2 \sin\phi. \end{aligned}$$

Thus we arrive at the formula

CHANGE OF VARIABLES—SPHERICAL COORDINATES

$$\iiint_D f(x, y, z)\, dx\, dy\, dz$$

$$= \iiint_{D^*} f(\rho \sin\phi \cos\theta, \rho \sin\phi \sin\theta, \rho \cos\phi) \rho^2 \sin\phi\, d\rho\, d\theta\, d\phi. \qquad (10)$$

To prove formula (10), one must show that the transformation S on the set D^* is one-to-one except on a set that is the union of finitely many graphs of continuous functions. We shall leave this verification as Exercise 20.

EXAMPLE 6   Evaluate $\iiint_D \exp(x^2 + y^2 + z^2)^{3/2} \, dV$, where D is the unit ball in $\mathbb{R}^3$.

SOLUTION   First note that we cannot *easily* integrate this function using iterated integrals (try it!). Hence, let us try a change of variables. The transformation into spherical coordinates seems appropriate, since then the entire quantity $x^2 + y^2 + z^2$ can be replaced by one variable, namely, ρ^2. If D^* is the region such that

$$0 \leq \rho \leq 1, \qquad 0 \leq \theta \leq 2\pi, \qquad 0 \leq \phi \leq \pi.$$

We may apply formula (10) and write

$$\iiint_D \exp(x^2 + y^2 + z^2)^{3/2} \, dV = \iiint_{D^*} \rho^2 e^{\rho^3} \sin\phi \, d\rho \, d\theta \, d\phi.$$

This integral equals the iterated integral

$$\int_0^1 \int_0^\pi \int_0^{2\pi} e^{\rho^3} \rho^2 \sin\phi \, d\theta \, d\phi \, d\rho = 2\pi \int_0^1 \int_0^\pi e^{\rho^3} \rho^2 \sin\phi \, d\phi \, d\rho$$

$$= -2\pi \int_0^1 \rho^2 e^{\rho^3} [\cos\phi]_0^\pi \, d\rho$$

$$= 4\pi \int_0^1 e^{\rho^3} \rho^2 \, d\rho = \frac{4}{3}\pi \int_0^1 e^{\rho^3} (3\rho^2) \, d\rho$$

$$= \left[\frac{4}{3}\pi e^{\rho^3} \right]_0^1 = \frac{4}{3}\pi(e - 1). \quad \blacktriangle$$

EXAMPLE 7   Let D be the ball of radius R and center $(0, 0, 0)$ in $\mathbb{R}^3$. Find the volume of D.

SOLUTION   The volume of D is $\iiint_D dx \, dy \, dz$. This integral may be evaluated by reducing it to iterated integrals or by regarding D as a volume of revolution, but let us evaluate it here by using spherical coordinates. We get

$$\iiint_D dx \, dy \, dz = \int_0^\pi \int_0^{2\pi} \int_0^R \rho^2 \sin\phi \, d\rho \, d\theta \, d\phi = \frac{R^3}{3} \int_0^\pi \int_0^{2\pi} \sin\phi \, d\theta \, d\phi$$

$$= \frac{2\pi R^3}{3} \int_0^\pi \sin\phi \, d\phi = \frac{2\pi R^3}{3} \{-[\cos(\pi) - \cos(0)]\} = \frac{4\pi R^3}{3},$$

which is the standard formula for the volume of a solid sphere. $\blacktriangle$

EXERCISES

1.   Let D be the unit disk: $x^2 + y^2 \leq 1$. Evaluate

$$\iint_D \exp(x^2 + y^2) \, dx \, dy$$

by making a change of variables to polar coordinates.

2. Let D be the region $0 \leq y \leq x$ and $0 \leq x \leq 1$. Evaluate

$$\iint_D (x + y) \, dx \, dy$$

by making the change of variables $x = u + v$, $y = u - v$. Check your answer by evaluating the integral directly by using an iterated integral.

3. Let $T(u, v) = (x(u, v), y(u, v))$ be the mapping defined by $T(u, v) = (4u, 2u + 3v)$. Let D^* be the rectangle $[0, 1] \times [1, 2]$. Find $D = T(D^*)$ and evaluate

(a) $\iint_D xy \, dx \, dy$ (b) $\iint_D (x - y) \, dx \, dy$

by making a change of variables to evaluate them as integrals over D^*.

4. Repeat Exercise 3 for $T(u, v) = (u, v(1 + u))$.

5. Evaluate

$$\iint_D \frac{dx \, dy}{\sqrt{1 + x + 2y}},$$

where $D = [0, 1] \times [0, 1]$, by setting $T(u, v) = (u, v/2)$ and evaluating an integral over D^*, where $T(D^*) = D$.

6. Define $T(u, v) = (u^2 - v^2, 2uv)$. Let D^* be the set of (u, v) with $u^2 + v^2 \leq 1, u \geq 0, v \geq 0$. Find $T(D^*) = D$. Evaluate $\iint_D dx \, dy$.

7. Let $T(u, v)$ be as in Exercise 6. By making a change of variables, "formally" evaluate the "improper" integral

$$\iint_D \frac{dx \, dy}{\sqrt{x^2 + y^2}}.$$

[*Note*: This integral (and the one in the next exercise) is *improper*, since the integrand $1/\sqrt{x^2 + y^2}$ is neither continuous nor bounded on the domain of integration. (The theory of improper integrals is discussed in Section 6.4.)]

8. Calculate $\iint_R \frac{1}{x+y} \, dy \, dx$, where R is the region bounded by $x = 0, y = 0, x + y = 1, x + y = 4$, by using the mapping $T(u, v) = (u - uv, uv)$.

9. Evaluate $\iint_D (x^2 + y^2)^{3/2} \, dx \, dy$ where D is the disk $x^2 + y^2 \leq 4$.

10. Let D^* be a region of type 1 in the uv plane bounded by $v = g(u)$ and $v = h(u) \leq g(u)$ for $a \leq u \leq b$. Let $T: \mathbb{R}^2 \to \mathbb{R}^2$ be the transformation given by $x = u$ and $y = \psi(u, v)$, where ψ is of class C^1 and $\partial\psi/\partial v$ is never zero. Assume that $T(D^*) = D$ is a region of type 1; show that if $f: D \to \mathbb{R}$ is continuous, then

$$\iint_D f(x, y) \, dx \, dy = \iint_{D^*} f(u, \psi(u, v)) \left| \frac{\partial\psi}{\partial v} \right| du \, dv.$$

11. Use double integrals to find the area inside the curve $r = 1 + \sin\theta$.

12. (a) Express $\int_0^1 \int_0^{x^2} xy\,dy\,dx$ as an integral over the triangle D^*, which is the set of (u, v) where $0 \le u \le 1, 0 \le v \le u$. (HINT: Find a one-to-one mapping T of D^* onto the given region of integration.)
(b) Evaluate this integral directly and as an integral over D^*.

13. Integrate $ze^{x^2+y^2}$ over the cylinder $x^2 + y^2 \le 4, 2 \le z \le 3$.

14. Let D be the unit disk. Express $\iint_D (1 + x^2 + y^2)^{3/2}\,dx\,dy$ as an integral over $[0, 1] \times [0, 2\pi]$ and evaluate.

15. Using polar coordinates, find the area bounded by the *lemniscate* $(x^2 + y^2)^2 = 2a^2 (x^2 - y^2)$.

16. Redo Exercise 11 of Section 5.3 using a change of variables and compare the effort involved in each method.

17. Calculate $\iint_R (x+y)^2 e^{x-y}\,dx\,dy$ where R is the region bounded by $x+y = 1, x+y = 4, x-y = -1$, and $x - y = 1$.

18. Let $T: \mathbb{R}^3 \to \mathbb{R}^3$ be defined by

$$T(u, v, w) = (u\cos v\cos w, u\sin v\cos w, u\sin w).$$

(a) Show that T is onto the unit sphere; that is, every (x, y, z) with $x^2 + y^2 + z^2 = 1$ can be written as $(x, y, z) = T(u, v, w)$ for some (u, v, w).
(b) Show that T is not one-to-one.

19. Integrate $x^2 + y^2 + z^2$ over the cylinder $x^2 + y^2 \le 2, -2 \le z \le 3$.

20. Show that $S(\rho, \theta, \phi) = (\rho\sin\phi\cos\theta, \rho\sin\phi\sin\theta, \rho\cos\phi)$, the spherical change-of-coordinate mapping, is one-to-one except on a set that is a union of finitely many graphs of continuous functions.

21. Let B be the unit ball. Evaluate

$$\iiint_B \frac{dx\,dy\,dz}{\sqrt{2 + x^2 + y^2 + z^2}}$$

by making the appropriate change of variables.

22. Evaluate $\iint_A [1/(x^2 + y^2)^2]\,dx\,dy$ where A is determined by the conditions $x^2 + y^2 \le 1$ and $x + y \ge 1$.

23. Evaluate $\iiint_S \frac{dx\,dy\,dz}{(x^2 + y^2 + z^2)^{3/2}}$, where S is the solid bounded by the two spheres $x^2+y^2+z^2 = a^2$ and $x^2 + y^2 + z^2 = b^2$, where $0 < b < a$.

24. Evaluate $\iint_D x^2\,dx\,dy$ where D is determined by the two conditions $0 \le x \le y$ and $x^2 + y^2 \le 1$.

25. Integrate $\sqrt{x^2 + y^2 + z^2}\,e^{-(x^2+y^2+z^2)}$ over the region described in Exercise 23.

26. Evaluate the following by using cylindrical coordinates.

(a) $\iiint_B z \, dx \, dy \, dz$ where B is the region within the cylinder $x^2 + y^2 = 1$ above the xy plane and below the cone $z = (x^2 + y^2)^{1/2}$.

(b) $\iiint_D (x^2 + y^2 + z^2)^{-1/2} \, dx \, dy \, dz$ where D is the region determined by the conditions $\frac{1}{2} \leq z \leq 1$ and $x^2 + y^2 + z^2 \leq 1$.

27. Evaluate $\iint_B (x + y) \, dx \, dy$ where B is the rectangle in the xy plane with vertices at $(0, 1)$, $(1, 0)$, $(3, 4)$, and $(4, 3)$.

28. Evaluate $\iint_D (x + y) \, dx \, dy$ where D is the square with vertices at $(0, 0)$, $(1, 2)$, $(3, 1)$, and $(2, -1)$.

29. Let E be the ellipsoid $(x^2/a^2) + (y^2/b^2) + (z^2/c^2) \leq 1$, where a, b, and c are positive.

(a) Find the volume of E.

(b) Evaluate $\iiint_E [(x^2/a^2) + (y^2/b^2) + (z^2/c^2)] \, dx \, dy \, dz$. (HINT: Change variables and then use spherical coordinates.)

30. Using spherical coordinates, compute the integral of $f(\rho, \phi, \theta) = 1/\rho$ over the region in the first octant of $\mathbb{R}^3$ which is bounded by the cones $\phi = \pi/4$, $\phi = \arctan 2$ and the sphere $\rho = \sqrt{6}$.

31. The mapping $T(u, v) = (u^2 - v^2, 2uv)$ transforms the rectangle $1 \leq u \leq 2$, $1 \leq v \leq 3$ of the uv plane into a region R of the xy plane.

(a) Show that T is one-to-one.

(b) Find the area of R using the change of variables formula.

32. Let R denote the region inside $x^2 + y^2 = 1$ but outside $x^2 + y^2 = 2y$ with $x \geq 0$, $y \geq 0$.

(a) Sketch this region.

(b) Let $u = x^2 + y^2$, $v = x^2 + y^2 - 2y$. Sketch the region D in the uv plane which corresponds to R under this change of coordinates.

(c) Compute $\iint_R x e^y \, dx \, dy$ using this change of coordinates.

33. Let D be the region bounded by $x^{3/2} + y^{3/2} = a^{3/2}$, for $x \geq 0$, $y \geq 0$, and the coordinate axes $x = 0$, $y = 0$. Express $\iint_D f(x, y) \, dx \, dy$ as an integral over the triangle D^*, which is the set of points $0 \leq u \leq a$, $0 \leq v \leq a - u$. (Do not attempt to evaluate.)

6.3 APPLICATIONS OF DOUBLE AND TRIPLE INTEGRALS

In this section we shall discuss average values, centers of mass, moments of inertia, and the gravitational potential as applications.

If $x_1, \ldots, x_n$ are n numbers, their **average** is defined by

$$[x_i]_{\text{av}} = \frac{x_1 + \cdots + x_n}{n} = \frac{1}{n} \sum_{i=1}^{n} x_i.$$

This concept leads one to define average values more generally.

AVERAGE VALUES The *average value* of a function of one variable on the interval $[a, b]$ is defined by

$$[f]_{av} = \frac{\int_a^b f(x)\,dx}{b-a}.$$

Likewise, for functions of two variables, the ratio of the integral to the area of D,

$$[f]_{av} = \frac{\iint_D f(x, y)\,dx\,dy}{\iint_D dx\,dy}, \tag{1}$$

is called the *average value* of f over D. Similarly, the *average value* of a function f on a region W in three space is defined by

$$[f]_{av} = \frac{\iiint_W f(x, y, z)\,dx\,dy\,dz}{\iiint_W dx\,dy\,dz}.$$

EXAMPLE 1 Find the average value of $f(x, y) = x\sin^2(xy)$ on the region $D = [0, \pi] \times [0, \pi]$.

SOLUTION First we compute

$$\iint_D f(x, y)\,dx\,dy = \int_0^\pi \int_0^\pi x\sin^2(xy)\,dx\,dy$$

$$= \int_0^\pi \left[\int_0^\pi \frac{1 - \cos(2xy)}{2} x\,dy \right] dx = \int_0^\pi \left[\frac{y}{2} - \frac{\sin(2xy)}{4x} \right] x \Big|_{y=0}^\pi dx$$

$$= \int_0^\pi \left[\frac{\pi x}{2} - \frac{\sin(2\pi x)}{4} \right] dx = \left[\frac{\pi x^2}{4} + \frac{\cos(2\pi x)}{8\pi} \right] \Big|_0^\pi$$

$$= \frac{\pi^3}{4} + \frac{\cos(2\pi^2) - 1}{8\pi}.$$

Thus the average value of f, by formula (1), is

$$\frac{\pi^3/4 + [\cos(2\pi^2) - 1]/8\pi}{\pi^2} = \frac{\pi}{4} + \frac{\cos(2\pi^2) - 1}{8\pi^3} \approx 0.7839. \quad \blacktriangle$$

If masses $m_1, \ldots, m_n$ are placed at points $x_1, \ldots, x_n$ on the x axis, their *center of mass* is defined to be

$$\bar{x} = \frac{\sum m_i x_i}{\sum m_i}. \tag{2}$$

This definition arises from the following observation: If one is balancing masses on a lever (Figure 6.3.1), the balance point $\bar{x}$ occurs where the total moment (mass times distance from the balance point) is zero, that is, where $\sum m_i(x_i - \bar{x}) = 0$. A physical principle, going back first to Archimedes and then in this generality to Newton, states that this condition means there is no tendency for the lever to rotate.

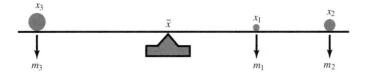

Figure 6.3.1   The lever is balanced if $\sum (x_i - \bar{x}) m_i = 0$.

For a continuous mass density $\rho(x)$ along the lever, the analog of formula (2) is

$$\bar{x} = \frac{\int x\rho(x)\,dx}{\int \rho(x)\,dx}. \tag{3}$$

For two-dimensional plates, this generalizes (see Figure 6.3.2) to:

Figure 6.3.2   The plate balances when supported at its center of mass.

THE CENTER OF MASS OF TWO-DIMENSIONAL PLATES

$$\bar{x} = \frac{\iint_D x\rho(x, y)\,dx\,dy}{\iint_D \rho(x, y)\,dx\,dy} \quad \text{and} \quad \bar{y} = \frac{\iint_D y\rho(x, y)\,dx\,dy}{\iint_D \rho(x, y)\,dx\,dy}, \tag{4}$$

where again $\rho(x, y)$ is the mass density.

EXAMPLE 2   Find the center of mass of the rectangle $[0, 1] \times [0, 1]$ if the mass density is e^{x+y}.

SOLUTION   First we compute the total mass:

$$\iint_D e^{x+y}\,dx\,dy = \int_0^1 \int_0^1 e^{x+y}\,dx\,dy = \int_0^1 \left(e^{x+y}\big|_{x=0}^1 \right) dy = \int_0^1 (e^{1+y} - e^y)\,dy$$
$$= (e^{1+y} - e^y)\big|_{y=0}^1 = e^2 - e - (e - 1) = e^2 - 2e + 1.$$

The numerator in formula (4) for $\bar{x}$ is

$$\int_0^1 \int_0^1 xe^{x+y}\,dx\,dy = \int_0^1 (xe^{x+y} - e^{x+y})|_{x=0}^1\,dy = \int_0^1 [e^{1+y} - e^{1+y} - (0e^y - e^y)]\,dy$$

$$= \int_0^1 e^y\,dy = e^y|_{y=0}^1 = e - 1,$$

so that

$$\bar{x} = \frac{e-1}{e^2 - 2e + 1} = \frac{e-1}{(e-1)^2} = \frac{1}{e-1} \approx 0.582.$$

The roles of x and y may be interchanged in all these calculations, so that $\bar{y} = 1/(e-1) \approx 0.582$ as well. ▲

For a region W in space with mass density $\rho(x, y, z)$ we know that

$$\text{volume} = \iiint_W dx\,dy\,dz, \tag{5}$$

$$\text{mass} = \iiint_W \rho(x, y, z)\,dx\,dy\,dz. \tag{6}$$

If one denotes the coordinates of the center of mass by $(\bar{x}, \bar{y}, \bar{z})$, then one can generalize the formulas in the preceding box.

COORDINATES FOR THE CENTER OF MASS OF THREE-DIMENSIONAL REGIONS

$$\bar{x} = \frac{\iiint_W x\rho(x, y, z)\,dx\,dy\,dz}{\text{mass}},$$

$$\bar{y} = \frac{\iiint_W y\rho(x, y, z)\,dx\,dy\,dz}{\text{mass}}, \tag{7}$$

$$\bar{z} = \frac{\iiint_W z\rho(x, y, z)\,dx\,dy\,dz}{\text{mass}}.$$

EXAMPLE 3   The cube $[1, 2] \times [1, 2] \times [1, 2]$ has mass density given by $\rho(x, y, z) = (1 + x)e^z y$. Find the total mass of the box.

SOLUTION   The mass of the box is, by formula (6),

$$\int_1^2 \int_1^2 \int_1^2 (1 + x)e^z y\,dx\,dy\,dz = \int_1^2 \int_1^2 \left[\left(x + \frac{x^2}{2}\right)e^z y\right]_{x=1}^{x=2}\,dy\,dz$$

$$= \int_1^2 \int_1^2 \frac{5}{2}e^z y\,dy\,dz = \int_1^2 \frac{15}{4}e^z\,dz = \left[\frac{15}{4}e^z\right]_{z=1}^{z=2} = \frac{15}{4}(e^2 - e). ▲$$

If a region and its mass density are reflection-symmetric across a plane, then the center of mass lies on that plane. For example, in formula (7) for $\bar{x}$, if the region and mass density are symmetric in the yz plane, then the integrand is odd in x, and so $\bar{x} = 0$. This sort of use of symmetry is illustrated in the next example.

EXAMPLE 4   Find the center of mass of the hemispherical region W defined by the inequalities $x^2 + y^2 + z^2 \leq 1, z \geq 0$. (Assume that the density is constant.)

SOLUTION   By symmetry, the center of mass must lie on the z axis, and so $\bar{x} = \bar{y} = 0$. To find $\bar{z}$, we must compute, by formula (7), $I = \iiint_W z\, dx\, dy\, dz$. The hemisphere is an elementary region and thus the integral becomes

$$I = \int_0^1 \int_{-\sqrt{1-z^2}}^{\sqrt{1-z^2}} \int_{-\sqrt{1-y^2-z^2}}^{\sqrt{1-y^2-z^2}} z\, dx\, dy\, dz.$$

Since z is a constant for the x and y integrations, we can remove it from the integral signs, to obtain

$$I = \int_0^1 z \left(\int_{-\sqrt{1-z^2}}^{\sqrt{1-z^2}} \int_{-\sqrt{1-y^2-z^2}}^{\sqrt{1-y^2-z^2}} dx\, dy \right) dz.$$

Instead of calculating the inner two integrals explicitly, we observe that they equal the double integral $\iint_D dx\, dy$ over the disk $x^2 + y^2 \leq 1 - z^2$, considered as a type 2 region in the plane. The area of this disk $\pi(1 - z^2)$, and so

$$I = \pi \int_0^1 z(1 - z^2)\, dz = \pi \int_0^1 (z - z^3)\, dz = \pi \left[\frac{z^2}{2} - \frac{z^4}{4} \right]_0^1 = \frac{\pi}{4}.$$

The volume of the hemisphere is $\frac{2}{3}\pi$, and so $\bar{z} = (\pi/4)/(\frac{2}{3}\pi) = \frac{3}{8}$.   ▲

EXAMPLE 5   The temperature at points in the cube $W = [-1, 1] \times [-1, 1] \times [-1, 1]$ is proportional to the square of the distance from the origin.

(a)   What is the average temperature?

(b)   At which points of the cube is the temperature equal to the average temperature?

SOLUTION   (a) Let c be the constant of proportionality so $T = c(x^2 + y^2 + z^2)$ and the average temperature is $[T]_{av} = \frac{1}{8} \iiint_W T\, dx\, dy\, dz$, since the volume of the cube is 8. Thus,

$$[T]_{av} = \frac{c}{8} \int_{-1}^1 \int_{-1}^1 \int_{-1}^1 (x^2 + y^2 + z^2)\, dx\, dy\, dz.$$

The triple integral is the sum of the integrals of x^2, y^2, and z^2. Since x, y, and z enter symmetrically into the description of the cube, the three integrals will be equal, so that

$$[T]_{\text{av}} = \frac{3c}{8} \int_{-1}^{1}\int_{-1}^{1}\int_{-1}^{1} z^2\, dx\, dy\, dz = \frac{3c}{8}\int_{-1}^{1} z^2\left(\int_{-1}^{1}\int_{-1}^{1} dx\, dy\right) dz.$$

The inner integral is equal to the area of the square $[-1,1]\times[-1,1]$. The area of that square is 4, and so

$$[T]_{\text{av}} = \frac{3c}{8}\int_{-1}^{1} 4z^2\, dz = \frac{3c}{2}\left(\frac{z^3}{3}\right)\Big|_{-1}^{1} = c.$$

(b) The temperature is equal to the average temperature when $c(x^2 + y^2 + z^2) = c$, that is, on the sphere $x^2 + y^2 + z^2 = 1$, which is inscribed in the cube W. ▲

Another important concept in mechanics, one that is needed in studying the dynamics of a rotating rigid body, is that of *moment of inertia*. If the solid W has uniform density ρ, the **moments of inertia** I_x, I_y, and I_z about the x, y, and z axes, respectively, are defined by:

MOMENTS OF INERTIA ABOUT THE COORDINATE AXES

$$I_x = \iiint_W \rho(y^2 + z^2)\, dx\, dy\, dz, \quad I_y = \iiint_W \rho(x^2 + z^2)\, dx\, dy\, dz,$$
$$I_z = \iiint_W \rho(x^2 + y^2)\, dx\, dy\, dz. \tag{8}$$

The moment of inertia measures a body's response to efforts to spin it; it is analogous to the mass of a body, which measures its response to efforts to translate it. In contrast to translational motion, however, the moments of inertia depend on the shape and not just the total mass. (It is harder to spin up a large plate than a compact ball of the same mass.)

EXAMPLE 6   Compute the moment of inertia I_z for the solid above the xy plane bounded by the paraboloid $z = x^2 + y^2$ and the cylinder $x^2 + y^2 = a^2$, assuming a and the mass density to be constants.

SOLUTION   The paraboloid and cylinder intersect at the plane $z = a^2$. Using cylindrical coordinates, we find from (8),

$$I_z = \int_0^a\int_0^{2\pi}\int_0^{r^2} \rho r^2 \cdot r\, dz\, d\theta\, dr = \rho \int_0^a\int_0^{2\pi}\int_0^{r^2} r^3\, dz\, d\theta\, dr = \frac{\pi\rho a^6}{3}. \quad ▲$$

Another interesting physical application of triple integration is the determination of the gravitational fields of solid objects. Example 6, Section 2.6, showed that the gravitational force field $\mathbf{F}(x, y, z)$ of a particle is the negative of the gradient of a function $V(x, y, z)$ called the ***gravitational potential***. If there is a point mass M at (x, y, z), then the gravitational potential acting on a mass m at (x_1, y_1, z_1) due to this mass is $GmM[(x - x_1)^2 + (y - y_1)^2 + (z - z_1)^2]^{-1/2}$, where G is the universal gravitational constant.

If our attracting object occupies a domain W with mass density $\rho(x, y, z)$, we may think of it as made of infinitesimal box-shaped regions with masses $dM = \rho(x, y, z)\, dx\, dy\, dz$ located at points (x, y, z). The total gravitational potential V for W is then obtained by "summing" the potentials from the infinitesimal masses—that is, as a triple integral (see Figure 6.3.3):

$$V(x_1, y_1, z_1) = Gm \iiint_W \frac{\rho(x, y, z)\, dx\, dy\, dz}{\sqrt{(x - x_1)^2 + (y - y_1)^2 + (z - z_1)^2}}. \qquad (9)$$

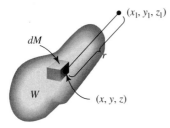

Figure 6.3.3 The gravitational potential acting on a mass m at (x_1, y_1, z_1) arising from the mass $dM = \rho(x, y, z)\, dx\, dy\, dz$ at (x, y, z) is $[Gm\rho(x, y, z)\, dx\, dy\, dz]/r$.

HISTORICAL NOTE

The theory of gravitational force fields and gravitational potentials was developed by Sir Isaac Newton (1642–1727). Newton withheld publication of his gravitational theories for quite some time. The result that a spherical planet has the same gravitational field as it would have if its mass were all concentrated at the planet's center first appeared in his famous *Philosophia Naturalis, Principia Mathematica*, the first edition of which appeared in 1687. Using multiple integrals and spherical coordinates, we shall solve Newton's problem below; remarkably, Newton's published solution used only Euclidean geometry.

EXAMPLE 7 Let W be a region of constant density and total mass M. Show that the gravitational potential is given by

$$V(x_1, y_1, z_1) = \left[\frac{1}{r}\right]_{\text{av}} GMm,$$

where $[1/r]_{av}$ is the average over W of

$$f(x, y, z) = \frac{1}{\sqrt{(x - x_1)^2 + (y - y_1)^2 + (z - z_1)^2}}.$$

SOLUTION   According to formula (9),

$$V(x_1, y_1, z_1) = Gm \iiint_W \frac{\rho \, dx \, dy \, dz}{\sqrt{(x - x_1)^2 + (y - y_1)^2 + (z - z_1)^2}}$$

$$= Gm\rho \iiint_W \frac{dx \, dy \, dz}{\sqrt{(x - x_1)^2 + (y - y_1)^2 + (z - z_1)^2}}$$

$$= Gm[\rho \text{ volume } (W)] \frac{\displaystyle\iiint_W \frac{dx \, dy \, dz}{\sqrt{(x - x_1)^2 + (y - y_1)^2 + (z - z_1)^2}}}{\text{volume } (W)}$$

$$= GmM \left[\frac{1}{r} \right]_{av},$$

are required.   ▲

Let us now use formula (9) and spherical coordinates to find the gravitational potential $V(x_1, y_1, z_1)$ for the region W between the concentric spheres $\rho = \rho_1$ and $\rho = \rho_2$, assuming the density is constant. Before evaluating the integral in formula (9), we make some observations which will simplify the computation. Since G, m, and the density are constants, we may ignore them at first. Since the attracting body W is symmetric with respect to rotations about the origin, the potential $V(x_1, y_1, z_1)$ should itself be symmetric—thus $V(x_1, y_1, z_1)$ depends only on the distance $R = \sqrt{x_1^2 + y_1^2 + z_1^2}$ from the origin. Our computation will be simplest if we look at the point $(0, 0, R)$ on the z axis (see Figure 6.3.4). Thus our integral is

$$V(0, 0, R) = \iiint_W \frac{dx \, dy \, dz}{\sqrt{x^2 + y^2 + (z - R)^2}}.$$

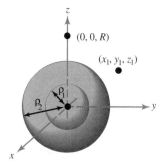

Figure 6.3.4   The gravitational potential at (x_1, y_1, z_1) is the same as at $(0, 0, R)$, where $R = \sqrt{x_1^2 + y_1^2 + z_1^2}$.

In spherical coordinates, W is described by the inequalities $\rho_1 \leq \rho \leq \rho_2, 0 \leq \theta \leq 2\pi$, and $0 \leq \phi \leq \pi$, and so

$$V(0, 0, R) = \int_{\rho_1}^{\rho_2} \int_0^\pi \int_0^{2\pi} \frac{\rho^2 \sin\phi \, d\theta \, d\phi \, d\rho}{\sqrt{\rho^2 \sin^2 \phi (\cos^2 \theta + \sin^2 \theta) + (\rho \cos\phi - R)^2}}.$$

Replacing $\cos^2 \theta + \sin^2 \theta$ by 1, so that the integrand no longer involves θ, we may integrate over θ to get

$$V(0, 0, R) = 2\pi \int_{\rho_1}^{\rho_2} \int_0^\pi \frac{\rho^2 \sin\phi \, d\phi \, d\rho}{\sqrt{\rho^2 \sin^2 \phi + (\rho \cos\phi - R)^2}}$$

$$= 2\pi \int_{\rho_1}^{\rho_2} \rho^2 \left(\int_0^\pi \frac{\sin\phi \, d\phi}{\sqrt{\rho^2 - 2R\rho \cos\phi + R^2}} \right) d\rho.$$

The inner integral is easily evaluated using the substitution $u = -2R\rho \cos\phi$. We get

$$\frac{1}{2R\rho} \int_{-2R\rho}^{2R\rho} (\rho^2 + u + R^2)^{-1/2} \, du = \frac{2}{2R\rho} (\rho^2 + u + R^2)^{1/2} \Big|_{-2R\rho}^{2R\rho}$$

$$= \frac{1}{R\rho} [(\rho^2 + 2R\rho + R^2)^{1/2} - (\rho^2 - 2R\rho + R^2)^{1/2}$$

$$= \frac{1}{R\rho} \{[(\rho + R)^2]^{1/2} - [(\rho - R)^2]^{1/2}\}$$

$$= \frac{1}{R\rho} (\rho + R - |\rho - R|).$$

The expression $\rho + R$ is always positive, but $\rho - R$ may not be, so we must keep the absolute value sign. Substituting into the formula for V, we get

$$V(0, 0, R) = 2\pi \int_{\rho_1}^{\rho_2} \frac{\rho^2}{R\rho} (\rho + R - |\rho - R|) \, d\rho = \frac{2\pi}{R} \int_{\rho_1}^{\rho_2} \rho(\rho + R - |\rho - R|) \, d\rho.$$

We will now consider two possibilities for R, corresponding to the gravitational potential for objects outside and inside the hollow ball W.

If $R \geq \rho_2$ [that is, if (x_1, y_1, z_1) is outside W], then $|\rho - R| = R - \rho$ for all ρ in the interval $[\rho_1, \rho_2]$, so that

$$V(0, 0, R) = \frac{2\pi}{R} \int_{\rho_1}^{\rho_2} \rho[\rho + R - (R - \rho)] \, d\rho = \frac{4\pi}{R} \int_{\rho_1}^{\rho_2} \rho^2 \, d\rho = \frac{1}{R} \frac{4\pi}{3} (\rho_2^3 - \rho_1^3).$$

The factor $(4\pi/3)(\rho_2^3 - \rho_1^3)$ equals the volume of W. Putting back the constants G, m, and the mass density, we find that *the gravitational potential is GmM/R, where M is the mass of W. Thus, V is just as it would be if all the mass of W were concentrated at the central point.*

If $R \leq \rho_1$ [that is, if (x_1, y_1, z_1) is inside the hole], then $|\rho - R| = \rho - R$ for ρ in $[\rho_1, \rho_2]$, and so

$$V(0, 0, R) = (Gm) \frac{2\pi}{R} \int_{\rho_1}^{\rho_2} \rho[\rho + R - (\rho - R)] \, d\rho = (Gm)4\pi \int_{\rho_1}^{\rho_2} \rho \, d\rho$$

$$= (Gm)2\pi(\rho_2^2 - \rho_1^2).$$

The result is independent of R, and so the potential V is *constant* inside the hole. Since the gravitational force is minus the gradient of V, we conclude that *there is no gravitational force inside a uniform hollow planet*!

We leave it to the reader to compute $V(0, 0, R)$ for the case $\rho_1 < R < \rho_2$.

A similar argument shows that the gravitational potential outside any spherically symmetric body of mass M (even if the density is variable) is $V = GMm/R$, where R is the distance to its center (which is its center of mass).

EXAMPLE 8   Find the gravitational potential acting on a unit mass of a spherical star with a mass $M = 3.02 \times 10^{30}$ kg at a distance of 2.25×10^{11} m from its center $(G = 6.67 \times 10^{-11}$ N $\cdot$ m^2/kg$^2)$.

SOLUTION   The potential is

$$V = \frac{GM}{R} = \frac{6.67 \times 10^{-11} \times 3.02 \times 10^{30}}{2.25 \times 10^{11}} = 8.95 \times 10^8 \text{ m}^2/\text{s}^2. \quad \blacktriangle$$

EXERCISES

1.   Find the average of $f(x, y) = y \sin xy$ over $D = [0, \pi] \times [0, \pi]$.

2.   Find the average of $f(x, y) = e^{x+y}$ over the triangle with vertices $(0, 0)$, $(0, 1)$, and $(1, 0)$.

3.   Find the center of mass of the region between $y = x^2$ and $y = x$ if the density is $x + y$.

4.   Find the center of mass of the region between $y = 0$ and $y = x^2$, where $0 \le x \le \frac{1}{2}$.

5.   A sculptured gold plate D is defined by $0 \le x \le 2\pi$ and $0 \le y \le \pi$ (centimeters) and has mass density $\rho(x, y) = y^2 \sin^2 4x + 2$ (grams per square centimeter). If gold sells for \$7 per gram, how much is the gold in the plate worth?

6.   In Exercise 5, what is the average mass density in grams per square centimeter?

7.   (a)   Find the mass of the box $[0, \frac{1}{2}] \times [0, 1] \times [0, 2]$, assuming the density to be uniform.
     (b)   Same as part (a), but with a mass density $\rho(x, y, z) = x^2 + 3y^2 + z + 1$.

8.   Find the mass of the solid bounded by the cylinder $x^2 + y^2 = 2x$ and the cone $z^2 = x^2 + y^2$ if the density is $\rho = \sqrt{x^2 + y^2}$.

9.   Find the center of mass of the region bounded by $x + y + z = 2$, $x = 0$, $y = 0$, and $z = 0$.

10.   Find the center of mass of the cylinder $x^2 + y^2 \le 1$, $1 \le z \le 2$ if the density is $\rho = (x^2 + y^2)z^2$.

11.   Find the average value of $\sin^2 \pi z \cos^2 \pi x$ over the cube $[0, 2] \times [0, 4] \times [0, 6]$.

12. Find the average value of e^{-z} over the ball $x^2 + y^2 + z^2 \leq 1$.

13. A solid with constant density is bounded above by the plane $z = a$ and below by the cone described in spherical coordinates by $\phi = k$, where k is a constant $0 < k < \pi/2$. Set up an integral for its moment of inertia about the z axis.

14. Find the moment of inertia around the y axis for the ball $x^2 + y^2 + z^2 \leq R^2$ if the mass density is a constant ρ.

15. Find the gravitational potential on a mass m of a spherical planet with mass $M = 3 \times 10^{26}$ kg, at a distance of 2×10^8 m from its center.

16. Find the gravitational force exerted on a 70-kg object at the position in Exercise 15.

17. A body W in xyz coordinates is called *symmetric with respect to a given plane* if for every particle on one side of the plane there is a particle of equal mass located at its mirror image through the plane.

   (a)   Discuss the planes of symmetry for an automobile shell.

   (b)   Let the plane of symmetry be the xy plane, and denote by W^+ and W^- the portions of W above and below the plane, respectively. By our assumption, the mass density $\rho(x, y, z)$ satisfies $\rho(x, y, -z) = \rho(x, y, z)$. Justify the following steps:

$$\bar{z} \cdot \iiint_W \rho(x, y, z)\, dx\, dy\, dz = \iiint_W z\rho(x, y, z)\, dx\, dy\, dz$$

$$= \iiint_{W^+} z\rho(x, y, z)\, dx\, dy\, dz + \iiint_{W^-} z\rho(x, y, z)\, dx\, dy\, dz$$

$$= \iiint_{W^+} z\rho(x, y, z)\, dx\, dy\, dz + \iiint_{W^+} -z\rho(u, v, -w)\, du\, dv\, dw$$

$$= 0.$$

   (c)   Explain why part (b) proves that if a body is symmetrical with respect to a plane, then its center of mass lies in that plane.

   (d)   Derive this law of mechanics: *If a body is symmetric with respect to two planes, then its center of mass lies on their line of intersection.*

18. A uniform rectangular steel plate of sides a and b rotates about its center of gravity with constant angular velocity ω.

   (a)   Kinetic energy equals $\frac{1}{2}$(mass)(velocity)^2. Argue that the kinetic energy of any element of mass $\rho\, dx\, dy$ ($\rho = $ constant) is given by $\rho(\omega^2/2)(x^2 + y^2)\, dx\, dy$, provided the origin $(0, 0)$ is placed at the center of gravity of the plate.

   (b)   Justify the formula for kinetic energy:

$$\text{K.E.} = \iint_{\text{plate}} \rho \frac{\omega^2}{2}(x^2 + y^2)\, dx\, dy.$$

   (c)   Evaluate the integral, assuming that the plate is described by the inequalities $-a/2 \leq x \leq a/2, -b/2 \leq y \leq b/2$.

19. As is well known, the density of a typical planet is not constant throughout the planet. Assume that planet C.M.W. has a radius of 5×10^8 cm and a mass density (in grams per cubic centimeter)

$$\rho(x, y, z) = \begin{cases} \dfrac{3 \times 10^4}{r}, & r \geq 10^4 \text{ cm,} \\ 3, & r \leq 10^4 \text{ cm,} \end{cases}$$

where $r = \sqrt{x^2 + y^2 + z^2}$. Find a formula for the gravitational potential outside C.M.W.

*6.4 IMPROPER INTEGRALS

In Chapter 5 we defined the integral of functions of two and three variables, and stated criteria guaranteeing that f was indeed integrable over a set D. Recall that one of the hypotheses was that f be bounded. The following example shows how the sum S_n may fail to converge if f is not bounded.

Let R be the unit square $[0, 1] \times [0, 1]$ and let $f \colon R \to \mathbb{R}$ be defined by $f(x, y) = 1/\sqrt{x}$ if $x \neq 0$ and $f(0, y) = 0$. Clearly, f is not bounded on R, since, as x gets close to zero, f gets arbitrarily large. Let R_{ij} be a regular partition of R and form the Riemann sum (see Figure 6.4.1)

$$S_n = \sum_{i=0}^{n-1} \sum_{j=0}^{n-1} f(\mathbf{c}_{ij}) \, \Delta x \, \Delta y;$$

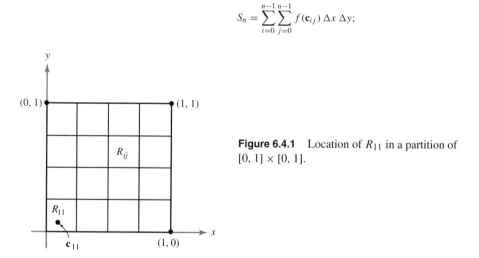

Figure 6.4.1   Location of R_{11} in a partition of $[0, 1] \times [0, 1]$.

Let R_{11} be the subrectangle that contains $(0, 0)$ (see Figure 6.4.1), and choose some $\mathbf{c}_{11} \in R_{11}$. For a fixed n, we can make S_n as large as we please by picking $\mathbf{c}_{11}$ closer and closer to $(0, 0)$; hence, $\lim_{n \to \infty} S_n$ cannot be *independent* of the choice of the $\mathbf{c}_{ij}$.

However, let us formally evaluate the iterated integral of f, following the rules for integrating a function of a single variable:

$$\int_0^1 \int_0^1 f(x, y) \, dx \, dy = \int_0^1 \int_0^1 \frac{dx}{\sqrt{x}} \, dy = \int_0^1 [2\sqrt{x}]_0^1 \, dy = \int_0^1 2 \, dy = 2.$$

Moreover, if we reverse the order of integration, we obtain the same value. Thus, in some sense this function is integrable. The question is, in what sense?

Recall from one-variable calculus how the ***improper*** integral $\int_0^1 dx/\sqrt{x}$ is treated: $1/\sqrt{x}$ is unbounded on the interval $(0, 1)$, yet $\text{limit}_{\delta \to 0} \int_\delta^1 dx/\sqrt{x} = 2$, and we *define* $\int_0^1 (dx/\sqrt{x})$ to be this limit. This example of an improper integral for functions of one variable is in fact rather simple. More complicated types of improper integrals can arise in multiple integrals.

In what follows we describe two types of improper integrals for functions of two variables for which we can determine whether they are convergent or not. We shall allow the function to be unbounded at certain points on the boundary of its domain and at one (or at most a finite number) of points in the interior of its domain of definition. As with functions of one variable such improper integrals arise naturally, as in Example 1 below where an improper integral occurs in computing the surface area of a hemisphere.

More specifically, suppose the region D is of type 1 and $f: D \to \mathbb{R}$ is continuous and bounded except at certain points on the boundary. *For simplicity of exposition, first assume that f is nonnegative.* Assume further that D is described by $a \leq x \leq b, \phi_1(x) \leq y \leq \phi_2(x)$. Choose numbers $\delta, \eta > 0$ such that $D_{\eta,\delta}$ is the subset of D consisting of points (x, y) with $a + \eta \leq x \leq b - \eta, \phi_1(x) + \delta \leq y \leq \phi_2(x) - \delta$ (Figure 6.4.2), where η and δ are chosen small enough that $D_{\eta,\delta} \subset D$. [If either $\phi_1(a) = \phi_2(a)$ or $\phi_1(b) = \phi_2(b)$, we must modify this slightly, because in this case $D_{\eta,\delta}$ may not be a subset of D (see Example 2).] Since f is continuous and bounded on $D_{\eta,\delta}$ the integral $\iint_{D_{\eta,\delta}} f\, dA$ exists. We can now ask what happens as the region $D_{\eta,\delta}$ expands to fill the region, D, that is, as $(\eta, \delta) \to (0, 0)$.

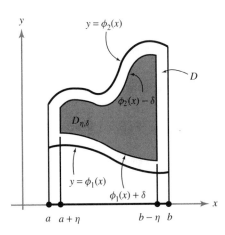

Figure 6.4.2 A shrunken domain $D_{\eta,\delta}$ for improper integrals.

Provided that

$$\underset{(\eta,\delta) \to (0,0)}{\text{limit}} \iint_{D_{\eta,\delta}} f\, dA$$

exists, we say that the integral of f over D is ***convergent*** or that f is ***integrable*** over D, and we define $\iint_D f\, dx\, dy$ to be equal to this limit.

Since f is integrable over $D_{\eta,\delta}$, we can apply Fubini's theorem to obtain

$$\iint_{D_{\eta,\delta}} f\, dA = \int_{a+\eta}^{b-\eta} \int_{\phi_1(x)+\delta}^{\phi_2(x)-\delta} f(x, y)\, dy\, dx. \tag{1}$$

Hence, if f is integrable over D,

$$\iint_D f \, dA = \lim_{(\eta, \delta) \to (0,0)} \int_{a+\eta}^{b-\eta} \int_{\phi_1(x)+\delta}^{\phi_2(x)-\delta} f(x, y) \, dy \, dx.$$

It may be convenient to work with the iterated limits

$$\lim_{\eta \to 0} \int_{a+\eta}^{b-\eta} \left[\lim_{\delta \to 0} \int_{\phi_1(x)+\delta}^{\phi_2(x)-\delta} f(x, y) \, dy \right] dx \qquad (2)$$

if these limits exist. If the limits do exist, then we denote expression (2) by

$$\int_a^b \int_{\phi_1(x)}^{\phi_2(x)} f(x, y) \, dy \, dx$$

and call it the **iterated improper integral** of f over D. Because we assumed $f \geq 0$, the existence of the limits shown in expression (2) implies the convergence of $\iint_D f \, dA$; therefore expression (2) equals $\iint_D f \, dA$ in this case. The definition of $\iint_D f \, dA$ when D is a region of type 2 is analogous, and in this case, if f is integrable over D, one can evaluate $\iint_D f \, dA$ as an iterated integral.

REMARK If f takes both positive and negative values, one can use a more advanced integration theory, called the Lebesgue integral to generalize the notion of convergent integral $\iint_D f \, dA$. Using this theory it is possible to show that if $\iint_D f \, dA$ exists, then it can be evaluated as an iterated integral. This latter fact is also known as Fubini's theorem.

EXAMPLE 1 Evaluate $\iint_D f(x, y) \, dy \, dx$ where $f(x, y) = 1/\sqrt{1 - x^2 - y^2}$ and D is the unit disk $x^2 + y^2 \leq 1$.

SOLUTION We can describe D as the set of points (x, y) satisfying $-1 \leq x \leq 1$, and $-\sqrt{1 - x^2} \leq y \leq \sqrt{1 - x^2}$. Because ∂D is the set of points (x, y) with $x^2 + y^2 = 1$, f is undefined at every point on ∂D, since at such points the denominator of f is 0. We calculate the iterated improper integrals and obtain

$$\int_{-1}^1 \int_{-\sqrt{(1-x^2)}}^{\sqrt{(1-x^2)}} \frac{dy \, dx}{\sqrt{(1 - x^2 - y^2)}} = \int_{-1}^1 \left[\sin^{-1} \left(\frac{y}{\sqrt{1 - x^2}} \right) \right]_{y=-\sqrt{(1-x^2)}}^{\sqrt{(1-x^2)}} dx$$

$$= \int_{-1}^1 [\sin^{-1}(1) - \sin^{-1}(-1)] \, dx = \int_{-1}^1 dx = 2\pi. \quad \blacktriangle$$

In this example we used the fact, stated above, that the iterated improper integral is equal to the improper integral of $1/(\sqrt{1 - x^2 - y^2})$ over the unit disk.

EXAMPLE 2 Let $f(x, y) = 1/(x - y)$ and let D be the set of (x, y) with $0 \leq x \leq 1$ and $0 \leq y \leq x$. Show that f is not integrable over D.

SOLUTION Since the denominator of f is zero on the line $y = x$, f is unbounded on part of the boundary of D. Let $0 < \eta < 1$ and $0 < \delta < \eta$ and let $D_{\eta,\delta}$ be the set of (x, y) with $\eta \leq x \leq 1 - \eta$ and $\delta \leq y \leq x - \delta$ (Figure 6.4.3). We choose $\delta < \eta$ to guarantee that $D_{\eta,\delta}$ is contained in D. Consider

$$
\begin{aligned}
\iint_{D_{\eta,\delta}} f \, dA &= \int_{\delta+\eta}^{1-\eta} \int_{\delta}^{x-\delta} \frac{1}{x-y} \, dy \, dx \\
&= \int_{\eta}^{1-\eta} [-\log (x - y)]|_{y=\delta}^{x-\delta} \, dx \\
&= \int_{\eta}^{1-\eta} [-\log (\delta) + \log (x - \delta)] \, dx \\
&= [-\log \delta] \int_{\eta}^{1-\eta} dx + \int_{\eta}^{1-\eta} \log (x - \delta) \, dx \\
&= -(1 - 2\eta) \log \delta + [(x - \delta) \log (x - \delta) - (x - \delta)]|_{\eta}^{1-\eta}.
\end{aligned}
$$

In the last step we used the fact that $\int \log u \, du = u \log u - u$. Continuing the above set of equalities, we have

$$
\begin{aligned}
\iint_{D_{\eta,\delta}} f \, dA = -(1 - 2\eta) \log \delta + (1 - \eta - \delta) \log(1 - \eta - \delta) \\
-(1 - \eta - \delta) - (\eta - \delta) \log(\eta - \delta) + (\eta - \delta).
\end{aligned}
$$

Figure 6.4.3 The shrunken domain $D_{\eta,\delta}$ for a triangular domain D.

As $(\eta, \delta) \to (0, 0)$, the second term converges to $1 \log 1 = 0$, and the third and fifth terms converge to -1 and 0, respectively. Let $\nu = \eta - \delta$. Since $\nu \log \nu \to 0$ as $\nu \to 0$ (a limit established by using l'Hôpital's rule from calculus), we see that the fourth term goes to zero as $(\eta, \delta) \to (0, 0)$. It is the first term which will give us trouble. Now

$$
-(1 - 2\eta) \log \delta = -\log \delta + 2\eta \log \delta, \tag{3}
$$

and it is not hard so see that this does not converge as $(\eta, \delta) \to (0, 0)$. For example, let $\eta = 2\delta$; then expression (3) becomes $-\log \delta + 4\delta \log \delta$. As before, $4\delta \log \delta \to 0$ as $\delta \to 0$ but $-\log \delta \to +\infty$ as $\delta \to 0$, which shows that expression (3) does not converge. Hence $\lim_{(\eta,\delta)\to(0,0)} \iint_{D_{\eta,\delta}} f \, dA$ does not exist and so f is not integrable. ▲

We now consider nonnegative functions f which become "infinite" or are undefined at isolated points in a region D of type 1 or 2. For example, consider the function $f(x, y) = 1/\sqrt{x^2 + y^2}$ on the unit disk $D = \{(x, y) \mid x^2 + y^2 \le 1\}$. Again $f \ge 0$, but f is not defined at the origin and is unbounded.

Let (x_0, y_0) be a point of a general region D where a nonnegative function f is undefined. Further, let $D_\delta = D_\delta(x_0, y_0)$ be the disk of radius δ centered at (x_0, y_0) and let $D \backslash D_\delta$ denote the region D with D_δ removed. Assume that f is continuous at every point of D except (x_0, y_0). Then $\iint_{D \backslash D_\delta} f \, dA$ is defined. We say that $\iint_D f \, dA$ is **convergent**, or that f is **integrable** over D if

$$\lim_{\delta \to 0} \iint_{D \backslash D_\delta} f \, dA$$

exists.

EXAMPLE 3   Show that $f(x, y) = 1/\sqrt{x^2 + y^2}$ is integrable over the unit disk D and evaluate $\iint_D f \, dA$.

SOLUTION   Let D_δ be the disk of radius δ centered at the origin. Then f is continuous everywhere on D except at $(0, 0)$. Thus $\iint_{D \backslash D_\delta} f \, dA$ exists. To evaluate this integral, we change variables to polar coordinates, $x = r \cos \theta$, $y = r \sin \theta$. Then $f(r \cos \theta, r \sin \theta) = 1/r$, and

$$\iint_{D \backslash D_\delta} f \, dA = \int_\delta^1 \int_0^{2\pi} \frac{1}{r} r \, d\theta \, dr = \int_\delta^1 \int_0^{2\pi} d\theta \, dr = 2\pi(1 - \delta).$$

Thus

$$\iint_D f \, dA = \lim_{\delta \to 0} \iint_{D \backslash D_\delta} f \, dA = 2\pi. \quad \blacktriangle$$

More generally, one can, in an analogous manner, define the integral of nonnegative functions f which are continuous except at a finite number of points in D. One can also combine both types of improper integrals; that is, one may consider functions which are continuous except at a finite number of points of D or at points on the boundary of D, and define $\iint_D f \, dA$ appropriately.

EXERCISES

1. Evaluate the following integrals if they exist (discuss how you define the integral if it was not given in the text).

(a) $\displaystyle\iint_D \frac{1}{\sqrt{xy}} \, dA$, where $D = [0, 1] \times [0, 1]$

(b) $\displaystyle\iint_D \frac{1}{\sqrt{|x - y|}} \, dx \, dy$, where $D = \{(x, y) \mid 0 \le x \le 1, 0 \le y \le 1, y \le x\}$

(c) $\displaystyle\iint_D (y/x) \, dx \, dy$, where D is bounded by $x = 1$, $x = y$, and $x = 2y$

(d) $\displaystyle\int_0^1 \int_0^{e^y} \log x \, dx \, dy$

2. (a) Discuss how you would define $\iint_D f\,dA$ if D is an unbounded region, for example, the set of (x, y) such that $a \leq x < \infty$ and $\phi_1(x) \leq y \leq \phi_2(x)$, where $\phi_1 \leq \phi_2$ are given (Figure 6.4.4).

(b) Evaluate $\iint_D xye^{-(x^2+y^2)}\,dx\,dy$ if $x \geq 0, 0 \leq y \leq 1$.

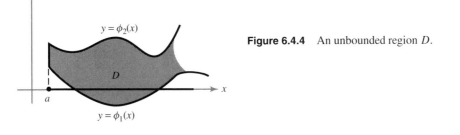

Figure 6.4.4 An unbounded region D.

3. (a) Evaluate

$$\iint_D \frac{dA}{(x^2 + y^2)^{2/3}},$$

where D is the unit disk in $\mathbb{R}^2$

(b) Determine the real numbers λ for which the integral

$$\iint_D \frac{dA}{(x^2 + y^2)^{\lambda}}$$

is convergent, where again D is the unit disk.

4. One can also consider improper integrals of functions which fail to be continuous on entire curves lying in some region D. For example, by breaking $D = [0, 1] \times [0, 1]$ into two regions, define and then discuss the convergence of the integral

$$\iint_D \frac{1}{\sqrt{|x - y|}}\,dx\,dy.$$

5. Using Exercise 2, integrate e^{-xy} for $x \geq 0, 1 \leq y \leq 2$ in two ways. Assuming Fubini's theorem can be used, show that

$$\int_0^\infty \frac{e^{-x} - e^{-2x}}{x}\,dx = \log 2.$$

6. Show that the integral $\int_0^1 \int_0^a (x/\sqrt{a^2 - y^2})\,dy\,dx$ exists, and compute its value.

7. Discuss whether the integral

$$\iint_D \frac{x + y}{x^2 + 2xy + y^2}\,dx\,dy$$

exists where $D = [0, 1] \times [0, 1]$. If it exists, compute its value.

8. Let f be a nonnegative function that may be unbounded and discontinuous on the boundary of an elementary region D. Let g be a similar function such that $f(x, y) \leq g(x, y)$ whenever both are defined. Suppose $\iint_D g(x, y)\, dA$ exists. Argue informally that this implies the existence of $\iint_D f(x, y)\, dA$.

9. Use Exercise 8 to show that

$$\iint_D \frac{\sin^2(x - y)}{\sqrt{1 - x^2 - y^2}}\, dy\, dx$$

exists where D is the unit disk $x^2 + y^2 \leq 1$.

10. Let f be as in Exercise 8 and let g be a function such that $0 \leq g(x, y) \leq f(x, y)$ whenever both are defined. Suppose that $\iint_D g(x, y)\, dA$ does not exist. Argue informally that $\iint_D f(x, y)\, dA$ cannot exist.

11. Use Exercise 10 to show that

$$\iint_D \frac{e^{x^2 + y^2}}{x - y}\, dy\, dx$$

does not exist, where D is the set of (x, y) with $0 \leq x \leq 1$ and $0 \leq y \leq x$.

12. (The Gaussian integral) A change of variables can help to find the value of an improper integral over the unbounded region $\mathbb{R}^2$. Evaluate

$$\iint_{\mathbb{R}^2} e^{-x^2 - y^2}\, dx\, dy$$

by changing to polar coordinates. Combining this with Fubini's theorem (assume its validity), conclude that

$$\int_{-\infty}^{\infty} e^{-x^2}\, dx = \sqrt{\pi}.$$

13. Let W be the first octant of the ball $x^2 + y^2 + z^2 \leq a^2$, where $x \geq 0, y \geq 0, z \geq 0$. Evaluate the improper integral

$$\iiint_W \frac{(x^2 + y^2 + z^2)^{1/4}}{\sqrt{z + (x^2 + y^2 + z^2)^2}}\, dx\, dy\, dz$$

by changing variables.

14. Let D be the unbounded region defined as the set of (x, y, z) with $x^2 + y^2 + z^2 \geq 1$. By making a change of variables, evaluate the improper integral

$$\iiint_D \frac{dx\, dy\, dz}{(x^2 + y^2 + z^2)^2}.$$

REVIEW EXERCISES FOR CHAPTER 6

1. (a) Find a linear transformation taking the square $S = [0, 1] \times [0, 1]$ to the parallelogram P with vertices $(0, 0), (2, 0), (1, 2), (3, 2)$.

(b) Write down a change of variables formula appropriate to the transformation you found in (a).

2. (a)   Find the image of the square $[0, 1] \times [0, 1]$ under the transformation $T(x, y) = (2x, x + 3y)$.
   (b)   Write down a change of variables formula appropriate to the transformation and the region you found in (a).

3.   Let B be the region in the first quadrant bounded by the curves $xy = 1$, $xy = 3$, $x^2 - y^2 = 1$, and $x^2 - y^2 = 4$. Evaluate $\iint_B (x^2 + y^2) \, dx \, dy$ using the change of variables $u = x^2 - y^2$, $v = xy$.

4.   In parts (a) to (d), make the indicated change of variables. (Do not evaluate.)

(a) $\displaystyle \int_0^1 \int_{-1}^1 \int_{-\sqrt{(1-y^2)}}^{\sqrt{(1-y^2)}} (x^2 + y^2)^{1/2} \, dx \, dy \, dz$, cylindrical coordinates

(b) $\displaystyle \int_{-1}^1 \int_{-\sqrt{(1-y^2)}}^{\sqrt{(1-y^2)}} \int_{-\sqrt{(4-x^2-y^2)}}^{\sqrt{(4-x^2-y^2)}} xyz \, dz \, dx \, dy$, cylindrical coordinates

(c) $\displaystyle \int_{-\sqrt{2}}^{\sqrt{2}} \int_{-\sqrt{(2-y^2)}}^{\sqrt{(2-y^2)}} \int_{\sqrt{(x^2+y^2)}}^{\sqrt{(4-x^2-y^2)}} z^2 \, dz \, dx \, dy$, spherical coordinates

(d) $\displaystyle \int_0^1 \int_0^{\pi/4} \int_0^{2\pi} \rho^3 \sin 2\phi \, d\theta \, d\phi \, d\rho$, rectangular coordinates

5.   Find the volume inside the surfaces $x^2 + y^2 = z$ and $x^2 + y^2 + z^2 = 2$.

6.   Find the volume enclosed by the cone $x^2 + y^2 = z^2$ and the plane $2z - y - 2 = 0$.

7.   A cylindrical hole of diameter 1 is bored through a sphere of radius 2. Assuming that the axis of the cylinder passes through the center of the sphere, find the volume of the solid that remains.

8.   Let C_1 and C_2 be two cylinders of infinite extent, of diameter 2, and with axes on the x and y axes respectively. Find the volume of $C_1 \cap C_2$.

9.   Find the volume bounded by $x/a + y/b + z/c = 1$ and the coordinate planes.

10.   Find the volume determined by $z \le 6 - x^2 - y^2$ and $z \ge \sqrt{x^2 + y^2}$.

11.   The *tetrahedron* defined by $x \ge 0$, $y \ge 0$, $z \ge 0$, $x + y + z \le 1$ is to be sliced into n segments of equal volume by planes parallel to the plane $x + y + z = 1$. Where should the slices be made?

12.   Let E be the solid ellipsoid $E = \{(x, y, z) \mid (x^2/a^2) + (y^2/b^2) + (z^2/c^2) \le 1\}$ where $a > 0, b > 0$, and $c > 0$. Evaluate

$$\iiint xyz \, dx \, dy \, dz$$

(a)   over the whole ellipsoid; and
(b)   over that part of it in the first quadrant:

$$x \ge 0, \qquad y \ge 0, \quad \text{and} \quad z \ge 0, \quad \frac{x^2}{a^2} + \frac{y^2}{b^2} + \frac{z^2}{c^2} \le 1.$$

13.   Find the volume of the "ice cream cone" defined by the inequalities $x^2 + y^2 \le \frac{1}{5}z^2$, and $0 \le z \le 5 + \sqrt{5 - x^2 - y^2}$.

14. Let ρ, θ, ϕ be spherical coordinates in $\mathbb{R}^3$ and suppose that a surface surrounding the origin is described by a continuous positive function $\rho = f(\theta, \phi)$. Show that the volume enclosed by the surface is

$$V = \frac{1}{3} \int_0^{2\pi} \int_0^{\pi} [f(\theta, \phi)]^3 \sin \phi \, d\phi \, d\theta.$$

15. Using an appropriate change of variables, evaluate $\iint_B \exp[(y - x)/(y + x)] \, dx \, dy$ where B is the interior of the triangle with vertices at $(0, 0)$, $(0, 1)$, and $(1, 0)$.

16. Suppose the density of a solid of radius R is given by $(1 + d^3)^{-1}$ where d is the distance to the center of the sphere. Find the total mass of the sphere.

17. The density of the material of a spherical shell whose inner radius is 1 m and whose outer radius is 2 m is $0.4d^2$ g/cm^3, where d is the distance to the center of the sphere in meters. Find the total mass of the shell.

18. If the shell in Exercise 17 were dropped into a large tank of pure water, would it float? What if the shell leaked? (Assume that the density of water is exactly 1 g/cm^3.)

19. The temperature at points in the cube $C = \{(x, y, z) \mid -1 \le x \le 1, -1 \le y \le 1, \text{and } -1 \le z \le 1\}$ is $32d^2$, where d is the distance to the origin.

   (a) What is the average temperature?

   (b) At what points of the cube is the temperature equal to the average temperature?

20. Use cylindrical coordinates to find the center of mass of the region defined by

$$y^2 + z^2 \le \frac{1}{4}, \qquad (x - 1)^2 + y^2 + z^2 \le 1, \qquad x \ge 1.$$

21. Find the center of mass of the solid hemisphere

$$V = \{(x, y, z) \mid x^2 + y^2 + z^2 \le a^2 \text{ and } z \ge 0\}$$

if the density is constant.

22. Evaluate $\iint_B e^{-x^2 - y^2} \, dx \, dy$ where B consists of those (x, y) satisfying $x^2 + y^2 \le 1$ and $y \le 0$.

23. Evaluate

$$\iiint_S \frac{dx \, dy \, dz}{(x^2 + y^2 + z^2)^{3/2}},$$

where S is the solid bounded by the spheres $x^2 + y^2 + z^2 = a^2$ and $x^2 + y^2 + z^2 = b^2$, where $a > b > 0$.

24. Evaluate $\iiint_D (x^2 + y^2 + z^2) xyz \, dx \, dy \, dz$ over each of the following regions.

   (a) the sphere $D = \{(x, y, z) \mid x^2 + y^2 + z^2 \le R^2\}$;

   (b) the hemisphere $D = \{(x, y, z) \mid x^2 + y^2 + z^2 \le R^2 \text{ and } z \ge 0\}$;

   (c) the octant $D = \{(x, y, z) \mid x \ge 0, y \ge 0, z \ge 0, \text{ and } z^2 + y^2 + z^2 \le R^2\}$.

25. Let C be the cone-shaped region $\{(x, y, z) \mid \sqrt{x^2 + y^2} \le z \le 1\}$ in $\mathbb{R}^3$ and evaluate the integral $\iiint_C (1 + \sqrt{x^2 + y^2}) \, dx \, dy \, dz$.

26. Find $\iiint_{\mathbb{R}^3} f(x, y, z) \, dx \, dy \, dz$ where $f(x, y, z) = \exp[-(x^2 + y^2 + z^2)^{3/2}]$.

27. The *flexural rigidity* EI of a uniform beam is the product of its Young's modulus of elasticity E and the moment of inertia I of the cross section of the beam with respect to a horizontal line l passing through the center of gravity of this cross section. Here

$$I = \iint_R [d(x, y)]^2 \, dx \, dy,$$

where $d(x, y) = $ the distance from (x, y) to l and $R = $ the cross section of the beam being considered.
  (a) Assume that the cross section R is the rectangle $-1 \le x \le 1, -1 \le y \le 2$ and l is the x axis. Find I.
  (b) Assume the cross section R is a circle of radius 4 and l is the x axis. Find I, using polar coordinates.

28. Find, $\iiint_{\mathbb{R}^3} f(x, y, z) \, dx \, dy \, dz$ where

$$f(x, y, z) = \frac{1}{[1 + (x^2 + y^2 + z^2)^{3/2}]^{3/2}}.$$

29. Suppose D is the unbounded region of $\mathbb{R}^2$ given by the set of (x, y) with $0 \le x < \infty, 0 \le y \le x$. Let $f(x, y) = x^{-3/2} e^{y-x}$. Does the improper integral $\iint_D f(x, y) \, dx \, dy$ exist?

30. If the world were two-dimensional, the laws of physics would predict that the gravitational potential of a mass point is proportional to the logarithm of the distance from the point. Using polar coordinates, write an integral giving the gravitational potential of a disk of constant density.

31. (a) Evaluate the improper integral

$$\int_0^\infty \int_0^y x e^{-y^3} \, dx \, dy.$$

  (b) Evaluate $\iint_B (x^4 + 2x^2 y^2 + y^4) \, dx \, dy$, where B is the portion of the disk of radius 2 [centered at $(0, 0)$ in the first quadrant].

32. Let f be a continuous nonnegative function on a region $D \subset \mathbb{R}^2$ of type 1 or 2 which is continuous except for points on the boundary of D and at most finitely many points interior to D. Give a suitable definition of $\iint_D f \, dA$.

33. Evaluate $\iint_{\mathbb{R}^2} f(x, y) \, dx \, dy$ where $f(x, y) = 1/(1 + x^2 + y^2)^{3/2}$. (HINT: You may assume that changing variables and Fubini's theorem are valid for improper integrals.)

7 INTEGRALS OVER PATHS AND SURFACES

I hold in fact: (1) That small portions of space are of a nature analogous to little hills on a surface which is on the average flat. (2) That this property of being curved or distorted is continually passed on from one portion of space to another after the manner of a wave. (3) That this variation of curvature of space is really what happens in that phenomenon which we call the motion of matter whether ponderable or ethereal. (4) That in this physical world nothing else takes place but this variation, subject, possibly, to the law of continuity.

W. K. CLIFFORD (1870)

In Chapter 5 we studied integration over regions in $\mathbb{R}^2$ and $\mathbb{R}^3$. In this chapter we study integration over paths and surfaces. This is basic to an understanding of Chapter 8; that chapter will discuss the basic relation between vector differential calculus (Chapter 4) and vector integral calculus (this chapter), a relation that generalizes the fundamental theorem of calculus to several variables. This generalization is summarized in the theorems of Green, Gauss, and Stokes.

7.1 THE PATH INTEGRAL

This section introduces the concept of a path integral; this is one of the several ways in which integrals of functions of one variable can be generalized to functions of several variables. Besides those in Chapter 5, there are other generalizations, to be discussed in later sections.

Suppose we are given a scalar function $f: \mathbb{R}^3 \to \mathbb{R}$, so that f sends points in $\mathbb{R}^3$ to real numbers. It will be useful to define the integral of a function f along a path $\mathbf{c}: I = [a, b] \to \mathbb{R}^3$, where $\mathbf{c}(t) = (x(t), y(t), z(t))$. To relate this notion to something tangible, suppose that the image of $\mathbf{c}$ represents a wire. We can let $f(x, y, z)$ denote the mass density at (x, y, z) and the integral of f will be the total mass of the wire. By letting $f(x, y, z)$ indicate temperature, we can also use the integral to determine the average temperature along the wire.

DEFINITION: PATH INTEGRALS The *path integral*, or the *integral of* $f(x, y, z)$ *along the path* $\mathbf{c}$, is defined when $\mathbf{c}: I = [a, b] \to \mathbb{R}^3$ is of class C^1 and when the composite function $t \mapsto f(x(t), y(t), z(t))$ is continuous on I. We define this integral by the equation

$$\int_{\mathbf{c}} f \, ds = \int_a^b f(x(t), y(t), z(t)) \|\mathbf{c}'(t)\| \, dt.$$

Sometimes $\int_{\mathbf{c}} f \, ds$ is denoted

$$\int_{\mathbf{c}} f(x, y, z) \, ds$$

or

$$\int_a^b f(\mathbf{c}(t)) \|\mathbf{c}'(t)\| \, dt.$$

If $\mathbf{c}(t)$ is only piecewise C^1 or $f(\mathbf{c}(t))$ is piecewise continuous, we define $\int_{\mathbf{c}} f \, ds$ by breaking $[a, b]$ into pieces over which $f(\mathbf{c}(t)) \|\mathbf{c}'(t)\|$ is continuous, and summing the integrals over the pieces.

When $f = 1$, we recover the definition of the arc length of $\mathbf{c}$. Also note that f need only be defined on the image curve C of $\mathbf{c}$ and not necessarily on the whole space in order for the above definition to make sense.

EXAMPLE 1 Let $\mathbf{c}$ be the helix $\mathbf{c}: [0, 2\pi] \to \mathbb{R}^3, t \mapsto (\cos t, \sin t, t)$ (see Figure 2.4.9), and let $f(x, y, z) = x^2 + y^2 + z^2$. Evaluate the integral $\int_{\mathbf{c}} f(x, y, z) \, ds$.

SOLUTION

$$\|\mathbf{c}'(t)\| = \sqrt{\left[\frac{d(\cos t)}{dt}\right]^2 + \left[\frac{d(\sin t)}{dt}\right]^2 + \left[\frac{dt}{dt}\right]^2} = \sqrt{\sin^2 t + \cos^2 t + 1} = \sqrt{2}.$$

We substitute for x, y, and z in terms of t to obtain

$$f(x, y, z) = x^2 + y^2 + z^2 = \cos^2 t + \sin^2 t + t^2 = 1 + t^2$$

along **c**. This yields

$$\int_{\mathbf{c}} f(x, y, z)\, ds = \int_0^{2\pi} (1 + t^2)\sqrt{2}\, dt = \sqrt{2}\left[t + \frac{t^3}{3}\right]_0^{2\pi} = \frac{2\sqrt{2}\pi}{3}(3 + 4\pi^2). \quad \blacktriangle$$

To motivate the definition of the path integral we shall consider "Riemann-like" sums S_N in the same general way we did to define arc length in Section 4.2. For simplicity, let **c** be of class C^1 on I. Subdivide the interval $I = [a, b]$ by means of a partition

$$a = t_0 < t_1 < \cdots < t_N = b.$$

This leads to a decomposition of **c** into paths $\mathbf{c}_i$ (Figure 7.1.1) defined on $[t_i, t_{i+1}]$ for $0 \le i \le N - 1$. Denote the arc length of $\mathbf{c}_i$ by Δs_i; thus,

$$\Delta s_i = \int_{t_i}^{t_{i+1}} \|\mathbf{c}'(t)\|\, dt.$$

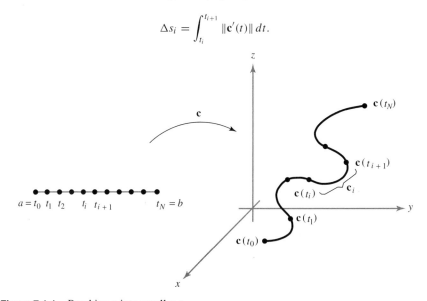

Figure 7.1.1   Breaking **c** into smaller $\mathbf{c}_i$.

When N is large, the arc length Δs_i is small and $f(x, y, z)$ is approximately constant for points on $\mathbf{c}_i$. We consider the sums

$$S_N = \sum_{i=0}^{N-1} f(x_i, y_i, z_i)\, \Delta s_i,$$

where $(x_i, y_i, z_i) = \mathbf{c}(t)$ for some $t \in [t_i, t_{i+1}]$. By the mean value theorem we know that $\Delta s_i = \|\mathbf{c}'(t_i^*)\|\Delta t_i$, where $t_i \le t_i^* \le t_{i+1}$ and $\Delta t_i = t_{i+1} - t_i$. Now, from the theory of Riemann sums, it can be shown that

$$\lim_{N \to \infty} S_N = \sum_{i=0}^{N-1} f(x_i, y_i, z_i)\|\mathbf{c}'(t_i^*)\|\Delta t_i = \int_I f(x(t), y(t), z(t))\|\mathbf{c}'(t)\|\, dt$$

$$= \int_{\mathbf{c}} f(x, y, z)\, ds.$$

An important special case of the path integral occurs when the path **c** describes a plane curve. Suppose that all points **c**(t) lie in the xy plane and f is a real-valued function of two variables. The path integral of f along **c** is

$$\int_{\mathbf{c}} f(x, y)\, ds = \int_{a}^{b} f(x(t), y(t))\sqrt{x'(t)^2 + y'(t)^2}\, dt.$$

When $f(x, y) \geq 0$, this integral has a geometric interpretation as the "area of a fence." We can construct a "fence" with base the image of **c** and with height $f(x, y)$ at (x, y) (Figure 7.1.2). If **c** moves only once along the image of **c**, the integral $\int_{\mathbf{c}} f(x, y)\, ds$ represents the area of a side of this fence. Readers should try to justify this interpretation for themselves, using an argument like the one used to justify the arc length formula.

Figure 7.1.2   The path integral as the area of a fence.

EXAMPLE 2   Tom Sawyer's aunt has asked him to whitewash both sides of the old fence shown in Figure 7.1.3. Tom estimates that for each 25 ft² of whitewashing he lets someone do for him, the willing victim will pay 5 cents. How much can Tom hope to earn, assuming his aunt will provide whitewash free of charge?

Figure 7.1.3   Tom Sawyer's fence.

SOLUTION From Figure 7.1.3, the base of the fence in the first quadrant is the path $\mathbf{c}\colon [0, \pi/2] \to \mathbb{R}^2, t \mapsto (30 \cos^3 t, 30 \sin^3 t)$, and the height of the fence at (x, y) is $f(x, y) = 1 + y/3$. The area of one side of the half of the fence is equal to the *integral* $\int_{\mathbf{c}} f(x, y)\, ds = \int_{\mathbf{c}} (1 + y/3)\, ds$. Since $\mathbf{c}'(t) = (-90 \cos^2 t \sin t, 90 \sin^2 t \cos t)$, we have $\|\mathbf{c}'(t)\| = 90 \sin t \cos t$. Thus the integral is

$$\int_{\mathbf{c}} \left(1 + \frac{y}{3}\right) ds = \int_0^{\pi/2} \left(1 + \frac{30 \sin^3 t}{3}\right) 90 \sin t \cos t \, dt$$

$$= 90 \int_0^{\pi/2} (\sin t + 10 \sin^4 t) \cos t \, dt$$

$$= 90 \left[\frac{\sin^2 t}{2} + 2 \sin^5 t\right]_0^{\pi/2} = 90 \left(\frac{1}{2} + 2\right) = 225,$$

which is the area in the first quadrant. Hence, the area of one side of the fence is 450 ft^2. Since both sides are to be whitewashed, we must multiply by 2 to find the total area, which is 900 ft^2. Dividing by 25 and then multiplying by 5, we find that Tom could realize as much as $1.80 for the job. ▲

This concludes our study of integration of *scalar* functions over paths. In the next section we shall turn our attention to the integration of *vector fields* over paths, and we shall see many further applications of the path integral in Chapter 8, when we study vector analysis.

EXERCISES

1. Let $f(x, y, z) = y$ and $\mathbf{c}(t) = (0, 0, t), 0 \le t \le 1$. Prove that $\int_{\mathbf{c}} f \, ds = 0$.

2. Evaluate the following path integrals $\int_{\mathbf{c}} f(x, y, z)\, ds$, where
   (a) $f(x, y, z) = x + y + z$ and $\mathbf{c}\colon t \mapsto (\sin t, \cos t, t), t \in [0, 2\pi]$
   (b) $f(x, y, z) = \cos z$, $\mathbf{c}$ as in part (a)
   (c) $f(x, y, z) = x \cos z$, $\mathbf{c}\colon t \mapsto t\mathbf{i} + t^2\mathbf{j}, t \in [0, 1]$

3. Evaluate the following path integrals $\int_{\mathbf{c}} f(x, y, z)\, ds$, where
   (a) $f(x, y, z) = \exp \sqrt{z}$, and $\mathbf{c}\colon t \mapsto (1, 2, t^2), t \in [0, 1]$
   (b) $f(x, y, z) = yz$, and $\mathbf{c}\colon t \mapsto (t, 3t, 2t), t \in [1, 3]$
   (c) $f(x, y, z) = (x + y)/(y + z)$, and $\mathbf{c}\colon t \mapsto \left(t, \frac{2}{3}t^{3/2}, t\right), t \in [1, 2]$

4. (a) Show that the path integral of $f(x, y)$ along a path given in polar coordinates by $r = r(\theta)$, $\theta_1 \le \theta \le \theta_2$, is

$$\int_{\theta_1}^{\theta_2} f(r \cos \theta, r \sin \theta) \sqrt{r^2 + \left(\frac{dr}{d\theta}\right)^2} \, d\theta.$$

   (b) Compute the arc length of $r = 1 + \cos \theta, 0 \le \theta \le 2\pi$.

5. Let $f\colon \mathbb{R}^3 \setminus \{xz \text{ plane}\} \to \mathbb{R}$ be defined by $f(x, y, z) = 1/y^3$. Evaluate $\int_{\mathbf{c}} f(x, y, z)\, ds$ where $\mathbf{c}\colon [1, e] \to \mathbb{R}^3$ is given by $\mathbf{c}(t) = (\log t)\mathbf{i} + t\mathbf{j} + 2\mathbf{k}$.

6. Write the following limit as a path integral of $f(x, y, z) = xy$ over some path **c** on $[0, 1]$ and evaluate:

$$\underset{N \to \infty}{\text{limit}} \sum_{i=1}^{N-1} (t_i^*)^2 \left(t_{i+1}^2 - t_i^2 \right)$$

(here $t_1, \ldots, t_N$ is a partition of $[0, 1]$ and $t_i \le t_i^* \le t_{i+1}$).

7. Let $f(x, y) = 2x - y$, $x = t^4$, $y = t^4$, $-1 \le t \le 1$.

    (a) Compute the integral of f along this path and interpret the answer geometrically.

    (b) Evaluate the arc length function $s(t)$ and redo part (a) in terms of s (you may wish to consult Exercise 2, Section 4.2).

Exercise 8 to 11 are concerned with the application of the path integral to the problem of defining the average value of a scalar function along a path. Define the number

$$\frac{\int_{\mathbf{c}} f(x, y, z)\, ds}{l(\mathbf{c})}$$

*to be the **average value** of f along* **c**. *Here $l(\mathbf{c})$ is the length of the path:*

$$l(\mathbf{c}) = \int_{\mathbf{c}} \|\mathbf{c}'(t)\|\, dt.$$

(This is analogous to the average of a function over a region defined in Section 6.3.)

8. (a) Justify the formula $[\int_{\mathbf{c}} f(x, y, z)\, ds]/l(\mathbf{c})$ for the average value of f along **c** using Riemann sums.

    (b) Show that the average value of f along **c** in Example 1 is $(1 + \frac{4}{3}\pi^2)$.

    (c) In Exercise 2(a) and (b) above, find the average value of f over the given curves.

9. Find the average y coordinate of the points on the semicircle parametrized by **c**: $[0, \pi] \to \mathbb{R}^3$, $\theta \mapsto (0, a \sin \theta, a \cos \theta)$; $a > 0$.

10. Suppose the semicircle in Exercise 9 is made of a wire with a uniform density of 2 grams per unit length.

    (a) What is the total mass of the wire?

    (b) Where is the center of mass of this configuration of wire? (Consult Section 6.3.)

11. Let **c** be the path given by $\mathbf{c}(t) = (t^2, t, 3)$ for $t \in [0, 1]$.

    (a) Find $l(\mathbf{c})$, the length of the path.

    (b) Find the average y coordinate along the path **c**.

12. If $f \colon [a, b] \to \mathbb{R}$ is piecewise continuously differentiable, let the *length of the graph* of f on $[a, b]$ be defined as the length of the path $t \mapsto (t, f(t))$ for $t \in [a, b]$.

    (a) Show that the length of the graph of f on $[a, b]$ is

$$\int_a^b \sqrt{1 + [f'(x)]^2}\, dx.$$

    (b) Find the length of the graph of $y = \log x$ from $x = 1$ to $x = 2$.

13. Find the mass of a wire formed by the intersection of the sphere $x^2 + y^2 + z^2 = 1$ and the plane $x + y + z = 0$ if the density at (x, y, z) is given by $\rho(x, y, z) = x^2$ grams per unit length of wire.

14. Evaluate $\int_{\mathbf{c}} f\, ds$ where $f(x, y, z) = z$ and $\mathbf{c}(t) = (t \cos t, t \sin t, t)$ for $0 \leq t \leq t_0$.

7.2 LINE INTEGRALS

If $\mathbf{F}$ is a force field in space, then a test particle (for example, a small unit charge in an electric force field or a unit mass in a gravitational field) will experience the force $\mathbf{F}$. Suppose the particle moves along the image of a path $\mathbf{c}$ while being acted upon by $\mathbf{F}$. A fundamental concept is the *work done* by $\mathbf{F}$ on the particle as it traces out the path $\mathbf{c}$. If $\mathbf{c}$ is a straight-line displacement given by the vector $\mathbf{d}$ and $\mathbf{F}$ is a constant force, then the work done by $\mathbf{F}$ in moving the particle along the path is $\mathbf{F} \cdot \mathbf{d}$:

$$\mathbf{F} \cdot \mathbf{d} = (\text{magnitude of force}) \times (\text{displacement in direction of force}).$$

If the path is curved, we can imagine that it is made up of a succession of infinitesimal straight-line displacements or that it is approximated by a finite number of straight-line displacements. Then (as in our derivation of the formulas for the path integral in the preceding section) we are led to the following formula for the work done by the force field $\mathbf{F}$ on a particle moving along a path $\mathbf{c} \colon [a, b] \to \mathbb{R}^3$:

$$\text{work done by } \mathbf{F} = \int_a^b \mathbf{F}(\mathbf{c}(t)) \cdot \mathbf{c}'(t)\, dt.$$

Without giving a full proof, we can justify this derivation as follows. As t ranges over a small interval t to $t + \Delta t$, the particle moves from $\mathbf{c}(t)$ to $\mathbf{c}(t + \Delta t)$, a vector displacement of $\Delta \mathbf{s} = \mathbf{c}(t + \Delta t) - \mathbf{c}(t)$ (see Figure 7.2.1).

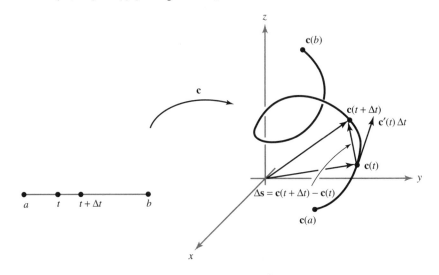

Figure 7.2.1   For small Δt, $\Delta \mathbf{s} = \mathbf{c}(t + \Delta t) - \mathbf{c}(t) \approx \mathbf{c}'(t)\Delta t$.

From the definition of the derivative, we get the approximation $\Delta \mathbf{s} \approx \mathbf{c}'(t)\Delta t$. The work done in going from $\mathbf{c}(t)$ to $\mathbf{c}(t + \Delta t)$ is therefore approximately

$$\mathbf{F}(\mathbf{c}(t)) \cdot \Delta \mathbf{s} \approx \mathbf{F}(\mathbf{c}(t)) \cdot \mathbf{c}'(t)\Delta t.$$

If we subdivide the interval $[a, b]$ into n equal parts $a = t_0 < t_1 < \cdots < t_n = b$, with $\Delta t = t_{i+1} - t_i$, then the work done by $\mathbf{F}$ is approximately

$$\sum_{i=0}^{n-1} \mathbf{F}(\mathbf{c}(t_i)) \cdot \Delta \mathbf{s} = \sum_{i=0}^{n-1} \mathbf{F}(\mathbf{c}(t_i)) \cdot \mathbf{c}'(\mathbf{t}_i)\Delta t.$$

As $n \to \infty$, this approximation becomes better and better, and so it is reasonable to take as our definition of work to be the limit of the above sum as $n \to \infty$. This limit is given by the integral

$$\int_a^b \mathbf{F}(\mathbf{c}(t)) \cdot \mathbf{c}'(t)\, dt.$$

This discussion leads us to make the following definition.

DEFINITION: LINE INTEGRALS Let $\mathbf{F}$ be a vector field on $\mathbb{R}^3$ which is continuous on the C^1 path $\mathbf{c}\colon [a, b] \to \mathbb{R}^3$. We define $\int_{\mathbf{c}} \mathbf{F} \cdot d\mathbf{s}$, the *line integral* of $\mathbf{F}$ along $\mathbf{c}$, by the formula

$$\int_{\mathbf{c}} \mathbf{F} \cdot d\mathbf{s} = \int_a^b \mathbf{F}(\mathbf{c}(t)) \cdot \mathbf{c}'(t)\, dt;$$

that is, we integrate the dot product of $\mathbf{F}$ with $\mathbf{c}'$ over the interval $[a, b]$.

As is the case with scalar functions, we can also define $\int_{\mathbf{c}} \mathbf{F} \cdot d\mathbf{s}$ if $\mathbf{F}(\mathbf{c}(t)) \cdot \mathbf{c}'(t)$ is only piecewise continuous.

For paths $\mathbf{c}$ that satisfy $\mathbf{c}'(t) \neq \mathbf{0}$, there is another useful formula for the line integral: namely, if $\mathbf{T}(t) = \mathbf{c}'(t)/\|\mathbf{c}'(t)\|$ denotes the unit tangent vector, we have

$$\int \mathbf{F} \cdot d\mathbf{s} = \int_a^b \mathbf{F}(\mathbf{c}(t)) \cdot \mathbf{c}'(t)\, dt \qquad \text{(by definition)}$$

$$= \int_a^b \left[\mathbf{F}(\mathbf{c}(t)) \cdot \frac{\mathbf{c}'(t)}{\|\mathbf{c}'(t)\|} \right] \|\mathbf{c}'(t)\|\, dt \qquad \text{(canceling } \|\mathbf{c}'(t)\|)$$

$$= \int_a^b [\mathbf{F}(\mathbf{c}(t)) \cdot \mathbf{T}(t)]\|\mathbf{c}'(t)\|\, dt \qquad (1)$$

This formula says that $\int_{\mathbf{c}} \mathbf{F} \cdot d\mathbf{s}$ is equal to something that looks like the path integral of the tangential component $\mathbf{F}(\mathbf{c}(t)) \cdot \mathbf{T}(t)$ of $\mathbf{F}$ along $\mathbf{c}$. In fact, the last part of formula (1) is analogous to the path integral of a scalar function f along $\mathbf{c}$. [NOTE: If $\mathbf{c}$ does not intersect itself (that is, if $\mathbf{c}(t_1) = \mathbf{c}(t_2)$ implies $t_1 = t_2$), then each point P on C (the image curve of $\mathbf{c}$) can be written uniquely as $\mathbf{c}(t)$ for some t. If we define $f(\mathrm{P}) = f(\mathbf{c}(t)) = \mathbf{F}(\mathbf{c}) \cdot \mathbf{T}(t)$,

f is a function on C and by definition its path integral along $\mathbf{c}$ is given by formula (1) and there is no difficulty in interpreting $\int_{\mathbf{c}} \mathbf{F} \cdot d\mathbf{s}$ as a path integral. If $\mathbf{c}$ intersects itself, we cannot define f as a function on C as above (why?); however, in this case it is still useful to think of the right side of formula (1) as a path integral.]

To compute a line integral in any particular case, one can either use the original definition or integrate the tangential component of $\mathbf{F}$ along $\mathbf{c}$, as prescribed by formula (1), whichever is easier or more appropriate.

EXAMPLE 1   Let $\mathbf{c}(t) = (\sin t, \cos t, t)$ with $0 \le t \le 2\pi$. Let the vector field $\mathbf{F}$ be defined by $\mathbf{F}(x, y, z) = x\mathbf{i} + y\mathbf{j} + z\mathbf{k}$. Compute $\int_{\mathbf{c}} \mathbf{F} \cdot d\mathbf{s}$.

SOLUTION   Here, $\mathbf{F}(\mathbf{c}(t)) = \mathbf{F}(\sin t, \cos t, t) = (\sin t)\mathbf{i} + (\cos t)\mathbf{j} + t\mathbf{k}$, and $\mathbf{c}'(t) = (\cos t)\mathbf{i} - (\sin t)\mathbf{j} + \mathbf{k}$. Therefore,

$$\mathbf{F}(\mathbf{c}(t)) \cdot \mathbf{c}'(t) = \sin t \cos t - \cos t \sin t + t = t,$$

and so

$$\int_{\mathbf{c}} \mathbf{F} \cdot d\mathbf{s} = \int_0^{2\pi} t \, dt = 2\pi^2. \quad \blacktriangle$$

Another common way of writing line integrals is

$$\int_{\mathbf{c}} \mathbf{F} \cdot d\mathbf{s} = \int_{\mathbf{c}} F_1 \, dx + F_2 \, dy + F_3 \, dz,$$

where F_1, F_2, and F_3 are the components of the vector field $\mathbf{F}$. We call the expression $F_1 \, dx + F_2 \, dy + F_3 \, dz$ a **differential form**.* By *definition*, the integral of a differential form along a path $\mathbf{c}$, where $\mathbf{c}(t) = (x(t), y(t), z(t))$, is

$$\int_{\mathbf{c}} F_1 \, dx + F_2 \, dy + F_3 \, dz = \int_a^b \left(F_1 \frac{dx}{dt} + F_2 \frac{dy}{dt} + F_3 \frac{dz}{dt} \right) dt = \int_{\mathbf{c}} \mathbf{F} \cdot d\mathbf{s}.$$

Note that we may think of $d\mathbf{s}$ as the differential form $d\mathbf{s} = dx\mathbf{i} + dy\mathbf{j} = dz\mathbf{k}$. Thus, the differential form $F_1 dx + F_2 dy + F_3 dz$ may be written as the dot product $\mathbf{F} \cdot d\mathbf{s}$.

EXAMPLE 2   Evaluate $\int_{\mathbf{c}} x^2 \, dx + xy \, dy + dz$, where $\mathbf{c}: [0, 1] \to \mathbb{R}^3$ is given by $\mathbf{c}(t) = (t, t^2, 1) = (x(t), y(t), z(t))$.

SOLUTION   We compute $dx/dt = 1$, $dy/dt = 2t$, $dz/dt = 0$; therefore,

$$\int_{\mathbf{c}} x^2 \, dx + xy \, dy + dz = \int_0^1 \left([x(t)]^2 \frac{dx}{dt} + [x(t)y(t)] \frac{dy}{dt} \right) dt$$

$$= \int_0^1 (t^2 + 2t^4) \, dt = \left[\frac{1}{3} t^3 + \frac{2}{5} t^5 \right]_0^1 = \frac{11}{15}. \quad \blacktriangle$$

---

*See Section 8.6 for a brief discussion of the general theory of differential forms.

EXAMPLE 3 Evaluate the integral $\int_{\mathbf{c}} \cos z \, dx + e^x \, dy + e^y \, dz$, where $\mathbf{c}(t) = (1, t, e^t)$ and $0 \leq t \leq 2$.

SOLUTION We compute $dx/dt = 0$, $dy/dt = 1$, $dz/dt = e^t$, and so

$$\int_{\mathbf{c}} \cos z \, dx + e^x \, dy + e^y \, dz = \int_0^2 (0 + e + e^{2t}) \, dt$$

$$= \left[et + \frac{1}{2}e^{2t} \right]_0^2 = 2e + \frac{1}{2}e^4 - \frac{1}{2}. \quad \blacktriangle$$

EXAMPLE 4 Let $\mathbf{c}$ be the path

$$x = \cos^3 \theta, \quad y = \sin^3 \theta, \quad z = \theta, \quad 0 \leq \theta \leq \frac{7\pi}{2}$$

(see Figure 7.2.2). Evaluate the integral $\int_{\mathbf{c}} (\sin z \, dx + \cos z \, dy - (xy)^{1/3} \, dz)$.

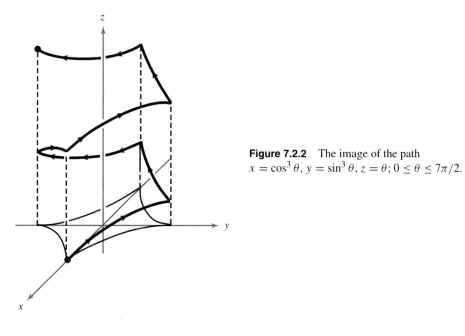

Figure 7.2.2 The image of the path $x = \cos^3 \theta$, $y = \sin^3 \theta$, $z = \theta$; $0 \leq \theta \leq 7\pi/2$.

SOLUTION In this case we have

$$\frac{dx}{d\theta} = -3\cos^2 \theta \sin \theta, \qquad \frac{dy}{d\theta} = 3\sin^2 \theta \cos \theta, \qquad \frac{dz}{d\theta} = 1,$$

and so the integral is

$$\int_{\mathbf{c}} \sin z \, dx + \cos z \, dy - (xy)^{1/3} \, dz$$

$$= \int_0^{7\pi/2} (-3 \cos^2 \theta \sin^2 \theta + 3 \sin^2 \theta \cos^2 \theta - \cos \theta \sin \theta) \, d\theta$$

$$= -\int_0^{7\pi/2} \cos \theta \sin \theta \, d\theta = -\left[\frac{1}{2} \sin^2 \theta \right]_0^{7\pi/2} = -\frac{1}{2}. \quad \blacktriangle$$

EXAMPLE 5   Suppose $\mathbf{F}$ is the vector force field $\mathbf{F}(x, y, z) = x^3\mathbf{i} + y\mathbf{j} + z\mathbf{k}$. Parametrize the circle of radius a in the yz plane by letting $\mathbf{c}(\theta)$ have components

$$x = 0, \qquad y = a \cos \theta, \qquad z = a \sin \theta, \qquad 0 \le \theta \le 2\pi.$$

Since $\mathbf{F}(\mathbf{c}(\theta)) \cdot \mathbf{c}'(\theta) = 0$, the force field $\mathbf{F}$ is normal to the circle at every point on the circle, so $\mathbf{F}$ will not do any work on a particle moving along the circle (Figure 7.2.3).

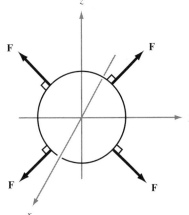

Figure 7.2.3   A vector field $\mathbf{F}$ normal to a circle in the yz plane.

We can verify by direct computation that the work done by $\mathbf{F}$ is zero.

$$W = \int_{\mathbf{c}} \mathbf{F} \cdot d\mathbf{s} = \int_{\mathbf{c}} x^3 \, dx + y \, dy + z \, dz$$

$$= \int_0^{2\pi} (0 - a^2 \cos \theta \sin \theta + a^2 \cos \theta \sin \theta) \, d\theta = 0. \quad \blacktriangle$$

EXAMPLE 6   If we consider the field and curve of Example 4, we see that the work done by the field is $-\frac{1}{2}$, a negative quantity. This means that the field impedes movement along the path. $\blacktriangle$

The line integral $\int_{\mathbf{c}} \mathbf{F} \cdot d\mathbf{s}$ depends not only on the field $\mathbf{F}$ but also on the path $\mathbf{c} \colon [a, b] \to \mathbb{R}^3$. In general, if $\mathbf{c}_1$ and $\mathbf{c}_2$ are two different paths in $\mathbb{R}^3$, $\int_{\mathbf{c}_1} \mathbf{F} \cdot d\mathbf{s} \ne \int_{\mathbf{c}_2} \mathbf{F} \cdot d\mathbf{s}$. On the other hand, we shall see that it is true that $\int_{\mathbf{c}_1} \mathbf{F} \cdot d\mathbf{s} = \pm \int_{\mathbf{c}_2} \mathbf{F} \cdot d\mathbf{s}$ for every vector field $\mathbf{F}$ if $\mathbf{c}_1$ is what we call a *reparametrization* of $\mathbf{c}_2$.

DEFINITION   Let $h: I \to I_1$ be a C^1 real-valued function that is a one-to-one map of an interval $I = [a, b]$ onto another interval $I_1 = [a_1, b_1]$. Let $\mathbf{c}: I_1 \to \mathbb{R}^3$ be a piecewise C^1 path. Then we call the composition

$$\mathbf{p} = \mathbf{c} \circ h: I \to \mathbb{R}^3$$

a ***reparametrization*** of $\mathbf{c}$.

   This means that $\mathbf{p}(t) = \mathbf{c}(h(t))$, and so h changes the variable; alternatively, one can think of h as changing the speed at which a point moves along the path. Indeed, observe that $\mathbf{p}'(t) = \mathbf{c}'(h(t))h'(t)$, so that the velocity vector for $\mathbf{p}$ is multiplied by the scalar factor $h'(t)$.

   It is implicit in the definition that h must carry endpoints to endpoints; that is, either $h(a) = a_1$ and $h(b) = b_1$, or $h(a) = b_1$ and $h(b) = a_1$. We thus distinguish two types of reparametrizations. If $\mathbf{c} \circ h$ is a reparametrization of $\mathbf{c}$, then either

$$(\mathbf{c} \circ h)(a) = \mathbf{c}(a_1) \qquad \text{and} \qquad (\mathbf{c} \circ h)(b) = \mathbf{c}(b_1)$$

or

$$(\mathbf{c} \circ h)(a) = \mathbf{c}(b_1) \qquad \text{and} \qquad (\mathbf{c} \circ h)(b) = \mathbf{c}(a_1).$$

In the first case, the reparametrization is said to be ***orientation-preserving***, and a particle tracing the path $\mathbf{c} \circ h$ moves in the same direction as a particle tracing $\mathbf{c}$. In the second case, the reparametrization is described as ***orientation-reversing***, and a particle tracing the path $\mathbf{c} \circ h$ moves in the opposite direction to that of a particle tracing $\mathbf{c}$ (Figure 7.2.4).

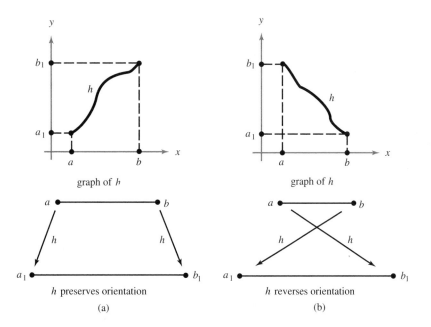

Figure 7.2.4   Illustrating an orientation-preserving reparametrization (a) and an orientation-reversing reparametrization (b).

For example, if C is the image of a path $\mathbf{c}$ as shown in Figure 7.2.5, that is, $C = \mathbf{c}([a_1, b_1])$, and if h is orientation-preserving, then $\mathbf{c} \circ h(t)$ will go from $\mathbf{c}(a_1)$ to $\mathbf{c}(b_1)$ as t goes from a to b; and if h is orientation-reversing, $\mathbf{c} \circ h(t)$ will go from $\mathbf{c}(b_1)$ to $\mathbf{c}(a_1)$ as t goes from a to b.

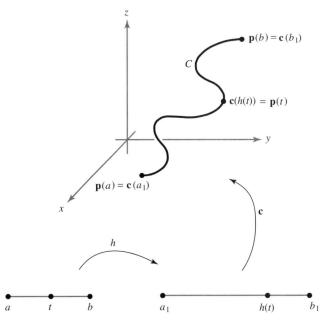

Figure 7.2.5   The path $\mathbf{p} = \mathbf{c} \circ h$ is a reparametrization of $\mathbf{c}$.

EXAMPLE 7   Let $\mathbf{c}$: $[a, b] \to \mathbb{R}^3$ be a piecewise C^1 path. Then:

(a) The path $\mathbf{c}_{\mathrm{op}}$: $[a, b] \to \mathbb{R}^3$, $t \mapsto \mathbf{c}(a + b - t)$, is reparametrization of $\mathbf{c}$ corresponding to the map h: $[a, b] \to [a, b]$, $t \mapsto a + b - t$; we call $\mathbf{c}_{\mathrm{op}}$ the ***opposite path*** to $\mathbf{c}$. This reparametrization is orientation-reversing.

(b) The path $\mathbf{p}$: $[0, 1] \to \mathbb{R}^3$, $t \mapsto \mathbf{c}(a + (b - a)t)$, is an orientation-preserving reparametrization of $\mathbf{c}$ corresponding to a change of coordinates h: $[0, 1] \to [a, b]$, $t \mapsto a + (b - a)t$.   ▲

THEOREM 1: CHANGE OF PARAMETRIZATION FOR LINE INTEGRALS   *Let* $\mathbf{F}$ *be a vector field continuous on the* C^1 *path* $\mathbf{c}$: $[a_1, b_1] \to \mathbb{R}^3$, *and let* $\mathbf{p}$: $[a, b] \to \mathbb{R}^3$ *be a reparametrization of* $\mathbf{c}$. *If* $\mathbf{p}$ *is orientation-preserving, then*

$$\int_{\mathbf{p}} \mathbf{F} \cdot d\mathbf{s} = \int_{\mathbf{c}} \mathbf{F} \cdot d\mathbf{s},$$

and if $\mathbf{p}$ *is orientation-reversing, then*

$$\int_{\mathbf{p}} \mathbf{F} \cdot d\mathbf{s} = -\int_{\mathbf{c}} \mathbf{F} \cdot d\mathbf{s}.$$

PROOF   By hypothesis, we have a map h such that $\mathbf{p} = \mathbf{c} \circ h$. By the chain rule,

$$\mathbf{p}'(t) = \mathbf{c}'(h(t))h'(t),$$

and so

$$\int_{\mathbf{p}} \mathbf{F} \cdot d\mathbf{s} = \int_a^b [\mathbf{F}(\mathbf{c}(h(t))) \cdot \mathbf{c}'(h(t))]h'(t)\, dt.$$

Changing variables with $s = h(t)$ this becomes

$$\int_{h(a)}^{h(b)} \mathbf{F}(\mathbf{c}(s)) \cdot \mathbf{c}'(s)\, ds$$

$$= \begin{cases} \displaystyle\int_{a_1}^{b_1} \mathbf{F}(\mathbf{c}(s)) \cdot \mathbf{c}'(s)\, ds = \int_{\mathbf{c}} \mathbf{F} \cdot d\mathbf{s} & \text{if } \mathbf{p} \text{ is orientation-} \\[2mm] & \text{preserving} \\[2mm] \displaystyle\int_{b_1}^{a_1} \mathbf{F}(\mathbf{c}(s)) \cdot \mathbf{c}'(s)\, ds = -\int_{\mathbf{c}} \mathbf{F} \cdot d\mathbf{s} & \text{if } \mathbf{p} \text{ is orientation-} \\[2mm] & \text{reversing.} \quad\blacksquare \end{cases}$$

Theorem 1 also holds for piecewise C^1 paths, as may be seen by breaking up the intervals into segments on which the paths are of class C^1 and summing the integrals over the separate intervals.

Thus, if it is convenient to reparametrize a path when evaluating an integral, Theorem 1 assures us that the value of the integral will not be affected, except possibly for the sign, depending on the orientation.

EXAMPLE 8   Let $\mathbf{F}(x, y, z) = yz\mathbf{i} + xz\mathbf{j} + xy\mathbf{k}$ and $\mathbf{c}\colon [-5, 10] \to \mathbb{R}^3$ be defined by $t \mapsto (t, t^2, t^3)$. Evaluate $\int_{\mathbf{c}} \mathbf{F} \cdot d\mathbf{s}$ and $\int_{\mathbf{c}_{\mathrm{op}}} \mathbf{F} \cdot d\mathbf{s}$.

SOLUTION   For $\mathbf{c}$, we have $dx/dt = 1$, $dy/dt = 2t$, $dz/dt = 3t^2$, and $\mathbf{F}(\mathbf{c}(t)) = t^5\mathbf{i} + t^4\mathbf{j} + t^3\mathbf{k}$. Therefore,

$$\int_{\mathbf{c}} \mathbf{F} \cdot d\mathbf{s} = \int_{-5}^{10} \left(F_1 \frac{dx}{dt} + F_2 \frac{dy}{dt} + F_3 \frac{dz}{dt} \right) dt$$

$$= \int_{-5}^{10} (t^5 + 2t^5 + 3t^5)\, dt = [t^6]_{-5}^{10} = 984{,}375.$$

On the other hand, for

$$\mathbf{c}_{\mathrm{op}}\colon [-5, 10] \to \mathbb{R}^3, t \mapsto \mathbf{c}(5 - t) = (5 - t, (5 - t)^2, (5 - t)^3),$$

we have $dx/dt = -1$, $dy/dt = -10 + 2t = -2(5 - t)$, $dz/dt = -75 + 30t - 3t^2 = -3(5 - t)^2$, and $\mathbf{F}(\mathbf{c}_{\mathrm{op}}(t)) = (5 - t)^5\mathbf{i} + (5 - t)^4\mathbf{j} + (5 - t)^3\mathbf{k}$. Therefore,

$$\int_{\mathbf{c}_{\mathrm{op}}} \mathbf{F} \cdot d\mathbf{s} = \int_{-5}^{10} [-(5 - t)^5 - 2(5 - t)^5 - 3(5 - t)^5]\, dt = [(5 - t)^6]_{-5}^{10} = -984{,}375. \quad\blacktriangle$$

We are interested in reparametrizations because if the image of a particular **c** can be represented in many ways, we want to be sure that path and line integrals depend only on the image curve and not on the particular parametrization. For example, for some problems the unit circle may be conveniently represented by the map **p** given by

$$x(t) = \cos 2t, \qquad y(t) = \sin 2t, \qquad 0 \leq t \leq \pi.$$

Theorem 1 guarantees that any integral computed for this representation will be the same as when we represent the circle by the map **c** given by

$$x(t) = \cos t, \qquad y(t) = \sin t, \qquad 0 \leq t \leq 2\pi,$$

since $\mathbf{p} = \mathbf{c} \circ h$, where $h(t) = 2t$, and thus **p** is a reparametrization of **c**. However, notice that the map γ given by

$$\gamma(t) = (\cos t, \sin t), \qquad y(t) = \sin t, \qquad 0 \leq t \leq 4\pi,$$

is *not* a reparametrization of **c**. Although it traces out the same image (the circle), it does so twice. (Why does this imply that γ is not a reparametrization of **c**?)

The line integral is an *oriented integral*, in that a change of sign occurs (as we have seen in Theorem 1) if the orientation of the curve is reversed. The *path integral* does not have this property. This follows from the fact that changing t to $-t$ (reversing orientation) just changes the sign of $\mathbf{c}'(t)$, not its length. This is one of the differences between line and path integrals. The following theorem, which is proved by the same method as Theorem 1, shows that path integrals are unchanged under reparametrizations—even orientation-reversing ones.

THEOREM 2: CHANGE OF PARAMETRIZATION FOR PATH INTEGRALS   Let **c** be piecewise C^1, f a continuous (real-valued) function on the image of **c**; and let **p** be any reparametrization of **c**. Then

$$\int_{\mathbf{c}} f(x, y, z)\, ds = \int_{\mathbf{p}} f(x, y, z)\, ds. \tag{2}$$

We next consider a useful technique for evaluating certain types of line integrals. Recall that a vector field **F** is a *gradient vector field* if $\mathbf{F} = \nabla f$ for some real-valued function f. Thus

$$\mathbf{F} = \frac{\partial f}{\partial x}\mathbf{i} + \frac{\partial f}{\partial y}\mathbf{j} + \frac{\partial f}{\partial z}\mathbf{k}.$$

Suppose g and G are real-valued continuous functions defined on a closed interval $[a, b]$, that G is differentiable on (a, b) and that $G' = g$. Then by the fundamental theorem of calculus

$$\int_a^b g(x)\, dx = G(b) - G(a).$$

Thus the value of the integral of g depends only on the value of G at the endpoints of the interval $[a, b]$. Since ∇f represents the derivative of f, one can ask whether $\int_{\mathbf{c}} \nabla f \cdot d\mathbf{s}$ is completely determined by the value of f at the endpoints $\mathbf{c}(a)$ and $\mathbf{c}(b)$. The answer is contained in the following *generalization of the fundamental theorem of calculus.*

THEOREM 3: LINE INTEGRALS FOR GRADIENT VECTOR FIELDS *Suppose that $f: \mathbb{R}^3 \to \mathbb{R}$ is of class C^1 and that $\mathbf{c}: [a, b] \to \mathbb{R}^3$ is a piecewise C^1 path. Then*

$$\int_{\mathbf{c}} \nabla f \cdot d\mathbf{s} = f(\mathbf{c}(b)) - f(\mathbf{c}(a)).$$

PROOF   Apply the chain rule to the composite function

$$F: t \mapsto f(\mathbf{c}(t))$$

to obtain

$$F'(t) = (f \circ \mathbf{c})'(t) = \nabla f(\mathbf{c}(t)) \cdot \mathbf{c}'(t).$$

The function F is a real-valued function of the variable t, and so, by the fundamental theorem of calculus,

$$\int_a^b F'(t)\, dt = F(b) - F(a) = f(\mathbf{c}(b)) - f(\mathbf{c}(a)).$$

Therefore,

$$\int_{\mathbf{c}} \nabla f \cdot d\mathbf{s} = \int_a^b \nabla f(\mathbf{c}(t)) \cdot \mathbf{c}'(t)\, dt = \int_a^b F'(t)\, dt = F(b) - F(a)$$
$$= f(\mathbf{c}(b)) - f(\mathbf{c}(a)). \quad \blacksquare$$

EXAMPLE 9   Let $\mathbf{c}$ be the path $\mathbf{c}(t) = (t^4/4, \sin^3(t\pi/2), 0), t \in [0, 1]$. Evaluate

$$\int_{\mathbf{c}} y\, dx + x\, dy$$

$\left(\text{which means } \int_{\mathbf{c}} y\, dx + x\, dy + 0\, dz\right).$

SOLUTION   We recognize $y\, dx + x\, dy$, or equivalently, the vector field $y\mathbf{i} + x\mathbf{j} + 0\mathbf{k}$, as the gradient of the function $f(x, y, z) = xy$. Thus,

$$\int_{\mathbf{c}} y\, dx + x\, dy = f(\mathbf{c}(1)) - f(\mathbf{c}(0)) = \frac{1}{4} \cdot 1 - 0 = \frac{1}{4}. \quad \blacktriangle$$

Obviously, if one can recognize the integrand as a gradient, then evaluation of the integral becomes much easier. For example, the reader should try to work out the integral in Example 9 directly. In one-variable calculus, every integral is, in principle, obtainable by finding an antiderivative. For vector fields, however, this is not always true, because a given vector field need not be a gradient. This point will be examined in detail in Section 8.3.

We have seen how to define path integrals (integrals of scalar functions) and line integrals (integrals of vector functions) over parametrized curves. We have also seen that our work is simplified if we make a judicious choice of parametrization. Since these integrals are independent of the parametrization (except possibly for the sign), it seems natural to express the theory in a way that is independent of the parametrization, and that is thereby more "geometric." We do this briefly and somewhat informally in the following discussion.

DEFINITION   We define a ***simple curve*** C to be the image of a piecewise C^1 map $\mathbf{c}: I \to \mathbb{R}^3$ that is one-to-one on an interval I; $\mathbf{c}$ is called a ***parametrization*** of C. Thus a simple curve is one that does not intersect itself (Figure 7.2.6). If $I = [a, b]$, we call $\mathbf{c}(a)$

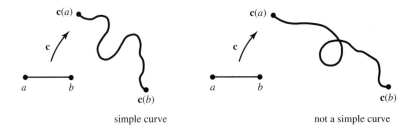

simple curve   not a simple curve

Figure 7.2.6   A simple curve that has no self-intersections is shown on the left. On the right we have a curve with a self-intersection, so it is not simple.

and $\mathbf{c}(b)$ ***endpoints*** of the curve. Each simple curve C has two orientations or directions associated with it. If P and Q are the endpoints of the curve, then we can consider C as directed either from P to Q or from Q to P. The simple curve C together with a sense of direction is called an ***oriented simple curve*** or ***directed simple curve*** (Figure 7.2.7).

Figure 7.2.7   There are two possible senses of direction on a curve joining P and Q.

DEFINITION: SIMPLE CLOSED CURVES By a *simple closed curve*
we mean the image of a piecewise C^1 map $\mathbf{c}$: $[a, b] \to \mathbb{R}^3$ that is one-to-one on $[a, b)$
and satisfies $\mathbf{c}(a) = \mathbf{c}(b)$ (Figure 7.2.8). If $\mathbf{c}$ satisfies the condition $\mathbf{c}(a) = \mathbf{c}(b)$ but is
not necessarily one-to-one on $[a, b)$, we call its image a *closed curve*. Simple closed
curves have two orientations, corresponding to the two possible directions of motion
along the curve (Figure 7.2.9).

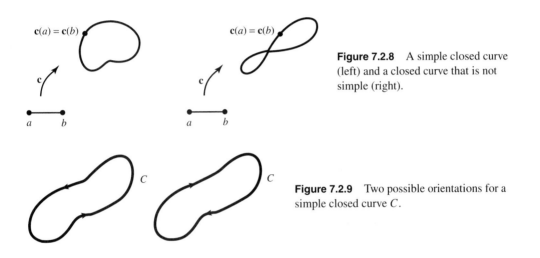

Figure 7.2.8 A simple closed curve
(left) and a closed curve that is not
simple (right).

Figure 7.2.9 Two possible orientations for a
simple closed curve C.

If C is an oriented simple curve or an oriented simple closed curve, we may unambiguously define:

**LINE INTEGRALS OVER ORIENTED SIMPLE CURVES AND
SIMPLE CLOSED CURVES** C:

$$\int_C \mathbf{F} \cdot d\mathbf{s} = \int_{\mathbf{c}} \mathbf{F} \cdot d\mathbf{s} \quad \text{and} \quad \int_C f \, ds = \int_{\mathbf{c}} f \, ds, \tag{3}$$

where $\mathbf{c}$ is any *orientation-preserving* parametrization of C.

These integrals do not depend on the choice of $\mathbf{c}$ as long as $\mathbf{c}$ is one-to-one (except
possibly at the endpoints) by virtue of Theorems 1 and 2. (We have not proved that any
two one-to-one paths $\mathbf{c}$ and $\mathbf{p}$ with the same image must be reparametrizations of each
other, but this technical point will be omitted.) The point we want to make here is that
although *a curve must be parametrized to make integration along it tractable, it is not
necessary to include the parametrization in our notation for the integral.*

EXAMPLE 10   If $I = [a, b]$ is a closed interval on the x axis, then I, as a curve, has two orientations: one corresponding to motion from a to b (left to right) and the other corresponding to motion from b to a (right to left). If f is a real-valued function continuous on I, then denoting I with the first orientation by I^+ and I with the second orientation by I^-, we have

$$\int_{I^+} f(x)\,dx = \int_a^b f(x)\,dx = -\int_b^a f(x)\,dx = -\int_{I^-} f(x)\,dx. \quad \blacktriangle$$

A given simple closed curve can be parametrized in many different ways. Figure 7.2.10 shows C represented as the image of a map $\mathbf{p}$, with $\mathbf{p}(t)$ progressing in a prescribed direction around an oriented curve C as t ranges from a to b. Note that $\mathbf{p}'(t)$ points in this direction also. The speed with which we traverse C may vary from parametrization to parametrization, but as long as the orientation is preserved the integral will not, according to Theorems 1 and 2.

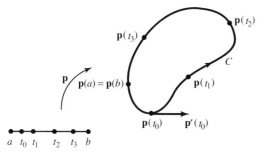

Figure 7.2.10   As t goes from a to b, $\mathbf{p}(t)$ moves around the curve C in some fixed direction.

The following precaution should be noted in regard to these remarks. It is possible to have two mappings $\mathbf{c}$ and $\mathbf{p}$ with the same image, and inducing the same orientation on the image, such that

$$\int_{\mathbf{c}} \mathbf{F} \cdot d\mathbf{s} \neq \int_{\mathbf{p}} \mathbf{F} \cdot d\mathbf{s}.$$

For an example, let $\mathbf{c}(t) = (\cos t, \sin t, 0)$ and $\mathbf{p}(t) = (\cos 2t, \sin 2t, 0), 0 \leq t \leq 2\pi$, with $\mathbf{F}(x, y, z) = (y, 0, 0)$. Then

$$\int_{\mathbf{c}} \mathbf{F} \cdot d\mathbf{s} = \int_0^{2\pi} F_1(\mathbf{c}(t)) \frac{dx}{dt}\,dt$$

(the terms containing F_2 and F_3 are zero)

$$= -\int_0^{2\pi} \sin^2 t\,dt = -\pi.$$

But $\int_{\mathbf{p}} \mathbf{F} \cdot d\mathbf{s} = -2 \int_0^{2\pi} \sin^2 2t\,dt = -2\pi$. Clearly $\mathbf{c}$ and $\mathbf{p}$ have the same image, namely, the unit circle in the xy plane, and, moreover, they traverse the unit circle in the same direction; yet $\int_{\mathbf{c}} \mathbf{F} \cdot d\mathbf{s} \neq \int_{\mathbf{p}} \mathbf{F} \cdot d\mathbf{s}$. The reason for this is that $\mathbf{c}$ is one-to-one but $\mathbf{p}$ is not ($\mathbf{p}$ traverses the unit circle twice in a counterclockwise direction); therefore $\mathbf{p}$ is not a parametrization of the unit circle as a simple closed curve.

As a consequence of Theorem 1 and generalizing the notation in Example 10, we introduce the following convention:

LINE INTEGRALS OVER CURVES WITH OPPOSITE ORIENTATIONS Let C^- be the same curve as C, but with the opposite orientation. Then

$$\int_C \mathbf{F} \cdot d\mathbf{s} = -\int_{C^-} \mathbf{F} \cdot d\mathbf{s}.$$

We also have

LINE INTEGRALS OVER CURVES CONSISTING OF SEVERAL COMPONENTS Let C be an oriented curve that is made up of several oriented component curves $C_i, i = 1, \ldots, k$, as in Figure 7.2.11. Then we shall write $C = C_1 + C_2 + \cdots + C_k$. Since we can parametrize C by parametrizing the pieces $C_1, \ldots, C_k$ separately, one can prove that

$$\int_C \mathbf{F} \cdot d\mathbf{s} = \int_{C_1} \mathbf{F} \cdot d\mathbf{s} + \int_{C_2} \mathbf{F} \cdot d\mathbf{s} + \cdots + \int_{C_k} \mathbf{F} \cdot d\mathbf{s}. \qquad (4)$$

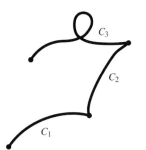

Figure 7.2.11   A curve can be made up of several components.

One reason for writing a curve as a sum of components is that it may be easier to parametrize the components C_i individually than it is to parametrize C as a whole. If that is the case, formula (4) provides a convenient way of evaluating $\int_C \mathbf{F} \cdot d\mathbf{s}$.

EXAMPLE 11   Consider C, the perimeter of the unit square in $\mathbb{R}^2$, oriented in the counterclockwise sense (see Figure 7.2.12). Evaluate $\int_C x^2 \, dx + xy \, dy$.

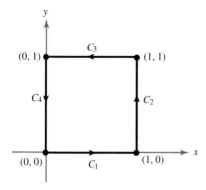

Figure 7.2.12   The perimeter of the unit square, parametrized in four pieces.

SOLUTION   We evaluate the integral using a convenient parametrization of C that induces the given orientation. For example:

$$\mathbf{c}\colon [0,4] \to \mathbb{R}^2, \qquad t \mapsto \begin{cases} (t,0) & 0 \le t \le 1 \\ (1, t-1) & 1 \le t \le 2 \\ (3-t, 1) & 2 \le t \le 3 \\ (0, 4-t) & 3 \le t \le 4. \end{cases}$$

Then

$$\int_C x^2\, dx + xy\, dy = \int_0^1 (t^2 + 0)\, dt + \int_1^2 [0 + (t-1)]\, dt$$
$$+ \int_2^3 [-(3-t)^2 + 0]\, dt + \int_3^4 (0+0)\, dt$$
$$= \frac{1}{3} + \frac{1}{2} + \left(-\frac{1}{3}\right) + 0 = \frac{1}{2}.$$

Now let us reevaluate this line integral, using formula (4) and parametrizing the C_i separately. Notice that $C = C_1 + C_2 + C_3 + C_4$, where C_i are the oriented curves pictured in Figure 7.2.12. These can be parametrized as follows:

$$C_1\colon \mathbf{c}_1(t) = (t, 0), 0 \le t \le 1$$
$$C_2\colon \mathbf{c}_2(t) = (1, t), 0 \le t \le 1$$
$$C_3\colon \mathbf{c}_3(t) = (1-t, 1), 0 \le t \le 1$$
$$C_4\colon \mathbf{c}_4(t) = (0, 1-t), 0 \le t \le 1$$

and so

$$\int_{C_1} x^2\, dx + xy\, dy = \int_0^1 t^2\, dt = \frac{1}{3}$$
$$\int_{C_2} x^2\, dx + xy\, dy = \int_0^1 t\, dt = \frac{1}{2}$$

$$\int_{C_3} x^2 \, dx + xy \, dy = \int_0^1 -(1-t)^2 \, dt = -\frac{1}{3}$$

$$\int_{C_4} x^2 \, dx + xy \, dy = \int_0^1 0 \, dt = 0.$$

Thus, again,

$$\int_C x^2 \, dx + xy \, dy = \frac{1}{3} + \frac{1}{2} - \frac{1}{3} + 0 = \frac{1}{2}. \quad \blacktriangle$$

EXAMPLE 12 An interesting application of the line integral is the mathematical formulation of Ampère's law, which relates electric currents to their magnetic effects.[*] Suppose $\mathbf{H}$ denotes a magnetic field in $\mathbb{R}^3$, and let C be a closed oriented curve in $\mathbb{R}^3$. In appropriate physical units Ampère's law states that

$$\int_C \mathbf{H} \cdot d\mathbf{s} = I,$$

where I is the net current that passes through any surface bounded by C (see Figure 7.2.13). $\blacktriangle$

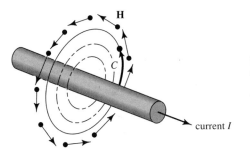

Figure 7.2.13 The magnetic field $\mathbf{H}$ surrounding a wire carrying a current I satisfies Ampère's law: $\int_C \mathbf{H} \cdot d\mathbf{s} = I$.

Finally, let us mention that the line integral has another important physical meaning, specifically, the interpretation of $\int_C \mathbf{V} \cdot d\mathbf{s}$ where $\mathbf{V}$ is the velocity field of a fluid which we shall discuss in Section 8.2. Thus, a wide variety of physical concepts, from the notion of work to electromagnetic fields and the motions of fluids, can be analyzed with the help of line integrals.

EXERCISES

1. Let $\mathbf{F}(x, y, z) = x\mathbf{i} + y\mathbf{j} + z\mathbf{k}$. Evaluate the integral of $\mathbf{F}$ along each of the following paths:

(a) $\mathbf{c}(t) = (t, t, t), \quad 0 \le t \le 1$

(b) $\mathbf{c}(t) = (\cos t, \sin t, 0), \quad 0 \le t \le 2\pi$

(c) $\mathbf{c}(t) = (\sin t, 0, \cos t), \quad 0 \le t \le 2\pi$

(d) $\mathbf{c}(t) = (t^2, 3t, 2t^3), \quad -1 \le t \le 2$

---

[*]The discovery that electric currents produce magnetic effects was made by Oersted about 1820. See any elementary physics text for discussions of the physical basis of these ideas.

2. Evaluate each of the following integrals:

(a) $\int_{\mathbf{c}} x\, dy - y\, dx$, $\mathbf{c}(t) = (\cos t, \sin t)$, $0 \le t \le 2\pi$

(b) $\int_{\mathbf{c}} x\, dx + y\, dy$, $\mathbf{c}(t) = (\cos \pi t, \sin \pi t)$, $0 \le t \le 2$

(c) $\int_{\mathbf{c}} yz\, dx + xz\, dy + xy\, dz$, where $\mathbf{c}$ consists of straight-line segments joining $(1, 0, 0)$ to $(0, 1, 0)$ to $(0, 0, 1)$

(d) $\int_{\mathbf{c}} x^2\, dx - xy\, dy + dz$, where $\mathbf{c}$ is the parabola $z = x^2$, $y = 0$ from $(-1, 0, 1)$ to $(1, 0, 1)$.

3. Consider the force $\mathbf{F}(x, y, z) = x\mathbf{i} + y\mathbf{j} + z\mathbf{k}$. Compute the work done in moving a particle along the parabola $y = x^2$, $z = 0$, from $x = -1$ to $x = 2$.

4. Let $\mathbf{c}$ be a smooth path.

(a) Suppose $\mathbf{F}$ is perpendicular to $\mathbf{c}'(t)$ at $\mathbf{c}(t)$. Show that

$$\int_{\mathbf{c}} \mathbf{F} \cdot d\mathbf{s} = 0.$$

(b) If $\mathbf{F}$ is parallel to $\mathbf{c}'(t)$ at $\mathbf{c}(t)$, show that

$$\int_{\mathbf{c}} \mathbf{F} \cdot d\mathbf{s} = \int_{\mathbf{c}} \|\mathbf{F}\|\, ds.$$

[By parallel to $\mathbf{c}'(t)$ we mean that $\mathbf{F}(\mathbf{c}(t)) = \lambda(t)\mathbf{c}'(t)$, where $\lambda(t) > 0$.]

5. Suppose $\mathbf{c}$ has length l, and $\|\mathbf{F}\| \le M$. Prove that

$$\left| \int_{\mathbf{c}} \mathbf{F} \cdot d\mathbf{s} \right| \le Ml.$$

6. Evaluate $\int_{\mathbf{c}} \mathbf{F} \cdot d\mathbf{s}$ where $\mathbf{F}(x, y, z) = y\mathbf{i} + 2x\mathbf{j} + y\mathbf{k}$ and $\mathbf{c}(t) = t\mathbf{i} + t^2\mathbf{j} + t^3\mathbf{k}$, $0 \le t \le 1$.

7. Evaluate $\int_{\mathbf{c}} y\, dx + (3y^3 - x)\, dy + z\, dz$ for each of the paths $\mathbf{c}(t) = (t, t^n, 0)$, $0 \le t \le 1$, where $n = 1, 2, 3, \ldots$.

8. This exercise refers to Example 12. Let L be a very long wire, a planar section of which (with the plane perpendicular to the wire) is shown in Figure 7.2.14. Suppose this plane is the xy plane. Experiments

Figure 7.2.14   A planar section of a long wire and a curve C about the wire.

show that **H** is tangent to every circle in the xy plane whose center is the axis of L, and that the magnitude of **H** is constant on every such circle C. Thus $\mathbf{H} = H\mathbf{T}$, where **T** is a unit tangent vector to C and H is some scalar. Using this information, show that $H = I/2\pi r$, where r is the radius of circle C and I is the current flowing in the wire.

9. The image of $t \mapsto (\cos^3 t, \sin^3 t), 0 \leq t \leq 2\pi$, in the plane is shown in Figure 7.2.15. Evaluate the integral of the vector field $\mathbf{F}(x, y) = x\mathbf{i} + y\mathbf{j}$ around this curve.

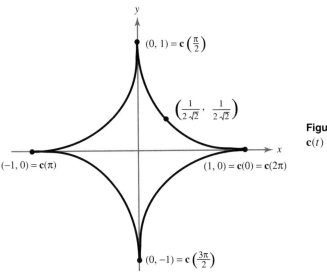

Figure 7.2.15 The hypocycloid $\mathbf{c}(t) = (\cos^3 t, \sin^3 t)$ (Exercise 9).

10. Suppose $\mathbf{c}_1$ and $\mathbf{c}_2$ are two paths with the same endpoints and **F** is a vector field. Show that $\int_{\mathbf{c}_1} \mathbf{F} \cdot d\mathbf{s} = \int_{\mathbf{c}_2} \mathbf{F} \cdot d\mathbf{s}$ is equivalent to $\int_c \mathbf{F} \cdot d\mathbf{s} = 0$, where C is the closed curve obtained by first moving along $\mathbf{c}_1$ and then moving along $\mathbf{c}_2$ in the opposite direction.

11. Let $\mathbf{c}(t)$ be a path and **T** the unit tangent vector. What is $\int_{\mathbf{c}} \mathbf{T} \cdot d\mathbf{s}$?

12. Let $\mathbf{F} = (z^3 + 2xy)\mathbf{i} + x^2\mathbf{j} + 3xz^2\mathbf{k}$. Show that the integral of **F** around the circumference of the unit square with vertices $(\pm 1, \pm 1)$ is zero.

13. Using the path in Exercise 9, observe that a C^1 map $\mathbf{c}$: $[a, b] \to \mathbb{R}^3$ can have an image that does not "look smooth." Do you think this could happen if $\mathbf{c}'(t)$ were always nonzero?

14. What is the value of the integral of a gradient field around a closed curve C?

15. Evaluate $\int_C 2xyz\,dx + x^2z\,dy + x^2y\,dz$, where C is an oriented simple curve connecting $(1, 1, 1)$ to $(1, 2, 4)$.

16. Suppose $\nabla f(x, y, z) = 2xyze^{x^2}\mathbf{i} + ze^{x^2}\mathbf{j} + ye^{x^2}\mathbf{k}$. If $f(0, 0, 0) = 5$, find $f(1, 1, 2)$.

17. Consider the gravitational force field (with $G = m = M = 1$) defined [for $(x, y, z) \neq (0, 0, 0)$] by

$$\mathbf{F}(x, y, z) = \frac{1}{(x^2 + y^2 + z^2)^{3/2}} (x\mathbf{i} + y\mathbf{j} + z\mathbf{k}).$$

Show that the work done by the gravitational force as a particle moves from (x_1, y_1, z_1) to (x_2, y_2, z_2) along any path depends only on the radii $R_1 = \sqrt{x_1^2 + y_1^2 + z_1^2}$ and $R_2 = \sqrt{x_2^2 + y_2^2 + z_2^2}$.

18. A cyclist rides up a mountain along the path shown in Figure 7.2.16. She makes one complete revolution around the mountain in reaching the top, while her vertical rate of climb is constant. Throughout the trip she exerts a force described by the vector field

$$\mathbf{F}(x, y, z) = y\mathbf{i} + x\mathbf{j} + \mathbf{k}$$

What is the work done by the cyclist in traveling from A to B? What is unrealistic about this model of a cyclist?

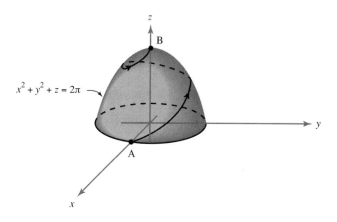

Figure 7.2.16   How much work is done in cycling up this mountain?

19. Let $\mathbf{c}: [a, b] \to \mathbb{R}^3$ be a path such that $\mathbf{c}'(t) \neq \mathbf{0}$. When this condition holds, $\mathbf{c}$ is said to be *regular*. Let the function f be defined by $f(x) = \int_a^x \|\mathbf{c}'(t)\| \, dt$.

(a) What is df/dx?

(b) Using the answer to part (a), prove that $f: [a, b] \to [0, L]$, where L is the length of $\mathbf{c}$, has a differentiable inverse $g: [0, L] \to [a, b]$ satisfying $f \circ g(s) = s$, $g \circ f(x) = x$. (You may use the one-variable inverse function theorem of Section 3.5.)

(c) Compute dg/ds.

(d) Recall that a path $s \mapsto \mathbf{b}(s)$ is of unit speed, or parametrized by arc length, if $\|\mathbf{c}'(s)\| = 1$. Show that the reparametrization of $\mathbf{c}$ given by $\mathbf{b}(s) = \mathbf{c} \circ g(s)$ is of unit speed. Conclude that any regular path can be reparametrized by arc length. (Thus, for example, the Frenet formulas in Exercise 17 of Section 4.2 can be applied to the reparametrization $\mathbf{b}$.)

20. Along a "thermodynamic path" C in (V, T, P) space,

(i) The heat gained is $\int_C \Lambda_V \, dV + K_V \, dT$; where Λ_V, K_V are functions of (V, T, P) depending on the particular physical system.

(ii) The work done is $\int_C P \, dV$.

For a van der Waals gas, we have

$$P(V, T) = \frac{RT}{V-b} - \frac{a}{V^2}, \qquad J\Lambda_V = \frac{RT}{V-b}, \qquad \text{and} \qquad K_V = \text{constant}$$

where R, b, a, and J are known constants. Initially the gas is at a temperature T_0 and volume V_0.

(a) An *adiabatic* process is a thermodynamic motion $(V(t), T(t), P(t))$ for which

$$\frac{dT}{dV} = \frac{dT/dt}{dV/dt} = -\frac{\Lambda_V}{K_V}.$$

If the van der Waals gas undergoes an adiabatic process in which the volume doubles to $2V_0$, compute

(1) the heat gained;
(2) the work done; and
(3) the final volume, temperature, and pressure.

(b)   After the process indicated in part (a), the gas is cooled (or heated) at constant volume until the original temperature T_0 is achieved. Compute

(1) the heat gained;
(2) the work done; and
(3) the final volume, temperature, and pressure.

(c)   After the process indicated in part (b), the gas is compressed until the gas returns to its original volume V_0. The temperature is held constant throughout the process. Compute

(1) the heat gained;
(2) the work done; and
(3) the final volume, temperature, and pressure.

(d)   For the cyclic process (a), (b), (c), compute

(1) the total heat gained and
(2) the total work done

7.3 PARAMETRIZED SURFACES

In Sections 7.1 and 7.2 we studied integrals of scalar and vector functions along curves. Now we turn to integrals over surfaces and begin by studying the geometry of surfaces themselves.

We are already used to one kind of surface, namely, the graph of a function $f(x, y)$. Graphs were extensively studied in Chapter 2, and we know how to compute their tangent planes. However, it would be unduly limiting to restrict ourselves to this case. For example, many surfaces arise as level surfaces of functions. Suppose our surface S is the set of points (x, y, z) where $x - z + z^3 = 0$. Here S is a sheet that (relative to the xy plane) doubles back on itself (see Figure 7.3.1). Obviously, we want to call S a surface, since it is just a plane with a wrinkle. However, S is *not* the graph of some function $z = f(x, y)$, because this means that for each $(x_0, y_0) \in \mathbb{R}^2$ there must be *one* z_0 with $(x_0, y_0, z_0) \in S$. As Figure 7.3.1 illustrates, this condition is violated.

Figure 7.3.1   A surface that is not the graph of a function $z = f(x, y)$.

Another example is the torus, or surface of a doughnut, which is depicted in Figure 7.3.2. Anyone would call a torus a surface; yet, by the same reasoning as above, a torus cannot be the graph of a differentiable function of two variables. These observations encourage us to extend our definition of a surface.

Figure 7.3.2   The torus is not the graph of a function of the form $z = f(x, y)$.

The motivation for the extended definition that follows is partly that a surface can be thought of as being obtained from the plane by "rolling," "bending," and "pushing." For example, to get a torus, we take a portion of the plane and roll it (see Figure 7.3.3), then take the two "ends" and bring them together until they meet (Figure 7.3.4).

Figure 7.3.3   The first step in obtaining a torus from a rectangle: making a cylinder.

ends glued

Figure 7.3.4 Bending the cylinder and gluing the ends to get a torus.

With surfaces, just as with curves, we want to distinguish a map (a parametrization) from its image (a geometric object). This leads us to the following definition.

DEFINITION: PARAMETRIZED SURFACES A *parametrized surface* is a function $\mathbf{\Phi}\colon D \subset \mathbb{R}^2 \to \mathbb{R}^3$, where D is some domain in $\mathbb{R}^2$. The *surface S* corresponding to the function $\mathbf{\Phi}$ is its image: $S = \mathbf{\Phi}(D)$. We can write

$$\mathbf{\Phi}(u, v) = (x(u, v), y(u, v), z(u, v)).$$

If $\mathbf{\Phi}$ is differentiable or is of class C^1 [which is the same as saying that $x(u, v)$, $y(u, v)$, and $z(u, v)$ are differentiable or C^1 functions of (u, v)], we call S a *differentiable or a C^1 surface*.

We can think of $\mathbf{\Phi}$ as twisting or bending the region D in the plane to yield the surface S (see Figure 7.3.5). Thus each point (u, v) in D becomes a label for a point $(x(u, v), y(u, v), z(u, v))$ on S.

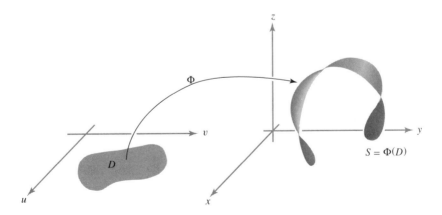

Figure 7.3.5 $\mathbf{\Phi}$ "twists" and "bends" D onto the surface $S = \mathbf{\Phi}(D)$.

Suppose that $\mathbf{\Phi}$ is differentiable at $(u_0, v_0) \in \mathbb{R}^2$. Fixing u at u_0, we get a map $\mathbb{R} \to \mathbb{R}^3$ given by $t \mapsto \mathbf{\Phi}(u_0, t)$, whose image is a curve on the surface (Figure 7.3.6). From Chapters 2 and 4 we know that the vector tangent to this curve at the point $\mathbf{\Phi}(u_0, v_0)$ is given by

$$\mathbf{T}_v = \frac{\partial x}{\partial v}(u_0, v_0)\mathbf{i} + \frac{\partial y}{\partial v}(u_0, v_0)\mathbf{j} + \frac{\partial z}{\partial v}(u_0, v_0)\mathbf{k}.$$

Similarly, if we fix v and consider the curve $t \mapsto \mathbf{\Phi}(t, v_0)$, we obtain the tangent vector to this curve at $\mathbf{\Phi}(u_0, v_0)$, given by

$$\mathbf{T}_u = \frac{\partial x}{\partial u}(u_0, v_0)\mathbf{i} + \frac{\partial y}{\partial u}(u_0, v_0)\mathbf{j} + \frac{\partial z}{\partial u}(u_0, v_0)\mathbf{k}.$$

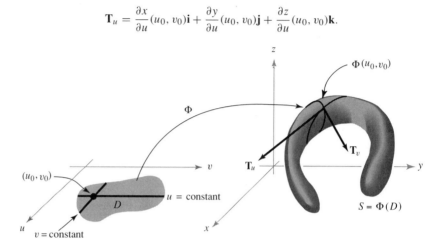

Figure 7.3.6   The tangent vectors $\mathbf{T}_u$ and $\mathbf{T}_v$ that are tangent to curve on a surface S, and hence tangent to S.

Since the vectors $\mathbf{T}_u$ and $\mathbf{T}_v$ are tangent to two curves on the surface at $\mathbf{\Phi}(u_0, v_0)$, they ought to determine the plane tangent to the surface at this point; that is, $\mathbf{T}_u \times \mathbf{T}_v$ ought to be normal to the surface.

We say that the surface S is **smooth**[*] at $\mathbf{\Phi}(u_0, v_0)$ if $\mathbf{T}_u \times \mathbf{T}_v \neq \mathbf{0}$ at (u_0, v_0); the surface is **smooth** if it is smooth at all points $\mathbf{\Phi}(u_0, v_0) \in S$. The nonzero vector $\mathbf{T}_u \times \mathbf{T}_v$ is *normal* to S (recall that the vector product of $\mathbf{T}_u$ and $\mathbf{T}_v$ is perpendicular to the plane spanned by $\mathbf{T}_u$ and $\mathbf{T}_v$); the fact that it is nonzero ensures that there will be a tangent plane. Intuitively, a smooth surface has no "corners."[†]

E X A M P L E   1   Consider the surface given by the equations

$$x = u \cos v, \quad y = u \sin v, \quad z = u, \quad u \geq 0.$$

Is this surface differentiable? Is it smooth?

---

[*]Strictly speaking, smoothness depends on the parametrization $\mathbf{\Phi}$ and not just on its image S. Therefore this terminology is somewhat imprecise; however, it is descriptive and should not cause confusion. (See Exercise 15.)

[†]In Section 3.5 we showed that level surfaces $f(x, y, z) = 0$ were in fact graphs of functions of two variables in some neighborhood of a point (x_0, y_0, z_0) satisfying $\nabla f(x_0, y_0, z_0) \neq \mathbf{0}$. This united two concepts of a surface. Again, using the implicit function theorem, it is likewise possible to show that the image of a parametrized surface $\mathbf{\Phi}$ in the neighborhood of a point (u_0, v_0) where $\mathbf{T}_u \times \mathbf{T}_v \neq \mathbf{0}$ is also the graph of a function of two variables. Thus all definitions of a surface are consistent. (See Exercise 16.)

SOLUTION These equations describe the surface $z = \sqrt{x^2 + y^2}$ (square the equations for x, y, and z to check this), which is shown in Figure 7.3.7. This surface is a cone with

Figure 7.3.7 The surface $z = \sqrt{x^2 + y^2}$ is a cone.

a "point" at $(0, 0, 0)$; it is a differentiable surface because each component function is differentiable as a function of u and v. However, *the surface is not smooth at* $(0, 0, 0)$. To see this, compute $\mathbf{T}_u$ and $\mathbf{T}_v$ at $(0, 0) \in \mathbb{R}^2$:

$$\mathbf{T}_u = \frac{\partial x}{\partial u}(0, 0)\mathbf{i} + \frac{\partial y}{\partial u}(0, 0)\mathbf{j} + \frac{\partial z}{\partial u}(0, 0)\mathbf{k} = (\cos 0)\mathbf{i} + (\sin 0)\mathbf{j} + \mathbf{k} = \mathbf{i} + \mathbf{k},$$

and similarly,

$$\mathbf{T}_v = 0(-\sin 0)\mathbf{i} + 0(\cos 0)\mathbf{j} + 0\mathbf{k} = \mathbf{0}.$$

Thus $\mathbf{T}_u \times \mathbf{T}_v = \mathbf{0}$, and so, by definition, the surface is not smooth at $(0, 0, 0)$. ▲

Let us summarize our conclusions in a formal definition:

DEFINITION: THE TANGENT PLANE TO A SURFACE If a parametrized surface $\mathbf{\Phi}: D \subset \mathbb{R}^2 \to \mathbb{R}^3$ is *smooth* at $\mathbf{\Phi}(u_0, v_0)$, that is, if $\mathbf{T}_u \times \mathbf{T}_v \neq \mathbf{0}$ at (u_0, v_0), we define the *tangent plane* of the surface at $\mathbf{\Phi}(u_0, v_0)$ to be the plane determined by the vectors $\mathbf{T}_u$ and $\mathbf{T}_v$. Thus $\mathbf{n} = \mathbf{T}_u \times \mathbf{T}_v$ is a normal vector, and an equation of the tangent plane at (x_0, y_0, z_0) on the surface is given by

$$(x - x_0, y - y_0, z - z_0) \cdot \mathbf{n} = 0, \tag{1}$$

where $\mathbf{n}$ is evaluated at (u_0, v_0); that is, the tangent plane is the set of (x, y, z) satisfying (1). If $\mathbf{n} = (n_1, n_2, n_3) = n_1\mathbf{i} + n_2\mathbf{j} + n_3\mathbf{k}$, then formula (1) becomes

$$n_1(x - x_0) + n_2(y - y_0) + n_3(z - z_0) = 0. \tag{1'}$$

EXAMPLE 2   Let $\Phi\colon \mathbb{R}^2 \to \mathbb{R}^3$ be given by

$$x = u \cos v, \quad y = u \sin v, \quad z = u^2 + v^2.$$

Where does a tangent plane exist? Find the tangent plane at $\Phi(1, 0)$.

SOLUTION   We compute

$$\mathbf{T}_u = (\cos v)\mathbf{i} + (\sin v)\mathbf{j} + 2u\mathbf{k} \qquad \text{and} \qquad \mathbf{T}_v = -u(\sin v)\mathbf{i} + u(\cos v)\mathbf{j} + 2v\mathbf{k}$$

and the tangent plane at $\Phi(u, v)$ is the set of vectors through $\Phi(u, v)$ perpendicular to

$$\mathbf{T}_u \times \mathbf{T}_v = (-2u^2 \cos v + 2v \sin v, -2u^2 \sin v - 2v \cos v, u)$$

if this vector is nonzero. Since $\mathbf{T}_u \times \mathbf{T}_v$ is equal to $\mathbf{0}$ at $(u, v) = (0, 0)$, we cannot find a tangent plane at $\Phi(0, 0) = (0, 0, 0)$. However, we can find an equation of the tangent plane at all the other points, where $\mathbf{T}_u \times \mathbf{T}_v \neq \mathbf{0}$. At the point $\Phi(0, 1) = (1, 0, 1)$,

$$\mathbf{n} = \mathbf{T}_u \times \mathbf{T}_v = (-2, 0, 1) = -2\mathbf{i} + \mathbf{k}.$$

Since we have the vector $\mathbf{n}$ normal to the surface and a point $(1, 0, 1)$ on the surface, we can use formula $(1')$ to obtain an equation of the tangent plane: $-2(x - 1) + (z - 1) = 0$; that is, $z = 2x - 1$.   ▲

EXAMPLE 3   Suppose a surface S is the graph of a differentiable function $g\colon \mathbb{R}^2 \to \mathbb{R}$. Show that the surface is smooth at all points $(u_0, v_0, g(u_0, v_0)) \in \mathbb{R}^3$.

SOLUTION   Write S in parametric form as follows:

$$x = u, \qquad y = v, \qquad z = g(u, v),$$

which is the same as $z = g(x, y)$. Then

$$\mathbf{T}_u = \mathbf{i} + \frac{\partial g}{\partial u}(u_0, v_0)\mathbf{k} \qquad \text{and} \qquad \mathbf{T}_v = \mathbf{j} + \frac{\partial g}{\partial v}(u_0, v_0)\mathbf{k},$$

and for $(u_0, v_0) \in \mathbb{R}^2$,

$$\mathbf{n} = \mathbf{T}_u \times \mathbf{T}_v = -\frac{\partial g}{\partial u}(u_0, v_0)\mathbf{i} - \frac{\partial g}{\partial v}(u_0, v_0)\mathbf{j} + \mathbf{k} \neq \mathbf{0}. \tag{2}$$

This is nonzero because the coefficient of $\mathbf{k}$ is 1; consequently the parametrization $(u, v) \mapsto (u, v, g(u, v))$ is smooth at all points. Moreover, the tangent plane at the point $(x_0, y_0, z_0) = (u_0, v_0, g(u_0, v_0))$ is given, by formula (1), as

$$(x - x_0, y - y_0, z - z_0) \cdot \left(-\frac{\partial g}{\partial u}, -\frac{\partial g}{\partial v}, 1\right) = 0,$$

where the partial derivatives are evaluated at (u_0, v_0). Remembering that $x = u$ and $y = v$, we can write this as

$$z - z_0 = \left(\frac{\partial g}{\partial x}\right)(x - x_0) + \left(\frac{\partial g}{\partial y}\right)(y - y_0), \tag{3}$$

where $\partial g/\partial x$ and $\partial g/\partial y$ are evaluated at (x_0, y_0). ▲

This example also shows that the definition of the tangent plane for parametrized surfaces agrees with the one for surfaces obtained as graphs, since equation (3) is the same formula we derived (in Chapter 2) for the plane tangent to S at the point $(x_0, y_0, z_0) \in S$.

It is also useful to consider piecewise smooth surfaces, that is, surfaces composed of a certain number of images of smooth parametrized surfaces. For example, the surface of a cube in $\mathbb{R}^3$ is such a surface. These surfaces are considered in Section 7.4.

EXAMPLE 4 Find a parametrization for the hyperboloid of one sheet:

$$x^2 + y^2 - z^2 = 1.$$

SOLUTION Since x and y appear in the combination $x^2 + y^2$, the surface is invariant under rotation about the z axis, and so it is natural to write

$$x = r\cos\theta, \ y = r\sin\theta.$$

Then $x^2 + y^2 - z^2 = 1$ becomes $r^2 - z^2 = 1$. This we can conveniently parametrize by

$$r = \cosh u, \quad z = \sinh u.$$

Thus a parametrization is

$$x = (\cosh u)(\cos\theta), \qquad y = (\cosh u)(\sin\theta), \qquad z = \sinh u,$$

where $0 \le \theta < 2\pi, -\infty < u < \infty$. ▲

EXERCISES

In Exercises 1 to 3, find an equation for the plane tangent to the given surface at the specified point.

1. $x = 2u, \qquad y = u^2 + v, \qquad z = v^2, \qquad at\ (0, 1, 1)$

2. $x = u^2 - v^2, \quad y = u + v, \quad z = u^2 + 4v, \qquad at\ (-\frac{1}{4}, \frac{1}{2}, 2)$

3. $x = u^2, \quad y = u\sin e^v, \quad z = \frac{1}{3}u\cos e^v, \qquad at\ (13, -2, 1)$

4. Are the surfaces in Exercises 1 and 2 smooth?

5. Find an expression for a unit vector normal to the surface

$$x = \cos v \sin u, \qquad y = \sin v \sin u, \qquad z = \cos u$$

at the image of a point (u, v) for u in $[0, \pi]$ and v in $[0, 2\pi]$. Identify this surface.

6. Repeat Exercise 5 for the surface

$$x = 3 \cos \theta \sin \phi, \qquad y = 2 \sin \theta \sin \phi, \qquad z = \cos \phi$$

for θ in $[0, 2\pi]$ and ϕ in $[0, \pi]$.

7. Repeat Exercise 5 for the surface

$$x = \sin v, \qquad y = u, \qquad z = \cos v$$

for $0 \le v \le 2\pi$ and $-1 \le u \le 3$.

8. Repeat Exercise 5 for the surface

$$x = (2 - \cos v) \cos u, \qquad y = (2 - \cos v) \sin u, \qquad z = \sin v$$

for $-\pi \le u \le \pi, -\pi \le v \le \pi$. Is this surface smooth?

9. (a) Develop a formula for the plane tangent to the surface $x = h(y, z)$.
   (b) Obtain a similar formula for $y = k(x, z)$.

10. Find the equation of the plane tangent to each surface at the indicated point.
   (a) $x = u^2, y = v^2, z = u^2 + v^2, u = 1, v = 1$
   (b) $z = 3x^2 + 8xy, x = 1, y = 0$
   (c) $x^3 + 3xy + z^2 = 2, x = 1, y = \frac{1}{3}, z = 0$

11. Consider the surface in $\mathbb{R}^3$ parametrized by

$$\Phi(r, \theta) = (r \cos \theta, r \sin \theta, \theta), \qquad 0 \le r \le 1 \qquad \text{and} \qquad 0 \le \theta \le 4\pi.$$

   (a) Sketch and describe the surface.
   (b) Find an expression for a unit normal to the surface.
   (c) Find an equation for the plane tangent to the surface at the point (x_0, y_0, z_0).
   (d) If (x_0, y_0, z_0) is a point on the surface, show that the horizontal line segment of unit length from the z axis through (x_0, y_0, z_0) is contained in the surface and in the plane tangent to the surface at (x_0, y_0, z_0).

12. Given a sphere of radius 2 centered at the origin, find the equation for the plane that is tangent to it at the point $(1, 1, \sqrt{2})$ by considering the sphere as:
   (a) a surface parametrized by $\Phi(\theta, \phi) = (2 \cos \theta \sin \phi, 2 \sin \theta \sin \phi, 2 \cos \phi)$;

(b) a level surface of $f(x, y, z) = x^2 + y^2 + z^2$; and
(c) the graph of $g(x, y) = \sqrt{4 - x^2 - y^2}$.

13. (a) Find a parametrization for the hyperboloid $x^2 + y^2 - z^2 = 25$.
(b) Find an expression for a unit normal to this surface.
(c) Find an equation for the plane tangent to the surface at $(x_0, y_0, 0)$, where $x_0^2 + y_0^2 = 25$.
(d) Show that the lines $(x_0, y_0, 0) + t(-y_0, x_0, 5)$ and $(x_0, y_0, 0) + t(y_0, -x_0, 5)$ lie in the surface *and* in the tangent plane found in part (c).

14. A parametrized surface is described by a differentiable function $\Phi: \mathbb{R}^2 \to \mathbb{R}^3$. According to Chapter 2, the derivative should give a linear approximation that yields a representation of the tangent plane. This exercise demonstrates that this is indeed the case.

(a) Assuming $\mathbf{T}_u \times \mathbf{T}_v \neq \mathbf{0}$, show that the range of the linear transformation $\mathbf{D}\Phi(u_0, v_0)$ is the plane spanned by $\mathbf{T}_u$ and $\mathbf{T}_v$. [Here $\mathbf{T}_u$ and $\mathbf{T}_v$ are evaluated at (u_0, v_0).]
(b) Show that $\mathbf{w} \perp (\mathbf{T}_u \times \mathbf{T}_v)$ if and only if $\mathbf{w}$ is in the range of $\mathbf{D}\Phi(u_0, v_0)$.
(c) Show that the tangent plane as defined in this section is the same as the "parametrized plane"

$$(u, v) \mapsto \Phi(u_0, v_0) + \mathbf{D}\Phi(u_0, v_0)\begin{bmatrix} u - u_0 \\ v - v_0 \end{bmatrix}.$$

15. Consider the surfaces $\Phi_1(u, v) = (u, v, 0)$ and $\Phi_2(u, v) = (u^3, v^3, 0)$.
(a) Show that the image of Φ_1 and of Φ_2 is the xy plane.
(b) Show that Φ_1 describes a smooth surface, yet Φ_2 does not. Conclude that the notion of smoothness of a surface S depends on the existence of at least one smooth parametrization for S.
(c) Prove that the tangent plane of S is well defined independently of the smooth (one-to-one) parametrization (you will need to use the inverse function theorem from Section 3.5).
(d) After these remarks, do you think you can find a smooth parametrization of the cone of Figure 7.3.7?

16. Let Φ be a smooth surface; that is, Φ is of class C^1 and $\mathbf{T}_u \times \mathbf{T}_v \neq \mathbf{0}$ at (u_0, v_0).
(a) Use the implicit function theorem (Section 3.5) to show that the image of Φ *near* (u_0, v_0) is the graph of a C^1 function of two variables. If the z component of $\mathbf{T}_u \times \mathbf{T}_v$ is nonzero, we can write it as $z = f(x, y)$.
(b) Show that the tangent plane at $\Phi(u_0, v_0)$ defined by the plane spanned by $\mathbf{T}_u$ and $\mathbf{T}_v$ coincides with the tangent plane of the graph of $z = f(x, y)$ at this point.

7.4 AREA OF A SURFACE

Before proceeding to general surface integrals, let us first consider the problem of computing the area of a surface, just as we considered the problem of finding the arc length of a curve before discussing path integrals.

In Section 7.3, we defined a parametrized surface S to be the *image* of a function $\Phi: D \subset \mathbb{R}^2 \to \mathbb{R}^3$, written as $\Phi(u, v) = (x(u, v), y(u, v), z(u, v))$. The map Φ was called the parametrization of S and S was said to be smooth at $\Phi(u, v) \in S$ if $\mathbf{T}_u \times \mathbf{T}_v \neq \mathbf{0}$, where

$$\mathbf{T}_u = \frac{\partial x}{\partial u}(u, v)\mathbf{i} + \frac{\partial y}{\partial u}(u, v)\mathbf{j} + \frac{\partial z}{\partial u}(u, v)\mathbf{k}$$

and

$$\mathbf{T}_v = \frac{\partial x}{\partial v}(u, v)\mathbf{i} + \frac{\partial y}{\partial v}(u, v)\mathbf{j} + \frac{\partial z}{\partial v}(u, v)\mathbf{k}.$$

Recall that a smooth surface (loosely speaking) is one that has no corners or breaks.

In the rest of this chapter and in the next one we shall consider only piecewise smooth surfaces that are unions of images of parametrized surfaces $\mathbf{\Phi}_i \colon D_i \to \mathbb{R}^3$ for which.

(i) D_i is an elementary region in the plane;

(ii) $\mathbf{\Phi}_i$ is of class C^1 and one-to-one, except possibly on the boundary of D_i; and

(iii) S_i, the image of $\mathbf{\Phi}_i$ is smooth, except possibly at a finite number of points.

DEFINITION: AREA OF A PARAMETRIZED SURFACE   We define the **surface area**[*] $A(S)$ of a parametrized surface by

$$A(S) = \iint_D \|\mathbf{T}_u \times \mathbf{T}_v\| \, du \, dv \tag{1}$$

where $\|\mathbf{T}_u \times \mathbf{T}_v\|$ is the norm of $\mathbf{T}_u \times \mathbf{T}_v$. If S is a union of surfaces S_i, its area is the sum of the areas of the S_i.

As the reader can easily verify, we have

$$\|\mathbf{T}_u \times \mathbf{T}_v\| = \sqrt{\left[\frac{\partial(x, y)}{\partial(u, v)}\right]^2 + \left[\frac{\partial(y, z)}{\partial(u, v)}\right]^2 + \left[\frac{\partial(x, z)}{\partial(u, v)}\right]^2} \tag{2}$$

where

$$\frac{\partial(x, y)}{\partial(u, v)} = \begin{vmatrix} \dfrac{\partial x}{\partial u} & \dfrac{\partial x}{\partial v} \\[2mm] \dfrac{\partial y}{\partial u} & \dfrac{\partial y}{\partial v} \end{vmatrix}$$

and so on. Thus formula (1) becomes

$$A(S) = \iint_D \sqrt{\left[\frac{\partial(x, y)}{\partial(u, v)}\right]^2 + \left[\frac{\partial(y, z)}{\partial(u, v)}\right]^2 + \left[\frac{\partial(x, z)}{\partial(u, v)}\right]^2} \, du \, dv. \tag{3}$$

We can justify this definition by analyzing the integral $\iint_D \|\mathbf{T}_u \times \mathbf{T}_v\| \, du \, dv$ in terms of Riemann sums. For simplicity, suppose D is a rectangle; consider the nth reg-

---

[*]As we have not yet discussed the independence of parametrization, it may seem that $A(S)$ depends on the parametrization $\mathbf{\Phi}$. We shall discuss independence of parametrization in Section 7.6; the use of this notation here should not cause confusion.

ular partition of D, and let R_{ij} be the ijth rectangle in the partition, with vertices $(u_i, v_j), (u_{i+1}, v_j), (u_i, v_{j+1})$, and $(u_{i+1}, v_{j+1}), 0 \leq i \leq n-1, 0 \leq j \leq n-1$. Denote the values of $\mathbf{T}_u$ and $\mathbf{T}_v$ at (u_i, v_j) by $\mathbf{T}_{u_i}$ and $\mathbf{T}_{v_j}$. We can think of the vectors $\Delta u \mathbf{T}_{u_i}$ and $\Delta v \mathbf{T}_{v_j}$ as tangent to the surface at $\boldsymbol{\Phi}(u_i, v_j) = (x_{ij}, y_{ij}, z_{ij})$, where $\Delta u = u_{i+1} - u_i, \Delta v = v_{j+1} - v_j$. Then these vectors form a parallelogram P_{ij} that lies in the plane tangent to the surface at (x_{ij}, y_{ij}, z_{ij}) (see Figure 7.4.1). We thus have a "patchwork cover" of the surface by the P_{ij}. For n large, the area of P_{ij} is a good approximation to the area of $\boldsymbol{\Phi}(R_{ij})$. Since the area of the parallelogram spanned by two vectors $\mathbf{v}_1$ and $\mathbf{v}_2$ is $\|\mathbf{v}_1 \times \mathbf{v}_2\|$ (see Chapter 1), we see that

$$A(P_{ij}) = \|\Delta u \mathbf{T}_{u_i} \times \Delta v \mathbf{T}_{v_j}\| = \|\mathbf{T}_{u_i} \times \mathbf{T}_{v_j}\| \Delta u \, \Delta v.$$

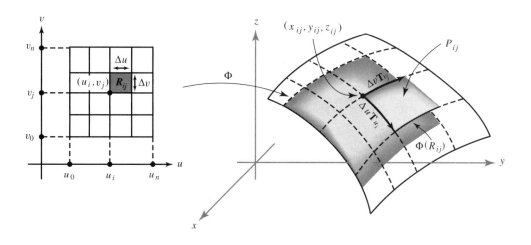

Figure 7.4.1 $\|\mathbf{T}_{u_i} \times \mathbf{T}_{v_j}\| \, \Delta u \, \Delta v$ is equal to the area of a parallelogram that approximates the area of a patch on a surface $S = \boldsymbol{\Phi}(D)$.

Therefore, the area of the patchwork cover is

$$A_n = \sum_{i=0}^{n-1} \sum_{j=0}^{n-1} A(P_{ij}) = \sum_{i=0}^{n-1} \sum_{j=0}^{n-1} \|\mathbf{T}_{u_i} \times \mathbf{T}_{v_j}\| \, \Delta u \, \Delta v.$$

As $n \to \infty$, the sums A_n converge to the integral

$$\iint_D \|\mathbf{T}_u \times \mathbf{T}_v\| \, du \, dv.$$

Since A_n should approximate the surface area better and better as $n \to \infty$, we are led to formula (1) for a reasonable definition of $A(S)$.

EXAMPLE 1   Let D be the region determined by $0 \leq \theta \leq 2\pi, 0 \leq r \leq 1$ and let the function $\boldsymbol{\Phi} \colon D \to \mathbb{R}^3$ defined by

$$x = r \cos \theta, \qquad y = r \sin \theta, \qquad z = r$$

be a parametrization of a cone S (see Figure 7.3.7). Find its surface area.

SOLUTION   In formula (3),

$$\frac{\partial(x, y)}{\partial(r, \theta)} = \begin{vmatrix} \cos \theta & -r \sin \theta \\ \sin \theta & r \cos \theta \end{vmatrix} = r,$$

$$\frac{\partial(y, z)}{\partial(r, \theta)} = \begin{vmatrix} \sin \theta & r \cos \theta \\ 1 & 0 \end{vmatrix} = -r \cos \theta,$$

and

$$\frac{\partial(x, z)}{\partial(r, \theta)} = \begin{vmatrix} \cos \theta & -r \sin \theta \\ 1 & 0 \end{vmatrix} = r \sin \theta,$$

so the area integrand is

$$\|\mathbf{T}_r \times \mathbf{T}_\theta\| = \sqrt{r^2 + r^2 \cos^2 \theta + r^2 \sin^2 \theta} = r\sqrt{2}.$$

Clearly, $\|\mathbf{T}_r \times \mathbf{T}_\theta\|$ vanishes for $r = 0$, but $\boldsymbol{\Phi}(0, \theta) = (0, 0, 0)$ for any θ. Thus $(0, 0, 0)$ is the only point where the surface is not smooth. We have

$$\iint_D \|\mathbf{T}_r \times \mathbf{T}_\theta\| \, dr \, d\theta = \int_0^{2\pi} \int_0^1 \sqrt{2} r \, dr \, d\theta = \int_0^{2\pi} \frac{1}{2} \sqrt{2} \, d\theta = \sqrt{2}\pi.$$

To confirm that this is the area of $\boldsymbol{\Phi}(D)$ we must verify that $\boldsymbol{\Phi}$ is one-to-one (for points not on the boundary of D). Let D^0 be the set of (r, θ) with $0 < r < 1$ and $0 < \theta < 2\pi$. Hence, D^0 is D without its boundary. To see that $\boldsymbol{\Phi} \colon D^0 \to \mathbb{R}^3$ is one-to-one, assume that $\boldsymbol{\Phi}(r, \theta) = \boldsymbol{\Phi}(r', \theta')$ for (r, θ) and $(r', \theta') \in D^0$. Then

$$r \cos \theta = r' \cos \theta', \qquad r \sin \theta = r' \sin \theta', \qquad r = r'.$$

From these equations it follows that $\cos \theta = \cos \theta'$ and $\sin \theta = \sin \theta'$. Thus either $\theta = \theta'$ or $\theta = \theta' + 2\pi n$. But the second case is impossible, since both θ and θ' belong to the open interval $(0, 2\pi)$ and thus cannot be 2π radians apart. This proves that off the boundary, $\boldsymbol{\Phi}$ is one-to-one. (Is $\boldsymbol{\Phi} \colon D \to \mathbb{R}^3$ one-to-one?) In future examples we shall not usually verify that the parametrization is one-to-one when it is intuitively clear.   ▲

EXAMPLE 2   A *helicoid* is defined by $\boldsymbol{\Phi} \colon D \to \mathbb{R}^3$, where

$$x = r \cos \theta, \qquad y = r \sin \theta, \qquad z = \theta$$

and D is the region where $0 \leq \theta \leq 2\pi$ and $0 \leq r \leq 1$ (Figure 7.4.2). Find its area.

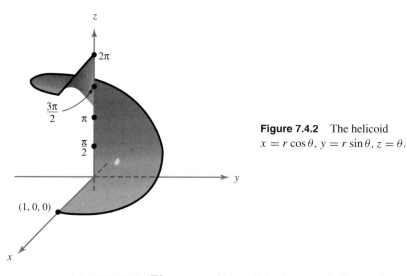

Figure 7.4.2 The helicoid
$x = r \cos \theta,\ y = r \sin \theta,\ z = \theta$.

SOLUTION We compute $\partial(x, y)/\partial(r, \theta) = r$ as before, and

$$\frac{\partial(y, z)}{\partial(r, \theta)} = \begin{vmatrix} \sin \theta & r \cos \theta \\ 0 & 1 \end{vmatrix} = \sin \theta,$$

$$\frac{\partial(x, z)}{\partial(r, \theta)} = \begin{vmatrix} \cos \theta & -r \sin \theta \\ 0 & 1 \end{vmatrix} = \cos \theta.$$

The area integrand is therefore $\sqrt{r^2 + 1}$, which never vanishes, so the surface is smooth. The area of the helicoid is

$$\iint_D \|\mathbf{T}_r \times \mathbf{T}_\theta\|\, dr\, d\theta = \int_0^{2\pi} \int_0^1 \sqrt{r^2 + 1}\, dr d\theta = 2\pi \int_0^1 \sqrt{r^2 + 1}\, dr.$$

After a little computation (using the table of integrals), we find that this integral is equal to

$$\pi[\sqrt{2} + \log(1 + \sqrt{2})]. \quad \blacktriangle$$

A surface S given in the form $z = f(x, y)$, where $(x, y) \in D$, admits the parametrization

$$x = u, \qquad y = v, \qquad z = f(u, v)$$

for $(u, v) \in D$. When f is of class C^1, this parametrization is smooth, and the formula for surface area reduces to

$$A(S) = \iint_D \left(\sqrt{\left(\frac{\partial f}{\partial x}\right)^2 + \left(\frac{\partial f}{\partial y}\right)^2 + 1} \right) dA \tag{4}$$

after applying the formulas

$$\mathbf{T}_u = \mathbf{i} + \frac{\partial f}{\partial u}\mathbf{k}, \qquad \mathbf{T}_v = \mathbf{j} + \frac{\partial f}{\partial v}\mathbf{k},$$

and

$$\mathbf{T}_u \times \mathbf{T}_v = -\frac{\partial f}{\partial u}\mathbf{i} - \frac{\partial f}{\partial v}\mathbf{j} + \mathbf{k} = -\frac{\partial f}{\partial x}\mathbf{i} - \frac{\partial f}{\partial y}\mathbf{j} + \mathbf{k},$$

as noted in Example 3 of Section 7.3.

In most books on one-variable calculus, it is shown that the lateral surface area generated by revolving the graph of a function $y = f(x)$ about the x axis is given by

$$A = 2\pi \int_a^b (|f(x)|\sqrt{1 + [f'(x)]^2})\,dx. \tag{5}$$

If the graph is revolved about the y axis, we have

$$A = 2\pi \int_a^b (|x|\sqrt{1 + [f'(x)]^2})\,dx. \tag{6}$$

We shall derive formula (5) by using the methods developed above; one can obtain formula (6) in a similar fashion (Exercise 10).

To derive formula (5) from formula (3), we must give a parametrization of S. Define the parametrization by

$$x = u, \qquad y = f(u)\cos v, \qquad z = f(u)\sin v$$

over the region D given by

$$a \le u \le b, \qquad 0 \le v \le 2\pi.$$

This is indeed a parametrization of S, because for fixed u,

$$(u,\, f(u)\cos v,\, f(u)\sin v)$$

traces out a circle of radius $|f(u)|$ with the center $(u, 0, 0)$ (Figure 7.4.3).

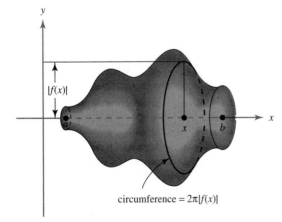

Figure 7.4.3   The curve $y = f(x)$ rotated about the x axis.

We calculate

$$\frac{\partial(x, y)}{\partial(u, v)} = -f(u)\sin v, \qquad \frac{\partial(y, z)}{\partial(u, v)} = f(u)f'(u), \qquad \frac{\partial(x, z)}{\partial(u, v)} = f(u)\cos v,$$

and so by formula (3)

$$
\begin{aligned}
A(S) &= \iint_D \sqrt{\left[\frac{\partial(x, y)}{\partial(u, v)}\right]^2 + \left[\frac{\partial(y, z)}{\partial(u, v)}\right]^2 + \left[\frac{\partial(x, z)}{\partial(u, v)}\right]^2} \, du \, dv \\
&= \iint_D \sqrt{[f(u)]^2 \sin^2 v + [f(u)]^2[f'(u)]^2 + [f(u)]^2 \cos^2 v} \, du \, dv \\
&= \iint_D |f(u)|\sqrt{1 + [f'(u)]^2} \, du \, dv \\
&= \int_a^b \int_0^{2\pi} |f(u)|\sqrt{1 + [f'(u)]^2} \, dv \, du \\
&= 2\pi \int_a^b |f(u)|\sqrt{1 + [f'u)]^2} \, du,
\end{aligned}
$$

which is formula (5).

If S is the surface of revolution, then $2\pi|f(x)|$ is the circumference of the vertical cross section to S at the point x (Figure 7.4.3). Observe that we can write

$$2\pi \int_a^b |f(x)|\sqrt{1 + [f'(x)]^2} \, dx = \int_{\mathbf{c}} 2\pi|f(x)| \, ds,$$

where the expression on the right is the path integral of $2\pi|f(x)|$ along the path given by $\mathbf{c} \colon [a, b] \to \mathbb{R}^2, t \mapsto (t, f(t))$. Therefore, the lateral surface of a solid of revolution is obtained by integrating the cross-sectional circumference along the path determined by the given function.

HISTORICAL NOTE

Calculus was invented (or discovered?) by Isaac Newton (1647–1727) about 1669 and by Gottfried Wilhelm Leibniz (1646–1716) about 1684. In the beginning of the eighteenth century, mathematicians were interested in the problem of finding paths of shortest length on a surface by using the methods of calculus. At this time, surfaces were regarded as boundaries of solids that were defined by inequalities (the ball $x^2 + y^2 + z^2 \le 1$ is bounded by the sphere $x^2 + y^2 + z^2 = 1$).

Christian Huygens (1629–1695) was the first person since Archimedes to give results on the areas of special surfaces beyond the sphere, and he obtained the areas of portions of surfaces of revolution, such as the paraboloid and hyperboloid.

The brilliant and prolific mathematician Leonhard Euler (1707–1783) presented the first fundamental work on the theory of surfaces in 1760 with "Recherches sur la courbure des surfaces," and it may have been in this work that a surface was first defined as a graph $z = f(x, y)$. Euler was interested in studying the curvature of surfaces, and in 1771 he introduced the notion of the parametric surfaces that are described in this section.

After the rapid development of calculus in the early eighteenth century, formulas for the lengths of curves and areas of surfaces were developed. Although we do not know when the area formulas presented in this section first appeared, they were certainly common by the end of the eighteenth century. The underlying concepts of the length of a curve and the area of a surface were understood intuitively before this time, and the use of formulas from calculus to compute areas was considered a great achievement.

Augustin Louis Cauchy (1789–1857) was the first to take the step of defining the quantities of length and surface areas by integrals as we have done in this book. The question of defining surface area independent of integrals was taken up somewhat later, but this posed many difficult problems that were not properly resolved until this century.

We end this section by describing the fascinating classical problem of Plateau, which has enjoyed a long history in mathematics. The Belgian physicist Joseph Plateau (1801–1883) carried out many experiments from 1830 to 1869 on surface tension and capillary phenomena, experiments that had enormous impact at the time and were repeated by notable nineteenth-century physicists, such as Michael Faraday (1791–1867).

If a wire is dipped into a soap or glycerine solution, then one usually withdraws a soap film spanning the wire. Some examples are given in Figure 7.4.4, although readers might like to perform the experiment for themselves. Plateau raised the mathematical question: For a given boundary (wire), how does one prove the existence of such a surface (soap film) and how many surfaces can there be? The underlying physical principle is that nature tends to minimize area; that is, the surface that forms should be a surface of least area among all possible surfaces that have the given curve as their boundary.

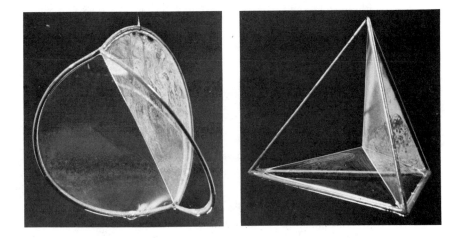

Figure 7.4.4   Two soap films spanning wires. (*Fritz Goro.*)

Plateau formulated the problem in a special way. Let $D \subset \mathbb{R}^2$ be the unit disk defined by $\{(x, y) \mid x^2 + y^2 \leq 1\}$; let ∂D be its boundary. Furthermore, suppose that the image Γ of $\mathbf{c}$: $[0, 2\pi] \rightarrow \mathbb{R}^3$ is a simple closed curve, Γ representing a wire in $\mathbb{R}^3$ (Figure 7.4.5).

Let $\mathcal{S}$ be the set of all maps $\mathbf{\Phi} \colon D \to \mathbb{R}^3$ such that $\mathbf{\Phi}(\partial D) = \Gamma$, $\mathbf{\Phi}$ is of class C^1, and $\mathbf{\Phi}$ is one-to-one on ∂D. Each $\mathbf{\Phi} \in \mathcal{S}$ represents a parametric C^1 "disklike" surface "spanning" the wire Γ.

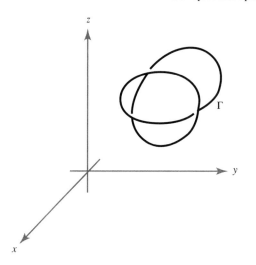

Figure 7.4.5 Γ is to be the boundary of a soap film.

The soap films in Figure 7.4.4 are not disklike but represent a system of disklike surfaces. Figure 7.4.6 shows a contour which bounds both a disklike and a nondisklike surface.

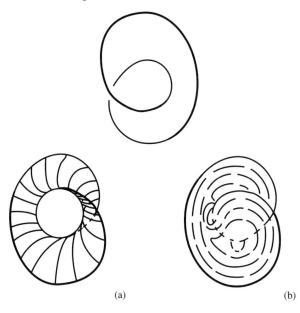

Figure 7.4.6 Soap film surfaces; (b) is a disklike surface, but (a) is not.

(a)   (b)

For each $\mathbf{\Phi} \in \mathcal{S}$ consider the area of the image surface, $A(\mathbf{\Phi}) = \iint_D \|\mathbf{T}_u \times \mathbf{T}_v\| \, du \, dv$. This area is a function that assigns to each parametric surface its area. Plateau asked whether A has a minimum on $\mathcal{S}$; that is, does there exist a $\mathbf{\Phi}_0$ such that $A(\mathbf{\Phi}_0) \leq A(\mathbf{\Phi})$ for all $\mathbf{\Phi} \in \mathcal{S}$? Unfortunately, the methods of this book are not adequate to solve this problem. We can tackle questions of finding minima of real-valued functions of several variables, but in no way can the set S be thought of as a region in $\mathbb{R}^n$ for *any* n! The set S is really a space of functions, and the problem of finding minima of functions like A on such sets is part of a subject called the "calculus of variations," a subject that is almost as old as the calculus itself. It is also a discipline that is intimately connected with partial differential equations.

Plateau showed that if a minimum

$$\Phi_0(u, v) = (x(u, v), y(u, v), z(u, v))$$

existed at all it would have to satisfy (after suitable normalizations) the partial differential equations

(i) $\nabla^2 \Phi_0 = 0$

(ii) $\dfrac{\partial \Phi_0}{\partial u} \cdot \dfrac{\partial \Phi_0}{\partial v} = 0$

(iii) $\left\| \dfrac{\partial \Phi_0}{\partial u} \right\| = \left\| \dfrac{\partial \Phi_0}{\partial v} \right\|$

where $\|\mathbf{w}\|$ denotes the "norm" or length of the vector $\mathbf{w}$.

For well over 70 years, mathematicians such as Riemann, Weierstrass, H. A. Schwarz, Darboux, and Lebesgue puzzled over the challenge posed by Plateau. In 1931 the question was finally settled when Jesse Douglas showed that such a Φ_0 existed. However, many questions about soap films remain unsolved, and this area of research is still active today.

For more information on this fascinating subject, the reader may consult S. Hildebrandt and A. Tromba, *The Parsimonious Universe*: *Shape and Form in the Natural World*, Springer-Verlag, New York/Heidelberg, 1995.

EXERCISES

1. Find the surface area of the unit sphere S represented parametrically by $\Phi: D \to S \subset \mathbb{R}^3$, where D is the rectangle $0 \le \theta \le 2\pi, 0 \le \phi \le \pi$ and Φ is given by the equations

$$x = \cos\theta \sin\phi, \qquad y = \sin\theta \sin\phi, \qquad z = \cos\phi.$$

Note that we can represent the entire sphere parametrically, but we cannot represent it in the form $z = f(x, y)$.

2. In Exercise 1, what happens if we allow ϕ to vary from $-\pi/2$ to $\pi/2$? From 0 to 2π? Why do we obtain different answers?

3. Find the area of the helicoid in Example 2 if the domain D is $0 \le r \le 1$ and $0 \le \theta \le 3\pi$.

4. The torus T can be represented parametrically by the function $\Phi: D \to \mathbb{R}^3$, where Φ is given by the coordinate functions $x = (R + \cos\phi)\cos\theta, y = (R + \cos\phi)\sin\theta, z = \sin\phi$; D is the rectangle $[0, 2\pi] \times [0, 2\pi]$, that is, $0 \le \theta \le 2\pi, 0 \le \phi \le 2\pi$; and $R > 1$ is fixed (see Figure 7.4.7). Show that $A(T) = (2\pi)^2 R$, first by using formula (3) and then by using formula (6).

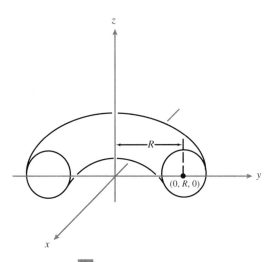

Figure 7.4.7   A cross section of a torus.

5.   Let $\Phi(u, v) = (u - v, u + v, uv)$ and let D be the unit disk in the uv plane. Find the area of $\Phi(D)$.

6.   Find the area of the portion of the unit sphere that is cut out by the cone $z \geq \sqrt{x^2 + y^2}$ (see Exercise 1).

7.   Show that the surface $x = 1/\sqrt{y^2 + z^2}$, where $1 \leq x < \infty$, can be filled but not painted!

8.   Find a parametrization of the surface $x^2 - y^2 = 1$, where $x > 0, -1 \leq y \leq 1$ and $0 \leq z \leq 1$. Use your answer to express the area of the surface as an integral.

9.   Represent the ellipsoid E:

$$\frac{x^2}{a^2} + \frac{y^2}{b^2} + \frac{z^2}{c^2} = 1$$

parametrically and write out the integral for its surface area $A(E)$. (Do not evaluate the integral.)

10.   Let the curve $y = f(x), a \leq x \leq b$, be rotated about the y axis. Show that the area of the surface swept out is

$$A = 2\pi \int_a^b |x|\sqrt{1 + [f'(x)]^2}\, dx.$$

Interpret using arc length and slant height.

11.   Find the area of the surface obtained by rotating the curve $y = x^2, 0 \leq x \leq 1$, about the y axis.

12.   Use formula (4) to compute the surface area of the cone in Example 1.

13.   Find the area of the surface defined by $x + y + z = 1, x^2 + 2y^2 \leq 1$.

14.   Show that for the vectors $\mathbf{T}_u$ and $\mathbf{T}_v$, we have the formula

$$\|\mathbf{T}_u \times \mathbf{T}_v\| = \sqrt{\left[\frac{\partial(x, y)}{\partial(u, v)}\right]^2 + \left[\frac{\partial(y, z)}{\partial(u, v)}\right]^2 + \left[\frac{\partial(x, z)}{\partial(u, v)}\right]^2}.$$

15. Compute the area of the surface given by

$$x = r \cos \theta, \qquad y = 2r \cos \theta, \qquad z = \theta, \qquad 0 \le r \le 1, \qquad 0 \le \theta \le 2\pi.$$

Sketch.

16. Prove *Pappus' theorem*: Let $\mathbf{c}$: $[a, b] \to \mathbb{R}^2$ be a C^1 path whose image lies in the right half plane and is a simple closed curve. The area of the lateral surface generated by rotating the image of $\mathbf{c}$ about the y axis is equal to $2\pi \bar{x} l(\mathbf{c})$, where $\bar{x}$ is the average value of x coordinates of points on $\mathbf{c}$ and $l(\mathbf{c})$ is the length of $\mathbf{c}$. (See Exercises 8 to 11, Section 7.1, for a discussion of average values.)

17. The cylinder $x^2 + y^2 = x$ divides the unit sphere S into two regions S_1 and S_2, where S_1 is inside the cylinder and S_2 outside. Find the ratio of areas $A(S_2)/A(S_1)$.

18. Suppose a surface S that is the graph of a function $z = f(x, y)$, where $(x, y) \in D \subset \mathbb{R}^2$ can also be described as the set of $(x, y, z) \in \mathbb{R}^3$ with $F(x, y, z) = 0$ (a level surface). Derive a formula for $A(S)$ that involves only F.

19. Calculate the area of the frustum shown in Figure 7.4.8 using (a) geometry alone and (b) a surface area formula.

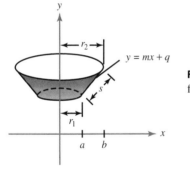

Figure 7.4.8 A line segment revolved around the y axis becomes a frustum of a cone.

20. A cylindrical hole of radius 1 is bored through a solid ball of radius 2 to form a ring coupler as shown in Figure 7.4.9. Find the volume and outer surface area of this coupler.

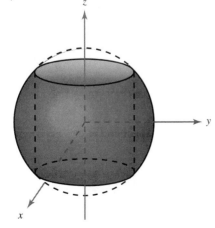

Figure 7.4.9 Find the surface area and volume of the shaded region.

21. Find the area of the graph of the function $f(x, y) = \frac{2}{3}(x^{3/2} + y^{3/2})$ which lies over the domain $[0, 1] \times [0, 1]$.

22. Express the surface area of the following graphs over the indicated region D as a double integral. Do not evaluate.

    (a) $(x + 2y)^2$; $D = [-1, 2] \times [0, 2]$

    (b) $xy + x/(y + 1)$; $D = [1, 4] \times [1, 2]$

    (c) $xy^3 e^{x^2 y^2}$; $D =$ unit circle centered at the origin.

    (d) $y^3 \cos^2 x$; $D =$ triangle with vertices $(-1, 1)$, $(0, 2)$, and $(1, 1)$.

23. Show that the surface area of the upper hemisphere of radius R, $z = \sqrt{R^2 - x^2 - y^2}$, can be computed by formula (4), evaluated as an improper integral.

7.5 INTEGRALS OF SCALAR FUNCTIONS OVER SURFACES

Now we are ready to define the integral of a *scalar* function f over a surface S. This concept is a natural generalization of the area of a surface, which corresponds to the integral over S of the scalar function $f(x, y, z) = 1$. This is quite analogous to considering the path integral as a generalization of arc length. In the next section we shall deal with the integral of a *vector* function $\mathbf{F}$ over a surface. These concepts will play a crucial role in the vector analysis treated in the final chapter.

Let us start with a surface S parametrized by a mapping $\mathbf{\Phi} \colon D \to S \subset \mathbb{R}^3$, which we write as $\mathbf{\Phi}(u, v) = (x(u, v), y(u, v), z(u, v))$.

DEFINITION: THE INTEGRAL OF A SCALAR FUNCTION OVER A SURFACE If $f(x, y, z)$ is a real-valued continuous function defined on a parametrized surface S, we define the **integral of f over S** to be

$$\iint_S f(x, y, z) \, dS = \iint_S f \, dS = \iint_D f(\mathbf{\Phi}(u, v)) \| \mathbf{T}_u \times \mathbf{T}_v \| \, du \, dv, \qquad (1)$$

Written out, equation (1) becomes

$$\iint_S f \, dS$$

$$= \iint_D f(x(u, v), y(u, v), z(u, v)) \sqrt{\left[\frac{\partial(x, y)}{\partial(u, v)}\right]^2 + \left[\frac{\partial(y, z)}{\partial(u, v)}\right]^2 + \left[\frac{\partial(x, z)}{\partial(u, v)}\right]^2} \, du \, dv. \, (2)$$

Thus if f is identically 1, we recover the area formula (3) of Section 7.4. Like surface area, the surface integral is independent of the particular parametrization used. This will be discussed in Section 7.6.

We can gain some intuitive knowledge about this integral by considering it as a limit of sums. Let D be a rectangle partitioned into n^2 rectangles R_{ij} with areas $\Delta u \, \Delta v$. Let $S_{ij} = \Phi(R_{ij})$ *be the portion of the surface* $\Phi(D)$ corresponding to R_{ij} (see Figure 7.5.1), and let $A(S_{ij})$ be the area of this portion of the surface. For large n, f will be approximately constant on S_{ij}, and we form the sum

$$S_n = \sum_{i=0}^{n-1} \sum_{j=0}^{n-1} f(\Phi(u_i, v_j)) A(S_{ij}), \tag{3}$$

where $(u_i, v_j) \in R_{ij}$. From Section 7.4 we have a formula for $A(S_{ij})$:

$$A(S_{ij}) = \iint_{R_{ij}} \|\mathbf{T}_u \times \mathbf{T}_v\| \, du \, dv,$$

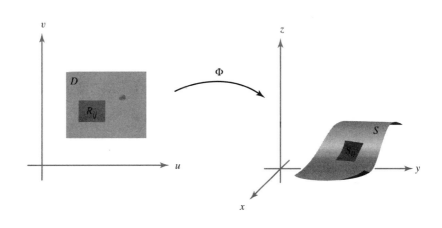

Figure 7.5.1 Φ takes a portion R_{ij} of D to a portion of S.

which, by the mean value theorem for integrals, equals $\|\mathbf{T}_{u_i^*} \times \mathbf{T}_{v_j^*}\| \, \Delta u \, \Delta v$ for some point (u_i^*, v_j^*) in R_{ij}. Hence our sum becomes

$$S_n = \sum_{i=0}^{n-1} \sum_{j=0}^{n-1} f(\Phi(u_i, v_j)) \|\mathbf{T}_{u_i^*} \times \mathbf{T}_{v_j^*}\| \Delta u \, \Delta v,$$

which is an approximating sum for the last integral in formula (1). Therefore,

$$\lim_{n \to \infty} S_n = \iint_S f \, dS.$$

Note that each term in the sum in formula (3) is the value of f at some point $\Phi(u_i, v_j)$ times the area of S_{ij}. Compare this with the Riemann-sum interpretation of the path integral in Section 7.1.

If S is a union of parametrized surfaces S_i, $i = 1, \ldots, N$, that do not intersect except possibly along curves defining their boundaries, then the integral of f over S is defined by

$$\iint_S f \, dS = \sum_{i=1}^N \iint_{S_i} f \, dS,$$

as we should expect. For example, the integral over the surface of a cube may be expressed as the sum of the integrals over the six sides.

EXAMPLE 1 Suppose a helicoid is described as in Example 2, Section 7.4, and let f be given by $f(x, y, z) = \sqrt{x^2 + y^2 + 1}$. Find $\iint_S f \, dS$.

SOLUTION As before,

$$\frac{\partial(x, y)}{\partial(r, \theta)} = r, \qquad \frac{\partial(y, z)}{\partial(r, \theta)} = \sin \theta, \qquad \frac{\partial(x, z)}{\partial(r, \theta)} = \cos \theta.$$

Also, $f(r \cos \theta, r \sin \theta, \theta) = \sqrt{r^2 + 1}$. Therefore

$$\iint_S f(x, y, z) \, dS = \iint_D f(\mathbf{\Phi}(r, \theta)) \|\mathbf{T}_r \times \mathbf{T}_\theta\| \, dr \, d\theta$$

$$= \int_0^{2\pi} \int_0^1 \sqrt{r^2 + 1} \sqrt{r^2 + 1} \, dr \, d\theta = \int_0^{2\pi} \frac{4}{3} \, d\theta = \frac{8}{3}\pi. \quad \blacktriangle$$

Suppose S is the graph of a C^1 function $z = g(x, y)$. Recall from Section 7.4 that we can parametrize S by

$$x = u, \qquad y = v, \qquad z = g(u, v).$$

and that in this case

$$\|\mathbf{T}_u \times \mathbf{T}_v\| = \sqrt{1 + \left(\frac{\partial g}{\partial u}\right)^2 + \left(\frac{\partial g}{\partial v}\right)^2},$$

and so

$$\iint_S f(x, y, z) \, dS = \iint_D f(x, y, g(x, y)) \sqrt{1 + \left(\frac{\partial g}{\partial x}\right)^2 + \left(\frac{\partial g}{\partial y}\right)^2} \, dx \, dy. \qquad (4)$$

EXAMPLE 2 Let S be the surface defined by $z = x^2 + y$, where D is the region $0 \leq x \leq 1$, $-1 \leq y \leq 1$. Evaluate $\iint_S x \, dS$.

SOLUTION   If we let $z = g(x, y) = x^2 + y$, formula (4) gives

$$\iint_S x \, dS = \iint_D x \sqrt{1 + \left(\frac{\partial g}{\partial x}\right)^2 + \left(\frac{\partial g}{\partial y}\right)^2} \, dx \, dy = \int_{-1}^{1} \int_{0}^{1} x \sqrt{1 + 4x^2 + 1} \, dx \, dy$$

$$= \frac{1}{8} \int_{-1}^{1} \left[\int_{0}^{1} (2 + 4x^2)^{1/2} (8x \, dx) \right] dy = \frac{2}{3} \cdot \frac{1}{8} \int_{-1}^{1} [(2 + 4x^2)^{3/2}]|_0^1 \, dy$$

$$= \frac{1}{12} \int_{-1}^{1} (6^{3/2} - 2^{3/2}) \, dy = \frac{1}{6}(6^{3/2} - 2^{3/2}) = \sqrt{6} - \frac{\sqrt{2}}{3} = \sqrt{2}\left(\sqrt{3} - \frac{1}{3}\right). \quad \blacktriangle$$

EXAMPLE 3   Evaluate $\iint_S z^2 \, dS$, where S is the unit sphere $x^2 + y^2 + z^2 = 1$.

SOLUTION   For this problem it is convenient to represent the sphere parametrically by the equations $x = \cos\theta \sin\phi$, $y = \sin\theta \sin\phi$, $z = \cos\phi$, over the region D in the $\theta\phi$ plane given by $0 \le \phi \le \pi$, $0 \le \theta \le 2\pi$. From equation (1) we get

$$\iint_S z^2 \, dS = \iint_D (\cos\phi)^2 \|\mathbf{T}_\theta \times \mathbf{T}_\phi\| \, d\theta \, d\phi.$$

Now a little computation [use formula (2) of Section 7.4] shows that

$$\|\mathbf{T}_\theta \times \mathbf{T}_\phi\| = |\sin\phi|,$$

so that

$$\iint_S z^2 \, dS = \int_0^{2\pi} \int_0^{\pi} \cos^2\phi |\sin\phi| \, d\phi \, d\theta = \int_0^{2\pi} \int_0^{\pi} \cos^2\phi \sin\phi \, d\phi \, d\theta$$

$$= \frac{1}{3} \int_0^{2\pi} [-\cos^3\phi]_0^{\pi} \, d\theta = \frac{2}{3} \int_0^{2\pi} d\theta = \frac{4\pi}{3}. \quad \blacktriangle$$

We now develop another formula for surface integrals when the surface can be represented as a graph. To do so, we let S be the graph of $z = g(x, y)$ and consider formula (4). We claim that

$$\iint_S f(x, y, z) \, dS = \iint_D \frac{f(x, y, g(x, y))}{\cos\theta} \, dx \, dy, \qquad (5)$$

where θ is the angle the normal to the surface makes with the unit vector $\mathbf{k}$ at the point $(x, y, g(x, y))$ (see Figure 7.5.2). Describing the surface by the equation $\phi(x, y, z) = z - g(x, y) = 0$, a normal vector is $\nabla\phi$; that is,

$$\mathbf{n} = -\frac{\partial g}{\partial x}\mathbf{i} - \frac{\partial g}{\partial y}\mathbf{j} + \mathbf{k}$$

[see Example 3, of Section 7.3, or recall that the normal to a surface with equation $g(x, y, z) = $ constant is given by ∇g]. Thus

$$\cos \theta = \frac{\mathbf{n} \cdot \mathbf{k}}{\|\mathbf{n}\|} = \frac{1}{\sqrt{(\partial g / \partial x)^2 + (\partial g / \partial y)^2 + 1}}.$$

Substitution of this formula into equation (4) gives equation (5).

The result is, in fact, obvious geometrically, for if a small rectangle in the xy plane has area ΔA, then the area of the portion above it on the surface is $\Delta S = \Delta A / \cos \theta$ (Figure 7.5.2). This approach can help us to remember formula (5) and to apply it in problems.

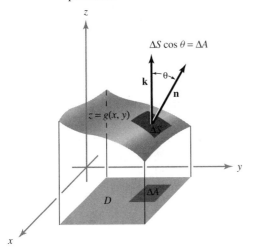

Figure 7.5.2 The area of a patch of area ΔS over a patch ΔA is $\Delta S = \Delta A / \cos \theta$, where θ is the angle the normal $\mathbf{n}$ makes with $\mathbf{k}$.

EXAMPLE 4 Compute $\iint_S x \, dS$, where S is the triangle with vertices $(1, 0, 0)$, $(0, 1, 0)$, $(0, 0, 1)$ (See Figure 7.5.3).

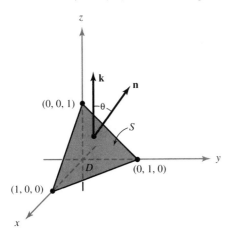

Figure 7.5.3 In computing a specific surface integral, one finds a formula for the normal $\mathbf{n}$ and computes the angle θ in preparation for formula (5).

SOLUTION This surface is the plane described by the equation $x + y + z = 1$. Since the surface is a plane, the angle θ is constant and a unit normal vector is $\mathbf{n} =$

$(1/\sqrt{3}, 1/\sqrt{3}, 1/\sqrt{3})$. Thus, $\cos\theta = \mathbf{n}\cdot\mathbf{k} = 1/\sqrt{3}$, and by equation (5),

$$\iint_S x\,dS = \sqrt{3}\iint_D x\,dx\,dy,$$

where D is the domain in the xy plane. But

$$\sqrt{3}\iint_D x\,dx\,dy = \sqrt{3}\int_0^1\int_0^{1-x} x\,dy\,dx = \sqrt{3}\int_0^1 x(1-x)\,dx = \frac{\sqrt{3}}{6}. \quad \blacktriangle$$

Integrals of functions over surfaces are useful for computing the mass of a surface when the mass density function m is known. The total mass of a surface with mass density (per unit area) m is given by

$$M(S) = \iint_S m(x, y, z)\,dS. \tag{6}$$

EXAMPLE 5   Let $\Phi\colon D \to \mathbb{R}^3$ be the parametrization of the helicoid $S = \Phi(D)$ of Example 2 of Section 7.4. Recall that $\Phi(r, \theta) = (r\cos\theta, r\sin\theta, \theta)$, where $0 \le \theta \le 2\pi$, and $0 \le r \le 1$. Suppose S has a mass density at $(x, y, z) \in S$ equal to twice the distance of (x, y, z) from the central axis (see Figure 7.4.2), that is, $m(x, y, z) = 2\sqrt{x^2 + y^2} = 2r$, in the cylindrical coordinate system. Find the total mass of the surface.

SOLUTION   Applying formula (6),

$$M(S) = \iint_S 2\sqrt{x^2 + y^2}\,dS = \iint_D 2r\,dS = \iint_D 2r\|\mathbf{T}_r \times \mathbf{T}_\theta\|\,dr\,d\theta.$$

From Example 2 of Section 7.4 we see that $\|\mathbf{T}_r \times \mathbf{T}_\theta\| = \sqrt{1 + r^2}$. Thus

$$M(S) = \iint_D 2r\sqrt{1 + r^2}\,dr\,d\theta = \int_0^{2\pi}\int_0^1 2r\sqrt{1 + r^2}\,dr\,d\theta$$

$$= \int_0^{2\pi}\left[\frac{2}{3}(1 + r^2)^{3/2}\right]_0^1 d\theta = \int_0^{2\pi}\frac{2}{3}(2^{3/2} - 1)\,d\theta = \frac{4\pi}{3}(2^{3/2} - 1). \quad \blacktriangle$$

EXERCISES

1.   Compute $\iint_S xy\,dS$, where S is the surface of the tetrahedron with sides $z = 0$, $y = 0$, $x + z = 1$, and $x = y$.

2.   Let $\Phi\colon D \subset \mathbb{R}^2 \to \mathbb{R}^3$ be a parametrization of a surface S defined by

$$x = x(u, v), \quad y = y(u, v), \quad z = z(u, v)$$

(a)   Let

$$\frac{\partial\Phi}{\partial u} = \left(\frac{\partial x}{\partial u}, \frac{\partial y}{\partial u}, \frac{\partial z}{\partial u}\right) \quad \text{and} \quad \frac{\partial\Phi}{\partial v} = \left(\frac{\partial x}{\partial v}, \frac{\partial y}{\partial v}, \frac{\partial z}{\partial v}\right),$$

that is, $\partial\Phi/\partial u = \mathbf{T}_u$ and $\partial\Phi/\partial v = \mathbf{T}_v$, and set

$$E = \left\| \frac{\partial\Phi}{\partial u} \right\|^2, \qquad F = \frac{\partial\Phi}{\partial u} \cdot \frac{\partial\Phi}{\partial v}, \qquad G = \left\| \frac{\partial\Phi}{\partial v} \right\|^2.$$

Show that the surface area of S is $\iint_D \sqrt{EG - F^2}\, du\, dv$. In this notation, how can we express $\iint_S f\, dS$ for a general function f?

(b) What does the formula become if the vectors $\partial\Phi/\partial u$ and $\partial\Phi/\partial v$ are orthogonal?

(c) Use parts (a) and (b) to compute the surface area of a sphere of radius a.

3. Evaluate $\iint_S z\, dS$, where S is the upper hemisphere of radius a, that is, the set of (x, y, z) with $z = \sqrt{a^2 - x^2 - y^2}$.

4. Evaluate $\iint_S (x + y + z)\, dS$, where S is the boundary of the unit ball B; that is, S is the set of (x, y, z) with $x^2 + y^2 + z^2 = 1$. (HINT: Use the symmetry of the problem.)

5. Evaluate $\iint_S xyz\, dS$, where S is the triangle with vertices $(1, 0, 0)$, $(0, 2, 0)$ and $(0, 1, 1)$.

6. Let a surface S be defined implicitly by $F(x, y, z) = 0$ for (x, y) in a domain D of $\mathbb{R}^2$. Show that

$$\iint_S \left| \frac{\partial F}{\partial z} \right| dS = \iint_D \sqrt{\left(\frac{\partial F}{\partial x} \right)^2 + \left(\frac{\partial F}{\partial y} \right)^2 + \left(\frac{\partial F}{\partial z} \right)^2}\, dx\, dy$$

Compare with Exercise 18 of Section 7.4.

7. Evaluate $\iint_S z\, dS$, where S is the surface $z = x^2 + y^2$, $x^2 + y^2 \le 1$.

8. Evaluate $\iint_S z^2\, dS$, where S is the boundary of the cube $C = [-1, 1] \times [-1, 1] \times [-1, 1]$. (HINT: Do each face separately and add the results.)

9. Find the mass of a spherical surface S of radius R such that at each point $(x, y, z) \in S$ the mass density is equal to the distance of (x, y, z) to some fixed point $(x_0, y_0, z_0) \in S$.

10. A metallic surface S is in the shape of a hemisphere $z = \sqrt{R^2 - x^2 - y^2}$, $0 \le x^2 + y^2 \le R^2$. The mass density at $(x, y, z) \in S$ is given by $m(x, y, z) = x^2 + y^2$. Find the total mass of S.

11. Let S be the sphere of radius R.

(a) Argue by symmetry that

$$\iint_S x^2\, dS = \iint_S y^2\, dS = \iint_S z^2\, dS.$$

(b) Use this fact and some clever thinking to evaluate, with very little computation, the integral

$$\iint_S x^2\, dS.$$

(c) Does this help in Exercise 10?

12. (a)   Use Riemann sums to justify the formula

$$\frac{1}{A(S)}\iint_S f(x, y, z)\, dS$$

for the *average value* of f over the surface S.

(b)   In Example 3 of this section, show that the average of $f(x, y, z) = z^2$ over the sphere is $\frac{1}{3}$.

(c)   Define the ***center of gravity*** $(\bar{x}, \bar{y}, \bar{z})$ of a surface S to be such that $\bar{x}$, $\bar{y}$, and $\bar{z}$ are the average values of the x, y, and z coordinates on S. Show that the center of gravity of the triangle in Example 4 of this section is $(\frac{1}{3}, \frac{1}{3}, \frac{1}{3})$.

13.   Find the x, y, and z coordinates of the center of gravity of the octant of the solid sphere of radius R and centered at the origin determined by $x \geq 0$, $y \geq 0$, $z \geq 0$. (HINT: Write this octant as a parametrized surface—see Example 3 of this section and Exercise 12.)

14.   Find the z coordinate of the center of gravity (the average z coordinate) of the surface of a hemisphere ($z \leq 0$) with radius r (see Exercise 12). Argue by symmetry that the average x and y coordinates are both zero.

15.   ***Dirichlet's functional*** for a parametrized surface $\boldsymbol{\Phi}: D \to \mathbb{R}^3$ is defined by[*]

$$J(\boldsymbol{\Phi}) = \frac{1}{2}\iint_D \left(\left\|\frac{\partial\boldsymbol{\Phi}}{\partial u}\right\|^2 + \left\|\frac{\partial\boldsymbol{\Phi}}{\partial v}\right\|^2 \right) du\, dv.$$

Using Exercise 15, Section 1.5, argue that the area $A(\boldsymbol{\Phi}) \leq J(\boldsymbol{\Phi})$ and equality holds if

(a) $\left\|\dfrac{\partial\boldsymbol{\Phi}}{\partial u}\right\|^2 = \left\|\dfrac{\partial\boldsymbol{\Phi}}{\partial v}\right\|^2$ and   (b) $\dfrac{\partial\boldsymbol{\Phi}}{\partial u}\cdot\dfrac{\partial\boldsymbol{\Phi}}{\partial v} = 0$.

Compare these equations with Exercise 2 and the remarks at the end of Section 7.4. A parametrization $\boldsymbol{\Phi}$ that satisfies conditions (a) and (b) is said to be ***conformal***.

16.   Let $D \subset \mathbb{R}^2$ and $\boldsymbol{\Phi}: D \to \mathbb{R}^2$ be a smooth function $\boldsymbol{\Phi}(u, v) = (x(u, v), y(u, v))$ satisfying conditions (a) and (b) of Exercise 15. Show that x and y satisfy the ***Cauchy-Riemann equations*** $\partial x/\partial u = \partial y/\partial v$, $\partial x/\partial v = -\partial y/\partial u$ or the ***conjugate Cauchy-Riemann equations*** $\partial x/\partial u = -\partial y/\partial v$, $\partial x/\partial v = \partial y/\partial u$. Conclude that $\nabla^2\boldsymbol{\Phi} = 0$ (i.e., each component of $\boldsymbol{\Phi}$ is harmonic).

17.   (a)   Compute the area of the portion of the cone $x^2 + y^2 = z^2$ with $z \geq 0$ that is inside the sphere $x^2 + y^2 + z^2 = 2Rz$, where R is a positive constant.

(b)   What is the area of that portion of the sphere that is inside the cone?

18.   Let S be a sphere of radius r and $\mathbf{p}$ be a point inside or outside the sphere (but not on it). Show that

$$\iint_S \frac{1}{\|\mathbf{x} - \mathbf{p}\|}\, dS = \begin{cases} 4\pi r & \text{if } \mathbf{p} \text{ is inside } S \\ 4\pi r^2/d & \text{if } \mathbf{p} \text{ is outside } S \end{cases}$$

where d is the distance from $\mathbf{p}$ to the center of the sphere and the integration is over x.

---

[*]Dirichlet's functional played a major role in the mathematics of the nineteenth century. The mathematician Georg Friedrich Bernhard Riemann (1826–1866) used it to develop his complex function theory and to give a proof of the famous Riemann mapping theorem. Today it is still used extensively as a tool in the study of partial differential equations.

19.   Find the surface area of that part of the cylinder $x^2 + z^2 = a^2$ that is inside the cylinder $x^2 + y^2 = 2ay$ and also in the positive octant ($x \geq 0$, $y \geq 0$, $z \geq 0$). Assume $a > 0$.

7.6   SURFACE INTEGRALS OF VECTOR FUNCTIONS

In this section we turn our attention to integrals of *vector* fields over surfaces. The definition we give here is a natural extension of that for scalar functions discussed in Section 7.5. The physical interpretation of surface integrals and a discussion in terms of Riemann sums is given below.

DEFINITION: THE SURFACE INTEGRAL OF VECTOR FIELDS
Let $\mathbf{F}$ be a vector field defined on S, the image of a parametrized surface $\boldsymbol{\Phi}$. The ***surface integral*** of $\mathbf{F}$ over $\boldsymbol{\Phi}$, denoted by

$$\iint_{\boldsymbol{\Phi}} \mathbf{F} \cdot d\mathbf{S},$$

is defined by

$$\iint_{\boldsymbol{\Phi}} \mathbf{F} \cdot d\mathbf{S} = \iint_{D} \mathbf{F} \cdot (\mathbf{T}_u \times \mathbf{T}_v)\, du\, dv.$$

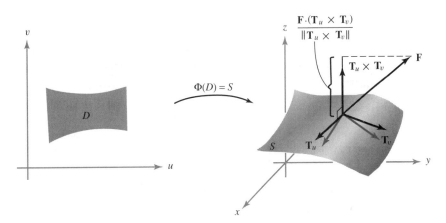

Figure 7.6.1   The geometric significance of $\mathbf{F} \cdot (\mathbf{T}_u \times \mathbf{T}_v)$.

EXAMPLE   1   Let D be the rectangle in the $\theta\phi$ plane defined by

$$0 \leq \theta \leq 2\pi, \qquad 0 \leq \phi \leq \pi,$$

and let the surface S be defined by the parametrization $\boldsymbol{\Phi}: D \to \mathbb{R}^3$ given by

$$x = \cos\theta \sin\phi, \qquad y = \sin\theta \sin\phi, \qquad z = \cos\phi.$$

(Thus θ and ϕ are the angles of spherical coordinates, and S is the unit sphere parametrized by Φ.) Let $\mathbf{r}$ be the position vector $\mathbf{r}(x, y, z) = x\mathbf{i} + y\mathbf{j} + z\mathbf{k}$. Compute $\iint_\Phi \mathbf{r} \cdot d\mathbf{S}$.

SOLUTION   First we find

$$\mathbf{T}_\theta = (-\sin\phi \sin\theta)\mathbf{i} + (\sin\phi \cos\theta)\mathbf{j}$$
$$\mathbf{T}_\phi = (\cos\theta \cos\phi)\mathbf{i} + (\sin\theta \cos\phi)\mathbf{j} - (\sin\phi)\mathbf{k}$$

and hence

$$\mathbf{T}_\theta \times \mathbf{T}_\phi = (-\sin^2\phi \cos\theta)\mathbf{i} - (\sin^2\phi \sin\theta)\mathbf{j} - (\sin\phi \cos\phi)\mathbf{k}.$$

Then we evaluate

$$\mathbf{r} \cdot (\mathbf{T}_\theta \times \mathbf{T}_\phi) = (x\mathbf{i} + y\mathbf{j} + z\mathbf{k}) \cdot (\mathbf{T}_\theta \times \mathbf{T}_\phi)$$
$$= [(\cos\theta \sin\phi)\mathbf{i} + (\sin\theta \sin\phi)\mathbf{j} + (\sin\phi)\mathbf{k}]$$
$$\cdot (-\sin\phi)[(\sin\phi \cos\theta)\mathbf{i} + (\sin\phi \sin\theta)\mathbf{j} + (\cos\phi)\mathbf{k}]$$
$$= (-\sin\phi)(\sin^2\phi \cos^2\theta + \sin^2\phi \sin^2\theta + \cos^2\phi) = -\sin\phi.$$

Thus

$$\iint_\Phi \mathbf{r} \cdot d\mathbf{S} = \iint_D -\sin\phi \, d\phi \, d\theta = \int_0^{2\pi} (-2)d\theta = -4\pi. \quad \blacktriangle$$

An analogy can be drawn between the surface integral $\iint_\Phi \mathbf{F} \cdot d\mathbf{S}$ and the line integral $\int_\mathbf{c} \mathbf{F} \cdot d\mathbf{s}$. Recall that the line integral is an oriented integral. We needed the notion of orientation of a curve to extend the definition of $\int_\mathbf{c} \mathbf{F} \cdot d\mathbf{s}$ to line integrals $\int_C \mathbf{F} \cdot d\mathbf{s}$ over oriented curves. We extend the definition of $\iint_\Phi \mathbf{F} \cdot d\mathbf{S}$ to *oriented surfaces* in a similar fashion; that is, given a surface S parametrized by a mapping Φ, we want to define $\iint_S \mathbf{F} \cdot d\mathbf{S} = \iint_\Phi \mathbf{F} \cdot d\mathbf{S}$ and show that it is independent of the parametrization, except possibly for the sign. To accomplish this, we need the notion of orientation of a surface.

DEFINITION: ORIENTED SURFACES   An *oriented surface* is a two-sided surface with one side specified as the *outside* or *positive side*; we call the other side the *inside* or *negative side*.* At each point $(x, y, z) \in S$ there are two unit normal vectors $\mathbf{n}_1$ and $\mathbf{n}_2$, where $\mathbf{n}_1 = -\mathbf{n}_2$ (see Figure 7.6.2). Each of these two normals can be associated with one side of the surface. Thus to specify a side of a surface S, at each point we choose a unit normal vector $\mathbf{n}$ that points away from the positive side of S at that point.

---

*We use the term "side" in an intuitive sense. This concept can be developed rigorously. Also, the choice of the side to be named the "outside" is often dictated by the surface itself, as, for example, is the case with a sphere. In other cases the naming is somewhat arbitrary (see the piece of surface depicted in Figure 7.6.2).

Figure 7.6.2 The two possible unit normals to a surface at a point.

This definition assumes that our surface does have two sides. We should give an example of a surface with only one side. The first known example of such a surface was the Möbius strip (named after the German mathematician and astronomer A. F. Möbius, who, along with the mathematician J. B. Listing, discovered it in 1858). Pictures of such a surface are given in Figures 7.6.3 and 7.6.4. At each point of M there are two unit normals, $\mathbf{n}_1$ and $\mathbf{n}_2$. However, $\mathbf{n}_1$ does not determine a unique side of M, and neither does $\mathbf{n}_2$. To see this intuitively, we can slide $\mathbf{n}_2$ around the closed curve C (Figure 7.6.3). When $\mathbf{n}_2$ returns to a fixed point p on C it will coincide with $\mathbf{n}_1$, showing that both $\mathbf{n}_1$ and $\mathbf{n}_2$ point away from the same side of M and, consequently, that M has only one side.

Figure 7.6.3 The Möbius strip: slide $\mathbf{n}_2$ around C once; when $\mathbf{n}_2$ returns to its initial point, $\mathbf{n}_2$ will coincide with $\mathbf{n}_1 = -\mathbf{n}_2$.

Figure 7.6.4 is a Möbius strip as drawn by the well-known twentieth-century mathematician and artist M. C. Escher. It depicts ants crawling along the Möbius band. After one trip around the band (without crossing an edge) they end up on the "opposite side" of the surface.

Figure 7.6.4 Ants walking on a Möbius strip. (*Moebius Strip II, 1963, by M. C. Escher. Escher Foundation, Haags Gemeentemuseum, The Hague.*)

Let $\mathbf{\Phi}\colon D \to \mathbb{R}^3$ be a parametrization of an oriented surface S and suppose S is smooth at $\mathbf{\Phi}(u_0, v_0)$, $(u_0, v_0) \in D$; that is, the unit normal vector $(\mathbf{T}_{u_0} \times \mathbf{T}_{v_0})/\|\mathbf{T}_{u_0} \times \mathbf{T}_{v_0}\|$ is defined. If $\mathbf{n}(\mathbf{\Phi}(u_0, v_0))$ denotes the unit normal to S at $\mathbf{\Phi}(u_0, v_0)$, it follows that $(\mathbf{T}_{u_0} \times \mathbf{T}_{v_0})/\|\mathbf{T}_{u_0} \times \mathbf{T}_{v_0}\| = \pm\mathbf{n}(\mathbf{\Phi}(u_0, v_0))$. The parametrization $\mathbf{\Phi}$ is said to be ***orientation-preserving*** if $(\mathbf{T}_u \times \mathbf{T}_v)/\|\mathbf{T}_u \times \mathbf{T}_v\| = \mathbf{n}(\mathbf{\Phi}(u, v))$ at all $(u, v) \in D$ for which S is smooth at $\mathbf{\Phi}(u, v)$. In other words, $\mathbf{\Phi}$ is orientation-preserving if the vector $\mathbf{T}_u \times \mathbf{T}_v$ points to the outside of the surface. If $\mathbf{T}_u \times \mathbf{T}_v$ points to the inside of the surface at all points $(u, v) \in D$ for which S is smooth at $\mathbf{\Phi}(u, v)$, then $\mathbf{\Phi}$ is said to be ***orientation-reversing***. Using the above notation, this condition corresponds to the choice $(\mathbf{T}_u \times \mathbf{T}_v)/\|\mathbf{T}_u \times \mathbf{T}_v\| = -\mathbf{n}(\mathbf{\Phi}(u, v))$.

It follows from this discussion that the Möbius band M cannot be parametrized by a single parametrization for which $\mathbf{n} = \mathbf{T}_u \times \mathbf{T}_v \neq \mathbf{0}$ (if there were such a parameterization, then M would indeed have two sides, one determined by $\mathbf{n}$ and one determined by $-\mathbf{n}$). The sphere in Example 1 can be parametrized by a single parametrization, but not by one that is everywhere one-to-one—see the discussion at the beginning of Section 7.4.

Thus, any parametrized surface for which $\mathbf{T}_u \times \mathbf{T}_v$ never vanishes can be considered as an oriented surface with a positive side determined by the direction of $\mathbf{T}_u \times \mathbf{T}_v$.

E X A M P L E   2   We can give the unit sphere $x^2 + y^2 + z^2 = 1$ in $\mathbb{R}^3$ (Figure 7.6.5) an orientation by selecting the unit vector $\mathbf{n}(x, y, z) = \mathbf{r}$, where $\mathbf{r} = x\mathbf{i} + y\mathbf{j} + z\mathbf{k}$, which points to the outside of the surface. This choice corresponds to our intuitive notion of outside for the sphere.

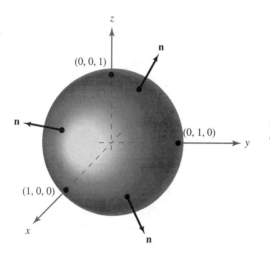

Figure 7.6.5   The unit sphere oriented by its outward normal $\mathbf{n}$.

Now that the sphere S is an oriented surface, consider the parametrization $\mathbf{\Phi}$ of S given in Example 1. The cross product of the tangent vectors $\mathbf{T}_\theta$ and $\mathbf{T}_\phi$—that is, a normal to S is given by

$$(-\sin\phi)[(\cos\theta \sin\phi)\mathbf{i} + (\sin\theta \sin\phi)\mathbf{j} + (\cos\phi)\mathbf{k}] = -\mathbf{r}\sin\phi.$$

Since $-\sin\phi \le 0$ for $0 \le \phi \le \pi$, this normal vector points inward from the sphere. Thus the given parametrization Φ is *orientation-reversing*. ▲

EXAMPLE 3 Let S be a surface described by $z = f(x, y)$. There are two unit normal vectors to S at $(x_0, y_0, f(x_0, y_0))$, namely, $\pm\mathbf{n}$, where

$$\mathbf{n} = \frac{-\dfrac{\partial f}{\partial x}(x_0, y_0)\mathbf{i} - \dfrac{\partial f}{\partial y}(x_0, y_0)\mathbf{j} + \mathbf{k}}{\sqrt{\left[\dfrac{\partial f}{\partial x}(x_0, y_0)\right]^2 + \left[\dfrac{\partial f}{\partial y}(x_0, y_0)\right]^2 + 1}}.$$

We can orient all such surfaces by taking the positive side of S to be the side away from which $\mathbf{n}$ points (Figure 7.6.6). Thus the positive side of such a surface is determined by the unit normal $\mathbf{n}$ with positive $\mathbf{k}$ component. If we parametrize this surface by $\Phi(u, v) = ((u, v, f(u, v)))$, then Φ will be orientation-preserving. ▲

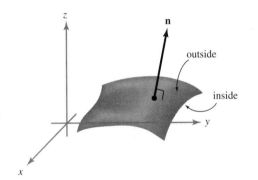

Figure 7.6.6 $\mathbf{n}$ points away from the outside of the surface.

We now state without proof a theorem showing that the integral over an oriented surface is independent of the parametrization. The proof of this theorem is analogous to that of Theorem 1 (Section 7.2); the heart of the proof is again the change of variables formula—this time applied to double integrals.

THEOREM 4: INDEPENDENCE OF SURFACE INTEGRALS ON PARAMETRIZATIONS *Let S be an oriented surface and let Φ_1 and Φ_2 be two smooth orientation-preserving parametrizations, with $\mathbf{F}$ a continuous vector field defined on S. Then*

$$\iint_{\Phi_1} \mathbf{F} \cdot d\mathbf{S} = \iint_{\Phi_2} \mathbf{F} \cdot d\mathbf{S}.$$

> If $\mathbf{\Phi}_1$ is orientation-preserving and $\mathbf{\Phi}_2$ orientation-reversing, then
>
> $$\iint_{\mathbf{\Phi}_1} \mathbf{F} \cdot d\mathbf{S} = -\iint_{\mathbf{\Phi}_2} \mathbf{F} \cdot d\mathbf{S}.$$
>
> If f is a real-valued continuous function defined on S, and if $\mathbf{\Phi}_1$ and $\mathbf{\Phi}_2$ are parametrizations of S, then
>
> $$\iint_{\mathbf{\Phi}_1} f\, dS = \iint_{\mathbf{\Phi}_2} f\, dS.$$

Note that if $f = 1$, we obtain

$$A(S) = \iint_{\mathbf{\Phi}_1} dS = \iint_{\mathbf{\Phi}_2} dS,$$

thus showing that area is independent of parametrization.

We can thus unambiguously use the notation

$$\iint_S \mathbf{F} \cdot d\mathbf{S} = \iint_{\mathbf{\Phi}} \mathbf{F} \cdot d\mathbf{S}$$

(or a sum of such integrals, if S is a union of parametrized surfaces that intersect only along their boundary curves) where $\mathbf{\Phi}$ is an orientation-preserving parametrization. Theorem 4 guarantees that the value of the integral does not depend on the selection of $\mathbf{\Phi}$.

Recall from formula (1) of Section 7.2 that a line integral $\int_{\mathbf{c}} \mathbf{F} \cdot d\mathbf{s}$ can be thought of as the path integral of the tangential component of $\mathbf{F}$ along $\mathbf{c}$ (although for the case in which $\mathbf{c}$ intersects itself, the integral obtained is technically not a path integral). A similar situation will hold for surface integrals in our setting, since we are assuming that the mappings $\mathbf{\Phi}$ defining the surface S are one-to-one except perhaps on the boundary of D, which can be ignored for the purposes of integration. Thus in defining integrals over surfaces, we assume in this book that the surfaces are nonintersecting.

For an oriented smooth surface S and an orientation-preserving parametrization $\mathbf{\Phi}$ of S we can express $\iint_S \mathbf{F} \cdot d\mathbf{S}$ as an integral of a real-valued function f over the surface. Let $\mathbf{n} = (\mathbf{T}_u \times \mathbf{T}_v)/\|\mathbf{T}_u \times \mathbf{T}_v\|$ be the unit normal pointing to the outside of S. Then

$$
\begin{aligned}
\iint_S \mathbf{F} \cdot d\mathbf{S} &= \iint_{\mathbf{\Phi}} \mathbf{F} \cdot d\mathbf{S} = \iint_D \mathbf{F} \cdot (\mathbf{T}_u \times \mathbf{T}_v)\, du\, dv \\
&= \iint_D \mathbf{F} \cdot \left(\frac{\mathbf{T}_u \times \mathbf{T}_v}{\|\mathbf{T}_u \times \mathbf{T}_v\|} \right) \|\mathbf{T}_u \times \mathbf{T}_v\|\, du\, dv \\
&= \iint_D (\mathbf{F} \cdot \mathbf{n})\|\mathbf{T}_u \times \mathbf{T}_v\|\, du\, dv = \iint_S (\mathbf{F} \cdot \mathbf{n})\, dS = \iint_S f\, dS,
\end{aligned}
$$

where $f = \mathbf{F} \cdot \mathbf{n}$. We have thus proved the following theorem.

THEOREM 5 $\iint_S \mathbf{F} \cdot d\mathbf{S}$, *the surface integral of* $\mathbf{F}$ *over* S, *is equal to the integral of the normal component of* $\mathbf{F}$ *over the surface. In short,*

$$\iint_S \mathbf{F} \cdot d\mathbf{S} = \iint_S \mathbf{F} \cdot \mathbf{n} \, dS.$$

This observation can often save computational effort, as Example 4 below demonstrates. The geometric and physical significance of the surface integral can be understood by expressing it as a limit of Riemann sums. For simplicity, we assume D is a rectangle. Fix a parametrization $\mathbf{\Phi}$ of S that preserves orientation and partition the region D into n^2 pieces D_{ij}, $0 \le i \le n-1$, $0 \le j \le n-1$. We let Δu denote the length of the horizontal side of D_{ij} and Δv denote the length of the vertical side of D_{ij}. Let (u, v) be a point in D_{ij}, and $(x, y, z) = \mathbf{\Phi}(u, v)$ the corresponding point on the surface. We consider the parallelogram with sides $\Delta u \, \mathbf{T}_u$ and $\Delta v \, \mathbf{T}_v$ lying in the plane tangent to S at (x, y, z) and the parallelepiped formed by $\mathbf{F}$, $\Delta u \, \mathbf{T}_u$, and $\Delta v \, \mathbf{T}_v$. The volume of the parallelepiped is the absolute value of the triple product

$$\mathbf{F} \cdot (\Delta u \, \mathbf{T}_u \times \Delta v \, \mathbf{T}_v) = \mathbf{F} \cdot (\mathbf{T}_u \times \mathbf{T}_v) \, \Delta u \, \Delta v.$$

The vector $\mathbf{T}_u \times \mathbf{T}_v$ is normal to the surface at (x, y, z) and points away from the outside of the surface. Thus the number $\mathbf{F} \cdot (\mathbf{T}_u \times \mathbf{T}_v)$ is positive when the parallelepiped lies on the outside of the surface (Figure 7.6.7).

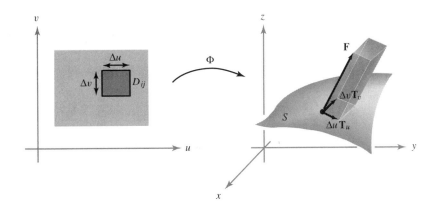

Figure 7.6.7 $\mathbf{F} \cdot (\mathbf{T}_u \times \mathbf{T}_v) > 0$ when the parallelepiped formed by $\Delta v \, \mathbf{T}_v$, $\Delta u \, \mathbf{T}_u$, and $\mathbf{F}$ lies to the "outside" of the surface S.

In general, the parallelepiped lies on that side of the surface away from which $\mathbf{F}$ is pointing. If we think of $\mathbf{F}$ as the velocity field of a fluid, $\mathbf{F}(x, y, z)$ is pointing in the direction in which fluid is moving across the surface near (x, y, z). Moreover, the number

$$|\mathbf{F} \cdot (\mathbf{T}_u \, \Delta u \times \mathbf{T}_v \, \Delta v)|$$

measures the amount of fluid that passes through the tangent parallelogram per unit time. Since the sign of $\mathbf{F} \cdot (\Delta u\, \mathbf{T}_u \times \Delta v\, \mathbf{T}_v)$ is positive if the vector $\mathbf{F}$ is pointing outward at (x, y, z) and negative if $\mathbf{F}$ is pointing inward, $\sum_{i,j} \mathbf{F} \cdot (\mathbf{T}_u \times \mathbf{T}_v)\, \Delta u\, \Delta v$ is an approximate measure of the net quantity of fluid to flow outward across the surface per unit time. (Remember that "outward" or "inward" depends on our choice of parametrization. Figure 7.6.8 illustrates $\mathbf{F}$ directed outward and inward, given $\mathbf{T}_u$ and $\mathbf{T}_v$.) Hence, *the integral $\iint_S \mathbf{F} \cdot d\mathbf{S}$ is the net quantity of fluid to flow across the surface per unit time, that is, the rate of fluid flow.* This integral is also called the ***flux*** of $\mathbf{F}$ across the surface.

Figure 7.6.8   When $\mathbf{F} \cdot (\mathbf{T}_u \times \mathbf{T}_v) > 0$ (left), $\mathbf{F}$ points outward; when $\mathbf{F} \cdot (\mathbf{T}_u \times \mathbf{T}_v) < 0$ (right), $\mathbf{F}$ points inward.

In the case where $\mathbf{F}$ represents an electric or a magnetic field, $\iint_S \mathbf{F} \cdot d\mathbf{S}$ is also commonly known as the flux. The reader may be familiar with physical laws (such as Faraday's law) that relate flux of a vector field to a circulation (or current) in a bounding loop. This is the historical and physical basis of Stokes' theorem, which we shall discuss in Section 8.2. The corresponding principle in fluid mechanics is called Kelvin's circulation theorem.

Surface integrals also apply to the study of heat flow. Let $T(x, y, z)$ be the temperature at a point $(x, y, z) \in W \subset \mathbb{R}^3$, where W is some region and T is a C^1 function. Then

$$\nabla T = \frac{\partial T}{\partial x}\mathbf{i} + \frac{\partial T}{\partial y}\mathbf{j} + \frac{\partial T}{\partial z}\mathbf{k}$$

represents the temperature gradient, and heat "flows" with the vector field $-k\, \nabla T = \mathbf{F}$, where k is a positive constant (see Section 8.5). Therefore $\iint_S \mathbf{F} \cdot d\mathbf{S}$ is the total rate of heat flow or flux across the surface S.

EXAMPLE 4   Suppose a temperature function is given as $T(x, y, z) = x^2 + y^2 + z^2$, and let S be the unit sphere $x^2 + y^2 + z^2 = 1$ oriented with the outward normal (see Example 2). Find the heat flux across the surface S if $k = 1$.

SOLUTION   We have

$$\mathbf{F} = -\nabla T(x, y, z) = -2x\mathbf{i} - 2y\mathbf{j} - 2z\mathbf{k}.$$

On S, $\mathbf{n}(x, y, z) = x\mathbf{i} + y\mathbf{j} + z\mathbf{k}$ is the unit "outward" normal to S at (x, y, z), and $f(x, y, z) = \mathbf{F} \cdot \mathbf{n} = -2x^2 - 2y^2 - 2z^2 = -2$ is the normal component of $\mathbf{F}$. From Theorem 5 we can see that the surface integral of $\mathbf{F}$ is equal to the integral of its normal component $f = \mathbf{F} \cdot \mathbf{n}$ over S. Thus

$$\iint_S \mathbf{F} \cdot d\mathbf{S} = \iint_S f \, dS = -2 \iint_S dS = -2A(S) = -2(4\pi) = -8\pi.$$

The flux of heat is directed toward the center of the sphere (why toward?). Clearly, our observation that $\iint_S \mathbf{F} \cdot d\mathbf{S} = \iint_S f \, dS$ has saved us considerable computational time.

In this example, $\mathbf{F}(x, y, z) = -2x\mathbf{i} - 2y\mathbf{j} - 2z\mathbf{k}$ could also represent an electric field, in which case $\iint_S \mathbf{F} \cdot d\mathbf{S} = -8\pi$ would be the electric flux across S. ▲

EXAMPLE 5 There is an important physical law, due to the great mathematician and physicist K. F. Gauss, that relates the flux of an electric field $\mathbf{E}$ over a "closed" surface S (for example, a sphere or an ellipsoid) to the net charge Q enclosed by the surface, namely (in suitable units),

$$\iint_S \mathbf{E} \cdot d\mathbf{S} = Q \tag{1}$$

(see Figure 7.6.9). Gauss' law will be discussed in detail in Chapter 8. This law is analogous to Ampère's law (see Example 12, Section 7.2).

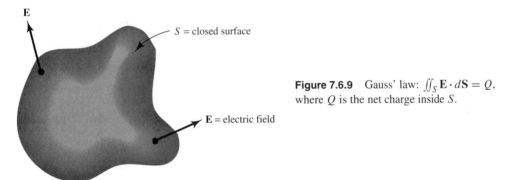

S = closed surface

$\mathbf{E}$ = electric field

Figure 7.6.9 Gauss' law: $\iint_S \mathbf{E} \cdot d\mathbf{S} = Q$, where Q is the net charge inside S.

Suppose that $\mathbf{E} = E\mathbf{n}$; that is, $\mathbf{E}$ is a constant scalar multiple of the unit normal to S. Then Gauss' law, equation (1) above, becomes

$$\iint_S \mathbf{E} \cdot d\mathbf{S} = \iint_S E \, dS = E \iint_S dS = Q$$

since $E = \mathbf{E} \cdot \mathbf{n}$. Thus,

$$E = \frac{Q}{A(S)}. \tag{2}$$

In the case where S is the sphere of radius R, equation (2) becomes

$$E = \frac{Q}{4\pi R^2} \qquad (3)$$

(see Figure 7.6.10).

Figure 7.6.10   The field $\mathbf{E}$ due to a point charge Q is $\mathbf{E} = Q\mathbf{n}/4\pi R^2$.

Now suppose that $\mathbf{E}$ arises from an isolated point charge Q. From symmetry it is reasonable that $\mathbf{E} = E\mathbf{n}$, where $\mathbf{n}$ is the unit normal to any sphere centered at Q. Hence equation (3) holds. Consider a second point charge Q_0 located at a distance R from Q. The force $\mathbf{F}$ that acts on this second charge Q_0 is given by

$$\mathbf{F} = \mathbf{E}Q_0 = EQ_0\mathbf{n} = \frac{QQ_0}{4\pi R^2}\mathbf{n}.$$

If F is the magnitude of $\mathbf{F}$, we have

$$F = \frac{QQ_0}{4\pi R^2},$$

which is the well-known ***Coulomb's law*** for the force between two point charges.*   ▲

Finally, let us derive the surface-integral formulas for vector fields $\mathbf{F}$ over surfaces S which are graphs of functions. Consider the surface S described by $z = f(x, y)$, where $(x, y) \in D$, where S is oriented so that

$$\mathbf{n} = \frac{-\dfrac{\partial f}{\partial x}\mathbf{i} - \dfrac{\partial f}{\partial y}\mathbf{j} + \mathbf{k}}{\sqrt{\left(\dfrac{\partial f}{\partial x}\right)^2 + \left(\dfrac{\partial f}{\partial y}\right)^2 + 1}}$$

---

*Sometimes one sees the formula $F = (1/4\pi\varepsilon_0)QQ_0/R^2$. The extra constant ε_0 appears when MKS units are used for measuring charge. We are using CGS, or Gaussian, units.

points outward. We have seen that we can parametrize S by $\mathbf{\Phi} \colon D \to \mathbb{R}^3$ given by $\mathbf{\Phi}(x, y) = (x, y, f(x, y))$. In this case, $\iint_S \mathbf{F} \cdot d\mathbf{S}$ can be written in a particularly simple form. We have

$$\mathbf{T}_x = \mathbf{i} + \frac{\partial f}{\partial x}\mathbf{k}, \quad \mathbf{T}_y = \mathbf{j} + \frac{\partial f}{\partial y}\mathbf{k}.$$

Thus $\mathbf{T}_x \times \mathbf{T}_y = -(\partial f/\partial x)\mathbf{i} - (\partial f/\partial y)\mathbf{j} + \mathbf{k}$. If $\mathbf{F} = F_1\mathbf{i} + F_2\mathbf{j} + F_3\mathbf{k}$ is a continuous vector field, then we get

---

THE SURFACE INTEGRAL OF A VECTOR FIELD OVER A GRAPH S

$$\iint_S \mathbf{F} \cdot d\mathbf{S} = \iint_D \mathbf{F} \cdot (\mathbf{T}_x \times \mathbf{T}_y)\, dx\, dy$$

$$= \iint_D \left[F_1\left(-\frac{\partial f}{\partial x}\right) + F_2\left(-\frac{\partial f}{\partial y}\right) + F_3 \right] dx\, dy. \tag{4}$$

EXAMPLE 6   The equations

$$z = 12, \quad x^2 + y^2 \le 25$$

describe a disk of radius 5 lying in the plane $z = 12$. Suppose $\mathbf{r}$ is the vector field

$$\mathbf{r}(x, y, z) = x\mathbf{i} + y\mathbf{j} + z\mathbf{k}.$$

Compute $\iint_S \mathbf{r} \cdot d\mathbf{S}$.

SOLUTION   We shall do this in three ways. First, we have $\partial z/\partial x = \partial z/\partial y = 0$, since $z = 12$ is constant on the disk and so

$$\mathbf{r}(x, y, z) \cdot (\mathbf{T}_x \times \mathbf{T}_y) = \mathbf{r}(x, y, z) \cdot (\mathbf{i} \times \mathbf{j}) = \mathbf{r}(x, y, z) \cdot \mathbf{k} = z$$

and so using the original definition at the beginning of this section, the integral becomes

$$\iint_S \mathbf{r} \cdot d\mathbf{S} = \iint_D z\, dx\, dy = \iint_D 12\, dx\, dy = 12(\text{ area of } D) = 300\pi.$$

A second solution: Since the disk is parallel to the xy plane, the outward unit normal is $\mathbf{k}$. Hence $\mathbf{n}(x, y, z) = \mathbf{k}$ and $\mathbf{r} \cdot \mathbf{n} = z$. However, $\|\mathbf{T}_x \times \mathbf{T}_y\| = \|\mathbf{k}\| = 1$, and so we know from the discussion preceding Theorem 5 that

$$\iint_S \mathbf{r} \cdot d\mathbf{S} = \iint_S \mathbf{r} \cdot \mathbf{n}\, dS = \iint_S z\, dS = \iint_D 12\, dx\, dy = 300\pi.$$

Third, we may solve this problem by using formula (4) directly, with $f(x, y) = 12$ and D the disk $x^2 + y^2 \leq 25$:

$$\iint_S \mathbf{r} \cdot d\mathbf{S} = \iint_D (x \cdot 0 + y \cdot 0 + 12) \, dx \, dy = 12(\text{area of } D) = 300\pi. \quad \blacktriangle$$

EXERCISES

1. Let the temperature of a point in $\mathbb{R}^3$ be given by $T(x, y, z) = 3x^2 + 3z^2$. Compute the heat flux across the surface $x^2 + z^2 = 2, 0 \leq y \leq 2$, if $k = 1$.

2. Compute the heat flux across the unit sphere S if $T(x, y, z) = x$ (see Example 4). Can you interpret your answer physically?

3. Let S be the closed surface that consists of the hemisphere $x^2 + y^2 + z^2 = 1, z \geq 0$, and its base $x^2 + y^2 \leq 1, z = 0$. Let $\mathbf{E}$ be the electric field defined by $\mathbf{E}(x, y, z) = 2x\mathbf{i} + 2y\mathbf{j} + 2z\mathbf{k}$. Find the electric flux across S. [HINT: Break S into two pieces S_1 and S_2 and evaluate $\iint_{S_1} \mathbf{E} \cdot d\mathbf{S}$ and $\iint_{S_2} \mathbf{E} \cdot d\mathbf{S}$ separately.]

4. Let the velocity field of a fluid be described by $\mathbf{F} = \sqrt{y}\mathbf{j}$ (measured in meters per second). Compute how many cubic meters of fluid per second are crossing the surface $x^2 + z^2 = 1, 0 \leq y \leq 1$, in the direction of increasing y.

5. Evaluate $\iint_S (\nabla \times \mathbf{F}) \cdot d\mathbf{S}$, where S is the surface $x^2 + y^2 + 3z^2 = 1, z \leq 0$, and $\mathbf{F} = y\mathbf{i} - x\mathbf{j} + zx^3y^2\mathbf{k}$. (Let $\mathbf{n}$, the unit normal, be upward pointing.)

6. Evaluate $\iint_S (\nabla \times \mathbf{F}) \cdot d\mathbf{S}$ where $\mathbf{F} = (x^2 + y - 4)\mathbf{i} + 3xy\mathbf{j} + (2xz + z^2)\mathbf{k}$ and S is the surface $x^2 + y^2 + z^2 = 16, z \geq 0$. (Let $\mathbf{n}$, the unit normal, be upward pointing.)

7. Calculate the integral $\iint_S \mathbf{F} \cdot d\mathbf{S}$, where S is the surface of the half ball $x^2 + y^2 + z^2 \leq 1, z \geq 0$, and $\mathbf{F} = (x + 3y^5)\mathbf{i} + (y + 10xz)\mathbf{j} + (z - xy)\mathbf{k}$. (Let $\mathbf{n}$, the unit normal, be upward pointing.)

8. A restaurant is being built on the side of a mountain. The architect's plans are shown in Figure 7.6.11.

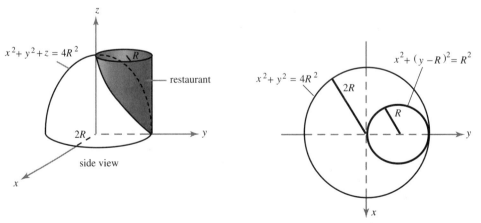

Figure 7.6.11   Restaurant plans.

(a)   The vertical curved wall of the restaurant is to be built of glass. What will be the surface area of this wall?

(b)   To be large enough to be profitable, the consulting engineer informs the developer that the volume of the interior must exceed $\pi R^4/2$.   For what R does the proposed structure satisfy this requirement?

(c)   During a typical summer day, the environs of the restaurant are subject to a temperature field given by

$$T(x, y, z) = 3x^2 + (y - R)^2 + 16z^2.$$

A heat flux density $\mathbf{V} = -k\,\nabla T$ (k is a constant depending on the grade of insulation to be used) through all sides of the restaurant (including the top and the contact with the hill) produces a heat flux. What is this total heat flux? (Your answer will depend on R and k.)

9.   Find the flux of $\mathbf{\Phi}(x, y, z) = 3xy^2\mathbf{i} + 3x^2y\mathbf{j} + z^3\mathbf{k}$ out of the unit sphere.

10.   Evaluate the surface integral $\iint_S \mathbf{F} \cdot \mathbf{n}\,dA$, where $\mathbf{F}(x, y, z) = \mathbf{i} + \mathbf{j} + z(x^2 + y^2)^2\mathbf{k}$ and S is the surface of the cylinder $x^2 + y^2 \leq 1, 0 \leq z \leq 1$.

11.   Let S be the surface of the unit sphere. Let $\mathbf{F}$ be a vector field and F_r its radial component. Prove that

$$\iint_S \mathbf{F} \cdot d\mathbf{S} = \int_{\theta=1}^{2\pi} \int_{\phi=0}^{\pi} F_r \sin \phi \, d\phi \, d\theta.$$

What is the corresponding formula for real-valued functions f?

12.   Prove the mean value theorem for surface integrals: If $\mathbf{F}$ is a continuous vector field, then

$$\iint_S \mathbf{F} \cdot \mathbf{n} \, dS = [\mathbf{F}(Q) \cdot \mathbf{n}(Q)]A(S)$$

for some point $Q \in S$, where $A(S)$ is the area of S. [HINT: Prove it for real functions first, by reducing the problem to one of a double integral: show that if $g \geq 0$, then

$$\iint_D fg \, dA = f(Q) \iint_D g \, dA$$

for some $Q \in D$ (do it by considering $(\iint_D fg \, dA)/(\iint_D g \, dA)$ and using the intermediate value theorem).]

13.   Work out a formula like that in Exercise 11 for integration over the surface of a cylinder.

14.   Let S be a surface in $\mathbb{R}^3$ that is actually a subset D of the xy plane. Show that the integral of a scalar function $f(x, y, z)$ over S reduces to the double integral of $f(x, y, z)$ over D. What does the surface integral of a vector field over S become? (Make sure your answer is compatible with Example 6.)

15.   Let the velocity field of a fluid be described by $\mathbf{F} = \mathbf{i} + x\mathbf{j} + z\mathbf{k}$ (measured in meters per second). Compute how many cubic meters of fluid per second are crossing the surface described by $x^2 + y^2 + z^2 = 1, z \geq 0$.

16.   (a)   A uniform fluid that flows vertically downward (heavy rain) is described by the vector field $\mathbf{F}(x, y, z) = (0, 0, -1)$. Find the total flux through the cone $z = (x^2 + y^2)^{1/2}, x^2 + y^2 \leq 1$.

(b)   The rain is driven sideways by a strong wind so that it falls at a $45°$ angle, and it is described by $\mathbf{F}(x, y, z) = -(\sqrt{2}/2, 0, \sqrt{2}/2)$. Now what is the flux through the cone?

17.   For $a > 0, b > 0, c > 0$, let S be the upper half ellipsoid

$$S = \left\{(x, y, z) \,\middle|\, \frac{x^2}{a^2} + \frac{y^2}{b^2} + \frac{y^2}{c^2} = 1, z \geq 0\right\}$$

with orientation determined by the upward normal. Compute $\iint_S \mathbf{F} \cdot d\mathbf{S}$ where $\mathbf{F}(x, y, z) = (x^3, 0, 0)$.

18.   If S is the upper hemisphere $\{(x, y, z) \mid x^2 + y^2 + z^2 = 1, z \geq 0\}$ oriented by the normal pointing out of the sphere, compute $\iint_S \mathbf{F} \cdot d\mathbf{S}$ for parts (a) and (b).

(a) $\mathbf{F}(x, y, z) = x\mathbf{i} + y\mathbf{j}$
(b) $\mathbf{F}(x, y, z) = y\mathbf{i} + x\mathbf{j}$
(c)   For each of the vector fields above, compute $\iint_S (\nabla \times \mathbf{F}) \cdot d\mathbf{S}$ and $\int_C \mathbf{F} \cdot d\mathbf{S}$ where C is the unit circle in the xy-plane traversed in the counterclockwise direction (as viewed from the positive z axis). (Notice that C is the boundary of S. The phenomenon illustrated here will be studied more thoroughly in the next chapter, using Stokes' theorem.)

REVIEW EXERCISES FOR CHAPTER 7

1.   Integrate $f(x, y, z) = xyz$ along the following paths:

(a) $\mathbf{c}(t) = (e^t \cos t, e^t \sin t, 3), 0 \leq t \leq 2\pi$
(b) $\mathbf{c}(t) = (\cos t, \sin t, t), 0 \leq t \leq 2\pi$
(c) $\mathbf{c}(t) = \frac{3}{2}t^2\mathbf{i} + 2t^2\mathbf{j} + t\mathbf{k}, 0 \leq t \leq 1$
(d) $\mathbf{c}(t) = t\mathbf{i} + (1/\sqrt{2})t^2\mathbf{j} + \frac{1}{3}t^3\mathbf{k}, 0 \leq t \leq 1$

2.   Compute the integral of f along the path $\mathbf{c}$ in each of the following cases:

(a) $f(x, y, z) = x + y + yz; \mathbf{c}(t) = (\sin t, \cos t, t), 0 \leq t \leq 2\pi$
(b) $f(x, y, z) = x + \cos^2 z; \mathbf{c}(t) = (\sin t, \cos t, t), 0 \leq t \leq 2\pi$
(c) $f(x, y, z) = x + y + z; \mathbf{c}(t) = (t, t^2, \frac{2}{3}t^3), 0 \leq t \leq 1$

3.   Compute each of the following line integrals

(a) $\int_C (\sin \pi x) \, dy - (\cos \pi y) \, dz$, where C is the triangle whose vertices are $(1, 0, 0), (0, 1, 0)$, and $(0, 0, 1)$ in that order
(b) $\int_C (\sin z) \, dx + (\cos z) \, dy - (xy)^{1/3} \, dz$, where C is the path $\mathbf{c}(\theta) = (\cos^3 \theta, \sin^3 \theta, \theta)$, $0 \leq \theta \leq 7\pi/2$.

4.   If $\mathbf{F}(\mathbf{x})$ is orthogonal to $\mathbf{c}'(t)$ at each point on the curve $\mathbf{x} = \mathbf{c}(t)$, what can you say about $\int_c \mathbf{F} \cdot d\mathbf{s}$?

5.   Find the work done by the force $\mathbf{F}(x, y) = (x^2 - y^2)\mathbf{i} + 2xy\mathbf{j}$ in moving a particle counterclockwise around the square with corners $(0, 0), (a, 0), (a, a), (0, a), a > 0$.

6.   A ring in the shape of the curve $x^2 + y^2 = a^2$ is formed of thin wire weighing $|x| + |y|$ grams per unit length at (x, y). Find the mass of the ring.

7. Find a parametrization for each of the following surfaces;

    (a) $x^2 + y^2 + z^2 - 4x - 6y = 12$

    (b) $2x^2 + y^2 + z^2 - 8x = 1$

    (c) $4x^2 + 9y^2 - 2z^2 = 8$

8. Find the area of the surface defined by Φ: $(u, v) \mapsto (x, y, z)$ where

$$x = h(u, v) = u + v, \qquad y = g(u, v) = u, \qquad z = f(u, v) = v;$$

$0 \le u \le 1, 0 \le v \le 1$. Sketch.

9. Write a formula for the surface area of Φ: $(r, \theta) \mapsto (x, y, z)$ where

$$x = r \cos \theta, \qquad y = 2r \sin \theta, \qquad z = r;$$

$0 \le r \le 1, 0 \le \theta \le 2\pi$. Describe the surface.

10. Suppose $z = f(x, y)$ and $(\partial f/\partial x)^2 + (\partial f/\partial y)^2 = c, c > 0$. Show that the area of the graph of f lying over a region D in the xy plane is $\sqrt{1 + c}$ times the area of D.

11. Compute the integral of $f(x, y, z) = x^2 + y^2 + z^2$ over the surface in Review Exercise 8.

12. Find $\iint_S f \, dS$ in each of the following cases:

    (a) $f(x, y, z) = x$; S is the part of the plane $x + y + z = 1$ in the positive octant defined by $x \ge 0, y \ge 0, z \ge 0$

    (b) $f(x, y, z) = x^2$; S is the part of the plane $x = z$ inside the cylinder $x^2 + y^2 = 1$

    (c) $f(x, y, z) = x$; S is the part of the cylinder $x^2 + y^2 = 2x$ with $0 \le z \le \sqrt{x^2 + y^2}$

13. Compute the integral of $f(x, y, z) = xyz$ over the rectangle with vertices $(1, 0, 1)$, $(2, 0, 0)$, $(1, 1, 1)$, and $(2, 1, 0)$.

14. Compute the integral of $x + y$ over the surface of the unit sphere.

15. Compute the surface integral of x over the triangle with vertices $(1, 1, 1)$, $(2, 1, 1)$, and $(2, 0, 3)$.

16. A paraboloid of revolution S is parametrized by $\Phi(u, v) = (u \cos v, u \sin v, u^2), 0 \le u \le 2, 0 \le v \le 2\pi$.

    (a) Find an equation in x, y, and z describing the surface.

    (b) What are the geometric meanings of the parameters u and v?

    (c) Find a unit vector orthogonal to the surface at $\Phi(u, v)$.

    (d) Find the equation for the tangent plane at $\Phi(u_0, v_0) = (1, 1, 2)$ and express your answer in the following two ways:

        (i) parametrized by u and v; and

        (ii) in terms of x, y, and z.

    (e) Find the area of S.

17. Let $f(x, y, z) = xe^y \cos \pi z$.

  (a) Compute $\mathbf{F} = \nabla f$.

  (b) Evaluate $\int_C \mathbf{F} \cdot d\mathbf{s}$ where $\mathbf{c}(t) = (3 \cos^4 t, 5 \sin^7 t, 0), 0 \le t \le \pi$.

18. Let $\mathbf{F}(x, y, z) = x\mathbf{i} + y\mathbf{j} + z\mathbf{k}$. Evaluate $\iint_S \mathbf{F} \cdot d\mathbf{S}$ where S is the upper hemisphere of the unit sphere $x^2 + y^2 + z^2 = 1$.

19. Let $\mathbf{F}(x, y, z) = x\mathbf{i} + y\mathbf{j} + z\mathbf{k}$. Evaluate $\int_{\mathbf{c}} \mathbf{F} \cdot d\mathbf{s}$ where $\mathbf{c}(t) = (e^t, t, t^2), 0 \le t \le 1$.

20. Let $\mathbf{F} = \nabla f$ for a given scalar function. Let $\mathbf{c}(t)$ be a closed curve, that is, $\mathbf{c}(b) = \mathbf{c}(a)$. Show that $\int_{\mathbf{c}} \mathbf{F} \cdot d\mathbf{s} = 0$.

21. Consider the surface $\mathbf{\Phi}(u, v) = (u^2 \cos v, u^2 \sin v, u)$. Compute the unit normal at $u = 1, v = 0$. Compute the equation of the tangent plane at this point.

22. Let S be the part of the cone $z^2 = x^2 + y^2$ with z between 1 and 2 oriented by the normal pointing out of the cone. Compute $\iint_S \mathbf{F} \cdot d\mathbf{S}$ where $\mathbf{F}(x, y, z) = (x^2, y^2, z^2)$.

23. Let $\mathbf{F} = x\mathbf{i} + x^2\mathbf{j} + yz\mathbf{k}$ represent the velocity field of a fluid (velocity measured in meters per second). Compute how many cubic meters of fluid per second are crossing the xy plane through the square $0 \le x \le 1, 0 \le y \le 1$.

24. Show that the surface area of the part of the sphere $x^2 + y^2 + z^2 = 1$ lying above the rectangle $[-a, a] \times [-a, a]$, where $2a^2 < 1$, in the xy plane is

$$A = 2 \int_{-a}^{a} \sin^{-1}\left(\frac{a}{\sqrt{1 - x^2}}\right) dx.$$

25. Let S be a surface and C a closed curve bounding S. Verify the equality

$$\iint_S (\nabla \times \mathbf{F}) \cdot d\mathbf{S} = \int_C \mathbf{F} \cdot d\mathbf{S}$$

if $\mathbf{F}$ is a gradient field (use Review Exercise 20).

26. Calculate $\iint_S \mathbf{F} \cdot d\mathbf{S}$ where $\mathbf{F}(x, y, z) = (x, y, -y)$ and S is the cylindrical surface defined by $x^2 + y^2 = 1, 0 \le z \le 1$, with normal pointing out of the cylinder.

27. Let S be the portion of the cylinder $x^2 + y^2 = 4$ between the planes $z = 0$ and $z = x + 3$. Compute the following:

  (a) $\iint_S x^2 \, dS$

  (b) $\iint_S y^2 \, dS$

  (c) $\iint_S z^2 \, dS$

28. Let Γ be the curve of intersection of the plane $z = ax + by$ with the cylinder $x^2 + y^2 = 1$. Find all values of the real numbers a and b such that $a^2 + b^2 = 1$ and

$$\int_{\Gamma} y \, dx + (z - x) \, dy - y \, dz = 0.$$

29. A circular helix that lies on the cylinder $x^2 + y^2 = R^2$ with pitch p may be described parametrically by

$$x = R \cos \theta, \qquad y = R \sin \theta, \qquad z = p\theta, \qquad \theta \geq 0.$$

A particle slides under the action of gravity (which acts parallel to the z axis) without friction along the helix. If the particle starts out at the height $z_0 > 0$, then when it reaches the height $z, 0 \leq z < z_0$, along the helix, its speed is given by

$$\frac{ds}{dt} = \sqrt{(z_0 - z)2g},$$

where s is arc length along the helix, g is the constant of gravity, and t is time.

(a) Find the length of the part of the helix between the planes $z = z_0$ and $z = z_1, 0 \leq z_1 < z_0$.
(b) Compute the time T_0 it takes the particle to reach the plane $z = 0$.

8 THE INTEGRAL THEOREMS OF VECTOR ANALYSIS

All the theory of the motion of fluids has just been reduced to the solution of analytic formulas.

L. EULER

We are now prepared to tie together vector differential calculus and vector integral calculus. This will be done by means of the important theorems of Green, Gauss, and Stokes. We shall also point out some of the physical applications of these theorems to the study of electricity and magnetism, hydrodynamics, heat conduction, and differential equations.

The basic integral theorems in vector analysis had their origins in physics. For example, Green's theorem, discovered about 1828, arose in connection with potential theory (this includes gravitational and electrical potentials). Gauss' theorem—the divergence theorem—arose in connection with electrostatics (this theorem should actually be jointly credited to the Russian mathematician Ostrogradsky). Stokes' theorem was first suggested in a letter to Stokes from the physicist Lord Kelvin in 1850 and was used by Stokes on the examination for the Smith Prize in 1854.

8.1 GREEN'S THEOREM

Green's theorem relates a line integral along a closed curve C in the plane $\mathbb{R}^2$ to a double integral over the region enclosed by C. This important result will be generalized in the following sections to curves and surfaces in $\mathbb{R}^3$. We shall be referring to line integrals around curves that are the boundaries of elementary regions of type 1, 2, or 3 (see Section 5.3). To understand the ideas in this section you may also need to refer to Section 7.2.

A simple closed curve C that is the boundary of a region of type 1, 2, or 3 has two orientations—counterclockwise (positive) and clockwise (negative). We denote C with the counterclockwise orientation as C^+, and with the clockwise orientation as C^- (Figure 8.1.1).

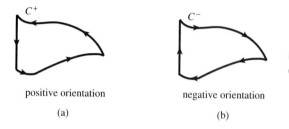

positive orientation

(a)

negative orientation

(b)

Figure 8.1.1 (a) Positive orientation of C and (b) negative orientation of C.

The boundary C of a region of type 1 can be decomposed into bottom and top portions, C_1 and C_2, and (if applicable) left and right vertical portions, B_1 and B_2. Following Figure 8.1.2, we write,

$$C^+ = C_1^+ + B_2^+ + C_2^- + B_1^-,$$

where the pluses denote the curves oriented in the direction of left to right or bottom to top, and the minuses denote the curves oriented from right to left or from top to bottom.

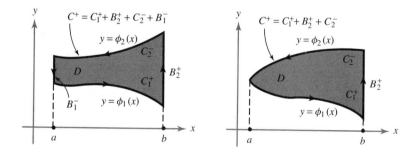

Figure 8.1.2 Two examples showing how to break the positively oriented boundary of a region D of type 1 into oriented components.

We may make a similar decomposition of the boundary of a region of type 2 into left and right portions, and upper and lower horizontal portions (if applicable) (Figure 8.1.3).

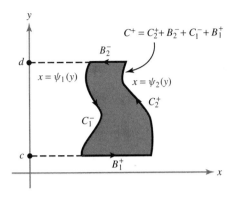

Figure 8.1.3 An example showing how to break the positively oriented boundary of a region D of type 2 into oriented components.

Similarly, the boundary of a region of type 3 has *two* decompositions—one into upper and lower halves, the other into left and right halves.

We shall now prove two lemmas in preparation for Green's theorem.

LEMMA 1 Let D be a region of type 1 and let C be its boundary. Suppose $P: D \to \mathbb{R}$ is of class C^1. Then

$$\int_{C^+} P\, dx = -\iint_D \frac{\partial P}{\partial y}\, dx\, dy.$$

(The left-hand side denotes the line integral $\int_{C^+} P\, dx + Q\, dy + R\, dz$ where $Q = 0$ and $R = 0$.)

PROOF Suppose D is described by

$$a \le x \le b \qquad \phi_1(x) \le y \le \phi_2(x).$$

We decompose C^+ by writing $C^+ = C_1^+ + B_2^+ + C_2^- + B_1^-$ (see Figure 8.1.2). By Fubini's theorem, we may evaluate the double integral as an iterated integral and then use the fundamental theorem of calculus:

$$\iint_D \frac{\partial P}{\partial y}(x, y)\, dx\, dy = \int_a^b \int_{\phi_1(x)}^{\phi_2(x)} \frac{\partial P}{\partial y}(x, y)\, dy\, dx = \int_a^b [P(x, \phi_2(x)) - P(x, \phi_1(x))]\, dx.$$

However, since C_1^+ can be parametrized by $x \mapsto (x, \phi_1(x))$, $a \le x \le b$, and C_2^+ can be parametrized by $x \mapsto (x, \phi_2(x))$, $a \le x \le b$, we have

$$\int_a^b P(x, \phi_1(x))\, dx = \int_{C_1^+} P(x, y)\, dx$$

and

$$\int_a^b P(x, \phi_2(x))\, dx = \int_{C_2^+} P(x, y)\, dx.$$

Thus, by reversing orientations,

$$-\int_a^b P(x, \phi_2(x))\, dx = \int_{C_2^-} P(x, y)\, dx.$$

Hence,

$$\iint_D \frac{\partial P}{\partial y}\, dx\, dy = -\int_{C_1^+} P\, dx - \int_{C_2^-} P\, dx.$$

Since x is constant on B_2^+ and B_1^-, we have

$$\int_{B_2^+} P\, dx = 0 = \int_{B_1^-} P\, dx,$$

and so

$$\int_{C^+} P\, dx = \int_{C_1^+} P\, dx + \int_{B_2^+} P\, dx + \int_{C_2^-} P\, dx + \int_{B_1^-} P\, dx = \int_{C_1^+} P\, dx + \int_{C_2^-} P\, dx.$$

Thus,

$$\iint_D \frac{\partial P}{\partial y}\, dx\, dy = -\int_{C_1^+} P\, dx - \int_{C_2^-} P\, dx = -\int_{C^+} P\, dx. \quad \blacksquare$$

We now prove the analogous lemma with the roles of x and y interchanged.

LEMMA 2 Let D be a region of type 2 with boundary C. Then if $Q: D \to \mathbb{R}$ is C^1,

$$\int_{C^+} Q\, dy = \iint_D \frac{\partial Q}{\partial x}\, dx\, dy.$$

The negative sign does not occur here, because reversing the role of x and y corresponds to a change of orientation for the plane.

PROOF Suppose D is given by

$$\psi_1(y) \le x \le \psi_2(y), \qquad c \le y \le d.$$

Using the notation of Figure 8.1.3, and noting that y is constant on B_1^+ and B_2^-, we have

$$\int_{C^+} Q\, dy = \int_{C_1^- + B_1^+ + C_2^+ + B_2^-} Q\, dy = \int_{C_2^+} Q\, dy + \int_{C_1^-} Q\, dy,$$

where C_2^+ is the curve parametrized by $y \mapsto (\psi_2(y), y), c \le y \le d$, and C_1^+ is the curve $y \mapsto (\psi_1(y), y), c \le y \le d$. Applying Fubini's theorem and the fundamental theorem of calculus, we obtain

$$\iint_D \frac{\partial Q}{\partial x}\, dx\, dy = \int_c^d \int_{\psi_1(y)}^{\psi_2(y)} \frac{\partial Q}{\partial x}\, dx\, dy = \int_c^d [Q(\psi_2(y), y) - Q(\psi_1(y), y)]\, dy$$

$$= \int_{C_2^+} Q\, dy - \int_{C_1^+} Q\, dy = \int_{C_2^+} Q\, dy + \int_{C_1^-} Q\, dy = \int_{C^+} Q\, dy. \quad \blacksquare$$

Adding the results of Lemmas 1 and 2 proves the following important theorem.

THEOREM 1: GREEN'S THEOREM *Let D be a region of type 3 and let C be its boundary. Suppose $P\colon D \to \mathbb{R}$ and $Q\colon D \to \mathbb{R}$ are of class C^1. Then*

$$\int_{C^+} P\, dx + Q\, dy = \iint_D \left(\frac{\partial Q}{\partial x} - \frac{\partial P}{\partial y} \right) dx\, dy.$$

The correct (positive) orientation for the boundary curves of region D can be remembered by the following device: *If you walk along the curve C with the correct orientation, the region D will be on your left* (see Figure 8.1.4).

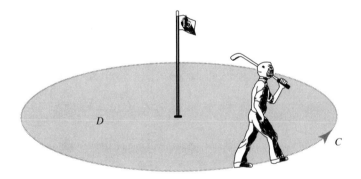

Figure 8.1.4 The correct orientation for the boundary of a region D.

Green's theorem actually applies to any "decent" region in $\mathbb{R}^2$. For instance, Green's theorem applies to regions that are not of type 3, but that can be broken up into pieces, each of which is of type 3. An example is shown in Figure 8.1.5. The region D is an annulus; its boundary consists of two curves $C = C_1 + C_2$ with the indicated orientations. (Note that for the inner region the correct orientation to ensure the validity of Green's theorem is *clockwise*; the device in Figure 8.1.4 still works for remembering the orientation!) If Theorem 1 is applied to each of the regions D_1, D_2, D_3, and D_4 and the results are summed, the equality of Green's theorem will be obtained for D and its boundary curve C.

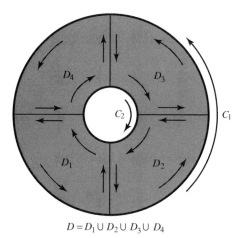

$D = D_1 \cup D_2 \cup D_3 \cup D_4$

Figure 8.1.5 Green's theorem applies to $D = D_1 \cup D_2 \cup D_3 \cup D_4$.

Let us use the notation ∂D for the oriented curve C^+, that is, the boundary curve of D oriented in the sense as described by the device in Figure 8.1.4. Then we may write Green's theorem as

$$\int_{\partial D} P\, dx + Q\, dy = \iint_D \left(\frac{\partial Q}{\partial x} - \frac{\partial P}{\partial y} \right) dx\, dy.$$

Green's theorem is very useful because it relates a line integral around the boundary of a region to an area integral over the interior of the region, and in many cases it is easier to evaluate the line integral than the area integral or vice versa. For example, if we know that P vanishes on the boundary, we can immediately conclude that $\iint_D (\partial P/\partial y)\, dx\, dy = 0$ even though $\partial P/\partial y$ need not vanish on the interior. (Can you construct such a P on the unit square?)

EXAMPLE 1   Verify Green's theorem for $P(x, y) = x$ and $Q(x, y) = xy$ where D is the unit disk $x^2 + y^2 \leq 1$.

SOLUTION   We do this by evaluating both sides in Green's theorem directly. The boundary of D is the unit circle parametrized by $x = \cos t, y = \sin t, 0 \leq t \leq 2\pi$, and so

$$\int_{\partial D} P\, dx + Q\, dy = \int_0^{2\pi} [(\cos t)(-\sin t) + \cos t \sin t \cos t]\, dt$$

$$= \left[\frac{\cos^2 t}{2} \right]_0^{2\pi} + \left[-\frac{\cos^3 t}{3} \right]_0^{2\pi} = 0.$$

On the other hand,

$$\iint_D \left(\frac{\partial Q}{\partial x} - \frac{\partial P}{\partial y} \right) dx\, dy = \iint_D y\, dx\, dy,$$

which is also zero by symmetry. Thus Green's theorem is verified in this case.   ▲

We can use Green's theorem to obtain a formula for the area of a region bounded by a simple closed curve.

THEOREM 2: AREA OF A REGION   *If C is a simple closed curve that bounds a region to which Green's theorem applies, then the area of the region D bounded by C = ∂D is*

$$A = \frac{1}{2} \int_{\partial D} x\, dy - y\, dx.$$

PROOF   Let $P(x, y) = -y$, $Q(x, y) = x$; then by Green's theorem we have

$$\frac{1}{2} \int_{\partial D} x\, dy - y\, dx = \frac{1}{2} \iint_D \left[\frac{\partial x}{\partial x} - \frac{\partial (-y)}{\partial y} \right] dx\, dy$$

$$= \frac{1}{2} \iint_D [1 + 1]\, dx\, dy = \iint_D dx\, dy = A. \quad \blacksquare$$

EXAMPLE 2   Compute the area (see Figure 8.1.6) of the region enclosed by the hypocycloid defined by $x^{2/3} + y^{2/3} = a^{2/3}$ using the parametrization

$$x = a \cos^3 \theta, \qquad y = a \sin^3 \theta, \qquad 0 \le \theta \le 2\pi.$$

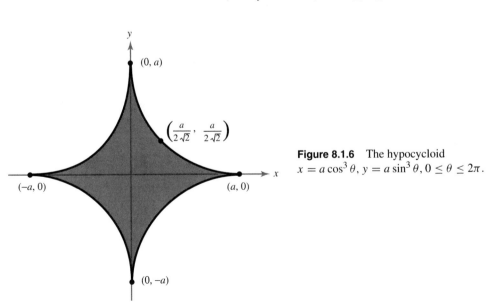

Figure 8.1.6   The hypocycloid $x = a \cos^3 \theta,\ y = a \sin^3 \theta,\ 0 \le \theta \le 2\pi$.

SOLUTION From the preceding box,

$$A = \frac{1}{2} \int_{\partial D} x \, dy - y \, dx$$

$$= \frac{1}{2} \int_0^{2\pi} [(a \cos^3 \theta)(3a \sin^2 \theta \cos \theta) - (a \sin^3 \theta)(-3a \cos^2 \theta \sin \theta)] \, d\theta$$

$$= \frac{3}{2} a^2 \int_0^{2\pi} (\sin^2 \theta \cos^4 \theta + \cos^2 \theta \sin^4 \theta) \, d\theta = \frac{3}{2} a^2 \int_0^{2\pi} \sin^2 \theta \cos^2 \theta \, d\theta$$

$$= \frac{3}{8} a^2 \int_0^{2\pi} \sin^2 2\theta \, d\theta = \frac{3}{8} a^2 \int_0^{2\pi} \left(\frac{1 - \cos 4\theta}{2} \right) d\theta$$

$$= \frac{3}{16} a^2 \int_0^{2\pi} d\theta - \frac{3}{16} a^2 \int_0^{2\pi} \cos 4\theta \, d\theta = \frac{3}{8} \pi a^2. \quad \blacktriangle$$

The statement of Green's theorem can be neatly rewritten in the language of vector fields. As we will see, this points the way to one possible generalization to $\mathbb{R}^3$.

THEOREM 3: VECTOR FORM OF GREEN'S THEOREM *Let $D \subset \mathbb{R}^2$ be a region to which Green's theorem applies, let ∂D be its boundary (oriented counterclockwise), and let $\mathbf{F} = P\mathbf{i} + Q\mathbf{j}$ be a C^1 vector field on D. Then*

$$\int_{\partial D} \mathbf{F} \cdot d\mathbf{s} = \iint_D (\text{curl } \mathbf{F}) \cdot \mathbf{k} \, dA = \iint_D (\nabla \times \mathbf{F}) \cdot \mathbf{k} \, dA$$

(see Figure 8.1.7).

Figure 8.1.7 The vector form of Green's theorem.

This result follows easily from Theorem 1 after we interpret the various symbols. We ask the reader to supply the details in Exercise 14.

EXAMPLE 3   Let $\mathbf{F} = (xy^2, y + x)$. Integrate $(\nabla \times \mathbf{F}) \cdot \mathbf{k}$ over the region in the first quadrant bounded by the curves $y = x^2$ and $y = x$.

SOLUTION   *Method 1.*   We first compute the curl

$$\nabla \times \mathbf{F} = \left(0, 0, \frac{\partial F_2}{\partial x} - \frac{\partial F_1}{\partial y}\right) = (1 - 2xy)\mathbf{k}.$$

Thus $(\nabla \times \mathbf{F}) \cdot \mathbf{k} = 1 - 2xy$. This can be integrated over the given region D (see Figure 8.1.8) using an iterated integral as follows:

$$\iint_D (\nabla \times \mathbf{F}) \cdot \mathbf{k}\, dx\, dy = \int_0^1 \int_{x^2}^x (1 - 2xy)\, dy\, dx = \int_0^1 [y - xy^2]\big|_{x^2}^x\, dx$$

$$= \int_0^1 [x - x^3 - x^2 + x^5]\, dx = \frac{1}{2} - \frac{1}{4} - \frac{1}{3} + \frac{1}{6} = \frac{1}{12}.$$

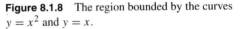

Figure 8.1.8   The region bounded by the curves $y = x^2$ and $y = x$.

Method 2.   Here we use Theorem 3 to obtain

$$\iint_D (\nabla \times \mathbf{F}) \cdot \mathbf{k}\, dx\, dy = \int_{\partial D} \mathbf{F} \cdot d\mathbf{s}.$$

The line integral of $\mathbf{F}$ along the curve $y = x$ from left to right is

$$\int_0^1 F_1\, dx + F_2\, dy = \int_0^1 (x^3 + 2x)\, dx = \frac{1}{4} + 1 = \frac{5}{4}.$$

Along the curve $y = x^2$ we get

$$\int_0^1 F_1\, dx + F_2\, dy = \int_0^1 x^5\, dx + (x + x^2)(2x\, dx) = \frac{1}{6} + \frac{2}{3} + \frac{1}{2} = \frac{4}{3}.$$

Thus, remembering that the integral along $y = x$ is to be taken from right to left, as in Figure 8.1.8,

$$\int_{\partial D} \mathbf{F} \cdot d\mathbf{s} = \frac{4}{3} - \frac{5}{4} = \frac{1}{12}. \quad \blacktriangle$$

There is another form of Green's theorem that is capable of being generalized to $\mathbb{R}^3$.

THEOREM 4: DIVERGENCE THEOREM IN THE PLANE *Let $D \subset \mathbb{R}^2$ be a region to which Green's theorem applies and let ∂D be its boundary. Let $\mathbf{n}$ denote the outward unit normal to ∂D. If $\mathbf{c}$: $[a, b] \to \mathbb{R}^2, t \mapsto \mathbf{c}(t) = (x(t), y(t))$ is a positively oriented parametrization of ∂D, $\mathbf{n}$ is given by*

$$\mathbf{n} = \frac{(y'(t) - x'(t))}{\sqrt{[x'(t)]^2 + [y'(t)]^2}},$$

(see Figure 8.1.9). Let $\mathbf{F} = P\mathbf{i} + Q\mathbf{j}$ be a C^1 vector field on D. Then

$$\int_{\partial D} \mathbf{F} \cdot \mathbf{n} \, ds = \iint_D \operatorname{div} \mathbf{F} \, dA.$$

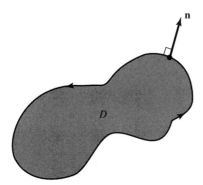

Figure 8.1.9 $\mathbf{n}$ is the outward unit normal to ∂D.

PROOF   Since $\mathbf{c}'(t) = (x'(t), y'(t))$ is tangent to ∂D, it is clear that $\mathbf{n} \cdot \mathbf{c}' = 0$, and so $\mathbf{n}$ is normal to the boundary. The sign of $\mathbf{n}$ is chosen to make it correspond to the *outward* (rather than the inward) direction. By the definition of the line integral (see Section 7.2),

$$\int_{\partial D} \mathbf{F} \cdot \mathbf{n} \, ds = \int_a^b \frac{P(x(t), y(t))y'(t) - Q(x(t), y(t))x'(t)}{\sqrt{[x'(t)]^2 + [y'(t)]^2}} \sqrt{[x'(t)]^2 + [y'(t)]^2} \, dt$$

$$= \int_a^b [P(x(t), y(t))y'(t) - Q(x(t), y(t))x'(t)] \, dt$$

$$= \int_{\partial D} P \, dy - Q \, dx.$$

By Green's theorem, this equals

$$\iint_D \left(\frac{\partial P}{\partial x} + \frac{\partial Q}{\partial y} \right) dx \, dy = \iint_D \operatorname{div} \mathbf{F} \, dA. \quad \blacksquare$$

EXAMPLE 4   Let $\mathbf{F} = y^3\mathbf{i} + x^5\mathbf{j}$. Compute the integral of the normal component of $\mathbf{F}$ around the unit square.

SOLUTION   This can be done using the divergence theorem. Indeed,

$$\int_{\partial D} \mathbf{F} \cdot \mathbf{n} \, ds = \iint_D \operatorname{div} \mathbf{F} \, dA.$$

But div $\mathbf{F} = 0$, and so the integral is zero.   ▲

EXERCISES

1.   Evaluate $\int_C y \, dx - x \, dy$ where C is the boundary of the square $[-1, 1] \times [-1, 1]$ oriented in the counterclockwise direction (use Green's theorem).

 2.   Find the area of the disk D of radius R using Green's theorem.

3.   Verify Green's theorem for the disk D with center $(0, 0)$ and radius R and the functions:

    (a) $P(x, y) = xy^2, Q(x, y) = -yx^2$
    (b) $P(x, y) = x + y, Q(x, y) = y$
    (c) $P(x, y) = xy = Q(x, y)$
    (d) $P(x, y) = 2y, Q(x, y) = x$

4.   Using the divergence theorem, show that $\int_{\partial D} \mathbf{F} \cdot \mathbf{n} \, ds = 0$ where $\mathbf{F}(x, y) = y\mathbf{i} - x\mathbf{j}$ and D is the unit disk. Verify this directly.

5.   Find the area bounded by one arc of the cycloid $x = a(\theta - \sin \theta)$, $y = a(1 - \cos \theta)$, where $a > 0$, and $0 \le \theta \le 2\pi$, and the x axis (use Green's theorem).

6.   Under the conditions of Green's theorem, prove that

    (a) $\displaystyle \int_{\partial D} PQ \, dx + PQ \, dy = \iint_D \left[Q \left(\frac{\partial P}{\partial x} - \frac{\partial P}{\partial y} \right) + P \left(\frac{\partial Q}{\partial x} - \frac{\partial Q}{\partial y} \right) \right] dx \, dy$

    (b) $\displaystyle \int_{\partial D} \left(Q \frac{\partial P}{\partial x} - P \frac{\partial Q}{\partial x} \right) dx + \left(P \frac{\partial Q}{\partial y} - Q \frac{\partial P}{\partial y} \right) dy = 2 \iint_D \left(P \frac{\partial^2 Q}{\partial x \, \partial y} - Q \frac{\partial^2 P}{\partial x \, \partial y} \right) dx \, dy$

7.   Evaluate $\int_C (2x^3 - y^3) \, dx + (x^3 + y^3) \, dy$, where C is the unit circle, and verify Green's theorem for this case.

8.   Prove the following generalization of Green's theorem: Let D be a region in the xy plane with boundary a finite number of oriented simple closed curves. Suppose that by means of a finite number of line segments parallel to the coordinate axes, D can be decomposed into a finite number of regions D_i of type 3 with the boundary of each D_i oriented counterclockwise (see Figure 8.1.5). Then if P and Q are of class C^1 on D,

$$\iint_D \left(\frac{\partial Q}{\partial x} - \frac{\partial P}{\partial y} \right) dx \, dy = \int_{\partial D} P \, dx + Q \, dy,$$

where ∂D is the oriented boundary of D. (HINT: Apply Green's theorem to each D_i.)

9.   Verify Green's theorem for the integrand of Exercise 7 (with $P = 2x^3 - y^3$, $Q = x^3 + y^3$) and the annular region D described by $a \le x^2 + y^2 \le b$, with boundaries oriented as in Figure 8.1.5.

10. Let D be a region for which Green's theorem holds. Suppose f is harmonic; that is,

$$\frac{\partial^2 f}{\partial x^2} + \frac{\partial^2 f}{\partial y^2} = 0$$

on D. Prove that

$$\int_{\partial D} \frac{\partial f}{\partial y}\, dx - \frac{\partial f}{\partial x}\, dy = 0.$$

11. (a) Verify the divergence theorem for $\mathbf{F} = x\mathbf{i} + y\mathbf{j}$ and D the unit disk $x^2 + y^2 \le 1$.
 (b) Evaluate the integral of the normal component of $2xy\mathbf{i} - y^2\mathbf{j}$ around the ellipse defined by $x^2/a^2 + y^2/b^2 = 1$.

12. Let $P(x, y) = -y(x^2 + y^2)$ and $Q(x, y) = x/(x^2 + y^2)$. Assuming D is the unit disk, investigate why Green's theorem fails for this P and Q.

13. Use Green's theorem to evaluate $\int_{C^+} (y^2 + x^3)\, dx + x^4\, dy$, where C^+ is the perimeter of the square $[0, 1] \times [0, 1]$ in the counterclockwise direction.

14. Verify Theorem 3.

15. Use Theorem 2 to compute the area inside the ellipse $x^2/a^2 + y^2/b^2 = 1$.

16. Use Theorem 2 to recover the formula $A = \frac{1}{2} \int_a^b r^2\, d\theta$ for a region in polar coordinates.

17. Sketch the proof of Green's theorem for the region shown in Figure 8.1.10.

Figure 8.1.10 Prove Green's theorem for this region.

18. Prove the identity

$$\int_{\partial D} \phi \nabla \phi \cdot \mathbf{n}\, ds = \iint_D (\phi \nabla^2 \phi + \nabla \phi \cdot \nabla \phi)\, dA.$$

19. Use Green's theorem to find the area of one loop of the four-leafed rose $r = 3 \sin 2\theta$. (HINT: $x\, dy - y\, dx = r^2 d\theta$).

20. Show that if C is a simple closed curve that bounds a region to which Green's theorem applies, then the area of the region D bounded by C is

$$A = \int_{\partial D} x\, dy = -\int_{\partial D} y\, dx.$$

Show how this implies Theorem 2.

Exercises 21 to 29 illustrate the application of Green's theorem to partial differential equations. They are particularly concerned with solutions to Laplace's equation, that is, with harmonic functions. (See Section 8.5 for additional results.) For these exercises, let D be an open region in $\mathbb{R}^2$ with boundary ∂D. Let $u: D \cup \partial D \to \mathbb{R}$ be a continuous function that is of class C^2 on D. Suppose $\mathbf{p} \in D$ and the closed disks $B_\rho = B_\rho(\mathbf{p})$ of radius ρ centered at $\mathbf{p}$ are contained in D for $0 < \rho \leq R$. Define $I(\rho)$ by

$$I(\rho) = \frac{1}{\rho} \int_{\partial B_\rho} u \, ds.$$

21. Show that $\text{limit}_{\rho \to 0} \, I(\rho) = 2\pi u(\mathbf{p})$.

22. Let $\mathbf{n}$ denote the outward unit normal to ∂B_ρ and $\partial u / \partial n = \nabla u \cdot \mathbf{n}$. Show that

$$\int_{\partial B_\rho} \frac{\partial u}{\partial n} \, ds = \iint_{B_\rho} \nabla^2 u \, dA.$$

23. Using Exercise 22, show that $I'(\rho) = \frac{1}{\rho} \iint_{B_\rho} \nabla^2 u \, dA$.

24. Suppose u satisfies Laplace's equation: $\nabla^2 u = 0$ on D. Use the preceding exercises to show that

$$u(\mathbf{p}) = \frac{1}{2\pi R} \int_{\partial B_R} u \, ds.$$

(This expresses the fact that the value of a harmonic function at a point is the average of its values on the circumference of any disk centered about it.)

25. Use Exercise 24 to show that if u is harmonic (i.e., if $\nabla^2 u = 0$), then $u(\mathbf{p})$ can be expressed as an area integral

$$u(\mathbf{p}) = \frac{1}{\pi R^2} \iint_{B_R} u \, dA.$$

26. Suppose u is a harmonic function defined on D (i.e., $\nabla^2 u = 0$ on D) and that u has a local maximum (or minimum) at a point $\mathbf{p}$ in D.
    (a)   Show that u must be constant on some disk centered at $\mathbf{p}$. (HINT: Use the results of Exercise 25.)
    (b)   Suppose that D is path-connected [i.e., for any points $\mathbf{p}$ and $\mathbf{q}$ in D there is a continuous path $\mathbf{c}: [0, 1] \to D$ such that $\mathbf{c}(0) = \mathbf{p}$ and $\mathbf{c}(1) = \mathbf{q}$] and that for some $\mathbf{p}$ the maximum or minimum at $\mathbf{p}$ is absolute; thus $u(\mathbf{q}) \leq u(\mathbf{p})$ or $u(\mathbf{q}) \geq u(\mathbf{p})$ for every $\mathbf{q}$ in D. Show that u must be constant on D.

(The result in this Exercise is called a ***strong maximum*** or ***minimum principle*** for harmonic functions. Compare this with Exercises 36 to 40 in Section 3.3.)

27. A function is said to be ***subharmonic*** on D if $\nabla^2 u \geq 0$ everywhere in D. It is said to be ***superharmonic*** if $\nabla^2 u \leq 0$.
    (a)   Derive a strong maximum principal for subharmonic functions.
    (b)   Derive a strong minimum principal for superharmonic functions.

28. Suppose D is the disk $\{(x, y) \mid x^2 + y^2 < 1\}$ and C is the circle $\{(x, y) \mid x^2 + y^2 = 1\}$. In Section 8.5 we shall show that if f is a continuous real-valued function on C, then there is a continuous function u on $D \cup C$ that agrees with f on C and is harmonic on D. That is, f has a harmonic extension to the disk. Assuming this, show the following:

(a)   If q is a nonconstant continuous function on $D \cup C$ that is subharmonic (but not harmonic) on D, then there is a continuous function u on $D \cup C$ that is harmonic on D such that u agrees with q on C and $q < u$ everywhere on D.

(b)   The same assertion holds if "subharmonic" is replaced by "superharmonic" and "$q < u$" by "$q > u$."

29.   Let D be as in Exercise 28. Let $f : D \to \mathbb{R}$ be continuous. Show that a solution to the equation $\nabla^2 u = 0$ satisfying $u(\mathbf{x}) = f(\mathbf{x})$ for all $\mathbf{x} \in \partial D$ is unique.

30.   Use Green's theorem to prove the change of variables formula in the following special case:

$$\iint_D dx\, dy = \iint_{D^*} \left| \frac{\partial(x, y)}{\partial(u, v)} \right| du\, dv$$

for a transformation $(u, v) \mapsto (x(u, v), y(u, v))$.

8.2   STOKES' THEOREM

Stokes' theorem relates the line integral of a vector field around a simple closed curve C in $\mathbb{R}^3$ to an integral over a surface S for which C is the boundary. In this regard it is very much like Green's theorem.

Let us begin by recalling a few facts from Chapter 7. Consider a surface S that is the graph of a function $f(x, y)$, so that S is parametrized by

$$\begin{cases} x = u \\ y = v \\ z = f(u, v) = f(x, y) \end{cases}$$

for (u, v) in some domain D. The integral of a vector function $\mathbf{F}$ over S was developed in Section 7.6 as

$$\iint_S \mathbf{F} \cdot d\mathbf{S} = \iint_D \left[F_1 \left(-\frac{\partial z}{\partial x} \right) + F_2 \left(-\frac{\partial z}{\partial y} \right) + F_3 \right] dx\, dy, \tag{1}$$

where $\mathbf{F} = F_1 \mathbf{i} + F_2 \mathbf{j} + F_3 \mathbf{k}$.

In Section 8.1 we first assumed that the regions D under consideration were of type 3; while this was used in our proof of Green's theorem, we noted there that the theorem is valid for a wider class of regions. In this section we assume that D is a region whose boundary is a simple closed curve and to which Green's theorem applies. Green's theorem involves choosing an orientation on the boundary of D, as was explained in Section 8.1. The choice of orientation that validates Green's theorem will be called **positive**. Recall that if D is of type 3, then the positive orientation is the counterclockwise one.

Suppose that $\mathbf{c} \colon [a, b] \to \mathbb{R}^2$, $\mathbf{c}(t) = (x(t), y(t))$ is a parametrization of ∂D in the positive direction. Then we define the **boundary curve** ∂S to be the oriented simple closed curve that is the image of the mapping $\mathbf{p} \colon t \mapsto (x(t), y(t), f(x(t), y(t)))$ with the orientation induced by $\mathbf{p}$ (Figure 8.2.1).

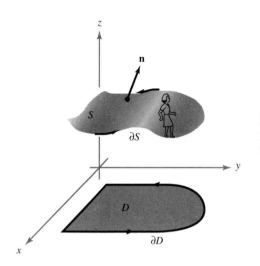

Figure 8.2.1 The induced orientation on ∂S: As you walk around the boundary, the surface should be on your left.

To remember this orientation (that is, the positive direction) on ∂S, imagine that you are an "observer" walking along the boundary of the surface with the normal as your upright direction; you are moving in the positive direction if the surface is on your left. This orientation on ∂S is often called the ***orientation induced by an upward normal*** **n**.

We are now ready to state and prove one of the primary results of this section.

THEOREM 5: STOKES' THEOREM FOR GRAPHS *Let S be the oriented surface defined by a C^2 function $z = f(x, y)$, where $(x, y) \in D$, and let* **F** *be a C^1 vector field on S. Then if ∂S denotes the oriented boundary curve of S as defined above, we have*

$$\iint_S \text{curl } \mathbf{F} \cdot d\mathbf{S} = \iint_S (\nabla \times \mathbf{F}) \cdot d\mathbf{S} = \int_{\partial S} \mathbf{F} \cdot d\mathbf{s}.$$

Remember that $\int_{\partial S} \mathbf{F} \cdot d\mathbf{s}$ is the integral around ∂S of the tangential component of **F**, while $\iint_S \mathbf{G} \cdot d\mathbf{S}$ is the integral over S of $\mathbf{G} \cdot \mathbf{n}$, the normal component of **G** (see Sections 7.2 and 7.6). Thus Stokes' theorem says that the integral of *the normal component of the curl of a vector field* **F** *over a surface S is equal to the integral of the tangential component of* **F** *around the boundary ∂S.*

PROOF   If $\mathbf{F} = F_1\mathbf{i} + F_2\mathbf{j} + F_3\mathbf{k}$, then

$$\text{curl } \mathbf{F} = \left(\frac{\partial F_3}{\partial y} - \frac{\partial F_2}{\partial z} \right)\mathbf{i} + \left(\frac{\partial F_1}{\partial z} - \frac{\partial F_3}{\partial x} \right)\mathbf{j} + \left(\frac{\partial F_2}{\partial x} - \frac{\partial F_1}{\partial y} \right)\mathbf{k}.$$

Therefore we use formula (1) to write

$$\iint_S \text{curl } \mathbf{F} \cdot d\mathbf{S} = \iint_D \left[\left(\frac{\partial F_3}{\partial y} - \frac{\partial F_2}{\partial z} \right) \left(-\frac{\partial z}{\partial x} \right) \right.$$
$$\left. + \left(\frac{\partial F_1}{\partial z} - \frac{\partial F_3}{\partial x} \right) \left(-\frac{\partial z}{\partial y} \right) + \left(\frac{\partial F_2}{\partial x} - \frac{\partial F_1}{\partial y} \right) \right] dA. \tag{2}$$

On the other hand,

$$\int_{\partial S} \mathbf{F} \cdot d\mathbf{s} = \int_{\mathbf{p}} \mathbf{F} \cdot d\mathbf{s} = \int_{\mathbf{p}} F_1 \, dx + F_2 \, dy + F_3 \, dz,$$

where $\mathbf{p}: [a, b] \to \mathbb{R}^3, \mathbf{p}(t) = (x(t), y(t), f(x(t), y(t)))$ is the orientation-preserving parametrization of the oriented simple closed curve ∂S discussed above. Thus,

$$\int_{\partial S} \mathbf{F} \cdot d\mathbf{s} = \int_a^b \left(F_1 \frac{dx}{dt} + F_2 \frac{dy}{dt} + F_3 \frac{dz}{dt} \right) dt. \tag{3}$$

By the chain rule

$$\frac{dz}{dt} = \frac{\partial z}{\partial x} \frac{dx}{dt} + \frac{\partial z}{\partial y} \frac{dy}{dt}.$$

Substituting this expression into equation (3), we obtain

$$\int_{\partial S} \mathbf{F} \cdot d\mathbf{s} = \int_a^b \left[\left(F_1 + F_3 \frac{\partial z}{\partial x} \right) \frac{dx}{dt} + \left(F_2 + F_3 \frac{\partial z}{\partial y} \right) \frac{dy}{dt} \right] dt$$
$$= \int_{\mathbf{c}} \left(F_1 + F_3 \frac{\partial z}{\partial x} \right) dx + \left(F_2 + F_3 \frac{\partial z}{\partial y} \right) dy$$
$$= \int_{\partial D} \left(F_1 + F_3 \frac{\partial z}{\partial x} \right) dx + \left(F_2 + F_3 \frac{\partial z}{\partial y} \right) dy. \tag{4}$$

Applying Green's theorem to equation (4) yields (we are assuming that Green's theorem applies to D)

$$\iint_D \left[\frac{\partial (F_2 + F_3 \, \partial z/\partial y)}{\partial x} - \frac{\partial (F_1 + F_3 \, \partial z/\partial x)}{\partial y} \right] dA.$$

Now we use the chain rule, remembering that $F_1, F_2,$ and F_3 are functions of x, y, and z and that z is a function of x and y, to obtain

$$\iint_D \left[\left(\frac{\partial F_2}{\partial x} + \frac{\partial F_2}{\partial z} \frac{\partial z}{\partial x} + \frac{\partial F_3}{\partial x} \frac{\partial z}{\partial y} + \frac{\partial F_3}{\partial z} \frac{\partial z}{\partial x} \frac{\partial z}{\partial y} + F_3 \frac{\partial^2 z}{\partial x \partial y} \right) \right.$$
$$\left. - \left(\frac{\partial F_1}{\partial y} + \frac{\partial F_1}{\partial z} \frac{\partial z}{\partial y} + \frac{\partial F_3}{\partial y} \frac{\partial z}{\partial x} + \frac{\partial F_3}{\partial z} \frac{\partial z}{\partial y} \frac{\partial z}{\partial x} + F_3 \frac{\partial^2 z}{\partial y \partial x} \right) \right] dA.$$

The last two terms in each parenthesis cancel each other, and we can rearrange terms to obtain the integral of equation (2), which completes the proof. ∎

EXAMPLE 1   Let $\mathbf{F} = ye^z\mathbf{i} + xe^z\mathbf{j} + xye^z\mathbf{k}$. Show that the integral of $\mathbf{F}$ around an oriented simple closed curve C that is the boundary of a surface S is 0. (Assume S is the graph of a function, as in Theorem 5.)

SOLUTION   Indeed, $\int_C \mathbf{F} \cdot d\mathbf{s} = \iint_S (\nabla \times \mathbf{F}) \cdot d\mathbf{S}$, by Stokes' theorem. But we compute

$$\nabla \times \mathbf{F} = \begin{vmatrix} \mathbf{i} & \mathbf{j} & \mathbf{k} \\ \dfrac{\partial}{\partial x} & \dfrac{\partial}{\partial y} & \dfrac{\partial}{\partial z} \\ ye^z & xe^z & xye^z \end{vmatrix} = \mathbf{0},$$

and so $\int_C \mathbf{F} \cdot d\mathbf{s} = 0$. (Alternatively, we can observe that $\mathbf{F} = \nabla(xye^z)$, so its integral around a closed curve is zero.) ▲

EXAMPLE 2   Use Stokes' theorem to evaluate the line integral

$$\int_C -y^3\,dx + x^3\,dy - z^3\,dz,$$

where C is the intersection of the cylinder $x^2 + y^2 = 1$ and the plane $x + y + z = 1$, and the orientation on C corresponds to counterclockwise motion in the xy plane.

SOLUTION   The curve C bounds the surface S defined by $z = 1 - x - y = f(x, y)$ for (x, y) in $D = \{(x, y) \mid x^2 + y^2 \le 1\}$ (Figure 8.2.2). We set $\mathbf{F} = -y^3\mathbf{i} + x^3\mathbf{j} - z^3\mathbf{k}$, which has curl $\nabla \times \mathbf{F} = (3x^2 + 3y^2)\mathbf{k}$. Then by Stokes' theorem, the line integral is equal to the surface integral

$$\iint_S (\nabla \times \mathbf{F}) \cdot d\mathbf{S}.$$

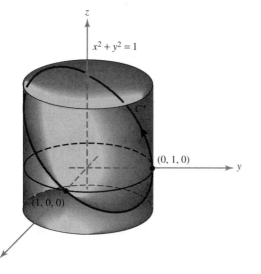

Figure 8.2.2   The curve C is the intersection of the cylinder $x^2 + y^2 = 1$ and the plane $x + y + z = 1$.

But $\nabla \times \mathbf{F}$ has only a $\mathbf{k}$ component. Thus, by formula (1) we have

$$\iint_S (\nabla \times \mathbf{F}) \cdot d\mathbf{S} = \iint_D (3x^2 + 3y^2)\, dx\, dy.$$

This integral can be evaluated by changing to polar coordinates. Doing this, we get

$$3 \iint_D (x^2 + y^2)\, dx\, dy = 3 \int_0^1 \int_0^{2\pi} r^2 \cdot r\, d\theta\, dr = 6\pi \int_0^1 r^3\, dr = \frac{6\pi}{4} = \frac{3\pi}{2}.$$

Let us verify this result by directly evaluating the line integral

$$\int_C -y^3\, dx + x^3\, dy - z^3\, dz.$$

We can parametrize the curve ∂D by the equations

$$x = \cos t, \qquad y = \sin t, \qquad z = 0, \qquad 0 \le t \le 2\pi.$$

The curve C is therefore parametrized by the equations

$$x = \cos t, \qquad y = \sin t, \qquad z = 1 - \sin t - \cos t, \qquad 0 \le t \le 2\pi.$$

Thus

$$\begin{aligned}
\int_C &-y^3\, dx + x^3\, dy - z^3\, dz \\
&= \int_0^{2\pi} [(-\sin^3 t)(-\sin t) + (\cos^3 t)(\cos t) \\
&\qquad - (1 - \sin t - \cos t)^3(-\cos t + \sin t)]\, dt \\
&= \int_0^{2\pi} (\cos^4 t + \sin^4 t)\, dt - \int_0^{2\pi} (1 - \sin t - \cos t)^3 (-\cos t + \sin t)\, dt.
\end{aligned}$$

The second integrand is of the form $u^3\, du$, where $u = 1 - \sin t - \cos t$, and thus the integral is equal to

$$\tfrac{1}{4}[(1 - \sin t - \cos t)^4]_0^{2\pi} = 0.$$

Hence we are left with

$$\int_0^{2\pi} (\cos^4 t + \sin^4 t)\, dt.$$

This integral can be evaluated using formulas (18) and (19) of the table of integrals. We can also proceed as follows. Using the trigonometric identities

$$\sin^2 t = \frac{1 - \cos 2t}{2}, \qquad \cos^2 t = \frac{1 + \cos 2t}{2},$$

we reduce the above integral to

$$\frac{1}{2} \int_0^{2\pi} (1 + \cos^2 2t)\, dt = \pi + \frac{1}{2} \int_0^{2\pi} \cos^2 2t\, dt.$$

Again using the identity $\cos^2 2t = (1 + \cos 4t)/2$, we find

$$\pi + \frac{1}{4} \int_0^{2\pi} (1 + \cos 4t)\, dt = \pi + \frac{1}{4} \int_0^{2\pi} dt + \frac{1}{4} \int_0^{2\pi} \cos 4t\, dt = \pi + \frac{\pi}{2} + 0 = \frac{3\pi}{2}. \quad \blacktriangle$$

To simplify the proof of Stokes' theorem above, we assumed that the surface S could be described as the graph of a function $z = f(x, y)$, $(x, y) \in D$, where D is some region to which Green's theorem applies. However, without too much more effort we can obtain a more general theorem for oriented parametrized surfaces S. The main complication is in the definition of ∂S.

Suppose $\Phi\colon D \to \mathbb{R}^3$ is a parametrization of a surface S and $\mathbf{c}(t) = (u(t), v(t))$ is a parametrization of ∂D. We might be tempted to define ∂S as the curve parametrized by $t \mapsto \mathbf{p}(t) = \Phi(u(t), v(t))$. However, with this definition, ∂S might not be the boundary of S in any reasonable geometric sense.

For example, we would conclude that the boundary of the unit sphere S parametrized by spherical coordinates in $\mathbb{R}^3$ is half of the great circle on S lying in the xz plane, but clearly in a geometric sense S is a smooth surface (no points or cusps) with no boundary or edge at all (see Figure 8.2.3 and Exercise 20). Thus this great circle is in some sense the "mistaken" boundary of S.

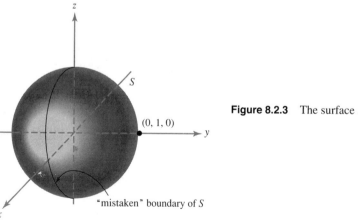

Figure 8.2.3   The surface S is a portion of a sphere.

We can get around this difficulty by assuming that Φ is one-to-one on all of D. Then the image of ∂D under Φ, namely, $\Phi(\partial D)$, will be the geometric boundary of $S = \Phi(D)$. If $\mathbf{c}(t) = (u(t), v(t))$ is a parametrization of ∂D in the positive direction, we define ∂S to be the oriented simple closed curve that is the image of the mapping $\mathbf{p}\colon t \mapsto \Phi(u(t), v(t))$ with the orientation of ∂S induced by $\mathbf{p}$ (see Figure 8.2.1).

> **THEOREM 6: STOKES' THEOREM: PARAMETRIZED SUR-**
> **FACES** *Let S be an oriented surface defined by a one-to-one parametrization*
> $\Phi: D \subset \mathbb{R}^2 \to S$. *Let ∂S denote the oriented boundary of S and let $\mathbf{F}$ be a C^1*
> *vector field on S. Then*
>
> $$\iint_S (\nabla \times \mathbf{F}) \cdot d\mathbf{S} = \int_{\partial S} \mathbf{F} \cdot d\mathbf{s}.$$

This is proved in the same way as Theorem 5.

EXAMPLE 3 Let S be the surface shown in Figure 8.2.4, with the indicated orientation. Let $\mathbf{F} = y\mathbf{i} - x\mathbf{j} + e^{xz}\mathbf{k}$. Evaluate $\iint_S (\nabla \times \mathbf{F}) \cdot d\mathbf{S}$.

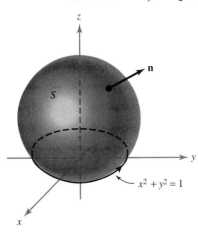

Figure 8.2.4 The boundary of a surface S that is parametrized by $\Phi: D \to \mathbb{R}^3$ is the image of the boundary of D only if Φ is one-to-one on D.

SOLUTION This surface could be parametrized using spherical coordinates based at the center of the sphere. However, we need not explicitly find Φ in order to solve this problem. By Theorem 6, $\iint_S (\nabla \times \mathbf{F}) \cdot d\mathbf{S} = \int_{\partial S} \mathbf{F} \cdot d\mathbf{s}$, and so if we parametrize ∂S by $x(t) = \cos t$, $y(t) = \sin t$, $0 \le t \le 2\pi$, we determine

$$\int_{\partial S} \mathbf{F} \cdot d\mathbf{s} = \int_0^{2\pi} \left(y \frac{dx}{dt} - x \frac{dy}{dt} \right) dt = \int_0^{2\pi} (-\sin^2 t - \cos^2 t)\, dt = -\int_0^{2\pi} dt = -2\pi$$

and therefore $\iint_S (\nabla \times \mathbf{F}) \cdot d\mathbf{S} = -2\pi$. ▲

Let us now use Stokes' theorem to justify the physical interpretation of $\nabla \times \mathbf{F}$ in terms of paddle wheels that was proposed in Chapter 4. Paraphrasing Theorem 6, we have

$$\iint_S (\text{curl } \mathbf{F}) \cdot \mathbf{n}\, dS = \iint_S (\text{curl } \mathbf{F}) \cdot d\mathbf{S} = \int_{\partial S} \mathbf{F} \cdot d\mathbf{s} = \int_{\partial S} F_T\, ds,$$

where F_T is the tangential component of $\mathbf{F}$. This says that the integral of the normal component of the curl of a vector field over an oriented surface S is equal to the line integral of **F** along ∂S, which in turn is equal to the path integral of the tangential component of **F** over ∂S.

Suppose **V** represents the velocity vector field of a fluid. Consider a point P and a unit vector **n**. Let S_ρ denote the disk of radius ρ and center P, which is perpendicular to **n**. By Stokes' theorem,

$$\iint_{S_\rho} \text{curl } \mathbf{V} \cdot d\mathbf{S} = \iint_{S_\rho} \text{curl } \mathbf{V} \cdot \mathbf{n} \, dS = \int_{\partial S_\rho} \mathbf{V} \cdot d\mathbf{s},$$

where ∂S_ρ has the orientation induced by **n** (see Figure 8.2.5).

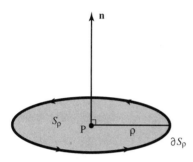

Figure 8.2.5   A normal **n** induces an orientation on the boundary ∂S_ρ of the disk S_ρ.

By the mean value theorem for integrals (Exercise 12, Section 7.6), there is a point Q in S_ρ such that

$$\iint_{S_\rho} \text{curl } \mathbf{V} \cdot \mathbf{n} \, dS = [\text{curl } \mathbf{V}(Q) \cdot \mathbf{n}] A(S_\rho)$$

where $A(S_\rho) = \pi\rho^2$ is the area of S_ρ, curl **V**(Q) is the value of curl **V** at Q, and **n** is evaluated at Q as well. Thus

$$\lim_{\rho \to 0} \frac{1}{A(S_\rho)} \int_{\partial S_\rho} \mathbf{V} \cdot d\mathbf{s} = \lim_{\rho \to 0} \frac{1}{A(S_\rho)} \iint_{S_\rho} (\text{curl } \mathbf{V}) \cdot d\mathbf{S}$$

$$= \lim_{\rho \to 0} \text{curl } \mathbf{V}(Q) \cdot \mathbf{n}(Q) = \text{curl } \mathbf{V}(P) \cdot \mathbf{n}(P).$$

Thus,[*]

$$\text{curl } \mathbf{V}(P) \cdot \mathbf{n}(P) = \lim_{\rho \to 0} \frac{1}{A(S_\rho)} \int_{\partial S_\rho} \mathbf{V} \cdot d\mathbf{s}. \tag{5}$$

Let us pause to consider the physical meaning of $\int_C \mathbf{V} \cdot d\mathbf{s}$ when **V** is the velocity field of a fluid. Suppose, for example, that **V** points in the direction tangent to the oriented curve C (Figure 8.2.6). Then clearly $\int_C \mathbf{V} \cdot d\mathbf{s} > 0$, and particles on C tend to rotate

---

[*]Some informal texts adopt equation (5) as the *definition* of the curl, and use it to "prove" Stokes' theorem. However, this raises the danger of circular reasoning, for to show that equation (5) really defines a vector "curl **V**(P)" requires Stokes' theorem, or some similar argument.

counterclockwise. If **V** is pointing in the opposite direction, then $\int_C \mathbf{V} \cdot d\mathbf{s} < 0$. If **V** is perpendicular to C, then particles don't rotate on C at all and $\int_C \mathbf{V} \cdot d\mathbf{s} = 0$. In general, $\int_C \mathbf{V} \cdot d\mathbf{s}$, being the integral of the tangential component of **V**, represents the net amount of turning of the fluid in a counterclockwise direction around C. One therefore refers to $\int_C \mathbf{V} \cdot d\mathbf{s}$ as the ***circulation*** of **V** around C (see Figure 8.2.7).

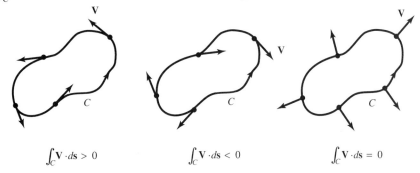

$$\int_C \mathbf{V} \cdot d\mathbf{s} > 0 \qquad \int_C \mathbf{V} \cdot d\mathbf{s} < 0 \qquad \int_C \mathbf{V} \cdot d\mathbf{s} = 0$$

Figure 8.2.6 The intuitive meaning of the possible signs of $\int_C \mathbf{V} \cdot d\mathbf{s}$.

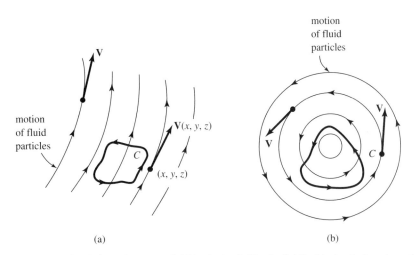

(a) (b)

Figure 8.2.7 Circulation of a vector field (velocity field of a fluid): (a) circulation about C is zero; (b) nonzero circulation about C ("whirlpool").

These results allow us to see just what curl **V** means for the motion of a fluid. The circulation $\int_{\partial S_\rho} \mathbf{V} \cdot d\mathbf{s}$ is the net velocity of the fluid around ∂S_ρ, so that curl $\mathbf{V} \cdot \mathbf{n}$ represents the turning or rotating effect of the fluid around the axis **n**.

CIRCULATION AND CURL curl $\mathbf{V}(P) \cdot \mathbf{n}(P)$ is the circulation of **V** per unit area at P on a surface perpendicular to $\mathbf{n}(P)$.

Observe that the magnitude of curl $\mathbf{V} \cdot \mathbf{n}$ is maximized when $\mathbf{n} = \text{curl } \mathbf{V}/\|\text{curl } \mathbf{V}\|$. There-fore, the rotating effect at P is greatest about the axis which is parallel to curl $\mathbf{V}/\|\text{curl } \mathbf{V}\|$. Thus, curl $\mathbf{V}$ is aptly called the ***vorticity vector.***

We can use these ideas to compute the curl in cylindrical coordinates.

EXAMPLE 4   Let the unit vectors $\mathbf{e}_r$, $\mathbf{e}_\theta$, $\mathbf{e}_z$ associated to cylindrical coordinates be as shown in Figure 8.2.8. Let $\mathbf{F} = F_r \mathbf{e}_r + F_\theta \mathbf{e}_\theta + F_z \mathbf{e}_z$. (The subscripts here denote components of $\mathbf{F}$, not partial derivatives.) Find a formula for the $\mathbf{e}_r$ component of $\nabla \times \mathbf{F}$ in cylindrical coordinates.

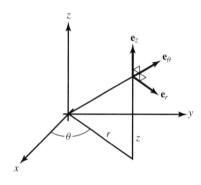

Figure 8.2.8   Orthonormal vectors $\mathbf{e}_r$, $\mathbf{e}_\theta$, and $\mathbf{e}_z$ associated with cylindrical coordinates. The vector $\mathbf{e}_r$ is parallel to the line labeled r.

SOLUTION   Let S be the surface shown in Figure 8.2.9.

Figure 8.2.9   A surface element in cylindrical coordinates.

The area of S is $r \, d\theta \, dz$ and the unit normal is $\mathbf{e}_r$. The integral of $\mathbf{F}$ around the edges of S is approximately

$$[F_\theta(r, \theta, z) - F_\theta(r, \theta, z + dz)]r \, d\theta + [F_z(r, \theta + d\theta, z) - F_z(r, \theta, z)] \, dz$$

$$\approx -\frac{\partial F_\theta}{\partial z} \, dz \, r \, d\theta + \frac{\partial F_z}{\partial \theta} \, d\theta \, dz.$$

Thus, the circulation per unit area is this expression divided by $r\,d\theta\,dz$, namely,

$$\frac{1}{r}\frac{\partial F_z}{\partial \theta} - \frac{\partial F_\theta}{\partial z}.$$

According to the previous box, this must be the $\mathbf{e}_r$ component of the curl.   ▲

By similar arguments, one finds that the curl in cylindrical coordinates is given by

$$\nabla \times \mathbf{F} = \frac{1}{r}\begin{vmatrix} \mathbf{e}_r & r\mathbf{e}_\theta & \mathbf{e}_z \\ \dfrac{\partial}{\partial r} & \dfrac{\partial}{\partial \theta} & \dfrac{\partial}{\partial z} \\ F_r & rF_\theta & F_z \end{vmatrix}.$$

We can find other important vector quantities expressed in different coordinate systems. For example, the chain rule shows that the gradient in cylindrical coordinates is

$$\nabla f = \frac{\partial f}{\partial r}\mathbf{e}_r + \frac{1}{r}\frac{\partial f}{\partial \theta}\mathbf{e}_\theta + \frac{\partial f}{\partial z}\mathbf{e}_z$$

and in Section 8.4 we establish techniques that give the following formula for the divergence in cylindrical coordinates:

$$\nabla \cdot \mathbf{F} = \frac{1}{r}\left[\frac{\partial}{\partial r}(rF_r) + \frac{\partial F_\theta}{\partial \theta} + \frac{\partial}{\partial z}(rF_z)\right].$$

Corresponding formulas for gradient, divergence, and curl in spherical coordinates are

$$\nabla f = \frac{\partial f}{\partial \rho}\mathbf{e}_\rho + \frac{1}{\rho}\frac{\partial f}{\partial \phi}\mathbf{e}_\phi + \frac{1}{\rho\sin\phi}\frac{\partial f}{\partial \theta}\mathbf{e}_\theta$$

$$\nabla \cdot \mathbf{F} = \frac{1}{\rho^2}\frac{\partial}{\partial \rho}(\rho^2 F_\rho) + \frac{1}{\rho\sin\phi}\frac{\partial}{\partial \phi}(\sin\phi F_\phi) + \frac{1}{\rho\sin\phi}\frac{\partial F_\theta}{\partial \theta}$$

and

$$\nabla \times \mathbf{F} = \left[\frac{1}{\rho\sin\phi}\frac{\partial}{\partial \phi}(\sin\phi F_\theta) - \frac{1}{\rho\sin\phi}\frac{\partial F_\phi}{\partial \theta}\right]\mathbf{e}_\rho$$

$$+ \left[\frac{1}{\rho\sin\phi}\frac{\partial F_\rho}{\partial \theta} - \frac{1}{\rho}\frac{\partial}{\partial \rho}(\rho F_\theta)\right]\mathbf{e}_\phi + \left[\frac{1}{\rho}\frac{\partial}{\partial \rho}(\rho F_\phi) - \frac{1}{\rho}\frac{\partial F_\rho}{\partial \phi}\right]\mathbf{e}_\theta$$

where $\mathbf{e}_\rho, \mathbf{e}_\phi, \mathbf{e}_\theta$ are as shown in Figure 8.2.10 and where $\mathbf{F} = F_\rho\mathbf{e}_\rho + F_\phi\mathbf{e}_\phi + F_\theta\mathbf{e}_\theta$.

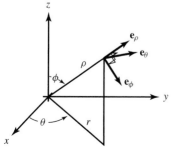

Figure 8.2.10 Orthonormal vectors $\mathbf{e}_\rho, \mathbf{e}_\phi$, and $\mathbf{e}_\theta$ associated with spherical coordinates.

Vector calculus plays an essential role in the theory of electromagnetism. Here is how Stokes' theorem applies.

EXAMPLE 5   Let $\mathbf{E}$ and $\mathbf{H}$ be time-dependent electric and magnetic fields, respectively, in space. Let S be a surface with boundary C. We define

$$\int_C \mathbf{E} \cdot d\mathbf{s} = \text{voltage around } C,$$

$$\iint_S \mathbf{H} \cdot d\mathbf{S} = \text{magnetic flux across } S.$$

Faraday's law (see Figure 8.2.11) states that the *voltage around C equals the negative rate of change of magnetic flux through S.* Show that Faraday's law follows from the following differential equation (one of the Maxwell equations):

$$\nabla \times \mathbf{E} = -\frac{\partial \mathbf{H}}{\partial t}.$$

Figure 8.2.11   Faraday's law.

SOLUTION   Assume that $-\partial \mathbf{H}/\partial t = \nabla \times \mathbf{E}$ holds. By Stokes' theorem,

$$\int_C \mathbf{E} \cdot d\mathbf{s} = \iint_S (\nabla \times \mathbf{E}) \cdot d\mathbf{S}.$$

Assuming that we can move $\partial/\partial t$ under the integral sign, we get

$$-\frac{\partial}{\partial t} \iint_S \mathbf{H} \cdot d\mathbf{S} = \iint_S -\frac{\partial \mathbf{H}}{\partial t} \cdot d\mathbf{S} = \iint_S (\nabla \times \mathbf{E}) \cdot d\mathbf{S} = \int_C \mathbf{E} \cdot d\mathbf{s}$$

and so

$$\int_C \mathbf{E} \cdot d\mathbf{s} = -\frac{\partial}{\partial t} \iint_S \mathbf{H} \cdot d\mathbf{S},$$

which is Faraday's law.   ▲

EXERCISES

1. Redo Exercise 5 of Section 7.6 using Stokes' theorem.

2. Redo Exercise 6 of Section 7.6 using Stokes' theorem.

3. Verify Stokes' theorem for the upper hemisphere $z = \sqrt{1 - x^2 - y^2}, z \geq 0$, and the radial vector field $\mathbf{F}(x, y, z) = x\mathbf{i} + y\mathbf{j} + z\mathbf{k}$.

4. Let S be a surface with boundary ∂S, and suppose $\mathbf{E}$ is an electric field that is perpendicular to ∂S. Show that the induced magnetic flux across S is constant in time. (HINT: Use Faraday's law.)

5. Let S be the capped cylindrical surface shown in Figure 8.2.12. S is the union of two surfaces S_1 and S_2, where S_1 is the set of (x, y, z) with $x^2 + y^2 = 1, 0 \leq z \leq 1$, and S_2 is the set of (x, y, z) with $x^2 + y^2 + (z - 1)^2 = 1, z \geq 1$. Set $\mathbf{F}(x, y, z) = (zx + z^2 y + x)\mathbf{i} + (z^3 yx + y)\mathbf{j} + z^4 x^2 \mathbf{k}$. Compute $\iint_S (\nabla \times \mathbf{F}) \cdot d\mathbf{S}$. (HINT: Stokes' theorem holds for this surface.)

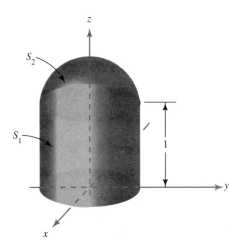

Figure 8.2.12 The capped cylinder is the union of S_1 and S_2.

6. Let c consist of straight lines joining $(1, 0, 0)$, $(0, 1, 0)$, and $(0, 0, 1)$ and let S be the triangle with these vertices. Verify Stokes' theorem directly with $\mathbf{F} = yz\mathbf{i} + xz\mathbf{j} + xy\mathbf{k}$.

7. Evaluate the integral $\iint_S (\nabla \times \mathbf{F}) \cdot d\mathbf{S}$, where S is the portion of the surface of a sphere defined by $x^2 + y^2 + z^2 = 1$ and $x + y + z \geq 1$, and where $\mathbf{F} = \mathbf{r} \times (\mathbf{i} + \mathbf{j} + \mathbf{k}), \mathbf{r} = x\mathbf{i} + y\mathbf{j} + z\mathbf{k}$.

8. Show that the calculation in Exercise 7 can be simplified by observing that $\int_{\partial S} \mathbf{F} \cdot d\mathbf{r} = \int_{\partial \Sigma} \mathbf{F} \cdot d\mathbf{r}$ for any other surface Σ. By picking Σ appropriately, $\iint_\Sigma (\nabla \times \mathbf{F}) \cdot d\mathbf{S}$ may be easy to compute. Show that this is the case if Σ is taken to be the portion of the plane $x + y + z = 1$ inside the circle ∂S.

9. Calculate the surface integral $\iint_S (\nabla \times \mathbf{F}) \cdot d\mathbf{S}$ where S is the hemisphere $x^2 + y^2 + z^2 = 1, x \geq 0$ and $\mathbf{F} = x^3\mathbf{i} - y^3\mathbf{j}$.

10. Find $\iint_S (\nabla \times \mathbf{F}) \cdot d\mathbf{S}$, where S is the ellipsoid $x^2 + y^2 + 2z^2 = 10$ and $\mathbf{F} = (\sin xy)\mathbf{i} + e^x\mathbf{j} - yz\mathbf{k}$.

11. Let $\mathbf{F} = y\mathbf{i} - x\mathbf{j} + zx^3 y^2 \mathbf{k}$. Evaluate $\iint_S (\nabla \times \mathbf{F}) \cdot \mathbf{n} \, dA$, where S is the surface defined by $x^2 + y^2 + z^2 = 1, z \leq 0$.

12. A hot-air balloon has the truncated spherical shape shown in Figure 8.2.13. The hot gases escape through the porous envelope with a velocity vector field

$$\mathbf{V}(x, y, z) = \nabla \times \mathbf{\Phi}(x, y, z) \qquad \text{where} \qquad \mathbf{\Phi}(x, y, z) = -y\mathbf{i} + x\mathbf{j}.$$

If $R = 5$, compute the volume flow rate of the gases through the surface.

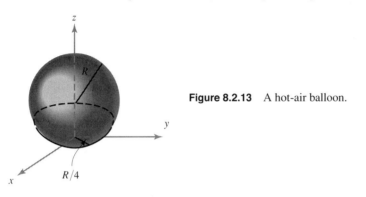

Figure 8.2.13   A hot-air balloon.

13.   Prove that Faraday's law implies $\nabla \times \mathbf{E} = -\partial \mathbf{H}/\partial t$.

14.   Let S be a surface and let $\mathbf{F}$ be perpendicular to the tangent to the boundary of S. Show that

$$\iint_S (\nabla \times \mathbf{F}) \cdot d\mathbf{S} = 0.$$

What does this mean physically if $\mathbf{F}$ is an electric field?

15.   Consider two surfaces S_1, S_2 with the same boundary ∂S. Describe with sketches how S_1 and S_2 must be oriented to ensure that

$$\iint_{S_1} (\nabla \times \mathbf{F}) \cdot d\mathbf{S} = \iint_{S_2} (\nabla \times \mathbf{F}) \cdot d\mathbf{S}.$$

16.   For a surface S and a fixed vector $\mathbf{v}$, prove that

$$2 \iint_S \mathbf{v} \cdot \mathbf{n}\, dS = \int_{\partial S} (\mathbf{v} \times \mathbf{r}) \cdot d\mathbf{s},$$

where $\mathbf{r}(x, y, z) = (x, y, z)$.

17.   Argue informally that if S is a closed surface then

$$\iint_S (\nabla \times \mathbf{F}) \cdot d\mathbf{S} = 0$$

(see Exercise 15). (A *closed surface* is one that forms the boundary of a region in space; thus, for example, a sphere is a closed surface.)

18.   If C is a closed curve which is the boundary of a surface S and f and g are C^2 functions, show that
(a) $\int_C f \nabla g \cdot d\mathbf{s} = \iint_S (\nabla f \times \nabla g) \cdot d\mathbf{S}$
(b) $\int_C (f \nabla g + g \nabla f) \cdot d\mathbf{s} = 0$.

19. (a) If C is a closed curve that is the boundary of a surface S and $\mathbf{v}$ is a constant vector, show that

$$\int_C \mathbf{v} \cdot d\mathbf{s} = 0.$$

(b) Show that this is true even if C is not the boundary of a surface S.

20. Show that $\boldsymbol{\Phi}: D \to \mathbb{R}^3$, $D = [0, \pi] \times [0, 2\pi]$, $\boldsymbol{\Phi}(\phi, \theta) = (\cos\theta \sin\phi, \sin\theta \sin\phi, \cos\phi)$, which parametrizes the unit sphere, takes the boundary of D to half of a great circle on S.

21. Verify Theorem 6 for the helicoid $\boldsymbol{\Phi}(r, \theta) = (r\cos\theta, r\sin\theta, \theta)$, $(r, \theta) \in [0, 1] \times [0, \pi/2]$, and the vector field $\mathbf{F}(x, y, z) = (z, x, y)$.

22. Prove Theorem 6.

23. Let $\mathbf{F} = x^2\mathbf{i} + (2xy + x)\mathbf{j} + z\mathbf{k}$. Let C be the circle $x^2 + y^2 = 1$ and S the disk $x^2 + y^2 \leq 1$ within the plane $z = 0$.
 (a) Determine the flux of $\mathbf{F}$ out of S.
 (b) Determine the circulation of $\mathbf{F}$ around C.
 (c) Find the flux of $\nabla \times \mathbf{F}$. Verify Stokes' theorem directly in this case.

24. Let S be a surface with boundary ∂S, and suppose that $\mathbf{E}$ is an electric field that is perpendicular to ∂S. Use Faraday's law to show that the induced magnetic flux across S is constant in time.

25. Integrate $\nabla \times \mathbf{F}$, $\mathbf{F} = (3y, -xz, -yz^2)$ over the portion of the surface $2z = x^2 + y^2$ below the plane $z = 2$, both directly and by using Stokes' theorem.

26. *Ampere's law* states that if the electric current density is described by a vector field $\mathbf{J}$ and the induced magnetic field is $\mathbf{H}$, then the circulation of $\mathbf{H}$ around the boundary C of a surface S equals the integral of $\mathbf{J}$ over S (i.e., the total current crossing S). See Figure 8.2.14. Show that this is implied by the steady-state *Maxwell equation* $\nabla \times \mathbf{H} = \mathbf{J}$.

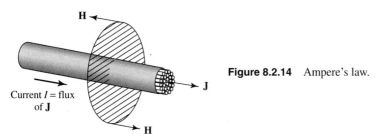

Figure 8.2.14 Ampere's law.

27. Faraday's law relates the line integral of the electric field around a loop C to the surface integral of the rate of change of the magnetic field over a surface S with boundary C. Regarding the equation $\nabla \times \mathbf{E} = -\partial\mathbf{H}/\partial t$ as the basic equation, Faraday's law is a consequence of Stokes' theorem, as we have seen in Example 4.

Suppose we are given electric and magnetic fields in space that satisfy $\nabla \times \mathbf{E} = -\partial\mathbf{H}/\partial t$. Suppose C is the boundary of the Möbius band shown in Figures 7.6.3 and 7.6.4. Since the Möbius band cannot be oriented, Stokes' theorem does not apply. What becomes of Faraday's law? What do you guess $\int_C \mathbf{E} \cdot d\mathbf{s}$ equals?

8.3 CONSERVATIVE FIELDS

We saw in Section 7.2 that for a gradient force field $\mathbf{F} = \nabla f$, line integrals of $\mathbf{F}$ were evaluated as follows:

$$\int_{\mathbf{c}} \mathbf{F} \cdot d\mathbf{s} = f(\mathbf{c}(b)) - f(\mathbf{c}(a)).$$

The value of the integral depends only on the endpoints $\mathbf{c}(b)$ and $\mathbf{c}(a)$ of the path. In other words, if we used another path with the same endpoints, we would still get the same answer. This leads us to say that the integral is *path-independent*.

Gradient fields are important in many physical problems. For example, if $V = -f$ represents a potential energy (gravitational, electrical, and so on), then $\mathbf{F}$ represents a force.* Consider the example of a particle of mass m in the field of the earth; in this case one takes f to be GmM/r or $V = -GmM/r$, where G is the gravitational constant, M is the mass of the earth, and r is the distance from the center of the earth. The corresponding force is $\mathbf{F} = (GmM/r^3)\mathbf{r} = (GmM/r^2)\mathbf{n}$, where $\mathbf{n}$ is the unit radial vector. Note that $\mathbf{F}$ fails to be defined at the point $r = 0$.

We wish to characterize those vector fields that can be written as a gradient. Our task is simplified considerably by Stokes' theorem.

THEOREM 7: CONSERVATIVE FIELDS *Let $\mathbf{F}$ be a C^1 vector field defined on $\mathbb{R}^3$ except possibly for a finite number of points. The following conditions on $\mathbf{F}$ are all equivalent*:

(i) *For any oriented simple closed curve C, $\int_C \mathbf{F} \cdot d\mathbf{s} = 0$.*

(ii) *For any two oriented simple curves C_1 and C_2 which have the same endpoints,*

$$\int_{C_1} \mathbf{F} \cdot d\mathbf{s} = \int_{C_2} \mathbf{F} \cdot d\mathbf{s}.$$

(iii) *$\mathbf{F}$ is the gradient of some function f; that is, $\mathbf{F} = \nabla f$ (and if $\mathbf{F}$ has an exceptional point where it fails to be defined, f is also undefined there).*

(iv) *$\nabla \times \mathbf{F} = \mathbf{0}$.*

A vector field satisfying one (and, hence, all) of the conditions (i)–(iv) is called a ***conservative vector field.***[†]

---

*If the minus sign is used, then V is decreasing in the direction $\mathbf{F}$. Thus a particle acted on by $\mathbf{F}$ moves in a direction that decreases the potential.

[†] In the plane $\mathbb{R}^2$, exceptional points are not allowed (see Exercise 12). Theorem 7 can be proved in the same way if $\mathbf{F}$ is defined and is of class C^1 only on an open convex set in $\mathbb{R}^2$ or $\mathbb{R}^3$. (A set D is convex if P, Q $\in D$ implies the line joining P and Q belongs to D.)

PROOF We shall establish the following chain of implications, which will prove the theorem:

$$(i) \Rightarrow (ii) \Rightarrow (iii) \Rightarrow (iv) \Rightarrow (i).$$

First we show that condition (i) implies condition (ii). Suppose $\mathbf{c}_1$ and $\mathbf{c}_2$ are parametrizations representing C_1 and C_2, with the same endpoints. Construct the closed curve $\mathbf{c}$ obtained by first traversing $\mathbf{c}_1$ and then $-\mathbf{c}_2$ (Figure 8.3.1), or, symbolically, the curve $\mathbf{c} = \mathbf{c}_1 - \mathbf{c}_2$. Assuming $\mathbf{c}$ is simple, condition (i) gives

$$\int_{\mathbf{c}} \mathbf{F} \cdot d\mathbf{s} = \int_{\mathbf{c}_1} \mathbf{F} \cdot d\mathbf{s} - \int_{\mathbf{c}_2} \mathbf{F} \cdot d\mathbf{s} = 0,$$

and so condition (ii) holds. (If $\mathbf{c}$ is not simple, an additional argument, omitted here, is needed.)

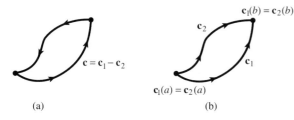

(a)

(b)

Figure 8.3.1 Constructing an oriented simple closed curve $\mathbf{c}_1 - \mathbf{c}_2$ (a) from two oriented simple curves (b).

Next we prove that condition (ii) implies condition (iii). Let C be any oriented simple curve joining a point such as $(0, 0, 0)$ to (x, y, z), and suppose C is represented by the parametrization $\mathbf{c}$ (if $(0, 0, 0)$ is the exceptional point of $\mathbf{F}$, we can choose a different starting point for $\mathbf{c}$ without affecting the argument). Define $f(x, y, z)$ to be $\int_{\mathbf{c}} \mathbf{F} \cdot d\mathbf{s}$. By hypothesis (ii), $f(x, y, z)$ is independent of C. We shall show that $\mathbf{F} = \operatorname{grad} f$. Indeed, choose $\mathbf{c}$ to be the path shown in Figure 8.3.2, so that

$$f(x, y, z) = \int_0^x F_1(t, 0, 0)\, dt + \int_0^y F_2(x, t, 0)\, dt + \int_0^z F_3(x, y, t)\, dt,$$

where $\mathbf{F} = (F_1, F_2, F_3)$.

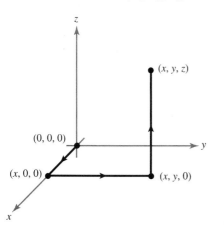

Figure 8.3.2 A path joining $(0, 0, 0)$ to (x, y, z).

It follows from the fundamental theorem of calculus that $\partial f/\partial z = F_3$. We can repeat this processing using two other paths from $(0, 0, 0)$ to (x, y, z) (for example, by drawing the lines from $(0, 0, 0)$ to $(0, y, 0)$ to $(x, y, 0)$ to (x, y, z)), we can similarly show that $\partial f/\partial x = F_1$ and $\partial f/\partial y = F_2$ (see Exercise 22). Thus, that is, $\nabla f = \mathbf{F}$.

Third, condition (iii) implies condition (iv), because, as proved in Section 4.4,

$$\nabla \times \nabla f = \mathbf{0}.$$

Finally, let $\mathbf{c}$ represent a closed curve C and let S be any surface whose boundary is $\mathbf{c}$ (if $\mathbf{F}$ has exceptional points, choose S to avoid them). Figure 8.3.3 indicates that we can probably always find such a surface; however, a formal proof of this would require the development of more sophisticated mathematical ideas than we can present here. By Stokes' theorem,

$$\int_C \mathbf{F} \cdot d\mathbf{s} = \int_{\mathbf{c}} \mathbf{F} \cdot d\mathbf{s} = \iint_S (\nabla \times \mathbf{F}) \cdot \mathbf{n} \, dS = \iint_S (\text{curl } \mathbf{F}) \cdot \mathbf{n} \, dS.$$

Since $\nabla \times \mathbf{F} = \mathbf{0}$, this integral vanishes, so that condition (iv) $\Rightarrow$ condition (i). ∎

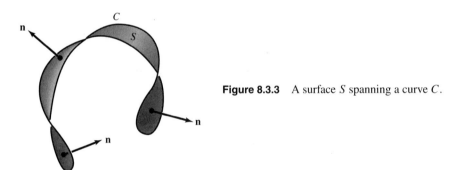

Figure 8.3.3   A surface S spanning a curve C.

There are several useful physical interpretations of $\int_C \mathbf{F} \cdot d\mathbf{s}$. We have already seen that one is the work done by $\mathbf{F}$ in moving a particle along C. A second interpretation is the notion of circulation, which we encountered at the end of the last section. In this case we think of $\mathbf{F}$ as the velocity field of a fluid; that is, to each point P in space, $\mathbf{F}$ assigns the velocity vector of the fluid at P. Take C to be a closed curve, and let $\Delta\mathbf{s}$ be a small directed chord of C. Then $\mathbf{F} \cdot \Delta\mathbf{s}$ is approximately the tangential component of $\mathbf{F}$ times $\|\Delta\mathbf{s}\|$. The circulation $\int_C \mathbf{F} \cdot d\mathbf{s}$ is the net tangential component around C. A small paddle wheel placed in the fluid would rotate if it is centered at a point where $\mathbf{F}$ vanishes and if the circulation of the fluid is nonzero, or $\int_C \mathbf{F} \cdot d\mathbf{s} \neq 0$ for small loops C (see Figure 8.3.4). Additional interpretations of circulation were discussed in Section 4.4.

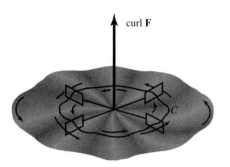

Figure 8.3.4 $\int_C \mathbf{F} \cdot d\mathbf{s} \neq 0$ implies that a paddle wheel in a fluid with velocity field $\mathbf{F}$ will rotate around its axis.

There is a similar interpretation in electromagnetic theory: If $\mathbf{F}$ represents an electric field, then a current will flow around a loop C if $\int_C \mathbf{F} \cdot d\mathbf{s} \neq 0$.

By Theorem 7, a field $\mathbf{F}$ has no circulation if and only if curl $\mathbf{F} = \nabla \times \mathbf{F} = \mathbf{0}$. Hence, a vector field $\mathbf{F}$ with curl $\mathbf{F} = \mathbf{0}$ is called *irrotational*. We have therefore proved that *a vector field in $\mathbb{R}^3$ is irrotational if and only if it is a gradient field for some function, that is, if and only if $\mathbf{F} = \nabla f$*. The function f is called a *potential* for $\mathbf{F}$.

EXAMPLE 1 Consider the vector field $\mathbf{F}$ on $\mathbb{R}^3$ defined by

$$\mathbf{F}(x, y, z) = y\mathbf{i} + (z \cos yz + x)\mathbf{j} + (y \cos yz)\mathbf{k}.$$

Show that $\mathbf{F}$ is irrotational and find a scalar potential for $\mathbf{F}$.

SOLUTION We compute $\nabla \times \mathbf{F}$:

$$\nabla \times \mathbf{F} = \begin{vmatrix} \mathbf{i} & \mathbf{j} & \mathbf{k} \\ \dfrac{\partial}{\partial x} & \dfrac{\partial}{\partial y} & \dfrac{\partial}{\partial z} \\ y & x + z \cos yz & y \cos yz \end{vmatrix}$$

$$= (\cos yz - yz \sin yz - \cos yz + yz \sin yz)\mathbf{i} + (0 - 0)\mathbf{j} + (1 - 1)\mathbf{k}$$

$$= 0\mathbf{i} + 0\mathbf{j} + 0\mathbf{k} = \mathbf{0},$$

and so $\mathbf{F}$ is irrotational. Thus, a potential exists by Theorem 7. We can find it in several ways.

Method 1. By the technique used to prove that condition (ii) implies condition (iii) in Theorem 7, we may set

$$f(x, y, z) = \int_0^x F_1(t, 0, 0)\, dt + \int_0^y F_2(x, t, 0)\, dt + \int_0^z F_3(x, y, t)\, dt$$

$$= \int_0^x 0\, dt + \int_0^y x\, dt + \int_0^z y \cos yt\, dt$$

$$= 0 + xy + \sin yz = xy + \sin yz.$$

One easily verifies that $\nabla f = \mathbf{F}$, as required:

$$\nabla f = \frac{\partial f}{\partial x}\mathbf{i} + \frac{\partial f}{\partial y}\mathbf{j} + \frac{\partial f}{\partial z}\mathbf{k} = y\mathbf{i} + (x + z\cos yz)\mathbf{j} + (y\cos yz)\mathbf{k}.$$

Method 2. Because we know that f exists, we know that we can solve the system of equations

$$\frac{\partial f}{\partial x} = y, \quad \frac{\partial f}{\partial y} = x + z\cos yz, \quad \frac{\partial f}{\partial z} = y\cos yz,$$

for $f(x, y, z)$. These are equivalent to the simultaneous equations

(a) $f(x, y, z) = xy + h_1(y, z)$

(b) $f(x, y, z) = \sin yz + xy + h_2(x, z)$

(c) $f(x, y, z) = \sin yz + h_3(x, y)$

for functions h_1, h_2, h_3 independent of x, y, and z (respectively). When $h_1(y, z) = \sin yz$, $h_2(x, z) = 0$, and $h_3(x, y) = xy$, the three equations agree and so yield a potential for $\mathbf{F}$. However, we have only guessed at the values of h_1, h_2, and h_3. To derive the formula for f more systematically we note that since $f(x, y, z) = xy + h_1(y, z)$ and $\partial f/\partial z = y\cos yz$, we find that

$$\frac{\partial h_1(y, z)}{\partial z} = y\cos yz$$

or

$$h_1(y, z) = \int y\cos yz\, dz + g(y) = \sin yz + g(y).$$

Therefore, substituting this back into equation (a), we get

$$f(x, y, z) = xy + \sin yz + g(y);$$

but by equation (b),

$$g(y) = h_2(x, z).$$

Since the right side of this equation is a function of x and z and the left side is a function of y alone, we conclude that they must equal some constant C. Thus

$$f(x, y, z) = xy + \sin yz + C$$

and we have determined f up to a constant.   ▲

EXAMPLE 2   A mass M at the origin in $\mathbb{R}^3$ exerts a force on a mass m located at $\mathbf{r} = (x, y, z)$ with magnitude GmM/r^2 and directed toward the origin. Here G is the gravitational constant, which depends on the units of measurement, and $r = \|\mathbf{r}\| = \sqrt{x^2 + y^2 + z^2}$. If we remember that $-\mathbf{r}/r$ is a unit vector directed toward the origin, then we can write the force field as

$$\mathbf{F}(x, y, z) = -\frac{GmM\mathbf{r}}{r^3}.$$

Show that $\mathbf{F}$ is irrotational and find a scalar potential for $\mathbf{F}$. (Notice that $\mathbf{F}$ is not defined at the origin, but Theorem 7 still applies, since it allows an exceptional point.)

SOLUTION   First let us verify that $\nabla \times \mathbf{F} = \mathbf{0}$. Referring to formula 11 in the table of vector identities in Section 4.4, we get

$$\nabla \times \mathbf{F} = -GmM\left[\nabla\left(\frac{1}{r^3}\right) \times \mathbf{r} + \frac{1}{r^3}\nabla \times \mathbf{r}\right].$$

But $\nabla(1/r^3) = -3\mathbf{r}/r^5$ (see Exercise 30, Section 4.4), and so the first term vanishes, since $\mathbf{r} \times \mathbf{r} = \mathbf{0}$. The second term vanishes, because

$$\nabla \times \mathbf{r} = \begin{vmatrix} \mathbf{i} & \mathbf{j} & \mathbf{k} \\ \dfrac{\partial}{\partial x} & \dfrac{\partial}{\partial y} & \dfrac{\partial}{\partial z} \\ x & y & z \end{vmatrix} = \left(\frac{\partial z}{\partial y} - \frac{\partial y}{\partial z}\right)\mathbf{i} + \left(\frac{\partial x}{\partial z} - \frac{\partial z}{\partial x}\right)\mathbf{j} + \left(\frac{\partial y}{\partial x} - \frac{\partial x}{\partial y}\right)\mathbf{k} = \mathbf{0}.$$

Hence $\nabla \times \mathbf{F} = \mathbf{0}$ (for $\mathbf{r} \neq \mathbf{0}$).

If we recall the formula $\nabla(r^n) = nr^{n-2}\mathbf{r}$ (Exercise 30, Section 4.4), then we can read off a scalar potential for $\mathbf{F}$ by inspection. We have $\mathbf{F} = -\nabla\phi$, where $\phi(x, y, z) = -GmM/r$ is called the ***gravitational potential energy.***

[We observe in passing that by Theorem 3 of Section 7.2, the work done by $\mathbf{F}$ in moving a particle of mass m from a point P_1 to a point P_2 is given by

$$\phi(P_1) - \phi(P_2) = GmM\left(\frac{1}{r_2} - \frac{1}{r_1}\right),$$

where r_1 is the radial distance of P_1 from the origin, with r_2 similarly defined.]   ▲

By the same proof, Theorem 7 is also true for C^1 vector fields $\mathbf{F}$ on $\mathbb{R}^2$. In this case we require that $\mathbf{F}$ has *no* exceptional points; that is, $\mathbf{F}$ is smooth everywhere (see Exercise 12). Notice, however, that the conclusion *might* still hold even if there are exceptional points, an example being $(x\mathbf{i} + y\mathbf{j})/(x^2 + y^2)^{3/2}$.

If $\mathbf{F} = P\mathbf{i} + Q\mathbf{j}$, then

$$\nabla \times \mathbf{F} = \left(\frac{\partial Q}{\partial x} - \frac{\partial P}{\partial y}\right)\mathbf{k}.$$

Sometimes $\partial Q/\partial x - \partial P/\partial y$ is called the ***scalar curl*** of $\mathbf{F}$. Thus, the condition $\nabla \times \mathbf{F} = \mathbf{0}$ reduces to

$$\frac{\partial P}{\partial y} = \frac{\partial Q}{\partial x}.$$

Thus we have:

COROLLARY 1   If $\mathbf{F}$ is a C^1 vector field on $\mathbb{R}^2$ of the form $P\mathbf{i}+Q\mathbf{j}$ with $\partial P/\partial y = \partial Q/\partial x$, then $\mathbf{F}=\nabla f$ for some f on $\mathbb{R}^2$.

We emphasize that this corollary can be false if $\mathbf{F}$ fails to be of class C^1 at even a single point (an example is given in Exercise 12). In $\mathbb{R}^3$, however, as already noted exceptions at single points are allowed (see Theorem 7).

EXAMPLE 3   (a)   Determine whether the vector field

$$\mathbf{F} = e^{xy}\mathbf{i} + e^{x+y}\mathbf{j}$$

is a gradient field.

(b)   Repeat part (a) for

$$\mathbf{F} = (2x\cos y)\mathbf{i} - (x^2\sin y)\mathbf{j}.$$

SOLUTION   (a)   Here $P(x, y) = e^{xy}$ and $Q(x, y) = e^{x+y}$, and so we compute

$$\frac{\partial P}{\partial y} = xe^{xy}, \qquad \frac{\partial Q}{\partial x} = e^{x+y}.$$

These are not equal, and so $\mathbf{F}$ cannot have a potential function.

(b)   In this case, we find

$$\frac{\partial P}{\partial y} = -2x\sin y = \frac{\partial Q}{\partial x},$$

and so $\mathbf{F}$ has a potential function f. To compute f we solve the equations

$$\frac{\partial f}{\partial x} = 2x\cos y, \qquad \frac{\partial f}{\partial y} = -x^2\sin y.$$

Thus, $f(x, y) = x^2\cos y + h_1(y)$ and $f(x, y) = x^2\cos y + h_2(x)$. If h_1 and h_2 are the same constant, then both equations are satisfied, and so $f(x, y) = x^2\cos y$ is a potential for $\mathbf{F}$.   ▲

EXAMPLE 4   Let $\mathbf{c}: [1, 2] \to \mathbb{R}^2$ be given by $x = e^{t-1}$, $y = \sin(\pi/t)$. Compute the integral

$$\int_{\mathbf{c}} \mathbf{F} \cdot d\mathbf{s} = \int_{\mathbf{c}} 2x\cos y\, dx - x^2\sin y\, dy,$$

where $\mathbf{F} = (2x\cos y)\mathbf{i} - (x^2\sin y)\mathbf{j}$.

SOLUTION   The endpoints are $\mathbf{c}(1) = (1, 0)$ and $\mathbf{c}(2) = (e, 1)$. Since $\partial(2x\cos y)/\partial y = \partial(-x^2\sin y)/\partial x$, $\mathbf{F}$ is irrotational and hence a gradient vector field (as we saw in

Example 3). Thus, by Theorem 7, we can replace $\mathbf{c}$ by any piecewise C^1 curve having the same endpoints, in particular, by the polygonal path from $(1, 0)$ to $(e, 0)$ to $(e, 1)$. Thus the line integral must be equal to

$$\int_{\mathbf{c}} \mathbf{F} \cdot d\mathbf{s} = \int_1^e 2t \cos 0 \, dt + \int_0^1 -e^2 \sin t \, dt = (e^2 - 1) + e^2(\cos 1 - 1) = e^2 \cos 1 - 1.$$

Alternatively, using Theorem 3 of Section 7.2, we have

$$\int_{\mathbf{c}} 2x \cos y \, dx - x^2 \sin y \, dy = \int_{\mathbf{c}} \nabla f \cdot d\mathbf{s} = f(\mathbf{c}(2)) - f(\mathbf{c}(1)) = e^2 \cos 1 - 1,$$

since $f(x, y) = x^2 \cos y$ is a potential function for $\mathbf{F}$. Evidently this technique is simpler than computing the integral directly. ▲

We conclude this section with a theorem that is quite similar in spirit to Theorem 7. Theorem 7 was motivated partly as a converse to the result that curl $\nabla f = \mathbf{0}$ for any C^1 function $f \colon \mathbb{R}^3 \to \mathbb{R}$—or, if curl $\mathbf{F} = \mathbf{0}$, then $\mathbf{F} = \nabla f$. We also know (formula 10 in the table of vector identities in Section 4.4) that div (curl $\mathbf{G}$) $= 0$ for any C^2 vector field $\mathbf{G}$. We may ask about the converse statement: If div $\mathbf{F} = 0$, is $\mathbf{F}$ the curl of a vector field $\mathbf{G}$? The following theorem answers this in the affirmative.

THEOREM 8   *If $\mathbf{F}$ is a C^1 vector field on $\mathbb{R}^3$ with div $\mathbf{F} = 0$, then there exists a C^1 vector field $\mathbf{G}$ with $\mathbf{F} = $ curl $\mathbf{G}$.*

The proof is outlined in Exercise 16. We should warn the reader at this point that, unlike the $\mathbf{F}$ in Theorem 7, the vector field $\mathbf{F}$ in Theorem 8 is not allowed to have an exceptional point. For example, the gravitational force field $\mathbf{F} = -(Gm M\mathbf{r}/r^3)$ has the property that div $\mathbf{F} = 0$, and yet there is no $\mathbf{G}$ for which $\mathbf{F} = $ curl $\mathbf{G}$ (see Exercise 25). Theorem 8 does not apply, because the gravitational force field $\mathbf{F}$ is not defined at $\mathbf{0} \in \mathbb{R}^3$.

EXERCISES

1. Show that any two potential functions for a vector field differ at most by a constant.

2. (a) Let $\mathbf{F}(x, y) = (xy, y^2)$ and let $\mathbf{c}$ be the path $y = 2x^2$ joining $(0, 0)$ to $(1, 2)$ in $\mathbb{R}^2$. Evaluate $\int_{\mathbf{c}} \mathbf{F} \cdot d\mathbf{s}$.
   (b) Does the integral in part (a) depend on the path joining $(0, 0)$ to $(1, 2)$?

3. Let $\mathbf{F}(x, y, z) = (2xyz + \sin x)\mathbf{i} + x^2 z\mathbf{j} + x^2 y\mathbf{k}$. Find a function f such that $\mathbf{F} = \nabla f$.

4. Evaluate $\int_{\mathbf{c}} \mathbf{F} \cdot d\mathbf{s}$, where $\mathbf{c}(t) = (\cos^5 t, \sin^3 t, t^4)$, $0 \le t \le \pi$, and $\mathbf{F}$ is as in Exercise 3.

5. What is the work done by the force $\mathbf{F} = -\mathbf{r}/\|\mathbf{r}\|^3$ in moving a particle from a point $\mathbf{r}_0 \in \mathbb{R}^3$ "to ∞," where $\mathbf{r}(x, y, z) = (x, y, z)$?

6. In Exercise 5, show that $\mathbf{F} = \nabla(1/r)$, $r \neq 0$, $r = \|\mathbf{r}\|$. In what sense is the integral of $\mathbf{F}$ independent of the path?

7. Let $\mathbf{F}(x, y, z) = xy\mathbf{i} + y\mathbf{j} + z\mathbf{k}$. Can there exist a function f such that $\mathbf{F} = \nabla f$?

8. Let $\mathbf{F} = F_1\mathbf{i} + F_2\mathbf{j} + F_3\mathbf{k}$ and suppose each F_k satisfies the homogeneity condition

$$F_k(tx, ty, tz) = t F_k(x, y, z), \quad k = 1, 2, 3.$$

Suppose also $\nabla \times \mathbf{F} = \mathbf{0}$. Prove that $\mathbf{F} = \nabla f$ where

$$2 f(x, y, z) = x F_1(x, y, z) + y F_2(x, y, z) + z F_3(x, y, z).$$

(HINT: Use Review Exercise 23, Chapter 2.)

9. Let $\mathbf{F}(x, y, z) = (e^x \sin y)\mathbf{i} + (e^x \cos y)\mathbf{j} + z^2\mathbf{k}$. Evaluate the integral $\int_{\mathbf{c}} \mathbf{F} \cdot d\mathbf{s}$, where $\mathbf{c}(t) = (\sqrt{t}, t^3, \exp \sqrt{t})$, $0 \leq t \leq 1$.

10. Let a fluid have the velocity field $\mathbf{F}(x, y, z) = xy\mathbf{i} + yz\mathbf{j} + xz\mathbf{k}$. What is the circulation around the unit circle in the xy plane? Interpret your answer.

11. The mass of the earth is approximately 6×10^{27} g and that of the sun is 330,000 times as much. The gravitational constant is 6.7×10^{-8} cm^3/s$^2 \cdot$ g. The distance of the earth from the sun is about 1.5×10^{12} cm. Compute, approximately, the work necessary to increase the distance of the earth from the sun by 1 cm.

12. (a) Show that $\int_C (x \, dy - y \, dx)/(x^2 + y^2) = 2\pi$, where C is the unit circle.
   (b) Conclude that the associated vector field $[-y/(x^2+y^2)]\mathbf{i}+[x/(x^2+y^2)]\mathbf{j}$ is not a conservative field.
   (c) Show, however, that $\partial P/\partial y = \partial Q/\partial x$. Does this contradict the corollary to Theorem 7? If not, why not?

13. Determine which of the following vector fields $\mathbf{F}$ in the plane is the gradient of a scalar function f. If such an f exists, find it.

   (a) $\mathbf{F}(x, y) = x\mathbf{i} + y\mathbf{j}$
   (b) $\mathbf{F}(x, y) = xy\mathbf{i} + xy\mathbf{j}$
   (c) $\mathbf{F}(x, y) = (x^2 + y^2)\mathbf{i} + 2xy\mathbf{j}$

14. Repeat Exercise 13 for the following vector fields:

   (a) $\mathbf{F}(x, y) = (\cos xy - xy \sin xy)\mathbf{i} - (x^2 \sin xy)\mathbf{j}$
   (b) $\mathbf{F}(x, y) = (x\sqrt{x^2 y^2 + 1})\mathbf{i} + (y\sqrt{x^2 y^2 + 1})\mathbf{j}$
   (c) $\mathbf{F}(x, y) = (2x \cos y + \cos y)\mathbf{i} - (x^2 \sin y + x \sin y)\mathbf{j}$

15. Show that the following vector fields are conservative. Calculate $\int_C \mathbf{F} \cdot d\mathbf{s}$ for the given curve.

   (a) $\mathbf{F} = (xy^2 + 3x^2 y)\mathbf{i} + (x + y)x^2\mathbf{j}$; C is the curve consisting of line segments from $(1, 1)$ to $(0, 2)$ to $(3, 0)$.
   (b) $\mathbf{F} = \dfrac{2x}{y^2 + 1}\mathbf{i} - \dfrac{2y(x^2 + 1)}{(y^2 + 1)^2}\mathbf{j}$; C is parametrized by $x = t^3 - 1$, $y = t^6 - t$, $0 \leq t \leq 1$.
   (c) $\mathbf{F} = [\cos(xy^2) - xy^2 \sin(xy^2)]\mathbf{i} - 2x^2 y \sin(xy^2)\mathbf{j}$; C is the curve (e^t, e^{t+1}), $-1 \leq t \leq 0$.

16. Prove Theorem 8. (HINT: Define $\mathbf{G} = G_1\mathbf{i} + G_2\mathbf{j} + G_3\mathbf{k}$ by

$$G_1(x, y, z) = \int_0^z F_2(x, y, t)\, dt - \int_0^y F_3(x, t, 0)\, dt$$

$$G_2(x, y, z) = -\int_0^z F_1(x, y, t)\, dt$$

and $G_3(x, y, z) = 0$.)

17. Is each of the following vector fields the curl of some other vector field? If so, find the vector field.

    (a) $\mathbf{F} = x\mathbf{i} + y\mathbf{j} + z\mathbf{k}$
    (b) $\mathbf{F} = (x^2 + 1)\mathbf{i} + (z - 2xy)\mathbf{j} + y\mathbf{k}$

18. Let $\mathbf{F} = xz\mathbf{i} - yz\mathbf{j} + y\mathbf{k}$. Verify that $\nabla \cdot \mathbf{F} = 0$. Find a $\mathbf{G}$ such that $\mathbf{F} = \nabla \times \mathbf{G}$.

19. Repeat Exercise 18 for $\mathbf{F} = y^2\mathbf{i} + z^2\mathbf{j} + x^2\mathbf{k}$.

20. Let $\mathbf{F} = xe^y\mathbf{i} - (x\cos z)\mathbf{j} - ze^y\mathbf{k}$. Find a $\mathbf{G}$ such that $\mathbf{F} = \nabla \times \mathbf{G}$.

21. Let $\mathbf{F} = (x\cos y)\mathbf{i} - (\sin y)\mathbf{j} + (\sin x)\mathbf{k}$. Find a $\mathbf{G}$ such that $\mathbf{F} = \nabla \times \mathbf{G}$.

22. By using different paths from $(0, 0, 0)$ to (x, y, z), show that the function f defined in the proof of Theorem 7 for "condition (ii) implies condition (iii)" satisfies $\partial f/\partial x = F_1$ and $\partial f/\partial y = F_2$.

23. Let $\mathbf{F}$ be the vector field on $\mathbb{R}^3$ given by $\mathbf{F} = -y\mathbf{i} + x\mathbf{j}$.

    (a) Show that $\mathbf{F}$ is rotational, that is, $\mathbf{F}$ is not irrotational.
    (b) Suppose $\mathbf{F}$ represents the velocity vector field of a fluid. Show that if we place a cork in this fluid it will revolve in a plane parallel to the xy plane, in a circular trajectory about the z axis.
    (c) In what direction does the cork revolve?

24. Let $\mathbf{G}$ be the vector field on $\mathbb{R}^3 \backslash \{z \text{ axis}\}$ defined by

$$\mathbf{G} = \frac{-y}{x^2 + y^2}\mathbf{i} + \frac{x}{x^2 + y^2}\mathbf{j}.$$

    (a) Show that $\mathbf{G}$ is irrotational.
    (b) Show that the result of Exercise 23(b) holds for $\mathbf{G}$ also.
    (c) How can we resolve the fact that the trajectories of $\mathbf{F}$ and $\mathbf{G}$ are both the same (circular about the z axis) yet $\mathbf{F}$ is rotational and $\mathbf{G}$ is not? (HINT: The property of being rotational is a local condition, that is, a property of the fluid in the neighborhood of a point.)

25. Let $\mathbf{F} = -(GmM\mathbf{r}/r^3)$ be the gravitational force field defined on $\mathbb{R}^3 \backslash \{\mathbf{0}\}$.

    (a) Show that div $\mathbf{F} = 0$.
    (b) Show that $\mathbf{F} \neq$ curl $\mathbf{G}$ for any C^1 vector field $\mathbf{G}$ on $\mathbb{R}^3 \backslash \{\mathbf{0}\}$.

8.4 GAUSS' THEOREM

Gauss' theorem states that the flux of a vector field out of a closed surface equals the integral of the divergence of that vector field over the volume enclosed by the surface.

The result parallels Stokes' theorem and Green's theorem in that it relates an integral over a closed geometrical object (curve or surface) to an integral over a contained region (surface or volume).

We shall begin by asking the reader to review the various elementary regions in space that were introduced when we considered the volume integral; these regions are illustrated in Figure 5.6.1. As that figure indicates, the boundary of an elementary region in $\mathbb{R}^3$ is a surface made up of a finite number (at most six, at least two) of surfaces that can be described as graphs of functions from $\mathbb{R}^2$ to $\mathbb{R}$. This kind of surface is called a *closed surface*. The surfaces $S_1, S_2, \ldots, S_N$ composing such a closed surface are called its *faces*.

EXAMPLE 1   The cube in Figure 8.4.1(a) is an elementary region, and in fact a symmetric elementary region, with six rectangles composing its boundary. The sphere in Figure 8.4.1(b) is the boundary of a solid ball, which is also a symmetric elementary region.   ▲

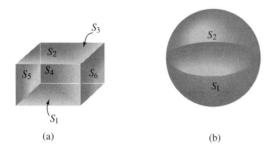

Figure 8.4.1   (a) Symmetric elementary regions and (b) the surface S_i composing their boundaries.

(a)   (b)

Closed surfaces can be oriented in two ways. The outward orientation makes the normal point outward into space, and the inward orientation makes the normal point into the bounded region (Figure 8.4.2).

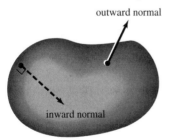

Figure 8.4.2   Two possible orientations for a closed surface.

Suppose S is a closed surface oriented in one of these two ways and $\mathbf{F}$ is a vector field on S. Then, as we defined it in Section 7.6,

$$\iint_S \mathbf{F} \cdot d\mathbf{S} = \sum_i \iint_{S_i} \mathbf{F} \cdot d\mathbf{S}.$$

If S is given the outward orientation, the integral $\iint_S \mathbf{F} \cdot d\mathbf{S}$ measures the total flux of $\mathbf{F}$ outward across S. That is, if we think of $\mathbf{F}$ as the velocity field of a fluid, $\iint_S \mathbf{F} \cdot d\mathbf{S}$

indicates the amount of fluid leaving the region bounded by S per unit time. If S is given the inward orientation, the integral $\iint_S \mathbf{F} \cdot d\mathbf{S}$ measures the total flux of $\mathbf{F}$ inward across S.

We recall another common way of writing these surface integrals, a way that explicitly specifies the orientation of S. Let the orientation of S be given by a unit normal vector $\mathbf{n}(x, y, z)$ at each point of S. Then we have the oriented integral

$$\iint_S \mathbf{F} \cdot d\mathbf{S} = \iint_S (\mathbf{F} \cdot \mathbf{n}) dS,$$

that is, the integral of the normal component of $\mathbf{F}$ over S. In the remainder of this section, if S is a closed surface enclosing a region Ω, we adopt the default convention that $S = \partial\Omega$ is given the outward orientation, with outward unit normal $\mathbf{n}(x, y, z)$ at each point $(x, y, z) \in S$. Furthermore, we denote the surface with the opposite (inward) orientation by $\partial\Omega_{op}$. Then the associated unit normal direction for this orientation is $-\mathbf{n}$. Thus

$$\iint_{\partial\Omega} \mathbf{F} \cdot d\mathbf{S} = \iint_S (\mathbf{F} \cdot \mathbf{n}) \, dS = -\iint_S [\mathbf{F} \cdot (-\mathbf{n})] \, dS = -\iint_{\partial\Omega_{op}} \mathbf{F} \cdot d\mathbf{S}.$$

EXAMPLE 2   The unit cube Ω given by

$$0 \le x \le 1, \qquad 0 \le y \le 1, \qquad 0 \le z \le 1$$

is a symmetric elementary region in space (see Figures 8.4.3 and 5.6.5).

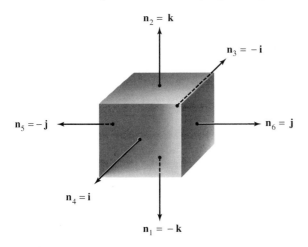

Figure 8.4.3 The outward orientation on the cube.

We write the faces as

$$S_1: z = 0, \qquad 0 \le x \le 1, \qquad 0 \le y \le 1$$

$$S_2: z = 1, \qquad 0 \le x \le 1, \qquad 0 \le y \le 1$$

$$S_3: x = 0, \qquad 0 \le y \le 1, \qquad 0 \le z \le 1$$

$$S_4: x = 1, \qquad 0 \le y \le 1, \qquad 0 \le z \le 1$$

$$S_5: y = 0, \qquad 0 \le x \le 1, \qquad 0 \le z \le 1$$

$$S_6: y = 1, \qquad 0 \le x \le 1, \qquad 0 \le z \le 1$$

From Figure 8.4.3 we see that

$$\mathbf{n}_2 = \mathbf{k} = -\mathbf{n}_1,$$

$$\mathbf{n}_4 = \mathbf{i} = -\mathbf{n}_3,$$

$$\mathbf{n}_6 = \mathbf{j} = -\mathbf{n}_5,$$

and so for a continuous vector field $\mathbf{F} = F_1\mathbf{i} + F_2\mathbf{j} + F_3\mathbf{k}$,

$$\iint_{\partial\Omega} \mathbf{F} \cdot d\mathbf{S} = \iint_S \mathbf{F} \cdot \mathbf{n}\, dS = - \iint_{S_1} F_3\, dS + \iint_{S_2} F_3\, dS - \iint_{S_3} F_1\, dS$$

$$+ \iint_{S_4} F_1\, dS - \iint_{S_5} F_2\, dS + \iint_{S_6} F_2\, dS. \quad \blacktriangle$$

We have now come to the last of the three central theorems of this chapter. This theorem relates surface integrals to volume integrals; in other words, the theorem states that if Ω is a region in $\mathbb{R}^3$, then the flux of a vector field $\mathbf{F}$ outward across the closed surface $\partial\Omega$ is equal to the integral of div $\mathbf{F}$ over Ω. We begin by assuming Ω is a symmetric elementary region (Figure 5.6.5).

THEOREM 9: GAUSS' DIVERGENCE THEOREM   *Let Ω be a symmetric elementary region in space. Denote by $\partial\Omega$ the oriented closed surface that bounds Ω. Let $\mathbf{F}$ be a smooth vector field defined on Ω. Then*

$$\iiint_\Omega (\nabla \cdot \mathbf{F})\, dV = \iint_{\partial\Omega} \mathbf{F} \cdot d\mathbf{S}$$

or, alternatively,

$$\iiint_\Omega (\text{div } \mathbf{F})\, dV = \iint_{\partial\Omega} (\mathbf{F} \cdot \mathbf{n})\, dS.$$

PROOF   If $\mathbf{F} = P\mathbf{i} + Q\mathbf{j} + R\mathbf{k}$, then by definition, div $\mathbf{F} = \partial P/\partial x + \partial Q/\partial y + \partial R/\partial z$, so we can write (using additivity of the volume integral)

$$\iiint_\Omega \text{div } \mathbf{F}\, dV = \iiint_\Omega \frac{\partial P}{\partial x}\, dV + \iiint_\Omega \frac{\partial Q}{\partial y}\, dV + \iiint_\Omega \frac{\partial R}{\partial z}\, dV.$$

On the other hand, the surface integral in question is

$$\iint_{\partial\Omega} \mathbf{F} \cdot \mathbf{n}\, dS = \iint_{\partial\Omega} (P\mathbf{i} + Q\mathbf{j} + R\mathbf{k}) \cdot \mathbf{n}\, dS$$

$$= \iint_{\partial\Omega} P\mathbf{i} \cdot \mathbf{n}\, dS + \iint_{\partial\Omega} Q\mathbf{j} \cdot \mathbf{n}\, dS + \iint_{\partial\Omega} R\mathbf{k} \cdot \mathbf{n}\, dS.$$

The theorem will follow if we establish the three equalities

$$\iint_{\partial\Omega} P\mathbf{i} \cdot \mathbf{n}\, dS = \iiint_{\Omega} \frac{\partial P}{\partial x}\, dV, \tag{1}$$

$$\iint_{\partial\Omega} Q\mathbf{j} \cdot \mathbf{n}\, dS = \iiint_{\Omega} \frac{\partial Q}{\partial y}\, dV, \tag{2}$$

and

$$\iint_{\partial\Omega} R\mathbf{k} \cdot \mathbf{n}\, dS = \iiint_{\Omega} \frac{\partial R}{\partial z}\, dV. \tag{3}$$

We shall prove equation (3); the other two equalities can be proved in an analogous fashion.

Since Ω is a symmetric elementary region, there is a pair of functions

$$z = f_1(x, y), \qquad z = f_2(x, y),$$

with common domain an elementary region D in the xy plane, such that Ω is the set of all points (x, y, z) satisfying

$$f_1(x, y) \leq z \leq f_2(x, y), \qquad (x, y) \in D.$$

By reduction to iterated integrals, we have

$$\iiint_{\Omega} \frac{\partial R}{\partial z}\, dV = \iint_{D} \left(\int_{z=f_1(x,y)}^{z=f_2(x,y)} \frac{\partial R}{\partial z}\, dz \right) dx\, dy,$$

and so

$$\iiint_{\Omega} \frac{\partial R}{\partial z}\, dV = \iint_{D} [R(x, y, f_2(x, y)) - R(x, y, f_1(x, y))]\, dx\, dy. \tag{4}$$

The boundary of Ω is a closed surface whose top S_2 is the graph of $z = f_2(x, y)$, where $(x, y) \in D$, and whose bottom S_1 is the graph of $z = f_1(x, y)$, $(x, y) \in D$. The four other sides of $\partial\Omega$ consist of surfaces S_3, S_4, S_5, and S_6, whose normals are always perpendicular to the z axis. (See Figure 8.4.4. Note that some of the other four sides might be absent—for instance, if Ω is a solid ball and $\partial\Omega$ is a sphere.) By definition,

$$\iint_{\partial\Omega} R\mathbf{k} \cdot \mathbf{n}\, dS = \iint_{S_1} R\mathbf{k} \cdot \mathbf{n}_1\, dS + \iint_{S_2} R\mathbf{k} \cdot \mathbf{n}_2\, dS + \sum_{i=3}^{6} \iint_{S_i} R\mathbf{k} \cdot \mathbf{n}_i\, dS.$$

$z = f_2(x, y)$

S_2

S_4

S_5

S_6

S_3

S_1

$z = f_1(x, y)$

Figure 8.4.4 A symmetric elementary region Ω for which $\iint_{\partial\Omega} R\mathbf{k} \cdot d\mathbf{S} = \iiint_{\Omega} (\partial R/\partial z)\, dV$. The four sides of $\partial\Omega$, which are S_3, S_4, S_5, S_6, have normals perpendicular to the z axis.

Since on each of S_3, S_4, S_5, and S_6 the normal $\mathbf{n}_i$ is perpendicular to $\mathbf{k}$, we have $\mathbf{k} \cdot \mathbf{n} = 0$ along these faces, and so the integral reduces to

$$\iint_{\partial\Omega} R\mathbf{k} \cdot \mathbf{n}\, ds = \iint_{S_1} R\mathbf{k} \cdot \mathbf{n}_1\, dS + \iint_{S_2} R\mathbf{k} \cdot \mathbf{n}_2\, dS. \tag{5}$$

The surface S_1 is defined by $z = f_1(x, y)$, and so

$$\mathbf{n}_1 = \frac{\dfrac{\partial f_1}{\partial x}\mathbf{i} + \dfrac{\partial f_1}{\partial y}\mathbf{j} - \mathbf{k}}{\sqrt{\left(\dfrac{\partial f_1}{\partial x}\right)^2 + \left(\dfrac{\partial f_1}{\partial y}\right)^2 + 1}}$$

(since S_1 is the bottom portion of Ω, for $\mathbf{n}_1$ to point outward it must have a negative $\mathbf{k}$ component; see Example 2). Thus

$$\mathbf{n}_1 \cdot \mathbf{k} = \frac{-1}{\sqrt{\left(\dfrac{\partial f_1}{\partial x}\right)^2 + \left(\dfrac{\partial f_1}{\partial y}\right)^2 + 1}}$$

and so

$$\iint_{S_1} R(\mathbf{k} \cdot \mathbf{n}_1)\, dS$$

$$= \iint_D R(x, y, f_1(x, y)) \frac{-1}{\sqrt{\left(\dfrac{\partial f_1}{\partial x}\right)^2 + \left(\dfrac{\partial f_1}{\partial y}\right)^2 + 1}} \left(\sqrt{\left(\dfrac{\partial f_1}{\partial x}\right)^2 + \left(\dfrac{\partial f_1}{\partial y}\right)^2 + 1}\right) dA$$

$$= -\iint_D R(x, y, f_1(x, y))\, dx\, dy. \tag{6}$$

This equation also follows from formula (4), Section 7.5.

Similarly, on the top face S_2 we have

$$\mathbf{k} \cdot \mathbf{n}_2 = \frac{1}{\sqrt{\left(\dfrac{\partial f_2}{\partial x}\right)^2 + \left(\dfrac{\partial f_2}{\partial y}\right)^2 + 1}},$$

and so

$$\iint_{S_1} R(\mathbf{k} \cdot \mathbf{n}_2)\, dS = \iint_D R(x, y, f_2(x, y))\, dx\, dy. \tag{7}$$

Substituting equations (6) and (7) into equation (5) and then comparing with equation (4), we obtain

$$\iiint_\Omega \frac{\partial R}{\partial z}\, dV = \iint_{\partial \Omega} R(\mathbf{k} \cdot \mathbf{n})\, dS.$$

The remaining equalities, (1) and (2), can be established in the same way to complete the proof. ■

The reader should note that the proof is similar to that of Green's theorem. By the procedure used in Exercise 8 of Section 8.1, we can extend Gauss' theorem to any region that can be broken up into symmetric elementary regions. This includes all regions of interest to us. As an example, consider the region between two closed surfaces, one inside the other. The surface of this region consists of two pieces oriented as shown in Figure 8.4.5. We shall apply the divergence theorem to such a region when we prove Gauss' law in Theorem 10, below.

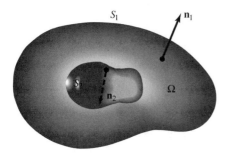

Figure 8.4.5 A more general region to which Gauss' theorem applies.

EXAMPLE 3   Consider $\mathbf{F} = 2x\mathbf{i} + y^2\mathbf{j} + z^2\mathbf{k}$. Let S be the unit sphere defined by $x^2 + y^2 + z^2 = 1$. Evaluate $\iint_S \mathbf{F} \cdot \mathbf{n}\, dS$.

SOLUTION   By Gauss' theorem,

$$\iint_S \mathbf{F} \cdot \mathbf{n}\, dS = \iiint_\Omega (\operatorname{div} \mathbf{F})\, dV,$$

where Ω is the ball bounded by the sphere. The integral on the right is

$$2 \iiint_{\Omega} (1 + y + z) \, dV = 2 \iiint_{\Omega} dV + 2 \iiint_{\Omega} y \, dV + 2 \iiint_{\Omega} z \, dV.$$

By symmetry, we can argue that $\iiint_{\Omega} y \, dV = \iiint_{\Omega} z \, dV = 0$ (see Exercise 17, Section 6.3). Thus, since a sphere of radius R has volume $4\pi R^3/3$,

$$2 \iiint_{\Omega} (1 + y + z) \, dV = 2 \iiint_{\Omega} dV = \frac{8\pi}{3}.$$

Readers can convince themselves that direct computation of $\iint_S \mathbf{F} \cdot \mathbf{n} \, dS$ proves unwieldy.   ▲

EXAMPLE 4   Use the divergence theorem to evaluate

$$\iint_{\partial W} (x^2 + y + z) \, dS,$$

where W is the solid ball $x^2 + y^2 + z^2 \le 1$.

SOLUTION   To apply Gauss' divergence theorem, we find a vector field $\mathbf{F} = F_1 \mathbf{i} + F_2 \mathbf{j} + F_3 \mathbf{k}$ on W with $\mathbf{F} \cdot \mathbf{n} = x^2 + y + z$. At any point $(x, y, z) \in \partial W$, the outward unit normal $\mathbf{n}$ to ∂W is

$$\mathbf{n} = x\mathbf{i} + y\mathbf{j} + z\mathbf{k},$$

since on ∂W, $x^2 + y^2 + z^2 = 1$ and the radius vector $\mathbf{r} = x\mathbf{i} + y\mathbf{j} + z\mathbf{k}$ is normal to the sphere ∂W (Figure 8.4.6).

Figure 8.4.6 $\mathbf{n}$ is the unit normal to ∂W, the boundary of the ball W.

Therefore, if $\mathbf{F}$ is the desired vector field, then

$$\mathbf{F} \cdot \mathbf{n} = F_1 x + F_2 y + F_2 z.$$

We set

$$F_1 x = x^2, \qquad F_2 y = y, \qquad F_3 z = z$$

and solve for F_1, F_2, and F_3 to find that $\mathbf{F} = x\mathbf{i} + \mathbf{j} + \mathbf{k}$. Computing div $\mathbf{F}$, we get

$$\text{div } \mathbf{F} = 1 + 0 + 0 = 1.$$

Thus, by Gauss' divergence theorem,

$$\iint_{\partial W} (x^2 + y + z) \, dS = \iiint_W dV = \text{volume } (W) = \frac{4}{3}\pi. \quad \blacktriangle$$

The physical meaning of divergence is that at a point P, div $\mathbf{F}(P)$ is the rate of net outward flux at P per unit volume. This follows from Gauss' theorem and the mean value theorem for integrals (as well as from the supplement to Section 4.4): If Ω_ρ is a ball in $\mathbb{R}^3$ of radius ρ centered at P, then there is a point $Q \in \Omega_\rho$ such that

$$\iint_{\partial \Omega_\rho} \mathbf{F} \cdot \mathbf{n} \, dS = \iiint_{\Omega_\rho} \text{div } \mathbf{F} \, dV = \text{div } \mathbf{F}(Q) \cdot \text{volume } (\Omega_\rho)$$

and so

$$\text{div } \mathbf{F}(P) = \underset{\rho \to 0}{\text{limit}} \, \text{div } \mathbf{F}(Q) = \underset{\rho \to 0}{\text{limit}} \, \frac{1}{V(\Omega_\rho)} \iint_{\partial \Omega_\rho} \mathbf{F} \cdot \mathbf{n} \, dS.$$

This is analogous to the limit formulation of the curl given at the end of Section 8.2. Thus, if div $\mathbf{F}(P) > 0$, we consider P to be a ***source***, for there is a net outward flow near P. If div $\mathbf{F}(P) < 0$, P is called a ***sink*** for $\mathbf{F}$.

A C^1 vector field $\mathbf{F}$ defined on $\mathbb{R}^3$ is said to be ***divergence free*** if div $\mathbf{F} = 0$. If $\mathbf{F}$ is divergence-free, we have $\iint_S \mathbf{F} \cdot d\mathbf{S} = 0$ for all closed surfaces S. The converse can also be demonstrated readily using Gauss' theorem: If $\iint_S \mathbf{F} \cdot d\mathbf{S} = 0$ for all closed surfaces S, then $\mathbf{F}$ is divergence-free. If $\mathbf{F}$ is divergence-free, we thus see that the flux of $\mathbf{F}$ across any closed surface S is 0, so that if $\mathbf{F}$ is the velocity field of a fluid, the net amount of fluid which flows out of any region will be 0. Thus, exactly as much fluid must flow into the region as flows out (in unit time). A fluid with this property is therefore described as ***incompressible.*** (Additional justification for this terminology is given in Exercise 22.)

EXAMPLE 5 Evaluate $\iint_S \mathbf{F} \cdot d\mathbf{S}$, where $\mathbf{F}(x, y, z) = xy^2\mathbf{i} + x^2 y\mathbf{j} + y\mathbf{k}$ and S is the surface of the cylinder $x^2 + y^2 = 1$, bounded by the planes $z = 1$ and $z = -1$ and including the portions $x^2 + y^2 \leq 1$ when $z = \pm 1$.

SOLUTION   One can compute this integral directly but it is easier to use the divergence theorem.

Now S is the boundary of the region Ω given by $x^2 + y^2 \leq 1, -1 \leq z \leq 1$. Thus $\iint_S \mathbf{F} \cdot d\mathbf{S} = \iiint_\Omega (\text{div } \mathbf{F}) \, dV$. Moreover,

$$\iiint_\Omega (\text{div } \mathbf{F}) \, dV = \iiint_\Omega (x^2 + y^2) \, dx \, dy \, dz = \int_{-1}^{1} \left(\iint_{x^2+y^2 \leq 1} (x^2 + y^2) \, dx \, dy \right) dz$$

$$= 2 \iint_{x^2+y^2 \leq 1} (x^2 + y^2) \, dx \, dy.$$

Before evaluating the double integral, we note that the surface integral satisfies

$$\iint_{\partial \Omega} \mathbf{F} \cdot \mathbf{n} \, dS = 2 \iint_{x^2+y^2 \leq 1} (x^2 + y^2) \, dx \, dy > 0.$$

This means that $\iint_{\partial \Omega} \mathbf{F} \cdot d\mathbf{S}$, the net flux of $\mathbf{F}$ out of the cylinder, is positive.

We change variables to polar coordinates to evaluate the double integral:

$$x = r \cos \theta, \qquad y = r \sin \theta, \qquad 0 \leq r \leq 1, \qquad 0 \leq \theta \leq 2\pi.$$

Hence, we have $\partial(x, y)/\partial(r, \theta) = r$ and $x^2 + y^2 = r^2$. Thus

$$\iint_{x^2+y^2 \leq 1} (x^2 + y^2) \, dx \, dy = \int_0^{2\pi} \left(\int_0^1 r^3 \, dr \right) d\theta = \frac{1}{2} \pi.$$

Therefore, $\iiint_\Omega \text{div } \mathbf{F} \, dV = \pi$.   ▲

As we remarked above, Gauss' divergence theorem can be applied to regions in space more general than symmetric elementary regions. To conclude this section, we shall use this observation *in the proof* of the following important results.

THEOREM 10: GAUSS' LAW   *Let M be a symmetric elementary region in $\mathbb{R}^3$. Then if $(0, 0, 0) \notin \partial M$, we have*

$$\iint_{\partial M} \frac{\mathbf{r} \cdot \mathbf{n}}{r^3} \, dS = \begin{cases} 4\pi & \text{if } (0, 0, 0) \in M \\ 0 & \text{if } (0, 0, 0) \notin M \end{cases}$$

where

$$\mathbf{r}(x, y, z) = x\mathbf{i} + y\mathbf{j} + z\mathbf{k}$$

and

$$r(x, y, z) = \|\mathbf{r}(x, y, z)\| = \sqrt{x^2 + y^2 + z^2}.$$

PROOF OF GAUSS' LAW   First suppose $(0, 0, 0) \notin M$. Then $\mathbf{r}/r^3$ is a C^1 vector field on M and ∂M, and so by the divergence theorem,

$$\iint_{\partial M} \frac{\mathbf{r} \cdot \mathbf{n}}{r^3} \, dS = \iiint_M \nabla \cdot \left(\frac{\mathbf{r}}{\mathbf{r}^3} \right) dV.$$

But $\nabla \cdot (\mathbf{r}/r^3) = 0$ for $r \neq 0$, as the reader can easily verify (see Exercise 30, Section 4.4). Thus

$$\iint_{\partial M} \frac{\mathbf{r} \cdot \mathbf{n}}{r^3} \, dS = 0.$$

Now let us suppose $(0, 0, 0) \in M$. We can no longer use the above method, because $\mathbf{r}/r^3$ is not smooth on M, on account of the zero denominator at $\mathbf{r} = (0, 0, 0)$. Since $(0, 0, 0) \in M$ and $(0, 0, 0) \notin \partial M$, there is an $\varepsilon > 0$ such that the ball N of radius ε centered at $(0, 0, 0)$ is contained completely inside M. Now let Ω be the region between M and N. Then Ω has boundary $\partial N \cup \partial M = S$. But the orientation on ∂N induced by the *outward* normal on Ω is the opposite of that obtained from N (see Figure 8.4.7).

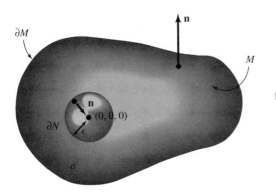

Figure 8.4.7   Induced outward orientation on S.

Now $\nabla \cdot (\mathbf{r}/r^3) = 0$ on Ω, and so, by the divergence theorem applied to the (nonelementary) region Ω,

$$\iint_S \frac{\mathbf{r} \cdot \mathbf{n}}{r^3} \, dS = \iiint_\Omega \nabla \cdot \left(\frac{\mathbf{r}}{r^3} \right) dV = 0.$$

Since

$$\iint_S \frac{\mathbf{r} \cdot \mathbf{n}}{r^3} \, dS = \iint_{\partial M} \frac{\mathbf{r} \cdot \mathbf{n}}{r^3} \, dS + \iint_{\partial N} \frac{\mathbf{r} \cdot \mathbf{n}}{r^3} \, dS,$$

where $\mathbf{n}$ is the outward normal to S, we have

$$\iint_{\partial M} \frac{\mathbf{r} \cdot \mathbf{n}}{r^3} \, dS = - \iint_{\partial N} \frac{\mathbf{r} \cdot \mathbf{n}}{r^3} \, dS.$$

However, on ∂N, $\mathbf{n} = -\mathbf{r}/r$ and $r = \varepsilon$, since ∂N is a sphere of radius ε, so that

$$-\iint_{\partial N} \frac{\mathbf{r} \cdot \mathbf{n}}{r^3} \, dS = \iint_{\partial N} \frac{\varepsilon^2}{\varepsilon^4} \, dS = \frac{1}{\varepsilon^2} \iint_{\partial N} \, dS.$$

But $\iint_{\partial N} dS = 4\pi \varepsilon^2$, the surface area of the sphere of radius ε. This proves the result. ∎

EXAMPLE 6   Gauss' law has the following physical interpretation. The potential due to a point charge Q at $(0, 0, 0)$ is given by

$$\phi(x, y, z) = \frac{Q}{4\pi r} = \frac{Q}{4\pi \sqrt{x^2 + y^2 + z^2}}$$

and the corresponding electric field is

$$\mathbf{E} = -\nabla \phi = \frac{Q}{4\pi} \left(\frac{\mathbf{r}}{r^3} \right).$$

Thus Theorem 10 states that the total electric flux $\iint_{\partial M} \mathbf{E} \cdot d\mathbf{S}$ (that is, the flux of $\mathbf{E}$ out of a closed surface ∂M) equals Q if the charge lies inside M and zero otherwise. Note that even if $(0, 0, 0) \notin M$, $\mathbf{E}$ will still be *nonzero* on M.

For a continuous charge distribution described by a charge density ρ, the field $\mathbf{E}$ is related to the density ρ by

$$\operatorname{div} \mathbf{E} = \nabla \cdot \mathbf{E} = \rho.$$

Thus, by Gauss' theorem,

$$\iint_S \mathbf{E} \cdot d\mathbf{S} = \iiint_\Omega \rho \, dV = Q,$$

or the flux out of a surface is equal to the total charge inside. ▲

We next use Gauss' theorem to derive the formula

$$\operatorname{div} \mathbf{F} = \frac{1}{\rho^2} \frac{\partial}{\partial \rho} (\rho^2 F_\rho) + \frac{1}{\rho \sin \phi} \frac{\partial}{\partial \phi} (\sin \phi \, F_\phi) + \frac{1}{\rho \sin \phi} \frac{\partial F_\theta}{\partial \theta} \tag{8}$$

for the divergence of a vector field $\mathbf{F}$ in spherical coordinates, which was stated in Section 8.2. (Again, the subscripts here denote *components*, *not* partial derivatives.) The method is to use the formula

$$\operatorname{div} \mathbf{F}(\mathrm{P}) = \lim_{\Omega \to \mathrm{P}} \frac{1}{V(\Omega)} \iint_{\partial \Omega} \mathbf{F} \cdot \mathbf{n} \, dS, \tag{9}$$

where Ω is a region with volume $V(\Omega)$ which shrinks down to a point P (in the main text we took a ball Ω_ρ, but one can use regions of any shape). Let Ω be the shaded region in Figure 8.4.8.

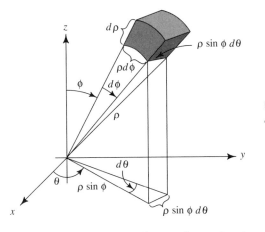

Figure 8.4.8 Infinitestimal volume determined by $d\rho, d\theta, d\phi$ at (ρ, θ, ϕ).

For the two faces orthogonal to the radial direction, the surface integral in equation (9) is, approximately,

$$F_\rho(\rho + d\rho, \phi, \theta) \times \text{ (area of outer face)} - F_\rho(\rho, \phi, \theta) \times \text{ (area of inner face)}$$

$$\approx F_\rho(\rho + d\rho, \phi, \theta)(\rho + d\phi)^2 \sin\phi \, d\phi \, d\theta - F_\rho(\rho, \phi, \theta)\rho^2 \sin\phi \, d\phi \, d\theta$$

$$\approx \frac{\partial}{\partial\rho}(F_\rho \rho^2 \sin\phi) \, d\rho \, d\phi \, d\theta \qquad (10)$$

by the one-variable mean value theorem. Dividing by the volume of the region Ω, namely, $\rho^2 \sin\phi \, d\rho \, d\phi \, d\theta$, we see that the contribution to the right-hand side of equation (9) is

$$\frac{1}{\rho^2} \frac{\partial}{\partial\rho}(\rho^2 F_\rho) \qquad (11)$$

for these faces. Likewise, the contribution from the faces orthogonal to the ϕ direction is

$$\frac{1}{\rho \sin\phi} \frac{\partial}{\partial\phi}(\sin\phi \, F_\phi), \quad \text{and for the } \theta \text{ direction,} \quad \frac{1}{\rho \sin\phi} \frac{\partial F_\theta}{\partial\theta}.$$

Substituting (11) and these expressions in equation (9) and taking the limit gives equation (8).

EXERCISES

1. Let S be a closed surface. Use Gauss' theorem to show that if $\mathbf{F}$ is a C^2 vector field, then we have $\iint_S (\nabla \times \mathbf{F}) \cdot d\mathbf{S} = 0$. (Compare with Exercise 14 of Section 8.2.)

2. Let $\mathbf{F} = x^3\mathbf{i} + y^3\mathbf{j} + z^3\mathbf{k}$. Evaluate the surface integral of $\mathbf{F}$ over the unit sphere.

3. Evaluate $\iint_{\partial\Omega} \mathbf{F} \cdot d\mathbf{S}$, where $\mathbf{F} = x\mathbf{i} + y\mathbf{j} + z\mathbf{k}$ and Ω is the unit cube (in the first octant). Perform the calculation directly and check by using the divergence theorem.

4. Repeat Exercise 3 for

(a) $\mathbf{F} = \mathbf{i} + \mathbf{j} + \mathbf{k}$

(b) $\mathbf{F} = x^2\mathbf{i} + x^2\mathbf{j} + z^2\mathbf{k}$

5. Let $\mathbf{F} = y\mathbf{i} + z\mathbf{j} + xz\mathbf{k}$. Evaluate $\iint_{\partial\Omega} \mathbf{F} \cdot d\mathbf{S}$ for each of the following regions Ω:

    (a) $x^2 + y^2 \leq z \leq 1$
    (b) $x^2 + y^2 \leq z \leq 1$ and $x \geq 0$
    (c) $x^2 + y^2 \leq z \leq 1$ and $x \leq 0$

6. Repeat Exercise 5 for $\mathbf{F} = (x - y)\mathbf{i} + (y - z)\mathbf{j} + (z - x)\mathbf{k}$. [The solution to part (b) only is in the Study Guide to this text.]

7. Let S be the surface of region Ω. Show that

$$\iint_S \mathbf{r} \cdot \mathbf{n}\, dS = 3 \text{ volume } (\Omega).$$

Attempt to explain this geometrically. (HINT: Assume $(0, 0, 0) \in \Omega$ and consider the skew cone with its vertex at $(0, 0, 0)$ with base ΔS and altitude $\|\mathbf{r}\|$. Its volume is $\frac{1}{3}(\Delta S)(\mathbf{r} \cdot \mathbf{n})$.)

8. Evaluate $\iint_S \mathbf{F} \cdot d\mathbf{S}$, where $\mathbf{F} = 3xy^2\mathbf{i} + 3x^2y\mathbf{j} + z^3\mathbf{k}$ and S is the surface of the unit sphere.

9. Evaluate $\iint_{\partial W} \mathbf{F} \cdot \mathbf{n}\, dA$, where $\mathbf{F}(x, y, z) = x\mathbf{i} + y\mathbf{j} - z\mathbf{k}$ and W is the unit cube in the first octant. Perform the calculation directly and check by using the divergence theorem.

10. Evaluate the surface integral $\iint_{\partial S} \mathbf{F} \cdot \mathbf{n}\, dA$, where $\mathbf{F}(x, y, z) = \mathbf{i} + \mathbf{j} + z(x^2 + y^2)^2\mathbf{k}$ and ∂S is the surface of the cylinder $x^2 + y^2 \leq 1, 0 \leq z \leq 1$.

11. Prove that

$$\iiint_W (\nabla f) \cdot \mathbf{F}\, dx\, dy\, dz = \iint_{\partial W} f\mathbf{F} \cdot \mathbf{n}\, dS - \iiint_W f\nabla \cdot \mathbf{F}\, dx\, dy\, dz.$$

12. Prove the identity

$$\nabla \cdot (\mathbf{F} \times \mathbf{G}) = \mathbf{G} \cdot (\nabla \times \mathbf{F}) - \mathbf{F} \cdot (\nabla \times \mathbf{G}).$$

13. Show that $\iiint_\Omega (1/r^2)\, dx\, dy\, dz = \iint_{\partial\Omega} (\mathbf{r} \cdot \mathbf{n}/r^2)\, dS$ where $\mathbf{r} = x\mathbf{i} + y\mathbf{j} + z\mathbf{k}$.

14. Fix vectors $\mathbf{v}_1, \ldots, \mathbf{v}_k \in \mathbb{R}^3$ and numbers ("charges") $q_1, \ldots, q_k$. Define the function ϕ by $\phi(x, y, z) = \sum_{i=1}^k q_i/(4\pi \|\mathbf{r} - \mathbf{v}_i\|)$, where $\mathbf{r} = (x, y, z)$. Show that for a closed surface S and $\mathbf{E} = -\nabla\phi$,

$$\iint_S \mathbf{E} \cdot d\mathbf{S} = Q,$$

where Q is the total charge inside S. (Assume that Gauss' law from Theorem 10 applies and that none of the charges are on S.)

15. Prove *Green's identities*

$$\iint_{\partial\Omega} f\nabla g \cdot \mathbf{n}\, dS = \iiint_\Omega (f\nabla^2 g + \nabla f \cdot \nabla g)\, dV$$

and

$$\iint_{\partial\Omega} (f\nabla g - g\nabla f) \cdot \mathbf{n}\, dS = \iiint_\Omega (f\nabla^2 g - g\nabla^2 f)\, dV.$$

16. Suppose $\mathbf{F}$ satisfies div $\mathbf{F} = 0$ and curl $\mathbf{F} = \mathbf{0}$ on all of $\mathbb{R}^3$. Show that we can write $\mathbf{F} = \nabla f$, where $\nabla^2 f = 0$.

17. Let ρ be a continuous function on $\mathbb{R}^3$ such that $\rho(\mathbf{q}) = 0$ except for $\mathbf{q}$ in some region Ω. Let $\mathbf{q} \in \Omega$ be denoted by $\mathbf{q} = (x, y, z)$. The ***potential*** of ρ is defined as the function

$$\phi(\mathbf{p}) = \iiint_\Omega \frac{\rho(\mathbf{q})}{4\pi \|\mathbf{p} - \mathbf{q}\|} \, dV(\mathbf{q}),$$

where $\|\mathbf{p} - \mathbf{q}\|$ is the distance between $\mathbf{p}$ and $\mathbf{q}$.

(a) Using the method of Theorem 10, show that $\iint_{\partial W} \nabla \phi \cdot \mathbf{n} \, dS = -\iiint_W \rho \, dV$ for those regions W that can be partitioned into a finite union of symmetric elementary regions.

(b) Show that ϕ satisfies ***Poisson's equation***

$$\nabla^2 \phi = -\rho.$$

[HINT: Use part (a).] (Notice that if ρ is a charge density, then the integral defining ϕ may be thought of as the sum of the potential at $\mathbf{p}$ caused by point charges distributed over Ω according to the density ρ.)

18. Suppose $\mathbf{F}$ is tangent to the closed surface $S = \partial \Omega$ of a region Ω. Prove that

$$\iiint_\Omega (\text{div } \mathbf{F}) \, dV = 0.$$

19. Use Gauss' law and symmetry to prove that the electric field due to a charge Q evenly spread over the surface of a sphere is the same outside the surface as the field from a point charge Q located at the center of the sphere. What is the field inside the sphere?

20. Reformulate Exercise 19 in terms of gravitational fields.

21. Show how Gauss' law can be used to solve part (b) of Exercise 25 in Section 8.3.

22. *(Transport theorem.)* Let $\phi(\mathbf{x}, t)$ be the flow of the vector field $\mathbf{F}$ on $\mathbb{R}^3$ (see Section 4.4), and let $J(\mathbf{x}, t)$ be the Jacobian of the map $\phi_t \colon \mathbf{x} \mapsto \phi(\mathbf{x}, t)$ for t fixed.

(a) Using the proof of Theorem 3, Section 4.4, show that

$$\frac{\partial}{\partial t} J(\mathbf{x}, t) = [\text{div } \mathbf{F}(\mathbf{x})] J(\mathbf{x}, t).$$

(b) Using the change of variables theorem and part (a), show that if $f(x, y, z, t)$ is a given function and $\Omega \subset \mathbb{R}^3$ is any region, then the ***transport equation*** holds:

$$\frac{d}{dt} \iiint_{\Omega_t} f(x, y, z, t) \, dx \, dy \, dz = \iiint_{\Omega_t} \left(\frac{Df}{Dt} + f \text{div } \mathbf{F} \right) dx \, dy \, dz$$

where $\Omega_t = \phi_t(\Omega)$, which is the region moving with the flow, and where $Df/Dt = \partial f/\partial t + \mathbf{D_x} f \cdot \mathbf{F}$ is the material derivative.

(c) Taking $f = 1$ in part (b), show that the following assertions are equivalent (which justifies the use of the term incompressible):

  (i) div $\mathbf{F} = 0$

  (ii) volume (Ω_t) = volume (Ω)

  (iii) $J(\mathbf{x}, t) = 1$

23. Let ϕ, J, $\mathbf{F}$, f be as in Exercise 22. Prove the vector form of the transport theorem, namely,

$$\frac{d}{dt} \iiint_{\Omega_t} (f\mathbf{F})\, dx\, dy\, dz = \iiint_{\Omega_t} \left[\frac{\partial}{\partial t}(f\mathbf{F}) + \mathbf{F} \cdot \nabla(f\mathbf{F}) + (f\mathbf{F})\, \text{div } \mathbf{F} \right] dx\, dy\, dz,$$

where $\mathbf{F} \cdot \nabla(f\mathbf{F})$ denotes the 3×3 derivative matrix $\mathbf{D}(f\mathbf{F})$ operating on the column vector $\mathbf{F}$; in cartesian coordinates, $\mathbf{F} \cdot \nabla \mathbf{G}$ is the vector whose ith component is

$$\sum_{j=1}^{3} F_j \frac{\partial G^i}{\partial x_j}$$

[labeling (x, y, z) as (x_1, x_2, x_3)].

8.5 APPLICATIONS TO PHYSICS, ENGINEERING, AND DIFFERENTIAL EQUATIONS

We apply the concepts developed in this chapter to the formulation of some physical theories.

Conservation Laws

Let us first discuss an important equation that is referred to as a ***conservation*** equation. For fluids, it expresses the conservation of mass, and for electromagnetic theory, the conservation of charge. We shall apply these ideas to the equation for heat conduction and to electromagnetism.

Let $\mathbf{V}(t, x, y, z)$ be a C^1 vector field on $\mathbb{R}^3$ for each t and let $\rho(t, x, y, z)$ be a C^1 real-valued function. By the ***law of conservation of mass*** for $\mathbf{V}$ and ρ, we shall mean that the condition

$$\frac{d}{dt} \iiint_{\Omega} \rho\, dV = -\iint_{\partial\Omega} \mathbf{J} \cdot \mathbf{n}\, dS$$

holds for all regions Ω in $\mathbb{R}^3$, where $\mathbf{J} = \rho\mathbf{V}$; see Figure 8.5.1.

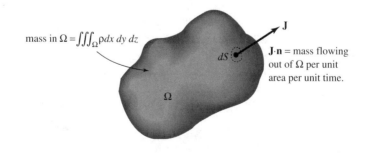

mass in $\Omega = \iiint_{\Omega} \rho\, dx\, dy\, dz$

dS

$\mathbf{J}$

$\mathbf{J} \cdot \mathbf{n}$ = mass flowing out of Ω per unit area per unit time.

Ω

Figure 8.5.1   The rate of change of mass in Ω equals the rate at which mass crosses $\partial\Omega$.

If we think of ρ as a mass density (ρ could also be charge density), that is, the mass per unit volume, and $\mathbf{V}$ as the velocity field of a fluid, the condition just says that the rate of change of total mass in Ω equals the rate at which mass flows *into* Ω. Recall that $\iint_{\partial\Omega} \mathbf{J} \cdot \mathbf{n} \, dS$ is called the flux of $\mathbf{J}$. We need the following result.

THEOREM 11 *For $\mathbf{V}$ and ρ defined on $\mathbb{R}^3$, the law of conservation of mass for $\mathbf{V}$ and ρ is equivalent to the condition*

$$\operatorname{div} \mathbf{J} + \frac{\partial \rho}{\partial t} = 0, \tag{1}$$

that is,

$$\rho \operatorname{div} \mathbf{V} + \mathbf{V} \cdot \nabla \rho + \frac{\partial \rho}{\partial t} = 0. \tag{1'}$$

Here, div $\mathbf{J}$ means that we compute div $\mathbf{J}$ for t held fixed, and $\partial\rho/\partial t$ means we differentiate ρ with respect to t for x, y, z fixed.

PROOF   First, observe that $(d/dt) \iiint_\Omega \rho \, dx \, dy \, dz = \iiint_\Omega (\partial\rho/\partial t) \, dx \, dy \, dz$, and

$$\iint_{\partial\Omega} \mathbf{J} \cdot \mathbf{n} \, dS = \iiint_\Omega \operatorname{div} \mathbf{J} \, dV,$$

by the divergence theorem. Thus conservation of mass is equivalent to the condition

$$\iiint_\Omega \left(\operatorname{div} \mathbf{J} + \frac{\partial\rho}{\partial t} \right) dx \, dy \, dz = 0.$$

Since this is to hold for all regions Ω, it is equivalent to div $\mathbf{J} + \partial\rho/\partial t = 0$. ∎

The equation div $\mathbf{J} + \partial\rho/\partial t = 0$ is called **the equation of continuity**. This is not the only equation governing fluid motion and it does not determine the motion of the fluid; it is just one equation that must hold. Later in this section we shall obtain the additional equations needed to determine the flow. Using the transport theorem (Exercise 22, Section 8.4), the reader can check that the equation of continuity is equivalent to the condition

$$\frac{d}{dt} \iiint_{\Omega_t} \rho \, dx \, dy \, dz = 0,$$

which says that the mass of a region moving with the fluid is constant in time. (The notation Ω_t is explained in the same exercise.)

The fluids that the continuity equation governs can be compressible. If div $\mathbf{V} = 0$ (incompressible case) and ρ is constant, equation (1') follows automatically. But in general, even for incompressible fluids the equation is not automatic, because ρ can depend on (x, y, z) and t. Thus, even if div $\mathbf{V} = 0$ holds, div$(\rho\mathbf{V}) \neq 0$ may still be true.

Perfect Fluids

Next we discuss Euler's equation for a perfect fluid. Consider a nonviscous fluid moving in space with a velocity field $\mathbf{V}$. When we say that the fluid is **perfect**, we mean that if Ω is any portion of the fluid, forces of pressure act on the boundary of Ω along its normal. We assume that the force per unit area acting on $\partial\Omega$ is $-p\mathbf{n}$, where $p(x, y, z, t)$ is some function called the **pressure**; see Figure 8.5.2. Thus the total pressure force acting on Ω is

$$\mathbf{F}_{\partial\Omega} = \text{force} = -\iint_{\partial\Omega} p\mathbf{n}\, dS.$$

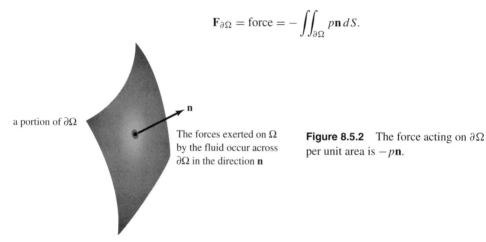

a portion of $\partial\Omega$

The forces exerted on Ω by the fluid occur across $\partial\Omega$ in the direction $\mathbf{n}$

Figure 8.5.2   The force acting on $\partial\Omega$ per unit area is $-p\mathbf{n}$.

This is a *vector* quantity; the ith component of $\mathbf{F}_{\partial\Omega}$ is the integral of the ith component of $p\mathbf{n}$ over the surface $\partial\Omega$ (this is therefore the surface integral of a real-valued function). If $\mathbf{e}$ is any fixed vector in space, we have

$$\mathbf{F}_{\partial\Omega} \cdot \mathbf{e} = -\iint_{\partial\Omega} p\mathbf{e}\cdot\mathbf{n}\, dS,$$

which is the integral of a scalar over $\partial\Omega$. By the divergence theorem and identity 8 in the table of vector identities (Section 4.4),

$$\mathbf{e}\cdot\mathbf{F}_{\partial\Omega} = -\iiint_{\Omega} \operatorname{div}(p\mathbf{e})\, dx\, dy\, dz = -\iiint_{\Omega} (\operatorname{grad} p)\cdot\mathbf{e}\, dx\, dy\, dz,$$

so that

$$\mathbf{F}_{\partial\Omega} = -\iiint_{\Omega} \nabla p\, dx\, dy\, dz.$$

Now we apply Newton's second law to a moving region Ω_t. Here $\Omega_t = \phi_t(\Omega)$, where $\phi_t(\mathbf{x}) = \phi(\mathbf{x}, t)$ denotes the flow of $\mathbf{V}$. The rate of change of momentum of Ω_t equals the force acting on it:

$$\frac{d}{dt}\iiint_{\Omega_t} \rho\mathbf{V}\, dx\, dy\, dz = \mathbf{F}_{\partial\Omega_t} = \iiint_{\Omega_t} \nabla p\, dx\, dy\, dz.$$

We apply the vector form of the transport theorem to the left-hand side (Exercise 23, Section 8.4) to get

$$\iiint_{\Omega_t} \left[\frac{\partial}{\partial t}(\rho \mathbf{V}) + \mathbf{V} \cdot \nabla(\rho \mathbf{V}) + \rho \mathbf{V} \operatorname{div} \mathbf{V} \right] dx\, dy\, dz = - \iiint_{\Omega_t} \nabla p\, dx\, dy\, dz.$$

Since Ω_t is arbitrary, this is equivalent to

$$\frac{\partial}{\partial t}(\rho \mathbf{V}) + \mathbf{V} \cdot \nabla(\rho \mathbf{V}) + \rho \mathbf{V} \operatorname{div} \mathbf{V} = -\nabla p.$$

Simplification using the equation of continuity formula (1′), gives

$$\rho \left(\frac{\partial \mathbf{V}}{\partial t} + \mathbf{V} \cdot \nabla \mathbf{V} \right) = -\nabla p. \tag{2}$$

This is **Euler's equation for a perfect fluid**. For compressible fluids, p is a given function of ρ (for instance, for many gases, $p = A\rho^\gamma$ for constants A and γ). If the fluid is incompressible, on the other hand, p is to be determined from the condition $\operatorname{div} \mathbf{V} = 0$. Equations (1) and (2) then govern the motion of the fluid.

HISTORICAL NOTE

The equations describing the motion of a fluid were first derived by Leonhard Euler in 1755, in a paper entitled "General Principles of the Motion of Fluids." Euler did basic work in mechanics as well as in pure mathematics; he essentially began the subject of analytical mechanics (as opposed to the Euclidean geometric methods used by Newton). He is responsible for the equations of a rigid body (the equations that apply, for example, to a tumbling satellite) and the formulation of many basic equations of mechanics in terms of variational principles; that is, by the methods of maxima and minima of real-valued functions. Euler wrote the first comprehensive textbook on calculus and contributed to virtually all branches of mathematics. He wrote several books and hundreds of research papers after he became totally blind, and he was working on a new treatise on fluid mechanics at the time of his death in 1783. Euler's equations for a fluid were eventually modified to include viscous effects by Navier and Stokes; the resulting Navier-Stokes equations are described in virtually every textbook on fluid mechanics. Stokes is, of course, also responsible for Stokes' theorem, one of the main results in this book.

The Heat Equation

If $T(t, x, y, z)$ (a C^2 function) denotes the temperature in a body at time t, then ∇T represents the temperature gradient: heat "flows" with the vector field $-\nabla T = \mathbf{F}$. Note that ∇T points in the direction of *increasing* T. Since heat flows from hot to cold, we have inserted a minus sign. The energy density, that is, the energy per unit volume, is $c\rho_0 T$, where c is a constant (specific heat) and ρ_0 is the mass density, assumed constant.

(We accept these assertions from elementary physics.) The *energy flux vector* is defined to be $\mathbf{J} = k\mathbf{F}$, where k is a constant called the *conductivity.*

We propose that energy be conserved. This means that $\mathbf{J}$ and $\rho = c\rho_0 T$ should obey the law of conservation of mass, with ρ playing the role of "mass" (note that it is *energy density*, not mass); that is,

$$\frac{d}{dt} \iiint_\Omega \rho\, dV = - \iint_{\partial\Omega} \mathbf{J} \cdot \mathbf{n}\, dS.$$

By Theorem 11, this assertion is equivalent to

$$\operatorname{div} \mathbf{J} + \frac{\partial \rho}{\partial t} = 0.$$

But $\operatorname{div} \mathbf{J} = \operatorname{div}(-k\nabla T) = -k\nabla^2 T$. (Recall that $\nabla^2 T = \partial^2 T/\partial x^2 + \partial^2 T/\partial y^2 + \partial^2 T/\partial z^2$ and ∇^2 is the Laplace operator.) Continuing, we have $\partial \rho/\partial t = \partial (c\rho_0 T)/\partial t = c\rho_0 (\partial T/\partial t)$. Thus equation (1) becomes, in this case,

$$\frac{\partial T}{\partial t} = \frac{k}{c\rho_0}\nabla^2 T = \kappa \nabla^2 T, \tag{3}$$

where $\kappa = k/c\rho_0$ is called the *diffusivity.* Equation (3) is the important *heat equation.*

Just as equations (1) and (2) govern the flow of an ideal fluid, equation (3) governs the conduction of heat, in the following sense. If $T(0, x, y, z)$ is a given initial temperature distribution, then a unique $T(t, x, y, z)$ is determined that satisfies equation (3). In other words the initial condition, at $t = 0$, gives us the result for $t > 0$. Notice that if T does not change with time (the steady-state case), then we must have $\nabla^2 T = 0$ (Laplace's equation).

Maxwell's Equations

We now discuss *Maxwell's equations*, governing the propagation of electromagnetic fields. The form of these equations depends on the physical units one is employing, and changing units introduces factors like 4π, the velocity of light, and so on. We shall choose the system in which Maxwell's equations are simplest.

Let $\mathbf{E}$ and $\mathbf{H}$ be C^1 functions of (t, x, y, z) that are vector fields for each t. They satisfy (by definition) *Maxwell's equations with charge density* $\rho(t, x, y, z)$ and *current density* $\mathbf{J}(t, x, y, z)$ when the following hold:

$$\nabla \cdot \mathbf{E} = \rho \qquad \text{(Gauss' law)}, \tag{4}$$

$$\nabla \cdot \mathbf{H} = 0 \qquad \text{(no magnetic sources)}, \tag{5}$$

$$\nabla \times \mathbf{E} + \frac{\partial \mathbf{H}}{\partial t} = \mathbf{0} \qquad \text{(Faraday's law)}, \tag{6}$$

and

$$\nabla \times \mathbf{H} - \frac{\partial \mathbf{E}}{\partial t} = \mathbf{J} \qquad \text{(Ampère's law)}. \tag{7}$$

Of these laws, equations (4) and (6) were discussed earlier in Sections 8.4 and 8.2 in integral form; historically, they arose in these forms as physically observed laws. Ampère's law was mentioned for a special case in Example 12, Section 7.2.

Physically, one interprets $\mathbf{E}$ as the ***electric field*** and $\mathbf{H}$ as the ***magnetic field***. As time t progresses, these fields interact according to the above equations with each other and with any charges and currents that are present. For example, the propagation of electromagnetic waves (TV signals, radio waves, light from the sun, etc.) in a vacuum is governed by these equations with $\mathbf{J} = \mathbf{0}$ and $\rho = 0$.

Since $\nabla \cdot \mathbf{H} = 0$, we can apply Theorem 8 of Section 8.3 to conclude that $\mathbf{H} = \nabla \times \mathbf{A}$ for some vector field $\mathbf{A}$. (We are assuming that $\mathbf{H}$ is defined on all of $\mathbb{R}^3$ for each time t.) This vector field $\mathbf{A}$ is not unique, and we can equally well use $\mathbf{A}' = \mathbf{A} + \nabla f$ for any function $f(t, x, y, z)$, since $\nabla \times \nabla f = \mathbf{0}$. (This freedom in the choice of $\mathbf{A}$ is called ***gauge freedom***.) For any such choice of $\mathbf{A}$, we have, by equation (6),

$$\mathbf{0} = \nabla \times \mathbf{E} + \frac{\partial \mathbf{H}}{\partial t} = \nabla \times \mathbf{E} + \frac{\partial}{\partial t} \nabla \times \mathbf{A} = \nabla \times \mathbf{E} + \nabla \times \frac{\partial \mathbf{A}}{\partial t} = \nabla \times \left(\mathbf{E} + \frac{\partial \mathbf{A}}{\partial t} \right).$$

Applying Theorem 7, Section 8.3, there is a real-valued function ϕ on $\mathbb{R}^3$ such that

$$\mathbf{E} + \frac{\partial \mathbf{A}}{\partial t} = -\nabla \phi.$$

Substituting this equation and $\mathbf{H} = \nabla \times \mathbf{A}$ into equation (7) and using the following vector identity

$$\nabla \times (\nabla \times \mathbf{A}) = \nabla(\nabla \cdot \mathbf{A}) - \nabla^2 \mathbf{A},$$

we get

$$\mathbf{J} = \nabla \times \mathbf{H} - \frac{\partial \mathbf{E}}{\partial t} = \nabla \times (\nabla \times \mathbf{A}) - \frac{\partial}{\partial t} \left(-\frac{\partial \mathbf{A}}{\partial t} - \nabla \phi \right)$$

$$= \nabla(\nabla \cdot \mathbf{A}) - \nabla^2 \mathbf{A} + \frac{\partial^2 \mathbf{A}}{\partial t^2} + \frac{\partial}{\partial t}(\nabla \phi).$$

Thus

$$\nabla^2 \mathbf{A} - \frac{\partial^2 \mathbf{A}}{\partial t^2} = -\mathbf{J} + \nabla(\nabla \cdot \mathbf{A}) + \frac{\partial}{\partial t}(\nabla \phi),$$

that is,

$$\nabla^2 \mathbf{A} - \frac{\partial^2 \mathbf{A}}{\partial t^2} = -\mathbf{J} + \nabla \left(\nabla \cdot \mathbf{A} + \frac{\partial \phi}{\partial t} \right). \tag{8}$$

Again using the equation $\mathbf{E} + \partial \mathbf{A}/\partial t = -\nabla \phi$ and the equation $\nabla \cdot \mathbf{E} = \rho$, we obtain

$$\rho = \nabla \cdot \mathbf{E} = \nabla \cdot \left(-\nabla \phi - \frac{\partial \mathbf{A}}{\partial t} \right) = -\nabla^2 \phi - \frac{\partial(\nabla \cdot \mathbf{A})}{\partial t},$$

that is,

$$\nabla^2 \phi = -\rho - \frac{\partial(\nabla \cdot \mathbf{A})}{\partial t}. \tag{9}$$

Now let us exploit the freedom in our choice of **A**. We impose the "condition"

$$\nabla \cdot \mathbf{A} + \frac{\partial \phi}{\partial t} = 0. \tag{10}$$

We must be sure we can do this. Supposing we have a given $\mathbf{A}_0$ and a corresponding ϕ_0, can we choose a new $\mathbf{A} = \mathbf{A}_0 + \nabla f$ and then a new ϕ such that $\nabla \cdot \mathbf{A} + \partial \phi / \partial t = 0$? With this new **A**, the new ϕ is $\phi_0 - \partial f / \partial t$; we leave verification as an exercise for the reader. Condition (10) on f then becomes

$$0 = \nabla \cdot (\mathbf{A}_0 + \nabla f) = \frac{\partial(\phi_0 - \partial f/\partial t)}{\partial t} = \nabla \cdot \mathbf{A}_0 + \nabla^2 f + \frac{\partial \phi_0}{\partial t} - \frac{\partial^2 f}{\partial t^2}$$

or

$$\nabla^2 f - \frac{\partial^2 f}{\partial t^2} = -\left(\nabla \cdot \mathbf{A}_0 + \frac{\partial \phi_0}{\partial t}\right). \tag{11}$$

Thus, to be able to choose **A** and ϕ satisfying $\nabla \cdot \mathbf{A} + \partial \phi / \partial t = 0$, we must be able to solve equation (11) for f. One can indeed do this under general conditions, although we do not prove it here. Equation (11) is called the ***inhomogeneous wave equation.***

If we accept that **A** and ϕ can be chosen to satisfy $\nabla \cdot \mathbf{A} + \partial \phi / \partial t = 0$, then equations (8) and (9) for **A** and ϕ become

$$\nabla^2 \mathbf{A} - \frac{\partial^2 \mathbf{A}}{\partial t^2} = -\mathbf{J}; \tag{8'}$$

$$\nabla^2 \phi - \frac{\partial^2 \phi}{\partial t^2} = -\rho. \tag{9'}$$

Equation (9′) follows from equation (9) by substituting $-\partial \phi / \partial t$ for $\nabla \cdot \mathbf{A}$. Thus the wave equation appears again.

Conversely, if **A** and ϕ satisfy the equations $\nabla \cdot \mathbf{A} + \partial \phi / \partial t = 0$, $\nabla^2 \phi - \partial^2 \phi / \partial t^2 = -\rho$, and $\nabla^2 \mathbf{A} - \partial^2 \mathbf{A} / \partial t^2 = -\mathbf{J}$, then $\mathbf{E} = -\nabla \phi - \partial \mathbf{A} / \partial t$ and $\mathbf{H} = \nabla \times \mathbf{A}$ satisfy Maxwell's equations. *This procedure then "reduces" Maxwell's equations to study of the wave equation.*[*]

Solutions to the wave equation have been well studied (one learns how to solve it in most courses in differential equations). To indicate the wavelike nature of the solutions, for example, observe that for any function f

$$\phi(t, x, y, z) = f(x - t)$$

solves the wave equation $\nabla^2 \phi - (\partial^2 \phi / \partial t^2) = 0$. This solution just propagates the graph of f like a wave; thus one might conjecture that solutions of Maxwell's equations are wavelike in nature. Historically, all of this was Maxwell's great achievement, and it soon

---

[*]There are variations on this procedure. For further details see, for example, G. F. D. Duff and D. Naylor, *Differential Equations of Applied Mathematics*, Wiley, New York, 1966, or books on electromagnetic theory, such as J. D. Jackson, *Classical Electrodynamics*, Wiley, New York, 1962.

led to Hertz' discovery of radio waves. To quote from the Feynman *Lectures on Physics* (Vol. II): "From a long view of the history of mankind—seen from, say, ten thousand years from now—there can be little doubt that the most significant event of the 19th century will be judged as Maxwell's discovery of the laws of electrodynamics. The American Civil War will pale into provincial insignificance in comparison with this important scientific event of the same decade."

Green's Function

Next we show how vector analysis can be used to solve differential equations by a method called ***potential theory*** or the ***Green's-function method.*** The presentation will be quite informal; the reader may consult the aforementioned references (see preceding footnote) for further information.

Suppose we wish to solve Poisson's equation $\nabla^2 u = \rho$ for $u(x, y, z)$, where $\rho(x, y, z)$ is a given function (this equation arises from Gauss' law if $\mathbf{E} = \nabla u$).

A function $G(\mathbf{x}, \mathbf{y})$ that has the properties

$$G(\mathbf{x}, \mathbf{y}) = G(\mathbf{y}, \mathbf{x}) \qquad \text{and} \qquad \nabla^2 G(\mathbf{x}, \mathbf{y}) = \delta(\mathbf{x} - \mathbf{y}) \tag{12}$$

(in this expression $\mathbf{y}$ is *held fixed*), that is, which solves the differential equations with ρ replaced by δ, is called ***Green's function*** for this differential equation. Here $\delta(\mathbf{x} - \mathbf{y})$ represents the Dirac delta function, "defined" by*

(i) $\delta(\mathbf{x} - \mathbf{y}) = 0$ for $\mathbf{x} \neq \mathbf{y}$

and

(ii) $\iiint_{\mathbb{R}^3} \delta(\mathbf{x} - \mathbf{y}) \, d\mathbf{y} = 1$.

It has the following operational property that formally follows from conditions (i) and (ii): For any continuous function $f(\mathbf{x})$,

$$\iiint_{\mathbb{R}^3} f(\mathbf{y}) \delta(\mathbf{x} - \mathbf{y}) \, d\mathbf{y} = f(\mathbf{x}). \tag{13}$$

This is sometimes called the ***sifting property*** of δ.

THEOREM 12   *If $G(\mathbf{x}, \mathbf{y})$ satisfies the differential equation $\nabla^2 u = \rho$ with ρ replaced by $\delta(\mathbf{x} - \mathbf{y})$, then*

$$u(\mathbf{x}) = \iiint_{\mathbb{R}^3} G(\mathbf{x}, \mathbf{y}) \rho(\mathbf{y}) \, d\mathbf{y} \tag{14}$$

is a solution to $\nabla^2 u = \rho$.

---

*This is not a precise definition; nevertheless, it is enough here to assume that δ is a symbolic expression with the operational property shown in equation (13). See the references in the preceding footnote for a more careful definition of δ.

PROOF   To see this, note that

$$\nabla^2 \iiint_{\mathbb{R}^3} G(\mathbf{x}, \mathbf{y}) \rho(\mathbf{y}) \, d\mathbf{y} = \iiint_{\mathbb{R}^3} (\nabla^2 G(\mathbf{x}, \mathbf{y})) \rho(\mathbf{y}) \, d\mathbf{y}$$

$$= \iiint_{\mathbb{R}^3} \delta(\mathbf{x} - \mathbf{y}) \rho(\mathbf{y}) \, d\mathbf{y} \qquad \text{by (12)}$$

$$= \rho(\mathbf{x}) \qquad \text{by (13).} \quad \blacksquare$$

The "function" $\rho(\mathbf{x}) = \delta(\mathbf{x})$ represents a unit charge concentrated at a single point [see conditions (i) and (ii), above]. Thus $G(\mathbf{x}, \mathbf{y})$ *represents the potential at* $\mathbf{x}$ *due to a charge placed at* $\mathbf{y}$.

We claim that equation (12) is satisfied if we choose

$$G(\mathbf{x}, \mathbf{y}) = -\frac{1}{4\pi \|\mathbf{x} - \mathbf{y}\|}. \qquad (15)$$

Clearly $G(\mathbf{x}, \mathbf{y}) = G(\mathbf{y}, \mathbf{x})$. To check the second part of equation (12), we must verify that $\nabla^2 G(\mathbf{x}, \mathbf{y})$ has the following two properties of the δ function:

(i) $\nabla^2 G(\mathbf{x}, \mathbf{y}) = 0$ for $\mathbf{x} \neq \mathbf{y}$

and

(ii) $\iiint_{\mathbb{R}^3} \nabla^2 G(\mathbf{x}, \mathbf{y}) \, d\mathbf{y} = 1$.

We will explain the meaning of (ii) in the course of the following discussion. Property (i) is true because the gradient of G is

$$\nabla G(\mathbf{x}, \mathbf{y}) = \frac{\mathbf{r}}{4\pi r^3},$$

where $\mathbf{r} = \mathbf{x} - \mathbf{y}$ is the vector from $\mathbf{y}$ to $\mathbf{x}$ and $r = \|\mathbf{r}\|$ (see Exercise 30, Section 4.4), and therefore for $r \neq 0$, $\nabla \cdot \nabla G(\mathbf{x}, \mathbf{y}) = 0$ (as in the aforementioned exercise). For property (ii), let B be a ball about $\mathbf{x}$; by property (i),

$$\iiint_{\mathbb{R}^3} \nabla^2 G(\mathbf{x}, \mathbf{y}) \, d\mathbf{y} = \iiint_B \nabla^2 G(\mathbf{x}, \mathbf{y}) \, d\mathbf{y}.$$

This, in turn, equals

$$\iint_{\partial B} \nabla G(\mathbf{x}, \mathbf{y}) \cdot \mathbf{n} \, dS,$$

by Gauss' theorem. Thus, by Theorem 10,

$$\iint_{\partial B} \nabla G(\mathbf{x}, \mathbf{y}) \cdot \mathbf{n} \, dS = \iint_{\partial B} \frac{\mathbf{r} \cdot \mathbf{n}}{4\pi r^3} \, dS = 1,$$

which proves property (ii).

Thus, *a solution of* $\nabla^2 u = \rho$ *is*

$$u(\mathbf{x}) = \iiint_{\mathbb{R}^3} \frac{-\rho(\mathbf{y})}{4\pi \|\mathbf{x} - \mathbf{y}\|} \, d\mathbf{y}, \tag{16}$$

by Theorem 12.

In two dimensions, one can similarly show that

$$G(\mathbf{x}, \mathbf{y}) = \frac{1}{2\pi} \log \|\mathbf{x} - \mathbf{y}\|, \tag{17}$$

and so a solution of $\nabla^2 u = \rho$ is

$$u(\mathbf{x}) = \frac{1}{2\pi} \iint_{\mathbb{R}^2} \rho(\mathbf{y}) \log \|\mathbf{x} - \mathbf{y}\| \, d\mathbf{y}. \tag{18}$$

We now turn to the problem of using Green's functions to solve Poisson's equation in a bounded region with given boundary conditions. To do this, we need Green's first and second identities, which can be obtained from the divergence theorem (see Exercise 15, Section 8.4). We start with the identity

$$\iiint_V \nabla \cdot \mathbf{F} \, dV = \iint_S \mathbf{F} \cdot \mathbf{n} \, dS,$$

where V is a region in space, S is its boundary, and $\mathbf{n}$ is the outward unit normal vector. Replacing $\mathbf{F}$ by $f \nabla g$, where f and g are scalar functions, we obtain

$$\iiint_V \nabla f \cdot \nabla g \, dV + \iiint_V f \nabla^2 g \, dV = \iint_S f \frac{\partial g}{\partial n} \, dS, \tag{19}$$

where $\partial g / \partial n = \nabla g \cdot \mathbf{n}$. This is ***Green's first identity***. If we simply permute f and g and subtract the result from equation (19), we obtain ***Green's second identity***,

$$\iiint_V (f \nabla^2 g - g \nabla^2 f) \, dV = \iint_S \left(f \frac{\partial g}{\partial n} - g \frac{\partial f}{\partial n} \right) dS, \tag{20}$$

which we shall use shortly.

Consider Poisson's equation $\nabla^2 u = \rho$ in some region V, and the corresponding equations for Green's function

$$G(\mathbf{x}, \mathbf{y}) = G(\mathbf{y}, \mathbf{x}) \qquad \text{and} \qquad \nabla^2 G(\mathbf{x}, \mathbf{y}) = \delta(\mathbf{x} - \mathbf{y}).$$

Inserting u and G into Green's second identity (20), we obtain

$$\iiint_V (u \nabla^2 G - G \nabla^2 u) \, dV = \iint_S \left(u \frac{\partial G}{\partial n} - G \frac{\partial u}{\partial n} \right) dS.$$

Choosing our integration variable to be $\mathbf{y}$ and using $G(\mathbf{x}, \mathbf{y}) = G(\mathbf{x}, \mathbf{y})$, this becomes

$$\iiint_V [u(\mathbf{y})\delta(\mathbf{x} - \mathbf{y}) - G(\mathbf{x}, \mathbf{y})\rho(\mathbf{y})] \, d\mathbf{y} = \iint_S \left(u \frac{\partial G}{\partial n} - G \frac{\partial u}{\partial n} \right) dS;$$

and by equation (13),

$$u(\mathbf{x}) = \iiint_V G(\mathbf{x}, \mathbf{y})\rho(\mathbf{y})\,d\mathbf{y} + \iint_S \left(u\frac{\partial G}{\partial n} - G\frac{\partial u}{\partial n} \right) dS. \tag{21}$$

Note that for an unbounded region, this becomes identical to our previous result, equation (14), for all of space. Equation (21) enables us to solve for u in a bounded region where $\rho = 0$ by incorporating the conditions that u must obey on S.

If $\rho = 0$, equation (21) reduces to

$$u = \iint_S \left(u\frac{\partial G}{\partial n} - G\frac{\partial u}{\partial n} \right) dS,$$

or, fully,

$$u(\mathbf{x}) = \iint_S \left[u(\mathbf{y})\frac{\partial G}{\partial n}(\mathbf{x}, \mathbf{y}) - G(\mathbf{x}, \mathbf{y})\frac{\partial u}{\partial n}(\mathbf{y}) \right] dS, \tag{22}$$

where u appears on both sides of the equation and the integration is over $\mathbf{y}$. The crucial point is that evaluation of the integral requires only that we know the behavior of u on S. Commonly, either u is given on the boundary (for a **Dirichlet problem**) or $\partial u/\partial n$ is given on the boundary (for a **Neumann problem**). If we know u on the boundary, we want to make $G\partial u/\partial n$ vanish on the boundary so we can evaluate the integral. Therefore if u is given on S we must find a G such that $G(\mathbf{x}, \mathbf{y})$ vanishes whenever $\mathbf{y}$ lies on S. This is called the **Dirichlet Green's function for the region** V. Conversely, if $\partial u/\partial n$ is given on S we must find a G such that $\partial G/\partial n$ vanishes on S. This is the **Neumann Green's function.**

Thus, a Dirichlet Green's function $G(\mathbf{x}, \mathbf{y})$ is defined for $\mathbf{x}$ and $\mathbf{y}$ in the volume V and satisfies these three conditions:

(a) $G(\mathbf{x}, \mathbf{y}) = G(\mathbf{y}, \mathbf{x})$,

(b) $\nabla^2 G(\mathbf{x}, \mathbf{y}) = \delta(\mathbf{x} - \mathbf{y})$,

(c) $G(\mathbf{x}, \mathbf{y}) = 0$ when $\mathbf{y}$ lies on S, the boundary of the region V.

[Note that by condition (a), in conditions (b) and (c) the variables $\mathbf{x}$ and $\mathbf{y}$ can be interchanged without changing the condition.]

It is perhaps surprising that *condition* (a) *is actually a consequence of conditions* (b) *and* (c), *provided* (b) *and* (c) *also hold with* $\mathbf{x}$ *and* $\mathbf{y}$ *interchanged.*

To see this, we fix $\mathbf{y}$ and use equation (20) with $f(\mathbf{x}) = G(\mathbf{x}, \mathbf{y})$ and $g(\mathbf{x}) = G(\mathbf{y}, \mathbf{x})$. By condition (b), $\nabla^2 f(\mathbf{x}) = \delta(\mathbf{x} - \mathbf{y})$ and $\nabla^2 g(\mathbf{x}) = \delta(\mathbf{x} - \mathbf{y})$, and by condition (c), f and g vanish on S. Thus, equation (20) becomes

$$\iiint_V [f(\mathbf{x})\delta(\mathbf{x} - \mathbf{y}) - g(\mathbf{x})\delta(\mathbf{x} - \mathbf{y})]\,dV = 0$$

or

$$f(\mathbf{x}) - g(\mathbf{x}) = 0.$$

Thus $G(\mathbf{x}, \mathbf{y}) = G(\mathbf{y}, \mathbf{x})$. This means, in effect, that it is not necessary to check condition (a). (This result is sometimes called the ***principle of reciprocity***.)

Solving any particular Dirichlet or Neumann problem thus reduces to the task of finding the appropriate Green's function. We shall do this by modifying the Green's function for Laplace's equations on all of $\mathbb{R}^2$ or $\mathbb{R}^3$, namely, equations (16) and (17).

As an example, we shall now use the two-dimensional Green's function method to construct the Dirichlet Green's function for the disk of radius R (see Figure 8.5.3). This will enable us to solve $\nabla^2 u = 0$ (or $\nabla^2 u = \rho$) with u given on the boundary circle.

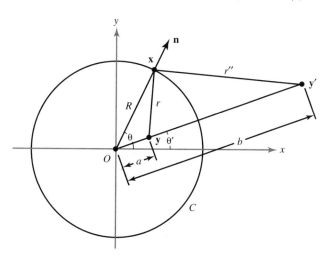

Figure 8.5.3 Geometry of the construction of Green's function for a disk.

In Figure 8.5.3 we have drawn the point $\mathbf{x}$ on the circumference because that is where we want G to vanish. [According to the procedure above, $G(\mathbf{x}, \mathbf{y})$ is supposed to vanish when either $\mathbf{x}$ or $\mathbf{y}$ is on C. We have chosen $\mathbf{x}$ on C to begin with.] The Green's function $G(\mathbf{x}, \mathbf{y})$ that we shall find will, of course, be valid for all $\mathbf{x}, \mathbf{y}$ in the disk. The point $\mathbf{y}'$ represents the "reflection" of the point $\mathbf{y}$ into the region outside the circle, such that $ab = R^2$. Now when $\mathbf{x} \in C$, by the similarity of the triangles $\mathbf{x}O\mathbf{y}$ and $\mathbf{x}O\mathbf{y}'$,

$$\frac{r}{R} = \frac{r''}{b}, \qquad \text{that is,} \qquad r = \frac{r''R}{b} = \frac{r''a}{R}.$$

Hence, if we choose our Green's function to be

$$G(\mathbf{x}, \mathbf{y}) = \frac{1}{2\pi}\left(\log r - \log \frac{r''a}{R} \right), \tag{23}$$

we see that G is zero if $\mathbf{x}$ is on C. Since $r''a/R$ reduces to r when $\mathbf{y}$ is on C, G also vanishes when $\mathbf{y}$ is on C. If we can show that G satisfies $\nabla^2 G = \delta(\mathbf{x} - \mathbf{y})$ in the circle, then we will have proved that G is indeed the Dirichlet Green's function. From equation (17) we know that $\nabla^2(\log r)/2\pi = \delta(\mathbf{x} - \mathbf{y})$, so that

$$\nabla^2 G(\mathbf{x}, \mathbf{y}) = \delta(\mathbf{x} - \mathbf{y}) - \delta(\mathbf{x} - \mathbf{y}'),$$

but $\mathbf{y}'$ is always outside the circle, and so $\mathbf{x}$ can never be equal to $\mathbf{y}'$ and $\delta(\mathbf{x} - \mathbf{y}')$ is always zero. Hence

$$\nabla^2 G(\mathbf{x}, \mathbf{y}) = \delta(\mathbf{x} - \mathbf{y}),$$

and thus G is the Dirichlet Green's function for the circle.

Now we shall consider the problem of solving

$$\nabla^2 u = 0$$

in this circle if $u(R, \theta) = f(\theta)$ is the given boundary condition. By equation (22) we have a solution

$$u = \int_C \left(u \frac{\partial G}{\partial n} - G \frac{\partial u}{\partial n} \right) ds.$$

But $G = 0$ on C, and so we are left with the integral

$$u = \int_C u \frac{\partial G}{\partial n} ds,$$

where we can replace u by $f(\theta)$, since the integral is around C. Thus the task of solving the Dirichlet problem in the circle is reduced to finding $\partial G/\partial n$. From equation (23) we can write

$$\frac{\partial G}{\partial n} = \frac{1}{2\pi} \left(\frac{1}{r} \frac{\partial r}{\partial n} - \frac{1}{r''} \frac{\partial r''}{\partial n} \right).$$

Now

$$\frac{\partial r}{\partial n} = \nabla r \cdot \mathbf{n} \qquad \text{and} \qquad \nabla r = \frac{\mathbf{r}}{r},$$

where $\mathbf{r} = \mathbf{x} - \mathbf{y}$, and so

$$\frac{\partial r}{\partial n} = \frac{\mathbf{r} \cdot \mathbf{n}}{r} = \frac{r \cos(nr)}{r} = \cos(nr),$$

where (nr) represents the angle between $\mathbf{n}$ and $\mathbf{r}$. Likewise,

$$\frac{\partial r''}{\partial n} = \cos(nr'').$$

In triangle $\mathbf{x}\mathbf{y}O$, we have, by the cosine law, $a^2 = r^2 + R^2 - 2rR\cos(nr)$, and in triangle $\mathbf{x}\mathbf{y}'O$, we get $b^2 = (r'')^2 + R^2 - 2r''R\cos(nr'')$, and so

$$\frac{\partial r}{\partial n} = \cos(nr) = \frac{R^2 + r^2 - a^2}{2rR} \qquad \text{and} \qquad \frac{\partial r''}{\partial n} = \cos(nr'') = \frac{R^2 + (r'')^2 - b^2}{2r''R}.$$

Hence

$$\frac{\partial G}{\partial n} = \frac{1}{2\pi} \left[\frac{R^2 + r^2 - a^2}{2r^2 R} - \frac{R^2 - (r'')^2 - b^2}{2(r'')^2 R} \right].$$

Using the relationship between r and r'' when $\mathbf{x}$ is on C, we get

$$\left.\frac{\partial G}{\partial n}\right|_{\mathbf{x} \in C} = \frac{1}{2\pi}\left(\frac{R^2 - a^2}{Rr^2}\right).$$

Thus the solution can be written as

$$u = \frac{1}{2\pi}\int_C f(\theta)\frac{R^2 - a^2}{Rr^2}\,ds.$$

Let us write this in a more useful form. Note that in triangle $\mathbf{xy}O$, we can write

$$r = [a^2 + R^2 - 2aR\cos(\theta - \theta')]^{1/2},$$

where θ and θ' are the polar angles in $\mathbf{x}$ and $\mathbf{y}$ space, respectively. Second, our solution must be valid for all $\mathbf{y}$ in the circle; hence the distance of $\mathbf{y}$ from the origin must now become a variable, which we shall call r'. Finally, note that $ds = R\,d\theta$ on C, so we can write the solution in polar coordinates as

$$u(r', \theta') = \frac{R^2 - (r')^2}{2\pi}\int_0^{2\pi}\frac{f(\theta)\,d\theta}{(r')^2 + R^2 - 2r'R\cos(\theta - \theta')}.$$

This is known as ***Poisson's formula in two dimensions.**** As an exercise, the reader should use this to write down the solution of $\nabla^2 u = \rho$ with u a given function $f(\theta)$ on the boundary.

EXERCISES

1. (a) Supply the details for the assertion in this section that

$$\frac{d}{dt}\iiint_{\Omega_t}\rho\,dx\,dy\,dz = 0$$

is equivalent to the law of conservation of mass.

  (b) Using part (a) and the change of variables theorem, show that $\rho(\mathbf{x}, t)$ can be expressed in terms of the Jacobian $J(\mathbf{x}, t)$ of the flow map $\phi(\mathbf{x}, t)$ and $\rho(\mathbf{x}, 0)$ by the equation

$$\rho(\mathbf{x}, t)J(\mathbf{x}, t) = \rho(\mathbf{x}, 0).$$

  (c) What can you conclude from part (b) for incompressible flow?

2. Let $\mathbf{V}$ be a vector field with flow $\phi(\mathbf{x}, t)$ and let $\mathbf{V}$ and ρ satisfy the law of conservation of mass. Let Ω_t be the region transported with the flow. Prove the following version of the transport theorem (see Exercise 22, Section 8.4):

$$\frac{d}{dt}\iiint_{\Omega_t}\rho f\,dx\,dy\,dz = \iiint_{\Omega_t}\rho\frac{Df}{Dt}\,dx\,dy\,dz.$$

---

*There are several other ways of deriving this famous formula. For the method of complex variables, see J. Marsden and M. Hoffman, *Basic Complex Analysis*, 2d ed., Freeman, New York, 1987, p. 195. For the method of Fourier series, see J. Marsden, *Elementary Classical Analysis*, Freeman, New York, 1974, p. 466.

3. (*Bernoulli's law*) (a) Let $\mathbf{V}$, ρ satisfy the law of conservation of mass and equation (2) (Euler's equation for a perfect fluid). Suppose $\mathbf{V}$ is irrotational and hence that $\mathbf{V} = \nabla\phi$ for a function ϕ. Show that if C is a path connecting two points P_1 and P_2, then

$$\left(\frac{\partial\phi}{\partial t} + \frac{1}{2}\|\mathbf{V}\|^2\right)\Bigg|_{P_1}^{P_2} + \int_C \frac{dp}{\rho} = 0.$$

[HINT: You may need the vector identity, $(\mathbf{V}\cdot\nabla)\mathbf{V} = \frac{1}{2}\nabla(\|\mathbf{V}\|^2) + (\nabla\times\mathbf{V})\times\mathbf{V}$.]
  (b) If in part (a), $\mathbf{V}$ is stationary—that is, $\partial\mathbf{V}/\partial t = 0$—and ρ is constant, show that

$$\frac{1}{2}\|\mathbf{V}\|^2 + \frac{p}{\rho}$$

is constant in space. Deduce that, in this situation, *higher pressure is associated with lower fluid speed.*

4. Using Exercise 3, show that if ϕ satisfies Laplace's equation $\nabla^2\phi = 0$, then $\mathbf{V} = \nabla\phi$ is a stationary solution to Euler's equation for a perfect *incompressible* fluid with constant density.

5. Verify that Maxwell's equations imply the equation of continuity for $\mathbf{J}$ and ρ.

6. Let H denote the upper half space $z \geq 0$. For a point $\mathbf{x} = (x, y, z)$ in H, let $R(\mathbf{x}) = (x, y, -z)$, the reflection of $\mathbf{x}$ in the xy plane. Let $G(\mathbf{x}, \mathbf{y}) = -1/(4\pi\|\mathbf{x} - \mathbf{y}\|)$ be the Green's function for all of $\mathbb{R}^3$.
  (a) Verify that the function $\tilde{G}$ defined by

$$\tilde{G}(\mathbf{x}, \mathbf{y}) = G(\mathbf{x}, \mathbf{y}) - G(R(\mathbf{x}), \mathbf{y})$$

is the Green's function for the Laplacian in H.
  (b) Write down a formula for the solution u of the problem

$$\nabla^2 u = \rho \quad \text{in} \quad H, \quad \text{and} \quad u(x, y, 0) = \phi(x, y).$$

7. (a) With notation as in Figure 8.5.3, show that the Dirichlet problem for the sphere of radius R in three dimensions has Green's function

$$G(\mathbf{x}, \mathbf{y}) = \frac{1}{4\pi}\left(\frac{R}{ar''} - \frac{1}{r}\right).$$

  (b) Prove Poisson's formula in three dimensions:

$$u(\mathbf{y}) = \frac{R(R^2 - a^2)}{4\pi}\int_0^{2\pi}\int_0^\pi \frac{f(\theta, \phi)\sin\phi\, d\theta\, d\phi}{(R^2 + a^2 - 2Ra\cos\gamma)^{3/2}}$$

where $f(\theta, \phi)$ is the function given on the boundary sphere, $\|\mathbf{y}\| = a$, and γ is the angle between $\mathbf{y}$ and the point $\mathbf{x} = (R\cos\theta\sin\phi, R\sin\theta\sin\phi, R\cos\phi)$ on the surface of the sphere.

Exercises 8 to 14 give some sample applications of vector calculus to shock waves.[*]

8. Consider the equation

$$u_t + uu_x = 0$$

---

[*]For additional details, consult A. J. Chorin and J. E. Marsden, *A Mathematical Introduction to Fluid Mechanics*, 3rd ed., Springer-Verlag, New York, 1992, and P. D. Lax, "The Formation and Decay of Shock Waves," *Am. Math. Monthly* 79 (1972): 227–241. We are grateful to Joel Smoller for suggesting this series of exercises.

for a function $u(x, t)$, $-\infty < x < \infty$, $t \geq 0$, where $u_t = \partial u/\partial t$ and $u_x = \partial u/\partial x$. Let $u(x, 0) = u_0(x)$ be the given value of u at $t = 0$. The curves $(x(s), t(s))$ in the xt plane defined by

$$\dot{x} = u, \quad \dot{t} = 1$$

are called **characteristic curves** (the overdot denotes the derivative with respect to s).

(a) Show that u is constant along each characteristic curve by showing that $\dot{u} = 0$.

(b) Show that the slopes of the characteristic curves are given by $dt/dx = 1/u$, and use it to prove that the characteristic curves are straight lines determined by the initial data.

(c) Suppose that $x_1 < x_2$ and $u_0(x_1) > u_0(x_2) > 0$. Show that the two characteristics through the points $(x_1, 0)$ and $(x_2, 0)$ intersect at a point $P = (\bar{x}, \bar{t})$, with $\bar{t} > 0$. Show that this together with the result in part (a) implies that the solution cannot be continuous at P (see Figure 8.5.4).

Figure 8.5.4 Characteristics of the equation $u_t + uu_x = 0$.

(d) Calculate $\bar{t}$.

9. Repeat Exercise 8 for the equation

$$u_t + f(u)_x = 0, \tag{24}$$

where $f'' > 0$ and $f'(u_0(x_2)) > 0$. The characteristics are now defined by $\dot{x} = f'(u)$, $\dot{t} = 1$. We call equation (24) an equation in **divergence form**. (This exercise shows that a continuous solution is generally impossible—irrespective of the smoothness of f!)

10. (*Weak solutions*) Since equations of the form in Exercise 9 arise in many physical applications [gas dynamics, magnetohydrodynamics, nonlinear optics (lasers)] and because it would be nice for a solution to exist for all time (t), it is desirable to make sense out of the equation by reinterpreting it when discontinuities develop. To this end, let $\phi = \phi(x, t)$ be a C^1 function. Let D be a rectangle in the xt plane determined by $-a \leq x \leq a$ and $0 \leq t \leq T$, such that $\phi(x, t) = 0$ for $x = \pm a$, $x = T$, and for all (x, t) in the upper half plane outside D. Let u be a "genuine" solution of equation (24).

(a) Show that

$$\iint_{t \geq 0} [u\phi_t + f(u)\phi_x] \, dx \, dt + \int_{t=0} u_0(x)\phi(x, 0) \, dx = 0. \tag{25}$$

(HINT: Start with $\iint_D [u_t + f(u)_x]\phi \, dx \, dt = 0$.) Thus, if u is a smooth solution, then equation (25) holds for all ϕ as above. We call the function u a **weak solution** of equation (24) if equation (25) holds for all such ϕ.

(b) Show that if u is a weak solution that is C^1 in an open set Ω in the upper half of the xt plane, then u is a genuine solution of equation (24) in Ω.

11. (*The **jump condition**, which is also known in gas dynamics as the **Rankine-Hugoniot condition**.*) The definition of a weak solution given in Exercise 10 clearly allows discontinuous solutions. However, the reader shall now determine that not every type of discontinuity is admissible, for there is a connection between the discontinuity curve and the values of the solution on both sides of the discontinuity.

Let u be a (weak) solution of equation (24) and suppose Γ is a smooth curve in the xt plane such that u "jumps" across a curve Γ; that is, u is of class C^1 except for a jump discontinuity across Γ. We call Γ a **shock wave**. Choose a point $P \in \Gamma$ and construct, near P, a "rectangle" $D = D_1 \cup D_2$, as shown in Figure 8.5.5. Choose ϕ to vanish on D and outside D.

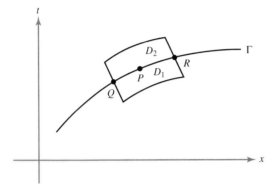

Figure 8.5.5   The solution u jumps in value from u_1 to u_2 across Γ.

(a) Show that

$$\iint_D [u\phi_t + f(u)\phi_x]\,dx\,dt = 0$$

and

$$\iint_{D_i} [u\phi_t + f(u)\phi_x]\,dx\,dt = \iint_{D_i} [(u\phi)_t + (f(u)\phi)_x]\,dx\,dt.$$

(b) Suppose that u jumps in value from u_1 to u_2 across Γ so that when (x,t) approaches a point (x_0, t_0) on Γ from ∂D_i, $u(x,t)$ approaches the value $u_i(x_0, t_0)$. Show that

$$0 = \int_{\partial D_1} \phi\{-u\,dx + f(u)\,dt\} + \int_{\partial D_2} \phi\{-u\,dx + f(u)\,dt\}$$

and deduce that

$$0 = \int_\Gamma \phi\{[-u]\,dx + [f(u)]\,dt\}$$

where $[\alpha(u)] = \alpha(u_2) - \alpha(u_1)$ denotes the jump in the quantity $\alpha(u)$ across Γ.

(c) If the curve Γ defines x implicitly as a function of t and ∂D intersects Γ at $Q = (x(t_1), t_1)$ and $R = (x(t_2), t_2)$, show that

$$0 = \int_Q^R \phi\{[-u]\,dx + [f(u)]\,dt\} = \int_{t_1}^{t_2} \phi\left\{[-u]\frac{dx}{dt} + [f(u)]\right\}dt.$$

(d)   Show that at the point P on Γ,

$$[u] \cdot s = [f(u)], \tag{26}$$

where $s = dx/dt$ at P. The number s is called the **speed** of the discontinuity. Equation (26) is called the **jump condition**; it is the relation that any discontinuous solution will satisfy.

12.   *(Loss of uniqueness)* One drawback of accepting weak solutions is loss of uniqueness. (In gas dynamics, some mathematical solutions are extraneous and rejected on physical grounds. For example, discontinuous solutions of rarefaction shock waves are rejected because they indicate that entropy *decreases* across the discontinuity.)

Consider the equation

$$u_t + \left(\frac{u^2}{2}\right)_x = 0, \qquad \text{with initial data} \qquad u(x, 0) = \begin{cases} -1, & x \geq 0 \\ 1, & x < 0. \end{cases}$$

Show that for every $\alpha \geq 1$, u_α is a weak solution, where u_α is defined by

$$u_\alpha(x, t) = \begin{cases} 1, & x \leq \dfrac{1-\alpha}{2}t \\[2mm] -\alpha, & \dfrac{1-\alpha}{2}t \leq x \leq 0 \\[2mm] \alpha, & 0 \leq x \leq \dfrac{\alpha-1}{2}t \\[2mm] -1, & \dfrac{\alpha-1}{2}t < x. \end{cases}$$

[It can be shown that if $f'' > 0$, uniqueness can be recovered by imposing an additional constraint on the solutions. Thus, there is a unique solution satisfying the "entropy" condition

$$\frac{u(x+a, t) - u(x, t)}{a} \leq \frac{E}{t}$$

for some $E > 0$ and all $a \neq 0$. Hence for fixed t, $u(x, t)$ can only "jump down" as x increases. In our example, this holds only for the solution with $\alpha = 1$.]

13.   [The solution of equation (24) depends on the particular divergence form used.]   The equation $u_t + uu_x = 0$ can be written in the two divergence forms

$$u_t + (\tfrac{1}{2}u^2)_x = 0, \tag{i}$$

$$(\tfrac{1}{2}u^2)_t + (\tfrac{1}{3}u^3)_x = 0. \tag{ii}$$

Show that a weak solution of equation (i) need not be a weak solution of equation (ii). [HINT: The equations have different jump conditions: In equation (i), $s = \tfrac{1}{2}(u_2 + u_1)$, while in equation (ii), $s = \tfrac{2}{3}(u_2^2 + u_1 u_2 + u_1^2)/(u_2 + u_1)$.]

14.   *(Noninvariance of weak solutions under nonlinear transformation)* Consider equation (24) where $f'' > 0$.

(a)   Show that the transformation $v = f'(u)$ takes this equation into

$$v_t + vv_x = 0. \tag{27}$$

(b)   Show that the above transformation *does not* necessarily map discontinuous solutions of equation (24) into discontinuous solutions of equation (27). (HINT: Check the jump conditions; for equation (27), $s[v] = \frac{1}{2}[v^2]$ implies $s[f'(u)] = \frac{1}{2}[f'(u)^2]$; for equation (24), $s[u] = [f(u)]$.)

15.   *(Requires a knowledge of complex numbers.)* Show that Poisson's formula in two dimensions may be written as

$$u(r', \theta') = \frac{1}{2\pi} \int_0^{2\pi} u(re^{i\theta}) \frac{r^2 - |z'|^2}{|re^{i\theta} - z'|^2} \, d\theta,$$

where $z' = r'e^{i\theta'}$.

16.   For a steady-state charge distribution and divergence-free current distribution, the electric and magnetic fields $\mathbf{E}(x, y, z)$ and $\mathbf{H}(x, y, z)$ satisfy

$$\nabla \times \mathbf{E} = \mathbf{0}, \quad \nabla \cdot \mathbf{H} = 0, \quad \nabla \cdot \mathbf{J} = 0, \quad \nabla \cdot \mathbf{E} = \rho, \quad \text{and} \quad \nabla \times \mathbf{H} = \mathbf{J}.$$

Here $\rho = \rho(x, y, z)$ and $\mathbf{J}(x, y, z)$ are assumed to be known. (The constants ε and μ which usually appear have been taken equal to unity by choice of units.) The radiation which the fields produce through a surface S is determined by a radiation flux density vector field, called the ***Poynting*** vector field,

$$\mathbf{P} = \mathbf{E} \times \mathbf{H}.$$

(a)   If S is a *closed* surface, show that the radiation flux through S given by the above field is given by

$$\iint_S \mathbf{P} \cdot d\mathbf{S} = - \iiint_V \mathbf{E} \cdot \mathbf{J} \, dV,$$

where V is the region enclosed by S.
(b)   Examples of such fields are

$$\mathbf{E}(x, y, z) = z\mathbf{j} + y\mathbf{k} \quad \text{and} \quad \mathbf{H}(x, y, z) = -xy\mathbf{i} + x\mathbf{j} + yz\mathbf{k}.$$

Find the flux of the Poynting vector through the hemispherical shell shown in Figure 8.5.6. (Notice that it is an *open* surface.)

Figure 8.5.6   The surface for Exercise 16(b).

(c)   The fields of part (b) produce a Poynting vector field passing through the toroidal surface shown in Figure 8.5.7. What is the flux through this torus?

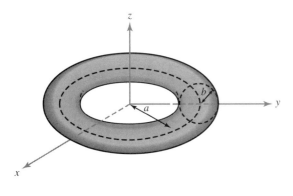

Figure 8.5.7 The surface for Exercise 16(c).

8.6 DIFFERENTIAL FORMS

The theory of differential forms provides a convenient and elegant way of phrasing Green's, Stokes', and Gauss' theorems. In fact, the use of differential forms shows that these theorems are all manifestations of a single underlying mathematical theory and provides the necessary language to generalize them to n dimensions. In this section we shall give a very elementary exposition of the theory of forms. Since our primary goal is to show that the theorems of Green, Stokes, and Gauss can be unified under a single theorem, we shall be satisfied with less than the strongest possible version of these theorems. Moreover, we shall introduce forms in a purely axiomatic and nonconstructive manner, thereby avoiding the tremendous number of formal algebraic preliminaries that are usually required for their construction. To the purist our approach will be far from complete, but to the student it may be comprehensible. We hope that this will motivate some students to delve further into the theory of differential forms.

We shall begin by introducing the notion of a 0-form.

DEFINITION Let K be an open set in $\mathbb{R}^3$. A *0-form* on K is a real-valued function $f: K \to \mathbb{R}$. When we differentiate f once, it is assumed to be of class C^1, and C^2 when we differentiate twice.

Given two 0-forms f_1 and f_2 on K, we can add them in the usual way to get a new 0-form $f_1 + f_2$ or multiply them to get a 0-form $f_1 f_2$.

EXAMPLE 1 $f_1(x, y, z) = xy + yz$ and $f_2(x, y, z) = y \sin xz$ are 0-forms on $\mathbb{R}^3$:

$$(f_1 + f_2)(x, y, z) = xy + yz + y \sin xz$$

and

$$(f_1 f_2)(x, y, z) = y^2 x \sin xz + y^2 z \sin xz. \quad \blacktriangle$$

DEFINITION   The *basic 1-forms* are the expressions dx, dy, and dz. At present we consider these to be only formal symbols. A *1-form* ω on an open set K is a formal linear combination

$$\omega = P(x, y, z)\, dx + Q(x, y, z)\, dy + R(x, y, z)\, dz,$$

or simply

$$\omega = P\, dx + Q\, dy + R\, dz,$$

where P, Q, and R are real-valued functions on K. By the expression $P\, dx$ we mean the 1-form $P\, dx + 0 \cdot dy + 0 \cdot dz$, and similarly for $Q\, dy$ and $R\, dz$. Also, the order of $P\, dx$, $Q\, dy$, and $R\, dz$ is immaterial, and so

$$P\, dx + Q\, dy + R\, dz = R\, dz + P\, dx + Q\, dy, \text{etc.}$$

Given two 1-forms $\omega_1 = P_1\, dx + Q_1\, dy + R_1\, dz$ and $\omega_2 = P_2\, dx + Q_2\, dy + R_2\, dz$, we can add them to get a new 1-form $\omega_1 + \omega_2$ defined by

$$\omega_1 + \omega_2 = (P_1 + P_2)\, dx + (Q_1 + Q_2)\, dy + (R_1 + R_2)\, dz,$$

and given a 0-form f, we can form the 1-form $f\omega_1$ defined by

$$f\omega_1 = (fP_1)\, dx + (fQ_1)\, dy + (fR_1)\, dz.$$

EXAMPLE 2   Let $\omega_1 = (x + y^2)\, dx + (zy)\, dy + (e^{xyz})\, dz$ and $\omega_2 = \sin y\, dx + \sin x\, dy$ be 1-forms. Then

$$\omega_1 + \omega_2 = (x + y^2 + \sin y)\, dx + (zy + \sin x)\, dy + (e^{xyz})\, dz.$$

If $f(x, y, z) = x$, then

$$f\omega_2 = x \sin y\, dx + x \sin x\, dy. \quad \blacktriangle$$

DEFINITION   The *basic 2-forms* are the formal expressions $dx\, dy$, $dy\, dz$, and $dz\, dx$. These expressions should be thought of as products of dx and dy, dy and dz, and dz and dx.

A *2-form* η on K is a formal expression

$$\eta = F\, dx\, dy + G\, dy\, dz + H\, dz\, dx,$$

where F, G, and H are real-valued functions on K. The order of $F\, dx\, dy$, $G\, dy\, dz$, and $H\, dz\, dx$ is immaterial; for example,

$$F\, dx\, dy + G\, dy\, dz + H\, dz\, dx = H\, dz\, dx + F\, dx\, dy + G\, dy\, dz, \text{etc.}$$

At this point it is useful to note that in a 2-form the basic 1-forms dx, dy, and dz always appear in cyclic pairs (see Figure 8.6.1), that is, $dx\,dy$, $dy\,dz$, and $dz\,dx$.

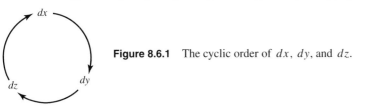

Figure 8.6.1   The cyclic order of dx, dy, and dz.

By analogy with 0-forms and 1-forms, we can add two 2-forms

$$\eta_i = F_i\,dx\,dy + G_i\,dy\,dz + H_i\,dz\,dx,$$

$i = 1$ and 2, to obtain a new 2-form,

$$\eta_1 + \eta_2 = (F_1 + F_2)\,dx\,dy + (G_1 + G_2)\,dy\,dz + (H_1 + H_2)\,dz\,dx.$$

Similarly, if f is a 0-form and η is a 2-form, we can take the product

$$f\eta = (fF)\,dx\,dy + (fG)\,dy\,dz + (fH)\,dz\,dx.$$

Finally, by the expression $F\,dx\,dy$ we mean the 2-form $F\,dx\,dy + 0 \cdot dy\,dz + 0 \cdot dz\,dx$.

EXAMPLE 3   The expressions

$$\eta_1 = x^2\,dx\,dy + y^3x\,dy\,dz + \sin zy\,dz\,dx$$

and

$$\eta_2 = y\,dy\,dz$$

are 2-forms. Their sum is

$$\eta_1 + \eta_2 = x^2\,dx\,dy + (y^3x + y)\,dy\,dz + \sin zy\,dz\,dx.$$

If $f(x, y, z) = xy$, then

$$f\eta_2 = xy^2\,dy\,dz. \quad \blacktriangle$$

DEFINITION   A *basic 3-form* is a formal expression $dx\,dy\,dz$ (in cyclic order, Figure 8.6.1). A *3-form* ν on an open set $K \subset \mathbb{R}^3$ is an expression of the form $\nu = f(x, y, z)\,dx\,dy\,dz$, where f is a real-valued function on K.

We can add two 3-forms and we can multiply them by 0-forms in the obvious way. There seems to be little difference between a 0-form and a 3-form, since both involve a single real-valued function. But we distinguish them for a purpose that will become clear when we multiply and differentiate forms.

EXAMPLE 4   Let $v_1 = y\,dx\,dy\,dz$, $v_2 = e^{x^2}\,dx\,dy\,dz$, and $f(x, y, z) = xyz$. Then $v_1 + v_2 = (y + e^{x^2})\,dx\,dy\,dz$ and $fv_1 = y^2xz\,dx\,dy\,dz$.   ▲

Although we can add two 0-forms, two 1-forms, two 2-forms, or two 3-forms, we do not need to add a k-form and a j-form if $k \neq j$. For example, we shall not need to write

$$f(x, y, z)\,dx\,dy + g(x, y, z)\,dz.$$

Now that we have defined these formal objects (forms), one can legitimately ask what they are good for, how they are used, and, perhaps most important, what they mean. The answer to the first question will become clear as we proceed, but we can immediately describe how to use and interpret them.

A real-valued function on a domain K in $\mathbb{R}^3$ is a rule that assigns to each point in K a real number. Differential forms are, in some sense, generalizations of the real-valued functions we have studied in calculus. In fact, 0-forms on an open set K are just functions on K. Thus a 0-form f takes points in K to real numbers.

We should like to interpret differential k-forms (for $k \geq 1$) not as functions on points in K, but as functions on geometric objects such as curves and surfaces. Many of the early Greek geometers viewed lines and curves as being made up of infinitely many points, and planes and surfaces as being made up of infinitely many curves. Consequently there is at least some historical justification for applying this geometric hierarchy to the interpretation of differential forms.

Given an open subset $K \subset \mathbb{R}^3$, we shall distinguish four types of subsets of K (see Figure 8.6.2).

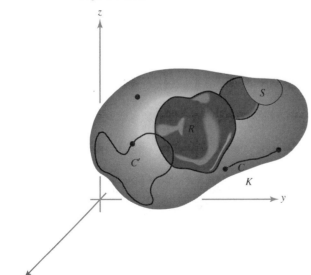

Figure 8.6.2   The four geometric types of subsets of an open set $K \subset \mathbb{R}^3$ to which the theory of forms applies.

(i)   points in K,

(ii)   oriented simple curves and oriented simple closed curves C in K,

(iii)   oriented surfaces $S \subset K$,

(iv)   elementary subregions $R \subset K$.

We shall begin with 1-forms. Let

$$\omega = P(x, y, z)\,dx + Q(x, y, z)\,dy + R(x, y, z)\,dz$$

be a 1-form on K and let C be an oriented simple curve as in Figure 8.6.2. The real number that ω assigns to C is given by the formula

$$\int_C \omega = \int_C P(x, y, z)\,dx + Q(x, y, z)\,dy + R(x, y, z)\,dz. \qquad (1)$$

Recall (see Section 7.2) that this integral is evaluated as follows. Let $\mathbf{c}\colon [a, b] \to K, \mathbf{c}(t) = (x(t), y(t), z(t))$ be an orientation-preserving parametrization of C. Then

$$\int_C \omega = \int_{\mathbf{c}} \omega = \int_a^b \left[P(x(t), y(t), z(t)) \cdot \frac{dx}{dt} \right.$$
$$\left. + Q(x(t), y(t), z(t)) \cdot \frac{dy}{dt} + R(x(t), y(t), z(t)) \cdot \frac{dz}{dt} \right] dt.$$

Theorem 1 of Section 7.2 guarantees that $\int_C \omega$ does not depend on the choice of the parametrization $\mathbf{c}$.

We can thus interpret a 1-form ω on K as a rule assigning a real number to each oriented curve $C \subset K$; a 2-form η will similarly be seen to be a rule assigning a real number to each oriented surface $S \subset K$; and a 3-form v will be a rule assigning a real number to each elementary subregion of K. The rules for associating real numbers with curves, surfaces, and regions are completely contained in the formal expressions we have defined.

EXAMPLE 5   Let $\omega = xy\,dx + y^2\,dy + dz$ be a 1-form on $\mathbb{R}^3$ and let C be the oriented simple curve in $\mathbb{R}^3$ described by the parametrization $\mathbf{c}(t) = (t^2, t^3, 1), 0 \le t \le 1$. C is oriented by choosing the positive direction of C to be the direction in which $\mathbf{c}(t)$ traverses C as t goes from 0 to 1. Then, by formula (1),

$$\int_C \omega = \int_0^1 [t^5(2t) + t^6(3t^2) + 0]\,dt = \int_0^1 (2t^6 + 3t^8)\,dt = \frac{13}{21}.$$

Thus this 1-form ω assigns to each oriented simple curve and each oriented simple closed curve C in $\mathbb{R}^3$ the number $\int_C \omega$.   ▲

A 2-form η on an open set $K \subset \mathbb{R}^3$ may similarly be interpreted as a function that associates with each oriented surface $S \subset K$ a real number. This is accomplished by means of the notion of integration of 2-forms over surfaces. Let

$$\eta = F(x, y, z)\, dx\, dy + G(x, y, z)\, dy\, dz + H(x, y, z)\, dz\, dx$$

be a 2-form on K, and let $S \subset K$ be an oriented surface parametrized by a function $\Phi: D \to \mathbb{R}^3$, $D \subset \mathbb{R}^2$, $\Phi(u, v) = (x(u, v), y(u, v), z(u, v))$ (see Section 7.3).

DEFINITION   If S is such a surface and η is a 2-form on K, we define $\iint_S \eta$ by the formula

$$\iint_S \eta = \iint_S F\, dx\, dy + G\, dy\, dz + H\, dz\, dx$$

$$= \iint_D \left[F(x(u, v), y(u, v), z(u, v)) \cdot \frac{\partial(x, y)}{\partial(u, v)} + G(x(u, v), y(u, v), z(u, v)) \cdot \frac{\partial(y, z)}{\partial(u, v)} \right.$$

$$\left. + H(x(u, v), y(u, v), z(u, v)) \cdot \frac{\partial(z, x)}{\partial(u, v)} \right] du\, dv, \tag{2}$$

where

$$\frac{\partial(x, y)}{\partial(u, v)} = \begin{vmatrix} \dfrac{\partial x}{\partial u} & \dfrac{\partial x}{\partial v} \\[2mm] \dfrac{\partial y}{\partial u} & \dfrac{\partial y}{\partial v} \end{vmatrix}, \qquad \frac{\partial(y, z)}{\partial(u, v)} = \begin{vmatrix} \dfrac{\partial y}{\partial u} & \dfrac{\partial y}{\partial v} \\[2mm] \dfrac{\partial z}{\partial u} & \dfrac{\partial z}{\partial v} \end{vmatrix}, \qquad \frac{\partial(z, x)}{\partial(u, v)} = \begin{vmatrix} \dfrac{\partial z}{\partial u} & \dfrac{\partial z}{\partial v} \\[2mm] \dfrac{\partial x}{\partial u} & \dfrac{\partial x}{\partial v} \end{vmatrix}.$$

If S is composed of several pieces S_i, $i = 1, \ldots, k$, as in Figure 8.4.4, each with its own parametrization Φ_i, we define

$$\iint_S \eta = \sum_{i=1}^{k} \iint_{S_i} \eta.$$

One must verify that $\iint_S \eta$ does not depend on the choice of parametrization Φ. This result is essentially (but not obviously) contained in Theorem 4, Section 7.6.

EXAMPLE 6   Let $\eta = z^2\, dx\, dy$ be a 2-form on $\mathbb{R}^3$, and let S be the upper unit hemisphere in $\mathbb{R}^3$. Find $\iint_S \eta$.

SOLUTION   Let us parametrize S by

$$\Phi(u, v) = (\sin u \cos v, \sin u \sin v, \cos u),$$

where $(u, v) \in D = [0, \pi/2] \times [0, 2\pi]$. By formula (2),

$$\iint_S \eta = \iint_D \cos^2 u \left[\frac{\partial(x, y)}{\partial(u, v)} \right] du\, dv,$$

where

$$\frac{\partial(x, y)}{\partial(u, v)} = \begin{vmatrix} \cos u \cos v & -\sin u \sin v \\ \cos u \sin v & \sin u \cos v \end{vmatrix}$$

$$= \sin u \cos u \cos^2 v + \cos u \sin u \sin^2 v = \sin u \cos u.$$

Therefore

$$\iint_S \eta = \iint_D \cos^2 u \cos u \sin u \, du \, dv$$

$$= \int_0^{2\pi} \int_0^{\pi/2} \cos^3 u \sin u \, du \, dv = \int_0^{2\pi} \left[-\frac{\cos^4 u}{4} \right]_0^{\pi/2} dv = \frac{\pi}{2}. \quad \blacktriangle$$

EXAMPLE 7   Evaluate $\iint_S x \, dy \, dz + y \, dx \, dy$, where S is the oriented surface described by the parametrization $x = u + v$, $y = u^2 - v^2$, $z = uv$, where $(u, v) \in D = [0, 1] \times [0, 1]$.

SOLUTION   By definition, we have

$$\frac{\partial(y, z)}{\partial(u, v)} = \begin{vmatrix} 2u & -2v \\ v & u \end{vmatrix} = 2(u^2 + v^2);$$

$$\frac{\partial(x, y)}{\partial(u, v)} = \begin{vmatrix} 1 & 1 \\ 2u & -2v \end{vmatrix} = -2(u + v).$$

Consequently,

$$\iint_S x \, dy \, dz + y \, dx \, dy = \iint_D [(u + v)(2)(u^2 + v^2) + (u^2 - v^2)(-2)(u + v)] \, du \, dv$$

$$= 4 \iint_D (v^3 + uv^2) \, du \, dv = 4 \int_0^1 \int_0^1 (v^3 + uv^2) \, du \, dv$$

$$= 4 \int_0^1 \left[uv^3 + \frac{u^2 v^2}{2} \right]_0^1 dv = 4 \int_0^1 \left(v^3 + \frac{v^2}{2} \right) dv$$

$$= \left[v^4 + \frac{2v^3}{3} \right]_0^1 = 1 + \frac{2}{3} = \frac{5}{3}. \quad \blacktriangle$$

Finally, we must interpret 3-forms as functions on the elementary subregions of K. Let $v = f(x, y, z) \, dx \, dy \, dz$ be a 3-form and let $R \subset K$ be an elementary subregion of K. Then to each such $R \subset K$ we assign the number

$$\iiint_R v = \iiint_R f(x, y, z) \, dx \, dy \, dz, \tag{3}$$

which is just the ordinary triple integral of f over R, as described in Section 5.6.

EXAMPLE 8   Suppose $v = (x + z)\,dx\,dy\,dz$ and $R = [0, 1] \times [0, 1] \times [0, 1]$. Evaluate $\iiint_R v$.

SOLUTION   We compute:

$$\iiint_R v = \iiint_R (x + z)\,dx\,dy\,dz = \int_0^1 \int_0^1 \int_0^1 (x + z)\,dx\,dy\,dz$$

$$= \int_0^1 \int_0^1 \left[\frac{x^2}{2} + zx\right]_0^1 dy\,dz = \int_0^1 \int_0^1 \left(\frac{1}{2} + z\right)dy\,dz = \int_0^1 \left(\frac{1}{2} + z\right)dz$$

$$= \left[\frac{z}{2} + \frac{z^2}{2}\right]_0^1 = 1. \quad \blacktriangle$$

We now discuss the algebra (or rules of multiplication) of forms, which, together with differentiation of forms, will enable us to state Green's, Stokes', and Gauss' theorems in terms of differential forms.

If ω is a k-form and η is an l-form on K, $0 \le k + l \le 3$, there is a product called the **wedge product** $\omega \wedge \eta$ of ω and η which is a $k + l$ form on K. The wedge product satisfies the following laws:

(i)   For each k there is a zero k-form 0 with the property that $0 + \omega = \omega$ for all k-forms ω and $0 \wedge \eta = 0$ for all l-forms η if $0 \le k + l \le 3$.

(ii)   (*Distributivity*) If f is a 0-form, then

$$(f\omega_1 + \omega_2) \wedge \eta = f(\omega_1 \wedge \eta) + (\omega_2 \wedge \eta).$$

(iii) (*Anticommutativity*) $\omega \wedge \eta = (-1)^{kl}(\eta \wedge \omega)$.

(iv)   (*Associativity*) If $\omega_1, \omega_2, \omega_3$ are k_1, k_2, k_3 forms, respectively, with $k_1 + k_2 + k_3 \le 3$, then

$$\omega_1 \wedge (\omega_2 \wedge \omega_3) = (\omega_1 \wedge \omega_2) \wedge \omega_3.$$

(v)   (*Homogeneity with respect to functions*) If f is a 0-form, then

$$\omega \wedge (f\eta) = (f\omega) \wedge \eta = f(\omega \wedge \eta).$$

Notice that rules (ii) and (iii) actually imply rule (v).

(vi)   The following multiplication rules for 1-forms hold:

$$dx \wedge dy = dx\,dy$$
$$dy \wedge dx = -dx\,dy = (-1)(dx \wedge dy)$$
$$dy \wedge dz = dy\,dz = (-1)(dz \wedge dy)$$
$$dz \wedge dx = dz\,dx = (-1)(dx \wedge dz)$$
$$dx \wedge dx = 0, \ dy \wedge dy = 0, \ dz \wedge dz = 0$$
$$dx \wedge (dy \wedge dz) = (dx \wedge dy) \wedge dz = dx\,dy\,dz.$$

(vii) If f is a 0-form and ω is any k-form, then $f \wedge \omega = f\omega$.

Using laws (i) to (vii), we can now find a unique product of any l-form η and any k-form ω, if $0 \le k + l \le 3$.

EXAMPLE 9   Show that $dx \wedge dy\,dz = dx\,dy\,dz$.

SOLUTION   By rule (vi), $dy\,dz = dy \wedge dz$. Therefore

$$dx \wedge dy\,dz = dx \wedge (dy \wedge dz) = dx\,dy\,dz. \quad \blacktriangle$$

EXAMPLE 10   If $\omega = x\,dx + y\,dy$ and $\eta = zy\,dx + xz\,dy + xy\,dz$, find $\omega \wedge \eta$.

SOLUTION   Computing $\omega \wedge \eta$, to get

$$
\begin{aligned}
\omega \wedge \eta &= (x\,dx + y\,dy) \wedge (zy\,dx + xz\,dy + xy\,dz) \\
&= [(x\,dx + y\,dy) \wedge (zy\,dx)] + [(x\,dx + y\,dy) \wedge (xz\,dy)] \\
&\quad + [(x\,dx + y\,dy) \wedge (xy\,dz)] \\
&= xyz(dx \wedge dx) + zy^2(dy \wedge dx) + x^2z(dx \wedge dy) + xyz(dy \wedge dy) \\
&\quad + x^2y(dx \wedge dz) + xy^2(dy \wedge dz) \\
&= -zy^2\,dx\,dy + x^2z\,dx\,dy - x^2y\,dz\,dx + xy^2\,dy\,dz \\
&= (x^2z - y^2z)\,dx\,dy - x^2y\,dz\,dx + xy^2\,dy\,dz. \quad \blacktriangle
\end{aligned}
$$

EXAMPLE 11   If $\omega = x\,dx - y\,dy$ and $\eta = x\,dy\,dz + z\,dx\,dy$, find $\omega \wedge \eta$.

SOLUTION

$$
\begin{aligned}
\omega \wedge \eta &= (x\,dx - y\,dy) \wedge (x\,dy\,dz + z\,dx\,dy) \\
&= [(x\,dx - y\,dy) \wedge (x\,dy\,dz)] + [(x\,dx - y\,dy) \wedge (z\,dx\,dy)] \\
&= (x^2\,dx \wedge dy\,dz) - (xy\,dy \wedge dy\,dz) + (xz\,dx \wedge dx\,dy) - (yz\,dy \wedge dx\,dy) \\
&= [x^2\,dx \wedge (dy \wedge dz)] - [xy\,dy \wedge (dy \wedge dz)] \\
&\quad + [xz\,dx \wedge (dx \wedge dy)] - [yz\,dy \wedge (dx \wedge dy)] \\
&= x^2\,dx\,dy\,dz - [xy(dy \wedge dy) \wedge dz] \\
&\quad + [xz(dx \wedge dx) \wedge dy] - [yz(dy \wedge dx) \wedge dy] \\
&= x^2\,dx\,dy\,dz - xy(0 \wedge dz) + xz(0 \wedge dy) + [yz(dy \wedge dy) \wedge dx] \\
&= x^2\,dx\,dy\,dz. \quad \blacktriangle
\end{aligned}
$$

The last major step in the development of this theory is to show how to differentiate forms. The derivative of a k-form is a $(k + 1)$-form if $k < 3$, and the derivative of a 3-form is always zero. If ω is a k-form, we shall denote the derivative of ω by $d\omega$. The operation d has the following properties:

(1)   If $f: K \to \mathbb{R}$ is a 0-form, then

$$df = \frac{\partial f}{\partial x}\,dx + \frac{\partial f}{\partial y}\,dy + \frac{\partial f}{\partial z}\,dz.$$

(2)   *(Linearity)* If ω_1 and ω_2 are k-forms, then

$$d(\omega_1 + \omega_2) = d\omega_1 + d\omega_2.$$

(3)   If ω is a k-form and η is an l-form,

$$d(\omega \wedge \eta) = (d\omega \wedge \eta) + (-1)^k(\omega \wedge d\eta).$$

(4) $d(d\omega) = 0$ and $d(dx) = d(dy) = d(dz) = 0$ or, simply, $d^2 = 0$.

Properties (1) to (4) provide enough information to allow us to uniquely differentiate any form.

EXAMPLE 12   Let $\omega = P(x, y, z)\,dx + Q(x, y, z)\,dy$ be a 1-form on some open set $K \subset \mathbb{R}^3$. Find $d\omega$.

SOLUTION

$d[P(x, y, z)\,dx + Q(x, y, z)\,dy]$

$\quad = d[P(x, y, z) \wedge dx] + d[Q(x, y, z) \wedge dy]$ (using 2)

$\quad = (dP \wedge dx) + [P \wedge d(dx)] + (dQ \wedge dy) + [Q \wedge d(dy)]$ (using 3)

$\quad = (dP \wedge dx) + (dQ \wedge dy)$ (using 4)

$\quad = \left(\dfrac{\partial P}{\partial x}\,dx + \dfrac{\partial P}{\partial y}\,dy + \dfrac{\partial P}{\partial z}\,dz\right) \wedge dx + \left(\dfrac{\partial Q}{\partial x}\,dx + \dfrac{\partial Q}{\partial y}\,dy + \dfrac{\partial Q}{\partial z}\,dz\right) \wedge dy$ (using 1)

$\quad = \left(\dfrac{\partial P}{\partial x}\,dx \wedge dx\right) + \left(\dfrac{\partial P}{\partial y}\,dy \wedge dx\right) + \left(\dfrac{\partial P}{\partial z}\,dz \wedge dx\right)$

$\quad\quad + \left(\dfrac{\partial Q}{\partial x}\,dx \wedge dy\right) + \left(\dfrac{\partial Q}{\partial y}\,dy \wedge dy\right) + \left(\dfrac{\partial Q}{\partial z}\,dz \wedge dy\right)$

$\quad = -\dfrac{\partial P}{\partial y}\,dx\,dy + \dfrac{\partial P}{\partial z}\,dz\,dx + \dfrac{\partial Q}{\partial x}\,dx\,dy - \dfrac{\partial Q}{\partial z}\,dy\,dz$

$\quad = \left(\dfrac{\partial Q}{\partial x} - \dfrac{\partial P}{\partial y}\right)dx\,dy + \dfrac{\partial P}{\partial z}\,dz\,dx - \dfrac{\partial Q}{\partial z}\,dy\,dz.$ ▲

EXAMPLE 13   Let f be a 0-form. Using only differentiation rules (1) to (3) and the fact that $d(dx) = d(dy) = d(dz) = 0$, show that $d(df) = 0$.

SOLUTION By rule (1),

$$df = \frac{\partial f}{\partial x}\,dx + \frac{\partial f}{\partial y}\,dy + \frac{\partial f}{\partial z}\,dz$$

and so

$$d(df) = d\left(\frac{\partial f}{\partial x}\,dx\right) + d\left(\frac{\partial f}{\partial y}\,dy\right) + d\left(\frac{\partial f}{\partial z}\,dz\right).$$

Working only with the first term, using rule (3), we get

$$d\left(\frac{\partial f}{\partial x}\,dx\right) = d\left(\frac{\partial f}{\partial x} \wedge dx\right) = d\left(\frac{\partial f}{\partial x}\right) \wedge dx + \frac{\partial f}{\partial x} \wedge d(dx)$$

$$= \left(\frac{\partial^2 f}{\partial x^2}\,dx + \frac{\partial^2 f}{\partial y \partial x}\,dy + \frac{\partial^2 f}{\partial z \partial x}\,dz\right) \wedge dx + 0$$

$$= \frac{\partial^2 f}{\partial y \partial x}\,dy \wedge dx + \frac{\partial^2 f}{\partial z \partial x}\,dz \wedge dx$$

$$= -\frac{\partial^2 f}{\partial y \partial x}\,dx\,dy + \frac{\partial^2 f}{\partial z \partial x}\,dz\,dx.$$

Similarly, we find that

$$d\left(\frac{\partial f}{\partial y}\,dy\right) = \frac{\partial^2 f}{\partial x \partial y}\,dx\,dy - \frac{\partial^2 f}{\partial z \partial y}\,dy\,dz$$

and

$$d\left(\frac{\partial f}{\partial z}\,dz\right) = -\frac{\partial^2 f}{\partial x \partial z}\,dz\,dx + \frac{\partial^2 f}{\partial y \partial z}\,dy\,dz.$$

Adding these up, we get $d(df) = 0$ by the equality of mixed partial derivatives. ▲

EXAMPLE 14 Show that $d(\,dx\,dy)$, $d(\,dy\,dz)$, and $d(\,dz\,dx)$ are all zero.

SOLUTION To prove the first case, we use property (3):

$$d(dx\,dy) = d(dx \wedge dy) = [d(dx) \wedge dy - dx \wedge d(dy)] = 0.$$

The other cases are similar. ▲

EXAMPLE 15 If $\eta = F(x, y, z)\,dx\,dy + G(x, y, z)\,dy\,dz + H(x, y, z)\,dz\,dx$, find $d\eta$.

SOLUTION By property (2),

$$d\eta = d(F\,dx\,dy) + d(G\,dy\,dz) + d(H\,dz\,dx).$$

We shall compute $d(F\,dx\,dy)$. Using property (3) again, we get

$$d(F\,dx\,dy) = d(F \wedge dx\,dy) = dF \wedge (dx\,dy) + F \wedge d(dx\,dy).$$

By Example 14, $d(dx\,dy) = 0$, so we are left with

$$dF \wedge (dx\,dy) = \left(\frac{\partial F}{\partial x}\,dx + \frac{\partial F}{\partial y}\,dy + \frac{\partial F}{\partial z}\,dz \right) \wedge (dx \wedge dy)$$

$$= \left[\frac{\partial F}{\partial x}dx \wedge (dx \wedge dy) \right] + \left[\frac{\partial F}{\partial y}dy \wedge (dx \wedge dy) \right] + \left[\frac{\partial F}{\partial z}dz \wedge (dx \wedge dy) \right].$$

Now

$$dx \wedge (dx \wedge dy) = (dx \wedge dx) \wedge dy = 0 \wedge dy = 0,$$

$$dy \wedge (dx \wedge dy) = -dy \wedge (dy \wedge dx)$$

$$= -(dy \wedge dy) \wedge dx = 0 \wedge dx = 0,$$

and

$$dz \wedge (dx \wedge dy) = (-1)^2 (dx \wedge dy) \wedge dz = dx\,dy\,dz.$$

Consequently,

$$d(F\,dx\,dy) = \frac{\partial F}{\partial z}\,dx\,dy\,dz.$$

Analogously, we find that

$$d(G\,dy\,dz) = \frac{\partial G}{\partial x}\,dx\,dy\,dz \quad \text{and} \quad d(H\,dz\,dx) = \frac{\partial H}{\partial y}\,dx\,dy\,dz.$$

Therefore

$$d\eta = \left(\frac{\partial F}{\partial z} + \frac{\partial G}{\partial x} + \frac{\partial H}{\partial y} \right) dx\,dy\,dz. \quad \blacktriangle$$

We have now developed all the concepts needed to reformulate Green's, Stokes', and Gauss' theorems in the language of forms.

THEOREM 13: GREEN'S THEOREM   *Let D be an elementary region in the xy plane, with ∂D given the counterclockwise orientation. Suppose $\omega = P(x, y)\,dx + Q(x, y)\,dy$ is a 1-form on some open set K in $\mathbb{R}^3$ that contains D. Then*

$$\int_{\partial D} \omega = \iint_D d\omega.$$

Here $d\omega$ is a 2-form on K and D is in fact a surface in $\mathbb{R}^3$ parametrized by $\boldsymbol{\Phi}: D \to \mathbb{R}^3$, $\boldsymbol{\Phi}(x, y) = (x, y, 0)$. Since P and Q are explicitly *not* functions of z, then $\partial P / \partial z$

and $\partial Q / \partial z = 0$, and by Example 12, $d\omega = (\partial Q / \partial x - \partial P / \partial y) \, dx \, dy$. Consequently, Theorem 13 means nothing more than that

$$\int_{\partial D} P \, dx + Q \, dy = \iint_D \left(\frac{\partial Q}{\partial x} - \frac{\partial P}{\partial y} \right) dx \, dy,$$

which is precisely Green's theorem. Hence Theorem 13 holds. Likewise, we have the following theorems.

THEOREM 14: STOKES' THEOREM *Let S be an oriented surface in $\mathbb{R}^3$ with a boundary consisting of a simple closed curve ∂S (Figure 8.6.3) oriented as the boundary of S (see Figure 8.2.1). Suppose that ω is a 1-form on some open set K that contains S. Then*

$$\int_{\partial S} \omega = \iint_S d\omega.$$

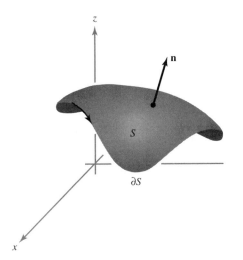

Figure 8.6.3 An oriented surface to which Stokes' theorem applies.

THEOREM 15: GAUSS' THEOREM *Let $\Omega \subset \mathbb{R}^3$ be an elementary region with $\partial \Omega$ given the outward orientation (see Section 8.4). If η is a 2-form on some region K containing Ω, then*

$$\iint_{\partial \Omega} \eta = \iiint_{\Omega} d\eta.$$

The reader has probably noticed the strong similarity in the statements of these theorems. In the vector-field formulations, we used divergence for regions in $\mathbb{R}^3$ (Gauss' theorem) and the curl for surfaces in $\mathbb{R}^3$ (Stokes' theorem) and regions in $\mathbb{R}^2$ (Green's theorem). Here we just use the unified notion of the derivative of a differential form for all three theorems; and, in fact, we can state all theorems as one by introducing a little more terminology.

By an ***oriented 2-manifold with boundary*** in $\mathbb{R}^3$ we mean a surface in $\mathbb{R}^3$ whose boundary is a simple closed curve with orientation as described in Section 8.2. By an ***oriented 3-manifold*** in $\mathbb{R}^3$ we mean an elementary region in $\mathbb{R}^3$ (we assume its boundary, which is a surface, is given the outward orientation discussed in Section 8.4). We call the following unified theorem "Stokes' theorem," according to the current convention.

THEOREM 16: GENERAL STOKES' THEOREM   *Let M be an oriented k-manifold in $\mathbb{R}^3$ ($k = 2$ or 3) contained in some open set K. Suppose ω is $(k-1)$-form on K. Then*

$$\int_{\partial M} \omega = \int_M d\omega.$$

Here the integral is interpreted as a single, double, or triple integral as is appropriate. In fact, it is this form of Stokes' theorem that generalizes to spaces of arbitrary dimension.

EXERCISES

1. Evaluate $\omega \wedge \eta$ if

(a) $\omega = 2x\,dx + y\,dy$
      $\eta = x^3\,dx + y^2\,dy$

(b) $\omega = x\,dx - y\,dy$
      $\eta = y\,dx + x\,dy$

(c) $\omega = x\,dx + y\,dy + z\,dz$
      $\eta = z\,dx\,dy + x\,dy\,dz + y\,dz\,dx$

(d) $\omega = xy\,dy\,dz + x^2\,dx\,dy$
      $\eta = dx + dz$

(e) $\omega = e^{xyz}\,dx\,dy$
      $\eta = e^{-xyz}\,dz$

2. Prove that

$$(a_1\,dx + a_2\,dy + a_3\,dz) \wedge (b_1\,dy\,dz + b_2\,dz\,dx + b_3\,dx\,dy) = \left(\sum_{i=1}^{3} a_i b_i\right) dx\,dy\,dz.$$

3. Find $d\omega$ in the following examples:

(a) $\omega = x^2 y + y^3$
(c) $\omega = xy\,dy + (x+y)^2\,dx$
(e) $\omega = (x^2 + y^2)\,dy\,dz$
(g) $\omega = \dfrac{-x}{x^2 + y^2}\,dx + \dfrac{y}{x^2 + y^2}\,dy$

(b) $\omega = y^2 \cos x\,dy + xy\,dx + dz$
(d) $\omega = x\,dx\,dy + z\,dy\,dz + y\,dz\,dx$
(f) $\omega = (x^2 + y^2 + z^2)\,dz$
(h) $\omega = x^2 y\,dy\,dz$

4. Let $\mathbf{V}: K \to \mathbb{R}^3$ be a vector field defined by $\mathbf{V}(x, y, z) = G(x, y, z)\mathbf{i} + H(x, y, z)\mathbf{j} + F(x, y, z)\mathbf{k}$, and let η be the 2-form on K given by

$$\eta = F\, dx\, dy + G\, dy\, dz + H\, dz\, dx.$$

Show that $d\eta = (\text{div } \mathbf{V})\, dx\, dy\, dz$.

5. If $\mathbf{V} = A(x, y, z)\mathbf{i} + B(x, y, z)\mathbf{j} + C(x, y, z)\mathbf{k}$ is a vector field on $K \subset \mathbb{R}^3$, define the operation Form$_2$: Vector Fields $\to$ 2-forms by

$$\text{Form}_2(\mathbf{V}) = A\, dy\, dz + B\, dz\, dx + C\, dx\, dy.$$

(a) Show that $\text{Form}_2(\alpha \mathbf{V}_1 + \mathbf{V}_2) = \alpha\, \text{Form}_2(\mathbf{V}_1) + \text{Form}_2(\mathbf{V}_2)$, where α is a real number.
(b) Show that $\text{Form}_2(\text{curl } \mathbf{V}) = d\omega$, where $\omega = A\, dx + B\, dy + C\, dz$.

6. Using the differential form version of Stokes' theorem, prove the vector field version in Section 8.2. Repeat for Gauss' theorem.

7. Interpret Theorem 16 in the case $k = 1$.

8. Let $\omega = (x + y)\, dz + (y + z)\, dx + (x + z)\, dy$, and let S be the upper part of the unit sphere; that is, S is the set of (x, y, z) with $x^2 + y^2 + z^2 = 1$ and $z \geq 0$. ∂S is the unit circle in the xy plane. Evaluate $\int_{\partial S} \omega$ both directly and by Stokes' theorem.

9. Let T be the triangular solid bounded by the xy plane, the xz plane, the yz plane, and the plane $2x + 3y + 6z = 12$. Compute

$$\iint_{\partial T} F_1\, dx\, dy + F_2\, dy\, dz + F_3\, dz\, dx$$

directly and by Gauss' theorem, if
   (a) $F_1 = 3y$, $F_2 = 18z$, $F_3 = -12$; and
   (b) $F_1 = z$, $F_2 = x^2$, $F_3 = y$.

10. Evaluate $\iint_S \omega$ where $\omega = z\, dx\, dy + x\, dy\, dz + y\, dz\, dx$ and S is the unit sphere, directly and by Gauss' theorem.

11. Let R be an elementary region in $\mathbb{R}^3$. Show that the volume of R is given by the formula

$$v(R) = \frac{1}{3} \iint_{\partial R} x\, dy\, dz + y\, dz\, dx + z\, dx\, dy.$$

12. In Section 4.2 we saw that the length $l(\mathbf{c})$ of a curve $\mathbf{c}(t) = (x(t), y(t), z(t))$, $a \leq t \leq b$, was given by the formula

$$l(\mathbf{c}) = \int ds = \int_a^b \left(\frac{ds}{dt}\right) dt$$

where, loosely speaking, $(ds)^2 = (dx)^2 + (dy)^2 + (dz)^2$, that is,

$$\frac{ds}{dt} = \sqrt{\left(\frac{dx}{dt}\right)^2 + \left(\frac{dy}{dt}\right)^2 + \left(\frac{dz}{dt}\right)^2}.$$

Now suppose a surface S is given in parametrized form by $\Phi(u, v) = (x(u, v), y(u, v), z(u, v))$, where $(u, v) \in D$. Show that the area of S can be expressed as

$$A(S) = \iint_D dS$$

where formally $(dS)^2 = (dx \wedge dy)^2 + (dy \wedge dz)^2 + (dz \wedge dx)^2$, a formula requiring interpretation. (HINT:

$$dx = \frac{\partial x}{\partial u} du + \frac{\partial x}{\partial v} dv,$$

and similarly for dy and dz. Use the law of forms for the basic 1-forms du and dv. Then dS turns out to be a function times the basic 2-form $du\, dv$, which we can integrate over D.)

REVIEW EXERCISES FOR CHAPTER 8

1. Let $\mathbf{F} = 2yz\mathbf{i} + (-x + 3y + 2)\mathbf{j} + (x^2 + z)\mathbf{k}$. Evaluate $\iint_S (\nabla \times \mathbf{F}) \cdot d\mathbf{S}$, where S is the cylinder $x^2 + y^2 = a^2, 0 \leq z \leq 1$ (without the top and bottom). What if the top and bottom are included?

2. Let Ω be a region in $\mathbb{R}^3$ with boundary $\partial\Omega$. Prove the identity

$$\iint_{\partial\Omega} [\mathbf{F} \times (\nabla \times \mathbf{G})] \cdot d\mathbf{S} = \iiint_{\Omega} (\nabla \times \mathbf{F}) \cdot (\nabla \times \mathbf{G}) \, dV - \iiint_{\Omega} \mathbf{F} \cdot (\nabla \times \nabla \times \mathbf{G}) \, dV.$$

3. Let $\mathbf{F} = x^2 y\mathbf{i} + z^8\mathbf{j} - 2xyz\mathbf{k}$. Evaluate the integral of $\mathbf{F}$ over the surface of the unit cube.

4. Verify Green's theorem for the line integral

$$\int_C x^2 y \, dx + y \, dy$$

when C is the boundary of the region between the curves $y = x$ and $y = x^3$, $0 \leq x \leq 1$.

5. (a) Show that $\mathbf{F} = (x^3 - 2xy^3)\mathbf{i} - 3x^2 y^2\mathbf{j}$ is a gradient vector field.
(b) Evaluate the integral of $\mathbf{F}$ along the path $x = \cos^3\theta, y = \sin^3\theta, 0 \leq \theta \leq \pi/2$.

6. Can you derive Green's theorem in the plane from Gauss' theorem?

7. (a) Show that $\mathbf{F} = 6xy(\cos z)\mathbf{i} + 3x^2(\cos z)\mathbf{j} - 3x^2 y(\sin z)\mathbf{k}$ is conservative (see Section 8.3).
(b) Find f such that $\mathbf{F} = \nabla f$.
(c) Evaluate the integral of $\mathbf{F}$ along the curve $x = \cos^3\theta, y = \sin^3\theta, z = 0, 0 \leq \theta \leq \pi/2$.

8. Let $\mathbf{r}(x, y, z) = (x, y, z), r = \|\mathbf{r}\|$. Show that $\nabla^2(\log r) = 1/r^2$ and $\nabla^2(r^n) = n(n + 1)r^{n-2}$.

9. Let the velocity of a fluid be described by $\mathbf{F} = 6xz\mathbf{i} + x^2 y\mathbf{j} + yz\mathbf{k}$. Compute the rate at which fluid is leaving the unit cube.

10. Let $\mathbf{F} = x^2\mathbf{i} + (x^2 y - 2xy)\mathbf{j} - x^2 z\mathbf{k}$. Does there exist a $\mathbf{G}$ such that $\mathbf{F} = \nabla \times \mathbf{G}$?

11. Let **a** be a constant vector and $\mathbf{F} = \mathbf{a} \times \mathbf{r}$ [as usual, $\mathbf{r}(x, y, z) = (x, y, z)$]. Is **F** conservative? If so, find a potential for it.

12. Consider the case of incompressible fluid flow with velocity field **F** and density ρ.
   (a) If ρ is constant for each fixed t, then show that ρ is constant in t as well.
   (b) If ρ is constant in t, then show that $\mathbf{F} \cdot \nabla \rho = 0$.

13. (a) Let $f(x, y, z) = 3xye^{z^2}$. Compute ∇f.
   (b) Let $\mathbf{c}(t) = (3\cos^3 t, \sin^2 t, e^t), 0 \le t \le \pi$. Evaluate

$$\int_{\mathbf{c}} \nabla f \cdot d\mathbf{s}.$$

   (c) Verify directly Stokes' theorem for gradient vector fields $\mathbf{F} = \nabla f$.

14. Using Green's theorem, or otherwise, evaluate $\int_C x^3 \, dy - y^3 \, dx$, where C is the unit circle $(x^2 + y^2 = 1)$.

15. Evaluate the integral $\iint_S \mathbf{F} \cdot d\mathbf{S}$ where $\mathbf{F} = x\mathbf{i} + y\mathbf{j} + 3\mathbf{k}$ and where S is the surface of the unit sphere $x^2 + y^2 + z^2 = 1$.

16. (a) State Stokes' theorem for surfaces in $\mathbb{R}^3$.
   (b) Let **F** be a vector field on $\mathbb{R}^3$ satisfying $\nabla \times \mathbf{F} = \mathbf{0}$. Use Stokes' theorem to show that $\int_C \mathbf{F} \cdot d\mathbf{s} = 0$ where C is a closed curve.

17. Use Green's theorem to find the area of the loop of the curve $x = a \sin \theta \cos \theta$, $y = a \sin^2 \theta$, for $a > 0$ and $0 \le \theta \le \pi$.

18. Evaluate $\int_C yz \, dx + xz \, dy + xy \, dz$ where C is the curve of intersection of the cylinder $x^2 + y^2 = 1$ and the surface $z = y^2$.

19. Evaluate $\int_C (x+y) \, dx + (2x-z) \, dy + (y+z) \, dz$ where C is the perimeter of the triangle connecting $(2, 0, 0)$, $(0, 3, 0)$, and $(0, 0, 6)$ in that order.

20. Which of the following are conservative fields on $\mathbb{R}^3$? For those that are, find a function f such that $\mathbf{F} = \nabla f$.
   (a) $\mathbf{F}(x, y, z) = 3x^2 y\mathbf{i} + x^3\mathbf{j} + 5\mathbf{k}$
   (b) $\mathbf{F}(x, y, z) = (x + z)\mathbf{i} - (y + z)\mathbf{j} + (x - y)\mathbf{k}$
   (c) $\mathbf{F}(x, y, z) = 2xy^3\mathbf{i} + x^2 z^3\mathbf{j} + 3x^2 yz^2\mathbf{k}$

21. Consider the following two vector fields in $\mathbb{R}^3$:
   (i) $\mathbf{F}(x, y, z) = y^2\mathbf{i} - z^2\mathbf{j} + x^2\mathbf{k}$
   (ii) $\mathbf{G}(x, y, z) = (x^3 - 3xy^2)\mathbf{i} + (y^3 - 3x^2 y)\mathbf{j} + z\mathbf{k}$
   (a) Which of these fields (if any) are conservative on $\mathbb{R}^3$? (That is, which are gradient fields?) Give reasons for your answer.
   (b) Find potential for the fields that are conservative.
   (c) Let α be the path that goes from $(0, 0, 0)$ to $(1, 1, 1)$ by following edges of the cube $0 \le x \le 1$, $0 \le y \le 1, 0 \le z \le 1$ from $(0, 0, 0)$ to $(0, 0, 1)$ to $(0, 1, 1)$ to $(1, 1, 1)$. Let β be the path from $(0, 0, 0)$ to

$(1, 1, 1)$ directly along the diagonal of the cube. Find the values of the line integrals

$$\int_\alpha \mathbf{F} \cdot d\mathbf{s}, \qquad \int_\alpha \mathbf{G} \cdot d\mathbf{s}, \qquad \int_\beta \mathbf{F} \cdot d\mathbf{s}, \qquad \int_\beta \mathbf{G} \cdot d\mathbf{s}.$$

22. Consider the *constant* vector field $\mathbf{F}(x, y, z) = \mathbf{i} + 2\mathbf{j} - \mathbf{k}$ in $\mathbb{R}^3$.
   (a) Find a scalar field $\phi(x, y, z)$ in $\mathbb{R}^3$ such that $\nabla\phi = \mathbf{F}$ in $\mathbb{R}^3$ and $\phi(0, 0, 0) = 0$.
   (b) On the sphere Σ of radius 2 about the origin find all the points at which
      (i) ϕ is a maximum and
      (ii) ϕ is a minimum.
   (c) Compute the maximum and minimum values of ϕ on Σ.

23. Suppose $\nabla \cdot \mathbf{F}(x_0, y_0, z_0) > 0$. Show that for a sufficiently small sphere S centered at (x_0, y_0, z_0), the flux of $\mathbf{F}$ out of S is positive.

ANSWERS TO ODD-NUMBERED EXERCISES

Some solutions requiring proofs may be incomplete or be omitted.

Chapter 1

SECTION 1.1

1. 4; 17

3. $(-104 + 16a, -24 - 4b, -22 + 26c)$

5.

7.

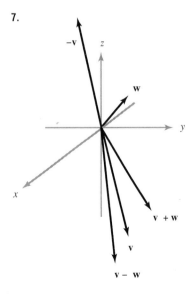

9. $x = 0, z = 0, y \in \mathbb{R}; x = 0, y = 0, z \in \mathbb{R}; y = 0, x, z \in \mathbb{R}; x = 0, y, z \in \mathbb{R}$

11. $\{(2s, 7s + 2t, 7t) \mid s \in \mathbb{R}, t \in \mathbb{R}\}$ **13.** $\mathbf{l}(t) = -\mathbf{i} + (t - 1)\mathbf{j} - \mathbf{k}$

15. $\mathbf{l}(t) = (2t - 1)\mathbf{i} - \mathbf{j} + (3t - 1)\mathbf{k}$ **17.** $\{s\mathbf{i} + 3s\mathbf{k} - 2t\mathbf{j} \mid 0 \le s \le 1, 0 \le t \le 1\}$

19. If (x, y, z) lies on the line, then $x = 2 + t$, $y = -2 + t$, and $z = -1 + t$. Therefore $2x - 3y + z - 2 = 4 + 2t + 6 - 3t - 1 + t - 2 = 7$, which is not zero. Hence no (x, y, z) satisfies both conditions.

21. The set of vectors of the form

$$\mathbf{v} = p\mathbf{a} + q\mathbf{b} + r\mathbf{c}$$

where $0 \le p \le 1, 0 \le q \le 1$, and $0 \le r \le 1$.

23. All points of the form

$$(x_0 + t(x_1 - x_0) + s(x_2 - x_0), y_0 + t(y_1 - y_0) + s(y_2 - y_0), z_0 + t(z_1 - z_0) + s(z_2 - z_0))$$

for real numbers t and s.

25. If one vertex is placed at the origin and the two adjacent sides are $\mathbf{u}$ and $\mathbf{v}$, the new triangle has sides $b\mathbf{u}, b\mathbf{v}$, and $b(\mathbf{u} - \mathbf{v})$.

27. $(1, 0, 1) + (0, 2, 1) = (0, 2, 0) + (1, 0, 2)$

29. Two such lines (there are many others) are $x = 1, y = t, z = t$ and $x = 1, y = t, z = -t$.

SECTION 1.2

1. 6

3. $99°$

5. No, it is 75.7; it would be zero only if the vectors were parallel.

7. $\|\mathbf{u}\| = \sqrt{5}, \|\mathbf{v}\| = \sqrt{2}, \mathbf{u} \cdot \mathbf{v} = -3$ **9.** $\|\mathbf{u}\| = \sqrt{11}, \|\mathbf{v}\| = \sqrt{62}, \mathbf{u} \cdot \mathbf{v} = -14$

11. $\|\mathbf{u}\| = \sqrt{14}, \|\mathbf{v}\| = \sqrt{26}, \mathbf{u} \cdot \mathbf{v} = -17$

13. In Exercise 9, $\cos^{-1}(-14/\sqrt{11}\sqrt{62})$; in Exercise 10, $\pi/2$; and in 11, $\cos^{-1}(-17/2\sqrt{3}\sqrt{26})$.

15. $-4(-\mathbf{i} + \mathbf{j} + \mathbf{k})/3$ **17.** Any (x, y, z) with $x + y + z = 0$; for example, $(1, -1, 0)$ and $(0, 1, -1)$.

19. $\mathbf{i} + 4\mathbf{j}, \theta \approx 0.24$ radian east of north

21. (a) 12:03 P.M. (b) 4.95 km

23.
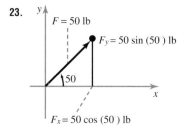

25. $(4.9, 4.9, 4.9)$ and $(-4.9, -4.9, 4.9)$ N.

27. (a) $\mathbf{F} = (3\sqrt{2}\mathbf{i} + 3\sqrt{2}\mathbf{j})$ (b) ≈ 0.322 radian (c) $18\sqrt{2}$

SECTION 1.3

1. $\begin{vmatrix} 1 & 2 & 1 \\ 3 & 0 & 1 \\ 2 & 0 & 2 \end{vmatrix} = -8; \quad \begin{vmatrix} 3 & 0 & 1 \\ 1 & 2 & 1 \\ 2 & 0 & 2 \end{vmatrix} = 8; \quad \begin{vmatrix} 2 & 1 & 1 \\ 0 & 3 & 1 \\ 0 & 2 & 2 \end{vmatrix} = 8; \quad$ etc.

3. $-3\mathbf{i} + \mathbf{j} + 5\mathbf{k}$

5. $\sqrt{35}$

7. 10

9. $\pm\mathbf{k}$

11. $\pm(113\mathbf{i} + 17\mathbf{j} - 103\mathbf{k})/\sqrt{23,667}$

13. $\mathbf{u} + \mathbf{v} = 3\mathbf{i} - 3\mathbf{j} + 3\mathbf{k}; \mathbf{u} \cdot \mathbf{v} = 6; \|\mathbf{u}\| = \sqrt{6}; \|\mathbf{v}\| = 3; \mathbf{u} \times \mathbf{v} = -3\mathbf{i} + 3\mathbf{k}$

15. (a) $x + y + z - 1 = 0$ (b) $x + 2y + 3z - 6 = 0$
     (c) $5x + 2z = 25$ (d) $x + 2y - 3z = 13$

17. (a) Do the first by working out each side in coordinates, and then use that and $\mathbf{A} \times (\mathbf{B} \times \mathbf{C}) = -(\mathbf{B} \times \mathbf{C}) \times \mathbf{A}$ to get the second.
     (b) Use the identities in part (a) to write the quantity in terms of inner products.
     (c) Use the identities in part (a) and collect terms.

19. Compute the results of Cramer's rule and check that they satisfy the equation.

21. $x - 2y + 3z + 12 = 0$

23. $4x - 6y - 10z = 14$

25. $10x - 17y + z + 25 = 0$

27. For Exercise 19, note that $(2, -3, 1) \cdot (1, 1, 1) = 0$, and so the line and plane are parallel and $(2, -2, -1)$ does not lie in the plane. For Exercise 20, the line and plane are parallel and $(1, -1, 2)$ *does* lie in the plane.

29. $\sqrt{2}/13$

31. (a) Show that $\mathbf{M}$ satisfies the geometric properties of $\mathbf{R} \times \mathbf{F}$. (b) $2\sqrt{3}$

33. Show that $n_1(\mathbf{N} \times \mathbf{a})$ and $n_2(\mathbf{N} \times \mathbf{b})$ have the same magnitude and direction.

35. One method is to write out all terms in the left-hand side and see that the terms involving λ all cancel. Another method is to first observe that the determinant is linear in each row or column and that if any row or column is repeated, the answer is zero. Then

$$\begin{vmatrix} a_1 & b_1 & c_1 \\ a_2 + \lambda a_1 & b_2 + \lambda b_1 & c_2 + \lambda c_1 \\ a_3 & b_3 & c_3 \end{vmatrix} = \begin{vmatrix} a_1 & b_1 & c_1 \\ a_2 & b_2 & c_2 \\ a_3 & b_3 & c_3 \end{vmatrix} + \lambda \begin{vmatrix} a_1 & b_1 & c_1 \\ a_1 & b_1 & c_1 \\ a_3 & b_3 & c_3 \end{vmatrix} = \begin{vmatrix} a_1 & b_1 & c_1 \\ a_2 & b_2 & c_2 \\ a_3 & b_3 & c_3 \end{vmatrix}.$$

SECTION 1.4

1. (a)

Cylindrical		
r	θ	z
1	$45°$	1
2	$\pi/2$	-4
0	$45°$	10
3	$\pi/6$	4
1	$\pi/6$	0
2	$3\pi/4$	-2

Rectangular		
x	y	z
$\sqrt{2}/2$	$\sqrt{2}/2$	1
0	2	-4
0	0	10
$3\sqrt{3}/2$	$3/2$	4
$\sqrt{3}/2$	$\frac{1}{2}$	0
$-\sqrt{2}$	$\sqrt{2}$	-2

Spherical		
ρ	θ	ϕ
$\sqrt{2}$	$45°$	$45°$
$2\sqrt{5}$	$\pi/2$	$\pi - \arccos(2\sqrt{5}/5)$
10	$45°$	0
5	$\pi/6$	$\arccos \frac{4}{5}$
1	$\pi/6$	$\pi/2$
$2\sqrt{2}$	$3\pi/4$	$3\pi/4$

(b)

Rectangular		
x	y	z
2	1	-2
0	3	4
$\sqrt{2}$	1	1
$-2\sqrt{3}$	-2	3

Spherical		
ρ	θ	ϕ
3	$\arctan \frac{1}{2}$	$\pi/2 + \arccos\sqrt{5}/3$
5	$\pi/2$	$\arcsin \frac{3}{5}$
2	$\arcsin \sqrt{3}/3$	$\pi/3$
5	$7\pi/6$	$\arccos \frac{3}{5}$

Cylindrical		
r	θ	z
$\sqrt{5}$	$\arctan \frac{1}{2}$	-2
3	$\pi/2$	4
$\sqrt{3}$	$\arcsin \sqrt{3}/3$	1
4	$7\pi/6$	3

3. (a) Rotation by π around the z axis   (b) Reflection across the xy plane

(c) Rotation by $\pi/2$ about the z axis together with a radial expansion by a factor of 2

5. No; (r, θ, ϕ) and $(-r, \theta + \pi, \pi - \phi)$ represent the same point.

7. (a) $\mathbf{e}_\rho = (x\mathbf{i} + y\mathbf{j} + z\mathbf{k})/\sqrt{x^2 + y^2 + z^2} = (x\mathbf{i} + y\mathbf{j} + z\mathbf{k})/\rho$
$\mathbf{e}_\theta = (-y\mathbf{i} + x\mathbf{j})/\sqrt{x^2 + y^2} = (-y\mathbf{i} + x\mathbf{j})/r$
$\mathbf{e}_\phi = (-xz\mathbf{i} - yz\mathbf{j} + (x^2 + y^2)\mathbf{k})/r\rho$
(b) $\mathbf{e}_\theta \times \mathbf{j} = -y\mathbf{k}/\sqrt{x^2 + y^2}, \mathbf{e}_\phi \times \mathbf{j} = (xz/r\rho)\mathbf{k} \pm (r/\rho)\mathbf{i}$

9. (a) The length of $x\mathbf{i} + y\mathbf{j} + z\mathbf{k}$ is $(x^2 + y^2 + z^2)^{1/2} = \rho$
(b) $\cos\phi = z/(x^2 + y^2 + z^2)^{1/2}$ (c) $\cos\theta = x/(x^2 + y^2)^{1/2}$

11. $0 \le r \le a, 0 \le \theta \le 2\pi$ means that (r, θ, z) is inside the cylinder with radius a centered on the z axis, and $|z| \le b$ means that it is no more than a distance b from the xy plane.

13. $-d/(6\cos\phi) \le \rho \le d/2, 0 \le \theta \le 2\pi,$ and $\pi - \cos^{-1}(\frac{1}{3}) \le \phi \le \pi$

15. This is a surface whose cross section with each surface $z = c$ is four-petaled rose. The petals shrink to zero as $|c|$ changes from 0 to 1.

SECTION 1.5

1. 7

3. $|\mathbf{x} \cdot \mathbf{y}| = 10 = \sqrt{5}\sqrt{20} = \|\mathbf{x}\|\|\mathbf{y}\|,$ so $|\mathbf{x} \cdot \mathbf{y}| \le \|\mathbf{x}\|\|\mathbf{y}\|$ is true.
$\|\mathbf{x} + \mathbf{y}\| = 3\sqrt{5} = \|\mathbf{x}\| + \|\mathbf{y}\|,$ so $\|\mathbf{x} + \mathbf{y}\| \le \|\mathbf{x}\| + \|\mathbf{y}\|$ is true.

5. $|\mathbf{x} \cdot \mathbf{y}| = 5 < \sqrt{65} = \|\mathbf{x}\|\|\mathbf{y}\|,$ so $|\mathbf{x} \cdot \mathbf{y}| \le \|\mathbf{x}\|\|\mathbf{y}\|$ is true.
$\|\mathbf{x} + \mathbf{y}\| = \sqrt{28} < \sqrt{5} + \sqrt{13} = \|\mathbf{x}\| + \|\mathbf{y}\|,$ so $\|\mathbf{x} + \mathbf{y}\| \le \|\mathbf{x}\| + \|\mathbf{y}\|$ is true.

7. $AB = \begin{bmatrix} -1 & -1 & 3 \\ -1 & 11 & 3 \\ -6 & 5 & 8 \end{bmatrix},$ det $A = -5,$ det $B = -24,$
det $AB = 120(=$ det A det $B),$ det $(A + B) = -61(\neq$ det $A +$ det $B)$

9. HINT: For $k = 2$ use the triangle inequality to show that $\|\mathbf{x}_1 + \mathbf{x}_2\| \le \|\mathbf{x}_1\| + \|\mathbf{x}_2\|$; then for $k = i+1$ note that $\|\mathbf{x}_1 + \mathbf{x}_2 + \cdots + \mathbf{x}_{i+1}\| \le \|\mathbf{x}_1 + \mathbf{x}_2 + \cdots + \mathbf{x}_i\| + \|\mathbf{x}_{i+1}\|.$

11. (a) Check $n = 1$ and $n = 2$ directly. Then reduce an $n \times n$ determinant to a sum of $(n-1) \times (n-1)$ determinants and use induction.
(b) The argument is similar to that for part (a). Suppose the first row is multiplied by λ. The first term of the sum will be λa_{11} times an $(n - 1) \times (n - 1)$ determinant with no factors of λ. The other terms obtained (by expanding across the first row) are similar.

13. Not necessarily. Try $A = \begin{bmatrix} 0 & 1 \\ 0 & 0 \end{bmatrix}$ and $B = \begin{bmatrix} 1 & 0 \\ 0 & 0 \end{bmatrix}.$

15. (a) The sum of two continuous functions and a scalar multiple of a continuous function are continuous.

(b) (i) $(\alpha f + \beta g) \cdot k = \int_0^1 (\alpha f + \beta g)(x)h(x)\,dx$

$= \int_0^1 f(x)h(x)\,dx + \beta \int_0^1 g(x)h(x)\,dx$

$= \alpha f \cdot h + \beta g \cdot h.$

(ii) $f \cdot g = \int_0^1 f(x)g(x)\,dx = \int_0^1 g(x)f(x)\,dx = g \cdot f.$

In conditions (iii) and (iv), the integrand is a perfect square. Therefore the integral in nonnegative and can be 0 only if the integrand is 0 everywhere. If $f(x) \neq 0$ for some x, then it would be positive in a neighborhood of x by continuity, and the integral would be positive.

17. Compute the matrix product in both orders and check that you get the identity.

19. $(\det A)(\det A^{-1}) = \det(AA^{-1}) = \det(I) = 1$

REVIEW EXERCISES FOR CHAPTER 1

1. $\mathbf{v} + \mathbf{w} = 4\mathbf{i} + 3\mathbf{j} + 6\mathbf{k}$; $3\mathbf{v} = 9\mathbf{i} + 12\mathbf{j} + 15\mathbf{k}$; $6\mathbf{v} + 8\mathbf{w} = 26\mathbf{i} + 16\mathbf{j} + 38\mathbf{k}$; $-2\mathbf{v} = -6\mathbf{i} - 8\mathbf{j} - 10\mathbf{k}$; $\mathbf{v} \cdot \mathbf{w} = 4$; $\mathbf{v} \times \mathbf{w} = 9\mathbf{i} + 2\mathbf{j} - 7\mathbf{k}$. Your sketch should display $\mathbf{v}, \mathbf{w}, 3\mathbf{v}, 6\mathbf{v}, 8\mathbf{w}, 6\mathbf{v} + 8\mathbf{w}, \mathbf{v} \cdot \mathbf{w}$ as the projection of $\mathbf{v}$ along $\mathbf{w}$ and $\mathbf{v} \times \mathbf{w}$ as a vector perpendicular to both $\mathbf{v}$ and $\mathbf{w}$.

3. (a) $\mathbf{l}(t) = -\mathbf{i} + (2 + t)\mathbf{j} - \mathbf{k}$ (b) $\mathbf{l}(t) = (5t - 3)\mathbf{i} + (t + 1)\mathbf{j} - t\mathbf{k}$
(c) $-2x + y + 2z = 9$

5. (a) 0   (b) 5   (c) -10

7. (a) $\pi/2$ (b) $5/2\sqrt{15}$ (c) $-10/\sqrt{6}\sqrt{17}$

9. $\{st\mathbf{a} + s(1 - t)\mathbf{b} \mid 0 \leq t \leq 1 \text{ and } 0 \leq s \leq 1\}$

11. Let $\mathbf{v} = (a_1, a_2, a_3)$, $\mathbf{w} = (b_1, b_2, b_3)$, and apply the CBS inequality.

13. The area is the absolute value of

$$\begin{vmatrix} a_1 & a_2 \\ b_1 & b_2 \end{vmatrix} = \begin{vmatrix} a_1 & a_2 \\ b_1 + \lambda a_1 & b_2 + \lambda a_2 \end{vmatrix}.$$

(A multiple of one row of a determinant may be added to another row without changing its value.) Your sketch should show two parallelograms with the same base and height.

15. The cosines of the two parts of the angle are equal, since
$\mathbf{a} \cdot \mathbf{v}/\|\mathbf{a}\|\|\mathbf{v}\| = (\mathbf{a} \cdot \mathbf{b} + \|\mathbf{a}\|\|\mathbf{b}\|)/\|\mathbf{v}\| = \mathbf{b} \cdot \mathbf{v}/\|\mathbf{b}\|\|\mathbf{v}\|.$

17. $\mathbf{i} \times \mathbf{j} = \begin{vmatrix} \mathbf{i} & \mathbf{j} & \mathbf{k} \\ 1 & 0 & 0 \\ 0 & 1 & 0 \end{vmatrix} = \mathbf{k}$; etc.

19. (a) HINT: The length of the projection of the vector connecting any pair of points, one on each line, onto $(\mathbf{a}_1 \times \mathbf{a}_2)/\|\mathbf{a}_1 \times \mathbf{a}_2\|$ is d.

(b) $\sqrt{2}$

21. (a) Note that

$$\frac{1}{2}\begin{vmatrix} 1 & 1 & 1 \\ x_1 & x_2 & x_3 \\ y_1 & y_2 & y_3 \end{vmatrix} = \frac{1}{2}\begin{vmatrix} 1 & 0 & 0 \\ x_1 & x_2 - x_1 & x_3 - x_1 \\ y_1 & y_2 - y_1 & y_3 - y_1 \end{vmatrix} = \frac{1}{2}\begin{vmatrix} x_2 - x_1 & x_3 - x_1 \\ y_2 - y_1 & y_3 - y_1 \end{vmatrix}.$$

(b) $\frac{1}{2}$

23. Rectangular   Spherical   (plot omitted)

(a) $(\sqrt{2}/2, \sqrt{2}/2, 1)$ (a) $(\sqrt{2}, \pi/4, \pi/4)$

(b) $(3\sqrt{3}/2, \frac{3}{2}, -4)$ (b) $(5, \pi/6, \arccos(-\frac{4}{5}))$

(c) $(0, 0, 1)$ (c) $(1, \pi/4, 0)$

(d) $(0, -2, 1)$ (d) $(\sqrt{5}, 3\pi/2, \arccos(\sqrt{5}/5))$

(e) $(0, 2, 1)$ (e) $(\sqrt{5}, \pi/2, \arccos(\sqrt{5}/5))$

25. $z = r^2 \cos 2\theta$; $\cos\phi = \rho \sin^2\phi \cos 2\theta$

27. $|\mathbf{x} \cdot \mathbf{y}| = 6 < \sqrt{98} = \|\mathbf{x}\| \|\mathbf{y}\|$; $\|\mathbf{x} + \mathbf{y}\| = \sqrt{33} < \sqrt{14} + \sqrt{7} = \|\mathbf{x}\| + \|\mathbf{y}\|$

29. (a) The associative law for matrix multiplication may be checked as follows:

$$[(AB)C]_{ij} = \sum_{k=1}^{n}(AB)_{ik}C_{kj} = \sum_{k=1}^{n}\sum_{l=1}^{n}A_{il}B_{lk}C_{kj}$$

$$= \sum_{l=1}^{n}A_i(BC)_{lj} = [A(BC)]_{ij}.$$

Use this with C taken to be a column vector.

(b) The matrix for the composition is the product matrix.

31. $\mathbb{R}^n$ is spanned by the vectors $\mathbf{e}_1, \mathbf{e}_2, \ldots, \mathbf{e}_n$. If $\mathbf{v} \in \mathbb{R}^n$, then

$$T\mathbf{v} = T\left[\sum_{i=1}^{n}(\mathbf{v}\cdot\mathbf{e}_i)\mathbf{e}_i\right] = \sum_{i=1}^{n}(\mathbf{v}\cdot\mathbf{e}_i)T\mathbf{e}_i.$$

Let $a_{ij} = (T\mathbf{e}_j \cdot \mathbf{e}_i)$, so that

$$T\mathbf{e}_j = \sum_{j=1}^{n}a_{ij}\mathbf{e}_i.$$

Then

$$T\mathbf{v}\cdot\mathbf{e}_k = \sum_{i=1}^{n}(\mathbf{v}\cdot\mathbf{e}_i)a_{ki}.$$

That is, if

$$\mathbf{v} = \begin{bmatrix} v_1 \\ \vdots \\ v_n \end{bmatrix}, \qquad \text{then} \qquad T\mathbf{v} = \begin{bmatrix} a_{11} & \cdots & a_{1n} \\ \vdots & & \vdots \\ a_{n1} & \cdots & a_{nn} \end{bmatrix} \begin{bmatrix} v_1 \\ \vdots \\ v_n \end{bmatrix},$$

as desired.

33. (a) $70\cos\theta + 20\sin\theta$ (b) $(21\sqrt{3} + 6)$ ft · lb

35. Each side equals $2xy - 7yz + 5z^2 - 48x + 54y - 5z - 96$. (Or switch the first two columns and then subtract the first row from the second.)

37. Add the last row to the first and subtract it from the second.

39. (a) $\dfrac{1}{6}\begin{vmatrix} a_1 & a_2 & a_3 \\ b_1 & b_2 & b_3 \\ c_1 & c_2 & c_3 \end{vmatrix}$ (b) $\frac{1}{3}$

41. Use the fact that $\|\mathbf{a}\|^2 = \mathbf{a} \cdot \mathbf{a}$, expand both sides, and use the definition of $\mathbf{c}$.

43. $(1/\sqrt{38})\mathbf{i} - (6/\sqrt{38})\mathbf{j} + (1/\sqrt{38})\mathbf{k}$ **45.** $(2/\sqrt{5})\mathbf{i} - (1/\sqrt{5})\mathbf{j}$

47. $(\sqrt{3}/2)\mathbf{i} + (1/2\sqrt{2})\mathbf{j} + (1/2\sqrt{2})\mathbf{k}$

Chapter 2

SECTION 2.1

1. The level curves and graphs are sketched below. The graph in part (c) is a hyperbolic paraboloid like that of Example 4 but rotated $45°$ and vertically compressed by a factor of $\frac{1}{4}$. To see this, use the variables $u = x + y$ and $v = x - y$. Then $z = (v^2 - u^2)/4$.

(a)

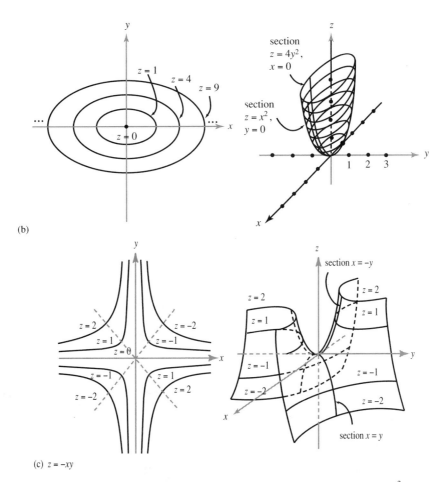

(b)

(c) $z = -xy$

3. For Example 2, $z = r(\cos\theta + \sin\theta) + 2$, shape depends on θ; for Example 3, $z = r^2$, shape is independent of θ; for Example 4, $z = r^2(\cos^2\theta - \sin^2\theta)$, shape depends on θ.

5. The level curves are circles $x^2 + y^2 = 100 - c^2$ when $c \le 10$. The graph is the upper hemisphere of $x^2 + y^2 + z^2 = 100$.

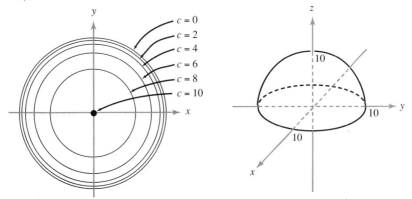

7. The level curves are circles, and the graph is a paraboloid revolution. See Example 3 of this section.

9. If $c = 0$, the level curve is the straight line $y = -x$ together with the line $x = 0$. If $c \neq 0$, then $y = -x + (c/x)$. The level curve is a hyperbola with the y axis and the line $y = -x$ as asymptotes. The graph is a hyperbolic paraboloid. Sections along the line $y = ax$ are the parabolas $z = (1 + a)x^2 = (1 + a)r^2/(1 + a^2)$.

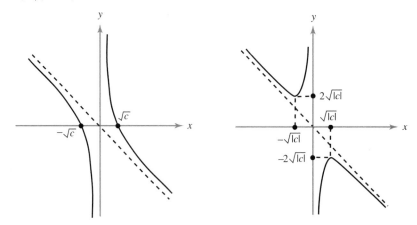

11. If $c > 0$, the level surface $f(x, y, z) = c$ is empty. If $c = 0$, the level surface is the point $(0, 0, 0)$. If $c < 0$, the level surface is the sphere of radius $\sqrt{c}$ centered at $(0, 0, 0)$. A section of the graph determined by $z = a$ is given by $t = -x^2 - y^2 - a^2$, which is a paraboloid of revolution opening down in xyt space.

13. If $c < 0$, the level surface is empty. If $c = 0$, the level surface is the z axis. If $c > 0$, it is the right-circular cylinder $x^2 + y^2 = c$ of radius $\sqrt{c}$ whose axis is the z axis. A section of the graph determined by $z = a$ is the paraboloid of revolution $t = x^2 + y^2$. A section determined by $x = b$ is a "trough" with parabolic cross section $t(y, z) = y^2 + b^2$.

15. Setting $u = (x - z)/\sqrt{2}$ and $v = (x + z)/\sqrt{2}$ gives u and v axes rotated $45°$ around the y axis from the x and z axes. Since $f = vy\sqrt{2}$, and the level surfaces $f = c$ are "cylinders" perpendicular to the vy plane ($z = -x$) whose cross sections are the hyperbolas $vy = c/\sqrt{2}$. The section $S_{x=a} \cap$ graph f is the hyperbolic paraboloid $t = (z+a)y$ in yzt space (see Exercise 1(c)). The section $S_{y=b} \cap$ graph f is the plane $t = bx+bz$ in xzt space. The section $S_{z=b} \cap$ graph f is the hyperbolic paraboloid $t = y(x+b)$ in xyt space.

17. If $c < 0$, the level curve is empty. If $c = 0$, the level curve is the x axis. If $c > 0$, it is the pair of parallel lines $|y| = c$. The sections of graph with x constant are V-shaped curves $x = |y|$ in yz space.

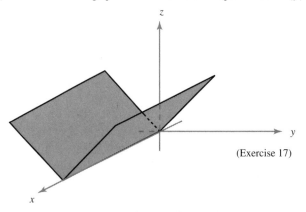

(Exercise 17)

19. The value of z does not matter, so we get a "cylinder" of elliptic cross section parallel to the z axis and intersecting the xy plane in the ellipse $4x^2 + y^2 = 16$.

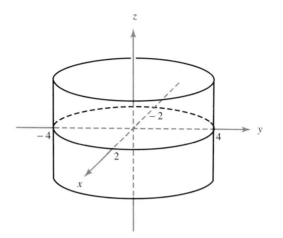

21. The value of x does not matter, so we get a "cylinder" parallel to the x axis of hyberbolic cross section intersecting the yz plane in the hyperbola $z^2 - y^2 = 4$.

23. An elliptic paraboloid with axis along the x axis.

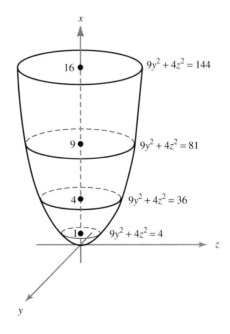

25. The value of y does not matter, so we get a "cylinder" of parabolic cross section.

27. This is a saddle surface similar to that of Example 4, but the hyperbolas, which are level curves, no longer have perpendicular asymptotes.

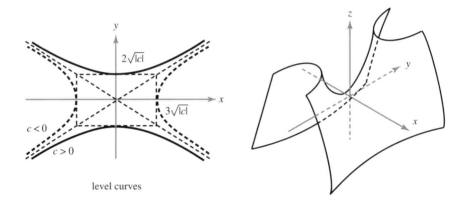

level curves

29. A double cone with axis along the y axis and elliptical cross sections

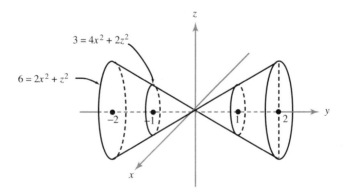

31. Complete the square to get $(x + 2)^2 + (y - b/2)^2 + (z + \frac{9}{2})^2 = (b^2 + 4b + 97)/4$. This is an ellipsoid with center at $(-2, b/2, -\frac{9}{2})$ and axes parallel to the coordinate axes.

33. Level curves are described by $\cos 2\theta = cr^2$. If $c > 0$, then $-\pi/4 \le \theta \le \pi/4$ or $3\pi/4 \le \theta \le 5\pi/4$. If $c < 0$, then $\pi/4 \le \theta \le 3\pi/4$ or $5\pi/4 \le \theta \le 7\pi/4$. In either case you get a figure-8 shape, called a *lemniscate*, through the origin. (Such shapes were first studied by Jacques Bernoulli and are sometimes called Bernoulli's lemniscates.)

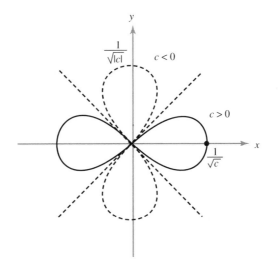

S E C T I O N 2.2

1.   (a)   If $(x_0, y_0) \in A$, then $|x_0| < 1$ and $|y_0| < 1$. Let $r < 1 - |x_0|$ and $r < 1 - |y_0|$. Prove that $D_r(x_0, y_0) \subset A$ either analytically or by drawing a figure.

      (b)   If $(x_0, y_0) \in B$ and $0 < r < y_0$ (e.g., if $r = y_0/2$), then $D_r(x_0, y_0) \subset B$ (prove analytically or by drawing a figure).

      (c)   Let $r = \min(4 - \sqrt{x_0^2 + y_0^2}, \sqrt{x_0^2 + y_0^2} - 2)$.

      (d)   Let r be the smallest of the three numbers used in parts (a), (b), and (c).

      (e)   Let $r = \min(|x_0|, |y_0|)$.

3.   For $|x - 2| < \delta = \sqrt{\varepsilon + 4} - 2$, we have $|x^2 - 4| = |x - 2||x + 2| < \delta(\delta + 4) = \varepsilon$. By Theorem 3(iii), $\lim_{x \to 2} x^2 = (\lim_{x \to 2} x)^2 = 2^2 = 4$.

5.   (a)   5   (b)   0   (c) $2x$ (d)   1   (e) $-\frac{1}{2}$.

7.   (a)   1   (b) $\|\mathbf{x}_0\|$ (c) $(1, e)$

      (d)   limit doesn't exist (look at the limits for $x = 0$ and $y = 0$ separately.)

9.   Compose $f(x, y) = xy$ with $g(t) = (\sin t)/t$ for $t \neq 0$ and $g(0) = 1$.

11.   0

13.   Use parts (ii) and (iii) of Theorem 4.

15.   (a)   Let the value of the function be 1 at $(0, 0)$.   (b)   No.

17.   Let $r = \|\mathbf{x} - \mathbf{y}\|/2$. If $\|\mathbf{z} - \mathbf{y}\| \leq r$, let $f(\mathbf{z}) = \|\mathbf{z} - \mathbf{y}\|/r$. If $\|\mathbf{z} - \mathbf{y}\| > r$, let $f(\mathbf{z}) = 1$.

19. (a) $\lim\limits_{x \to b^+} f(x) = L$ if for every $\varepsilon > 0$ there is a $\delta > 0$ such that $x > b$ and $0 < x - b < \delta$ imply $|f(x) - L| < \varepsilon$.

(b) $\lim\limits_{x \to 0^-} (1/x) = -\infty$, $\lim\limits_{t \to -\infty} e^t = 0$, and so $\lim\limits_{x \to 0^-} e^{1/x} = 0$. Hence $\lim\limits_{x \to 0^-} 1/(1 + e^{1/x}) = 1$. The other limit is 0.

(c)

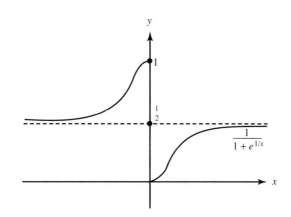

$$\frac{1}{1 + e^{1/x}}$$

21. If $\varepsilon > 0$ and $\mathbf{x}_0$ are given, let $\delta = (\varepsilon/K)^{1/\alpha}$. Then $\|f(\mathbf{x}) - f(\mathbf{x}_0)\| < K\delta^\alpha = \varepsilon$ whenever $\|\mathbf{x} - \mathbf{x}_0\| < \delta$. Notice that the choice of δ does not depend on $\mathbf{x}_0$. This means that f is *uniformly continuous*.

23. (a)   Choose $\delta < 1/500$.   (b)   Choose $\delta < 0.002$.

SECTION 2.3

1. (a) $\partial f/\partial x = y$; $\partial f/\partial y = x$

(b) $\partial f/\partial x = ye^{xy}$; $\partial f/\partial y = xe^{xy}$

(c) $\partial f/\partial x = \cos x \cos y - x \sin x \cos y$; $\partial f/\partial y = -x \cos x \sin y$

(d) $\partial f/\partial x = 2x[1 + \log(x^2 + y^2)]$; $\partial f/\partial y = 2y[1 + \log(x^2 + y^2)]$; $(x, y) \neq (0, 0)$

3. (a) $\partial w/\partial x = (1 + 2x^2) \exp(x^2 + y^2)$; $\partial w/\partial y = 2xy \exp(x^2 + y^2)$

(b) $\partial w/\partial x = -4xy^2/(x^2 - y^2)^2$; $\partial w/\partial y = 4yx^2/(x^2 - y^2)^2$

(c) $\partial w/\partial x = ye^{xy} \log(x^2 + y^2) + 2xe^{xy}/(x^2 + y^2)$;
      $\partial w/\partial y = xe^{xy} \log(x^2 + y^2) + 2ye^{xy}/(x^2 + y^2)$

(d) $\partial w/\partial x = 1/y$; $\partial w/\partial y = -x/y^2$

(e) $\partial w/\partial x = -y^2 e^{xy} \sin ye^{xy} \sin x + \cos ye^{xy} \cos x$; $\partial w/\partial y = (xye^{xy} + e^{xy})(-\sin ye^{xy} \sin x)$

5. $z = 6x + 3y - 11$

7. (a) $\begin{bmatrix} 1 & 0 \\ 0 & 1 \end{bmatrix}$;   (b) $\begin{bmatrix} e^y & xe^y - \sin y \\ 1 & 0 \\ 1 & e^y \end{bmatrix}$;

(c) $\begin{bmatrix} 1 & 1 & e^z \\ 2xy & x^2 & 0 \end{bmatrix}$ (d) $\begin{bmatrix} (y + xy^2)e^{xy} & (x + x^2 y)e^{xy} \\ \sin y & x \cos y \\ 5y^2 & 10xy \end{bmatrix}$

9.   At $z = 1$ **11.**   Both are xye^{xy}.

13. (a) $\nabla f = (e^{-x^2-y^2-z^2}(-2x^2+1), -2xye^{-x^2-y^2-z^2}, -2xze^{-x^2-y^2-z^2})$

(b) $\nabla f = (x^2+y^2+z^2)^{-2}(yz(y^2+z^2-x^2), xz(x^2+z^2-y^2), xy(x^2+y^2-z^2))$

(c) $\nabla f = (z^2e^x \cos y, -z^2e^x \sin y, 2ze^x \cos y)$

15. $2x + 6y - z = 5$ **17.** $-2\mathbf{k}$

19. They are constant. Show that the derivative is the zero matrix.

SECTION 2.4

1. This curve is the ellipse $(y^2/16) + x^2 = 1$:

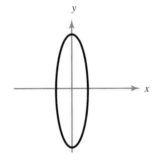

3. This curve is a straight line:

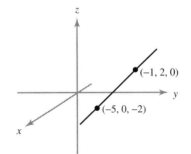

5. $6\mathbf{i} + 6t\mathbf{j} + 3t^2\mathbf{k}$ **7.** $(-2\cos t \sin t, 3 - 3t^2, 1)$

9. $\mathbf{c}'(t) = (e^t, -\sin t)$ **11.** $\mathbf{c}'(t) = (t \cos t + \sin t, 4)$

13. Horizontal when $t = (R/v)n\pi$, n an integer; speed is zero if n is even; speed is $2v$ if n is odd.

15. $(\sin 3, \cos 3, 2) + (3\cos 3, -3\sin 3, 5)(t - 1)$

17. $(8, 8, 0)$ **19.** $(8, 0, 1)$

SECTION 2.5

1. Use parts (i), (ii), and (iii) of Theorem 10. The derivative at $\mathbf{x}$ is $2(f(\mathbf{x})+1)\mathbf{D}f(\mathbf{x})$.

3. (a) $h(x,y)=f(x,u(x,y))=f(p(x),u(x,y))$. We use p here solely as notation: $p(x)=x$.

Written out: $\dfrac{\partial h}{\partial x}=\dfrac{\partial f}{\partial p}\dfrac{dp}{dx}+\dfrac{\partial f}{\partial u}\dfrac{\partial u}{\partial x}=\dfrac{\partial f}{\partial p}+\dfrac{\partial f}{\partial u}\dfrac{\partial u}{\partial x}$ since $\dfrac{dp}{dx}=\dfrac{dx}{dx}=1$

JUSTIFICATION: Call (p,u) the variables of f. To use the chain rule we must express h as a composition of functions; i.e., first find g such that $h(x,y)=f(g(x,y))$. Let $g(x,y)=(p(x),u(x,y))$. Therefore $\mathbf{D}h=(\mathbf{D}f)(\mathbf{D}g)$. Then

$$
\begin{bmatrix}\dfrac{\partial h}{\partial x} & \dfrac{\partial h}{\partial y}\end{bmatrix}=\begin{bmatrix}\dfrac{\partial f}{\partial p} & \dfrac{\partial f}{\partial u}\end{bmatrix}\begin{bmatrix}\dfrac{\partial g_1}{\partial x} & \dfrac{\partial g_1}{\partial y}\\[2mm] \dfrac{\partial g_2}{\partial x} & \dfrac{\partial g_2}{\partial y}\end{bmatrix}=\begin{bmatrix}\dfrac{\partial f}{\partial p} & \dfrac{\partial f}{\partial u}\end{bmatrix}\begin{bmatrix}1 & 0\\[2mm] \dfrac{\partial u}{\partial x} & \dfrac{\partial u}{\partial y}\end{bmatrix}
$$

$$
=\begin{bmatrix}\dfrac{\partial f}{\partial p}+\dfrac{\partial f}{\partial u}\dfrac{\partial u}{\partial x} & \dfrac{\partial f}{\partial u}\dfrac{\partial u}{\partial y}\end{bmatrix}
$$

and so $\dfrac{\partial h}{\partial x}=\dfrac{\partial f}{\partial p}+\dfrac{\partial f}{\partial u}\dfrac{\partial u}{\partial x}$. You may see $\dfrac{\partial h}{\partial x}=\dfrac{\partial f}{\partial x}+\dfrac{\partial f}{\partial u}\dfrac{\partial u}{\partial x}$ as an answer. This requires careful interpretation because of possible ambiguity about the meaning of $\partial f/\partial x$, which is why the name p was used.

(b) $\dfrac{\partial h}{\partial x}=\dfrac{\partial f}{\partial x}+\dfrac{\partial f}{\partial u}\dfrac{du}{dx}+\dfrac{\partial f}{\partial v}\dfrac{dv}{dx}$ (c) $\dfrac{\partial h}{\partial x}=\dfrac{\partial f}{\partial u}\dfrac{\partial u}{\partial x}+\dfrac{\partial f}{\partial v}\dfrac{\partial v}{\partial x}+\dfrac{\partial f}{\partial w}\dfrac{dw}{dx}$

5. Compute each in two ways; the answers are
(a) $(f\circ\mathbf{c})'(t)=e^{t}(\cos t-\sin t)$
(b) $(f\circ\mathbf{c})'(t)=15t^{4}\exp(3t^{5})$
(c) $(f\circ\mathbf{c})'(t)=(e^{2t}-e^{-2t})[1+\log(e^{2t}+e^{-2t})]$
(d) $(f\circ\mathbf{c})'(t)=(1+4t^{2})\exp(2t^{2})$

7. Use Theorem 10(iii) and replace matrices by vectors.

9. $(f\circ g)(x,y)=(\tan(e^{x-y}-1)-e^{x-y},e^{2(x-y)}-(x-y)^{2})$ and $\mathbf{D}(f\circ g)(1,1)=\begin{bmatrix}0 & 0\\ 2 & -2\end{bmatrix}$

11. $\tfrac{1}{2}\cos(1)\cos(\log\sqrt{2})$

13. $-2\cos t\sin t\,e^{\sin t}+\sin^{4}t+\cos^{3}t\,e^{\sin t}-3\cos^{2}t\sin^{2}t$.

15. $(2,0)$

17. (a) $G(x,y(x))=0$ and so $\dfrac{\partial G}{\partial x}+\dfrac{\partial G}{\partial y}\dfrac{dy}{dx}=0$.

(b) $\begin{bmatrix}\dfrac{dy_1}{dx}\\[3mm] \dfrac{dy_2}{dx}\end{bmatrix}=-\begin{bmatrix}\dfrac{\partial G_1}{\partial y_1} & \dfrac{\partial G_1}{\partial y_2}\\[3mm] \dfrac{\partial G_2}{\partial y_1} & \dfrac{\partial G_2}{\partial y_2}\end{bmatrix}^{-1}\begin{bmatrix}\dfrac{\partial G_1}{\partial x}\\[3mm] \dfrac{\partial G_2}{\partial x}\end{bmatrix}$ where^{-1} means the inverse matrix.

The first component of this equation reads

$$\frac{dy_1}{dx} = \frac{-\frac{\partial G_1}{\partial x}\frac{\partial G_2}{\partial y_2} + \frac{\partial G_2}{\partial x}\frac{\partial G_1}{\partial y_2}}{\frac{\partial G_1}{\partial y_1}\frac{\partial G_2}{\partial y_2} - \frac{\partial G_2}{\partial y_1}\frac{\partial G_1}{\partial y_2}}.$$

(c) $\dfrac{dy}{dx} = \dfrac{-2x}{3y^2 + e^y}.$

19. Apply the chain rule to $\partial G/\partial T$ where $G(t(T, P), P(T, P), V(T, P)) = P(V - b)e^{a/RVT} - RT$ is identically 0; $t(T, P) = T$; and $p(T, P) = P$.

21. Define $R_1(\mathbf{h}) = f(\mathbf{x}_0 + \mathbf{h}) - f(\mathbf{x}_0) - [\mathbf{D}f(\mathbf{x}_0)]\mathbf{h}.$

23. Let g_1 and g_2 be C^1 functions from $\mathbb{R}^3$ to $\mathbb{R}$ such that $g_1(\mathbf{x}) = 1$ for $\|\mathbf{x}\| < \sqrt{2}/3$; $g_1(\mathbf{x}) = 0$ for $\|\mathbf{x}\| > 2\sqrt{2}/3$; $g_2(\mathbf{x}) = 1$ for $\|\mathbf{x} - (1, 1, 0)\| < \sqrt{2}/3$; and $g_2(\mathbf{x}) = 0$ for $\|\mathbf{x} - (1, 1, 0)\| > 2\sqrt{2}/3$. (See Exercise 22.) Let

$$h_1(\mathbf{x}) = \begin{bmatrix} 1 & 0 & 0 \\ 0 & -1 & 0 \\ 0 & 0 & 0 \end{bmatrix}\begin{bmatrix} x_1 \\ x_2 \\ x_3 \end{bmatrix} + \begin{bmatrix} 1 \\ 1 \\ 0 \end{bmatrix} \quad \text{and} \quad h_2(\mathbf{x}) = \begin{bmatrix} 0 & 0 & -1 \\ 0 & 0 & 0 \\ 0 & 0 & 1 \end{bmatrix}\begin{bmatrix} x_1 \\ x_2 \\ x_3 \end{bmatrix}$$

and put $f(\mathbf{x}) = g_1(\mathbf{x})h_1(\mathbf{x}) + g_2(\mathbf{x})h_2(\mathbf{x}).$

25. Proof of rule (iii) follows:

$$\frac{|h(\mathbf{x}) - h(\mathbf{x}_0) - [f(\mathbf{x}_0)\mathbf{D}g(\mathbf{x}_0) + g(\mathbf{x}_0)\mathbf{D}f(\mathbf{x}_0)](\mathbf{x} - \mathbf{x}_0)|}{\|\mathbf{x} - \mathbf{x}_0\|}$$

$$\leq |f(\mathbf{x}_0)|\frac{|g(\mathbf{x}) - g(\mathbf{x}_0) - \mathbf{D}g(\mathbf{x}_0)(\mathbf{x} - \mathbf{x}_0)|}{\|\mathbf{x} - \mathbf{x}_0\|}$$

$$+ |g(\mathbf{x}_0)|\frac{|f(\mathbf{x}) - f(\mathbf{x}_0) - \mathbf{D}f(\mathbf{x}_0)(\mathbf{x} - \mathbf{x}_0)|}{\|\mathbf{x} - \mathbf{x}_0\|}$$

$$+ \frac{|f(\mathbf{x}) - f(\mathbf{x}_0)|}{\|\mathbf{x} - \mathbf{x}_0\|}\frac{|g(\mathbf{x}) - g(\mathbf{x}_0)|}{\|\mathbf{x} - \mathbf{x}_0\|}\|\mathbf{x} - \mathbf{x}_0\|$$

As $\mathbf{x} \to \mathbf{x}_0$, the first two terms go to 0 by the differentiability of f and g. The third does so because $|f(\mathbf{x}) - f(\mathbf{x}_0)|/\|\mathbf{x} - \mathbf{x}_0\|$ and $|g(\mathbf{x}) - g(\mathbf{x}_0)|/\|\mathbf{x} - \mathbf{x}_0\|$ are bounded by a constant, say M, on some ball $D_r(\mathbf{x}_0)$. To see, this, choose r small enough that $[f(\mathbf{x}) - f(\mathbf{x}_0)]/\|\mathbf{x} - \mathbf{x}_0\|$ is within 1 of $\mathbf{D}f(\mathbf{x}_0)(\mathbf{x} - \mathbf{x}_0)/\|\mathbf{x} - \mathbf{x}_0\|$ if $\|\mathbf{x} - \mathbf{x}_0\| < r$. Then we have $|f(\mathbf{x}) - f(\mathbf{x}_0)|/\|\mathbf{x} - \mathbf{x}_0\| \leq 1 + |\mathbf{D}f(\mathbf{x}_0)(\mathbf{x} - \mathbf{x}_0)|/\|\mathbf{x} - \mathbf{x}_0\| = 1 + |\nabla f(\mathbf{x}_0) \cdot (\mathbf{x} - \mathbf{x}_0)|/\|\mathbf{x} - \mathbf{x}_0\| \leq 1 + \|\nabla f(\mathbf{x}_0)\|$ by the Cauchy-Schwarz inequality.

The proof of rule (iv) follows from rule (iii) and the special case of the quotient rule, with f identically 1; that is, $\mathbf{D}(1/g)(\mathbf{x}_0) = [-1/g(\mathbf{x}_0)^2]\mathbf{D}g(\mathbf{x}_0)$. To obtain this answer, note that on some small ball $D_r(\mathbf{x}_0)$, $g(\mathbf{x}) > m > 0$. Use the triangle inequality and the Schwarz inequality to show that

$$\frac{\left|\frac{1}{g(\mathbf{x})} - \frac{1}{g(\mathbf{x})} + \frac{1}{g(\mathbf{x}_0)^2}\mathbf{D}g(\mathbf{x}_0)(\mathbf{x} - \mathbf{x}_0)\right|}{\|\mathbf{x} - \mathbf{x}_0\|}$$

$$\leq \frac{1}{|g(\mathbf{x})|}\frac{1}{|g(\mathbf{x}_0)|}\frac{|g(\mathbf{x}) - g(\mathbf{x}_0) - \mathbf{D}g(\mathbf{x}_0)(\mathbf{x} - \mathbf{x}_0)|}{\|\mathbf{x} - \mathbf{x}_0\|}$$

$$+ \frac{|g(\mathbf{x}) - g(\mathbf{x}_0)|}{|g(\mathbf{x})|g(\mathbf{x}_0)^2}\frac{|\mathbf{D}g(\mathbf{x}_0)(\mathbf{x} - \mathbf{x}_0)|}{\|\mathbf{x} - \mathbf{x}_0\|}$$

$$\leq \frac{1}{m^2}\frac{|g(\mathbf{x}) - g(\mathbf{x}_0) - \mathbf{D}g(\mathbf{x}_0)(\mathbf{x} - \mathbf{x}_0)|}{\|\mathbf{x} - \mathbf{x}_0\|} + \frac{\|\nabla g(\mathbf{x}_0)\|}{m^3}|g(\mathbf{x}) - g(\mathbf{x}_0)|$$

These last two terms both go to 0, since g is differentiable and continuous.

27. First find formula for $(\partial/\partial x)(F(x, x))$, using the chain rule. Let $F(x, z) = \int_0^x f(z, y)\, dy$ and use the fundamental theorem of calculus.

29. By Exercise 26 and Theorem 10(iii) (Exercise 25), each component of k is differentiable and $\mathbf{D}k_i(\mathbf{x}_0) = f(\mathbf{x}_0)\mathbf{D}g_i(\mathbf{x}_0) + g_i(\mathbf{x}_0)\mathbf{D}f(\mathbf{x}_0)$. Since $[\mathbf{D}g_i(\mathbf{x}_0)]\mathbf{y}$ is the ith component of $[\mathbf{D}g(\mathbf{x}_0)]\mathbf{y}$ and $[\mathbf{D}f(\mathbf{x}_0)]\mathbf{y}$ is the number $\nabla f(\mathbf{x}_0) \cdot \mathbf{y}$, we get $[\mathbf{D}k(\mathbf{x}_0)]\mathbf{y} = f(\mathbf{x}_0)[\mathbf{D}g(\mathbf{x}_0)]\mathbf{y} + [\mathbf{D}f(\mathbf{x}_0)]\mathbf{y}[g(\mathbf{x}_0)] = f(\mathbf{x}_0)[\mathbf{D}g(x_0)]\mathbf{y} + [\nabla f(\mathbf{x}_0) \cdot \mathbf{y}]g(\mathbf{x}_0)$.

SECTION 2.6

1. $\nabla f(1, 1, 2) \cdot \mathbf{v} = (4, 3, 4) \cdot (1/\sqrt{5}, 2/\sqrt{5}, 0) = 2\sqrt{5}$

3. (a) $17e^e/13$ (b) $e/\sqrt{3}$ (c) 0

5. (a) $z + 9x = 6$ (b) $z + y = \pi/2$ (c) $z = 1$

7. (a) $-\dfrac{1}{3\sqrt{3}}(\mathbf{i} + \mathbf{j} + \mathbf{k})$ (b) $2\mathbf{i} + 2\mathbf{j} + 2\mathbf{k}$ (c) $-\frac{2}{9}(\mathbf{i} + \mathbf{j} + \mathbf{k})$

9. $\mathbf{k}$

11. The graph of f is the level surface $0 = F(x, y, z) = f(x, y) - z$. Therefore the tangent plane is given by

$$0 = \nabla F(x_0, y_0, z_0) \cdot (x - x_0, y - y_0, z - z_0)$$

$$= \left(\frac{\partial f}{\partial x}(x_0, y_0), \frac{\partial f}{\partial y}(x_0, y_0), -1\right) \cdot (x - x_0, y - y_0, z - z_0)$$

Since $z_0 = f(x_0, y_0)$, this is $z = f(x_0, y_0) + (\partial f/\partial x)(x_0, y_0)(x - x_0) + (\partial f/\partial y)(x_0, y_0)(y - y_0)$.

13. (a) $\nabla f = (z + y, z + x, x + y)$, $\mathbf{g}'(t) = (e^t, -\sin t, \cos t)$,
$(f \circ \mathbf{g})'(1) = 2e \cos 1 + \cos^2 1 - \sin^2 1$
(b) $\nabla f = (yze^{xyz}, xze^{xyz}, xye^{xyz})$, $\mathbf{g}'(t) = [6, 6t, 3t^2]$, $(f \circ \mathbf{g})'(1) = 108e^{18}$
(c) $\nabla f = [1 + \log(x^2 + y^2 + z^2)](x\mathbf{i} + y\mathbf{j} + z\mathbf{k})$, $\mathbf{g}' = (e^t - e^{-t}, 1)$,
$(f \circ \mathbf{g})'(1) = [1 + \log(e^2 + e^{-2} + 1)](e^2 - e^{-2} + 1)$

15. Let $f(x, y, z) = 1/r = (x^2 + y^2 + z^2)^{-1/2}$; $\mathbf{r} = (x, y, z)$. Then we calculate $\nabla f = -(x^2 + y^2 + z^2)^{-3/2}(x, y, z) = -(1/r^3)\mathbf{r}$.

17. $\nabla f = (g'(x), 0)$.

19. $\mathbf{D}f(0, 0, \ldots, 0) = [0, \ldots, 0]$

21. $\mathbf{d}_1 = [-(0.03 + 2by_1)/2a]\mathbf{i} + y_1\mathbf{j}$, $\mathbf{d}_2 = [-(0.03 + 2by_2)/2a]\mathbf{i} + y_2\mathbf{j}$, where y_1 and y_2 are the solutions of $(a^2 + b^2)y^2 + 0.03by + \left(\frac{0.03^2}{4} - a^2\right) = 0$.

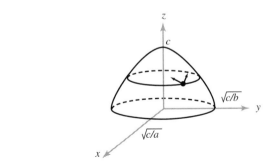

23. $\nabla V = \dfrac{\lambda}{2\pi\varepsilon_0}\left[\left(\dfrac{x + x_0}{r_1^2} - \dfrac{x - x_0}{r_2^2}\right)\mathbf{i} + 2y\left(\dfrac{1}{r_1^2} - \dfrac{1}{r_2^2}\right)\mathbf{j}\right]$

25. Crosses at $(2, 2, 0)$, $\sqrt{5}/10$ seconds later.

SECTION 2.7

1. $\mathbf{D}f(x, y, z) = \begin{bmatrix} e^x & 0 & 0 \\ 0 & -\sin y & 0 \\ 0 & 0 & \cos z \end{bmatrix}$;

$\mathbf{D}f$ is a diagonal matrix if each component function f_i depends only on x_i.

3. (a) Let $A = B = C = \mathbb{R}$ with $f(x) = 0$ and $g(x) = 0$ if $x \neq 0$ and $g(0) = 1$. Then $w = 0$ and $g(f(x)) = 1$ for all x.

(b) If $\varepsilon > 0$, let δ_1 and δ_2 be small enough that $D_{\delta_1}(\mathbf{y}_0) \subset B$ and $\|g(\mathbf{y}) - \mathbf{w}\| < \varepsilon$ whenever $\mathbf{y} \in B$ and $0 < \|\mathbf{y} - \mathbf{y}_0\| < \delta_2$. Since $g(\mathbf{y}_0) = \mathbf{w}$, the $0 < \|\mathbf{y} - \mathbf{y}_0\|$ restriction may be dropped. Let δ be small enough that $\|f(\mathbf{x}) - \mathbf{y}_0\| < \min(\delta_1, \delta_2)$ whenever $\mathbf{x} \in A$ and $0 < \|\mathbf{x} - \mathbf{x}_0\| < \delta$. Then for such $\mathbf{x}$, $\|f(\mathbf{x}) - \mathbf{y}_0\| < \delta_1$, so that $f(\mathbf{x}) \in B$ and $g(f(\mathbf{x}))$ is defined. Also, $\|f(\mathbf{x}) - \mathbf{y}_0\| < \delta_2$, and so $\|g(f(\mathbf{x})) - \mathbf{w}\| < \varepsilon$.

5. Let $\mathbf{x} = (x_1, \ldots, x_n)$ and fix an index k. Then

$$f(\mathbf{x}) = a_{kk}x_k^2 + \sum_{\substack{i=1 \\ i \neq k}}^{n} a_{kj}x_kx_j + \sum_{\substack{j=1 \\ j \neq k}}^{n} a_{kj}x_kx_j + [\text{terms not involving } x_k]$$

and therefore

$$\frac{\partial f}{\partial x_k} = 2a_{kk}x_k + \sum_{i \neq k} a_{ki}x_i + \sum_{i \neq k} a_{ki}x_i = 2\sum_{j=1}^{n} a_{kj}x_j = (2\mathbf{Ax})_k.$$

Since the kth components agree for each k, $\nabla f(\mathbf{x}) = 2\mathbf{Ax}$.

7. The matrix $\mathbf{T}$ of partial derivatives is formed by placing $\mathbf{D}g(x_0)$ and $\mathbf{D}h(y_0)$ next to each other in a matrix so that $\mathbf{T}(x_0, y_0)(x - x_0, y - y_0) = \mathbf{D}g(x_0)(x - x_0) + \mathbf{D}h(y_0)(y - y_0)$. Now use the triangle inequality and the fact that $\|(x - x_0, y - y_0)\|$ is larger than $|x - x_0|$ and $|y - y_0|$ to show that $\|f(x, y) - f(x_0, y_0) - \mathbf{T}(x_0, y_0)(x - x_0, y - y_0)\|/\|(x - x_0, y - y_0)\|$ goes to 0.

9. Use the limit theorems and the fact that the function $g(x) = \sqrt{|x|}$ is continuous. (Prove the last statement.)

11. For continuity at $(0, 0)$, use the fact that

$$\left| \frac{xy}{(x^2 + y^2)^{1/2}} \right| \leq \frac{|xy|}{(x^2)^{1/2}} = |y|$$

or that $|xy| \leq (x^2 + y^2)/2$.

13. 0, see Exercise 11.

15. Let $\mathbf{x}$ take the role of $\mathbf{x}_0$ and $\mathbf{x} + \mathbf{h}$ that of $\mathbf{x}$ in the definition.

17. The vector $\mathbf{a}$ takes the place of $\mathbf{x}_0$ in the definition of a limit or in Theorem 6. In either case, the limit depends only on values of $f(\mathbf{x})$ for $\mathbf{x}$ near $\mathbf{x}_0$, not for $\mathbf{x} = \mathbf{x}_0$. Therefore $f(\mathbf{x}) = g(\mathbf{x})$ for $\mathbf{x} \neq \mathbf{a}$ certainly suffices to make the limits equal.

19. (a) $\displaystyle\lim_{\mathbf{x} \to \mathbf{0}}(f_1 + f_2)(\mathbf{x})/\|\mathbf{x}\| = \lim_{\mathbf{x} \to \mathbf{0}} f_1(\mathbf{x})/\|\mathbf{x}\| + \lim_{\mathbf{x} \to \mathbf{0}} f_2(\mathbf{x})/\|\mathbf{x}\| = 0.$

(b) Let $\varepsilon > 0$. Since f is $o(\mathbf{x})$, there is a $\delta > 0$ such that $\|f(\mathbf{x})/\|\mathbf{x}\|\| < \varepsilon/c$ whenever $0 < \|\mathbf{x}\| < \delta$. Then $\|(gf)(\mathbf{x})/\|\mathbf{x}\|\| \leq |g(\mathbf{x})| \|f(\mathbf{x})/\|\mathbf{x}\|\| < \varepsilon$, so $\displaystyle\lim_{\mathbf{x} \to \mathbf{0}}(gf)(\mathbf{x})/\|\mathbf{x}\| = 0.$

(c) $\displaystyle\lim_{x \to 0} f(x)/|x| = \lim_{x \to 0} |x| = 0$, so that $f(x)$ is $o(x)$. But $\displaystyle\lim_{x \to 0} g(x)/|x|$ does not exist, since $g(x)/|x| = \pm 1$ (as x is positive or negative). Therefore $g(x)$ is not $o(x)$.

REVIEW EXERCISES FOR CHAPTER 2

1. (a) Elliptic paraboloid.

(b) Let $y' = y + 3$ and write $z = xy'$. This is a (shifted) hyperbolic paraboloid.

3. (a) $\mathbf{D}f(x, y) = \begin{bmatrix} 2xy & x^2 \\ -ye^{-xy} & -xe^{-xy} \end{bmatrix}$ (b) $\mathbf{D}f(x) = \begin{bmatrix} 1 \\ 1 \end{bmatrix}$

(c) $\mathbf{D}f(x, y, z)][e^x \quad e^y \quad e^z]$ (d) $\mathbf{D}f(x, y, z) = \begin{bmatrix} 1 & 0 & 0 \\ 0 & 1 & 0 \\ 0 & 0 & 1 \end{bmatrix}$

5. The plane tangent to a sphere at (x_0, y_0, z_0) is normal to the line from the center to (x_0, y_0, z_0).

7. (a) $z = x - y + 2$ (b) $z = 4x - 8y - 8$

(c) $x + y + z = 1$ (d) $10x + 6y - 4z = 6 - \pi$

(e) $2z = \sqrt{2}x + \sqrt{2}y$ (f) $x + 2y - z = 2$

9. (a) The level curves are hyperbolas $xy = 1/c$:

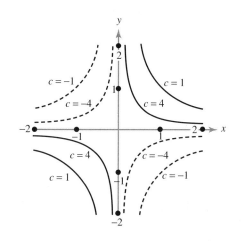

(b) $\mathbf{c} = x^2 - xy - y^2 = \left(x - \dfrac{1 + \sqrt{5}}{2} y \right)\left(x - \dfrac{1 - \sqrt{5}}{2} y \right)$

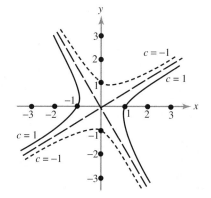

11. (a) 0   (b) Limit does not exist.

13. $(1 + 2x) \exp (1 + x^2 + y^2)$

15. (a) The line $\mathbf{L}(t) = (x_0, y_0, f(x_0, y_0)) + t(a, b, c)$ lies in the plane $z = f(x_0, y_0)$ if $c = 0$ and is perpendicular to $\nabla f (x_0, y_0)$ if $a(\partial f/\partial x)(x_0, y_0) + b(\partial b/\partial y)(x_0, y_0) = 0$. On $\mathbf{L}$, we have

$$f(x_0, y_0) + \left[\frac{\partial f}{\partial x}(x_0, y_0) \right](x - x_0) + \left[\frac{\partial f}{\partial y}(x_0, y_0) \right](y - y_0)$$

$$= f(x_0, y_0) + at\left[\frac{\partial f}{\partial x}(x_0, y_0) \right] + bt\left[\frac{\partial f}{\partial y}(x_0, y_0) \right]$$

$$= f(x_0, y_0) = z$$

Therefore **L** lies in the tangent plane. An upward unit normal to the tangent plane is $\mathbf{p} = (1 + \|\nabla f\|^{-1/2}(-(\partial f/\partial x)(x_0, y_0), -(\partial f/\partial y)(x_0, y_0), 1)$. Therefore $\cos\theta = \mathbf{p}\cdot\mathbf{k} = (1 + \|\nabla f\|^2)^{-1/2}$, and $\tan\theta = \sin\theta/\cos\theta = \{\|\nabla f\|^2/(1 + \|\nabla f\|^2)\}^{1/2}/(1 + \|\nabla f\|^2)^{-1/2} = \|\nabla f\|$ as claimed.

(b) The tangent plane contains the horizontal line through $(1, 0, 2)$ perpendicular to $\nabla f(1, 0) = (5, 0)$, that is, parallel to the y axis. It makes an angle of arctan $(\|\nabla f(1, 0)\|) = $ arctan $5 \approx 78.7°$ with respect to the xy plane.

17. $(1/\sqrt{2}, 1/\sqrt{2})$ or $(-1/\sqrt{2}, -1/\sqrt{2})$

19. A unit normal is $(\sqrt{2}/10)(3, 5, 4)$. The tangent plane is $3x + 5y + 4z = 18$.

21. $4\mathbf{i} + 16\mathbf{j}$

23. (a) Since g is the composition $\lambda \mapsto \lambda x \mapsto f(\lambda x)$, the chain rule gives

$$g'(\lambda) = \mathbf{D}f(\lambda\mathbf{x})\begin{bmatrix} x_1 \\ \vdots \\ x_n \end{bmatrix}.$$

Thus

$$g'(1) = \mathbf{D}f(\mathbf{x})\begin{bmatrix} x_1 \\ \vdots \\ x_n \end{bmatrix} = \nabla f(\mathbf{x})\cdot\mathbf{x}$$

But also $g(\lambda) = \lambda^p f(\mathbf{x})$, so $g'(\lambda) = p\lambda^{p-1}f(\mathbf{x})$ and $g'(1) = pf(\mathbf{x})$.

(b) $p = 1$.

25. Differentiate directly using the chain rule, or use Exercise 23(a) with $p = 0$.

27. (a) If $(x, y) \neq (0, 0)$, then one calculates for (i) that $\partial f/\partial x = (y^3 - yx^2)/(x^2 + y^2)^2$ and $\partial f/\partial y = (x^3 - xy^2)/(x^2 + y^2)^2$. If $x = y = 0$, use the definition directly to find that both partial derivatives are 0.

(b) Formula (i) is not continuous at $(0, 0)$; formula (ii) is differentiable but the derivative is not continuous.

29. (a) $\sqrt{2}\pi/8$ (b) $-\sin\sqrt{2}$ (c) $-2\sqrt{2}e^{-2}$

31. $(-4e^{-1}, 0)$

33. (a) See Theorem 11.

(b) $g(u) = (\sin 3u)^2 + \cos 8u$ $\nabla f = (2x, 1)$
$g'(u) = 6\sin 3u\cos 3u - 8\sin 8u$ $\nabla f(\mathbf{h}(0)) = \nabla f(0, 1) = (0, 1)$
$g'(0) = 0$ $\mathbf{h}'(u) = (3\cos 3u, -8\sin 8u)$
          $g'(0) = \nabla f(\mathbf{h}(0))\cdot\mathbf{h}'(0) = (0, 1)\cdot(3, 0) = 0$

35. $t = \sqrt{14}(-3 + \sqrt{359})/70$

37. $\partial z/\partial x = 4(e^{-2x-2y+2xy})(1 + y)/(e^{-2x-2y} - e^{2xy})^2$
$\partial z/\partial y = 4(e^{-2y-2x+2xy})(1 + x)/(e^{-2x-2y} - e^{2xy})^2$

45. Let $x = f(t)$, $y = t$, and use the chain rule to differentiate $u(x, y)$ with respect to t.

47. (a) $n = PV/RT$; $P = nRT/V$; $T = PV/nR$; $V = nRT/P$.

(b) $\partial V/\partial T = nR/P$; $\partial T/\partial P = V/nR$; $\partial P/\partial V = -nRT/V^2$. Multiply, remembering that $PV = nRT$.

49. (a) One may solve for any of the variables in terms of the other two.

(b) $\partial T/\partial P = (V - \beta)/R$;

$\partial P/\partial V = -RT/(V - \beta)^2 + 2\alpha/V^3$;

$\partial V/\partial T = R/[(V - \beta)(RT/(V - \beta)^2 - 2\alpha/V^3)]$

(c) Multiply and cancel factors.

51. (a) $(1/\sqrt{2}, 1/\sqrt{2})$

(b) The directional derivative is 0 in the direction

$$(x_0\mathbf{i} + y_0\mathbf{j})/\sqrt{x_0^2 + y_0^2}.$$

(c) The level curve through (x_0, y_0) must be tangent to the line through $(0, 0)$ and (x_0, y_0). The level curves are lines or half-lines emanating from the origin.

53. $G(x, y) = x - y$.

Chapter 3

SECTION 3.1

1. (a) $\dfrac{\partial^2 f}{\partial x^2} = 24\dfrac{x^3 y - xy^3}{(x^2 + y^2)^4}$, $\dfrac{\partial^2 f}{\partial y^2} = 24\dfrac{-x^3 y + xy^3}{(x^2 + y^2)^4}$,

$\dfrac{\partial^2 f}{\partial x \partial y} = \dfrac{\partial^2 f}{\partial y \partial x} = \dfrac{-6x^4 + 36x^2 y^2 - 6y^4}{(x^2 + y^2)^4}$

(b) $\dfrac{\partial^2 f}{\partial x^2} = \dfrac{2}{x^3}$, $\dfrac{\partial^2 f}{\partial y^2} = xe^{-y}$, $\dfrac{\partial^2 f}{\partial x \partial y} = \dfrac{\partial^2 f}{\partial y \partial x} = -e^{-y}$

(c) $\dfrac{\partial^2 f}{\partial x^2} = -y^4 \cos(xy^2)$, $\dfrac{\partial^2 f}{\partial y^2} = -2x \sin(xy^2) - 4x^2 y^2 \cos(xy^2)$,

$\dfrac{\partial^2 f}{\partial x \partial y} = \dfrac{\partial^2 f}{\partial y \partial x} = -2y \sin(xy^2) - 2xy^3 \cos(xy^2)$

(d) $\dfrac{\partial^2 f}{\partial x^2} = y^4 e^{-xy^2} + 12x^2 y^3,\ \dfrac{\partial^2 f}{\partial y^2} = -2xe^{-xy^2} + 4x^2 y^2 e^{-xy^2} + 6yx^4,$

$\dfrac{\partial^2 f}{\partial x \partial y} = \dfrac{\partial^2 f}{\partial y \partial x} = -2ye^{-xy^2} + 2xy^3 e^{-xy^2} + 12x^3 y^2$

(e) $\dfrac{\partial^2 f}{\partial x^2} = \dfrac{2(\cos^2 x + e^{-y})\cos 2x + 2\sin^2 2x}{(\cos^2 x + e^{-y})^3},$

$\dfrac{\partial^2 f}{\partial y^2} = \dfrac{e^{-y} - \cos^2 x}{e^y (\cos^2 x + e^{-y})^3}$

$\dfrac{\partial^2 f}{\partial x \partial y} = \dfrac{\partial^2 f}{\partial y \partial x} = \dfrac{2\sin 2x}{e^y (\cos^2 x + e^{-y})^3}$

3. (a) $\partial^2 z / \partial x^2 = 6,\ \partial^2 z / \partial y^2 = 4,$ (b) $\partial^2 z / \partial x^2 = 0,\ \partial^2 z / \partial y^2 = 4x/3y^3,$
$\partial^2 z / \partial x\, \partial y = \partial^2 z / \partial y\, \partial x = 0$ $\partial^2 z / \partial x\, \partial y = \partial^2 z / \partial y\, \partial x = -2/3y^2$

5. $f_{xy} = 2x + 2y,\ f_{yz} = 2z,\ f_{zx} = 0,\ f_{xyz} = 0$

7. Since f and $\partial f / \partial z$ are both of class C^2, we have

$$\frac{\partial^3 f}{\partial x\, \partial y\, \partial z} = \frac{\partial^2}{\partial x\, \partial y}\frac{\partial f}{\partial z} = \frac{\partial^2}{\partial y\, \partial x}\frac{\partial f}{\partial z} = \frac{\partial}{\partial y}\left(\frac{\partial^2 f}{\partial x\, \partial z}\right) = \frac{\partial}{\partial y}\left(\frac{\partial^2 f}{\partial z\, \partial x}\right) = \frac{\partial^3 f}{\partial y\, \partial z\, \partial x}.$$

9. $f_{xzw} = f_{zwx} = e^{xyz}[2xy\cos(xw) + x^2 y^2 z\cos(xw) - x^2 yw\sin(xw)]$

11. (a) $\dfrac{\partial f}{\partial x} = \arctan \dfrac{x}{y} + \dfrac{xy}{x^2 + y^2}$

$\dfrac{\partial f}{\partial y} = \dfrac{-x^2}{x^2 + y^2}$

$\dfrac{\partial^2 f}{\partial x^2} = \dfrac{2y^3}{(x^2 + y^2)^2},\ \dfrac{\partial^2 f}{\partial y^2} = \dfrac{2x^2 y^2}{(x^2 + y^2)^2}$

$\dfrac{\partial^2 f}{\partial x\, \partial y} = \dfrac{\partial^2 f}{\partial y\, \partial x} = \dfrac{-2xy^2}{(x^2 + y^2)^2}$

(b) $\dfrac{\partial f}{\partial x} = \dfrac{-x\sin\sqrt{x^2 + y^2}}{\sqrt{x^2 + y^2}},\ \dfrac{\partial f}{\partial y} = \dfrac{-y\sin\sqrt{x^2 + y^2}}{\sqrt{x^2 + y^2}}$

$\dfrac{\partial^2 f}{\partial x^2} = \dfrac{x^2\sin\sqrt{x^2 + y^2}}{(x^2 + y^2)^{3/2}} - \dfrac{x^2\cos\sqrt{x^2 + y^2}}{x^2 + y^2} - \dfrac{\sin\sqrt{x^2 + y^2}}{(x^2 + y^2)^{1/2}}$

$\dfrac{\partial^2 f}{\partial y^2} = \dfrac{y^2\sin\sqrt{x^2 + y^2}}{(x^2 + y^2)^{3/2}} - \dfrac{y^2\cos\sqrt{x^2 + y^2}}{x^2 + y^2} - \dfrac{\sin\sqrt{x^2 + y^2}}{(x^2 + y^2)^{1/2}}$

$\dfrac{\partial^2 f}{\partial x\, \partial y} = \dfrac{\partial^2 f}{\partial y\, \partial x} = xy\left[\dfrac{\sin\sqrt{x^2 + y^2}}{(x^2 + y^2)^{3/2}} - \dfrac{\cos\sqrt{x^2 + y^2}}{x^2 + y^2}\right]$

(c) $\dfrac{\partial f}{\partial x} = -2x\exp(-x^2 - y^2),\ \dfrac{\partial f}{\partial y} = -2y\exp(-x^2 - y^2),$

$\dfrac{\partial^2 f}{\partial x^2} = (4x^2 - 2)\exp(-x^2 - y^2),\ \dfrac{\partial^2 f}{\partial y^2} = (4y^2 - 2)\exp(-x^2 - y^2),$

$\dfrac{\partial^2 f}{\partial x\, \partial y} = \dfrac{\partial^2 f}{\partial y\, \partial x} = 4xy\exp(-x^2 - y^2)$

13. $\dfrac{\partial^2 f}{\partial x^2}\left(\dfrac{dx}{dt}\right)^2 + 2\dfrac{\partial^2 f}{\partial x\, \partial y}\dfrac{dx}{dt}\dfrac{dy}{dt} + \dfrac{\partial^2 f}{\partial y^2}\left(\dfrac{dy}{dt}\right)^2 + \dfrac{\partial f}{\partial x}\dfrac{d^2 x}{dt^2} + \dfrac{\partial f}{\partial y}\dfrac{d^2 y}{dt^2},$
where $\mathbf{c}(t) = (x(t), y(t))$

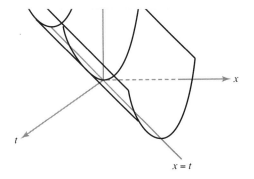

19. $V = -GmM/r = -GmM(x^2 + y^2 + z^2)^{-1/2}$. Check that
$$\frac{\partial^2 V}{\partial x^2} + \frac{\partial^2 V}{\partial y^2} + \frac{\partial^2 V}{\partial z^2} = GmM(x^2 + y^2 + z^2)^{-3/2}[3 - 3(x^2 + y^2 + z^2)(x^2 + y^2 + z^2)^{-2/2} = 0$$

SECTION 3.2

1. $f(h_1, h_2) = h_1^2 + 2h_1h_2 + h_2^2$ [$R_2(\mathbf{h}, \mathbf{0}) = 0$ in this case]

3. $f(h_1, h_2) = 1 + h_1 + h_2 + \dfrac{h_1^2}{2} + h_1h_2 + \dfrac{h_2^2}{2} + R_2(\mathbf{h}, 0)$

5. $f(h_1, h_2) = 1 + h_1h_2 + R_2(\mathbf{h}, \mathbf{0})$

7. (a)   Show that $|R_k(x, a)| \le AB^{k+1}/(k+1)!$ for constants A, B, and x in a fixed interval $[a, b]$. Prove that $R_k \to 0$ as $k \to \infty$. (Use convergence of the series $\sum c^k/k! = e^c$ and use Taylor's theorem.)

    (b)   The only possible trouble is at $x = 0$. Use L'Hôpital's rule to show that

$$\lim_{t \to \infty} p(t)e^t = \infty$$

for every polynomial $p(t)$. Using this, establish that $\lim_{x \to 0^+} p(x)e^{-1/x} = 0$ for every rational function $p(x)$, and conclude that $f^{(k)}(0) = 0$ for every k.

    (c) $f: \mathbb{R}^n \to \mathbb{R}$ is analytic at $\mathbf{x}_0$ if the series

$$f(\mathbf{x}_0) + \sum_{i=1}^{n} h_i \frac{\partial f}{\partial x_i}(\mathbf{x}_0) + \frac{1}{2} \sum_{i,j=1}^{n} h_i h_j \frac{\partial^2 f}{\partial x_i \, \partial x_j}(\mathbf{x}_0) + \cdots$$

$$+ \frac{1}{k!} \sum_{i_1,\dots,i_k=1}^{n} h_{i_1} h_{i_2} \cdots h_{i_k} \frac{\partial^k f}{\partial x_{i_1} \cdots \partial x_{i_k}}(\mathbf{x}_0) + \cdots$$

$$-\sum_{k=0}^{ } \frac{ }{k!}(x+y) .$$

SECTION 3.3

1. $(0, 0)$; saddle point.

3. The critical points are on the line $y = -x$; they are local minima, because $f(x, y) = (x + y)^2 \geq 0$, equaling zero only when $x = -y$.

5. $(0, 0)$; saddle point.

7. $(-\frac{1}{4}, -\frac{1}{4})$; local minimum.

9. $(0, 0)$; local maximum. (The tests fail, but use the fact that $\cos z \leq 1$).
$(\sqrt{\pi/2}, \sqrt{\pi/2})$, local minimum
$(0, \sqrt{\pi})$ local minimum.

11. No critical points,

13. $(1, 1)$ is a local minimum.

15. $(0, n\pi)$; critical points, no local maxima or minima.

17. Minimum at $(0, 0)$ and maxima at $(0, \pm 1)$.

19. (a) $\partial f/\partial x$ and $\partial f/\partial y$ vanish at $(0, 0)$.
(b) Show that $f(g(t)) = 0$ at $t = 0$ and that $f(g(t)) \geq 0$ if $|t| < |b|/3a^2$.
(c) f is negative on the parabola $y = 2x^2$.

21. The critical points are on the line $y = x$ and they are local minima (see Exercise 3).

23. Minimize $S = 2xy + 2yz + 2xz$ with $z = V/xy$, V the constant volume.

25. $40, 40, 40$

27. The only critical point is $(0, 0, 0)$. It is a minimum, since

$$f(x, y, z) \geq \frac{x^2 + y^2}{2} + z^2 + xy = \frac{1}{2}(x + y)^2 + z^2 \geq 0.$$

29. $(1, \frac{3}{2})$ is a saddle point; $(5, \frac{27}{2})$ is a local minimum.

31. $\frac{3}{2}$ is the absolute maximum and 0 is the absolute minimum.

33. -2 is the absolute minimum; 2 is the absolute maximum.

35. $(\frac{1}{2}, 4)$ is a local minimum.

37. If $u_n(x, y) = u(x, y) + (1/n)e^x$, then $\nabla^2 u_n = (1/n)e^x > 0$. Thus, u_n is strictly su[...] and can have its maximum only on ∂D, say, at $\mathbf{p}_n = (x_n, y_n)$. If $(x_0, y_0) \in D$, check that th[...] $u(x_n, y_n) > u(x_0, y_0) - e/n$. Thus there must be a point $\mathbf{q} = (x_\infty, y_\infty)$ on ∂D such that [...] close to $\mathbf{q}$ we can find an (x_n, y_n) for n as large as we like. Conclude from the continuity [...] $u(x_\infty, y_\infty) \geq u(x_0, y_0)$.

39. Follow the methods of Exercise 37.

41. (a) If there were an x_1 with $f(x_1) < f(x_0)$, then the maximum of f on the interval between x_0 and x would be another critical point.
(b) Verify (i) by the second-derivative test; for (ii), f goes to $-\infty$ as $y \to \infty$ and $x = -y$.

SECTION 3.4

1. Maximum at $\sqrt{\frac{2}{3}}(1, -1, 1)$, minimum at $\sqrt{\frac{2}{3}}(-1, 1, -1)$

3. Maximum at $(\sqrt{3}, 0)$, minimum at $(-\sqrt{3}, 0)$

5. Maximum at $(\frac{9}{\sqrt{70}}, \frac{4}{\sqrt{70}})$, minimum at $(-\frac{9}{\sqrt{70}}, -\frac{4}{\sqrt{70}})$

7. The minimum value 4 is attained at $(0, 2)$. Use a geometric picture rather than Lagrange multipliers.

9. $(0, 0, 2)$ is a minimum of f.

11. $\frac{3}{2}$ is the absolute maximum and 0 is the absolute minimum.

13. The diameter should equal the height, $20/\sqrt[3]{2\pi}$ cm.

15. Maximum value $\sqrt{3}$ at $(\frac{1}{\sqrt{3}}, \frac{1}{\sqrt{3}}, -\frac{1}{\sqrt{3}})$ and minimum value $-\sqrt{3}$ at $(-\frac{1}{\sqrt{3}}, -\frac{1}{\sqrt{3}}, \frac{1}{\sqrt{3}})$.

17. Horizontal length is $\sqrt{qA/p}$, vertical length is $\sqrt{pA/q}$.

19. For Exercise 1, the bordered Hessians required are

$$|\bar{H}_2| = \begin{vmatrix} 0 & 2x & 2y \\ 2x & -2\lambda & 0 \\ 2y & 0 & -2\lambda \end{vmatrix} = 8\lambda(x^2 + y^2),$$

$$|\bar{H}_3| = \begin{vmatrix} 0 & 2x & 2y & 2z \\ 2x & -2\lambda & 0 & 0 \\ 2y & 0 & -2\lambda & 0 \\ 2z & 0 & 0 & -2\lambda \end{vmatrix} = -16\lambda(x^2 + y^2 + z^2).$$

At $\sqrt{\frac{2}{3}}(1, -1, 1)$ the Lagrange multiplier is $\lambda = \sqrt{6}/4 > 0$, indicating a maximum at $\sqrt{\frac{2}{3}}(1, -1, 1)$, and

Theorem 9 applies. At an $\mathbf{x}$ where f is extreme, there is a $\lambda/2$ such that $\nabla f(\mathbf{x}) = (\lambda/2)/\nabla g(\mathbf{x})$. That is $A\mathbf{x} = \lambda\mathbf{x}$.

25. Minimum is $(-1/\sqrt{2}, 0)$ maximum is $(\frac{1}{4}, \pm\sqrt{7}/8)$, local minimum at $(+1/\sqrt{2}, 0)$.

SECTION 3.5

1. Use the special implicit function theorem with $n = 1$. (See Example 1.) Line (i) is given by
$0 = (x - x_0, y - y_0) \cdot \nabla F(x_0, y_0) = (x - x_0)(\partial F/\partial x)(x_0, y_0) + (y - y_0)(\partial F/\partial y)(x_0, y_0)$. For line (ii), Theorem 11 gives $dy/dx = -(\partial F/\partial x)(\partial F/\partial y)$, and so the lines agree and are given by

$$y = y_0 - \frac{(\partial F/\partial x)(x_0, y_0)}{(\partial F/\partial y)(x_0, y_0)}(x - x_0).$$

3. (a) If $x < -\frac{1}{4}$ we can solve for y in terms of x using the quadratic formula.
(b) $\partial F/\partial y = 2y + 1$ is nonzero for $\{y \mid y < -\frac{1}{2}\}$ and $\{y \mid > -\frac{1}{2}\}$. These regions correspond to the upper and lower halves of a horizontal parabola with vertex at $(-\frac{1}{4}, -\frac{1}{2})$ and to the choice of sign in the quadratic formula. The derivative $dy/dx = -3/(2y + 1)$ is negative on the top half of the parabola, positive on the bottom.

5. Let $F(x, y, z) = x^3 z^2 - z^3 yx$; $\partial F/\partial z = 2x^3 z - 3z^2 yz \neq 0$ at $(1, 1, 1)$. Near the origin, with $x = y \neq 0$, we get solutions $z = 0$ and $z = x$, and so there is no unique solution. At $(1, 1)$, $\partial z/\partial x = \frac{3}{4}$ and $\partial z/\partial y = -\frac{3}{4}$.

7. With $F_1 = y + x + uv$ and $F_2 = uxy + v$, the determinant in the general implicit function theorem is

$$\begin{vmatrix} \partial F_1/\partial u & \partial F_1/\partial v \\ \partial F_2/\partial u & \partial F_2/\partial v \end{vmatrix} = v - uxy,$$

which is 0 at $(0, 0, 0, 0)$. Thus the implicit function theorem does not apply. If we try directly, we find that $v = -uxy$, so $x + y = u^2 xy$. For a particular choice of (x, y) near $(0, 0)$, either there are no solutions for (u, v) or else there are two.

9. No. $f(x, y) = (-1, 0)$ has infinitely many solutions, namely, $(x, y) = (0, y)$ for any y.

11. (a) $x_0^2 + y_0^2 \neq 0$.
(b) $f'(z) = -z(x + 2y)/(x^2 + y^2)$; $\quad g'(z) = z(y - 2x)/(x^2 + y^2)$.

13. Multiply and equate coefficients to get $a_0, a_1,$ and a_2 as functions of $r_1, r_2,$ and r_3. Then compute the Jacobian determinant $\partial(a_0, a_1, a_2)/\partial(r_1, r_2, r_3) = (r_3 - r_2)(r_1 - r_2)(r_1 - r_3)$. This is not zero if the roots are distinct. Thus, the inverse function theorem shows that the roots may be found as functions of the coefficients in some neighborhood of any point at which the roots are distinct. That is, if the roots

r_1, r_2, r_3 of $x^3 + a_2 x^2 + a_1 x + a_0$ are all different, then there are neighborhoods V of (r_1, r_2, r_3) and W of (a_0, a_1, a_2) such that the roots in V are smooth functions of the coefficients in W.

SECTION 3.6

1. $(-\frac{1}{4}, -\frac{1}{4})$

3. $(2, 1)$ is an unstable equilibrium.

5. Stable equilibrium point $(2 + m^2 g^2)^{-1/2}(-1, -1, -mg)$

7. No critical points; no maximum or minimum

9. $(-1, 0, 1)$

11. At the optimum, $QK/\alpha = pL/(1 - \alpha)$.

REVIEW EXERCISES FOR CHAPTER 3

1. (a) Saddle point
   (b) Saddle point for any C

3. (a) 1   (b) $\sqrt{83/6}$

5. Use the second derivative test; $(0, 0)$ is a local maximum; $(-1, 0)$ is a saddle point; $(2, 0)$ is a local minumum.

7. Saddle points at $(n, 0)$, $n =$ integer.

9. $2, \frac{1}{2}$

11. maximum 1, minimum $\cos 1$

13. $z = \frac{1}{4}$

15. $(0, 0, \pm 1)$

17. If $b \geq 2$, the minimum distance is $2\sqrt{b - 1}$; if $b \leq 2$, the minimum distance is $|b|$.

19. Not stable

21. $f(-\frac{3}{2}, -\sqrt{3}/2) = 3\sqrt{3}/4$

23. $x = (20/3)\sqrt[3]{3}; y = 10\sqrt[3]{3}; z = 5\sqrt[3]{3}$

25. The determinant required in the general implicit function theorem is not zero, and so we can solve for u and v; $(\partial u/\partial x)(2, -1) = \frac{13}{32}$.

27. A new orthonormal basis may be found with respect to which the quadratic form given by the matrix

$$A = \begin{bmatrix} a & b \\ b & c \end{bmatrix}$$

takes diagonal form. This change of basis defines new variables ξ and η, which are linear functions of x and y. Manipulations of linear algebra and the chain rule show that $Lv = \lambda(\partial^2 v/\partial \xi^2) + \mu(\partial^2 v/\partial \eta^2)$. The numbers λ and μ are the eigenvalues of A and are positive, since the quadratic form in positive-definite. At a maximum, $\partial v/\partial \xi = \partial v/\partial \eta = 0$. Moreover, $\partial^2 v/\partial \xi^2 \le 0$ and $\partial^2 v/\partial \eta^2 \le 0$, since if either were greater than 0, the cross section of the graph in that direction would have a minimum. Then $Lv \le 0$, thus contradicting strict subharmonicity.

29. Reverse the inequalities in Exercises 19 and 20.

31. The equations for a critical point, $\partial s/\partial m = \partial s/\partial b = 0$, when solved for m and b give $m = (y_1 - y_2)/(x_1 - x_2)$ and $b = y_2 x_1 - y_1 x_2$. The line $y - mx + b$ then goes through (x_1, y_1) and (x_2, y_2).

33. At a minimum of s, we have $0 = \partial s/\partial b = -2 \sum_{i=1}^{n}(y_i - mx_i - b)$.

35. $y = \frac{9}{10}x + \frac{6}{5}$

Chapter 4

SECTION 4.1

1. $\mathbf{r}'(t) = -(\sin t)\mathbf{i} + 2(\cos 2t)\mathbf{j}, r'(0) = 2\mathbf{j}, \mathbf{a}(t) = -(\cos t)\mathbf{i} - 4(\sin 2t)\mathbf{j}, \mathbf{a}(0) = -\mathbf{i}, \mathbf{l}(t) = \mathbf{i} + 2t\mathbf{j}$

3. $\mathbf{r}'(t) = \sqrt{2}\mathbf{i} + e^t\mathbf{j} - e^{-t}\mathbf{k}, \mathbf{r}'(0) = \sqrt{2}\mathbf{i} + \mathbf{j} - \mathbf{k}, \mathbf{a}(t) = e^t\mathbf{j} + e^{-t}\mathbf{k}, \mathbf{a}(0) = \mathbf{j} + \mathbf{k}, \mathbf{l}(t) = \sqrt{2}t\mathbf{i} + (1+t)\mathbf{j} + (1-t)\mathbf{k}$

5. $(e^t - e^{-t}, \cos t - \sin t, -3t^2)$

7. $[-3t^2(2\sin t + \cos t) - t^3(2\cos t - \sin t)]\mathbf{i} + [3t^2(2e^t + e^{-t}) + t^3(2e^t - e^{-t})]\mathbf{j} + [e^t(\cos t - \sin t) - e^{-t}(-\sin t + \cos t)]\mathbf{k}$

9. $m(0, 6, 0)$

11. $-24\pi^2(\cos(2\pi t/5), \sin(2\pi t/5))/25$

13. $\dfrac{d}{dt}(\|\mathbf{v}\|^2) = \dfrac{d}{dt}(\mathbf{v} \cdot \mathbf{v}) = 2\mathbf{v} \cdot \dfrac{d\mathbf{v}}{dt} = 2\mathbf{v} \cdot \mathbf{a} = 0$

15. 6129 seconds

17. $\mathbf{c}(t) = \left(\dfrac{t^2}{2}, e^t - 6, \dfrac{t^3}{3} + 1\right)$

19. (a) $\mathbf{c}(t) = (t, e^t), -\infty < t < \infty$. The image of this path is the graph $y = e^x$.
(b) $\mathbf{c}(t) = (\frac{1}{2}\cos t, \sin t), 0 \le t \le 2\pi$, an ellipse.
(c) $\mathbf{c}(t) = (at, bt, ct)$
(d) $\mathbf{c}(t) = (\frac{2}{3}\cos t, \frac{1}{2}\sin t), 0 \le t \le 2\pi$, an ellipse.

21. $\mathbf{c}(t) \times \mathbf{c}'(t)$ is normal to the plane of the orbit at time t. As in Exercise 20, its derivative is 0, and so the orbital plane is constant.

SECTION 4.2

1. $2\sqrt{5}\pi$ **3.** $2(2\sqrt{2} - 1)$

5. $\dfrac{6 - \sqrt{3}}{\sqrt{2}} + \dfrac{1}{2}\log\left[\dfrac{2\sqrt{2} + 3}{\sqrt{2} + \sqrt{3}}\right]$ **7.** $2\pi(\sqrt{5} + \sqrt{2})$

9. $3 + \log 2$

11. (a) Since α is strictly increasing, it maps $[a, b]$ one-to-one onto $[\alpha(a), \alpha(b)]$. By definition, $\mathbf{v}$ is the image of $\mathbf{c}$ if and only if there is a t in $[a, b]$ with $\mathbf{c}(t) = \mathbf{v}$. There is one point s in $[\alpha(a), \alpha(b)]$ with $s = \alpha(t)$, so $\mathbf{d}(s) = \mathbf{c}(t) = \mathbf{v}$. Therefore the image of $\mathbf{c}$ is contained in that of $\mathbf{d}$. Use α^{-1} similarly for the opposite inclusion.
(b)
$$l_{\mathbf{d}} = \int_{\alpha(a)}^{\alpha(b)} \|\mathbf{d}'(s)\| \, ds = \int_{s=\alpha(a)}^{s=\alpha(b)} \|\mathbf{d}'(\alpha(t))\| \|\alpha'(t)\| \, dt$$
$$= \int_{t=a}^{t=b} \|\mathbf{d}'(\alpha(t))\alpha'(t)\| dt = \int_a^b \|\mathbf{c}'(t)\| \, dt = l_{\mathbf{c}}.$$

13. (a) $l_{\mathbf{c}} = \int_a^b \|\mathbf{c}'(s)\| \, ds = \int_a^b ds = b - a.$
(b) $\mathbf{T}(s) = \mathbf{c}'(s)/\|\mathbf{c}'(s)\| = \mathbf{c}'(s)$, so $\mathbf{T}'(s) = \mathbf{c}''(s)$. Then $k = \|\mathbf{T}'\| = \|\mathbf{c}''(s)\|$.
(c) Show that if $\mathbf{v}$ and $\mathbf{w}$ are in $\mathbb{R}^3$, $\|\mathbf{v} \times \mathbf{w}\| = \|\mathbf{w} - (\mathbf{v} \cdot \mathbf{w}/\|\mathbf{v}\|^2)\mathbf{v}\| \cdot \|\mathbf{v}\|$. Use this to show that if $\boldsymbol{\rho}(t) = (x(t), y(t), z(t))$ is never $(0, 0, 0)$ and $\mathbf{f}(t) = \boldsymbol{\rho}(t)/\|\boldsymbol{\rho}(t)\|$, then

$$\frac{d\mathbf{f}}{dt} = \frac{1}{\|\boldsymbol{\rho}(t)\|}\left[\boldsymbol{\rho}'(t) - \frac{\boldsymbol{\rho}(t) \cdot \boldsymbol{\rho}'(t)}{\|\boldsymbol{\rho}(t)\|^2}\boldsymbol{\rho}(t)\right] \quad \text{and} \quad \frac{d\mathbf{f}}{dt} = \frac{\|\boldsymbol{\rho}(t) \times \boldsymbol{\rho}'(t)\|}{\|\boldsymbol{\rho}(t)\|^2}.$$

With $\boldsymbol{\rho}(t) = \mathbf{c}'(t)$, this gives

$$\mathbf{T}'(t) = \frac{\mathbf{c}''(t)}{\|\mathbf{c}'(t)\|} - \frac{\mathbf{c}'(t) \cdot \mathbf{c}''(t)}{\|\mathbf{c}'(t)\|^3}\mathbf{c}'(t) \quad \text{and} \quad \|\mathbf{T}'(t)\| = \frac{\|\mathbf{c}'(t) \times \mathbf{c}''(t)\|}{\|\mathbf{c}'(t)\|^2}.$$

If s is the arc length of $\mathbf{c}$, $ds/dt = \|\mathbf{c}'(t)\|$, and therefore

$$\left\|\frac{d\mathbf{T}}{dt}\right\| = \left\|\frac{d\mathbf{T}}{ds}\frac{ds}{dt}\right\| = k\|\mathbf{c}'(t)\|.$$

Thus,

$$k = \frac{1}{\|\mathbf{c}'(t)\|}\frac{d\mathbf{T}}{dt} = \frac{\|\mathbf{c}'(t) \times \mathbf{c}''(t)\|}{\|\mathbf{c}'(t)\|^3}.$$

(This result is useful in Exercise 15.)

(d) $1/\sqrt{2}$

15. (a)   Since $\mathbf{c}$ is parametrized by arc length, $\mathbf{T}(s) = \mathbf{c}'(s)$, and $\mathbf{N}(s) = \mathbf{c}''(s)/\|\mathbf{c}''(s)\|$. Use Exercise 13 to show that

$$\frac{d\mathbf{B}}{ds} = \left(\mathbf{c}'' \times \frac{\mathbf{c}''}{\|\mathbf{c}''\|}\right) + \mathbf{c}' \times \left(\frac{\mathbf{c}'''}{\|\mathbf{c}''\|} - \frac{\mathbf{c}'' \cdot \mathbf{c}'''}{\|\mathbf{c}''\|^3}\mathbf{c}''\right)$$

and

$$\tau = -\frac{d\mathbf{B}}{ds} \cdot \mathbf{N} = -\frac{(\mathbf{c}' \times \mathbf{c}''') \cdot \mathbf{c}'''}{\|\mathbf{c}''\|^2} = \frac{(\mathbf{c}' \times \mathbf{c}'') \cdot \mathbf{c}'''}{\|\mathbf{c}''\|^2}.$$

(b)   Obtain $\mathbf{T}'(t)$ and $\|\mathbf{T}'(t)\|$ as in Exercise 13. $\mathbf{B}$ is a unit vector in the direction of $\mathbf{c}' \times \mathbf{T} = (\mathbf{c}' \times \mathbf{c}'')/\|\mathbf{c}'\|$, so $\mathbf{B} = (\mathbf{c}' \times \mathbf{c}'')/\|\mathbf{c}' \times \mathbf{c}''\|$. Use the solution of Exercise 13 with $\boldsymbol{\rho} = \mathbf{c}' \times \mathbf{c}''$ to obtain

$$d\mathbf{B}/dt = (\mathbf{c}' \times \mathbf{c}''')/\|\mathbf{c}' \times \mathbf{c}''\| - \{[(\mathbf{c}' \times \mathbf{c}'') \cdot (\mathbf{c}' \times \mathbf{c}''')]/\|\mathbf{c}' \times \mathbf{c}''\|^3\}(\mathbf{c}' \times \mathbf{c}''),$$

and the values of $\mathbf{T}'$ and $\|\mathbf{T}'\|$ to get

$$\mathbf{N} = (\|\mathbf{c}'\|/\|\mathbf{c}' \times \mathbf{c}''\|)(\mathbf{c}'' - (\mathbf{c}' \times \mathbf{c}'')/\|\mathbf{c}'\|^2).$$

Finally, use the chain rule and the inner product of these to obtain

$$\tau = -\left[\frac{d\mathbf{B}}{ds}(s(t))\right] \cdot \mathbf{N}(s(t)) = -\frac{1}{|ds/dt|}\frac{d\mathbf{B}}{dt} \cdot \mathbf{N} = \frac{(\mathbf{c}' \times \mathbf{c}'') \cdot \mathbf{c}'''}{\|\mathbf{c}' \times \mathbf{c}''\|^2}.$$

(c) $-\sqrt{2}/2$

17. (a) $\mathbf{N}$ is defined as $\mathbf{T}'/\|\mathbf{T}'\|$, so $\mathbf{T}' = \|\mathbf{T}'\|\mathbf{N} = k\mathbf{N}$. Since $\mathbf{T} \cdot \mathbf{T}' = 0$, $\mathbf{T}, \mathbf{N}$, and $\mathbf{B}$ are an orthonormal basis for $\mathbb{R}^3$. Differentiating $\mathbf{B}(s) \cdot \mathbf{B}(s) = 1$ and $\mathbf{B}(s) \cdot \mathbf{T}(s) = 0$ shows that $\mathbf{B}' \cdot \mathbf{B} = 0$ and $\mathbf{B}' \cdot \mathbf{T} + \mathbf{B} \cdot \mathbf{T}' = 0$. But $\mathbf{T}' \cdot \mathbf{B} = \|\mathbf{T}'\|\mathbf{N} \cdot \mathbf{B} = 0$, so $\mathbf{B}' \cdot \mathbf{T} = 0$ also. Thus $\mathbf{B}' = (\mathbf{B}' \cdot \mathbf{T}) + (\mathbf{B}' \cdot \mathbf{N})\mathbf{N} + (\mathbf{B}' \cdot \mathbf{B})\mathbf{B} = (\mathbf{B}' \cdot \mathbf{N})\mathbf{N} = -\tau\mathbf{N}$. Also, $\mathbf{N}' \cdot \mathbf{N} = 0$, since $\mathbf{N} \cdot \mathbf{N} = 1$. Thus $\mathbf{N}' = (\mathbf{N}' \cdot \mathbf{T})\mathbf{T} + (\mathbf{N}' \cdot \mathbf{B})\mathbf{B}$. But differentiating $\mathbf{N} \cdot \mathbf{T} = 0$ and $\mathbf{N} \cdot \mathbf{B} = 0$ gives $\mathbf{N}' \cdot \mathbf{T} = -\mathbf{N} \cdot \mathbf{T}' = -k$ and $\mathbf{N}' \cdot \mathbf{B} = -\mathbf{N} \cdot \mathbf{B}' = \tau$, and so the middle equation follows.

(b) $\boldsymbol{\omega} = -\tau\mathbf{i} - k\mathbf{k}$

SECTION 4.3

1.

3.

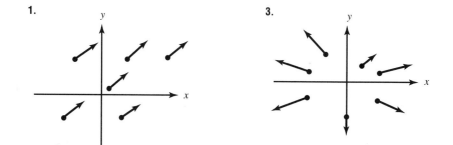

5. $\mathbf{F} = (2y, x)$ **7.**

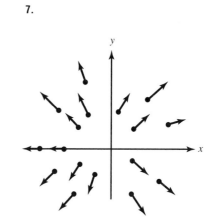

9.   The flow lines are concentric circles

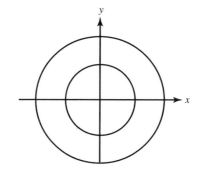

11.   The flow lines for $t > 0$:

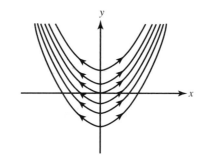

13. $\mathbf{c}'(t) = (2e^{2t}, 1/t, -1/t^2) = \mathbf{F}(\mathbf{c}(t))$

15. $\mathbf{c}'(t) = (\cos t, -\sin t, e^{-t}) = \mathbf{F}(\mathbf{c}(t))$

17. (a) $dE/dt = m\mathbf{r}''(t) \cdot \mathbf{r}(t) + \nabla V(\mathbf{r}(t)) \cdot \mathbf{r}'(t) = \mathbf{F} \cdot \mathbf{r}' - \mathbf{F} \cdot \mathbf{r}' = 0$

(b) $\dfrac{d}{dt} \|\mathbf{r}'(t)\|^2 = \dfrac{d}{dt}(\mathbf{r}'(t) \cdot \mathbf{r}'(t)) = 2\mathbf{r}''(t) \cdot \mathbf{r}'(t) = \dfrac{2}{m}\mathbf{F}(\mathbf{r}(t)) \cdot \mathbf{r}'(t)$

$= -\dfrac{2}{m}\nabla V(\mathbf{r}(t)) \cdot \mathbf{r}'(t) = 0$

19. Use the fact that $-\nabla T$ is perpendicular to the surface $T = $ constant.

21. If $\mathbf{x} = (x_1, x_2, x_3)$, $\boldsymbol{\phi}(\mathbf{x}, t) = (\phi_1, \phi_2, \phi_3)$, and $f = f(x_1, x_2, x_3, t)$, then by the chain rule,

$$\frac{d}{dt}(f(\boldsymbol{\phi}(\mathbf{x}, t), t)) = \frac{\partial f}{\partial t}(\mathbf{x}, t) + \sum_{i=1}^{3} \frac{\partial f}{\partial x_i}(\boldsymbol{\phi}(\mathbf{x}, t), t) \frac{\partial \phi_i}{\partial t}(\mathbf{x}, t)$$

$$= \frac{\partial f}{\partial t}(\mathbf{x}, t) + [\nabla f(\boldsymbol{\phi}(\mathbf{x}, t), t)] \cdot [\mathbf{F}(\boldsymbol{\phi}(\mathbf{x}, t))].$$

SECTION 4.4

1. $ye^{xy} - xe^{xy} + ye^{yz}$ **3.** 3

5. div $\mathbf{V} > 0$ in the first and third quadrants,
div $\mathbf{V} < 0$ in the second and fourth quadrants.

7. $\nabla \cdot \mathbf{F} = 0$; if $\mathbf{F}$ represents a fluid, there is neither expansion nor compression; the area of a small rectangle remains the same.

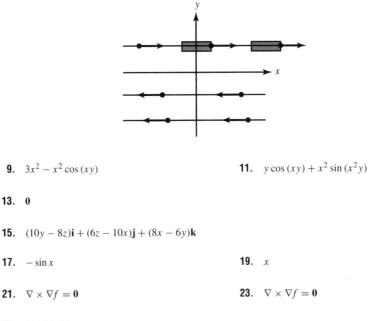

9. $3x^2 - x^2 \cos(xy)$ **11.** $y\cos(xy) + x^2 \sin(x^2 y)$

13. $\mathbf{0}$

15. $(10y - 8z)\mathbf{i} + (6z - 10x)\mathbf{j} + (8x - 6y)\mathbf{k}$

17. $-\sin x$ **19.** x

21. $\nabla \times \nabla f = \mathbf{0}$ **23.** $\nabla \times \nabla f = \mathbf{0}$

25. $\nabla \times \mathbf{F} \neq \mathbf{0}$

27. Let $\mathbf{F} = F_1 \mathbf{i} + F_2 \mathbf{j} + F_3 \mathbf{k}$ and compute both sides of the identity.

29. (a) $2xy\mathbf{i} + x^2\mathbf{j}$ (b) $(3y^2xz, 4xz - y^3z, 0)$
(c) $(-y^3zx^3, 2x^2y^4z, 2x^3z^2 - 2xy)$ (d) $4x^2yz^2 + x^2$

31. No

33. By the product rule,

$$\frac{d}{dt}\mathbf{v} \cdot \mathbf{w} = \frac{d\mathbf{v}}{dt} \cdot \mathbf{w} + \mathbf{v} \cdot \frac{d\mathbf{w}}{dt}.$$

Substitute equation (2) into this. For the last equality, use $A^T\mathbf{v} \cdot \mathbf{w} = \mathbf{v} \cdot A\mathbf{w}$ from Exercise 16 in Section 1.5.

35. By the choice of axes,

$$\mathbf{w} = \tfrac{1}{2}(\nabla \times \mathbf{F})(\mathbf{0}) = \omega\mathbf{k}.$$

By Example 2,

$$\mathbf{v} = -\omega y\mathbf{i} + \omega x\mathbf{j}$$

and therefore

$$\mathbf{Dv}(\mathbf{0}) = \begin{bmatrix} 0 & -\omega & 0 \\ \omega & 0 & 0 \\ 0 & 0 & 0 \end{bmatrix}.$$

On the other hand, by the definitions of W and $\nabla \times \mathbf{F}$,

$$W = \begin{bmatrix} w_{11} & w_{12} & w_{13} \\ w_{21} & w_{22} & w_{23} \\ w_{31} & w_{32} & w_{33} \end{bmatrix}$$

$$= \begin{bmatrix} 0 & \frac{1}{2}\left(\frac{\partial F_1}{\partial y} - \frac{\partial F_2}{\partial x}\right) & \frac{1}{2}\left(\frac{\partial F_1}{\partial z} - \frac{\partial F_3}{\partial x}\right) \\ \frac{1}{2}\left(\frac{\partial F_2}{\partial x} - \frac{\partial F_1}{\partial y}\right) & 0 & \frac{1}{2}\left(\frac{\partial F_2}{\partial z} - \frac{\partial F_3}{\partial y}\right) \\ \frac{1}{2}\left(\frac{\partial F_3}{\partial x} - \frac{\partial F_1}{\partial y}\right) & \frac{1}{2}\left(\frac{\partial F_3}{\partial y} - \frac{\partial F_2}{\partial z}\right) & 0 \end{bmatrix}.$$

$$= \frac{1}{2}\begin{bmatrix} 0 & -(\nabla \times \mathbf{F})_z & (\nabla \times \mathbf{F})_y \\ (\nabla \times \mathbf{F})_z & 0 & -(\nabla \times \mathbf{F})_x \\ -(\nabla \times \mathbf{F})_y & (\nabla \times \mathbf{F})_x & 0 \end{bmatrix}.$$

Our choice of coordinate axes gives

$$W = \begin{bmatrix} 0 & -\omega & 0 \\ \omega & 0 & 0 \\ 0 & 0 & 0 \end{bmatrix}.$$

To interpret the result, we note that the vector field $\mathbf{v}$ represents rotation around a fixed axis $\mathbf{w}$. The flow $\psi(\mathbf{x}, t)$ of $\mathbf{v}$ rotates points in this field, and, for fixed t, its derivative $\mathbf{D_x}\psi(\mathbf{x}, t)$ rotates vectors as well. Let $\mathbf{Y}$ be an arbitrary vector and set $\mathbf{Y}(t) = \mathbf{D_x}\psi(\mathbf{x}, t)\mathbf{Y}$. As t increases or decreases, $\mathbf{Y}(t)$ rotates around $\mathbf{w}$ and

$$\left.\frac{d\mathbf{Y}}{dt}\right|_{t=0} = \mathbf{D_x}\mathbf{v}(\mathbf{0})\mathbf{Y}.$$

This gives the rate of change of $\mathbf{Y}$ as it is transported (rotated) by $\mathbf{D_x}\psi$. By Exercise 33, the rate of change of any vector $\mathbf{X}$ at the origin under transport by the derivative of the flow $\phi(\mathbf{x}, t)$ of $\mathbf{F}$ is given by

$$\left.\frac{d\mathbf{X}}{dt}\right|_{t=0} = \mathbf{D_x F(0)X} = (S + W)\mathbf{X}.$$

Thus this rate of change of $\mathbf{X}$ has two components: the deformation matrix, which affects inner products, and the W matrix. Thus the W *matrix is precisely the rate of change of vectors as they undergo an infinitesimal rotation around the axis* (curl $\mathbf{F}$) $(\mathbf{0}) = (\nabla \times \mathbf{F})(\mathbf{0})$ by the mapping $\mathbf{D_x}\psi(\mathbf{x}, t)$.

[The deformation matrix S incorporates all the length and angle changes caused by the flow. In particular, volume changes are contained in S. In fact, the trace of S is the divergence; tr $S = $ div $\mathbf{F(x)}$. (The trace of a matrix is the sum of its diagonal entries). The trace-free part of S,

$$S' = S - \tfrac{1}{3}(\text{tr } S)^I$$

where I is the 3×3 identity, is called the *shear*.]

37. The line $\mathbf{x} + \lambda\mathbf{v}$ is carried to the curve $\lambda \mapsto \phi(\mathbf{x} + \lambda\mathbf{v}, t)$ after time t, which for λ small, is approximated by its tangent line, namely, $\lambda \mapsto \phi(\mathbf{x}, t) + \mathbf{D_x}\phi(\mathbf{x}, t) \cdot \lambda\mathbf{v}$.

REVIEW EXERCISES FOR CHAPTER 4

1. $\mathbf{v}(1) = (3, -e^{-1}, -\pi/2)$; $\mathbf{a}(1) = (6, e^{-1}, 0)$;
$s(1) = \sqrt{9 + e^{-2} + \frac{\pi^2}{4}}$; $\mathbf{l}(t) = (2, e^{-1}, 0) + (t - 1)(3, -e^{-1}, -\pi/2)$

3. $\mathbf{v}(0) = (1, 1, 0)$; $\mathbf{a}(0) = (1, 0, -1)$; $s = \sqrt{2}$; $l(t) = (1, 0, 1) + t(1, 1, 0)$

5. Tangent vector: $\mathbf{v} = -(1/\sqrt{2})\mathbf{i} + (1/\sqrt{2})\mathbf{j} + \mathbf{k}$
Acceleration vector: $\mathbf{a} = -(1/\sqrt{2})(\mathbf{i} + \mathbf{j})$

7. $m(2, 0, -1)$

9. $\displaystyle\int_1^4 \sqrt{1 + \tfrac{4}{9}t^{-2/3} + \tfrac{4}{25}t^{-6/5}}\, dt$

11. (a) $\mathbf{v} = (-2t\sin(t^2), 2t\cos(t^2), 0)$; $s = 2t$
(b) $\left(\dfrac{1}{2}, -\dfrac{\sqrt{3}}{2}, 0\right)$
(c) $\sqrt{5\pi/3}$
(d) $\mathbf{v} = 2\sqrt{5\pi/3}(\sqrt{3}/2, 1/2, 0)$; $s = 2\sqrt{5\pi/3}$
(e) $\left(\dfrac{3}{2} + \dfrac{5\pi}{\sqrt{3}}\right)\Big/\sqrt{5\pi}$

13. $x = 1 + t, \ y = -\dfrac{1}{2} + \dfrac{t}{2}, \ z = -\dfrac{2}{3} + \dfrac{t}{3}$

15. Compute $\mathbf{c}'(t)$ and check that it equals $\mathbf{F(c}(t))$.

17. 9; **0**

19. 3; $-\mathbf{i} - \mathbf{j} - \mathbf{k}$

21. 0; $-\mathbf{i} - \mathbf{j} - \mathbf{k}$

23. $\nabla f = (ye^{xy} - y\sin xy, xe^{xy} - x\sin xy, 0)$

25. $\nabla f = (2xe^{x^2} + y^2 \sin xy^2, 2xy \sin xy^2, 0)$

27. (a) $(yz^2, xz^2, 2xyz)$;   (b) $(z - y, 0, -x)$
  (c) $(2xyz^3 - 3xy^2z^2, 2x^2y^2z - y^2z^3, y^2z^3 - 2x^2yz^2)$

29. $\text{div}\,\mathbf{F} = 0$; $\text{curl}\,\mathbf{F} = (0, 0, 2(x^2 + y^2)f'(x^2 + y^2) + 2f(x^2 + y^2))$

31. (a) $[\partial P/\partial x)^2 + (\partial P/\partial y)^2]^{1/2}$
  (b) A small packet of air would obey $\mathbf{F} = m\mathbf{a}$.
  (c)   (d)

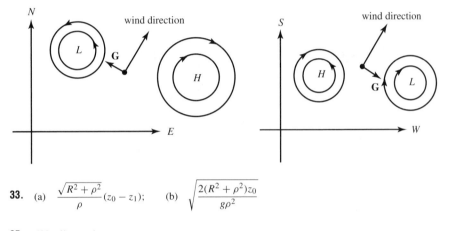

33. (a) $\dfrac{\sqrt{R^2 + \rho^2}}{\rho}(z_0 - z_1)$;   (b) $\sqrt{\dfrac{2(R^2 + \rho^2)z_0}{g\rho^2}}$

35. 680 miles per hour.

Chapter 5

SECTION 5.1

1. (a) $\frac{13}{15}$;   (b) $\pi + \frac{1}{2}$;   (c) 1;   (d) $\log 2 - \frac{1}{2}$.

3. To show that the volumes of the two cylinders are equal, show that their area functions are equal.

5. $2r^3(\tan\theta)/3$ **7.** $\frac{26}{9}$

9. $(2/\pi)(e^2 + 1)$ **11.** $\frac{196}{15}$

SECTION 5.2

1. (a) $\frac{7}{12}$;   (b) $e - 2$;   (c) $\frac{1}{9}\sin 1$;   (d) $2\ln 4 - 2$

3. If $f(x_0, y_0) > 0$, use continuity to show there is a small rectangle R_1 containing (x_0, y_0) with $f(x, y) > \frac{1}{2} f(x_0, y_0)$ on R_1. Let $g(x, y)$ be $\frac{1}{2} f(x_0, y_0)$ on R_1 and 0 elsewhere. By Theorem 2, g is integrable. Use properties (iii) and (iv) of the integral to conclude $\iint_R f \, dx \, dy > \frac{1}{2} f(x_0, y_0)$ area (R_1).

5. Use Fubini's theorem to write

$$\iint_R [f(x)g(y)] \, dx \, dy = \int_c^d g(y) \left[\int_a^b f(x) \, dx \right] dy,$$

and notice that $\int_a^b f(x) \, dx$ is a constant and so may be pulled out.

7. $\frac{11}{6}$

9. Since $\int_0^1 dy = \int_0^1 2y \, dy = 1$, we have $\int_0^1 [\int_0^1 f(x, y) \, dy] \, dx = 1$. In any partition of $R = [0, 1] \times [0, 1]$, each rectangle R_{jk} contains points $\mathbf{c}_{jk}^{(1)}$ with x rational and $\mathbf{c}_{jk}^{(2)}$ with x irrational. If in the regular partition of order n we choose $\mathbf{c}_{jk} = \mathbf{c}_{jk}^{(1)}$ in those rectangles with $0 \leq y \leq \frac{1}{2}$ and $\mathbf{c}_{jk} = \mathbf{c}_{jk}^{(2)}$ when $y > \frac{1}{2}$, the approximating sums are the same as those for

$$g(x, y) = \begin{cases} 1 & 0 \leq y \leq \frac{1}{2} \\ 2y & \frac{1}{2} < y < 1 \end{cases}$$

Since g is integrable, the approximating sums must converge to $\int_R g \, dA = \frac{7}{8}$. However, if we had picked all $\mathbf{c}_{ij} = \mathbf{c}_{jk}^{(1)}$, all approximating sums would have the value 1.

11. The function f is not bounded, since there must be a volume of -1 over each of the diagonal squares of area $1/[n(n+1)]^2$.

SECTION 5.3

1. (a) $\frac{1}{3}$, both;   (b) $\frac{5}{2}$, both;   (c) $(e^2 - 1)/4$, both;   (d) $\frac{1}{35}$, both.

3. $A = \int_{-r}^r \int_{-\sqrt{r^2-x^2}}^{\sqrt{r^2-x^2}} dy \, dz = 2 \int_{-r}^r \sqrt{r^2 - x^2} \, dx = r^2 [\arcsin 1 - \arcsin(-1)] = \pi r^2$.

5. 28,000 ft^3 **7.** 0

9. Type 1; $\pi/2$ **11.** 50π

13. $\pi/24$

15. Compute the integral with respect to y first. Split that into integrals over $[-\phi(x), 0]$ and $[0, \phi(x)]$ and change variables in the first integral, or use symmetry.

17. Let $\{R_{ij}\}$ be a partition of a rectangle R containing D and let f be 1 on D. Thus, f^* is 1 on D and 0 on $R \backslash D$. Let $\mathbf{c}_{jk} \in R \backslash D$ if R_{ij} is not wholly contained in D. The approximating Riemann sum is the sum of the areas of those rectangles of the partition that are contained in D.

SECTION 5.4

1. (a) $\frac{1}{8}$; (b) $\pi/4$; (c) $\frac{17}{12}$
(d) $G(b) - G(a)$, where $dG/dy = F(y, y) - F(a, y)$ and $\partial F/\partial x = f(x, y)$

3. Note that the maximum value of f on D is e and the minimum value of f on D is $1/e$. Use the ideas in the proof of Theorem 4 to show that

$$\frac{1}{e} \le \frac{1}{4\pi^2} \iint f(x, y)\, dA \le e.$$

5. The smallest value of $f(x, y) = 1/(x^2 + y^2 + 1)$ on D is $\frac{1}{6}$, at $(1, 2)$, and so

$$\iint_D f(x, y)\, dx\, dy \ge \frac{1}{6} \cdot \text{area } D = 1.$$

The largest value is 1, at $(0, 0)$, and so

$$\iint_D f(x, y)\, dx\, dy \le 1 \cdot \text{area } D = 6.$$

7. $\frac{4}{3}\pi abc$ **9.** $\pi(20\sqrt{10} - 52)/3$

11. $\sqrt{3}/4$

13. D looks like a slice of pie.

$$\int_0^1 \left[\int_0^x f(x, y)\, dy \right] dx + \int_1^{\sqrt{2}} \left[\int_0^{\sqrt{2-x^2}} f(x, y)\, du \right] dx.$$

15. Use the chain rule and the fundamental theorem of calculus.

SECTION 5.5

1. If $a \ne b$, let $\varepsilon = |a - b|/2$.

3. Let $e = 2d - c$, so that $d = (c+e)/2$. Consider the vertical "doubling" of R defined by $Q = R \cup R_1$, where $R_1 = [a, b] \times [d, e]$. If f is extended to Q by letting f be 0 on the added part, then f is integrable over Q by additivity. The nth regular partition of $[(a + b)/2, b] \times [c, d]$ is part of the $2n$th regular partition of Q. For large n, the Riemann sums for that $2n$th partition cannot vary by more than ε as we change the choice of points from the subrectangles, in particular if we change only those in $[(a + b)/2, b] \times [c, d]$. These changes correspond to the possible changes for the Riemann sums for the nth partition of $[(a + b)/2, b] \times [c, d]$. The argument for part (b) is similar.

5. Let $R = [a, b] \times [c, d]$ and $B = [e, f] \times [g, h]$. Since the rectangles of a partition of R intersect only along their edges, their areas can be added, and b_n is the area of the union of all subrectangles of the nth regular partition of R that intersect B. Since B is contained in this union, area $(B) \le b_n$. On the other hand, if (x, y) is in the union, then $e - (b-a)/n \le x \le f + (b-a)/n$ and $g - (d-c)/n \le y \le h + (d-c)/n$. This leads to $b_n \le$ area $(B) + 2[(b - a)(h - g) + (d - c)(f - e)]/n + 4(a - b)(d - c)/n^2$. Letting $n \to \infty$ and combining the inequalities proves the assertion.

7. (a)   The strategy is to go from point to point within $[a, b]$ by short steps, adding up the changes as you go. Given $\varepsilon > 0$, ϕ is uniformly continuous and therefore there exists a $\delta > 0$ such that $|\phi(x) - \phi(y)| \leq \varepsilon$ whenever $|x - y| < \delta$. Let $x \in [a, b]$ and introduce intermediate points $a = x_0 < x_1 < \cdots < x_{n-1} < x_n = x$ with $x_{i+1} - x_i < \delta$. This can be done with no more than $[(b - a)/\delta] + 1$ segments. By the triangle inequality,

$$|\phi(x) - \phi(a)| \leq \sum_{i=1}^{n} |\phi(x_i) - \phi(x_{i-1})| \leq \left(\frac{b - a}{\delta + 1}\right)\varepsilon.$$

Thus $|\phi(x)| \leq |\phi(a)| + [(b - a)/(\delta + 1)]\varepsilon$ for every x in $[a, b]$.

(b)   Use an argument like that for part (a), moving by short steps within the rectangle $[a, b] \times [c, d]$.

(c)   This is trickier, since D may be composed of many disconnected pieces so that the short steps cannot be taken within D. Nevertheless, given ε, there is a δ such that $|f(\mathbf{x}) - f(\mathbf{y})| \leq \varepsilon$ whenever $\mathbf{x}$ and $\mathbf{y}$ are in D and $\|\mathbf{x} - \mathbf{y}\| < \delta$, by the uniform boundedness principle. Since D is bounded, we may find a large "cube" R with sides of length L such that $D \subset R$. Partition R into subcubes by dividing each edge into m parts. The diagonal of each subcube has length $\sqrt{n}L/m$. If we take $m > \sqrt{n}L/\delta$, any two points in the same subcube are less than δ apart, and there are m^n subcubes. If $R_1, \ldots, R_n$ are those that intersect D, choose $\mathbf{x}_i \in D \cap R_i$. For any $\mathbf{x} \in D$, we have $|f(\mathbf{x})| < \varepsilon + \max(|f(\mathbf{x}_1)|, \ldots, |f(\mathbf{x})|)$.

SECTION 5.6

1. 1/3

3. 10

5. $x^2 + y^2 \leq z \leq \sqrt{x^2 + y^2}, -\sqrt{1 - y^2} \leq x \leq \sqrt{1 - y^2}, -1 \leq y \leq 1$

7. $0 \leq z \leq \sqrt{1 - x^2 - y^2}, -\sqrt{1 - y^2} \leq x \leq \sqrt{1 - y^2}, -1 \leq y \leq 1$

9. $50\pi/\sqrt{6}$

11. 1/2

13. 0

15. $a^5/20$

17. 0

19. 3/10

21. 1/6

23. $\int_{-1}^{1} \int_{-\sqrt{1-x^2}}^{\sqrt{1-x^2}} \int_{\sqrt{x^2+y^2}}^{1} f(x, y, z)\, dz\, dy\, dx$

25. $\iint_D \int_0^{f(x,y)} dz\, dx\, dy = \iint_D f(x, y)\, dx\, dy$

27.   Let M_ϵ and m_ϵ be the maximum and minimum of f on B_ϵ. Then $m_\epsilon \operatorname{vol}(B_\epsilon) \leq \iiint_{B_\epsilon} f\, dV \leq M_\epsilon \operatorname{vol}(B_\epsilon)$. Divide by $\operatorname{vol}(B_\epsilon)$, let $\epsilon \to 0$ and use continuity of f.

REVIEW EXERCISES FOR CHAPTER 5

1. (a) $\frac{81}{2}$;   (b) $\frac{29}{70}$;   (c) $\frac{1}{4}e^2 - e + \frac{9}{4}$

3. (a) $-\frac{1}{2}$;   (b) $\frac{1}{2}$

5. In the notation of Figure 5.3.1,

$$\iint_D dx\, dy = \int_a^b [\phi_2(x) - \phi_1(x)]\, dx.$$

7. (a) 0; (b) $\pi/24$; (c) 0;

9. Type 1; $2\pi + \pi^2$.

11. Type 2; 76/3.

13. Type 1; 33/140.

15. Type 1; 71/420.

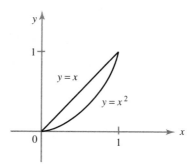

17. $\frac{1}{3}$

19. $\frac{19}{3}$

21. $\frac{11}{12}$

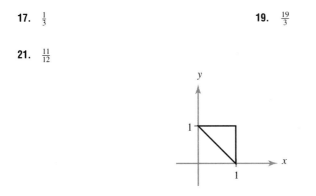

23. The function $f(x, y) = x^2 + y^2 + 1$ lies between 1 and $2^2 + 1 = 5$ on D, and so the integral lies between these values times 4π, the area of D.

25. Interchange the order of integration (the reader should draw a sketch in the (u, t) plane);

$$\int_0^x \int_0^t F(u)\, du\, dt = \int_0^x \int_u^x F(u)\, dt\, du$$
$$= \int_0^x (x - u) F(u)\, du.$$

27. $\dfrac{e^4}{4} - e + \dfrac{3}{4}$

29. The region is the shaded region W in the figure.

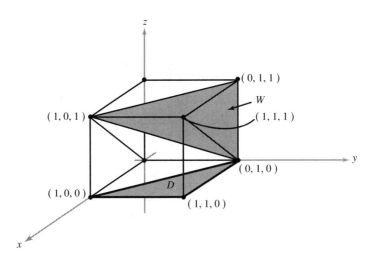

The integral in the order $dy\,dx\,dz$, for example, is

$$\int_0^1 \int_z^1 \int_{1-x}^1 f(x, y, z)\, dy\, dx\, dz$$

Chapter 6

SECTION 6.1

1. $S =$ the unit disk minus its center.

3. $D = [0, 3] \times [0, 1]$; yes

5. The image is the triangle with vertices $(0, 0)$, $(0, 1)$, and $(1, 1)$. T is not one-to-one, but becomes so if we eliminate the portion $x^* = 0$.

7. D is the set of (x, y, z) with $x^2 + y^2 + z^2 \leq 1$ (the unit ball). T is not one-to-one, but is one-to-one on $(0, 1] \times [0, \pi] \times (0, 2\pi]$.

9. Showing that T is onto, is equivalent in the 2×2 case, to showing that the system $ax + by = e$, $cx + dy = f$ can always be solved for x and y, where

$$A = \begin{bmatrix} a & b \\ c & d \end{bmatrix}.$$

When you do this by elimination or by Cramer's rule, the quantity by which you must divide is det A. Thus, if det $A \neq 0$, the equations can always be solved.

11. Since det $A \neq 0$, T maps $\mathbb{R}^2$ one-to-one onto $\mathbb{R}^2$. Let T^{-1} be the inverse transformation. Show that T^{-1} has matrix A^{-1} and det $(A^{-1}) = 1/\text{det } A$, where det $A \neq 0$. By Exercise 10, $P^* = T^{-1}(P)$ is a parallelogram.

SECTION 6.2

1. $\pi(e - 1)$

3. D is the region $0 \leq x \leq 4, \frac{1}{2}x + 3 \leq y \leq \frac{1}{2}x + 6$; (a) 140; (b) -42.

5. D^* is the region $0 \leq u \leq 1, 0 \leq v \leq 2$; $\frac{2}{3}(9 - 2\sqrt{2} - 3\sqrt{3})$.

7. π

9. $\dfrac{64\pi}{5}$

11. $3\pi/2$

13. $\dfrac{5\pi}{2}(e^4 - 1)$

15. $2a^2$

17. $\dfrac{21}{2}\left(e - \dfrac{1}{e}\right)$

19. $100\pi/3$

21. $4\pi[\sqrt{3}/2 - \log(1 + \sqrt{3}) + \log\sqrt{2}]$

23. $4\pi \ln(a/b)$

25. $2\pi[(b^2 + 1)e^{-b^2} - (a^2 + 1)e^{-a^2}]$

27. 24 (use the change of variables $x = 3u - v + 1$, $y = 3u + v$).

29. (a) $\frac{4}{3}\pi abc$;   (b) $\frac{4}{5}\pi abc$.

31. (b) $\frac{160}{3}$

33. $\frac{4}{9}a^{2/3} \iint_{D^*}[f((au^2)^{1/3}, (av^2)^{1/3})u^{-1/3}v^{-1/3}]\,du\,dv$

SECTION 6.3

1. $[\pi^2 - \sin(\pi^2)]/\pi^3$

3. $(\frac{11}{18}, \frac{65}{126})$

5. $503.06

7. ρ, where ρ is the (constant) mass density.   (b) $\frac{37}{12}$

9. $(\frac{1}{2}, \frac{1}{2}, \frac{1}{2})$

11. $\frac{1}{4}$

13. Letting d be density, the moment of inertia is $d \int_0^k \int_0^{2\pi} \int_0^{a\sec\phi}(\rho^4 \sin^3 \phi)\,d\rho\,d\theta\,d\phi$.

15. $(1.00 \times 10^8)m$

17. (a)   The only plane of symmetry for the body of an automobile is the one dividing the left and right sides of the car.

(b) $\bar{z} \cdot \iiint_W \rho(x, y, z)\,dx\,dy\,dz$ is the z coordinate of the center of mass times the mass of W. Rearrangement of the formula for $\bar{z}$ gives the first line of the equation. The next step is justified by the additivity property of integrals. By symmetry, we can replace z by $-z$ and integrate in the region above the xy plane. Finally, we can factor the minus sign outside the second integral, and since $\rho(x, y, z) = \rho(u, v, -w)$, we are subtracting the second integral from itself. Thus, the answer is 0.

(c)   In part (b), we showed that $\bar{z}$ times the mass of W is 0. Since the mass must be positive, $\bar{z}$ must be 0.

(d)   By part (c), the center of mass must lie in both planes.

19. $(4.71 \times 10^{22})G/R$

SECTION 6.4

1. (a) 4   (b) $\frac{8}{3}$ (c) $\frac{3}{16}$ (d) $2 - e$

3. (a) 3π (b) $\lambda < 1$

5. Integration of $\iint e^{-xy} \, dx \, dy$ with respect to x first and then y gives log 2. Reversing the order gives the integral on the left side of the inequality stated in the exercise.

7. Integrate over $[\varepsilon, 1] \times [\varepsilon, 1]$ and let $\varepsilon \to 0$ to show the improper integral exists and equals $2 \log 2$.

9. Use the fact that

$$\frac{\sin^2 (x - y)}{\sqrt{1 - x^2 - y^2}} \leq \frac{1}{\sqrt{1 - x^2 - y^2}}.$$

11. Use the fact that $e^{x^2 + y^2}/(x - y) \geq 1/(x - y)$ on the given region.

13. $\frac{2\pi}{9}[(1 + a^3)^{3/2} - a^{9/2} - 1]$

REVIEW EXERCISES FOR CHAPTER 6

1. (a) $T\begin{pmatrix} u \\ v \end{pmatrix} = \begin{pmatrix} 2 & 1 \\ 0 & 2 \end{pmatrix} \begin{pmatrix} u \\ v \end{pmatrix} = \begin{pmatrix} 2u + v \\ 2v \end{pmatrix} = \begin{pmatrix} x \\ y \end{pmatrix}$

    (b) $\iint_P f(x, y) \, dx \, dy = 4 \iint_S f(2u + v, 2v) \, du \, dv$

3. 3 (Use the change of variables $u = x^2 - y^2, v = xy$.)

5. $\frac{1}{3}\pi(4\sqrt{2} - \frac{7}{2})$ **7.** $(5\pi/2)\sqrt{15}$

9. $abc/6$

11. Cut with the planes $x + y + z = \sqrt[3]{k}/n, 1 \leq k \leq n - 1, k$ an integer.

13. $(25 + 10\sqrt{5})\pi/3$

15. $(e - e^{-1})/4$ (Use the change of variables $u = y - x, v = y + x$.)

17. $(9.92 \times 10^6)\pi$ grams

19. (a) 32.
    (b) This occurs at the point of the unit sphere $x^2 + y^2 + z^2 = 1$ inscribed in the cube.

21. $(0, 0, \frac{3}{8})$

23. $4\pi \ln(a/b)$

25. $\pi/2$

27. (a) $\frac{32}{3}$ (b) 64π

29. Work the integral with respect to y first on the region $D_{\varepsilon,L} = \{(x, y)|\varepsilon \leq x \leq L, 0 \leq y \leq x\}$ to obtain $I_{\varepsilon,L} = \iint_{D_{\varepsilon,L}} f\,dx\,dy = \int_{\varepsilon}^{L} x^{-3/2}(1 - e^{-x})\,dx$. The integrand is positive, and so $I_{\varepsilon,L}$ increases as $\varepsilon \to 0$ and $L \to \infty$. Bound $1 - e^{-x}$ above by x for $0 < x < 1$ and by 1 for $1 < x < \infty$ to see that $I_{\varepsilon,L}$ remains bounded and so must converge. The improper integral does exist.

31. (a) $\frac{1}{6}$ (b) $16\pi/3$

33. 2π

Chapter   7

SECTION   7.1

1. $\int_{\mathbf{c}} f(x, y, z)\,ds = \int_I f(x(t), y(t), z(t))\|\mathbf{c}'(t)\|\,dt = \int_0^1 0 \cdot 1\,dt = 0.$

3. (a) 2   (b) $52\sqrt{14}$ (c) $\frac{16}{3} - 2\sqrt{3}$

5. $-\frac{1}{3}(1 + 1/e^2)^{3/2} + \frac{1}{3}(2^{3/2})$

7. (a) The path follows the straight line from $(0, 0)$ to $(1, 1)$ and back to $(0, 0)$ in the xy plane. Over the path, the graph of f is a straight line from $(0, 0, 0)$ to $(1, 1, 1)$. The integral is the area of the resulting triangle covered twice and equals $\sqrt{2}$.

(b)
$$s(t) = \begin{cases} \sqrt{2}(1 - t^4) & \text{when} \quad -1 \leq t \leq 0 \\ \sqrt{2}(1 + t^4) & \text{when} \quad 0 < t \leq 1 \end{cases}$$

The path is

$$\mathbf{c}(s) = \begin{cases} (1 - s/\sqrt{2})(1, 1) & \text{when} \quad 0 \leq s \leq \sqrt{2} \\ (s/(\sqrt{2} - 1))(1, 1) & \text{when} \quad \sqrt{2} \leq s \leq 2\sqrt{2} \end{cases}$$

and $\int_{\mathbf{c}} f\,ds = \sqrt{2}$.

9. $2a/\pi$

11. (a) $[2\sqrt{5} + \log(2 + \sqrt{5})]/4$ (b) $(5\sqrt{5} - 1)/[6\sqrt{5} + 3\log(2 + \sqrt{5})]$

13. The path is a unit circle centered at $(0, 0, 0)$ in the plane $x + y + z = 0$ and so may be parametrized by $\mathbf{c}(\theta) = (\cos\theta)\mathbf{v} + (\sin\theta)\mathbf{w}$, where $\mathbf{v}$ and $\mathbf{w}$ are orthogonal unit vectors laying in that plane. For example, $\mathbf{v} = (1/\sqrt{2})(-1, 0, 1)$ and $\mathbf{w} = (1/\sqrt{6})(1, -2, 1)$ will do. The total mass is $2\pi/3$ grams.

SECTION   7.2

1. (a) $\frac{3}{2}$ (b) 0   (c) 0   (d) 147

3. 9

5. By the Cauchy-Schwarz inequality, $|\mathbf{F}(\mathbf{c}(t)) \cdot \mathbf{c}'(t)| \leq \|\mathbf{F}(\mathbf{c}(t))\| \|\mathbf{c}'(t)\|$ for every t. Thus

$$\left| \int_{\mathbf{c}} \mathbf{F} \cdot d\mathbf{s} \right| = \left| \int_a^b \mathbf{F}(\mathbf{c}(t)) \cdot \mathbf{c}'(t) \, dt \right| \leq \int_a^b |\mathbf{F}(\mathbf{c}(t)) \cdot \mathbf{c}'(t)| \, dt$$

$$\leq \int_a^b \|\mathbf{F}(\mathbf{c}(t))\| \|\mathbf{c}'(t)\| \, dt \leq M \int_a^b \|\mathbf{c}'(t)\| \, dt = Ml.$$

7. $\frac{3}{4} - (n-1)/(n+1)$

9. 0

11. The length of $\mathbf{c}$

13. If $\mathbf{c}'(t)$ is never 0, then the unit vector $\mathbf{T}(t) = \mathbf{c}'(t)/\|\mathbf{c}'(t)\|$ is a continuous function of t and so is a smoothly turning tangent to the curve. The answer is no.

15. 7

17. Let the path be given by $(x, y, z) = \mathbf{c}(t)$, where $\mathbf{c}(t_1) = (x_1, y_1, z_1)$ and $\mathbf{c}(t_2) = (x_2, y_2, z_2)$. Think of x, y, and z as functions of t. Then $\mathbf{c}'(t) = (dx/dt)\mathbf{i} + (dy/dt)\mathbf{j} + (dz/dt)\mathbf{k}$, and so the work is

$$W = \int_{t_1}^{t_2} \frac{-1}{(x^2 + y^2 + z^2)^{3/2}} (x\mathbf{i} + y\mathbf{j} + z\mathbf{k}) \cdot \left(\frac{dx}{dt}\mathbf{i} + \frac{dy}{dt}\mathbf{j} + \frac{dz}{dt}\mathbf{k} \right) dt$$

$$= -\int_{t_1}^{t_2} \frac{x(dx/dt) + y(dy/dt) + z(dz/dt)}{(x^2 + y^2 + z^2)^{3/2}} dt$$

$$= \int_{t_1}^{t_2} \frac{-\frac{1}{2}(d/dt)(x^2 + y^2 + z^2)}{(x^2 + y^2 + z^2)^{3/2}} dt$$

$$= \int_{t_1}^{t_2} \frac{d}{dt}(x^2 + y^2 + z^2)^{-1/2} dt = (x^2 + y^2 + z^2)^{-1/2} \Big|_{\mathbf{c}(t_1)}^{\mathbf{c}(t_2)}$$

$$= (x_2^2 + y_2^2 + z_2^2)^{-1/2} - (x_1^2 + y_1^2 + z_1^2)^{-1/2} = \frac{1}{R_2} - \frac{1}{R_1}.$$

Thus, the work done by the gravitational field when a particle moves from (x_1, y_1, z_1) to (x_2, y_2, z_2) is $1/R_2 - 1/R_1$. Notice that, in this case, the work is independent of the path taken between the two points.

19. (a) $\|\mathbf{c}'(x)\|$
(b) f has a positive derivative; it is one-to-one and onto $[0, L]$ by the mean value and intermediate value theorems. It has a differentiable inverse by the inverse function theorem.
(c) $g'(s) = 1/\|\mathbf{c}'(x)\|$ where $s = f(x)$.
(d) By the chain rule, $\mathbf{b}'(s) = \mathbf{c}'(x) \cdot g'(s)$, which has unit length by (c).

SECTION 7.3

1. $z = 2(y - 1) + 1$

3. $18(z - 1) - 4(y + 2) - (x - 13) = 0$ or $18z - 4y - x - 13 = 0$.

5. The vector $\mathbf{n} = (\cos v \sin u, \sin v \sin u, \cos u) = (x, y, z)$. The surface is the unit sphere centered at the origin.

7. $\mathbf{n} = -(\sin v)\mathbf{i} - (\cos v)\mathbf{k}$; the surface is a cylinder.

9. (a) $x = x_0 + (y - y_0)(\partial h/\partial y)(y_0, z_0) + (z - z_0)(\partial h/\partial z)(y_0, z_0)$ describes the plane tangent to $x = h(y, z)$ at (x_0, y_0, z_0), $x_0 = h(y_0, z_0)$.
(b) $y = y_0 + (x - x_0)(\partial k/\partial x)(x_0, z_0) + (z - z_0)(\partial k/\partial z)(x_0, z_0)$.

11. (a) The surfaces is a helicoid. It looks like a spiral ramp winding around the z axis. (See Figure 7.4.2) It winds twice around, since θ goes up to 4π.
(b) $\mathbf{n} = \pm(1/\sqrt{1 + r^2})(\sin\theta, -\cos\theta, r)$.
(c) $y_0 x - x_0 y + (x_0^2 + y_0^2)z = (x_0^2 + y_0^2)z_0$.
(d) If $(x_0, y_0, z_0) = (r_0 \cos\theta_0, r_0 \sin\theta_0, \theta_0)$, then representing the line segment in the form $\{(r \cos\theta_0, r \sin\theta_0, \theta_0) | 0 \leq r \leq 1\}$ shows that the line lies in the surface. Representing the line as $\{(x_0, t y_0, z_0) | 0 \leq t \leq 1/(x_0^2 + y_0^2)\}$ and substituting into the results of part (c) shows that it lies in the tangent plane at (x_0, y_0, z_0).

13. (a) Using cylindrical coordinates leads to the parametrization

$$\Phi(z, \theta) = ((25 + z^2)\cos\theta, (25 + z^2)\sin\theta, z), \qquad -\infty < z < \infty, 0 \leq \theta \leq 2\pi$$

as one possible solution.
(b) $\mathbf{n} = (1/\sqrt{1 + 4z^2})(\cos\theta, \sin\theta, -2z)$.
(c) $x_0 x + y_0 y = 25$.
(d) Substitute the coordinates along these lines into the defining equation of the surface and the result of part (c).

15. (a) $u \mapsto u, v \mapsto v, u \mapsto u^3$, and $v \mapsto v^3$ all map $\mathbb{R}$ onto $\mathbb{R}$.
(b) $\mathbf{T}_u \times \mathbf{T}_v = (0, 0, 1)$ for Φ_1, and this is never $\mathbf{0}$. For the surface Φ_2, $\mathbf{T}_u \times \mathbf{T}_v = 9u^2 v^2(0, 0, 1)$, and this is $\mathbf{0}$ along the u and v axes.
(c) We want to show that any two parametrizations of a surface that are smooth near a point will give the same tangent plane there. Thus, suppose $\Phi: D \subset \mathbb{R}^2 \to \mathbb{R}^3$ and $\Psi: B \subset \mathbb{R}^2 \to \mathbb{R}^3$ are parametrized surfaces such that

$$\Phi(u_0, v_0) = (x_0, y_0, z_0) = \Psi(s_0, t_0) \tag{i}$$

and

$$\left.\left(\mathbf{T}_u^{\Phi} \times \mathbf{T}_v^{\Phi}\right)\right|_{(u_0, v_0)} \neq \mathbf{0} \qquad \text{and} \qquad \left.\left(\mathbf{T}_s^{\Psi} \times \mathbf{T}_t^{\Psi}\right)\right|_{(s_0, t_0)} \neq \mathbf{0}, \tag{ii}$$

so that Φ and Ψ are smooth and one-to-one in neighborhoods of (u_0, v_0) and (s_0, t_0), which we may as well assume are D and B. Suppose further that they "describe the same surface," that is, $\Phi(D) = \Phi(B)$. To see that they give the same tangent plane at (x_0, y_0, z_0), show that they have parallel normal vectors. To do this, show that there is an open set C with $(u_0, v_0) \in C \subset D$ and a differentiable map $f: C \to B$ such that $\Phi(u, v) = \Psi(f(u, v))$ for $(u, v) \in C$. Once you have done this, computation shows that the normal vectors are related by $\mathbf{T}_u^{\Phi} \times \mathbf{T}_v^{\Phi} = [\partial(s, t)/\partial(u, v)]\mathbf{T}_s^{\Psi} \times \mathbf{T}_t^{\Psi}$.
To see that there is such an f, notice that since $\mathbf{T}_s^{\Psi} \times \mathbf{T}_t^{\Psi} \neq \mathbf{0}$, at least one of the 2×2 determinants in the cross product is not zero. Assume, for example, that

$$\begin{vmatrix} \dfrac{\partial x}{\partial s} & \dfrac{\partial y}{\partial s} \\ \dfrac{\partial x}{\partial t} & \dfrac{\partial y}{\partial t} \end{vmatrix} \neq 0.$$

Now use the inverse function theorem to write (s, t) as a differentiable function of (x, y) in some neighborhood of (x_0, y_0).

(d) No.

SECTION 7.4

1. 4π

3. $\frac{3}{2}\pi[\sqrt{2} + \log(1 + \sqrt{2})]$

5. $\frac{1}{3}\pi(6\sqrt{6} - 8)$

7. The integral for the volume converges, whereas that for the area diverges.

9. $A(E) = \int_0^{2\pi} \int_0^{\pi} \sqrt{a^2 b^2 \sin^2 \phi \cos^2 \phi + b^2 c^2 \sin^4 \phi \cos^2 \theta + a^2 c^2 \sin^4 \phi \sin^2 \theta} \, d\phi \, d\theta$

11. $(\pi/6)(5\sqrt{5} - 1)$

13. $(\pi/2)\sqrt{6}$

15. $4\sqrt{5}$; for fixed θ, (x, y, z) moves along the horizontal line segment $y = 2x, z = \theta$ from the z axis out to a radius of $\sqrt{5}|\cos\theta|$ into quadrant 1 if $\cos\theta > 0$ and into quadrant 3 if $\cos\theta < 0$.

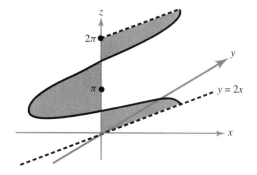

17. $(\pi + 2)/(\pi - 2)$

19. $\pi(a + b)\sqrt{1 + m^2}(b - a)$

21. $\frac{4}{15}(9\sqrt{3} - 8\sqrt{2} + 1)$

23. With $f(x, y) = \sqrt{R^2 - x^2 - y^2}$, (4) becomes

$$A(S') = \iint_D \sqrt{\frac{x^2 + y^2}{R^2 - x^2 - y^2} + 1} \, dx \, dy$$

$$= \iint_D \frac{R}{\sqrt{R^2 - x^2 - y^2}} \, dx \, dy$$

where D is the disk of radius R. Evaluate using polar coordinates, noting it is improper at the boundary, to get $2\pi R^2$.

SECTION 7.5

1. $\dfrac{5\sqrt{2}+3}{24}$

3. πa^3

5. $\sqrt{6}/30$

7. $\dfrac{\pi}{4}\left(\dfrac{5\sqrt{5}}{3}+\dfrac{1}{15}\right)$

9. $16\pi R^3/3$

11. (a) The sphere looks the same from all three axes, so these three integrals should be the same quantity with different labels on the axes.

   (b) $4\pi R^4/3$

   (c) $4\pi R^4/3$

13. $(R/2, R/2, R/2)$

15. Since $\|\partial\Phi/\partial u\|^2 - 2\|\partial\Phi/\partial u\|\|\partial\Phi/\partial v\| + \|\partial\Phi/\partial v\|^2 = (\|\partial\Phi/\partial u\| - \|\partial\Phi/\partial v\|)^2 \geq 0$, we have $\|\partial\Phi/\partial u \times \partial\Phi/\partial v\| = \|\partial\Phi/\partial u\|\|\partial\Phi/\partial v\||\sin\theta(u, v)| \leq \frac{1}{2}(\|\partial\Phi/\partial u\|^2 + \|\partial\Phi/\partial v\|^2)$, where θ is the angle between $\partial\Phi/\partial u$ and $\partial\Phi/\partial v$. This leads to $A(\Phi) \leq J(\Phi)$. Assuming continuity, equality can hold only if $\|\partial\Phi/\partial u\|\|\partial\Phi/\partial v\||\sin\theta| = \frac{1}{2}(\|\partial\Phi/\partial u\|^2 + \|\partial\Phi/\partial v\|^2) = \frac{1}{2}(\|\partial\Phi/\partial u\| - \|\partial\Phi/\partial v\|)^2 + \|\partial\Phi/\partial u\|\|\partial\Phi/\partial v\|$ at each point. This can occur if $|\sin\theta| = 1$ and if $\|\partial\Phi/\partial u\| = \|\partial\Phi/\partial v\|$. Thus $\cos\theta = 0$; that is, $\partial\Phi/\partial u \cdot \partial\Phi/\partial v = 0$.

17. (a) $\sqrt{2}\pi R^2$;   (b) $2\pi R^2$.

19. $2a^2$

SECTION 7.6

1. $\pm48\pi$ (the sign depends on orientation)

3. 4π

5. 2π (or -2π, if you choose a different orientation)

7. $-2\pi/3$

9. $12\pi/5$

11. With the usual spherical coordinate parametrization, $\mathbf{T}_\theta \times \mathbf{T}_\phi = -\sin\phi\,\mathbf{r}$ (see Example 1). Thus

$$\iint_S \mathbf{F} \cdot d\mathbf{S} = \iint \mathbf{F} \cdot (\mathbf{T}_\phi \times \mathbf{T}_\theta)\,d\phi\,d\theta = \iint (\mathbf{F} \cdot \mathbf{r})\sin\phi\,d\phi\,d\theta$$

$$= \int_0^{2\pi} \int_0^\pi F_r \sin\phi\,d\phi\,d\theta$$

and

$$\iint_S f\,dS = \int_0^{2\pi} \int_0^\pi f \sin\phi\,d\phi\,d\theta.$$

13. For a cylinder of radius $R = 1$ and normal component F_r,

$$\iint_S \mathbf{F} \cdot d\mathbf{S} = \int_a^b \int_0^{2\pi} F_r\,d\theta\,dz.$$

15. $2\pi/3$

17. $\frac{2}{3}a^3 bc\pi$

REVIEW EXERCISES FOR CHAPTER 7

1. (a) $3\sqrt{2}(1 - e^{6\pi})/13$ (b) $-\pi\sqrt{2}/2$
   (c) $(236, 158\sqrt{26} - 8)/35 \cdot (25)^3$ (d) $8\sqrt{2}/189$

3. (a) $2/\pi + 1$ (b) $-\frac{1}{2}$

5. $2a^3$

7. (a) A sphere of radius 5 centered at $(2, 3, 0)$; $\mathbf{\Phi}(\theta, \phi) = (2 + 5\cos\theta\sin\phi, 3 + 5\sin\theta\sin\phi, 5\cos\phi)$; $0 \le \theta \le 2\pi$; $0 \le \phi \le \pi$
   (b) An ellipsoid with center at $(2, 0, 0)$; $\mathbf{\Phi}(\theta, \phi) = (2 + (1/\sqrt{2})3\cos\theta\sin\phi, 3\sin\theta\sin\phi, 3\cos\phi)$; $0 \le \theta \le 2\pi, 0 \le \phi \le \pi$
   (c) An elliptic hyperboloid of one sheet; $\mathbf{\Phi}(\theta, z) = (\frac{1}{2}\sqrt{8 + 2z^2}\cos\theta, \frac{1}{3}\sqrt{8 + 2z^2}\sin\theta, z)$; $0 \le \theta \le 2\pi, -\infty < z < \infty$

9. $A(\mathbf{\Phi}) = \frac{1}{2}\int_0^{2\pi} \sqrt{3\cos^2\theta + 5}\,d\theta$; $\mathbf{\Phi}$ describes the upper nappe of a cone with elliptical horizontal cross sections.

11. $11\sqrt{3}/6$

13. $\sqrt{2}/3$

15. $5\sqrt{5}/6$

17. (a) $(e^y \cos\pi z, xe^y \cos\pi z, -\pi xe^y \sin\pi z)$ (b) 0

19. $\frac{1}{2}(e^2 + 1)$

21. $\mathbf{n} = (1/\sqrt{5})(-1, 0, 2)$, $2z - x = 1$

23. 0

25. If $\mathbf{F} = \nabla\phi$, then $\nabla \times \mathbf{F} = \mathbf{0}$ (at least if ϕ is of class C^2: see Theorem 1, Section 3.4). Theorem 3 of Section 7.2 shows that $\int_c \nabla\phi \cdot d\mathbf{s} = 0$ since $\mathbf{c}$ is a closed curve.

27. (a) 24π (b) 24π (c) 60π

29. (a) $\sqrt{R^2 + p^2}(z_0 - z_1)$ (b) $(R^2 + p^2)\sqrt{z_0/2g}$

Chapter 8

SECTION 8.1

1. -8

3. (a) 0   (b) $-\pi R^2$ (c) 0   (d) $-\pi R^2$

5. $3\pi a^2$ **7.** $3\pi/2$

9. $3\pi(b^2 - a^2)/2$ **11.** (a) both sides are 2π (b) 0

13. 0   **15.** πab

17. A horizontal line segment divides the region into three regions of which Green's theorem applies; now use Exercise 8 or the technique in Figure 8.1.5.

19. $9\pi/8$

21. If $\varepsilon > 0$, there is a $\delta > 0$ such that $|u(\mathbf{q}) - u(\mathbf{p})| < \varepsilon$ whenever $\|\mathbf{p} - \mathbf{q}\| = \rho < \delta$. Parametrize $\partial B_\rho(\mathbf{p})$ by $\mathbf{q}(\theta) = \mathbf{p} + \rho(\cos\theta, \sin\theta)$. Then

$$|I(\rho) - 2\pi u(\mathbf{p})| \leq \int_0^{2\pi} |u(\mathbf{q}(\theta)) - u(\mathbf{p})|\, d\theta \leq 2\pi\varepsilon.$$

23. Parametrize $\partial B_\rho(\mathbf{p})$ as in Exercise 21. If $\mathbf{p} = (p_1, p_2)$, then $I(\rho) = \int_0^{2\pi} u(p_1 + \rho\cos\theta, p_2 + \rho\sin\theta)\, d\theta$. Differentiation under the integral sign gives

$$\frac{dI}{d\rho} = \int_0^{2\pi} \nabla u \cdot (\cos\theta, \sin\theta)\, d\theta = \int_0^{2\pi} \nabla u \cdot \mathbf{n}\, d\theta = \frac{1}{\rho}\int_{\partial B_\rho} \frac{\partial u}{\partial \mathbf{n}}\, ds = \frac{1}{\rho}\int_{B_\rho} \nabla^2 u\, dA$$

(the last equality uses Exercise 22).

25.   Using Exercise 24,

$$\iint_{B_R} u \, dA = \int_0^R \int_0^{2\pi} u[\mathbf{p} + \rho(\cos\theta, \sin\theta)]\rho \, d\theta \, d\rho$$

$$= \int_0^R \left(\int_{\partial B_\rho} u \, ds \right) d\rho = \int_0^R 2\pi\rho u(\mathbf{p}) \, d\rho = \pi R^2 u(\mathbf{p})$$

27.   Suppose u is subharmonic. We establish the assertions corresponding to Exercise 26(a) and (b). The argument for superharmonic functions is similar, with inequalities reversed.

Suppose $\nabla^2 u \geq 0$ and $u(\mathbf{p}) \geq u(\mathbf{q})$ for all $\mathbf{q}$ in $B_R(\mathbf{p})$. By Exercise 23, $I'(\rho) \geq 0$ for $0 < \rho \leq R$, and so Exercise 24 shows that $2\pi u(\mathbf{p}) \leq I(\rho) \leq I(R)$ for $0 < \rho \leq R$. If $u(\mathbf{q}) > u(\mathbf{p})$ for some $\mathbf{q} = \mathbf{p} + \rho(\cos\theta_0, \sin\theta_0) \in B_R(\mathbf{p})$, then, by continuity, there is an arc $[\theta_0 - \delta, \theta_0 + \delta]$ on $\partial B_\rho(\mathbf{p})$ where $u < u(\mathbf{p}) - d$ for some $d > 0$. This would mean that

$$2\pi u(\mathbf{p}) \leq I(\rho) = \frac{1}{\rho}\int_0^{2\pi} u[\mathbf{p} + \rho(\cos\theta, \sin\theta)]\rho \, d\theta$$

$$\leq (2\pi - 2\delta)u(\mathbf{p}) + 2\delta[u(\mathbf{p}) - d] \leq 2\pi u(\mathbf{p}) - 2\delta d.$$

This contradiction shows that we must have $u(\mathbf{q}) = u(\mathbf{p})$ for every $\mathbf{q}$ in $B_B(\mathbf{p})$.

If the maximum at $\mathbf{p}$ is absolute for D, the last paragraph shows that $u(\mathbf{x}) = u(\mathbf{p})$ for all $\mathbf{x}$ in some disk around $\mathbf{p}$. If $\mathbf{c}: [0, 1) \rightarrow D$ is a path from $\mathbf{p}$ to $\mathbf{q}$, then $u(\mathbf{c}(t)) = u(\mathbf{p})$ for all t in some interval $[0, b)$. Let b_0 be the largest $b \in [0, 1]$ such that $u(\mathbf{c}(t)) = u(\mathbf{p})$ for all $t \in [0, b)$. (Strictly speaking, this requires the notion of the least upper bound from a good calculus text.) Since u is continuous, $u(\mathbf{c}(b_0)) = u(\mathbf{p})$. If $b_0 \neq 1$, then the last paragraph would apply at $\mathbf{c}(b_0)$ and u is constantly equal to $u(\mathbf{p})$ on a disk around $\mathbf{c}(b_0)$. In particular, there is a $\delta > 0$ such that $u(\mathbf{c}(t)) = u(\mathbf{c}(b_0)) = u(\mathbf{p})$ on $[0, b_0 + \delta)$. This contradicts the maximality of b_0, so we must have $b_0 = 1$. That is, $\mathbf{c}(\mathbf{q}) = \mathbf{c}(\mathbf{p})$. Since $\mathbf{q}$ was an arbitrary point in D, u is constant on D.

29.   Assume $\nabla^2 u_1 = 0$ and $\nabla^2 u_2 = 0$ are two solutions. Let $\phi = u_1 - u_2$. Then $\nabla^2 \phi = 0$ and $\phi(x) = 0$ for all $x \in \partial D$. Consider the integral $\iint_D \phi \nabla^2 \rho \, dA = -\iint_D \nabla \rho \cdot \mathbf{v}\rho \, dA$. Thus $\iint_D \nabla \phi \cdot \nabla \phi \, dA = 0$, which implies that $\nabla \phi = \mathbf{0}$ and so ϕ is a constant function and hence must be identically zero.

SECTION 8.2

1. -2π

3.   Each integral in Stokes' theorem is zero.

5.   0   **7.** $-4\pi/\sqrt{3}$

9.   0   **11.** $\pm 2\pi$

13.   Using Faraday's law, $\iint_S [\nabla \times \mathbf{E} - \partial\mathbf{H}/\partial t] \cdot d\mathbf{S} = 0$ for any surface S. If the integrand were a nonzero vector at some point, then by continuity the integral over some small disk centered at that point and lying perpendicular to that vector would be nonzero.

15.   The orientations of $\partial S_1 = \partial S_2$ must agree.

17. Suppose C is a closed loop on the surface drawn so that it divides the surface into two pieces S_1 and S_2. For the surface of a doughnut (torus) you must use two closed loops; can you see why? Then C bounds both S_1 and S_2 but with positive orientation with respect to one and negative with respect to the other. Therefore,

$$\iint_S \nabla \times \mathbf{F} \cdot d\mathbf{S} = \iint_{S_1} \nabla \times \mathbf{F} \cdot d\mathbf{S} + \iint_{S_2} \nabla \times \mathbf{F} \cdot d\mathbf{S} = \int_C \mathbf{F} \cdot d\mathbf{s} - \int_C \mathbf{F} \cdot d\mathbf{s} = 0.$$

19. (a) If $C = \partial S$, $\int_C \mathbf{v} \cdot d\mathbf{s} = \iint_S (\nabla \times \mathbf{v}) \cdot d\mathbf{S} = \iint_S \mathbf{0} \cdot d\mathbf{s} = 0$.
(b) $\int_C \mathbf{v} \cdot d\mathbf{s} = \int_a^b \mathbf{v} \cdot \mathbf{c}'(t) dt = \mathbf{v} \cdot \int_a^b \mathbf{c}'(t) dt = \mathbf{v} \cdot (\mathbf{c}(b) - \mathbf{c}(a))$, where $\mathbf{c}: [a, b] \to \mathbb{R}^3$ is a parametrization of C. (The vector integral is the vector whose components are the integrals of the component functions.) If C is closed, the last expression is 0.

21. Both integrals give $\pi/4$.   **23.** (a) 0   (b) π (c) π

25. 20π

27. One possible answer: The Möbius curve C is also the boundary of an *oriented* surface $\tilde{S}$; the equation in Faraday's law *is* valid for this new surface.

SECTION 8.3

1. If $\mathbf{F} = \nabla f = \nabla g$ and C is a curve from $\mathbf{v}$ to $\mathbf{w}$, then $(f - g)(\mathbf{w}) - (f - g)(\mathbf{v}) = \int_C \nabla (f - g) \cdot d\mathbf{s} = 0$ and so $f - g$ is constant.

3. $x^2 yz - \cos x + C$ **5.** $-1/\|\mathbf{r}_0\|$

7. No; $\nabla \times \mathbf{F} = (0, 0, -x) \neq \mathbf{0}$.   **9.** $e \sin 1 + \frac{1}{3}e^3 - \frac{1}{3}$

11. 4×10^{31} ergs

13. (a) $f = x^2/2 + y^2/2 + C$ (b) $\mathbf{F}$ is not a gradient field.
(c) $f = \frac{1}{3}x^3 + xy^2 + C$

15. Use Theorem 7 in each case.
(a) $-\frac{3}{2}$ (b) -1 (c) $\cos(e^2) - \cos(1/e)/e$

17. (a) No   (b) $(\frac{1}{2}z^2, xy - z, x^2 y)$ or $(\frac{1}{2}z^2 - 2xyz - \frac{1}{2}y^2, -x^2 z - z, 0)$

19. $\frac{1}{3}(z^3 \mathbf{i} + x^3 \mathbf{j} + y^3 \mathbf{k})$

21. $-(z \sin y + y \sin x, xz \cos y, 0)$ (Other answers are possible.)

23. (a) $\nabla \times \mathbf{F} = (0, 0, 2) \neq \mathbf{0}$.
(b) Let $\mathbf{c}(t)$ be the path of an object in the fluid. Then $\mathbf{F}(\mathbf{c}(t)) = \mathbf{c}'(t)$. Let $\mathbf{c}(t) = (x(t), y(t), z(t))$. Then $x' = -y$, $y' = x$, and $z' = 0$, and so z is constant and the motion is parallel to the xy-plane. Also, $x'' + x = 0$, $y'' + y = 0$. Thus $x = A \cos t + B \sin t$ and $y = C \cos t + D \sin t$. Substituting these values in $x' = -y$, $y' = x$, we get $C = -B$, $D = A$, so that $x^2 + y^2 = A^2 + B^2$ and we have a circle.
(c) Counterclockwise.

25. (a)
$$\mathbf{F} = -\frac{GmM}{(x^2 + y^2 + z^2)^{3/2}}(x, y, z);$$

$$\nabla \cdot \mathbf{F} = -GmM\left[\frac{x^2 + y^2 + z^2 - 3x^2}{(x^2 + y^2 + z^2)^{5/2}} + \frac{x^2 + y^2 + z^2 - 3y^2}{(x^2 + y^2 + z^2)^{5/2}} + \frac{x^2 + y^2 + z^2 - 3z^2}{(x^2 + y^2 + z^2)^{5/2}}\right]$$

$$= 0$$

(b) Let S be the unit sphere, S_1 the upper hemisphere, S_2 the lower hemisphere, and C the unit circle. If $\mathbf{F} = \nabla \times \mathbf{G}$, then

$$\iint_S \mathbf{F} \cdot d\mathbf{S} = \iint_{S_1} \mathbf{F} \cdot d\mathbf{S} + \iint_{S_2} \mathbf{F} \cdot d\mathbf{S} = \int_C \mathbf{G} \cdot d\mathbf{s} - \int_C \mathbf{G} \cdot d\mathbf{s} = 0.$$

But $\iint_S \mathbf{F} \cdot d\mathbf{S} = -GmM \iint_S (\mathbf{r}/\|\mathbf{r}\|^3) \cdot \mathbf{n} \, dS = -4\pi GmM$, since $\|\mathbf{r}\| = 1$ and $\mathbf{r} = \mathbf{n}$ on S. Thus $\mathbf{F} = \nabla \times \mathbf{G}$ is impossible. This does not contradict Theorem 8 since $\mathbf{F}$ is not smooth at the origin.

SECTION 8.4

1.
$$\iint_S \nabla \times \mathbf{F} \cdot d\mathbf{S} = \iiint_V \nabla \cdot (\nabla \times \mathbf{F}) \, dV = \iiint_V \left[\left(\frac{\partial^2 F_1}{\partial y \, \partial z} - \frac{\partial^2 F_1}{\partial z \, \partial y} \right) + \left(\frac{\partial^2 F_2}{\partial z \, \partial x} - \frac{\partial^2 F_2}{\partial x \partial z} \right) \right.$$
$$\left. + \left(\frac{\partial^2 F_3}{\partial x \, \partial y} - \frac{\partial^2 F_3}{\partial y \, \partial z} \right) \right] dV.$$

This is 0 if $\mathbf{F}$ is of class C^2, because the mixed second derivatives of its component functions are equal.

3. 3

5. (a) 0 (b) $\frac{4}{15}$ (c) $-\frac{4}{15}$

7. If $S = \partial\Omega$, then $\iint_S \mathbf{r} \cdot \mathbf{n} \, dS = \iiint_\Omega \nabla \cdot \mathbf{r} \, dV = \iiint_\Omega 3 \, dV = 3 \text{ volume}(\Omega)$.

9. 1

11. Apply the divergence theorem to $f\mathbf{F}$ using $\nabla \cdot (f\mathbf{F}) = \nabla f \cdot \mathbf{F} + f\nabla \cdot \mathbf{F}$.

13. If $\mathbf{F} = \mathbf{r}/r^2$, then $\nabla \cdot \mathbf{F} = 1/r^2$. If $(0, 0, 0) \notin \Omega$, the result follows from Gauss' theorem. If $(0, 0, 0) \in \Omega$, we compute the integral by deleting a small ball $B_\varepsilon = \{(x, y, z) | (x^2 + y^2 + z^2)^{1/2} < \varepsilon\}$ around the origin and then letting $\varepsilon \to 0$:

$$\iiint_\Omega \frac{1}{r^2} dV = \lim_{\varepsilon \to 0} \iiint_{\Omega \backslash B_\varepsilon} \frac{1}{r^2} dV = \lim_{\varepsilon \to 0} \iint_{\partial(\Omega \backslash B_\varepsilon)} \frac{\mathbf{r} \cdot \mathbf{n}}{r^2} dS$$

$$= \lim_{\varepsilon \to 0} \left(\iint_{\partial\Omega} \frac{\mathbf{r} \cdot \mathbf{n}}{r^2} dS - \iint_{\partial B_\varepsilon} \frac{\mathbf{r} \cdot \mathbf{n}}{r^2} dS \right) = \lim_{\varepsilon \to 0} \left(\iint_{\partial\Omega} \frac{\mathbf{r} \cdot \mathbf{n}}{r^2} dS - 4\pi\varepsilon \right)$$

$$= \iint_{\partial\Omega} \frac{\mathbf{r} \cdot \mathbf{n}}{r^2} dS.$$

The integral over ∂B_ε is obtained from Theorem 10 (Gauss' law), since $r = \varepsilon$ everywhere on B_ε.

15. Use the vector identity for $\operatorname{div}(f\mathbf{F})$ and the divergence theorem for part (a). Use the vector identity $\nabla\cdot(f\nabla g - g\nabla f) = f\nabla^2 g - g\nabla^2 f$ for part (b).

17. (a) If $\phi(\mathbf{p}) = \iiint_\Omega \rho(\mathbf{q})/(4\pi\|\mathbf{p}-\mathbf{q}\|)\,dV(\mathbf{q})$, then

$$\nabla\phi(\mathbf{p}) = \iiint_\Omega [\rho(\mathbf{q})/4\pi]\nabla_{\mathbf{p}}(1/\|\mathbf{p}-\mathbf{q}\|)\,dV(\mathbf{q})$$
$$= -\iiint_\Omega [\rho(\mathbf{q})/4\pi][(\mathbf{p}-\mathbf{q})/\|\mathbf{p}-\mathbf{q}\|^3]\,dV(\mathbf{q}),$$

where $\nabla_{\mathbf{p}}$ means the gradient with respect to the coordinates of $\mathbf{p}$ and the integral is the vector whose components are the three component integrals. If $\mathbf{p}$ varies in $V \cup \partial V$ and $\mathbf{n}$ is the outward unit normal to ∂V, we can take the inner product using these components and collect the pieces as

$$\nabla\phi(\mathbf{p})\cdot\mathbf{n} = -\iiint_\Omega \frac{\rho(\mathbf{q})}{4\pi}\frac{1}{\|\mathbf{p}-\mathbf{q}\|^3}(\mathbf{p}-\mathbf{q})\cdot\mathbf{n}\,dV(\mathbf{q}).$$

Thus

$$\iint_{\partial V}\nabla\phi(\mathbf{p})\cdot\mathbf{n}\,dV(\mathbf{p}) = -\iint_{\partial V}\left(\iiint_\Omega \frac{\rho(\mathbf{q})}{4\pi}\frac{1}{\|\mathbf{p}-\mathbf{q}\|^3}(\mathbf{p}-\mathbf{q})\cdot\mathbf{n}\,d\mathbf{q}\right)dV(\mathbf{p}).$$

There are essentially five variables of integration here, three placing $\mathbf{q}$ in Ω and two placing $\mathbf{p}$ on ∂V. Use Fubini's theorem to obtain

$$\iint_{\partial V}\nabla\phi\cdot\mathbf{n}\,dS = -\iiint_\Omega \frac{\rho(\mathbf{q})}{4\pi}\left[\iint_{\partial V}\frac{(\mathbf{p}-\mathbf{q})\cdot\mathbf{n}}{\|\mathbf{p}-\mathbf{q}\|^3}\,dS(\mathbf{p})\right]dV(\mathbf{q}).$$

If V is a symmetric elementary region, Theorem 10 says that the inner integral is 4π if $\mathbf{q} \in V$ and 0 if $\mathbf{q} \notin V$. Thus

$$\iint_{\partial V}\nabla\phi\cdot\mathbf{n}\,dS = -\iiint_{\Omega\cap V}\rho(\mathbf{q})\,dV(\mathbf{q}).$$

Since $\rho = 0$ outside Ω,

$$\iint_{\partial V}\nabla\phi\cdot\mathbf{n}\,dS = -\iiint_V \rho(\mathbf{q})\,dV(\mathbf{q}).$$

If V is not a symmetric elementary region, subdivide it into a sum of such regions. The equation holds on each piece, and, upon adding them together, the boundary integrals along appropriately oriented interior boundaries cancel, leaving the desired result.

(b) By Theorem 9, $\iint_{\partial V}\nabla\phi\cdot d\mathbf{S} = \iiint_V \nabla^2\phi\,dV$, and so $\iiint_V \nabla^2\phi\,dV = -\iiint_V \rho\,dV$. Since both ρ and $\nabla^2\phi$ are continuous and this holds for arbitrarily small regions, we must have $\nabla^2\phi = -\rho$.

19. If the charge Q is spread evenly over the sphere S of radius R centered at the origin, the density of charge per unit area must be $Q/4\pi R^2$. If $\mathbf{p}$ is a point not on S and $\mathbf{q} \in S$, then the contribution to the electric field at $\mathbf{p}$ due to charge near $\mathbf{q}$ is directed along the vector $\mathbf{p}-\mathbf{q}$. Since the charge is evenly distributed, the tangential component of this contribution will be canceled by that from a symmetric point on the other side of the sphere at the same distance from $\mathbf{p}$. (Draw the picture.) The total resulting field must be radial. Since S looks the same from any point at a distance $\|\mathbf{p}\|$ from the origin, the field must depend only on radius and be of the form $\mathbf{E} = f(r)\mathbf{r}$.

If we look at the sphere Σ of radius $\|\mathbf{p}\|$, we have

$$\text{(charge inside } \Sigma) = \iint_\Sigma \mathbf{E} \cdot d\mathbf{S} = \iint_\Sigma f(\|\mathbf{p}\|)\mathbf{r} \cdot \mathbf{n}\, dS$$
$$= f(\|\mathbf{p}\|)\|\mathbf{p}\| \text{ area } \Sigma = 4\pi \|\mathbf{p}\|^3 f(\|\mathbf{p}\|).$$

If $\|\mathbf{p}\| < R$, there is no charge inside Σ; if $\|\mathbf{p}\| > R$, the charge inside Σ is Q, and so

$$\mathbf{E}(\mathbf{p}) = \begin{cases} \dfrac{1}{4\pi}\dfrac{Q}{\|\mathbf{p}\|^3}\mathbf{p} & \text{if } \|\mathbf{p}\| > R \\ 0 & \text{if } \|\mathbf{p}\| < R \end{cases}$$

21. By Theorem 10, $\iint_{\partial M} \mathbf{F} \cdot d\mathbf{S} = 4\pi$ for any surface enclosing the origin. But if $\mathbf{F}$ were the curl of some field, then the integral over such a closed surface would have to be 0.

23. If v_i is the ith component of a vector $\mathbf{v}$, then Exercise 22(b) gives

$$\left[\frac{d}{dt}\iiint_{\Omega_t} f\mathbf{F}\,dx\,dy\,dz\right]_i = \frac{d}{dt}\iiint_{\Omega_t}(f\mathbf{F})_i\,dx\,dy\,dz = \frac{d}{dt}\iiint_{\Omega_t} fF_i\,dx\,dy\,dz$$

$$= \iiint_{\Omega_t}\left[\frac{D(fF_i)}{Dt} + (fF_i)\,\text{div }\mathbf{F}\right]dx\,dy\,dz$$

$$= \iiint_{\Omega_t}\left[\frac{\partial}{\partial t}(fF_i) + \mathbf{D}_x(fF_i) \cdot \mathbf{F} + (fF_i)\,\text{div }\mathbf{F}\right]dx\,dy\,dz$$

$$= \iiint_{\Omega_t}\left[\frac{\partial}{\partial t}(f\mathbf{F}_i) + \nabla(fF_i) \cdot \mathbf{F} + (fF_i)\,\text{div }\mathbf{F}\right]dx\,dy\,dz$$

$$= \iiint_{\Omega_t}\left\{\frac{\partial}{\partial t}(f\mathbf{F}) + [\mathbf{D}(f\mathbf{F})\mathbf{F}]_i + [(f\mathbf{F})\,\text{div }\mathbf{F}]_i\right\}dx\,dy\,dt$$

$$= \iiint_{\Omega_t}\left[\frac{\partial}{\partial t}(f\mathbf{F}) + \mathbf{D}(f\mathbf{F})\mathbf{F} + (f\mathbf{E})\,\text{div }\mathbf{F}\right]_i dx\,dy\,dz$$

$$= \left[\iiint_{\Omega_t}\frac{\partial}{\partial t}(f\mathbf{F}) + \mathbf{D}(f\mathbf{F})\mathbf{F} + (f\mathbf{F})\,\text{div }\mathbf{F}\,dx\,dy\,dz\right]_i$$

$$= \left[\iiint_{\Omega_t}\left(\frac{\partial}{\partial t}(f\mathbf{F}) + (\mathbf{F} \cdot \nabla)(f\mathbf{F}) + (f\mathbf{F})\,\text{div }\mathbf{F}\right)dx\,dy\,dt\right]_i.$$

SECTION 8.5

1. (a)   By Exercise 22, Section 8.4,

$$\frac{d}{dt}\iiint_{\Omega_t}\rho\,dV = \iiint_{\Omega_t}\left(\frac{\partial\rho}{\partial t} + \nabla\rho \cdot \mathbf{V} + \rho\,\text{div }\mathbf{V}\right)dV.$$

By Theorem 11, the conservation law for $\mathbf{V}$ is equivalent to stating that the integrand on the right is identically 0 and so implies that the left integral is 0. Conversely, if Ω is a small region, so is Ω_t. If the integrals on the right are 0 for all sufficiently small regions, then the integrand must be 0 (by continuity).

(b)   For each time t, the change of variables $(u, v, w) = \phi((x, y, z), t)$ gives

$$\iiint_{\Omega_t} \rho(u, v, w, 0)\, du\, dv\, dw = \iiint_{\Omega} \rho(x, y, z, t) \frac{\partial(u, v, w)}{\partial(x, y, z)}\, dx\, dy\, dz.$$

By part (a), the left side is constant in time and is therefore equal to its value at $t = 0$:

$$\iiint_{\Omega} \rho(x, y, z, 0)\, dx\, dy\, dz = \iiint_{\Omega} \rho(x, y, z, t) J(x, y, z, t)\, dx\, dy\, dz.$$

Since this holds for all small regions, the integrands must be equal by continuity.

(c)   By Exercise 22, Section 8.4, div $\mathbf{V} = 0$ implies $J(\mathbf{x}, t) = 1$. Applying this to part (b), $\rho(\mathbf{x}, t) = \rho(\mathbf{x}, 0)$. The density at each point is constant in time, and so $\partial\rho/\partial t = 0$ and the conservation law becomes $\mathbf{V} \cdot \nabla\rho = 0$. The flow is perpendicular to the gradient of ρ and so the flow lines lie in surfaces of constant density.

3.   (a)   Since $\mathbf{V} = \nabla\phi$, $\nabla \times \mathbf{V} = \mathbf{0}$, and therefore $(\mathbf{V} \cdot \nabla)\mathbf{V} = \frac{1}{2}\nabla(\|\mathbf{V}\|^2)$, Euler's equation becomes

$$-\frac{\nabla p}{\rho} = \frac{d\mathbf{V}}{dt} + \frac{1}{2}\nabla(\|\mathbf{V}\|^2) = \nabla\left(\frac{d\phi}{dt} + \frac{1}{2}\|\mathbf{V}\|^2\right)$$

If $\mathbf{c}$ is a path from P_1 to P_2, then

$$-\int_{\mathbf{c}} \frac{1}{\rho}\, dp = -\int \frac{1}{\rho}\nabla p \cdot \mathbf{c}'(t)\, dt = \int_{\mathbf{c}} \nabla\left(\frac{d\phi}{dt} + \frac{1}{2}\|\mathbf{V}\|^2\right) \cdot \mathbf{c}'(t)\, dt$$

$$= \left(\frac{d\phi}{dt} + \frac{1}{2}\|\mathbf{V}\|^2\right)\Big|_{P_1}^{P_2}$$

(b)   If $d\mathbf{V}/dt = \mathbf{0}$ and ρ is constant, then $\frac{1}{2}\nabla(\|\mathbf{V}\|^2) = -(\nabla p)/\rho = -\nabla(p/\rho)$ and therefore $\nabla(\frac{1}{2}\|\mathbf{V}\|^2 + p/\rho) = \mathbf{0}$.

5.   By Ampere's law, $\nabla \cdot \mathbf{J} = \nabla \cdot (\nabla \times \mathbf{H}) - \nabla \cdot (\partial\mathbf{E}/\partial t) = -\nabla \cdot (\partial\mathbf{E}/\partial t) = -(\partial/\partial t)(\nabla \cdot \mathbf{E})$. By Gauss' law this is $-\partial\rho/\partial t$. Thus $\nabla \cdot \mathbf{J} + \partial\rho/\partial t = 0$.

7.   (a)   If $\mathbf{x} \in S$, then $r''a/R = r$, so $G = 0$. In general, $r = \|\mathbf{x} - \mathbf{y}\|$ and $r'' = \|\mathbf{x} - \mathbf{y}\|$, and therefore

$$G = \frac{1}{4\pi}\left(\frac{R}{a}\frac{1}{\|\mathbf{x} - \mathbf{y}'\|} - \frac{1}{\|\mathbf{x} - \mathbf{y}\|}\right) \quad \text{and} \quad \nabla_{\mathbf{x}}G = \frac{1}{4\pi}\left(\frac{R}{a}\frac{\mathbf{y}' - \mathbf{x}}{\|\mathbf{x} - \mathbf{y}'\|^3} - \frac{\mathbf{y} - \mathbf{x}}{\|\mathbf{x} - \mathbf{y}\|^3}\right)$$

and $\nabla_{\mathbf{x}}^2 G = 0$ when $\mathbf{x} \neq \mathbf{y}$, just as in the analysis of equation (15) ($\mathbf{x} \neq \mathbf{y}'$, since $\mathbf{x}$ is inside and $\mathbf{y}'$ is outside the sphere). Theorem 10 gives $\nabla^2 G = \delta(\mathbf{x} - \mathbf{y}) - (R/a)\delta(\mathbf{x} - \mathbf{y}')$, but the second term is always 0, since $\mathbf{x}$ is never $\mathbf{y}'$. Therefore $\nabla^2 G = \delta(\mathbf{x} - \mathbf{y})$ for x and y in the sphere.

(b)   If $\mathbf{x}$ is on the surface of S, then $\mathbf{n} = \mathbf{x}/R$ is the outward unit normal, and

$$\frac{\partial G}{\partial \mathbf{n}} = \frac{1}{4\pi}\left[\frac{R}{a}\nabla\left(\frac{1}{r''}\right) \cdot \mathbf{n} - \nabla\left(\frac{1}{r}\right) \cdot \mathbf{n}\right] = \frac{1}{4\pi}\left(\frac{R}{a}\frac{\mathbf{y}' - \mathbf{x}}{\|\mathbf{y}' - \mathbf{x}\|^3} - \frac{\mathbf{y} - \mathbf{x}}{\|\mathbf{x} - \mathbf{y}\|^3}\right) \cdot \mathbf{n}.$$

If γ is the angle between $\mathbf{x}$ and $\mathbf{y}$, then $\|\mathbf{x} - \mathbf{y}\|^2 = r^2 = R^2 + a^2 - 2aR\cos\gamma$ and $\|\mathbf{x} - \mathbf{y}'\|^2 = r''^2 = R^2 + b^2 - 2bR\cos\gamma = (R^2/a^2)r^2$. Then

$$\frac{\partial G}{\partial \mathbf{n}} = \frac{1}{4\pi r^3}\left[\frac{R}{a}\frac{\mathbf{y}' - \mathbf{x}}{(R/a)^3} - (\mathbf{y} - \mathbf{x})\right] \cdot \mathbf{n}.$$

But $\mathbf{x} \cdot \mathbf{n} = R$ and $\mathbf{n} = \mathbf{x}/R$, and so this becomes

$$\frac{\partial G}{\partial \mathbf{n}} = \frac{1}{4\pi^3} \left(\frac{a^2}{R^3} \mathbf{y}' \cdot \mathbf{x} - \frac{a^2}{R} - \frac{\mathbf{y} \cdot \mathbf{x}}{R} + R \right)$$

$$= \frac{1}{4\pi^3 R} \left(\frac{a^2}{R^2} \|\mathbf{y}'\| R \cos\gamma - \|\mathbf{y}\| R \cos\gamma + R^2 - a^2 \right)$$

$$= \frac{R^2 - a^2}{4\pi R} \frac{1}{r^3} \quad \text{since} \quad \|\mathbf{y}'\| = R^2/a.$$

Integrating over the surface of the sphere,

$$u(\mathbf{y}) = \iint_S f \frac{\partial G}{\partial \mathbf{n}} \, dS = \int_0^{2\pi} \int_0^{\pi} \left[f(\theta, \phi) \frac{R^2 - a^2}{4\pi R} \frac{1}{r^3} R^2 \sin\phi \right] d\phi \, d\theta$$

$$= \frac{R(R^2 - a^2)}{4\pi} \int_0^{2\pi} \int_0^{\pi} \frac{f(\theta, \phi) \sin\phi \, d\phi \, d\theta}{(R^2 + a^2 - 2aR\cos\gamma)^{3/2}}.$$

9. (a) $\dot{u} = (d/ds)[u(x(s), t(s))] = u_x \dot{x} + u_t \dot{t} = u_x f'(u) + u_t = 0.$

(b) If the characteristic curve $u(x, t) = c$ (by part (a)), define t implicitly as a function of x; then $u_x + u_t(dt/dx) = 0$. But also $u_t + f(u)_x = 0$; that is, $u_t + f'(u)u_x = 0$. These two equations together give $dt/dx = 1/f'(u) = 1/f'(c)$. Therefore the curve is a straight line with slope $1/f'(c)$.

(c) If $x_1 < x_2, u_0(x_1) > u_0(x_2) > 0$, and $f'(u(x_2)) > 0$, then $f'(u_0(x_1)) > f'(u_0(x_2)) > 0$, since $f'' > 0$. The characteristic through $(x_1, 0)$ has slope $1/f'(u_0(x_1))$, which is less than $1/f'(u_0(x_2))$, (that of the characteristic through $(x_2, 0)$). So these lines must cross at a point $P = (\bar{x}, \bar{t})$ with $\bar{t} > 0$ and $\bar{x} > x_2$. The solution must be discontinuous at P, since these two crossing lines would give it different values there.

(d) $\bar{t} = (x_2 - x_1)/[f'(u_0(x_1)) - f'(u_0(x_2))].$

11. (a) Since the "rectangle" D does not touch the x axis and $\phi = 0$ on ∂D and outside D, equation (25) becomes

$$\iint [u\phi_t + f(u)\phi_x] \, dx \, dt = 0. \tag{i}$$

Since $(u\phi)_t + [f(u)\phi]_x = [u_t + f(u)_x]\phi + [u\phi_t + f(u)\phi_x]$, we have

$$\iint_{D_i} [u\phi_t + f(u)\phi_x] \, dx \, dt = \iint_{D_i} [(u\phi)_t + (f(u)\phi)_x] \, dx \, dt - \iint_{D_i} [u_t + f(u)_x]\phi \, dx \, dt.$$

But u is C^1 on the interior of D_i, and so Exercise 10(b) says $u_t + f(u)_x = 0$ there. Thus

$$\iint_{D_i} [u\phi_t + f(u)\phi_x] \, dx \, dt = \iint_{D_i} [(u\phi)_t + (f(u)\phi)_x] \, dx \, dt. \tag{ii}$$

(b) By Green's theorem,

$$\iint_{D_i} [(f(u)\phi)_x - (-u\phi)_t] \, dx \, dt = \int_{\partial D_i} [(-u\phi) \, dx + f(u)\phi \, dt],$$

and so expression (ii) becomes

$$\iint [u\phi_t + f(u)\phi_x] \, dx \, dt = \int_{\partial D_i} \phi[-u \, dx + f(u) \, dt]$$

Adding the above for $i = 1, 2$ and using expression (i) gives

$$0 = \int_{\partial D_1} \phi[-u\,dx + f(u)\,dt] + \int_{\partial D_2} \phi[-u\,dx + f(u)\,dt]$$

The union of these two boundaries traverses ∂D once and that portion of Γ within D once in each direction, once with the values u_1 and once with the values u_2. Since $\phi = 0$ outside D and on ∂D_2 this becomes $0 = \int_{\Gamma} \phi\{[-u]\,dx + [f(u)]\,dt\}$.

(c) Since $\phi = 0$ outside D, the first integral is the same as that of the second conclusion of part (b). The second integral results from parametrizing the portion of Γ by $\alpha(t) = (x(t), t), t_1 \leq t \leq t_2$.

(d) If $[-u]s + [f(u)] = c > 0$ at P, then we can choose a small disk B_ε centered at P contained in D (described above) such that $[-u](dx/dt) + [f(u)] > c/2$ on the part of Γ inside B_ε. Now take a slightly smaller disk $B_\delta \subset B_\varepsilon$ centered at P and pick ϕ such that $\phi \equiv 1$ on B_δ. $0 \leq \phi \leq 1$ on the annulus $B_\varepsilon \setminus B_\delta$ and $\phi \equiv 0$ outside B_ε. If $\alpha(t_0) = $ P, then there are t_3 and t_4 with $t_1 < t_3 < t_0 < t_4 < t_2$ and $\alpha(t) \in B_\delta$ for $t_3 < t < t_4$. But then

$$\int_{t_1}^{t_2} \phi\left\{[-u]\frac{dx}{dt} + [f(u)]\right\} dt > \frac{c}{2}(t_4 - t_3) > 0,$$

contradicting the result of part (c). A similar argument (reversing signs) works if $c < 0$.

13. Setting $P = g(u)\phi$ and $Q = -f(u)\phi$, applying Green's theorem on rectangular R, and using the function ϕ as in Exercise 10 shows that if u is a solution of $g(u)_t + f(u)_x = 0$, then

$$\iint_{t\geq 0} [g(u)\phi_t + f(u)\phi_x]\,dx\,dt + \int_{t=0} g(u_0(x))\phi(x, 0)\,dx = 0.$$

This is the appropriate analogy to equation (25), defining weak solutions of $g(u)_t + f(u)_x = 0$. Thus we want a u such that

$$\iint_{t\geq 0} \left(u\phi_t + \frac{1}{2}u^2\phi_x\right) dx\,dt + \int_{t=0} u_0(x)\phi(x, 0)\,dx = 0 \qquad \text{(i: } weak\text{)}$$

holds for all admissible ϕ but such that

$$\iint_{t\geq 0} \left(\frac{1}{2}u^2\phi_t + \frac{1}{3}u^3\phi_x\right) dx\,dt + \int_{t=0} \frac{1}{2}u_0^2(x)\phi(x, 0)\,dx = 0 \qquad \text{(ii: } weak\text{)}$$

fails for some admissible ϕ. The method of Exercise 11 produces the jump condition $s[g(u)] = [f(u)]$. For (a), this is $s(u_2 - u_1) = (\frac{1}{2}u_2^2 - \frac{1}{2}u_1^2)$ or

$$s = \frac{1}{2}(u_2 + u_1). \qquad \text{(i: } jump\text{)}$$

For (b), it is $s(\frac{1}{2}u_2^2 - \frac{1}{2}u_1^2) = (\frac{1}{3}u_2^3 - \frac{1}{3}u_1^3)$ or

$$s = \frac{2}{3}\frac{u_2^2 + u_1u_2 + u_1^2}{u_2 + u_1}. \qquad \text{(ii: } jump\text{)}$$

If we take for $u_0(x)$ a (Heaviside) function defined by $u_0(x) = 0$ for $x < 0$ and $u_0(x) = 1$ for $x > 0$, we are led to consider the function $u(x, t) = 0$ when $t > 2x$ and $u(x, t) = 1$ when $t \leq 2x$. Thus $u_1 = 1, u_2 = 0$, and the discontinuity curve Γ is given by $t = 2x$. Thus the jump condition (i: $jump$)(i.e., $dx/dt = \frac{1}{2}(u_1 + u_2)$) is satisfied.

For any particular ϕ, there are numbers T and a such that $\phi(x, t) = 0$ for $x \geq a$ and $t \geq T$. Letting Ω be the region $0 \leq x \leq a$ and $0 \leq t \leq T$, condition (i: *weak*) becomes

$$
\begin{aligned}
0 &= \iint_\Omega \left(\phi_t + \frac{1}{2}\phi_x \right) dx\, dt + \int_0^a \phi(x, 0)\, dx \\
&= \int_{\partial\Omega} \left(-\phi\, dx + \frac{\phi}{2}\, dt \right) + \int_0^a \phi(x, 0)\, dx \\
&= -\int_0^a \phi(x, 0)\, dx + \int_0^{T/2} \left[-\phi(x, 2x)(-dx) + \frac{1}{2}\phi(x, 2x)(-2\,dx) \right] \\
&\quad + \int_0^a \phi(x, 0)\, dx
\end{aligned}
$$

Thus (i: *weak*) is satisfied for every ϕ, and u is a weak solution of equation (i). However, (ii: *weak*) cannot be satisfied for every ϕ, since the jump condition (ii: *jump*) fails. Indeed, if we multiply (ii: *weak*) by 2 and insert u, (ii: *weak*) becomes

$$
0 = \iint_\Omega \left(\phi_t + \frac{2}{3}\phi_x \right) dx\, dt + \int_0^a \phi(x, 0)\, dx.
$$

The factor $\frac{1}{2}$ has changed to $\frac{2}{3}$, and the computation above now becomes

$$
0 = -\frac{1}{3} \int_0^{\pi/2} \phi(x, 2x)\, dx,
$$

which is certainly not satisfied for every admissible ϕ.

15. For example, write $|re^{i\theta} - z'|^2 = |re^{i\theta} - r'e^{i\theta'}|^2 = |re^{i(\theta-\theta')} - r'|^2$, use $e^{i\phi} = \cos\phi + i\sin\phi$, $|z|^2 = z\bar{z}$, and multiply out.

SECTION 8.6

1. (a) $(2xy^2 - yx^3)\, dx\, dy$ (b) $(x^2 + y^2)\, dx\, dy$

     (c) $(x^2 + y^2 + z^2)\, dx\, dy\, dz$ (d) $(xy + x^2)\, dx\, dy\, dz$

     (e) $dx\, dy\, dz$

3. (a) $2xy\, dx + (x^2 + 3y^2)\, dy$ (b) $-(x + y^2 \sin x)\, dx\, dy$

     (c) $-(2x + y)\, dx\, dy$ (d) $dx\, dy\, dz$

     (e) $2x\, dx\, dy\, dz$ (f) $2y\, dy\, dz - 2x\, dz\, dx$

     (g) $-\dfrac{4xy}{(x^2 + y^2)^2}\, dx\, dy$ (h) $2xy\, dx\, dy\, dz$

5. (a) $\text{Form}_2\,(\alpha \mathbf{V}_1 + \mathbf{V}_2) = \text{Form}_2\,(\alpha A_1 + A_2, \alpha B_1 + B_2, \alpha C_1 + C_2)$
$$= (\alpha A_1 + A_2)\,dy\,dz + (\alpha B_1 + B_2)\,dz\,dx$$
$$+ (\alpha C_1 + C_2)\,dx\,dy$$
$$= \alpha(A_1\,dy\,dz + B_1\,dz\,dx + C_1\,dx\,dy)$$
$$+ (A_2\,dy\,dz + B_2\,dz\,dx + C_2\,dx\,dy)$$
$$= \alpha\,\text{Form}_2\,(\mathbf{V}_1) + \text{Form}_2(\mathbf{V}_2)$$

(b) $d\omega = \left(\dfrac{\partial A}{\partial x}\,dx + \dfrac{\partial A}{\partial y}\,dy + \dfrac{\partial A}{\partial z}\,dz\right) \wedge dx - A(dx)^2$

$$+ \left(\dfrac{\partial B}{\partial x}\,dx + \dfrac{\partial B}{\partial y}\,dy + \dfrac{\partial B}{\partial z}\,dz\right) \wedge dy - B(dy)^2$$

$$+ \left(\dfrac{\partial C}{\partial x}\,dx + \dfrac{\partial C}{\partial y}\,dy + \dfrac{\partial C}{\partial z}\,dz\right) \wedge dz - C(dz)^2.$$

But $(dx)^2 = (dy)^2 = (dz)^2 = dx \wedge dx = dy \wedge dy = dz \wedge dz = 0$, $dy \wedge dx = -dx \wedge dy$, $dz \wedge dy = -dy \wedge dz$, and $dx \wedge dz = -dz \wedge dx$. Hence

$$d\omega = \left(\dfrac{\partial C}{\partial y} - \dfrac{\partial B}{\partial z}\right) dy\,dz + \left(\dfrac{\partial A}{\partial z} - \dfrac{\partial C}{\partial x)}\right) dz\,dx + \left(\dfrac{\partial B}{\partial x} - \dfrac{\partial A}{\partial y}\right) dx\,dy$$
$$= \text{Form}_2\,(\text{curl } \mathbf{V}).$$

7. An oriented 1-manifold is a curve. Its boundary is a pair of points that may be considered a 0-manifold. Therefore ω is a 0-form or function, and $\int_{\partial M} d\omega = \omega(b) - \omega(a)$ if the curve M runs from a to b. Furthermore, $d\omega$ is the 1-form $(\partial\omega/\partial x)\,dx + (\partial\omega/\partial y)\,dy$. Therefore $\int_M d\omega$ is the line integral $\int_M (\partial\omega/\partial x)\,d\omega + (\partial\omega/dy)\,dy = \int_M \nabla\omega \cdot d\mathbf{s}$. Thus we obtain Theorem 3 of Section 7.2, $\int_M \nabla\omega \cdot d\mathbf{s} = \omega(b) - \omega(a)$.

9. Put $\omega = F_1\,dx\,dy + f_2\,dy\,dz + F_3\,dz\,dx$. The integral becomes

$$\iint_{\partial T} \omega = \iiint_T d\omega = \iiint_T \left(\dfrac{\partial F_1}{\partial x} + \dfrac{\partial F_2}{\partial x} + \dfrac{\partial F_3}{\partial y}\right) dx\,dy\,dz.$$

(a) 0   (b) 40

11. Consider $\omega = x\,dy\,dz + y\,dz\,dx + z\,dx\,dy$. Compute that $d\omega = 3\,dx\,dy\,dz$, so that $\frac{1}{3} \iint_{\partial R} \omega = \frac{1}{3} \iiint_R d\omega = \iiint_R dx\,dy\,dz = v(R)$.

REVIEW EXERCISES FOR CHAPTER 8

1. (a) $2\pi a^2$ (b) 0

3. 0

5. (a) $f = x^4/4 - x^2 y^3$ (b) $-\frac{1}{4}$

7. (a) Check that $\nabla \times \mathbf{F} = \mathbf{0}$ (b) $f = 3x^2 y \cos z + C$ (c) 0

9. $\frac{23}{6}$

11. No: $\nabla \times (\mathbf{a} \times \mathbf{r}) = 2\mathbf{a}$

13. (a) $\nabla f = 3ye^{z^2}\mathbf{i} + 3xe^{z^2}\mathbf{j} + 6xyze^{z^2}\mathbf{k}$ (b) 0 (c) Both sides are 0

15. $8\pi/3$

17. $\pi a^2/4$

19. 21

21. (a) $\mathbf{G}$ is conservative; $\mathbf{F}$ is not
(b) $\mathbf{G} = \nabla\phi$ if $\phi = (x^4/4) + (y^4/4) - \frac{3}{2}x^2y^2 + \frac{1}{2}z^2 + C$, where C is any constant
(c) $\int_\alpha \mathbf{F} \cdot d\mathbf{s} = 0$; $\int_\alpha \mathbf{G} \cdot d\mathbf{s} = -\frac{1}{2}$; $\int_\beta \mathbf{F} \cdot d\mathbf{s} = \frac{1}{3}$; $\int_\beta \mathbf{G} \cdot d\mathbf{s} = -\frac{1}{2}$

23. Use $(\nabla \cdot \mathbf{F})(x_0, y_0, z_0) = \lim_{\rho \to 0} \frac{1}{V(\Omega_\rho)} \iint_{\partial\Omega_\rho} \mathbf{F} \cdot \mathbf{n} \, dS$ from Section 8.4.

INDEX

24. $\int \sinh x \, dx = \cosh x$

25. $\int \cosh x \, dx = \sinh x$

26. $\int \tanh x \, dx = \log |\cosh x|$

27. $\int \coth x \, dx = \log |\sinh x|$

28. $\int \operatorname{sech} x \, dx = \arctan (\sinh x)$

29. $\int \operatorname{csch} x \, dx = \log \left| \tanh \dfrac{x}{2} \right| = -\dfrac{1}{2} \log \dfrac{\cosh x + 1}{\cosh x - 1}$

30. $\int \sinh^2 x \, dx = \frac{1}{4} \sinh 2x - \frac{1}{2}x$

31. $\int \cosh^2 x \, dx = \frac{1}{4} \sinh 2x + \frac{1}{2}x$

32. $\int \operatorname{sech}^2 x \, dx = \tanh x$

33. $\int \sinh^{-1} \dfrac{x}{a} \, dx = x \sinh^{-1} \dfrac{x}{a} - \sqrt{x^2 + a^2} \quad (a > 0)$

34. $\int \cosh^{-1} \dfrac{x}{a} \, dx = \begin{cases} x \cosh^{-1} \dfrac{x}{a} - \sqrt{x^2 - a^2} & \left[\cosh^{-1} \left(\dfrac{x}{a} \right) > 0, a > 0 \right] \\ x \cosh^{-1} \dfrac{x}{a} + \sqrt{x^2 - a^2} & \left[\cosh^{-1} \left(\dfrac{x}{a} \right) < 0, a > 0 \right] \end{cases}$

35. $\int \tanh^{-1} \dfrac{x}{a} \, dx = x \tanh^{-1} \dfrac{x}{a} + \dfrac{a}{2} \log |a^2 - x^2|$

36. $\int \dfrac{1}{\sqrt{a^2 + x^2}} \, dx = \log (x + \sqrt{a^2 + x^2}) = \sinh^{-1} \dfrac{x}{a} \quad (a > 0)$

37. $\int \dfrac{1}{a^2 + x^2} \, dx = \dfrac{1}{a} \arctan \dfrac{x}{a} \quad (a > 0)$

38. $\int \sqrt{a^2 - x^2} \, dx = \dfrac{x}{2} \sqrt{a^2 - x^2} + \dfrac{a^2}{2} \arcsin \dfrac{x}{a} \quad (a > 0)$

39. $\int (a^2 - x^2)^{3/2} \, dx = \dfrac{x}{8} (5a^2 - 2x^2)\sqrt{a^2 - x^2} + \dfrac{3a^4}{8} \arcsin \dfrac{x}{a} \quad (a > 0)$

40. $\int \dfrac{1}{\sqrt{a^2 - x^2}} \, dx = \arcsin \dfrac{x}{a} \quad (a > 0)$

41. $\int \dfrac{1}{a^2 - x^2} \, dx = \dfrac{1}{2a} \log \left| \dfrac{a + x}{a - x} \right|$

42. $\int \dfrac{1}{(a^2 - x^2)^{3/2}} \, dx = \dfrac{x}{a^2 \sqrt{a^2 - x^2}}$

43. $\int \sqrt{x^2 \pm a^2} \, dx = \dfrac{x}{2} \sqrt{x^2 \pm a^2} \pm \dfrac{a^2}{2} \log |x + \sqrt{x^2 \pm a^2}|$

44. $\int \dfrac{1}{\sqrt{x^2 - a^2}} \, dx = \log |x + \sqrt{x^2 - a^2}| = \cosh^{-1} \dfrac{x}{a} \quad (a > 0)$

45. $\int \dfrac{1}{x(a + bx)} \, dx = \dfrac{1}{a} \log \left| \dfrac{x}{a + bx} \right|$

46. $\int x\sqrt{a + bx} \, dx = \dfrac{2(3bx - 2a)(a + bx)^{3/2}}{15b^2}$

47. $\int \dfrac{\sqrt{a + bx}}{x} \, dx = 2\sqrt{a + bx} + a \int \dfrac{1}{x\sqrt{a + bx}} \, dx$

48. $\displaystyle \int \frac{x}{\sqrt{a+bx}}\, dx = \frac{2(bx-2a)\sqrt{a+bx}}{3b^2}$

49. $\displaystyle \int \frac{1}{x\sqrt{a+bx}}\, dx = \begin{cases} \dfrac{1}{\sqrt{a}} \log \left| \dfrac{\sqrt{a+bx}-\sqrt{a}}{\sqrt{a+bx}+\sqrt{a}} \right| & (a>0) \\[3mm] \dfrac{2}{\sqrt{-a}} \arctan \left| \sqrt{\dfrac{a+bx}{-a}} \right| & (a<0) \end{cases}$

50. $\displaystyle \int \frac{\sqrt{a^2-x^2}}{x}\, dx = \sqrt{a^2-x^2} - a \log \left| \frac{a+\sqrt{a^2-x^2}}{x} \right|$

51. $\displaystyle \int x\sqrt{a^2-x^2}\, dx = -\tfrac{1}{3}(a^2-x^2)^{3/2}$

52. $\displaystyle \int x^2\sqrt{a^2-x^2}\, dx = \frac{x}{8}(2x^2-a^2)\sqrt{a^2-x^2} + \frac{a^4}{8} \arcsin \frac{x}{a} \qquad (a>0)$

53. $\displaystyle \int \frac{1}{x\sqrt{a^2-x^2}}\, dx = -\frac{1}{a} \log \left| \frac{a+\sqrt{a^2-x^2}}{x} \right|$

54. $\displaystyle \int \frac{x}{\sqrt{a^2-x^2}}\, dx = -\sqrt{a^2-x^2}$

55. $\displaystyle \int \frac{x^2}{\sqrt{a^2-x^2}}\, dx = -\frac{x}{2}\sqrt{a^2-x^2} + \frac{a^2}{2} \arcsin \frac{x}{a} \qquad (a>0)$

56. $\displaystyle \int \frac{\sqrt{x^2+a^2}}{x}\, dx = \sqrt{x^2+a^2} - a \log \left| \frac{a+\sqrt{x^2+a^2}}{x} \right|$

57. $\displaystyle \int \frac{\sqrt{x^2-a^2}}{x}\, dx = \sqrt{x^2-a^2} - a \arccos \frac{a}{|x|} \qquad (a>0)$

58. $\displaystyle \int \frac{x^2}{\sqrt{x^2+a^2}}\, dx = \frac{x\sqrt{x^2+a^2}}{2} - \frac{a^2}{2} \log \left(x + \sqrt{x^2+a^2} \right)$

59. $\displaystyle \int \frac{1}{x\sqrt{x^2+a^2}}\, dx = \frac{1}{a} \log \left| \frac{x}{a+\sqrt{x^2+a^2}} \right|$

60. $\displaystyle \int \frac{1}{x\sqrt{x^2-a^2}}\, dx = \frac{1}{a} \arccos \frac{a}{|x|} \qquad (a>0)$

61. $\displaystyle \int \frac{1}{x^2\sqrt{x^2\pm a^2}}\, dx = \mp \frac{\sqrt{x^2\pm a^2}}{a^2 x}$

62. $\displaystyle \int \frac{1}{\sqrt{x^2\pm a^2}}\, dx = \sqrt{x^2\pm a^2}$

63. $\displaystyle \int \frac{1}{ax^2+bx+c}\, dx = \begin{cases} \dfrac{1}{\sqrt{b^2-4ac}} \log \left| \dfrac{2ax+b-\sqrt{b^2-4ac}}{2ax+b+\sqrt{b^2-4ac}} \right| & (b^2>4ac) \\[4mm] \dfrac{2}{\sqrt{4ac-b^2}} \arctan \dfrac{2ax+b}{\sqrt{4ac-b^2}} & (b^2<4ac) \end{cases}$

64. $\displaystyle \int \frac{x}{ax^2+bx+c}\, dx = \frac{1}{2a} \log |ax^2+bx+c| - \frac{b}{2a} \int \frac{1}{ax^2+bx+c}\, dx$

65. $\displaystyle \int \frac{1}{\sqrt{ax^2+bx+c}}\, dx = \begin{cases} \dfrac{1}{\sqrt{a}} \log |2ax+b+2\sqrt{a}\sqrt{ax^2+bx+c}| & (a>0) \\[4mm] \dfrac{1}{\sqrt{-a}} \arcsin \dfrac{-2ax-b}{\sqrt{b^2-4ac}} & (a<0) \end{cases}$

66. $\displaystyle \int \sqrt{ax^2+bx+c}\, dx = \frac{2ax+b}{4a}\sqrt{ax^2+bx+c} + \frac{4ac-b^2}{8a} \int \frac{1}{\sqrt{ax^2+b+c}}\, dx$

67. $\int \dfrac{x}{\sqrt{ax^2 + bx + c}}\, dx = \dfrac{\sqrt{ax^2 + bx + c}}{a} - \dfrac{b}{2a}\int \dfrac{1}{\sqrt{ax^2 + bx + c}}\, dx$

68. $\int \dfrac{1}{x\sqrt{ax^2 + bx + c}}\, dx = \begin{cases} \dfrac{-1}{\sqrt{c}} \log \left| \dfrac{2\sqrt{c}\sqrt{ax^2 + bx + c} + bx + 2c}{x} \right| & (c > 0) \\[4mm] \dfrac{1}{\sqrt{-c}} \arcsin \dfrac{bx + 2c}{|x|\sqrt{b^2 - 4ac}} & (c < 0) \end{cases}$

69. $\int x^3 \sqrt{x^2 + a^2}\, dx = (\tfrac{1}{5}x^2 - \tfrac{2}{15}a^2)\sqrt{(a^2 + x^2)^3}$

70. $\int \dfrac{\sqrt{x^2 \pm a^2}}{x^4}\, dx = \dfrac{\mp \sqrt{(x^2 \pm a^2)^3}}{3a^2 x^3}$

71. $\int \sin ax \sin bx\, dx = \dfrac{\sin (a - b)x}{2(a - b)} - \dfrac{\sin (a + b)x}{2(a + b)} \quad (a^2 \neq b^2)$

72. $\int \sin ax \cos bx\, dx = -\dfrac{\cos (a - b)x}{2(a - b)} - \dfrac{\cos (a + b)x}{2(a + b)} \quad (a^2 \neq b^2)$

73. $\int \cos ax \cos bx\, dx = \dfrac{\sin (a - b)x}{2(a - b)} + \dfrac{\sin (a + b)x}{2(a + b)} \quad (a^2 \neq b^2)$

74. $\int \sec x \cot x \tan x\, dx = \sec x$

75. $\int \csc x\, dx = -\csc x$

76. $\int \cos^m x \sin^n x\, dx = \dfrac{\cos^{m-1} x \sin^{n+1} x}{m + n} + \dfrac{m - 1}{m + n}\int \cos^{m-2} x \sin^n x\, dx$

$\qquad = -\dfrac{\sin^{n-1} x \cos^{m+1} x}{m + n} + \dfrac{n - 1}{m + n}\int \cos^m x \sin^{n-2} x\, dx$

77. $\int x^n \sin ax\, dx = -\dfrac{1}{a} x^n \cos ax + \dfrac{n}{a}\int x^{n-1} \cos ax\, dx$

78. $\int x^n \cos ax\, dx = \dfrac{1}{a} x^n \sin ax - \dfrac{n}{a}\int x^{n-1} \sin ax\, dx$

79. $\int x^n e^{ax}\, dx = \dfrac{x^n e^{ax}}{a} - \dfrac{n}{a}\int x^{n-1} e^{ax}\, dx$

80. $\int x^n \log ax\, dx = x^{n+1}\left[\dfrac{\log ax}{n + 1} - \dfrac{1}{(n + 1)^2} \right]$

81. $\int x^n (\log ax)^m\, dx = \dfrac{x^{n+1}}{n + 1}(\log ax)^m - \dfrac{m}{n + 1}\int x^n (\log ax)^{m-1}\, dx$

82. $\int e^{ax} \sin bx\, dx = \dfrac{e^{ax}(a \sin bx - b \cos bx)}{a^2 + b^2}$

83. $\int e^{ax} \cos bx\, dx = \dfrac{e^{ax}(b \sin bx + a \cos bx)}{a^2 + b^2}$

84. $\int \operatorname{sech} x \tanh x\, dx = -\operatorname{sech} x$

85. $\int \operatorname{csch} x \coth x\, dx = -\operatorname{csch} x$